ANALYSIS OF HEAT AND MASS TRANSFER

McGRAW-HILL SERIES IN MECHANICAL ENGINEERING

KARL H. VESPER
University of Washington, Consulting Editor

BARRON Cryogenic Systems
CSANADY Theory of Turbomachines
ECKERT Introduction to Heat and Mass Transfer
ECKERT and DRAKE Analysis of Heat and Mass Transfer
ECKERT and DRAKE Heat and Mass Transfer
GRÖBER, ERK, and GRIGULL Fundamentals of Heat Transfer
HAM, CRANE, and ROGERS Mechanics of Machinery
HARTENBERG and DENAVIT Kinematic Synthesis of Linkages
HINZE Turbulence
JACOBSEN and AYRE Engineering Vibrations
JUVINALL Engineering Considerations of Stress, Strain, and Strength
KAYS Convective Heat and Mass Transfer
KAYS and LONDON Compact Heat Exchanges
LICHTY Combustion Engine Process
MARTIN Kinematics and Dynamics of Machines
PHELAN Dynamics of Machinery
PHELAN Fundamentals of Mechanical Design
RAVEN Automatic Control Engineering
SCHENCK Theories of Engineering Experimentation
SCHLICHTING Boundary-Layer Theory
SHIGLEY Dynamic Analysis of Machines
SHIGLEY Kinematic Analysis of Mechanics
SHIGLEY Mechanical Engineering Design
SHIGLEY Simulation of Mechanical Systems
SHIGLEY Theory of Machines
STOECKER Refrigeration and Air Conditioning
SUTTON and SHERMAN Engineering Magnetohydrodynamics
WILCOCK and BOOSER Bearing Design and Application

ANALYSIS OF HEAT AND MASS TRANSFER

E. R. G. ECKERT

Regents' Professor of Mechanical Engineering
and
Director of the Heat Transfer Laboratory
University of Minnesota

ROBERT M. DRAKE, JR.

Professor of Mechanical Engineering
and
Dean, College of Engineering
University of Kentucky

McGraw-Hill Book Company

New York St. Louis San Francisco Düsseldorf
Johannesburg Kuala Lumpur London Mexico Montreal
New Delhi Panama Rio de Janeiro
Singapore Sydney Toronto

ANALYSIS OF HEAT AND MASS TRANSFER

Library of Congress Catalog Card Number
73-159305

07-018925-0

3456789 KPKP 798765

This book was set in Modern and printed and bound by The Maple Press Company. The drawings were done by John Cordes, J & R. Technical Services, Inc. The editors were B. J. Clark and Madelaine Eichberg. John A. Sabella supervised production.

Contents

PART II HEAT TRANSFER BY CONVECTION

Nomenclature

CONDUCTION AND CONVECTION

a acceleration
b wall thickness; fin thickness
c specific heat
c_p specific heat at constant pressure
d diameter
d_h hydraulic diameter
e energy per unit mass
f friction coefficient for flow through a tube; dimensionless stream function
f_c drag coefficient of a cylinder
f_p friction coefficient of a plane plate
f_s velocity slip coefficient
g acceleration of gravity; body force per unit mass
h film heat-transfer coefficient
i enthalpy
i_{fg} heat of evaporation
j degrees of freedom
k thermal conductivity
$\mathbf{k}$ Boltzmann constant
l length, mixing length
m mass
$\dot{m}$ mass rate of flow
n normal; number of molecules
p pressure; stress
q heat flow per unit time and area (specific rate of heat flow)
r radius; recovery factor
s thickness; distance; molecular speed ratio
t temperature
t_B bulk temperature
u, U internal energy
u, v, w velocity components
v specific volume
v_i velocity components
v^* shearing-stress velocity
x, y, z coordinates
x_i coordinates
x_0 wavelength
A area
C constant; circumference
E voltage; energy
F force
$\mathbf{H}$ thermal field, enthalpy (in Sect. 11-3)
J turbulence intensity; electric current; integral
L length
L_e entrance length
M molecular weight
N_0 Avogadro's number
Q heat flow per unit time
R resistance; gas constant; radius
R_c thermal-conduction resistance
R_t thermal-convection resistance
T absolute temperature
U internal energy; overall transfer coefficient
V velocity
$\mathcal{V}$ volume
W weight; work
α angle; thermal diffusivity; parameter for properties
β angle; parameter for stream velocity; expansion coefficient
γ parameter for wall temperature; ratio of specific heats
δ boundary-layer thickness; phase lag

δ_i boundary-layer momentum thickness
δ_t thermal boundary-layer thickness
δ^* boundary-layer displacement thickness
ϵ porosity; energy flux
ϵ_H eddy diffusivity for heat
ϵ_m eddy diffusivity for momentum
ζ ratio of thermal and flow boundary-layer thickness
η dimensionless wall distance; fin effectiveness
ϑ temperature difference; parameter describing temperature
λ form parameter (for velocity profile); eigenvalue; mean-free-path length
μ dynamic viscosity
ν kinematic viscosity
ρ density; radius; electric resistivity
σ surface tension
τ shearing stress; time
τ_0 period; relaxation time
φ potential
Φ heat dissipation; heat source; fin effectiveness
ψ stream function

DIMENSIONLESS PARAMETERS

Bi Biot number
c_f friction factor
Ec Eckert number
f friction factor for tube
Fo Fourier number
Fr Froude number
Gr Grashof number
Kn Knudsen number
Le Lewis number
Le_t turbulent Lewis number
Ma Mach number
Nu Nusselt number
Pe Peclet number
Pr Prandtl number
Pr_t turbulent Prandtl number
Ra Rayleigh number
Re Reynolds number
Sc Schmidt number
Sc_t turbulent Schmidt number
Sh Sherwood number
St Stanton number

INDICES

b values on the border between laminar sublayer and turbulent boundary layer
cr critical values for transition to turbulent flow
d based on diameter d
dy dynamic values (dynamic pressure, etc.)
e exit
f fluid
i inlet; based on enthalpy
l liquid
m mean value
p partial
r recovery
s values in the stream outside the boundary layer
st static values (static pressure, etc.)
t turbulent
v vapor
w values on the wall surface
x based on length x
B bulk
M arithmetic mean value
0 values at a reference point
∞ values at great distance from a body
— mean value
′ fluctuating value; dimensionless

SUPERSCRIPTS

- ° total values (total pressure, . . .)
- * reference

ADDITIONALLY FOR MASS TRANSFER

- c mass concentration
- h_M mass-transfer coefficient
- i enthalpy
- j specific mass flux
- n mole density
- r relative humidity
- s specific humidity
- w mass ratio
- C_p specific heat per unit volume
- D mass diffusion coefficient
- M molecular weight
- N number of moles
- $\mathcal{R}$ universal gas constant
- α thermal diffusion ratio
- δ_D diffusion boundary-layer thickness
- ϵ_M eddy diffusivity for mass
- φ parameter describing mass fraction
- ψ mass generation rate
- ω mole fraction

INDICES

- a air
- m component m
- s saturation
- v vapor

RADIATION

- c wave velocity
- e emissive power
- h Planck's quantum
- i intensity of emitted flux
- j coefficient of emission
- $\mathbf{k}$ Boltzmann's gas constant
- n refractive index
- s path length
- u radiation density
- A area
- B radiosity, flux leaving a surface per unit area
- E energy
- F shape factor, angle factor
- H irradiance, flux arriving at a surface per unit area
- K intensity of radiant flux
- L length
- L_e equivalent beam length
- L_g geometric beam length
- S path length
- T absolute temperature
- V volume
- α absorptance, absorptivity
- β angle measured from the surface normal
- γ azimuth angle
- ϵ emittance, emissivity
- κ coefficient of absorption
- λ wavelength
- Λ mean molecular path length
- Λ_r mean photon path length
- ν frequency
- ρ density, reflectance, reflectivity
- σ scattering coefficient, Stefan-Boltzmann constant
- τ transmittance, transmissivity
- Φ radiative energy flux, energy source
- φ radiative energy flux per unit area
- ω solid angle

DEFINITIONS OF RADIATION PARAMETERS

	On surfaces directional	Hemispherical
Total flux	K	B, H
Emitted flux	i	e

	In space
Total flux	K
Emitted flux	j

	Dimensionless parameters			
On surfaces	ϵ	α	ρ	τ
In space		κ	σ	

Preface

The present book has evolved from a series of lectures to graduate students. The text is therefore envisioned as one which will enable the student to learn and understand the fundamental laws in depth and detail, to study and appreciate the power of analytical methods, and to recognize their limitations, after he has gained some familiarity with the phenomena of heat-transfer processes from such texts as Eckert, "Introduction to Heat and Mass Transfer" or Eckert and Drake, "Heat and Mass Transfer."

We have a strong conviction that the student should not be overburdened with numerous details, which may prevent him from developing a clear understanding of the fundamental principles required for application to specific situations as they occur in technological developments. Therefore no attempt is made to discuss all or even most of the newer developments. Instead, a few approaches which have broad applications are treated in sufficient detail for the student to follow the developments and derivations with minimum reference to other sources.

The literature of heat transfer has grown to such an extent that a clear separation between textbooks, like the present one, and reference books is desirable. In the latter category there are now available the annual *Advances in Heat Transfer,* edited by T. F. Irvine and J. P. Hartnett and the yearly reviews of heat-transfer literature in the *International Journal of Heat and Mass Transfer.* In preparation is the "Handbook of Heat Transfer," edited by W. M. Rohsenow and J. P. Hartnett. It is hoped that these volumes and others of a similar nature will provide a factual coverage not only of useful developments from the past but also of new developments as they occur.

Analysis in the field of heat transfer, as in many other subjects, has undergone radical changes in methodology in recent years because of the greater availability and use of large computers. The changes brought about by increasingly sophisticated computer technology are reflected in the field of education. The present text therefore emphasizes first the development and exposition of the basic conservation equations; then it treats the applications of the resulting equations to special situations, through simplification by similarity transformations, for example; and finally it discusses results. The necessary numerical computations are not treated, on the assumption that the graduate student of today is familiar with the programming techniques required to produce the results from the equations describing physical phenomena by the use of the digital or analog computer.

Modern computers make it possible to obtain numerical solutions for the complete system of conservation equations for quite complex situations, e.g., the determination of timewise unsteady three-dimensional flow and temperature fields. Extensive industrial applications have already been made of such calculations. Many groups, both industrial and academic, have developed computer programs of interest, and it is hoped that such programs will become generally available through computer libraries.

On the other hand, to gain an insight into the heat-transfer and flow processes involved and to develop an understanding of the trends in the phenomena, one must study simplified situations with a limited number of boundary conditions. This choice is quite analogous to the situation in experimental work, where basic understanding is sought by means of the selection and exploration of simple experiments with a restricted number of well-defined influencing parameters. It is hoped that the phenomena and examples discussed in this text are those best suited to improving and broadening the understanding of the processes of heat transfer, although we realize that such a selection will always be somewhat subjective.

Dean Drake again contributed Part I, Heat Conduction, Sec. 11-3,

Exchange Processes in Rarefied Gases, and Appendix B, Thermophysical Properties, and thus essentially helped round out the coverage in the text. Physical dimensions, including the thermophysical properties of matter, are given in *Système International* (SI) units, as it is expected that they will become more and more common in the United States, which is apparently destined to be the only country in the world using a nonmetric system in the near future. There seems to be a consensus today that a change to SI units must be made, and opinion differs only as to when the change will occur. Perhaps this book will help speed that day. The reader may find Appendix A helpful in acquainting himself with the SI system.

It is hoped that this book will be useful as a text in graduate courses, as well as in self-study by practicing engineers who need to solve problems in heat transfer or who want to look at the field of heat transfer from a more basic point of view and thus obtain a deeper understanding of the more advanced methods of analysis.

We wish to express here our thanks to our wives, Josefine (Eckert) and Jane (Drake), who patiently accepted the fact that considerable time had to be devoted to the preparation of the book and kept the distracting everyday chores and disturbances at a distance. We also thank our secretaries, Miss Ruth Fenton and Miss Marsha Harman, who typed the manuscript with their usual care and helped with the organization of the material. Thanks are expressed here to the many graduate students of the senior author for their assistance in carrying out the necessary computations and help in proofreading of the textual material. Professors Wendell DeMarcus and J. E. Funk graciously provided an invaluable review of Part I, for which we here note our graditude. We are indebteded to L. B. Davis, Jr. for the recalculation of the data in Appendix B into the *Système International d'Unités.* The help provided by McGraw-Hill Book Company through careful editing and in preparing the final drawings should also be acknowledged.

E. R. G. Eckert
Robert M. Drake, Jr.

ANALYSIS OF HEAT
AND MASS TRANSFER

part I

Heat Conduction

In the next several chapters we shall examine the physical phenomena, the fundamental laws, the thermophysical properties, and the characteristic mathematical formulations that are significant to the process of heat conduction in matter. The flow of heat in opaque solids takes place exclusively by the conduction process. In transparent or translucent solids conduction and radiation transfer can occur, and in liquids and gases the transfer processes of conduction, convection, and radiation can occur simultaneously. In the developments that follow the major interest will be with solids, and the results of the problems considered in the chapters on heat conduction can be applied to all solid bodies. In those specific cases where heat exchange by convection is prevented and exchange by radiation minimized, the principles of heat conduction can be applied to liquids and gases as well.

The treatment of heat conduction has been separated into five specific topics: (1) *theory of heat conduction and heat-conduction equations,* (2) *thermal conductivity and its measurement,* (3) *steady heat conduction,* (4) *unsteady heat conduction, and* (5) *heat conduction with moving boundaries.* These subdivisions are necessarily arbitrary but serve as an aid in the presentation of the material.

Heat conduction from the macroscopic, phenomenological point of view can be understood without a companion understanding of the mechanism of heat conduction as proposed by the theories of modern solid-state physics and kinetic theory. However, today, an acceptable understanding of the science of heat and heat transfer requires a familiarity with both the microscopic and macroscopic points of view. Many of the advances in technology from which we benefit today have been the result of intelligent application of knowledge of the behavior of materials on a microscopic as well as on a macroscopic scale. To note two examples: the effect of alloying on the reduction of thermal conductivity of metals has been used to advantage in the selection of materials for low-temperature work, and knowledge of the material behavior on a microscopic scale has been of major importance in the development of heat-shield materials for atmospheric reentry.

Within the section on heat conduction it is appropriate to treat in some detail, from both microscopic and macroscopic points of view, the important transport property *thermal conductivity.* In the discussion of heat conduction it is the thermal conductivity of solids that appears most important; however, heat conduction is an ever-present process in nominally heat-convection processes, and for this reason the discussion of thermal conductivity includes the liquid and gas as well as the solid phases of matter.

In the other chapters, where heat conduction is treated from the macroscopic, phenomenological sense, the thermophysical properties, such as the thermal conductivity, are presumed to be known. In such a treatment it is well to realize that although the physical notions may be simple, the mathematical techniques required to obtain usable results, i.e., temperature distributions, heat rates, temperature-time histories, etc., are generally complex and in many cases quite difficult. In the treatment here, the mathematical complexities have not been avoided; however, an attempt has been made to keep the presentation from becoming a mathematical treatise. For more complicated problems and problems with diverse boundary conditions the reader is referred to the excellent book by Carslaw and Jaeger.[1]

Certain elementary problems in heat conduction have been treated only briefly in this edition. The reader wishing more detail than shown here should consult the introductory volume by Eckert and Gross[2] or the more advanced work by Eckert and Drake.[3]

With some exceptions, the thermophysical properties which occur in the heat-conduction problems have been considered to be independent of temperature. Such practice not only simplifies the mathematical treatment but also is a reasonable approximation in many physical problems where the temperature variation is not large. In problems involving large temperature differences, chemical reactions, or phase changes, the neglect of temperature dependency of the thermophysical properties may be a serious omission. Therefore, each problem must be carefully considered physically before the assumption of constant properties is applied.

Certain heat-conduction problems involve the convection (or the radiation) mode of heat transfer, usually in the statement of some boundary condition. In the consideration of heat-conduction problems in which heat convection (or radiation) plays a part, it will be assumed that the heat-transfer coefficients are known. The nature of these heat-transfer coefficients as well as the methods for their determination are considered in the chapters on convection heat transfer and thermal radiation.

[1] H. S. Carslaw and J. C. Jaeger, "Conduction of Heat in Solids," 2d ed., Oxford University Press, New York, 1959.

[2] E. R. G. Eckert and J. F. Gross, "Introduction to Heat and Mass Transfer," McGraw-Hill, New York, 1963.

[3] E. R. G. Eckert and R. M. Drake, Jr., "Heat and Mass Transfer," 2d ed., McGraw-Hill, New York, 1959.

1

Theory of Heat Conduction and Heat-conduction Equations

1-1 THE CONCEPT OF HEAT CONDUCTION

The currently accepted theory of heat is closely associated with the internal energy of matter, which in thermodynamics is referred to as the energy related to the physical and chemical state of the body—the orientation and motion of the molecules and atoms within the body. Although incomplete, the dynamic theory of heat permits some important conclusions to be drawn which are quite generally confirmed by experiment:

1. Since heat as energy is associated with translational, rotational, and vibrational motions of the molecules, atoms, and their components, heat transfer by conduction must be strictly related to these motions.
2. Increased temperature increases the intensity and frequency of molecular and atomic motions; therefore, the conduction of heat should increase with increasing temperature. (There are exceptions to this statement.)

3. Changes from the denser solid phases to the liquid and gaseous phases result in lower thermal conductivities and thus smaller heat-conduction effects.

These general patterns of the behavior of materials are discussed in considerable detail in Chap. 2 in terms of the modern theories of the solid state and the kinetic theory of matter. The description of the heat-transfer mechanism by means of a dynamic theory and in terms of a meaningful molecular model has presented the physicist with some of the most complicated problems in theoretical physics. In recent years improvement of the fundamental theories and implementation of the theoretical treatment by means of accurate and detailed calculations made possible by modern computers have greatly advanced the knowledge and understanding of the transport properties. And while there remain large gaps in knowledge in certain areas, liquids in particular, the general understanding of these properties is fairly complete. In fact, in the case of gases at high temperatures the calculated results are superior to the experimental measurements.

While there remains some conjecture in regard to the precise physical model for the mechanism of the transfer of heat energy, for any event and for any theory the energy in transition is referred to as *heat* and the process of energy transfer is known as *conduction*.

It is not necessary to understand fully the mechanism of heat conduction in matter to proceed with the mathematical developments which lead to practical results. For even though the physicomathematical models representing the microscopic behavior are not yet perfect, one relies upon the fact that the practical developments of the heat-conduction processes resulted from a hypothesis based upon experimental observations. Subsequent use of this hypothesis as a basis for mathematical analysis to obtain results which have been experimentally verified is sufficient to establish the particular law which is characteristic of the transfer itself. The basic law so established is entirely consistent with the laws of thermodynamics.

1-2 THE FUNDAMENTAL LAW OF HEAT CONDUCTION

To be consistent thermodynamically is to require, by virtue of the second law of thermodynamics, that heat will be transferred from one body to another body (or from one part of a body to another part of the same body) only when the bodies are at different temperatures and that the heat will flow from the location of the highest temperature to the location of the lowest temperature; i.e., a temperature gradient exists, and the energy in the form of heat flows in the direction of decreasing temperature. The first law of thermodynamics states that the flowing thermal energy is conserved in the absence of heat sources or sinks. To this end a body may have a temperature distribution which is dependent upon the space coordinates and time of observation:

$$t = f(x,y,z,\tau)$$

One may suppose that within this body is a surface such that, when observed at a certain time, each point on it has an identical temperature. Such a surface is called an *isothermal surface.* One can further visualize other isothermal surfaces within this body which differ from one another by being hotter or colder by temperature increments $\pm\delta t$, respectively. These isothermal surfaces never intersect, because no point in the body can exist at two different temperatures at the same time. The body is thus visualized as being composed of a number of arbitrarily thin, isothermal shells that, of course, vary with time.

In the discussion which follows, unless stated differently, we shall consider only *isotropic solids,* i.e., solids whose properties and constitution in the neighborhood of any point are invariant with the direction from the point. In such a case, and because of the symmetry involved, the heat flow at a point is along a path perpendicular to the isothermal surface through the point. For a nonisotropic solid the heat flow is not necessarily in a direction perpendicular to the isothermal surface through the point, a situation which will be discussed in a subsequent paragraph.

The hypothesis which forms the basis for the mathematical formulation of the law of heat conduction had its foundation in the results of a simple experiment. A plate of some solid is bounded by planes sufficiently large to be supposed infinite in extent. The bounding planes are maintained experimentally at different but uniform temperatures, the temperature difference being not so great as to measurably change the thermophysical properties of the plate material. After some sufficient time the heat flow and temperature distribution in the plate become invariant with time, or *steady.* The heat flow can then be measured to be

$$Q = \frac{kA(t_1 - t_2)}{d} \tag{1-1}$$

where Q, the heat flow, is shown to be proportional to the area A of the plate surface and to the temperature difference $t_1 - t_2$ of the plate surface and inversely proportional to the plate thickness d. The constant of proportionality is the thermal conductivity k. Strictly speaking, the thermal conductivity is not a constant but depends upon temperature and to some extent pressure. Thermal conductivity is a *transport property* of the material. The nature of the thermal conductivity and its variation with physical parameters such as material composition, pressure, and temperature will be discussed in Chap. 2.

At a point in the interior of the plate, Eq. (1-1) can be written more generally as

$$\frac{Q}{A} = \frac{\partial \mathbf{H}}{\partial \tau} = -k\,\frac{\partial t}{\partial n} \tag{1-2}$$

where **H** is the heat-flow vector field. Equation (1-2) can be interpreted with the aid of Fig. 1-1. The heat flux Q/A flows along the normal n to the area A in the direction of the decreasing temperature, i.e., the negative thermal gradient. The negative sign in Eq. (1-2) indicates that the heat flow is in the direction of the negative gradient and serves to make the heat flux positive in that sense. Again, the proportionality factor is the thermal conductivity k, a property of the material through which the heat flows.

The form of Eq. (1-2), which in fact serves to define the thermal conductivity, implies that the process of heat conduction is a random process, a diffusion of energy. The heat energy does not enter one side of the plate and travel directly to the other side but moves randomly through the plate as the result of frequent collisions. If the heat were propagated through the plate without the random character, the expression for the heat flux in Eq. (1-2) would show a dependence only on the temperature difference and not on the temperature gradient, thus being independent of the thickness of the plate d. It is the random nature of the conduction process that brings the temperature gradient into the expression for the heat flux.

The random nature of the conduction process can be demonstrated in the following way: In the kinetic theory of gases under certain approximations, as will be shown elsewhere in this text, the thermal conductivity of certain solids can be shown as

$$k = \tfrac{1}{3}\rho\bar{w}\lambda c_v \tag{1-3}$$

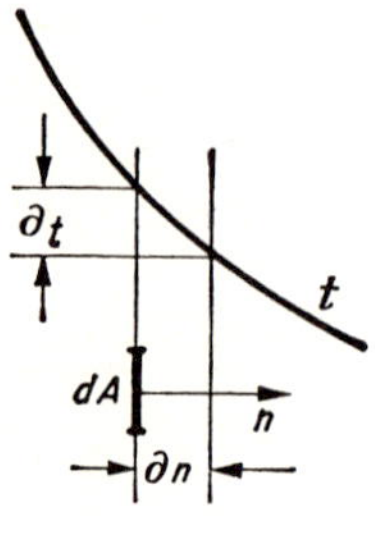

Fig. 1-1 Heat-conduction system.

where c_v = heat capacity at constant volume
$\bar{w}$ = average carrier velocity
ρ = mass density
λ = mean free path of carrier between collisions

If Eqs. (1-1) and (1-3) are combined,

$$q = \frac{Q}{A} = \frac{1}{3}\rho c_v(t_1 - t_2)\frac{\bar{w}\lambda}{d} \tag{1-4}$$

The expression for the thermal flux as shown by Eq. (1-4) can be interpreted as follows: $\rho c_v(t_1 - t_2)$ is the excess energy on one side of the plate over that of the other side. This energy is propagated across the plate at an effective transport velocity $\bar{w}\lambda/d$, which is just the carrier velocity reduced by the ratio of the mean free path to the significant dimension of the plate d. As will be shown in Chap. 2, the carriers are the individual molecules in the case of a gas; in the case of solids (and to a great extent liquids) the carriers are free electrons and phonons.

Equation (1-2) can be rewritten for the case of an infinitesimal area as

$$dQ = -k\,dA\,\frac{\partial t}{\partial n} \tag{1-5}$$

Equations (1-2) and (1-5) are generally attributed to the French mathematician *Jean Baptiste Fourier* and in his honor are designated the *Fourier heat-conduction equations.*

The heat flow per unit area per unit time across any surface is called the *heat flux* q and has units of watts per square meter. The heat flux is a *vector;* i.e., its magnitude and direction must be specified. The heat flux can be calculated for any point in reference to any arbitrary direction through the point if the area normal to the desired direction is considered.

In Fig. 1-2 are shown the isotherms t and $t + dt$ in a body. The normal to these isotherms is designated by the axis n, which is also normal to the differential area dA. The heat flux can be calculated in the direction of the normal n and in the arbitrary direction s as shown below:

$$q_n = \frac{dQ}{dA} = -k\frac{\partial t}{\partial n}$$

$$q_s = \frac{dQ}{dA\cos\alpha} = -k\frac{\partial t}{\partial s}$$

Since $n = s\cos\alpha$,

$$q_s = -k\frac{\partial t}{\partial n}\cos\alpha \tag{1-6}$$

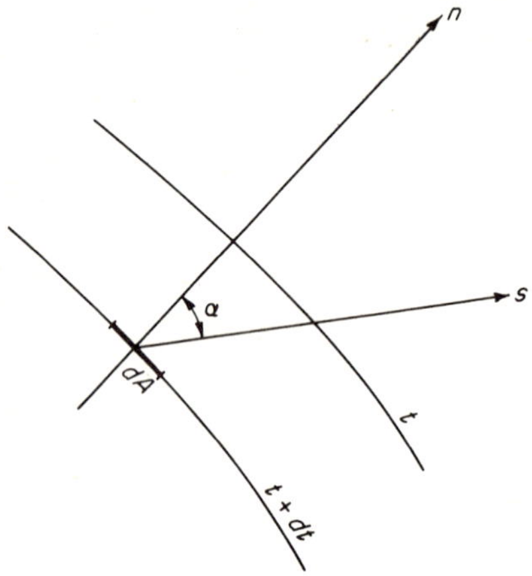

Fig. 1-2 Direction of heat flow.

or, in other words, q_s is a component of the heat-flux vector q_n. From Eq. (1-6) it can be seen that the greatest heat flux is that which is calculated along the normal to the isothermal surfaces. In particular, if the component fluxes are related to the planes of the (x,y,z) coordinate system, the heat fluxes are

$$q_x = -k\frac{\partial t}{\partial x} \qquad q_y = -k\frac{\partial t}{\partial y} \qquad q_z = -k\frac{\partial t}{\partial z} \tag{1-7}$$

The heat fluxes shown in Eq. (1-7) are components of the heat-flux vector

$$\mathbf{q} = i\mathbf{q}_x + j\mathbf{q}_y + k\mathbf{q}_z \tag{1-8}$$

Effect of variable thermal conductivity It should be noted here that the thermal conductivity k is not a constant but, in fact, is a function of the temperature for all phases and in liquids and gases depends also upon the pressure, especially when near the critical state. The thermal conductivity in wood and crystals also varies markedly in direction.

The dependence of thermal conductivity on temperature for small, select temperature ranges can be acceptably expressed in a linear form:

$$k = k_0(1 + at)$$

where k_0 is the value of the thermal conductivity at some reference condition and a is the temperature coefficient and is positive or negative depending upon the material in question. Figure 1-3 shows the effect on the temperature gradient in a body as a result of the positive or negative characteristic of a.

It can readily be seen that a linear temperature gradient exists only when the thermal conductivity is a constant.

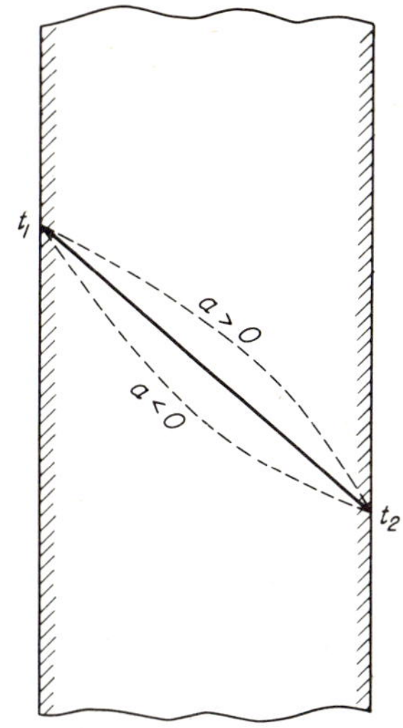

Fig. 1-3 Temperature profile in a simple slab.

It is interesting to note that the Fourier equation for heat conduction is exactly analogous to Ohm's law for an electric conductor. Ohm's law for a conductor of any shape can be expressed as

$$dI = -\sigma\, dA \frac{\partial E}{\partial n} \tag{1-9}$$

In Eq. (1-9) the electric current I corresponds to the heat flow Q, the electric potential E corresponds to the thermal potential t, and the electric conductivity σ ($\sigma = 1/\rho$, where ρ is the electric resistivity) corresponds to the thermal conductivity k. Since Eqs. (1-5) and (1-9) have the same form, the temperature field within a heated body and the field of electric potential in bodies of the same shape correspond to each other provided the temperature distribution on the surface corresponds to the surface distribution of the electric potential. This analogy enables one to study heat-conduction problems in accurate detail by means of similar electrical models.

1-3 THE HEAT–CONDUCTION EQUATION (DIFFERENTIAL FORM)

In the following section we discuss the establishment of the heat-conduction equation in differential form in the rectangular coordinate system. The differential form of the heat-conduction equation is most advantageous.

Heat-conduction equation for isotropic materials Consider an infinitesimal control volume of dimensions δx, δy, and δz which is oriented within a three-dimensional (x,y,z) coordinate system as in Fig. 1-4. The considerations here will include the nonsteady condition of temperature variation with time τ.

According to the Fourier heat-conduction law [Eq. (1-5)], the heat flowing into the leftmost face of the volume element in the x direction can be expressed as

$$dQ_x = -k(\delta y\ \delta z)\frac{\partial t}{\partial x}$$

The heat flow out of the right face of the volume element can be obtained by expanding dQ_x in a Taylor series and retaining only the first two terms as a reasonable approximation:

$$dQ_{x+\delta x} = dQ_x + \frac{\partial}{\partial x}dQ_x\ \delta x + \cdots$$

The net heat flow by conduction in the x direction is therefore

$$dQ_x - dQ_{x+\delta x} = \frac{\partial}{\partial x}\left(k\frac{\partial t}{\partial x}\right)\delta x\ \delta y\ \delta z \tag{1-10}$$

Two additional equations similar to Eq. (1-10) for the y and z directions can be written in the same way:

$$dQ_y - dQ_{y+\delta y} = \frac{\partial}{\partial y}\left(k\frac{\partial t}{\partial y}\right)\delta x\ \delta y\ \delta z$$

$$dQ_z - dQ_{z+\delta z} = \frac{\partial}{\partial z}\left(k\frac{\partial t}{\partial z}\right)\delta x\ \delta y\ \delta z$$

The sum of the net quantities of heat flow in the three coordinate directions is

$$\left[\frac{\partial}{\partial x}\left(k\frac{\partial t}{\partial x}\right)+\frac{\partial}{\partial y}\left(k\frac{\partial t}{\partial y}\right)+\frac{\partial}{\partial z}\left(k\frac{\partial t}{\partial z}\right)\right]\delta x\ \delta y\ \delta z \tag{1-11}$$

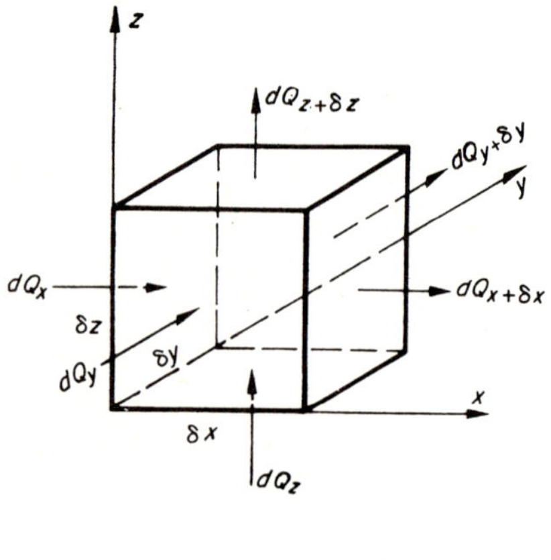

Fig. 1-4 The volume element for determination of the heat-conduction equation.

If per unit of time and space the quantity of heat $Q'(x,y,z,\tau)$ is generated, then the generation of heat in the volume element is

$$Q' \, \delta x \, \delta y \, \delta z \tag{1-12}$$

The net heat flow into the volume element due to conduction [Eq. (1-11)] and the heat generated within the volume element [Eq. (1-12)] together serve to increase the internal energy of the volume element. Such an increase in the internal energy is reflected in the time rate of change in the energy storage in the volume element and can be written

$$c\rho \, \delta x \, \delta y \, \delta z \, \frac{\partial t}{\partial \tau} \tag{1-13}$$

where c = specific heat
ρ = density
τ = time

An energy balance can be made on the volume element to equate the time rate of change of the energy stored to the sum of the net heat flow into the element due to conduction and the heat generated within the element, to yield the expression

$$c\rho \frac{\partial t}{\partial \tau} = \frac{\partial}{\partial x}\left(k \frac{\partial t}{\partial x}\right) + \frac{\partial}{\partial y}\left(k \frac{\partial t}{\partial y}\right) + \frac{\partial}{\partial z}\left(k \frac{\partial t}{\partial z}\right) + Q' \tag{1-14}$$

It should be noted here that $k = k(x,y,z,t)$, $c = c(x,y,z,t)$, and $\rho = \rho(x,y,z,t)$, so that Eq. (1-14) is valid for isotropic, heterogeneous media.

The heat-source term can be omitted where the body is free of sources, and Eq. (1-14) can be written in terms of the heat-flux components as

$$-c\rho \frac{\partial t}{\partial \tau} = \frac{\partial}{\partial x} q_x + \frac{\partial}{\partial y} q_y + \frac{\partial}{\partial z} q_z \tag{1-15}$$

This last formulation is somewhat more general and will be of use in the discussion of nonisotropic materials.

Equation (1-14) can be further simplified for an isotropic, homogeneous material where the thermal conductivity k can be considered a constant; thus

$$\frac{\partial t}{\partial \tau} = \frac{k}{\rho c}\left(\frac{\partial^2 t}{\partial x^2} + \frac{\partial^2 t}{\partial y^2} + \frac{\partial^2 t}{\partial z^2}\right) + \frac{Q'}{\rho c} = \alpha \nabla^2 t + \frac{Q'}{\rho c} \tag{1-16}$$

The parametric group $k/\rho c$, which has dimensions of length squared per time, is referred to as the *thermal diffusivity*, is designated by the symbol α, and is a property of the conducting material.

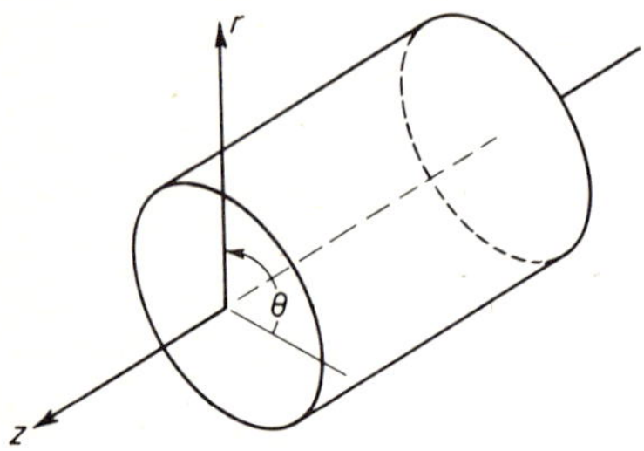

Fig. 1-5 Cylindrical coordinate system.

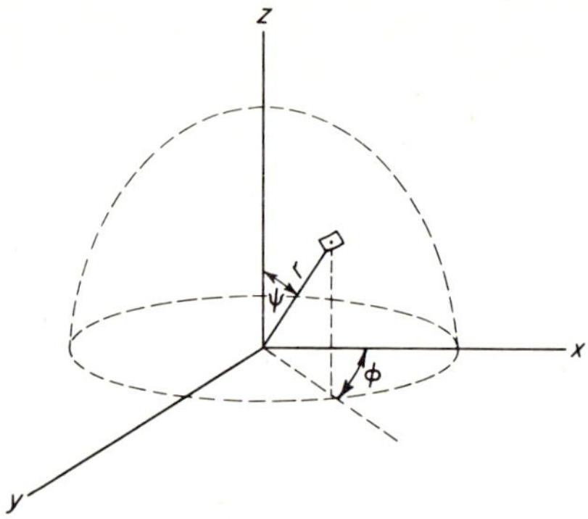

Fig. 1-6 Spherical coordinate system.

Heat-conduction equation in cylindrical coordinate system By means of a transformation of coordinates, Eq. (1-16) can be expressed in a form better suited for cylindrical systems. Thus from Fig. 1-5 for $x = r \cos \theta$, $y = r \sin \theta$, and $z = z$,

$$\frac{\partial t}{\partial \tau} = \alpha \left(\frac{\partial^2 t}{\partial r^2} + \frac{1}{r} \frac{\partial t}{\partial r} + \frac{1}{r^2} \frac{\partial^2 t}{\partial \theta^2} + \frac{\partial^2 t}{\partial z^2} \right) + \frac{Q'}{\rho c} \tag{1-17}$$

Heat-conduction equation in spherical coordinate system A similar transformation yields for the spherical system (Fig. 1-6) for

$$x = r \sin \psi \cos \phi$$

$y = r \sin \psi \sin \phi$; and $z = r \cos \psi$,

$$\frac{\partial t}{\partial \tau} = \alpha \left[\frac{1}{r} \frac{\partial^2 (rt)}{\partial r^2} + \frac{1}{r^2 \sin \psi} \frac{\partial}{\partial \psi} \left(\sin \psi \frac{\partial t}{\partial \psi} \right) + \frac{1}{r^2 \sin^2 \psi} \frac{\partial^2 t}{\partial \phi^2} \right] + \frac{Q'}{\rho c} \tag{1-18}$$

Heat conduction for nonisotropic materials In the preceding section the heat-conduction equation for isotropic media was derived. Certain technically important pure materials and laminates have thermal conductivities that vary markedly depending upon the direction of heat flow through the body. Included in this category of materials are crystalline substances, woods, laminated plastics and laminated metals, such as are used in transformer cores, and plywood. The heat-conduction equation requires certain modifications before these nonisotropic materials can be adequately treated.

The thermal-conductivity coefficients for nonisotropic materials k_{ij} form an ellipsoid. The major axes of the ellipsoid, ξ, η, ζ, are called the *principal axes of conductivity*, and the coefficients, k_ξ, k_η, k_ζ, are the *principal thermal conductivities*.

Retaining the assumption that the heat flux across any surface at a point in a solid body depends linearly upon the temperature gradients at the point, the heat fluxes in the (x,y,z) coordinate system can be written in terms of the principal conductivities.

The heat fluxes in the x, y, z direction in terms of the heat fluxes in the ξ, η, ζ directions are

$$\begin{aligned} q_x &= l_1 q_\xi + l_2 q_\eta + l_3 q_\zeta \\ q_y &= m_1 q_\xi + m_2 q_\eta + m_3 q_\zeta \\ q_z &= n_1 q_\xi + n_2 q_\eta + n_3 q_\zeta \end{aligned}$$

where the l_i, m_i, n_i are the appropriate direction cosines.

The heat fluxes q_ξ, q_η, and q_ζ can be written in terms of the temperature gradients and the principal thermal conductivities

$$\begin{aligned} q_\xi &= -k_\xi \frac{\partial t}{\partial \xi} \\ q_\eta &= -k_\eta \frac{\partial t}{\partial \eta} \\ q_\zeta &= -k_\zeta \frac{\partial t}{\partial \zeta} \end{aligned}$$

The temperature gradients in the ξ, η, ζ directions can be written in terms of the gradients in x, y, z

$$\begin{aligned} \frac{\partial t}{\partial \xi} &= l_1 \frac{\partial t}{\partial x} + m_1 \frac{\partial t}{\partial y} + n_1 \frac{\partial t}{\partial z} \\ \frac{\partial t}{\partial \eta} &= l_2 \frac{\partial t}{\partial x} + m_2 \frac{\partial t}{\partial y} + n_2 \frac{\partial t}{\partial z} \\ \frac{\partial t}{\partial \zeta} &= l_3 \frac{\partial t}{\partial x} + m_3 \frac{\partial t}{\partial y} + n_3 \frac{\partial t}{\partial z} \end{aligned}$$

A combination of the three sets of equations yields the expressions for the heat fluxes in the x, y, and z directions in terms of the temperature gradients in the x, y, and z directions, and the values of the principal thermal conductivities k_ξ, k_η, and k_ζ

$$\begin{aligned} -q_x = {} & (k_\xi l_1^2 + k_\eta l_2^2 + k_\zeta l_3^2) \frac{\partial t}{\partial x} + (k_\xi l_1 m_1 + k_\eta l_2 m_2 + k_\zeta l_3 m_3) \frac{\partial t}{\partial y} \\ & + (k_\xi l_1 n_1 + k_\eta l_2 n_2 + k_\zeta l_3 n_3) \frac{\partial t}{\partial z} \\ -q_y = {} & (k_\xi l_1 m_1 + k_\eta l_2 m_2 + k_\zeta l_3 m_3) \frac{\partial t}{\partial x} + (k_\xi m_1^2 + k_\eta m_2^2 + k_\zeta m_3^2) \frac{\partial t}{\partial y} \\ & + (k_\xi m_1 n_1 + k_\eta m_2 n_2 + k_\zeta m_3 n_3) \frac{\partial t}{\partial z} \\ -q_z = {} & (k_\xi l_1 n_1 + k_\eta l_2 n_2 + k_\zeta l_3 n_3) \frac{\partial t}{\partial x} + (k_\xi m_1 n_1 + k_\eta m_2 n_2 + k_\zeta m_3 n_3) \frac{\partial t}{\partial y} \\ & + (k_\xi n_1^2 + k_\eta n_2^2 + k_\zeta n_3^2) \frac{\partial t}{\partial z} \end{aligned}$$

The heat fluxes may be written

$$\begin{aligned}
-q_x &= k_{xx}\frac{\partial t}{\partial x} + k_{xy}\frac{\partial t}{\partial y} + k_{xz}\frac{\partial t}{\partial z} \\
-q_y &= k_{yx}\frac{\partial t}{\partial x} + k_{yy}\frac{\partial t}{\partial y} + k_{yz}\frac{\partial t}{\partial z} \\
-q_z &= k_{zx}\frac{\partial t}{\partial x} + k_{zy}\frac{\partial t}{\partial y} + k_{zz}\frac{\partial t}{\partial z}
\end{aligned} \tag{1-19}$$

where

$$\begin{aligned}
k_{xx} &= k_\xi l_1^2 + k_\eta l_2^2 + k_\zeta l_3^2 \\
k_{yy} &= k_\xi m_1^2 + k_\eta m_2^2 + k_\zeta m_3^2 \\
k_{zz} &= k_\xi n_1^2 + k_\eta n_2^2 + k_\zeta n_3^2 \\
k_{xy} &= k_{yx} = k_\xi l_1 m_1 + k_\eta l_2 m_2 + k_\zeta l_3 m_3 \\
k_{xz} &= k_{zx} = k_\xi l_1 n_1 + k_\eta l_2 n_2 + k_\zeta l_3 n_3 \\
k_{yz} &= k_{zy} = k_\xi m_1 n_1 + k_\eta m_2 n_2 + k_\zeta m_3 n_3
\end{aligned}$$

The heat-balance equation (1-15) remains valid for the nonisotropic case; thus

$$\begin{aligned}
\rho c \frac{\partial t}{\partial \tau} = k_{xx}\frac{\partial^2 t}{\partial x^2} + k_{yy}\frac{\partial^2 t}{\partial y^2} + k_{zz}\frac{\partial^2 t}{\partial z^2} + (k_{xy} + k_{yx})\frac{\partial^2 t}{\partial x\,\partial y} \\
+ (k_{yz} + k_{zy})\frac{\partial^2 t}{\partial y\,\partial z} + (k_{xz} + k_{zx})\frac{\partial^2 t}{\partial x\,\partial z}
\end{aligned} \tag{1-20}$$

if the medium is homogeneous and has no heat sources or heat sinks. Addition of heat sources or sinks is an easy extension of the equations.

If the axes of interest coincide with the axes of the principal thermal conductivities, Eq. (1-20) reduces to the form

$$\rho c \frac{\partial t}{\partial \tau} = k_\xi \frac{\partial^2 t}{\partial \xi^2} + k_\eta \frac{\partial^2 t}{\partial \eta^2} + k_\zeta \frac{\partial^2 t}{\partial \zeta^2} \tag{1-21}$$

An additional transformation is possible where

$$\xi_1 = \xi\sqrt{\frac{k_0}{k_\xi}} \qquad \eta_1 = \eta\sqrt{\frac{k_0}{k_\eta}} \qquad \zeta_1 = \zeta\sqrt{\frac{k_0}{k_\zeta}} \tag{1-22}$$

where k_0 is chosen arbitrarily. With this transformation Eq. (1-21) becomes

$$\frac{\partial t}{\partial \tau} = \frac{k_0}{\rho c}\left(\frac{\partial^2 t}{\partial \xi_1^2} + \frac{\partial^2 t}{\partial \eta_1^2} + \frac{\partial^2 t}{\partial \zeta_1^2}\right) \tag{1-23}$$

which now has the same form as Eq. (1-16) for the isotropic body. This transformation reduces the problem of the nonisotropic solid to the problem of corresponding isotropic solid when the solid is bounded by planes perpendicular to the principal axes of conductivity or when the solid is infinite.

In the particular case of wood, where the principal axes of conductivity coincide with the axial, radial, and circumferential coordinates of the cylindrical coordinate system, the heat fluxes are

$$q_z = -k_z \frac{\partial t}{\partial z} \qquad q_r = -k_r \frac{\partial t}{\partial r} \qquad q_\theta = -k_\theta \frac{1}{r} \frac{\partial t}{\partial \theta}$$

and the heat-conduction equation becomes, for the case of no heat sources or heat sinks,

$$\rho c \frac{\partial t}{\partial \tau} = \frac{k_r}{r} \frac{\partial}{\partial r}\left(r \frac{\partial t}{\partial r}\right) + \frac{k_\theta}{r^2} \frac{\partial^2 t}{\partial \theta^2} + k_z \frac{\partial^2 t}{\partial z^2} \tag{1-24}$$

It is informative to consider a detailed example of a nonisotropic material in two dimensions in which the principal axes of conductivity are not perpendicular to planes bounding the body.

Consider a material, as shown in Fig. 1-7, where the body is oriented in the (x,y) coordinate system which makes an angle β with the principal axes of thermal conductivity of the material. The coordinate system (ξ,η) is aligned with the principal axes of conductivity. The heat flux through the body can be considered to be made up of the component heat fluxes in the direction of the coordinates ξ and η:

$$q_\xi = -k_\xi \frac{\partial t}{\partial \xi} \qquad q_\eta = -k_\eta \frac{\partial t}{\partial \eta}$$

The fluxes in the x and y directions are therefore

$$\begin{aligned} q_x &= q_\xi \cos\beta - q_\eta \sin\beta = -k_\xi \cos\beta \frac{\partial t}{\partial \xi} + k_\eta \sin\beta \frac{\partial t}{\partial \eta} \\ q_y &= q_\xi \sin\beta + q_\eta \cos\beta = -k_\xi \sin\beta \frac{\partial t}{\partial \xi} - k_\eta \cos\beta \frac{\partial t}{\partial \eta} \end{aligned} \tag{1-25}$$

The temperature gradients can be transformed into gradients in x and y

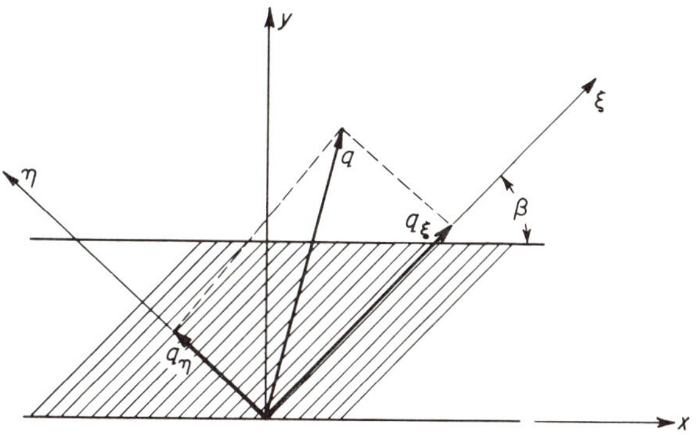

Fig. 1-7 Heat flux in nonisotropic medium.

by the relations

$$\frac{\partial t}{\partial \xi} = \frac{\partial t}{\partial x}\frac{\partial x}{\partial \xi} + \frac{\partial t}{\partial y}\frac{\partial y}{\partial \xi}$$
$$\frac{\partial t}{\partial \eta} = \frac{\partial t}{\partial x}\frac{\partial x}{\partial \eta} + \frac{\partial t}{\partial y}\frac{\partial y}{\partial \eta}$$

and by the geometry of the figure,

$$y = \xi \sin \beta = \eta \cos \beta \qquad x = \xi \cos \beta = -\eta \sin \beta$$

Substituting these values into Eq. (1-25) and rearranging, we obtain for the heat fluxes

$$\begin{aligned} q_x &= -(k_\xi \cos^2 \beta + k_\eta \sin^2 \beta)\frac{\partial t}{\partial x} - (k_\xi - k_\eta) \cos \beta \sin \beta \frac{\partial t}{\partial y} \\ q_y &= -(k_\xi - k_\eta) \sin \beta \cos \beta \frac{\partial t}{\partial x} - (k_\xi \sin^2 \beta + k_\eta \cos^2 \beta)\frac{\partial t}{\partial y} \end{aligned} \tag{1-26}$$

With the use of the general form of the conduction equation as shown in Eq. (1-15) and the heat fluxes as given in Eq. (1-26), the heat-conduction equation for a nonisotropic material in two dimensions can be expressed:

$$\begin{aligned} \rho c \frac{\partial t}{\partial \tau} = (k_\xi \cos^2 \beta + k_\eta \sin^2 \beta)\frac{\partial^2 t}{\partial x^2} + (k_\xi \sin^2 \beta + k_\eta \cos^2 \beta)\frac{\partial^2 t}{\partial y^2} \\ + (k_\xi - k_\eta) \sin 2\beta \frac{\partial^2 t}{\partial x\, \partial y} \end{aligned} \tag{1-27}$$

For an isotropic medium, $k_\xi = k_\eta$ and $\beta = 0$. For these specifications Eq. (1-27) reduces to the two-dimensional form of Eq. (1-16).

It is interesting to recognize here that if a sheet of nonisotropic material, such as a laminate, is held between the isothermal surface of a thermal-conductivity test apparatus, and if the laminated sample is prepared such that its principal axes make an angle of β with the isothermal surfaces, the thermal conductivities that are measured (depending on whether the measurements were made in the x or y direction) are given by the expressions

$$k_x = k_\xi \cos^2 \beta + k_\eta \sin^2 \beta$$
$$k_y = k_\xi \sin^2 \beta + k_\eta \cos^2 \beta$$

Example 1-1 A slab of laminated material is being used in a thermal-conductivity test. The laminations make an angle β with the smooth surfaces of the sample (Fig. 1-8). The surfaces A are kept at constant but different temperatures and thus are isothermal surfaces. It is required to calculate the angle that the heat-flux vector makes with the normal n to the isothermal surfaces. From Fig. 1-8 it may be seen that for $k_\xi > k_\eta$,

$$\tan \gamma = \frac{q_\eta}{q_\xi} = \frac{k_\eta\, \partial t/\partial_\eta}{k_\xi\, \partial t/\partial_\xi}$$

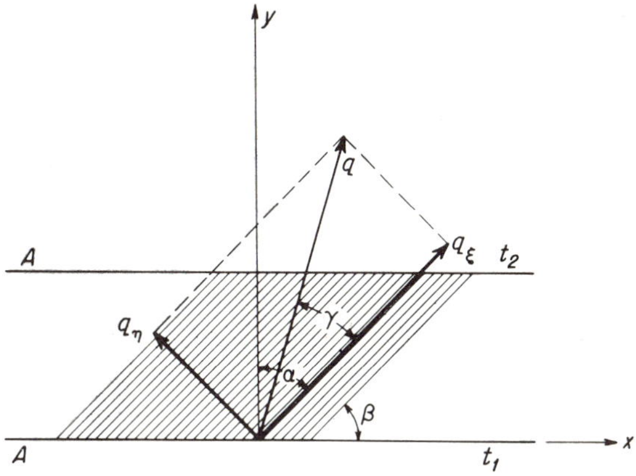

Fig. 1-8 Heat-flux vector in a laminated material.

but from earlier calculations,

$$\frac{\partial t}{\partial \eta} = -\sin\beta\frac{\partial t}{\partial x} + \cos\beta\frac{\partial t}{\partial y}$$

$$\frac{\partial t}{\partial \xi} = \cos\beta\frac{\partial t}{\partial x} + \sin\beta\frac{\partial t}{\partial y}$$

thus,
$$\tan\gamma = \frac{k_\eta[\cos\beta(\partial t/\partial y) - \sin\beta(\partial t/\partial x)]}{k_\xi[\cos\beta(\partial t/\partial x) + \sin\beta(\partial t/\partial y)]} \qquad (a)$$

In the system we are considering here, however, the bounding surfaces are kept isothermal; thus the coordinate axis y is in the direction of the normal n, and x lies in the isothermal surface. Thus $\partial t/\partial x = 0$, and expression ($a$) becomes

$$\tan\gamma = \frac{k_\eta}{k_\xi}\cot\beta = \frac{k_\eta}{k_\xi}\tan\alpha$$

There $\gamma < \alpha$, and the heat-flux vector is not normal to the isothermal surface as it would be if the material were isotropic. If $k_\xi = 2k_\eta$, as is possible in wood, and if $\beta = 45°$, then

$$\tan\gamma = \tfrac{1}{2}$$
$$\gamma = 26.6°$$

1-4 THE HEAT-CONDUCTION EQUATION (INTEGRATED FORM)

In recent years, and especially for problems of geometrical complexity or with nonlinear or otherwise complicated boundary conditions, resort has been made to approximate formulations of the heat-conduction equations. One such approximation is the integrated form of the heat-conduction equation that has been used so effectively by Goodman[1] and others. The

[1] T. R. Goodman, The Heat Balance Integral and Its Application to Problems Involving a Change in Phase, *Trans. ASME*, **8**:2 (1958); The Heating of Slabs with Arbitrary Heat Inputs, *J. Aerospace Sci.*, **26**:187 (1959).

technique and the resulting equation are similar to the formulations by T. von Kármán[1] which have had such wide use in the subfield of heat convection and the fluid mechanics of boundary layers. The resulting equations are approximate in that they are not accurate locally in terms of the specification of the heat-flux vector or the temperature but satisfy exactly the energy balance. The method has the virtue of simplicity, speed, and quite reasonable accuracy when compared with exact solutions.

The formulation of the integrated heat-conduction equation can easily be shown for the transient case in one space dimension, the semi-infinite body. When a semi-infinite body initially existing at some temperature T_∞ loses heat from the free surface, a temperature field is built up as is shown schematically in Fig. 1-9. This temperature field comprises a region within the body wherein the local temperature differs from T_∞. The depth to which the thermal effects are felt is called the *penetration depth* and following the boundary-layer practice is designated by δ. The penetration depth is a function of time.

The differential equation for the case of the semi-infinite solid with constant properties is

$$\frac{\partial t}{\partial \tau} = \alpha \frac{\partial^2 t}{\partial x^2} \tag{1-28}$$

Equation (1-28) can be integrated from $x = 0$ to $x = \delta(\tau)$

$$\int_0^\delta \frac{\partial t}{\partial \tau}\, dx = \int_0^\delta \alpha \frac{\partial^2 t}{\partial x^2}\, dx \tag{1-29}$$

which may be evaluated to yield[2]

$$\frac{d}{d\tau}\left[\int_0^\delta t(x,\tau)\, dx\right] - t(\delta,\tau)\frac{d\delta}{d\tau} = \alpha\left[\frac{\partial t}{\partial x}(\delta,\tau) - \frac{\partial}{\partial x}(0,\tau)\right] \tag{1-30}$$

[1] T. von Kármán, *Z. Angew. Math. Mech.*, **1**:235 (1921).

[2] $\frac{d}{d\beta}\int_a^b f(x,\beta)\, dx = f(b,\beta)\frac{db}{d\beta} - f(a,\beta)\frac{da}{d\beta} + \int_a^b \frac{\partial f}{\partial \beta}\, dx$

E. T. Whittaker and G. N. Watson, "Modern Analysis," p. 67, Macmillan, New York, (1946).

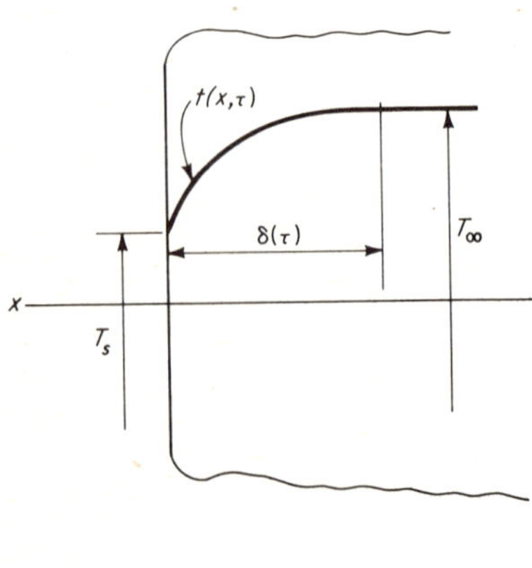

Fig. 1-9 Temperature distribution in a semi-infinite solid.

In Eq. (1-30), $t(\delta,\tau) = T_\infty$, in this case a constant, and

$$\frac{\partial t}{\partial x}(\delta,\tau) = 0$$

thus the equation becomes

$$\frac{d}{d\tau}\left[\int_0^\delta t(x,\tau)\,dx - T_\infty\delta(\tau)\right] = -\alpha\frac{\partial t}{\partial x}(0,\tau) \tag{1-31}$$

For cases in which the variation of the physical properties with temperature must be accounted for,

$$\rho c\frac{\partial t}{\partial \tau} = \frac{\partial}{\partial x}\left(k\frac{\partial t}{\partial x}\right) \tag{1-32}$$

where $\rho = \rho(t)$, $c = c(t)$, and $k = k(t)$.

A transformation in the dependent variable[1]

$$v = \int_0^t \rho c\,dt$$

will change Eq. (1-32) to read

$$\frac{\partial v}{\partial \tau} = \frac{\partial}{\partial x}\left[\alpha(v)\frac{\partial v}{\partial x}\right] \tag{1-33}$$

Integration of Eq. (1-33) from $x = 0$ to $x = \delta(\tau)$ produces

$$\frac{d}{d\tau}\left(\int_0^\delta v\,dx - V_\infty\delta\right) = -\alpha(v_0)\frac{\partial v}{\partial x}(0,\tau) \tag{1-34}$$

Both Eqs. (1-31) and (1-34) will be used in the following chapters to produce approximate solutions for transient heat conduction in semi-infinite bodies and slabs having both constant and variable thermophysical properties. In the integrated form of the heat-conduction equation it is necessary to describe the temperature distribution by means of a suitable analytical expression to evaluate the integrals and derivatives in Eqs. (1-31) and (1-34). The analytical temperature profile so selected must not only be physically representative of the actual temperature profile but also must satisfy the thermal boundary conditions under consideration.

1-5 THE HEAT-CONDUCTION EQUATION (VARIATIONAL FORM)

Variational methods have been used in mechanics for many years; however, such methods were until recently not much used in heat transfer.

[1] Other transformations have been suggested (see, for example, H. S. Carslaw and J. C. Jaeger, "Conduction of Heat in Solids," 2d ed., p. 89, Oxford University Press, New York, 1959); however, the transform here suggested by T. R. Goodman has the advantage in use with the integrated heat-conduction equation of requiring a specification of the thermal diffusivity at the surface $x = 0$ only.

Biot[1] in a series of papers beginning in 1955 formulated from the ideas of irreversible thermodynamics a variational equivalent of the heat-conduction equation that constituted a thermodynamical analogy to Hamilton's principle in mechanics and led to a lagrangian formulation of the heat-conduction problem in terms of generalized coordinates.

The methods formulated by Biot have seen some acceptance[2] as well as some sharply directed criticism,[3] which points out some of the theoretical weaknesses of the formulation. Despite the validity of the criticism, i.e., that the variational principle is not really stated and, as has been pointed out, does not exist in the classical sense, the Biot formulation is interesting and can be used to develop approximate solutions for a certain class of problems. It is for this reason that the method is included here.

For an isotropic body with thermal conductivity $k(x,y,z)$ and a heat capacity $c(x,y,z)$ in the absence of heat sources or sinks, the heat-conduction equation satisfied by the temperature $t(x,y,z,\tau)$ is given by Eq. (1-14) with Q' set to zero. Biot formulated a variational statement which is equivalent to Eq. (1-14).

The heat-flow vector field $\mathbf{H}$ is such that the heat-flux vector at every point is $\dfrac{\partial \mathbf{H}}{\partial \tau} = \mathbf{q}$. That energy is conserved is expressed by the statement

$$\rho c t = -\left(\frac{\partial \mathbf{H}}{\partial x} + \frac{\partial \mathbf{H}}{\partial y} + \frac{\partial \mathbf{H}}{\partial z}\right) = -\operatorname{div} \mathbf{H} \tag{1-35}$$

Biot defines a thermal potential

$$E = \frac{1}{2} \iiint\limits_V \rho c t^2 \, dV \tag{1-36}$$

[1] M. A. Biot, Theory of Stress-Strain Relations in Anisotropic Viscoelasticity and Relaxation Phenomena, *J. Appl. Phys.*, **25**:11 (1954); Variational Principles in Irreversible Thermodynamics with Applications to Viscoelasticity, *Phys. Rev.*, **97**:6 (1955); Thermoelasticity and Irreversible Thermodynamics, *J. Appl. Phys.*, **27**:3 (1956); New Methods in Heat Flow Analysis with Application to Flight Structures, *J. Aeron. Sci.*, **24**:12 (1957).

[2] T. J. Lardner, Biot's Variational Principle in Heat Conduction, *J. AIAA*, **1**:1 (1963); P. D. Richardson, Unsteady One-dimensional Heat Conduction with a Non-linear Boundary Condition, *J. Heat Transfer*, **86**:298–299 (1964); P. D. Richardson and W. W. Smith, Use of a Transcendental Approximation in Transient Conduction Analysis, private communication, 1965.

[3] B. A. Finlayson and L. E. Scriven, On the Search for Variational Principles, *Intern. J. Heat Mass Transfer*, **10**:6 (1967).

and a *variational invariant*

$$\delta D = \iiint_V \frac{1}{k} \frac{\partial \mathbf{H}}{\partial \tau} \cdot \delta \mathbf{H} \, dV \tag{1-37}$$

The thermal potential E plays a role analogous to a potential energy, while D is related to the concept of a dissipation function.

Using the specifications of Eqs. (1-36) and (1-37) the variational statement is expressed as

$$\delta E + \delta D = \iint_S t\mathbf{n} \cdot \delta \mathbf{H} \, dS \tag{1-38}$$

The surface integral is extended over the boundary surface S with the unit normal $\mathbf{n}$ to the boundary taken positive inward.

To show that Eq. (1-38) is equivalent to Eq. (1-14), it is necessary to evaluate the terms δE and δD

$$\delta E = \iiint_V \rho c t \, \delta t \, dV = -\iiint_V \delta t (\operatorname{div} \mathbf{H}) \, dV \tag{1-39}$$

and integrating by parts,

$$\delta E = \iiint_V \delta \mathbf{H} \cdot \operatorname{grad} t \, dV + \iint_S t\mathbf{n} \cdot \delta \mathbf{H} \, dS \tag{1-40}$$

Substituting Eqs. (1-40) and (1-37) into Eq. (1-38) yields

$$\iiint_V \left(\operatorname{grad} t + \frac{1}{k} \frac{\partial \mathbf{H}}{\partial \tau} \right) \cdot \delta \mathbf{H} \, dV = 0 \tag{1-41}$$

For arbitrary variations of $\delta \mathbf{H}$, Eq. (1-41) reduces to

$$\operatorname{grad} t + \frac{1}{k} \frac{\partial \mathbf{H}}{\partial \tau} = 0 \tag{1-42}$$

which is, using Eq. (1-35),

$$\operatorname{div} (k \operatorname{grad} t) = -\operatorname{div} \frac{\partial \mathbf{H}}{\partial \tau} = \rho c \frac{\partial t}{\partial \tau} \tag{1-43}$$

Equation (1-43) is identical with Eq. (1-14), and the equivalence is established.

As a further development Biot introduces the concept of generalized coordinates in conjunction with the variational statement. The heat-flow field $\mathbf{H}$ is shown as a function of certain generalized coordinates q_i,

$$\mathbf{H} = \mathbf{H}(q_1, \ldots, q_n, x, y, z) \tag{1-44}$$

where the generalized coordinates are a set of n parameters defining the configuration of the heat-flow field.

For an isotropic body the variations in $\delta\mathbf{H}$ are due entirely to the variations of q_i

$$\delta\mathbf{H} = \sum_i \frac{\partial \mathbf{H}}{\partial q_i}\,\delta q_i \tag{1-45}$$

also

$$\frac{\partial \mathbf{H}}{\partial \tau} = \sum_i \left(\frac{\partial \mathbf{H}}{\partial q_i}\right)\dot{q}_i \tag{1-46}$$

Hence,

$$\frac{\partial}{\partial \dot{q}_i}\frac{\partial \mathbf{H}}{\partial \tau} = \frac{\partial \mathbf{H}}{\partial q_i} \tag{1-47}$$

and the argument of Eq. (1-37) may be written

$$\frac{1}{k}\frac{\partial \mathbf{H}}{\partial \tau}\cdot\delta\mathbf{H} = \frac{1}{k}\sum_i \frac{\partial}{\partial \dot{q}_i}\left[\frac{1}{2}\left(\frac{\partial \mathbf{H}}{\partial \tau}\right)^2\right]\delta q_i \tag{1-48}$$

With the above results, Eq. (1-38) becomes

$$\delta E + \sum_i \left(\frac{\partial D}{\partial \dot{q}_i}\right)\delta q_i = \iint_S t\mathbf{n}\cdot\delta\mathbf{H}\,dS \tag{1-49}$$

where

$$D = \frac{1}{2}\iiint_V \frac{1}{k}\left(\frac{\partial \mathbf{H}}{\partial \tau}\right)^2 dV \tag{1-50}$$

and with

$$Q_i = \iint_S t\mathbf{n}\cdot\frac{\partial \mathbf{H}}{\partial q_i}\,\delta q_i\,dS \tag{1-51}$$

and

$$\delta E = \sum_i \left(\frac{\partial E}{\partial q_i}\right)\delta q_i \tag{1-52}$$

Equation (1-49) becomes

$$\frac{\partial E}{\partial q_i} + \frac{\partial D}{\partial \dot{q}_i} = Q_i \tag{1-53}$$

in analogy with the lagrangian equations for a dissipative mechanical system with a potential energy E and a dissipation function D. Biot refers to Q_i as a *generalized thermal force*, i.e., the work done by the temperature t on a virtual displacement of the heat-flow field $\delta\mathbf{H}$. D is referred to in purely thermal problems as the dissipation function.

In the case wherein the surface heat transfer by convection must be considered, with a heat-transfer coefficient h the heat flux through the boundary layer is (see Sec. 1-7)

$$\frac{\partial \mathbf{H}_n}{\partial \tau} = h(t_f - t_0) \tag{1-54}$$

where t_f is the fluid temperature and t_0 is the surface temperature. The normal component of the heat flow into the body is $\mathbf{H}_n$. The *thermal force* becomes

$$Q_i = \iint_S t \frac{\partial \mathbf{H}_n}{\partial q_i} dS = -\iint_S \frac{1}{h} \frac{\partial \mathbf{H}_n}{\partial \tau} \frac{\partial \mathbf{H}_n}{\partial q_i} dS + \iint t_f \frac{\partial \mathbf{H}_n}{\partial q_i} dS \quad (1\text{-}55)$$

The surface dissipation function becomes

$$D_s = \frac{1}{2} \iiint_V \frac{1}{k} \left(\frac{\partial \mathbf{H}}{\partial \tau} \right)^2 dV + \frac{1}{2} \iint_S \frac{1}{h} \left(\frac{\partial \mathbf{H}_n}{\partial \tau} \right)^2 dS \quad (1\text{-}56)$$

There then results the expression corresponding to Eq. (1-53)

$$\frac{\partial E}{\partial q_i} + \frac{\partial (D + D_s)}{\partial \dot{q}_i} = Q_i \quad (1\text{-}57)$$

For nonisotropic media Eqs. (1-53) and (1-57) will hold true provided

$$D = \frac{1}{2} \iiint_V \sum_{ij} \lambda_{ij} \frac{\partial \mathbf{H}_i}{\partial \tau} \frac{\partial \mathbf{H}_j}{\partial \tau} dV$$

where λ_{ij}, the inverse matrix of k_{ij}, is the *thermal resistivity*.

Applications of the Biot formulations will be discussed in later chapters.

1-6 THE HEAT-CONDUCTION EQUATION (HYPERBOLIC FORM)

Transient heat-conduction problems in the general technical literature relate to the solution of the Fourier heat-conduction equation, which is, for constant physical properties,

$$\frac{1}{\alpha} \frac{\partial t}{\partial \tau} = \nabla^2 t \quad (1\text{-}58)$$

Several investigators[1] have pointed out from time to time that the

[1] R. M. Morse and H. Feshbach, "Methods of Theoretical Physics," McGraw-Hill, New York, 1953; B. A. Boley, "The Analysis of Problems of Heat Conduction and Melting: High Temperature Structures and Materials," pp. 260–315, Pergamon, New York, 1964; C. R. Cattaneo, Sur une forme de l'équation de la chaleur éliminant le paradoxe d'une propagation instantanée, *Compt. Rend.*, **247**(4):431–433 (1958); P. Vernotte, Les Paradoxes de la theorie continue de l'équation de la chaleur, *Comp. Rend.*, **246**(22):3154–3155 (1958); R. E. Nettleton, Inertial Effects in the Phenomenological Theory of Thermal Diffusion in Liquids, *Nuovo Cimento*, **28**(5):952–969 (1963); U. S. Lindholm et al., Transient Heat Conduction at High Thermal Flux, *J. Heat Transfer*, **87**:49–52 (1965); S. Kaliski, Wave Equation in Heat Conduction, *Bull. Acad. Polon. Sci.*, **13**(4):211–219 (1965); A. V. Luikov, Applications of Irreversible Thermodynamics Methods to Investigations of Heat and Mass Transfer, *Intern. J. Heat Mass Transfer*, **9**(2):139–152 (1966); K. J. Baumeister and T. D. Hamill, Hyperbolic Heat Conduction Equation: A Solution for the Semi-infinite Body Problem, *J. Heat Transfer*, **91**(4):543–548 (1969).

parabolic form of Eq. (1-58) requires an infinite velocity of propagation of effect, i.e., that a portion of the solution extends to infinity. This implies that the effects of a thermal disturbance are instantaneously felt at a distance infinitely far from the location of the disturbance. That the effect at infinity is small does not remove this particular objection to the use of Eq. (1-58) on philosophical grounds or on practical grounds in certain instances, e.g., small elapsed time during transient behavior, where the effects may be important.

Morse and Feshbach[1] suggest that the governing differential equation for transient heat conduction should be the damped wave equation, often called the *telegraph equation*, which is *hyperbolic* instead of *parabolic* and admits to a velocity of propagation of heat C,

$$\frac{1}{C^2}\frac{\partial^2 t}{\partial \tau^2} + \frac{1}{\alpha}\frac{\partial t}{\partial \tau} = \nabla^2 t \tag{1-59}$$

$C^2 = \alpha/\tau_0$, where τ_0 is the *relaxation time.*

Boley[2] has pointed out that for most practical problems in heat conduction the effect of the finite speed of heat propagation in Eq. (1-59) is negligible; i.e., Eq. (1-58) is an acceptable approximation. Normally the thermal diffusivity α is 10 orders of magnitude smaller than the square of the velocity of propagation; however, for very low temperatures, or for very small times, the effect of the finite propagation speed could become important.

Tisza[3] in 1938 and Landau in 1941 predicted the possibility of wavelike heat flow with finite heat-propagation speeds in liquid helium II and called this effect *second sound* in analogy with the first sound of acoustic transmission. Peshkov[4] by means of a novel heat-transfer experiment later confirmed the prediction and obtained quantitative results. He found the speed of heat propagation to be 19 m/s at a temperature of 1.4 K, an order of magnitude smaller than the sound of speed in helium.

Chester[5] considers the possibility of second sound in solids. He justifies the existence of the propagation term in Eq. (1-59) by a hypothesis based on the results of the kinetic-theory calculations on the Boltzmann equations by Grad.[6]

[1] *Op. cit.*

[2] *Loc. cit.*

[3] K. R. Atkins, "Liquid Helium," Cambridge University Press, London, 1959.

[4] V. Peshkov, Second Sound in Helium II, *J. Phys. USSR*, **8**:381 (1944).

[5] M. Chester, Second Sound in Solids, *Phys. Rev.*, **131**(15):2013–2015 (1963).

[6] H. Grad, The Kinetic Theory of Rarefied Gases, *Commun. Pure Appl. Math.*, **2**(4): 331–407 (1949).

In most of the discussion herein the heat-flux vector is given as

$$\mathbf{q} = -k\,\nabla t \tag{1-60}$$

whereas to be consistent with Eq. (1-59) the heat-flux vector must be expressed as

$$\frac{\alpha}{C^2}\frac{\partial \mathbf{q}}{\partial \tau} + \mathbf{q} = -k\,\nabla t \tag{1-61}$$

Equation (1-61) was postulated by Chester as a truncation of a more general expression proposed by Grad for a rarefied gas. Consequently, for a solid or liquid, for example, Eq. (1-61) must be considered tentative, as it has not been established either experimentally or theoretically.

Baumeister and Hamill[1] and more recently Vujanovic and Strauss[2] have used Eq. (1-59) to obtain numerical and analytical results that show the expected differences from the inclusion of the propagation term. These are (1) the effect of finite speed correction is extremely small and will be negligible in most practical problems; (2) a step change in the temperature produces a thermal wave which travels at a finite speed C. The temperature discontinuity across the face of this wave is given by

$$\frac{t(C\tau,\tau) - t_0}{t_w - t_0} = e^{-C^2\tau/2\alpha}$$

and (3) the maximum heat flux in response to a step change in temperature in a heat-conduction system is finite.

Bertman and Sandiford[3] have reviewed past and recent work related to second sound in solids. Second sound in solid helium II has been determined, and indeed the effect might exist in other materials of great purity and at low temperatures; however, it is necessary to prepare materials of exceptional purity, otherwise trace impurities cause phonon scattering to such an extent that the transfer process is diffusional rather than wavelike. It thus seems that for most technical materials, which lack the requisite purity in the use state, the diffusional form of heat transfer will be the predominant one we observe physically, even at low temperatures. Thus the major practical importance of the wave-equation model will be in situations where there is interest in thermal effects taking place over very small time intervals.

Hyperbolic equation for heat conduction (variational form) The development by Vujanovic and Strauss[4] is of some importance in that it presents

[1] *Loc. cit.*

[2] B. Vujanovic and A. M. Strauss, Heat Transfer with Nonlinear Boundary Conditions via a Variational Principle, *J. AIAA*, **9**:2(327–330), 1971.

[3] B. Bertman and D. J. Sandiford, Second Sound in Solid Helium, *Sci. Amer.*, May, 1970.

[4] *Op. cit.*

a variational method for treating the heat-conduction equations which is free of the uncertainties associated with the Biot variational method. The variational method proposed by Vujanovic and Strauss corresponds to the general problem of heat conduction with finite wave speeds. The transition to the classical situation, i.e., where the speed of propagation of heat is infinite, is accomplished by allowing the relaxation time to approach zero.

The variational method developed is neither the quasi-variational formulation of the problem proposed by Biot[1] nor the restricted variational principle like that of Glansdorf and Prigogine.[2] By consideration of the heat-conduction equation with finite wave speeds [Eq. (1-59)] and making use of the *exact lagrangian,* the technique is applicable to nonlinear heat-conduction problems within the framework of the classical calculus of variations.

The main disadvantage of the method is that the error of approximation cannot be controlled; however, this is as well a disadvantage with other like methods, i.e., Biot's, Galerkin's, and the integral method.

The concept of generalized coordinates and the method of partial integration are basic to this method. By means of the variational method, nonlinear boundary-value problems can be reduced to linear boundary-value problems, the solutions to which can often be expressed in closed form.

By making use of the method of Rafalski and Zyszkowski[3] the boundary conditions are formulated from heat-balance equations at the body surface. These equations yield simple algebraic expressions rather than the differential expressions of Biot[4] and Lardner.[4]

Equation (1-59) can be written

$$\frac{\tau_0}{\alpha}\frac{\partial^2 t}{\partial \tau^2} + \frac{1}{\alpha}\frac{\partial t}{\partial \tau} = \nabla^2 t$$

where τ_0 is the *relaxation time* and $\alpha/\tau_0 = C^2$, the square of the speed of heat propagation.

It has been shown earlier that the heat-conduction equation possesses a lagrangian form. This lagrangian can be written as

$$L = \left[\tfrac{1}{2}\rho c\tau_0\left(\frac{\partial t}{\partial \tau}\right)^2 - \frac{k}{2}\sum_i\left(\frac{\partial t}{\partial x_i}\right)^2\right]e^{\tau/\tau_0} \tag{1-62}$$

[1] *Loc. cit.*

[2] Finlayson and Scriven, *loc. cit.*

[3] P. Rafalski and W. Zyszkowski, Lagrangian Approach to the Non-linear Boundary Heat Transfer Problem, *J. AIAA,* **6**:8 (1968).

[4] *Loc. cit.*

The heat-conduction equation can be obtained by substituting the lagrangian into the Euler-Lagrange equation

$$\frac{\partial L}{\partial t} - \frac{\partial}{\partial \tau} \frac{\partial L}{\partial(\partial t/\partial \tau)} - \frac{\partial}{\partial x_i} \frac{\partial L}{\partial(\partial t/\partial x_i)} = 0 \tag{1-63}$$

One can obtain solutions to Eq. (1-59) by making use of the lagrangian [Eq. (1-62)] and by performing the variation of the *action integral*

$$I = \int_{\tau}\int_{V} L\, dV\, d\tau \tag{1-64}$$

or solutions to Eq. (1-58) by following the same process, then letting $\tau_0 \rightarrow 0$.

It is tacitly assumed that no natural boundary conditions exist. This assumption is equivalent to the statement that the temperature variation δt vanishes on the boundary.

One advantage of the above formulation is immediately evident. The problem can be stated in terms of the lagrangian equations in generalized coordinates which define the heat-flow field. Initially, in a simple one-dimensional heat-conduction problem, for example, two generalized coordinates are defined: (1) the penetration distance δ and (2) the surface temperature t_s. The penetration distance is considered as an independent generalized coordinate to be determined by the variational procedure. The generalized coordinate t_s, the surface temperature, can be determined from the relationship

$$\int_S G(S,t,\tau) g(S)\, dS = 0 \tag{1-65}$$

The function $G(S,t,\tau)$ is determined from the prescribed boundary condition

$$G(S,t,\tau) = 0 \tag{1-66}$$

where S is the surface of the body and

$$g(S) \geq 0 \tag{1-67}$$

$g(S)$ can be interpreted as a statement expressing the accuracy desired of the approximation on the surface S.

The method of Vujanovic and Strauss will be applied in an example in Chap. 4.

1-7 HEAT CONVECTION FROM BOUNDING SURFACES

Although the study of heat conduction has as its primary objective the specification of the temperature distribution within the conducting body and the heat-flow rate within the body, surface conditions are most important. In the determination of the temperature distribution, the boundary conditions can be either a specification of the surface temperatures directly or a specification for the heat rate to or from the surface. In the cases where the heat-flow rate is desired, the convection conditions on the surface are often the controlling factors; i.e., the heat-flow resistance in the conducting body is much smaller than the resistance due to convection at the surface, with the result that conditions on the surface, and not those within the body itself, regulate the heat flow.

The general problem of convective heat transfer is rather complicated, with dependence on several variables, and as such will be treated in detail in Part II. However, it will be necessary in the developments that follow to make use of certain convection boundary conditions and for this reason to show here that certain similarities exist between the gross convection problem and pure conduction.

In the instance where a fluid flows over a solid surface, the temperature field in the fluid appears somewhat as shown in Fig. 1-10. The temperature gradient in the fluid is confined to a relatively narrow layer of thickness δ quite close to the wall, whereas at a greater distance from the wall in most cases only small temperature differences exist. Simplified, the temperature profile can be replaced by the dashed line. Physically this can be explained by the assumption that a thin film of fluid of thickness δ' adheres to the wall and through it the total temperature change takes place. Outside the film all temperature differences vanish

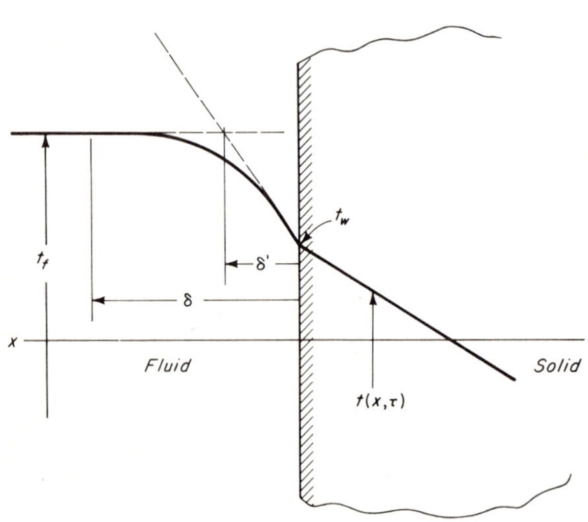

Fig. 1-10 Steady convective heat transfer and heat conduction at a bounding surface.

as a result of mixing motion of the fluid. Within the film the heat transfer is assumed to take place by pure conduction, as in the solid wall. This picture greatly oversimplifies the actual mechanism, but it does bring out the salient features.

The temperature distribution in the film is supposed linear, and the flow of heat follows from Eq. (1-1), into which the thermal conductivity k_f of the fluid and the thickness of the film δ' are to be inserted for k and d, respectively. Thus for the heat flow to the wall the equation

$$Q = \frac{k_f}{\delta'} A(t_f - t_w) \tag{1-68}$$

is obtained. From this equation the rate of heat flow Q can be calculated as soon as the film thickness δ' is known. The latter depends to a very great extent upon the external flow conditions, e.g., the velocity with which the fluid flows along the surface, the shape of the surface, the structure of the surface, and several other factors. In practice it has become convenient to calculate with the expression k_f/δ' rather than directly with the film thickness δ'. The ratio k_f/δ' is called the *convective-heat-transfer coefficient* h. Thus the expression

$$Q = hA(t_f - t_w) \tag{1-69}$$

as formulated by Isaac Newton (1643–1727) is obtained. For some time after its introduction the heat-transfer coefficient h was considered a property of the flowing fluid. It was not until the more recent developments in the theory of heat transfer that the truly intricate relationships that make the calculation of the heat-transfer coefficient possible became apparent. For the purpose of establishing convection boundary conditions for the problems in heat conduction, the heat-transfer coefficient as defined by Eq. (1-69) will be used.

2

Thermal Conductivity[1]

2-1 INTRODUCTION

One might anticipate from the formulations of the preceding chapter that the thermal conductivity is an important property of materials. Knowledge of the variation of the thermal conductivity with the physical structure of materials and its variation with the measurable thermodynamic variables, principally pressure, temperature, and composition, is of utmost importance to meaningful heat-transfer calculations and experiments.

Although thermal conductivity has been recognized, measured, and numerically tabulated for many years, only recently has the true nature of the property been understood in terms of the kinetic theory of matter. From a physical, microscopic point of view thermal conductivity is a

[1] L. S. Kowalczyk's survey, Thermal Conductivity and Its Variability with Temperature and Pressure, *Trans. ASME*, **77**:1021–1035 (1955), was used extensively in the preparation of this chapter, with the kind permission of Professor Kowalczyk; See also Max Jakob, "Heat Transfer," vol. I, chap. 6, Wiley, New York, 1949.

nonequilibrium property; therefore, its determination is based on measurements of the time necessary to restore thermal equilibrium, or *relaxation time.* Macroscopically, the thermal conductivity is defined in terms of the Fourier heat-conduction law

$$k = \frac{q_n}{-\partial t/\partial n} \tag{2-1}$$

which states that the thermal conductivity is the amount of heat conducted per unit time per unit area per unit negative temperature gradient. In the units of the Système International (SI) the dimensions of thermal conductivity become

$$\frac{\text{J}}{\text{s m K}} = \frac{\text{W}}{\text{m K}}$$

while in the English engineering system the dimensions are

$$\frac{\text{Btu}}{\text{hr ft F}}$$

In this book the values of the thermal conductivity will be given in SI units unless noted otherwise.

The magnitude of thermal conductivity for materials of technical interest at any given temperature varies over a range of a millionfold, from the value for chloroform vapor to the value for a natural crystal of copper. Some comparative values for the thermal conductivities of various technical materials for a temperature of 273 K are shown in Table 2-1.

Values of thermal conductivity much in excess of those shown in Table 2-1 have been measured in several investigations of single crystals of both metallic and nonmetallic substances. The highest thermal conductivity reported is 12,200 W/m K measured by Schott[1] for a natural copper crystal at a temperature of 20.8 K. Andrews, Webber, and Spohr[2] report the value 6,490 W/m K for a 99.996 percent pure crystal of aluminum at 14 K. Berman, Simon, and Wilks[3] report the value 6,000 W/m K for the thermal conductivity of a single crystal of synthetic sapphire, Al_2O_3, at a temperature of 50 K.

[1] H. Landolt, "Landolt-Bornstein, Physikalischchemische Tabellen, p. 1291, Springer-Verlag, Berlin, 1923.

[2] F. A. Andrews, R. T. Webber, and D. H. Spohr, Thermal Conductivities of Pure Metals at Low Temperatures. I. Aluminum, *Phys. Rev.*, **84**:994–996 (1951).

[3] R. Berman, F. E. Simon, and J. Wilks, Thermal Conductivity of Dielectric Crystals: The "Umklapp" Process, *Nature*, **168**:277–280 (1951).

Table 2-1 Thermal conductivities of selected materials at 273 K

Material	W/m K
Metallic solids:	
Silver (purest)	418
Copper (pure)	387
Aluminum (pure)	203
Zinc (pure)	112.7
Iron (pure)	73
Tin (pure)	66
Lead (pure)	34.7
Nonmetallic solids:	
Periclase, MgO	41.6
Quartz (parallel to axis)	19.1
Corundum, Al_2O_3	10.4
Marble	2.78
Ice, H_2O	2.22
Quartz (fused)	1.91
Pyrex glass	1.05
Liquids:	
Mercury	8.21
Water	0.552
Sulfur dioxide, SO_2	0.211
Methyl chloride, CH_3Cl	0.178
Carbon dioxide, CO_2	0.105
Freon, CCl_2F_2	0.0728
Gases:	
Hydrogen	0.175
Helium	0.141
Air	0.0243
Pentane	0.0128
Chloroform	0.0066

A comparison of the values reported for the thermal conductivity of various materials reveals several general facts about the nature of the property.

1. Material in a crystalline form, either metallic or nonmetallic, conducts heat better than material in an amorphous form.
2. In crystalline and other material of oriented structure, e.g., fibrous material like wood, the thermal conductivity has different values relative to the structural axes of the material, such that there are principal axes for the thermal conductivity (see Chap. 1).
3. Small structural differences in crystals related to their means of growth influence their thermal conductivity. For this reason pure, natural

crystals have higher conductivities than synthetic varieties. Also various types of mica[1] and graphite[2] show different conductivities. The variation in graphites is shown in Fig. 2-1.

[1] R. W. Powell and E. Griffiths, The Variation with Temperature of the Thermal Conductivities of Some Micas, *Proc. Royal Soc. London, Ser. A*, **163**:189–198 (1937).

[2] R. W. Powell, Thermal Conductivities of Metallic Conductors and Their Estimation, *Proc. Gen. Discussion Heat Transfer, ASME-IME, London, 1951*, pp. 290–295.

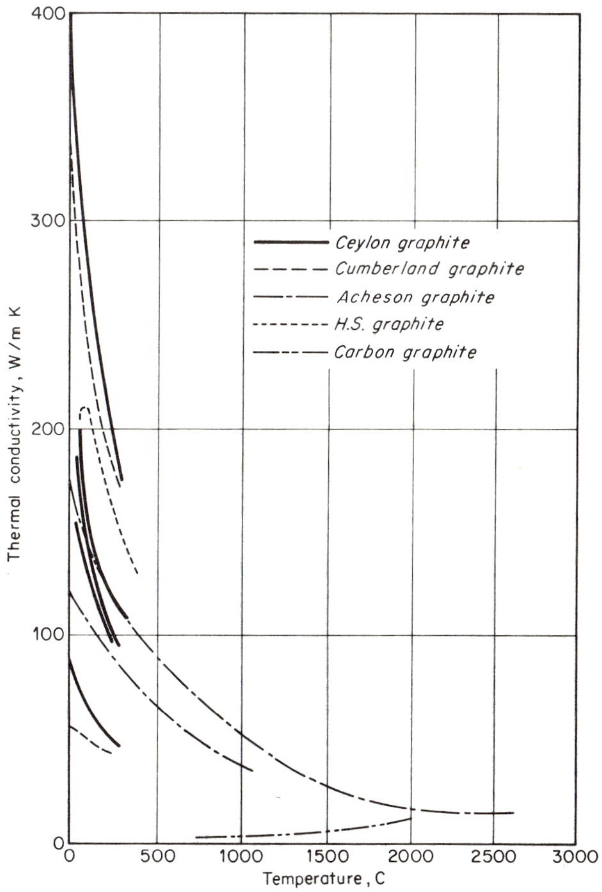

Fig. 2-1 Variation of thermal conductivity of graphites with temperature: ______ Ceylon graphite; _ _ _ _ _ Cumberland graphite; _._._._ Acheson graphite; .... Hilger "H.S." graphite; _.._.._ carbon graphite. (*R. W. Powell, Proc. Gen. Discussion Heat Transfer, ASME-IME, London, 1951.*)

4. Chemical impurities in crystalline substances result in lower thermal conductivities compared with the pure state. Pure metals have much higher thermal conductivities than their respective alloys.
5. Mechanical damage, such as cold-working and nuclear-irradiation damage, causes changes in the thermal conductivity of a material.
6. Generally speaking, metals are better heat conductors than nonmetals.
7. Solid phases of materials have higher conductivities than their respective liquid phases. For example, ice has a higher thermal conductivity than water; the solid phase of lead has a higher conductivity than the liquid phase of lead. (An exception may be noted in bismuth, where the liquid phase has a higher conductivity than the solid phase.)
8. Liquid phases show higher thermal conductivities than gaseous phases.

These facts serve to demonstrate that the transport property, thermal conductivity, is closely related to the physicochemical structure of matter and that the phenomena associated with the thermal conductivity may be explained on the basis of our knowledge of the nature of heat and the structure of matter.

2-2 THE NATURE OF THERMAL CONDUCTIVITY

It is worthwhile to discuss in some detail the mechanism of heat transfer by conduction in terms of the modern kinetic theory of heat and its relation to the structure of matter in the solid, liquid, and gaseous phases. The important points to be delineated are as follows:

1. The marked differences between metals and nonmetals
2. The differences between the solid, liquid, and gaseous phases
3. The relationship between the electric and thermal conductivities in metals, the *Wiedemann-Franz law*
4. The relationship between the viscosity and thermal conductivity of materials in the gaseous state
5. The effect on thermal conductivity of impurities, cold-working, radiation damage, and alloying
6. The dependence of thermal conductivity on temperature, pressure, and composition.

SOLIDS

The modern theory of heat conduction in solids makes a clear distinction between metallic and nonmetallic (dielectric) materials. In metallic conductors, and other electrically conducting solids, heat is carried simultaneously by means of lattice waves, *phonons*, and *free* (*or valence*) *electrons*.

In the nonmetallic, or dielectric material, heat is carried only by phonons. (A phonon is equivalent to one quantum of energy in the form of a thermoelastic wave of a fixed frequency. In this sense a phonon is analogous to the concept of a photon in electromagnetic radiation theory.)

Nonmetallic (dielectric) solids The thermal conductivity of nonmetallic electric nonconductors is due entirely to the thermal vibrations of the crystalline lattice, which can be regarded as a superposition of elastic (acoustic) waves. If a crystal has two faces at different temperatures, heat energy is transferred from the hotter face to the colder face by acoustic radiation in a manner analogous to the transmission of energy by electromagnetic waves through space.

If one resorts to classical kinetic theory and considers the discrete energy packets (phonons) as particles, an expression for the thermal conductivity can be derived. Consider the transfer of energy by particles crossing the xy plane, where the temperature gradient is along the z axis. A particle traveling a distance equal to the mean free path λ and striking the xy plane at an angle θ has a mean energy

$$e(0) + \lambda \cos \theta \frac{\partial e}{\partial z} \tag{2-2}$$

where $e(0)$ is the mean energy at the plane $z = 0$. From the kinetic theory of gases[1] the number of particles which cross a unit area of the plane xy per unit time in a direction making an angle $\theta + d\theta$ with the z axis is

$$\tfrac{1}{2} n\bar{w} \cos \theta \sin \theta \, d\theta \tag{2-3}$$

where n is the number of particles per unit volume and $\bar{w}$ is the average velocity. The net energy flux is

$$\tfrac{1}{2} n\bar{w} \frac{\partial e}{\partial z} \int_0^\pi \lambda \cos^2 \theta \sin \theta \, d\theta = \tfrac{1}{3} n\bar{w}\lambda \frac{\partial e}{\partial z} \tag{2-4}$$

The results of Eq. (2-4) when interpreted in terms of the definition of the thermal conductivity must be equal to

$$k \frac{\partial t}{\partial z} = \tfrac{1}{3} n\bar{w}\lambda \frac{\partial e}{\partial z} \tag{2-5}$$

from which the thermal conductivity k becomes

$$k = \tfrac{1}{3} n\bar{w}\lambda \frac{\partial e}{\partial t} = \tfrac{1}{3} \rho\bar{w}\lambda c_v \tag{2-6}$$

[1] See E. H. Kennard, "Kinetic Theory of Gases," McGraw-Hill, New York, 1938, or J. F. Lee, F. W. Sears, and D. L. Turcotte, "Statistical Thermodynamics," Addison-Wesley, Reading, Mass., 1963.

where ρ is the density of particles and c_v is the specific heat at constant volume.

In Debye's theory for the thermal conductivity of solid nonconductors the fundamental carriers of thermal energy are sound waves; thus the velocity $\bar{w}$ in Eq. (2-6) is identified with the velocity of sound in the material. (According to this theory, the specific heat of a body is the rate of change of the energy of the thermal-acoustic waves with temperature.)

The heat transfer may be visualized as taking place by the exchange of discrete packets of sound energy, which, if they were not attenuated, would result in an infinite value for the thermal conductivity. If, on the other hand, the packets of energy are scattered such that the attenuation in traversing some distance z is $1 - \exp^{-z/\lambda}$, then λ will be the effective mean-free-path length. The attenuation that takes place is the result of the sound waves scattering each other. One can interpret the attenuation resulting from the interaction of the thermal-acoustic waves in terms of a thermal resistivity R, which is the reciprocal of the thermal conductivity

$$R = \frac{3}{\rho \bar{w} \lambda c_v} \tag{2-7}$$

Equation (2-7) shows that the thermal resistance is inversely proportional to the mean-free-path length; thus strong scattering interaction among the phonons leads to small mean-free-path lengths and large resistance to the energy flow.

In *ideal crystals* of pure materials the displacement of an atom from its equilibrium position is proportional to the force causing the displacement; therefore, the atomic vibrations are harmonic, and any number of lattice waves can be superimposed upon one another without mutual interference. For this reason the phonons in *ideal crystals* are not scattered. From Eq. (2-7) the mean free path should be very large, the resistivity small, and the conductivity very large. The mean free path is limited solely by the collisions of phonons with the crystal boundaries. When λ becomes comparable with the smallest thickness of the test specimen, the value of λ is limited by that thickness and the conductivity becomes a function of the dimensions of the specimen. This effect was discovered by de Haas and Biermasz,[1] and the explanation was given by Casimir.[2] The effect can be seen from the data in Fig. 2-2, where at temperatures below about 7 K the resistivity becomes a function of the size of the specimen. The abrupt decrease in thermal conductivity of

[1] W. J. de Haas and T. Biermasz, Dependence on Thickness of the Thermal Resistance of Crystals at Low Temperatures, *Physica*, **5**:619–624 (1938).

[2] H. B. G. Casimir, Note on the Conduction of Heat in Crystals, *Physica*, **5**:495–500 (1938).

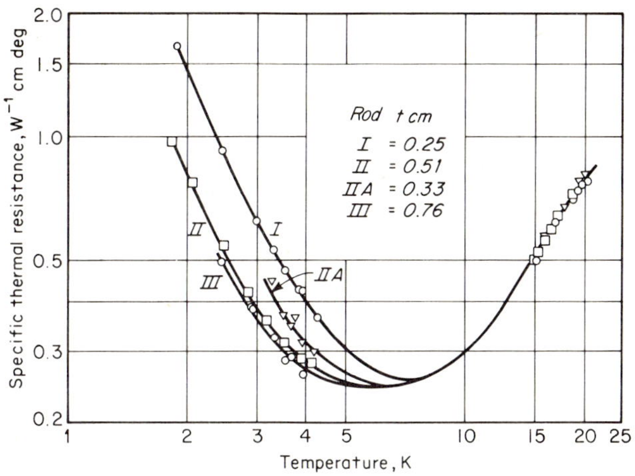

Fig. 2-2 Thermal resistivity of single crystal of potassium chloride as measured by Biermasz and de Haas. Below 5 K the resistivity is a function of the crystal thickness t. [*W. J. de Haas and T. Biermasz, Physica,* **5**:*619–624* (*1938*).]

pure crystals at low temperatures is caused by the size effect. One would expect that the phonon mean free path would be of the order of the smallest linear dimension d of the crystal; thus the conductivity k becomes

$$k = \tfrac{1}{3}\rho c_v \bar{w} d \tag{2-8}$$

The only temperature-dependent terms on the right are ρ and c_v, which as a product represents the heat capacity per unit volume. The heat capacity per unit volume varies as T^3 at low temperatures; thus it may be expected that the thermal conductivity will vary as T^3 at low temperatures. Experimental results generally confirm Eq. (2-8), but the exponent of T falls in the range between 2 and 2.5, and the value of λ is often much smaller than d. The smaller-than-expected value of the exponent of T is attributed to impurities. The sometimes small value of λ has no plausible explanation as yet.[1]

In a lattice in which the force on an atom is proportional to its displacement from its equilibrium position, any number of lattice waves can be superimposed without mutual interference. In an *actual crystal,* however, the force is not strictly proportional to the atomic displacement. Experimentally this can be shown by the fact that the compression of a body under a hydrostatically imposed pressure is less than its expansion caused by a hydrostatically imposed tension. The linear force-

[1] R. Berman, *Advan. Phys.,* **2**:103 (1953).

displacement relationship $F = Bx$ must be replaced by the expression $F = (B + Cx)x$, where C is the measure of the anharmonicity. With anharmonic lattice interactions there is a coupling between the different phonons which serves to limit the magnitude of the mean free path λ.

The theory of the effect of anharmonic coupling on thermal conductivity is one of the most complicated problems in solid-state physics. The problem has been treated in an approximate way by Debye,[1] and Peierls[2] has considered the problem in detail. Briefly, both found the scattering intensity to be proportional to the mean-square displacement, which was found proportional to the temperature T. The mean free path λ was found to be inversely proportional to T. At ordinary temperatures and above, the specific heat and the sound velocity are reasonably constant; thus the thermal conductivity k varies inversely with the temperature T. This dependence is confirmed by a number of experiments on nonmetallic, crystalline substances.

At low temperatures the mean free path λ is found to be approximately proportional to $\exp(-\Theta/2T)$, where Θ is the *Debye temperature.* Qualitatively we may follow the reasoning in Peierls' theory which lead to the prediction, $\lambda \sim \exp(-\Theta/2T)$. Peierls' theory admits of two types of phonon-phonon interaction. One type of collision process among phonons is that in which the total wave vector of the phonons is conserved; i.e., two phonons 1 and 2 collide to give rise to a single phonon 3

$$K_1 + K_2 = K_3$$

Collisions of this type do not change the direction of the energy flow. The second type of collision process has the form

$$K_1 + K_2 = K_3 + G$$

where G is 2π times a reciprocal lattice vector. This type of collision, which Peierls called an *Umklapp process*, gives rise to thermal resistance. The energy associated with suitable phonons K_1 and K_2 has been estimated to be of the order $\mathbf{k}\Theta/2$, where $\mathbf{k}$ is the Boltzmann constant. At low temperature the number of suitable phonons with the required energies as high as $\mathbf{k}\Theta/2$ has been estimated to be proportional to $\exp(-\Theta/2T)$. These results have also been confirmed generally by experiments.[3]

In addition to being an important mechanism in the determination

[1] P. Debye, "Vortrage über der kinetische Theorie der Materie und der Electrizität," Teubner, Berlin, 1941.

[2] R. Peierls, Zur kinetischen Theorie der Wärmeleitung in Kristallen, *Ann. Physik*, **3**:1055–1101 (1929); also Quelques propriétés typiques des corps solides, *Ann. Inst. Henri Poincaré*, **5**:177–222 (1935).

[3] Berman, Simon, and Wilks, *loc. cit.*

of the phonon mean-free-path length, anharmonic interaction is the only mechanism in solids by means of which thermal equilibrium can be established. In considering the transport property, thermal conductivity, it is necessary not only to have a means of limiting the mean free path but also a way of establishing an equilibrium distribution for the phonons at a given temperature.

In actual crystals, in addition to the *Umklapp process*, other scattering processes are identifiable. These are *elastic scattering*, *inelastic scattering*, *impurity scattering*, and *mosaic-structure scattering*—all essentially scattering phenomena which arise because of small structural anomalies in the crystal lattice.

Elastic and inelastic scattering result from natural or induced defects in the lattice. In elastic as compared with inelastic scattering the energy of the phonons is conserved.

It has been shown experimentally that irregularities in the position of the atoms in the lattice, *small-scale lattice defects*, serve to scatter phonons and decrease the thermal conductivity. The effect of fast neutron irradiation of quartz crystal is shown in Fig. 2-3. The radiation causes displacement of the atoms, gradually converting the ordered crystalline quartz into the disordered form of a dilute glass.

The effect of cold-working on specimens of 90/10 cupronickel alloys

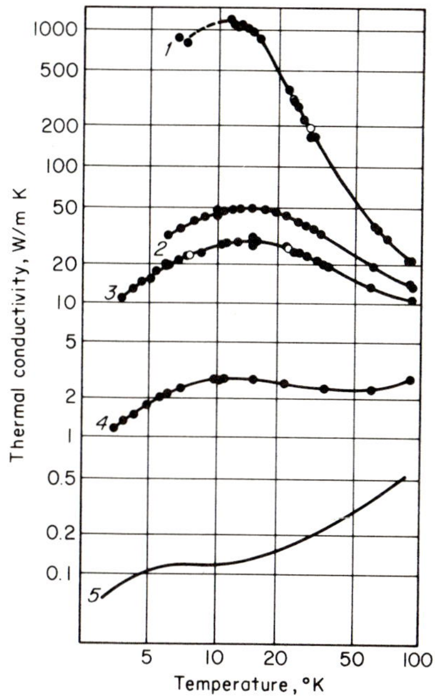

Fig. 2-3 Thermal conductivity of quartz crystal, irradiated quartz crystal, and quartz glass: (1) quartz crystal perpendicular to axis 5 mm² cross section; (2) concentration of displaced atoms 1×10^{-4} cm^{-3}; (3) 2×10^{-4}; (4) 2×10^{-3}; (5) quartz glass. [*R. Berman, P. G. Klemens, F. E. Simon, and T. M. Fry, Nature,* **166**:*864* (*1950*).]

in the temperature range of 10 to 20 K has been shown by experiment to have reduced the thermal conductivity to one-third the value of that for a single crystal.[1] These results may be attributed to the stronger scattering of the phonons by the dislocations and short-range lattice defects which were the result of the cold-working. On the other hand, after annealing, the specimen showed an increased thermal conductivity, indicating that the annealing process allowed a relaxation to a certain degree of the induced lattice defects.

Larger defects in the crystal lattice, such as mechanical breaks and cracks, contribute further to phonon scattering.

Chemical impurities in an otherwise pure material are known to be responsible for phonon scattering. Chemical impurities introduce foreign atoms into the crystal lattice, and these in turn scatter the phonons. Measurements of the thermal resistance induced by chromium atoms in an artificial sapphire in the temperature range between 5 and 100 K show that the impure crystal (ruby) has a lower thermal conductivity than the pure sapphire.[2] Klemens[3] has considered the problem of lattice-defect scattering in detail and concludes that the presence of impurities is the most important factor in determining the conductivity of potassium chloride and potassium bromide between 10 and 90 K. It has been suggested by Kittel that isotopic masses distributed in chemical elements might result in similar scattering phenomena.[4]

Phonon scattering of a higher order is caused by the *mosaic structure* of a crystal. The mosaic structure is a result of the division of the crystal into regions of slightly different orientations. It is a departure from the ideal regularity in the crystal lattice and may result in phonon scattering.[5]

In *glassy*, or *amorphous, materials* the atoms or molecules are distributed in a manner lacking both symmetry and periodicity. X-ray diffraction experiments reveal that the atomic arrangement in glassy materials shows short-range order but no long-range order as in crystals. Schematic analogs shown in Fig. 2-4 illustrate the structural differences between crystals and glasses. The present concept of the nature of the

[1] I. Estermann and J. E. Zimmerman, Effect of Lattice Defects on the Thermal Conductivity of Certain Alloys at Low Temperatures, "Problems Relating to Thermal Conductivity," *Annexe 1952–1 Bull. Inst. Intern. Froid,* p. 33.

[2] R. Berman, The Thermal Conductivity of Disordered Solids at Low Temperatures, in *ibid.*, p. 13.

[3] P. G. Klemens, The Thermal Conductivity of Dielectric Solids at Low Temperatures, *Proc. Roy. Soc. London, Ser. A,* **208**:108–133 (1951).

[4] C. Kittel, "Introduction to Solid State Physics," 2d ed., p. 145, Wiley, New York, 1956.

[5] J. Pomeranchuk, Thermal Conductivity of Dielectrics at Temperatures below the Debye Temperature, *J. Phys. USSR,* **6**:237–250 (1942).

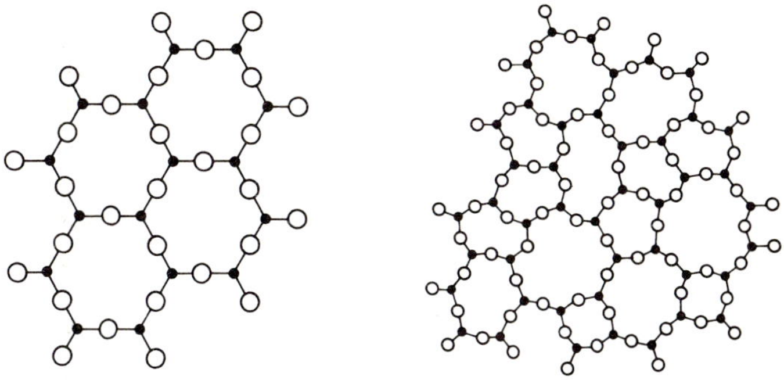

(a) (b)

Fig. 2-4 Schematic two-dimensional analogs illustrating (*a*) the regularly repeating structure of a crystal and (*b*) the random network of a glass. [*W. H. Zachariasen, J. Amer. Chem. Soc.*, **54**: *3841* (*1932*).]

glassy state pictures a glass, such as fused quartz, as a random but continuous network of silicon-oxygen bonds. The effective crystallite size is the order of magnitude of a single tetrahedron of the structure. Except at low temperatures the phonon mean free path will be limited essentially by the size of the crystallite. The decline of the thermal conductivity as the temperature is lowered may be attributed to the decline in the heat capacity.[1] As can be noted from Fig. 2-3, there is a distinct difference between the thermal conductivity of quartz crystal and quartz glass, the former being considerably the larger. Even at room temperature the difference is an order of magnitude.

Refractory materials may be considered as mixtures of crystalline and amorphous materials. The thermal conductivity of these materials will depend upon the thermal conductivities and the volume fractions of the components.

The remaining solid materials, *building and insulating materials, plastics, rubber, leather, fibrous and granular materials, powders*, etc., are heterogeneous in composition. In a broad sense they are mixtures of solid amorphous materials or amorphous solids with fluids. The observed thermal conductivity is an *apparent thermal conductivity* which in addition to the conductivities and volume fractions of the various components may depend upon convective and radiation energy exchange in the interstices.[2]

[1] C. Kittel, Interpretation of Thermal Conductivity of Glasses, *Phys. Rev.*, **75**:972–974 (1949); see also Klemens, *loc. cit.*

[2] E. A. Allcut, An Analysis of Heat Transfer through Thermal Insulating Materials, *Proc. Gen. Discussion Heat Transfer, ASME-IME, London, 1951*, pp. 232–235.

Metallic solids The thermal conductivity of metals and semiconductors depends upon two separate mechanisms: (1) energy transfer by means of lattice waves, *phonons*, and (2) energy transfer by means of *free* (*valence*) *electrons*.

The phonon contribution to the thermal conductivity has been discussed in relation to nonmetallic solids. The role played by the phonon transport of heat in metallic solids is essentially the same as that in nonmetallic solids; however, in metallic solids there exists the possibility of additional simultaneous and independent transport of energy by the valence electrons. According to the laws of parallel circuits, therefore, the total conductivity in a metallic solid is the sum of the phonon contribution and the electron contribution

$$k = k_p + k_e \tag{2-9}$$

The first point to decide in discussing the thermal conductivity of metals (or semiconductors) is the ratio in which the two means of energy transport, the phonons and the electrons, carry the heat current.

The scattering of lattice waves by various means as discussed for dielectric solids led to the concept of lattice resistance to heat flow. Similarly the scattering of the valence electrons leads to the concept of an electronic resistance to heat flow. The electrons are scattered by the thermal vibrations of the lattice, by impurities, and by other small-scale defects. In metals the phonons are also scattered by the electrons. It is clear that collisions between electrons and phonons are a matter of some importance.

A theoretical treatment of the electronic contribution to the thermal conductivity of metals is given by Sondheimer,[1] de Boer,[2] and others.

In pure metals the electronic contribution to the thermal conductivity is about 30 times the phonon contribution at room temperature. It can be seen from Table 2-1 that metallic solids tend to have thermal conductivities one to two orders of magnitude higher than dielectric solids.

In disordered alloys the mean free paths of both the phonons and the electrons are essentially limited by the scale of the disorder; therefore, the electron and phonon contributions to the thermal transport are the same order of magnitude. Measurements of thermal conductivities of three alloys, *german silver*, *stainless steel*, and *constantan*, between 2 and 90 K confirm the estimate that phonon and electron contributions are of

[1] E. H. Sondheimer, The Theory of the Transport Phenomena in Metals, *Proc. Roy. Soc. London, Ser. A*, **203**:75–98 (1950).

[2] J. de Boer, The Theory of the Heat Conductivity in Solids, "Problems Relating to Thermal Conductivity," *Annexe 1952–1 Bull. Inst. Intern. Froid*, p. 21.

equal order.[1] Makinson[2] has presented a careful treatment of the relative magnitudes of the electron and phonon heat currents for various metals over a wide range of temperature with results that generally confirm the above estimates.

There are several methods for the determination of the lattice conductivity in electric conductors independently of the electronic conductivity.

At room temperature and above, the thermal and electric conductivities of pure metals are related by the *Wiedemann-Franz law*, which came from the early electron-gas theory and which survives almost unchanged in more recent theories. The Wiedemann-Franz law is important because it relates the thermal conductivity to the electric conductivity, about which much more is known. When the phonon transport of energy is negligible,

$$k = \mathrm{Lo}\ \sigma T \tag{2-10}$$

where k = thermal conductivity
σ = electric conductivity
T = absolute temperature
Lo = Lorentz constant = 2.45×10^{-8} W Ω/K^2

If a similar mechanism for the transport of heat and electricity by the free electrons exists, then one concludes that

$$k_e = \mathrm{Lo}\ \sigma T \tag{2-11}$$

Substitution of Eq. (2-11) in Eq. (2-9) allows the calculation of the phonon contribution to the conductivity

$$k_p = k - \mathrm{Lo}\ \sigma T \tag{2-12}$$

This method is valid only in the temperature range for which the Wiedemann-Franz law is valid, i.e., room temperature and above.

Another method makes use of the fact that a strong magnetic field can considerably reduce the electric conductivity in metals and thus that part of the thermal conductivity. In the proper magnetic-field environment, therefore, the measured thermal conductivity will be only the lattice conductivity k_p.

Another feature of thermal conductivity that enables one to sepa-

[1] R. Berman, The Thermal Conductivity of Some Alloys at Low Temperatures, *Phil. Mag.*, **42**:642 (1951).

[2] R. E. B. Makinson, The Thermal Conductivity of Metals, *Proc. Cambridge Phil. Soc.*, **34**:474–497 (1938).

rate the electronic and lattice contributions is the fact that the conductivity of crystals depends upon the grain size, as earlier noted. Since k_e is independent of the grain size, i.e., the electrons pass grain boundaries without scattering, the grain boundary affects only k_p. A set of measurements on the same material but with different grain size allows the differential determination of k_p.

Alloys The thermal conductivity of alloys, like that of pure metals, is the sum of the lattice conductivity and the electronic conductivity; however, since alloys have different structural characteristics, and since phonons are sensitive to structural scattering and the electrons to impurity and small-scale-defect scattering, it is necessary to discuss the various alloys individually. Four common examples of alloys will be discussed: (1) *continuous solid solutions*, (2) *eutectics*, (3) *limited solid solutions*, and (4) *intermetallic compounds.*

In alloys which are *continuous solid solutions* the variation of thermal conductivity with composition is shown schematically in Fig. 2-5, where A and B represent two pure metals. In this class of alloy the smallest addition of the alloying metal into the solvent metal acts as an impurity and causes the scattering of both phonons and electrons, which results in the sharp drop down in the thermal conductivity. When the amount of alloying metal is increased, the phonon transport is completely surpressed and the alloy acts more like an amorphous material, where the scattering is largely inelastic. Further addition of alloying material has a negligible effect on the structure of the alloy, and the conductivity becomes independent of composition, as shown in Fig. 2-5. The greater the structural differences in the solvent metal and the solute metal, the greater the drop in conductivity.

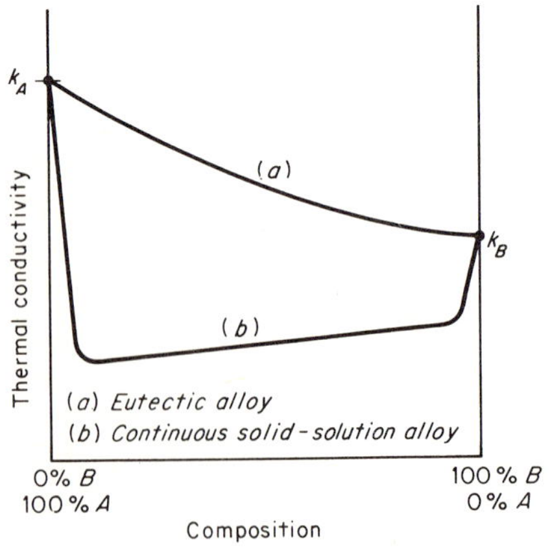

Fig. 2-5 Variation of thermal conductivity with composition in eutectic and continuous-solid-solution alloys.

In *eutectic alloys,* alloys which are soluble in the liquid phase but become mechanical mixtures in the solid phase, the change in thermal conductivity with composition may be roughly represented as shown in Fig. 2-5. The almost linear relationship indicates that the pure crystalline structure of each component is preserved for any composition and heat is conducted independently by each type of crystal. The slight deviation from linearity is no doubt due to scattering at the grain boundaries.

In alloys which are *limited solid solutions* the thermal-conductivity–composition curve varies according to the behavior of the solid phase; that is, k decreases rapidly with addition of the alloying metal to the solvent metal until the limit of solubility is reached and two different crystals are formed, after which the conductivity varies approximately linearly, as in the case of eutectic alloys. Certain maxima have been observed for certain compositions, which may be explained by the existence of an ordered structure of the alloy, *superstructure* or *superlattice,* which appears after certain heat treatments.

If the alloying metal forms *intermetallic compounds* with the solvent metal, their existence can be noticed by the appearance of maxima on the thermal-conductivity–composition curve. These compounds have a definite crystalline structure and thus offer less resistance to the phonons and electrons than do the continuous-solid-solution alloys. The difference between this kind of maximum in the thermal conductivity and that due to the superstructure effect is that the peaks of thermal conductivity in the intermetallic compounds are stable up to the melting point, whereas the superstructure may be destroyed before, e.g., by quenching.

Semiconductors In the broad class of semiconductors the heat transport by phonons is supplemented by heat transport by electrons. The number of free electrons in semiconductors is usually small. At low temperatures the electronic contribution to the transport of heat is fairly insignificant; however, at high temperatures the electronic contribution is considerable and may be larger than the phonon contribution.

Phase transition in solids Solids do not always exist in the same structural form but may have, in the case of crystalline, polymorphous solids, a transition point above which there exists one structural form and below which there exists another. The transition is the result of the rearrangement of the atoms in the lattice, and different physical properties ensue. In particular, there is a change in the thermal conductivity. An example of this type of phase transition is exhibited by the *rhombic* and *monoclinic* forms of sulfur. Below 368.5 K sulfur exists in the stable rhombic form, while above 368.5 K it is converted into the stable monoclinic form. The

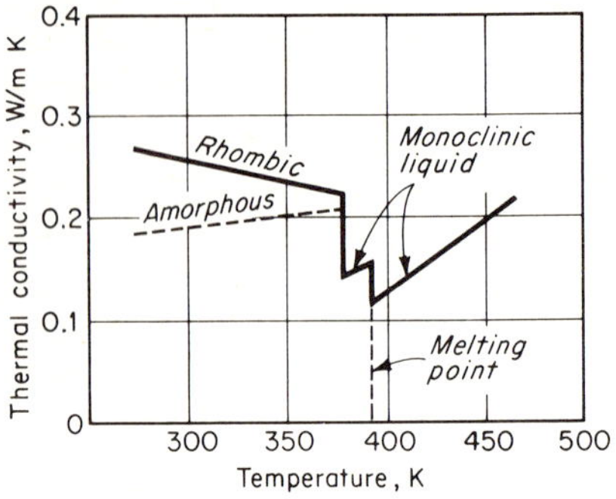

Fig. 2-6 Thermal conductivity of sulfur near the transition temperature. [*G. W. Kaye and W. F. Higgins, Proc. Roy. Soc. London, Ser. A,* **122**:*633–646* (*1929*).]

change in the thermal conductivity at and near the transition temperature is shown in Fig. 2-6.

The transition from one solid phase to another may not be marked by such abrupt changes in the thermal conductivity as was shown to be the case for sulfur. The transition from *alpha iron* to *gamma iron* results in no discontinuity in the thermal conductivity but does produce a change in the slope of the conductivity-temperature curve. Perhaps this effect is due to the fact that the electronic contribution to the transport of heat is not much affected by the structural changes in the lattice.

LIQUIDS

The situation in regard to the thermal conductivity of liquids is considerably less clear (but certainly no less important) than that of solids and gases. The process of melting is no doubt a change from a more orderly state to a less orderly one, and this in itself implies important changes in the molecular structure. The molecular bonds are loosened, and the possibility of free thermal movement is enhanced.

The fact that liquids will flow readily in most cases makes any measurement of thermal conductivity difficult because of the necessity of suppressing the convection currents caused by the buoyancy effects, which in turn result from the impressed temperature difference required by the experiment. Convective currents in thermal-conductivity experiments give rise to increased heat transport and spuriously high values of the conductivity.

Furthermore, the general macroscopic mobility of liquids led to the notion that liquids in their structure are more gaslike than solidlike. The major difference assumed between liquids and gases was that the molecules of the liquids are closer together and thus the intermolecular forces exert greater influence. It is now believed that the structure of liquids is more nearly like that of solids, at least near the phase-transition points, and retains short-range order to a great extent. This hypothesis has considerable support from the results of x-ray diffraction studies on liquids and the existence of *liquid crystals,* which exhibit some properties of anisotropic crystals.

Changes in the structure of matter strongly influence the thermal conductivity, which abruptly decreases at the melting point. More than likely the reduction in the thermal conductivity is the result of the degeneration of the order of the solid structure, precluding transport of heat energy by lattice waves. Experimental data for the thermal conductivity of solid and liquid phases near the melting point are available for water, sulfur, and a number of metals. These data confirm the abrupt decrease in the thermal conductivity as the solid phase melts, except for bismuth, for which the conductivity of the liquid phase is higher than the solid phase (Fig. 2-7). An explanation may lie in increased phonon transport in liquid bismuth since the speed of sound is higher in the liquid phase than in the solid phase.

Many attempts have been made to relate the thermal conductivity of liquids with other associated physical properties that are easier to measure. In general these efforts may be categorized into two groups: (1) correlation of the conductivity with the acoustic velocity and (2) correlation of the conductivity with properties other than the acoustic velocity. It would seem that the theories and correlations relating the conductivity in liquids to the acoustic velocity might be more accurate.

Early attempts by Bridgman[1] to relate the thermal conductivity of liquids to their velocity of sound resulted in the expression

$$k = \frac{2RC^{\text{liq}}}{z^2} \tag{2-13}$$

Here R is the universal gas constant, C^{liq} is the speed of sound in the liquid, and z is the mean distance between the centers of the molecules

[1] P. W. Bridgman, Thermal Conductivity of Liquids under Pressure, *Proc. Amer. Acad. Arts Sci.,* **59**:141–169 (1923).

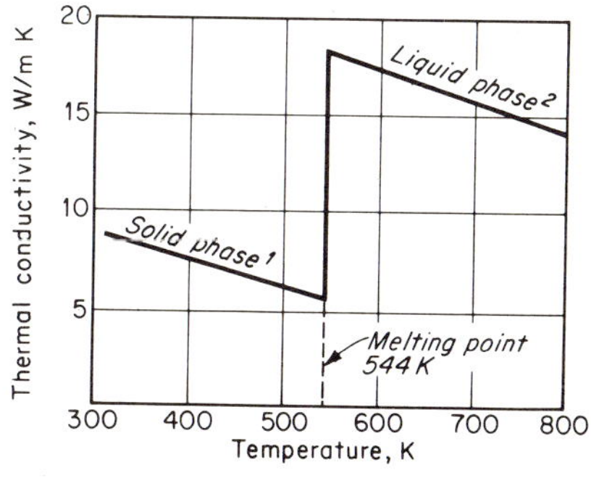

Fig. 2-7 Thermal conductivity of bismuth near the melting point. [(*1*) *J. H. Perry, "Chemical Engineers' Handbook, 3d ed., p. 458, McGraw-Hill, New York, 1950;* (*2*) *R. C. L. Bosworth, "Heat Transfer Phenomena," Wiley, New York, 1952.*]

in a cubical arrangement

$$z = \left(\frac{M}{\rho}\right)^{\frac{1}{3}} \tag{2-14}$$

where M is the molecular weight and ρ is the mass density. Equation (2-13) represents the available experimental data fairly well, though in some cases deviations are as large as 40 percent.

According to Eyring,[1] the thermal conductivity of a liquid (or a dense gas) depends upon the rapid transmission of the energy through the molecules themselves. The mechanism is best understood in terms of an analogy with the expression for the thermal conductivity for gases as derived from the elementary kinetic theory.

On the basis of the elementary kinetic theory the thermal conductivity for a monatomic gas is [see Eq. (2-6)]

$$k^{\text{gas}} = \tfrac{1}{3}\rho c_v \bar{w} \lambda \tag{2-15}$$

where ρ = density
c_v = specific heat at constant volume
λ = molecular-mean-free-path length

The mean molecular velocity $\bar{w}$ is defined as

$$\bar{w} = \sqrt{\frac{8RT}{\pi M}} = \sqrt{\frac{8}{\gamma\pi}}\, C^{\text{gas}}$$

where R = universal gas constant
T = absolute temperature
M = molecular weight
γ = ratio of specific heats at constant pressure and constant volume
C^{gas} = speed of sound in gas

It has already been shown that the thermal conductivity of some solids can be represented approximately by an expression similar to Eq. (2-15). That equation can be modified to be applicable in an approximate way for liquids. For a liquid the specific heat at constant volume is very nearly $3R/M$ rather than the value $3R/2M$ for a monatomic gas, since the molecules in a liquid may be thought to move in a simple three-dimensional harmonic potential. The molecular-mean-free-path length for a liquid may be approximated by the cube root of the volume per molecule in the liquid $v^{\frac{1}{3}}$, the distance over which the energy is transferred in one collision. The transfer velocity, the equivalent of the mean molecular velocity, may be written as $\bar{w}^{\text{liq}} = \sqrt{8/\gamma\pi}\, C^{\text{liq}}$. It is known

[1] R. E. Powell, W. E. Roseveare, and H. Eyring, Diffusion, Thermal Conductivity, and Viscous Flow of Liquids, *Ind. Eng. Chem.*, **33**:430–435 (1941); see also H. Eyring, Viscosity, Plasticity, and Diffusion as Examples of Absolute Reaction Rates, *J. Chem. Phys.*, **4**:283 (1936).

that the speed of sound in a liquid is 5 to 10 times larger than the speed of the molecules themselves.

The speed of sounds in liquids C^{liq} can be estimated from the speed of sound in a gas C^{gas} and the supposition that during a collision the energy is transferred instantly from the center of one molecule to the center of the other. Thus,

$$C^{\text{liq}} = \left(\frac{v}{v_f}\right)^{\frac{1}{3}} C^{\text{gas}} \tag{2-16}$$

where v_f is the *free volume* of the molecule in the liquid. The free volume may be visualized by considering an array of rigid-sphere molecules of diameter σ. If all but one of the molecules are fixed in a regular cubic lattice, the free molecule can move around in the cell formed by its nearest neighbors. The free volume within which the molecule moves has a complicated shape; however, the free volume can be estimated with the help of Fig. 2-8. The schematic diagram in the figure shows a row of molecules in the lattice. The black molecule may move a maximum distance $2(v^{\frac{1}{3}} - \sigma)$ along the row and has a similar freedom of movement along the other two coordinate axes. The free volume available for movement may be given approximately as

$$v_f = [2(v^{\frac{1}{3}} - \sigma)]^3 = 8(v^{\frac{1}{3}} - \sigma)^3 \tag{2-17}$$

The free volume of liquids can be calculated from the vapor pressure or the equation of state, for example. Equation (2-17) can be written as

$$v_f = 8\left[\left(\frac{V_0}{N_0}\right)^{\frac{1}{3}} - \left(\frac{3b}{2\pi N_0}\right)^{\frac{1}{3}}\right]^3 \tag{2-18}$$

where V_0 = volume per mole
N_0 = number of molecules per mole
b = van der Waals constant

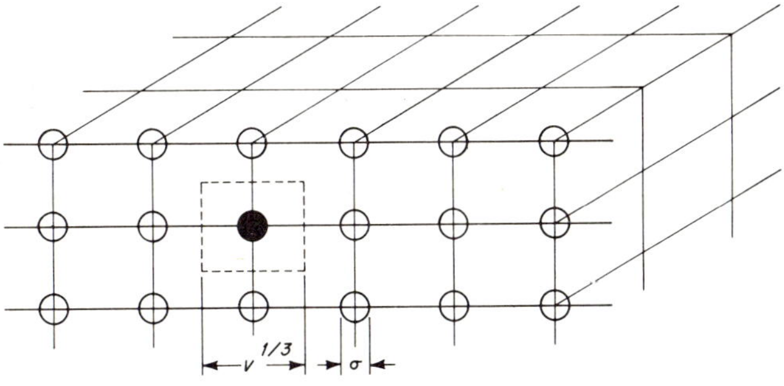

Fig. 2-8 One-row hard-sphere molecules in regular cubic lattice.

Equation (2-16) gives excellent agreement with the observed speed of sound for a large number of liquids.

An expression for the thermal conductivity of monatomic liquids in a form analogous to Eq. (2-15) can be written

$$k^{\text{liq}} = \rho \frac{R}{M} \sqrt{\frac{8}{\gamma\pi}}\, C^{\text{liq}} v^{\frac{1}{3}} \tag{2-19}$$

which, since $v = \rho/MN_0$, can be written finally as

$$k^{\text{liq}} = \left(\frac{\rho}{M}\right)^{\frac{2}{3}} R \sqrt{\frac{8}{\gamma\pi}}\, \frac{C^{\text{liq}}}{N_0^{\frac{1}{3}}} \tag{2-20}$$

The thermal conductivity for polyatomic molecules can be obtained by means of Eq. (2-20) and the Eucken correction[1]

$$\frac{9\gamma - 5}{15(\gamma - 1)} \tag{2-21}$$

which takes into account the internal degrees of freedom of the molecules. The question arises naturally as to what value should be used for the ratio of specific heat γ in the case of polyatomic molecules, as γ depends upon the number of degrees of freedom f in accordance with the expression

$$\gamma = 1 + \frac{2}{f} \tag{2-22}$$

Eyring proposes that only the translational and rotational degrees of freedom are effective in the transfer of energy if the resulting combination of Eqs. (2-20) and (2-21) is to agree with Bridgman's equation (2-13). Thus for six degrees of freedom, γ from Eq. (2-22) becomes $\frac{4}{3}$ for use in the Eucken correction term.

Combining Eqs. (2-20) and (2-21) as a product

$$k^{\text{liq}} = \sqrt{\frac{8}{\pi\gamma}} \left(\frac{\rho}{m}\right)^{\frac{2}{3}} R \frac{C^{\text{liq}}}{N_0^{\frac{1}{3}}} \frac{(9\gamma - 5)}{15(\gamma - 1)}$$

and using the value $\gamma = \frac{4}{3}$ in the correction term gives

$$k^{\text{liq}} = \frac{7}{5} \sqrt{\frac{8}{\pi\gamma}} \left(\frac{\rho}{M}\right)^{\frac{2}{3}} R \frac{C^{\text{liq}}}{N_0^{\frac{1}{3}}} \tag{2-23}$$

In Eq. (2-23) the value of γ should be determined from the speed-of-sound measurements. The results of Eq. (2-23) are very good for a large number of liquids, the mean deviation being about 10 percent. It has been noted by Bridgman[2] that Eq. (2-23) gives the correct variation of con-

[1] A. Eucken, Allgemeine Gesetzmässigkeiten für das Wärmeleitvermögen verschiedener Stoffarten und Aggregatzustande, *Forsch. Gebiete Ingenieurw.*, **11**:6–20 (1940).

[2] *Loc. cit.*

ductivity with temperature at 1 atm but overestimates the increase of thermal conductivities at high pressure by a factor of 2. The variation of conductivity with pressure is not large, however. The thermal conductivity of most liquids increases by only a factor of 2 for a pressure increase from 1 to 12,000 atm.

Liquid metals Liquid metals and electrolytes represent a separate class of liquids since the transport of heat energy is based on two different processes, the *atomic motions* and the *drift of free electrons.* The contribution to the heat current by the electrons increases the thermal conductivities of metallic liquids to values 10 to 1,000 times the values for nonmetallic liquids.

GASES

In the phase transition to the gaseous state the intermolecular bonds that existed in the liquid (or solid) state are greatly reduced, and the distance between molecules is so increased that they are free to move in any direction, their movements hindered only by collisions with boundary walls or other gaseous molecules. If one supposes a gas to exist at standard temperature and pressure, it is an easy calculation to show that the volume available per molecule is about 1,000 times the volume of the molecule.

The solid and liquid phases each have a type of order; the gaseous phase has none. The molecules of a gas are randomly distributed throughout the occupied volume. Because of this randomness in distribution, which lends itself to the methods of statistical mechanics and kinetic theory, all the properties of gases can be explained and determined to a satisfactory degree.

Calculations made on the basis of the arguments of the kinetic theory of gases lead to a simple but approximate representation for the thermal conductivity of gases. The form is identical with that of Eq. (2-6), and the derivation is accomplished in the same way; therefore it will not be repeated here. However, it is worth noting here that according to the theory of equipartition of energy, the specific heat at constant volume may be written for a gas in terms of the available and active degrees of freedom of the molecule, i.e., translational, rotational, and vibrational energy modes, such that the thermal conductivity becomes

$$k = \tfrac{1}{3}\rho\bar{w}\lambda\left(\frac{f}{2}\frac{R}{M}\right) \tag{2-24}$$

where f is the number of degrees of freedom and M is the molecular weight. Comparison with Eq. (2-6) shows that

$$c_v = \frac{f}{2}\frac{R}{M}$$

where c_v is the specific heat at constant volume

$$k = \tfrac{1}{3}\rho\bar{w}\lambda c_v \tag{2-25}$$

As might be anticipated, the viscosity μ, the property related to the transport of momentum in a gas by the same molecular-collision mechanism that gives rise to the thermal conductivity, can be related to the thermal conductivity. From the simple kinetic-theory model

$$\mu = \tfrac{1}{3}\rho\bar{w}\lambda$$

Thus

$$k = \mu c_v$$

for that model.

The results of the simple theory are correct in form, but the constant coefficient does not adequately represent the experimental values of either the thermal conductivity or the viscosity. It might be thought that a slightly more realistic molecular model would be sufficient to correct the constant, but the situation is more serious than that. Both the viscosity and the thermal conductivity are *transport properties;* as such they are *nonequilibrium properties* and must be treated by means of a *nonequilibrium theory.* Kinetic theory, the basis of Eq. (2-25), makes use of the maxwellian distribution function, fundamentally applicable only for an isentropic, equilibrium condition. Molecular collisions in the maxwellian sense do not disturb the equilibrium condition; thus transport properties, which are defined in terms of gradients, i.e., nonequilibrium conditions, cannot be obtained from the maxwellian model.

The rigorous development of the kinetic theory of gases is based upon the knowledge of a distribution function representing the number of molecules of some species which at some time τ lie in a unit volume of space around a point in space and have velocities lying in a unit velocity space around a point in velocity space. If there are no gradients in the composition, velocity, and temperature in the gas, the distribution function becomes the maxwellian distribution function. When equilibrium does not exist, the distribution function satisfies the Boltzmann equation. For conditions in the gas that vary only slightly from equilibrium, the distribution function is very nearly maxwellian, and the Boltzmann equation can be solved by a perturbation method developed by Chapman and

Enskog.[1] The solutions therefrom are used to obtain expressions for the mass, momentum, and energy fluxes and thus the coefficients, the transport properties.

The final result is that all the transport properties can be expressed in terms of a set of integrals, $\mathbf{\Omega}^{(l,s)}$ which involve explicitly the dynamics of molecular encounters and thus some intermolecular-force law.

The Chapman-Enskog kinetic theory is based upon several assumptions which limit the applicability of the final results:

1. Only binary collisions between molecules are considered; thus the results do not apply for densities high enough to result in any meaningful number of three-body collisions.
2. The theory is limited to the use of classical mechanics; thus it is invalid at low temperatures, where quantum effects may be important.
3. The perturbation method is valid only for nearly equilibrium conditions and thus only for those situations where the gradients of the properties are reasonably small.
4. The dimensions of the containing vessel are large compared with the molecular mean free path; i.e., the gas is in a continuum state.
5. The theory applies strictly to monatomic molecules, molecules with no internal degrees of freedom. Inelastic collisions occur between molecules with internal degrees of freedom. In such collisions momentum is conserved but energy is not. The viscosity is not affected, and the theory for monatomic gases may be applied to polyatomic molecules provided the molecules are reasonably spherical. The thermal conductivity is significantly affected by the presence of internal degrees of freedom, and the monatomic theory is not applicable to polyatomic molecules.

When the rigorous theory is applied to rigid-sphere molecules, the relationship between the viscosity and the thermal conductivity becomes

$$k = \tfrac{5}{2}\mu c_v \tag{2-26}$$

Equation (2-26) agrees fairly well with experimental data for *monatomic rigid-sphere molecules* but not for more complex molecules.

An extension of the theory to account for deviations from the hard-sphere molecules, i.e., to take into account for binary collisions some form

[1] The Chapman analysis is given in a very palatable form by G. N. Patterson, "Molecular Flow of Gases," Wiley, New York, 1956; see also S. Chapman and T. G. Cowling, "The Mathematical Theory of Non-uniform Gases," 2d ed., Cambridge University Press, London, 1958; or J. O. Hirschfelder, C. F. Curtiss, and R. B. Bird, "Molecular Theory of Gases and Liquids," Wiley, New York, 1954.

of attraction-repulsion law for the molecule, results in relationships in which Eq. (2-26) survives unchanged.

The first approximation for the viscosity becomes

$$\mu_1 = \frac{5}{16} \frac{\sqrt{\pi MRT}}{(\pi\sigma^2)\Omega^{(2,2)*}N_0} \tag{2-27}$$

where N_0 = Avogadro's number
σ = collision diameter
M = molecular weight
R = universal gas constant
T = absolute temperature

$\Omega^{(2,2)*}$ is the dimensionless function that indicates the deviation of any particular molecular model from the idealized rigid-sphere model. It is this last function which contains the information from the particular molecular potential function used.

With the help of Eq. (2-26) the expression for the thermal conductivity to the first approximation becomes

$$k_1 = \frac{25\sqrt{\pi MRT}\, c_v}{32(\pi\sigma^2)\Omega^{(2,2)*}N_0} \tag{2-28}$$

which is useful for monatomic molecules only. For polyatomic molecules Eq. (2-28) should be replaced by the relationship proposed by Eucken[1] to account for internal degrees of freedom in the molecule

$$k_1^{\text{Eucken}} = \mu_1\left(c_v + \frac{9}{4}\frac{R}{M}\right) = \frac{1}{4}(9\gamma - 5)\mu_1 c_v \tag{2-29}$$

In Eqs. (2-28) and (2-29) c_v is the usual specific heat at constant volume, and γ is the ratio of the specific heat at constant pressure to the specific heat at constant volume.

Equation (2-29) agrees quite well with the experimental data for simple and complex molecules. For monatomic gases, where $\gamma = \frac{5}{3}$, Eq. (2-29) reduces to Eq. (2-26).

Approximations higher than the first have been calculated[2] for both viscosity and thermal conductivity; however, in each case the correcting multiplier to the results of the first approximation is not much different from unity. In the case of viscosity (with the Lennard-Jones potential) the deviation of the multiplier from unity is less than 0.8 percent.

In anticipation of the means of measurement for the thermal conductivity of gases and the subsequent work in convective heat transfer, it is well to define at this time a dimensionless property of a gas (or

[1] A. Eucken, *Physik Z.*, **14**:324 (1913).

[2] Hirschfelder, Curtiss, and Bird, *op. cit.*

liquid) the *Prandtl number* (Pr) from Eq. (2-29)

$$\mathrm{Pr} = \frac{\mu c_p}{k} = \frac{4\gamma}{9\gamma - 5} \tag{2-30}$$

Equation (2-30) represents experimental data fairly well.

2-3 THE VARIATION OF THERMAL CONDUCTIVITY WITH TEMPERATURE

In our modern technology, use is made of materials over a truly impressive range of temperatures. Solid-state physicists have an increasing requirement for temperatures from $\frac{1}{2}$ to several degrees Kelvin, while plasma and magnetohydrodynamic researchers and astrophysicists are confronted with temperatures of many thousand degrees Kelvin, with the expectation of even higher temperatures. In the wide range between these extremes are myriad processes requiring knowledge of the effects of temperature on matter.

Thermal conductivity is a temperature-dependent property, the dependence largely determined by the structure of matter and its atomic or molecular motion. It is informative to explore in some detail the variability of the thermal conductivity with temperature for various materials.

SOLIDS

Nonmetallic (dielectric) solids The thermal conductivity of dielectric crystals is due to the lattice waves (*phonons*) which are attenuated by various scattering processes, as was discussed in the preceding section. In particular, the phonon-scattering processes increase the resistance to energy flow. As a first approximation the various thermal resistances arising from the scattering processes may be considered additive, so that the total thermal resistance, the reciprocal of the conductivity, becomes

$$\frac{1}{k} = R_{\mathrm{tot}} = R_U + R_B + R_D + R_I + R_M \tag{2-31}$$

where R_U = resistance due to *Umklapp* processes
R_B = resistance due to boundary scattering
R_D = resistance due to lattice defects
R_I = resistance due to chemical impurities
R_M = resistance due to mosaic structure

These various resistances depend upon the absolute temperature to various degrees. For low temperatures de Boer[1] gives the following relationships between the thermal resistances and temperature:

[1] de Boer, *loc. cit.*

1. For the *Umklapp* process (the mutual scattering of phonons) the resistance is

$$R_U = \frac{1}{AT} e^{-\Theta/2T} \tag{2-32}$$

where A = constant
T = temperature
Θ = Debye temperature

2. For boundary scattering,

$$R_B = \frac{1}{BLT^3} \tag{2-33}$$

where B is a constant and L is the smallest dimension of the sample.

3. For scattering by chemical impurities,

$$R_I = CT \tag{2-34}$$

where C is a numerical constant.

4. For mosaic-structure scattering, higher-order interaction,[1]

$$R_M = \frac{1}{DT} \tag{2-35}$$

where D is a numerical constant.

In well-formed, nearly perfect crystals free from chemical impurities, the phonon scattering occurs from processes (1) and (2) alone. Thus

$$R_T = R_U + R_B = \frac{1}{AT} e^{-\Theta/2T} + \frac{1}{B'LT^3} \tag{2-36}$$

where $B' < B$ to account for the fact that the relative contribution from boundary scattering to the total resistance is larger than indicated by the simple addition of the two resistances.

Equation (2-36) is the reciprocal of the thermal conductivity. Thus

$$k = \frac{1}{(1/AT) \exp(-\Theta/2T) + 1/B'LT^3} \tag{2-37}$$

Equation (2-37) is the basic equation for the thermal conductivity of regular dielectric crystals as a function of the absolute temperature, in the low-temperature range.

The theoretical curve on the basis of Eq. (2-37) reveals a maximum value in the conductivity at a temperature which depends upon the values of the constants A and B' and Θ. From the maximum value the conductivity falls gradually to its value at the melting point. At

[1] Pomeranchuk, *loc. cit.*

temperatures somewhat higher than the Debye temperature Θ the phonon wavelength becomes shorter and the *Umklapp* processes more significant. The thermal resistance in the higher-temperature case gradually shifts from the exponential form to a linear relation in T

$$R_U = A'T \tag{2-38}$$

At higher temperature, too, the contribution of boundary scattering becomes negligible, as does the contribution from lattice defects; thus

$$k = \frac{1}{A'T} \tag{2-39}$$

Pomeranchuk speculates that the dependence upon T is stronger than Eq. (2-39)—more like $T^{-\frac{3}{2}}$.

The existing experimental data on dielectric crystals for materials such as ice, quartz, mica, and sapphire show substantial agreement in form with the theories in Fig. 2-9. The numerical value of the maximum thermal conductivity is many times the value at room temperature. For example, the maximum value of thermal conductivity of a quartz crystal at 10 to 11 K as measured by Berman was 1,006 W/m K as compared with the value 19 W/m K for quartz crystal measured parallel to the axis at 273 K and a value measured perpendicular to the axis of 14 W/m K.

In *amorphous solids*, such as glasses, the less ordered structure in comparison with the orderly crystalline structure results directly in stronger inelastic scattering and distinct decreases in the thermal conductivity and a different kind of dependence upon temperature. The changes in temperature dependence may be seen from Fig. 2-3 by a comparison of quartz crystal and quartz glass shown there for reference.

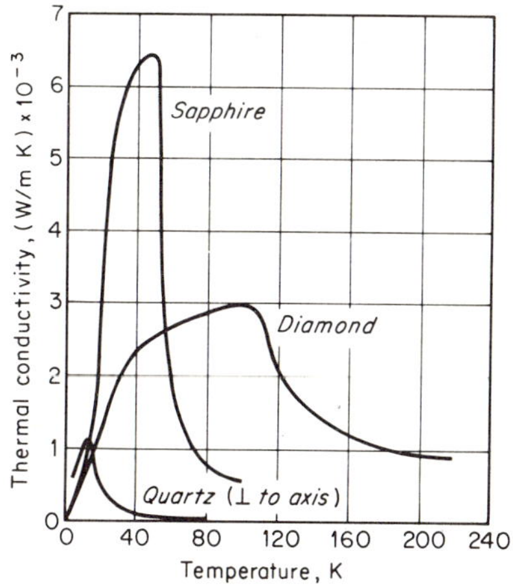

Fig. 2-9 Variation of thermal conductivity with temperature for dielectric crystals. [*R. Berman, F. E. Simon, and J. Wilks, Nature,* **168**:*277–280* (*1951*).]

If the only resistance-producing effect is the structural inelastic scattering, the thermal conductivity should be proportional to T

$$k = ET \tag{2-40}$$

where E is a numerical constant. Kittel[1] substituted known values of conductivities and specific heats into Eq. (2-6) and found that the mean free path λ is independent of temperature at room temperature and above; thus the thermal conductivity of glasses varies with the temperature in the same way as the specific heat c_v, or like Eq. (2-40).

The curve for quartz glass in Fig. 2-3 has been seen to vary in the region from 0 to 5 K about like $T^{1.3}$; in the temperature range from 5 to 15 K, conductivity is reasonably independent of temperature; above 15 K the conductivity varies with temperature linearly like Eq. (2-40). The shape of the curve may be partly explained by noting that at high temperatures the wavelengths of the phonons are short compared with the lattice dimensions and, as has been noted, the mean free path λ is independent of temperature. At low temperature the wavelengths of the phonons are much larger than the lattice dimensions, and the disorder in the structure has a smaller scattering effect; thus the mean free path increases.

Experimental data on amorphous and glassy materials follow, in form, the trace for quartz glass.

In *refractory materials* the variation of thermal conductivity with temperature depends mainly upon structural composition. Those containing mostly crystalline materials will show a decrease in conductivity with increasing temperature, while those containing mostly amorphous materials will show an increase with increasing temperature. The dependence of the conductivity upon the ratio of crystalline to amorphous components helps to explain the increase in conductivity of kaolin firebrick with temperature, while the conductivity of magnesite and silicon carbide bricks drops with temperature. In the same way one also can explain the variability with temperature of the thermal conductivity in refractory materials of the same chemical composition.

Metallic solids and semiconductors The thermal conductivity of pure metals depends upon the transport of heat energy by means of the phonons and valence electrons. Thus to determine the dependence of thermal conductivity upon temperature, the effect of temperature on the various scattering processes must be explored. To the resistances considered for the nonmetallic crystals it is necessary to add the phonon-

[1] Kittel, Interpretation of the Thermal Conductivity of Glasses, *loc. cit.*

electron and electron-phonon interactions. According to Laredo,[1] the resistance introduced from the scattering of phonons by the electrons at temperatures less than $\Theta/10$ is inversely proportional to the square of the absolute temperature

$$R_{p-e} = \frac{1}{FT^2} \tag{2-41}$$

where F is a numerical constant. In the same range of temperature, the resistance due to the scattering of the electrons by the phonons is proportional to the square of the absolute temperature

$$R_{e-p} = GT^2 \tag{2-42}$$

where G is a constant.

In metals the total conductivity may be written

$$k = k_p + k_e = \frac{1}{R_l} + \frac{1}{R_e} \tag{2-43}$$

where
$$R_l = R_U + R_B + R_{p-e} \tag{2-44}$$
and
$$R_e = R_{e-p} \tag{2-45}$$

Thus for low temperatures,

$$k = \frac{1}{R_U + R_B + R_{p-e}} + \frac{1}{R_{e-p}} \tag{2-46}$$

or
$$k = \frac{1}{(1/AT)\exp(-\Theta/2T) + 1/B'LT^3 + 1/FT^2} + \frac{1}{GT^2} \tag{2-47}$$

Since in pure metals the electronic contribution greatly outweighs the lattice contribution, a reasonable first approximation may be

$$k = \frac{1}{GT^2} \tag{2-48}$$

From Eq. (2-48) one might surmise that the thermal conductivity would become infinite at absolute zero; however, there are two effects that make it not so and in fact reduce the value to near zero at very low temperature: (1) the already discussed size effect for single crystals [Eq. (2-8)] and (2) the fact that at very low temperatures tiny amounts of impurities tend to scatter electrons according to T^{-1}; thus an additional resistance is present

$$R_{e,\,\mathrm{imp}} = \frac{M}{T} \tag{2-49}$$

[1] S. J. Laredo, Recent Work on Thermal Conductivity of Metals at Very Low Temperatures, "Problems Relating to Thermal Conductivity," *Annexe 1952–1 Bull. Inst. Intern. Froid,* p. 61.

where M is a measure of the amount of the impurity. Consequently, for nearly pure metals the conductivity is more properly written

$$k = \frac{1}{GT^2 + M/T} \tag{2-50}$$

Equation (2-50) shows a maximum, which for the same metal is shifted to lower temperatures as the amount of impurity is decreased.

In materials which are poor conductors of electricity the lattice conductivity cannot be neglected; thus for these materials

$$k = \frac{1}{(1/AT)\exp(-\Theta/2T) + 1/B'LT^3 + 1/FT^2} + \frac{1}{GT^2 + M/T} \tag{2-51}$$

The resistance term R_I (see Eq. 2-31) due to impurities has been omitted because at low temperatures and the resulting long wavelengths of the phonons the term is negligibly small.

Equation (2-51) shows a maximum value of the conductivity and a decrease thereafter as the temperature is increased. At higher temperatures the electric resistance becomes proportional to the temperature. Also the electric and thermal conductivity may be related through the Wiedemann-Franz law [Eq. (2-10)]. The electronic contribution, in this range, is independent of temperature; however, also in this temperature range the lattice contribution becomes proportional to T^{-1}. Consequently,

$$k = \frac{1}{A'T} + N \tag{2-52}$$

where A' and N are constants. Equation (2-52) indicates that the conductivity of pure metals will decrease with temperature slightly due to the lattice resistance. Equations (2-51) and (2-52) are confirmed fairly well by the existing experimental data. The position of the maximum [Eq. (2-51)] and the value of k are quite sensitive to the amount of impurity in the sample.

Alloys These may be treated from the standpoint of the scattering of phonons and electrons by chemical impurities. Impurities scatter phonons in a manner proportional to the absolute temperature, while electrons are scattered in a manner proportional to T^{-1}. Thus the dependence upon temperature for alloys depends mostly on the structure.

The effect of composition on the thermal conductivity of solid-solution alloys has been discussed earlier from the point of view of scattering of phonons and electrons by foreign atoms in the lattice (Fig. 2-5).

Since the mechanism of conduction in alloys is the same as that in pure metals (only smaller in magnitude), the equation representing the dependence of thermal conductivity of alloys on temperature will be essentially that of Eq. (2-51). If the amount of the alloying material is small, the thermal conductivity of solid-solution alloys will show a maximum value, as in the case of pure metals. Increases in the mixture of the alloying material depress the magnitude of the curve and cause the maximum to move toward higher temperatures.

A large amount of impurity or alloying material increases the resistance and lowers the thermal conductivity. In this case the scattering of electrons by impurities is the predominant factor in Eq. (2-51), which reduces, for low temperatures, to

$$k = \frac{T}{M} \tag{2-53}$$

Equation (2-53) shows that the conductivity increases with T and that for such materials no maximum occurs.

At high temperatures and where large amounts of alloying material are involved, the structure of the solid-solution alloys becomes very much disordered and the phonon activity is much like that of amorphous material; i.e., the lattice conductivity increases with temperature. At high temperatures the Wiedemann-Franz law is valid, and thus the electronic conductivity is independent of temperature; therefore

$$k = E'T + \text{const} \tag{2-54}$$

where E' is a constant. Thus the thermal conductivity of solid-solution alloys increases with temperature at high temperatures.

Figure 2-10 shows the schematic curves of the variation of thermal conductivity with temperature for pure metals, for metals with traces of impurities, and solid-solution alloys, as represented by Eqs. (2-51) to (2-54).

Binary, eutectic alloys are mechanical mixtures of different types of crystals. Heat is conducted independently by each type of crystal; thus the conductivity is the sum of the individual contributions. Inasmuch as each contribution will be characteristic of crystalline material, the total conductivity will be of the same character as that of pure metals, i.e., Eqs. (2-51) and (2-52).

Semiconductors In materials which are *semiconductors* heat energy is transported by both phonons and free electrons; however, the electronic conductivity is relatively small when compared with the phonon conductivity. The variation of thermal conductivity with temperature for semiconductors thus follows the trend for dielectric solids as shown in the case of graphite (Fig. 2-1). At high temperatures the number of free electrons in semiconductors increases rapidly, and at very high tem-

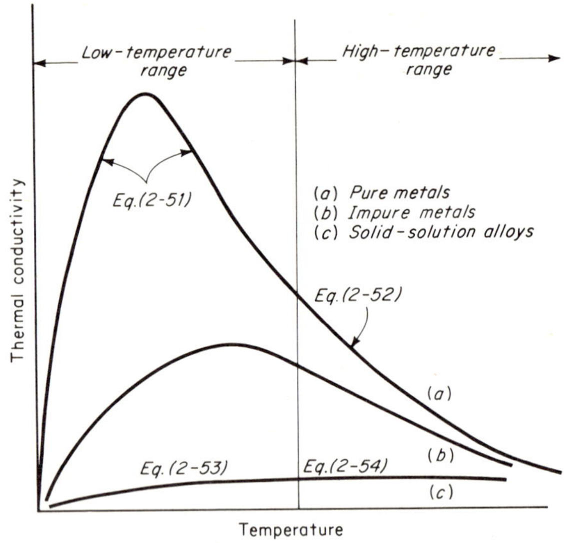

Fig. 2-10 Variation of thermal conductivity with temperature for (*a*) pure, (*b*) impure, and (*c*) solid-solution alloys.

peratures the electronic contribution to the conductivity may be larger than the phonon contribution, and the semiconductor will act as a metal. As can be seen from a comparison of Figs. 2-9 and 2-10, the variation of thermal conductivity with temperature is essentially the same for metals and dielectrics in the high-temperature range.

LIQUIDS

In solids and in gases the availability of adequate, or nearly adequate, theories permits reasonable estimates to be made concerning the dependence of thermal conductivity upon other properties, notably temperature and pressure. In the case of liquids no such theories are presently available, and the explanation of the variation of thermal conductivity with temperature (or pressure) is difficult.

In general, the volume of liquids increases with temperature, which indicates in turn that the degree of order is to some extent destroyed as the temperature rises. According to the order-disorder arguments used heretofore in this discussion, the thermal conductivity should decrease with an increase in temperature. From available experimental data it can be shown that the thermal conductivity of most liquids does decrease with temperature; however, some liquids show an increase of conductivity with temperature.

At 1 atm the Eyring-Bridgman formulation [Eqs. (2-20) and (2-23)] is useful. These equations indicate that the thermal conductivity depends directly upon the speed of sound and upon the $\frac{2}{3}$ power of the density for a given material. At 1-atm pressure the density of all liquids decreases as the temperature increases. The velocity of sound, however, decreases with temperature only for organic liquids. For water the velocity of sound increases with temperature. In organic liquids, therefore, both variables change with temperature in a way that decreases the thermal conductivity. In water, the velocity of sound increases faster than the density to the $\frac{2}{3}$ power decreases; thus the conductivity increases with temperature for this particular range of variables.

Experimental data available for liquids in a saturated state, i.e., water, ammonia, carbon dioxide, sulfur dioxide, and some organic liquids (see Appendix Table B-3), show thermal conductivities that decrease as the temperature increases toward the critical point. This trend is achieved either by passing through a maximum value which occurs near the saturated state at 1-atm pressure, in the instance of water, ammonia, and carbon dioxide, or by a monotonic decrease in thermal conductivity after melting, as in methyl chloride. The graphical representation for water is shown in Fig. 2-11. The thermal conductivity has a finite value at the critical state.

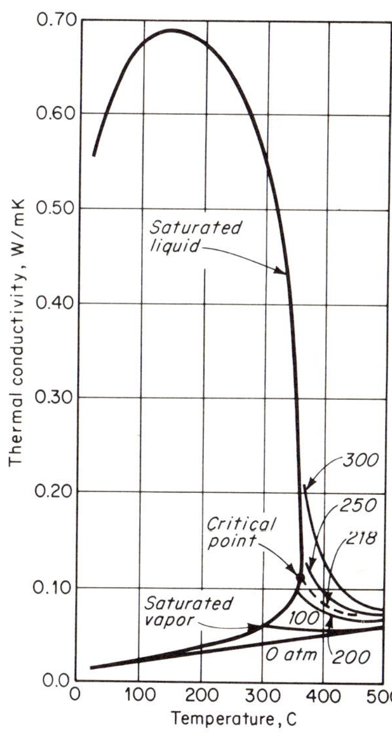

Fig. 2-11 The thermal conductivity k of water and water vapor in superheated and saturated state. [*D. L. Timrot and M. B. Vargaftik, The Thermal Conductivity and Viscosity of Steam at High Temperature and Pressure, J. Phys. USSR,* **2**:*101 (1940); and J. Hilsenrath et al., Natl. Bur. Std. U.S. Circ. 564 (1955).*]

In *metallic liquids* the majority of pure metals show a decrease in thermal conductivity with an increase in temperature; however, mercury, cadmium, and four eutectic alloys, lead-bismuth, tin-lead, tin-zinc, and lead-antimony, have positive temperature coefficients of thermal conductivity. There is no currently available theory for the variation of thermal conductivity of metallic liquids with temperature.

GASES

According to Eqs. (2-27) and (2-28), the determination of the thermal conductivity of gases depends upon the knowledge of σ and $\mathbf{\Omega}^{(2,2)*}$, knowledge which in turn requires a molecular model and an appropriate interaction potential. The interaction potentials generally take the form

$$\varphi(r) = \epsilon \frac{\sigma}{r} \tag{2-55}$$

where ϵ = maximum energy of attraction
σ = collision diameter
r = distance of separation of molecules

Of several such models that have been used for calculations, only two will be discussed here.

1. *The Sutherland model* represents molecules as rigid spheres of diameter σ surrounded by an inverse-power attractive force, with maximum energy of attraction ϵ. If the energy of attraction varies as $1/r^6$, the potential function can be expressed as

$$\varphi(r) = \begin{cases} \infty & r < \sigma \\ -\epsilon\left(\dfrac{\sigma}{r}\right)^6 & r > \sigma \end{cases} \tag{2-56}$$

2. *The Lennard-Jones 6-12 model* is a somewhat more realistic potential function obeying the equation

$$\varphi(r) = 4\epsilon\left[\left(\frac{\sigma}{r}\right)^{12} - \left(\frac{\sigma}{r}\right)^{6}\right] \tag{2-57}$$

where r = separation between molecules
ϵ = depth of potential well (maximum energy of attraction)
σ = collision diameter for low-energy collisions

that is, σ is the value of r for which $\varphi(r) = 0$. Equation (2-57) has been used widely for the calculation of transport properties of nonpolar gases. The potential function for polar molecules must contain an angular-dependent term proportional to the inverse third power of the intermolecular separation; thus the Lennard-Jones model is inappropriate.

The force constants ϵ and σ appropriate to any gas are obtained from experimental data for viscosity and self-diffusion and from second virial coefficient and Joule-Thomson coefficients. According to Hirschfelder et al., it is advisable in the calculation of transport properties to use the force constants obtained from viscosity data wherever possible.

Kihara and Kotani[1] have calculated the integral $\Omega^{(2,2)*}$ for the Sutherland model with the inverse-sixth-power attractive force, and calculations for the same integral for the Lennard-Jones 6-12 model have been made by several investigators.[2] With these data and calculations available, calculations of the viscosity and thermal conductivity follow from Eqs. (2-27) and (2-28).

The variation of the thermal conductivity with temperature can be seen to depend upon the temperature dependence of the viscosity–specific-heat product [Eq. (2-26)].

If the Sutherland model is used, the viscosity can be expressed as a function of the temperature

$$\mu = \frac{\mu_0 \sqrt{T}}{1 + C/T} \tag{2-58}$$

where μ_0 and C are constants for the gas in question and T is the absolute temperature.

The specific heat at constant volume generally increases with the temperature, and for a short interval in temperature may be represented approximately as a linear function of the temperature

$$c_v = c_{v_0}(1 + BT) \tag{2-59}$$

where c_{v_0} is a reference value of the specific heat and B is a numerical constant.

If Eqs. (2-58) and (2-59) are combined with Eq. (2-26),

$$\begin{aligned} k &= \frac{5}{2} \frac{\mu_0 \sqrt{T}}{1 + C/T} c_{v_0}(1 + BT) \\ &= \tfrac{5}{2} k_0 \frac{(1 + BT)T^{\frac{3}{2}}}{C + T} \end{aligned} \tag{2-60}$$

Equation (2-60) with the proper constants represents the experimental data within about 6 percent in the temperature range 95 to 475 K. For higher and lower temperatures Eq. (2-60) is not valid, and a more realistic molecular model is required.

[1] T. Kihara and M. Kotani, *Proc. Phys. Math. Soc. Japan*, **24**:76 (1942). The results are tabulated as $\Omega^{(l,s)*}$ as a function of $T^* = \mathbf{k}T/\epsilon$. **k** is the Boltzmann constant.

[2] *Ibid.*; J. de Boer and J. van Kranendonk, *Physica*, **14**:442 (1948); J. S. Rowlinson, *J. Chem. Phys.*, **17**:101 (1949); J. O. Hirschfelder, R. B. Bird, and E. L. Spotz, *J. Chem. Phys.*, **16**:968 (1948); **17**:149 (1950).

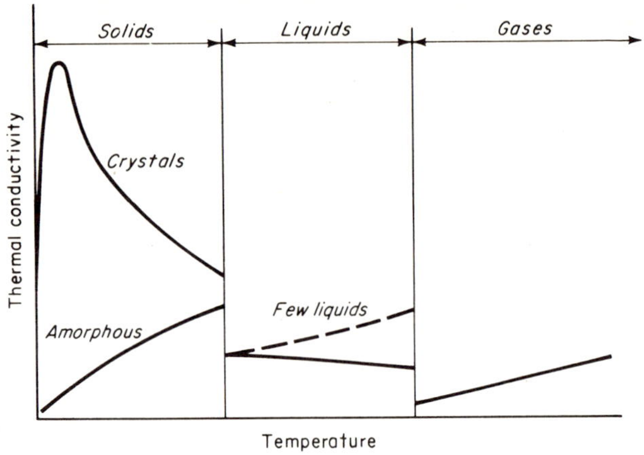

Fig. 2-12 Thermal conductivity as a function of temperature throughout three physical states.

In the *Lennard-Jones model* the value of the integral $\Omega^{(2,2)*}$ decreases as the temperature increases; thus if c_v is considered a linear function of the temperature, as before, the variation of the thermal conductivity with temperature may be between a T^2 and T^3 relationship. The experimental data of Stops[1] on air, nitrogen, and carbon dioxide in the temperature range of 273 to 1273 K tends to confirm this speculation.

The variation of the thermal conductivity with temperature for three phases of matter is shown schematically in Fig. 2-12. As a general rule, the thermal conductivity of all materials increases with pressure, the exception to the rule being a few metals.

2-4 THE VARIATION OF THERMAL CONDUCTIVITY WITH PRESSURE

In *metallic solids* it might be expected that an increase in pressure would distort the crystalline lattice and induce resistances to phonon and electron transport, resulting in a decrease in thermal conductivity. Experiments by Bridgman[2] on 11 metals with pressures up to 12,000 kg/cm² showed that the thermal conductivity of iron, copper, silver, nickel, platinum, bismuth, and antimony decreases linearly with pressure at various rates, while the conductivities for lead, tin, cadmium, and zinc increase in a similar way. The pressures used by Bridgman caused only elastic

[1] D. W. Stops, Effect of Temperature upon the Thermal Conductivity of Gases, *Nature*, **64**:966 (1949).

[2] P. W. Bridgman, "The Physics of High Pressures," pp. 307–326, G. Bell, London, 1949.

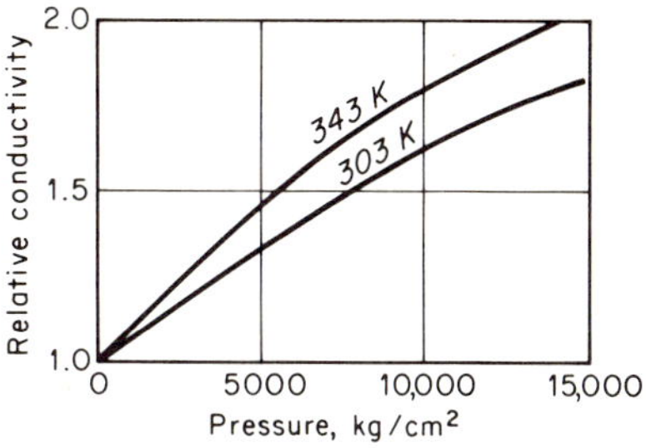

Fig. 2-13 Influence of pressure on thermal conductivity of liquids. [*P. W. Bridgman, "The Physics of High Pressures," pp. 307–326, G. Bell, London, 1949.*]

deformation of the metallic lattice and as such may have made the conductivity better or poorer, depending upon the structure of the lattice.

For *amorphous solids* the increase in pressure should increase the molecular contact and thus should promote an increase in the thermal conductivity. This situation was confirmed by the experiments of Bridgman, who has shown that for amorphous solids the thermal-conductivity dependence upon pressure can be represented by the equation

$$k(p) = k(1 \text{ atm}) + a \frac{p}{1{,}000} \tag{2-61}$$

where p is the pressure in kilograms per square centimeter and a is a numerical constant.

Liquids behave under the influence of pressure in a manner similar to that of amorphous solids. The thermal conductivity of liquids may increase by a factor of 2 in the pressure range 0 to 12,000 kg/cm^2 (Fig. 2-13).

The thermal conductivity of *gases* at moderate pressures is essentially independent of pressure. At low pressures, however, where the mean-free-molecular-path length becomes the same order as the dimensions of the container, i.e., the mean free path is determined by the container dimensions, the conductivity becomes pressure-dependent. (This situation is quite analogous to the crystal-size effect in solids.) Vacuum gages of the *Pirani* type take advantage of the pressure dependence of a contained low-pressure gas.

At high pressures both the viscosity and the specific heat at constant volume increase with pressure; thus, from Eq. (2-26), so does the thermal conductivity. The positive pressure coefficients for the thermal conductivity of gases are large for heavy polyatomic gases and vapors.

3
Steady Heat Conduction

3-1 SIMPLE CONDUCTION-EQUATION SOLUTIONS FOR STEADY CONDITIONS

Whenever the geometrical form of the conducting body is simple and the flow of heat is in one direction only and invariant with time, the heat-conduction equation can be greatly simplified. Some very important practical cases of heat-conduction problems fall into these simple categories, e.g., the slab, the hollow tube, and the hollow sphere.

The slab For a slab of constant thermal conductivity and a thickness l (Fig. 3-1) which extends to infinity in the other dimensions so that in effect the heat flow in the region considered is truly one-dimensional, it is convenient to treat the problem in the rectangular coordinate system. If there are no heat sources (or sinks) in the slab and the heat flow is steady and one-dimensional, Eq. (1-16) becomes

$$\frac{d^2t}{dx^2} = 0 \tag{3-1}$$

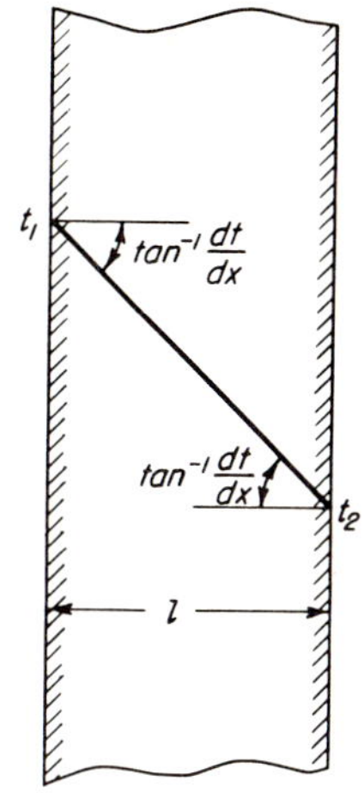

Fig. 3-1 Steady heat conduction in a slab.

which is readily solved to yield

$$t = C_1x + C_2$$

The constants C_1 and C_2 can be evaluated from the boundary conditions which prescribe the temperature of the surfaces $x = 0$ and $x = l$. Application of these conditions results in an expression for the temperature distribution in the slab:

$$\frac{t - t_1}{t_2 - t_1} = \frac{x}{l} \tag{3-2}$$

The heat flow through the slab can be obtained from the Fourier conduction law:

$$Q = -kA\,\frac{dt}{dx} = -kA\,\frac{t_2 - t_1}{l} = \frac{t_1 - t_2}{l/kA} \tag{3-3}$$

Again it is well to note here the similarity of Eq. (3-3) to the usual statement of Ohm's law. The term l/kA is the equivalent of the *electric resistance* and is called appropriately the *thermal resistance.*

The composite slab The analogy with Ohm's law and the concept of the thermal resistance provides a means of treating the problem of heat flow through a slab made up of material of differing thicknesses and thermophysical properties. Such a slab is represented schematically in Fig. 3-2. The treatment is that of resistances in series, such that if t_5 and t_2, the temperatures of each outer surface of the slab, the physical dimensions, and thermal properties of the slab are known, the heat flow can be written

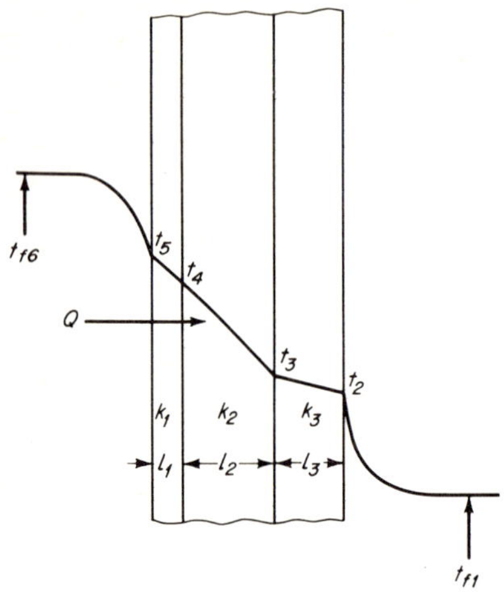

Fig. 3-2 Steady heat conduction in a composite slab.

immediately

$$Q = \frac{t_5 - t_2}{l_1/k_1A + l_2/k_2A + l_3/k_3A} \tag{3-4}$$

Should there be fluid adjacent to the slab surfaces wherein the fluid temperatures and not the surface temperatures are known, the thermal resistances due to the fluid must be included in the expression, or

$$Q = \frac{t_{f6} - t_{f1}}{1/h_1A + l_1/k_1A + l_2/k_2A + l_3/k_3A + 1/h_2A} \tag{3-5}$$

Intermediate temperatures can be determined by the inclusion of those resistances between a known temperature and that to be determined.

The tube The tube (Fig. 3-3) or the hollow cylinder is more conveniently treated from the cylindrical coordinate system. A case in point is one of

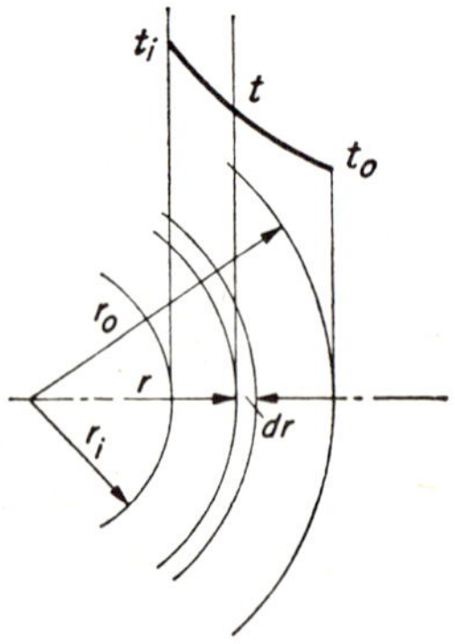

Fig. 3-3 Steady heat conduction in a thick-walled tube.

steady state, constant properties, and absence of heat sinks or sources. With these restrictions and another which limits the heat flow to the radial direction (this presupposes the tube to be very long axially), Eq. (1-17) reduces to the ordinary differential equation

$$\frac{d^2t}{dr^2} + \frac{1}{r}\frac{dt}{dr} = 0 \tag{3-6}$$

the solution of which becomes

$$t = C_1 \ln r + C_2$$

The appropriate boundary conditions are those specifying the temperatures t_i and t_o at the radii r_i and r_o, from which the constants C_1 and C_2 can be evaluated to yield the expression for the radial temperature distribution in the tube

$$\frac{t - t_o}{t_i - t_o} = \frac{\ln (r/r_o)}{\ln (r_i/r_o)} \tag{3-7}$$

The heat flow across the tube wall is determined from the Fourier conduction law; however, the area normal to the heat-flow vector now changes with radius and must be properly accounted for. Thus

$$Q = -kA(r)\frac{dt}{dr} = -k(2\pi rL)\frac{dt}{dr} \tag{3-8}$$

where L is the axial length of the tube. Differentiating Eq. (3-7) once and using that result in Eq. (3-8) gives the specification of the heat flow

$$Q = \frac{t_i - t_o}{(1/2\pi kL) \ln (r_o/r_i)} \tag{3-9}$$

Equation (3-9) has a form similar to Eq. (3-3) for the slab, except that here the thermal resistance is

$$\frac{1}{2\pi kL} \ln \frac{r_o}{r_i}$$

The composite tube As in the case of the composite slab, it is possible to treat a composite tube (Fig. 3-4). Physically this may be a pipe with covers of various sorts of insulation. The treatment is that of resistances in series as for the composite slab

$$Q = \frac{t_1 - t_4}{\dfrac{1}{2\pi k_1 L} \ln \dfrac{r_2}{r_1} + \dfrac{1}{2\pi k_2 L} \ln \dfrac{r_3}{r_2} + \dfrac{1}{2\pi k_3 L} \ln \dfrac{r_4}{r_3}} \tag{3-10}$$

To include resistances for convective effects of fluids flowing on the inner and outer surface of the tube it is necessary to include the correct

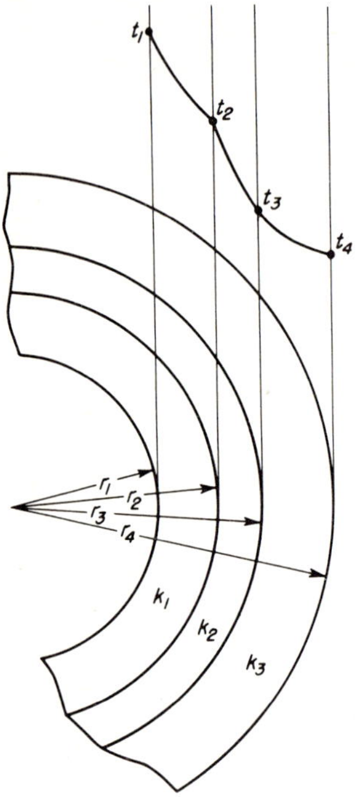

Fig. 3-4 Steady heat conduction in a composite tube.

areas swept by the flow, or

$$Q = \frac{t_{fi} - t_{fo}}{\dfrac{1}{2\pi r_4 L h_o} + \dfrac{1}{2\pi k_1 L} \ln \dfrac{r_2}{r_1} + \dfrac{1}{2\pi k_2 L} \ln \dfrac{r_3}{r_2} + \dfrac{1}{2\pi k_3 L} \ln \dfrac{r_4}{r_3} + \dfrac{1}{2\pi r_1 L h_i}} \tag{3-11}$$

Also, as in the case of the slab, the intermediate temperatures can be readily determined.

The sphere The temperature distribution and heat flow through the walls of a hollow sphere can be calculated from Eq. (1-18) and the Fourier conduction law by using assumptions similar to those for the slab and tube. The temperature distribution so obtained is

$$\frac{t - t_o}{t_i - t_o} = \frac{1/r - 1/r_o}{1/r_i - 1/r_o} \tag{3-12}$$

The heat flow is thence found to be

$$Q = \frac{t_i - t_o}{(1/4\pi k)[1/r_i - 1/r_o]} \tag{3-13}$$

The composite-sphere case can be readily synthesized from Eq. (3-13) in the manner used to obtain Eqs. (3-10) and (3-11).

3-2 CRITICAL THICKNESS OF INSULATION

In the case of small-diameter circular tubes the application of insulating material to the outer surface may in special instances increase the heat loss from the surface. This interesting phenomenon can be treated as follows. The heat loss from a tube with insulation to a surrounding fluid is given by

$$Q = \frac{2\pi L(t_i - t_f)}{(1/k)\ \ln\ (r_o/r_i) + 1/h_o r_o} \tag{3-14}$$

The heat rate will be a maximum when the denominator becomes a minimum. The minimum value for the denominator can be calculated by taking the derivative of the denominator with respect to r_o while r_i is held as a constant parameter and setting the result equal to zero. This gives

$$\frac{1}{kr_o} - \frac{1}{h_o r_o^2} = 0 \tag{3-15}$$

from which

$$r_{o,\text{crit}} = \frac{k}{h_o} \tag{3-16}$$

It can be seen that this result is independent of r_i. The heat-transfer coefficient was considered constant in this calculation. This approximation is reasonable for many practical cases where the variation in r_o is small.

Physically the result of Eq. (3-16) can be explained in the following way. In Eq. (3-14) the term $1/k \ln (r_o/r_i)$ is the thermal resistance due to the insulation; the term $1/h_o r_o$ is the thermal resistance due to the film of fluid. The former increases with r_o, while the latter decreases with r_o. At the critical radius $r_{o,\text{crit}}$ the rate of increase of the insulation resistance is equal to the rate of decrease in the film resistance, thus giving the minimum value of the sum of the resistances as shown by Eq. (3-15). The conclusion is that tubes whose outside radii (in this case r_i) are smaller than the critical radius $r_{o,\text{crit}}$ as calculated here, can have their heat losses increased by adding insulation up to the value of the critical thickness. This usually requires small tube radii, relatively large thermal conductivities in the insulation, and small heat-transfer coefficients. A practical application is the problem of insulating electrical wires where the objective would be the provision of adequate electrical insulation at the same time providing for maximum wire cooling.

In the preceding analysis the heat-transfer coefficient h is taken to be a constant, independent of both the radius and the temperature differences. For the necessary small radii the heat-transfer coefficient depends upon the radius. For free convection the heat-transfer coefficient depends upon the temperature difference to a fractional power.

Sparrow[1] has calculated critical radii for both cylinders and spheres, taking into account the dependence of the heat-transfer coefficient on the radius and the temperature difference as expressed by $h \sim r_o^{-m}|t_o - t_\infty|^n$, where $m \geq 0$ and $n \geq 0$. For a cylinder the critical radius for these specifications becomes

$$r_{o,\text{crit}} = \frac{1 - m}{1 + n}\frac{k}{h}$$

In comparison of this result with Eq. (3-16) the quantity $(1 - m)/(1 + n)$ appears as a correction factor with a value equal to or less than unity. For forced flow over a cylinder in a range of Reynolds numbers from 4,000 to 40,000, $m = 0.382$ and $n = 0$. The correction factor in this case will equal 0.618.

For free convection about a horizontal cylinder for which $m = n = \frac{1}{4}$, the correction factor equals 0.6.

A similar calculation for a sphere yields an expression for the critical radius,

$$r_{o,\text{crit}} = \left[\frac{(1 - m/2)}{1 + n}\right]\frac{2k}{h}$$

And as the expression for the critical radius corresponding to the calculation which yielded Eq. (3-16) is $2k/h$, the quantity $(1 - m/2)/(1 + n)$ is again shown as a correction factor similar to that obtained for the cylinder.

3-3 THE THIN ROD

Another simple but important solution of Eq. (1-16) is that for the physical system of a thin rod which is transferring heat at its surface to a surrounding fluid and which is connected at its base to a heated wall. The system is shown in Fig. 3-5. The temperature of the base at the wall is t_1; the cross-sectional area of the rod, A; its circumference, C; and its length, l. The convection on the surface is assumed to provide a constant value of the heat-transfer coefficient over the entire surface. The area A and circumference C are constant along the length of the rod. If the diameter of the rod is small compared with its length, and if the

[1] E. M. Sparrow: Reexamination and Correction of the Critical Radius for Radial Heat Conduction, *J. AIChE.*, **16**(1):149 (1970).

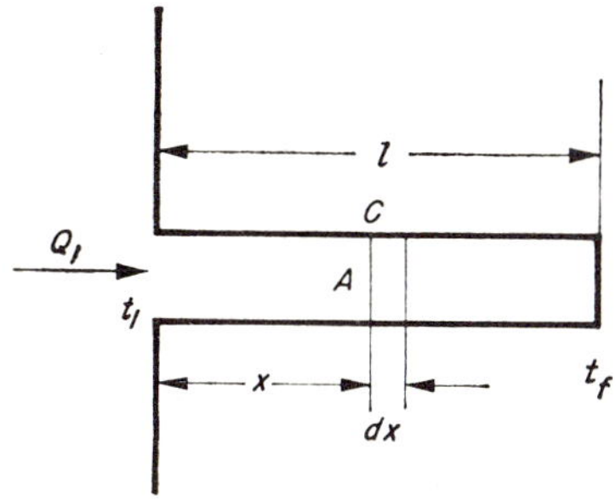

Fig. 3-5 Steady heat conduction in a thin rod.

convection film essentially controls the heat flow, there will be no radial temperature distribution in the rod, but there will be a large axial temperature distribution. This, then, is a case of one-dimensional steady heat conduction in the rod. The fact that the heat conducted along the rod from the base is lost to the surrounding fluid by convection suggests that the problem can be solved by reducing Eq. (1-16) to the terms describing axial conduction and a distributed heat sink which is equal to the convection loss. Such a reduction becomes

$$\frac{d^2t}{dx^2} = \frac{Q'}{k} \tag{3-17}$$

The heat sink Q' must be evaluated in terms of the convection loss, where Q' is the heat sink per unit volume:

$$\frac{d^2t}{dx^2} = \frac{hC\ \delta x\ (t - t_f)}{kA\ \delta x} \tag{3-18}$$

Equation (3-18) can be readily recognized as the same result that would be obtained by equating the net heat conduction in a volume to the convection loss by that same volume. If the symbol $\vartheta = t - t_f$ is used for the temperature difference, Eq. (3-18) becomes

$$\frac{d^2\vartheta}{dx^2} - \frac{hC}{kA}\vartheta = 0 \tag{3-19}$$

Equation (3-19) can be solved by routine methods[1] to yield a general solution in the form

$$\vartheta = C_1e^{mx} + C_2e^{-mx} \tag{3-20}$$

where $$m = \sqrt{\frac{hC}{kA}}$$

[1] I. S. Sokolnikoff and R. M. Redheffer, "Mathematics of Physics and Modern Engineering," 2d ed., McGraw-Hill, New York, 1966.

The constants C_1 and C_2 must be calculated with the aid of suitable boundary conditions. The end of the rod fastened to the wall is at the wall temperature t_1. The other end may transmit heat to the surrounding fluid. Mathematically this end-loss process can be expressed by

$$-kA\left(\frac{d\vartheta}{dx}\right)_{x=l} = hA\vartheta_{x=l} \tag{3-21}$$

If the rod is long and thin, as we have so far allowed, the heat loss from the end may be zero. This condition can be written as

$$\left(\frac{d\vartheta}{dx}\right)_{x=l} = 0 \tag{3-22}$$

Using the boundary condition in Eq. (3-22) for the moment and introducing it into Eq. (3-20) together with the wall condition gives

$$\begin{aligned} \vartheta_1 &= t_1 - t_f = C_1 + C_2 \\ \left(\frac{d\vartheta}{dx}\right)_{x=l} &= 0 = mC_1e^{ml} - mC_2e^{-ml} \end{aligned} \tag{3-23}$$

The solution of Eqs. (3-23) for C_1 and C_2 and the introduction of these values into Eq. (3-20) gives

$$\frac{\vartheta}{\vartheta_1} = \frac{e^{m(l-x)} + e^{-m(l-x)}}{e^{ml} + e^{-ml}} = \frac{\cosh m(l-x)}{\cosh ml} \tag{3-24}$$

The excess temperature of the end of the rod ($x = l$) is

$$\vartheta_2 = \frac{\vartheta_1}{\cosh ml} \tag{3-25}$$

The heat flow through the base of the rod ($x = 0$) is

$$Q_1 = -kA\left(\frac{d\vartheta}{dx}\right)_{x=0} = mkA\vartheta_1\left[\frac{\sinh m(l-x)}{\cosh ml}\right]_{x=0} = \sqrt{hCkA}\,\vartheta_1 \tanh ml \tag{3-26}$$

The two functions $\cosh^{-1} ml$ and $\tanh ml$ are given in Fig. 3-6 and in Table 3-1. As can be seen in Fig. 3-6, when the length l increases beginning at zero, the heat flow increases rapidly at first but the incremental

Table 3-1 Calculation functions for heat conduction in a rod

ml	0	0.5	1	1.5	2	3	4	5	6
$\cosh ml$	1	1.1276	1.543	2.352	3.762	10.07	27.31	74.21	201.7
$\tanh ml$	0	0.4621	0.7616	0.9052	0.9640	0.9951	0.9993	0.9999	1

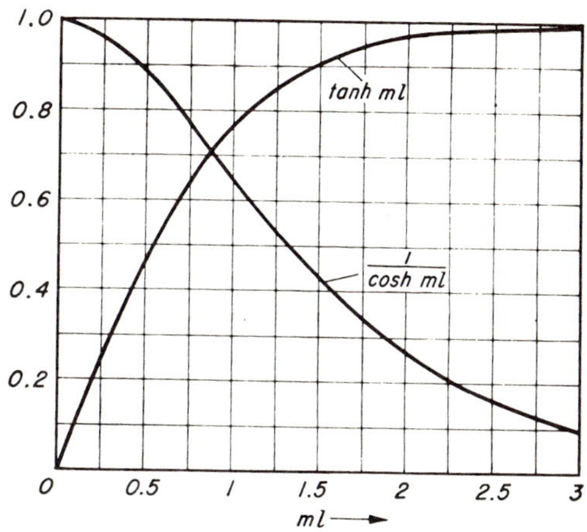

Fig. 3-6 Functions for the determination of heat flow and temperature distribution in a thin rod.

increase becomes smaller and smaller and finally the heat flow approaches an asymptotic value. The excess temperature at the end of a very long rod is zero.

The solution of Eq. (3-19) with the boundary condition expressed by Eq. (3-21), i.e., the case of heat loss from the end of the rod, becomes somewhat lengthy but results in the expression for the temperature distribution along the rod

$$\frac{\vartheta}{\vartheta_1} = \frac{\cosh m(l - x) + (h_2/mk) \sinh m(l - x)}{\cosh ml + (h_2/mk) \sinh ml} \tag{3-27}$$

The temperature excess at the end of the rod ($x = l$) becomes

$$\frac{\vartheta_2}{\vartheta_1} = \frac{1}{\cosh ml + (h_2/mk) \sinh ml} \tag{3-28}$$

The heat flow through the base of the rod ($x = 0$) becomes

$$Q_1 = mkA\vartheta_1 \frac{h_2/mk + \tanh ml}{1 + (h_2/mk) \tanh ml} \tag{3-29}$$

In the last three equations the value h_2 is the heat-transfer coefficient at the end of the rod; this value is generally different from the heat-transfer coefficient along the rod surface. Equation (3-29) will reduce to Eq. (3-26) for small enough values of h_2/mk.

The following calculated example has an important technical application. The temperature of a fluid flowing in a tube is usually measured by a thermometer or thermocouple put into a well which is welded into the tube wall, as shown in Fig. 3-7. If the fluid temperature is higher and differs greatly from the ambient temperature, the tube wall will have a lower temperature than the fluid and heat will flow by conduction from

Fig. 3-7 Temperature measurement in flow in a tube.

the well to the tube wall. The end of the well where the thermometer bulb or thermocouple junction is placed may become colder than the fluid, and the indicated temperature will not be the true fluid temperature. This error can be calculated by Eq. (3-25) or (3-28), whichever is deemed appropriate. Figure 3-6 and Table 3-1 give the necessary length of the tube when the error must be confined within a certain limit.

Example 3-1 In a tube of 10-cm diameter in which flows superheated steam, there is placed an iron thermometer well (Fig. 3-7) with a diameter $d = 1.50$ cm for a thermometer. The steam has a pressure of 1.00 kg/cm² and a temperature of 315 C. The flow velocity is 20 m/s. It is required to determine the length of the well which gives an error in temperature measurement of less than 0.5 percent of the difference between the steam temperature and the tube-wall temperature. The film heat-transfer coefficient between the steam and the tube wall is $h = 105$ W/m² K as calculated. The thermal conductivity of iron $k = 55$ W/m K is to be found in Appendix Table B-1. If the wall thickness of the well is $s = 0.1$ cm, the cross section for the heat flow in the well is $A = \pi ds$, and the circumference is $C = \pi d$, then

$$m = \sqrt{\frac{hC}{kA}} = \sqrt{\frac{h}{ks}} = \sqrt{\frac{105(100)}{55(0.1)}} = 45.7 \text{ m}^{-1}$$

With the signs used in the text the error is $\vartheta_2/\vartheta_1 = 0.005$. For this value the product ml must be 6, as found in Table 3-1. Therefore, the length of the well, $l = 6/m = 6/0.457 = 13.1$ cm. As this length is greater than the tube diameter, it is necessary to place the well obliquely in the tube (Fig. 3-7).

Thermal radiation between the well and the tube wall may cause an additional error in the temperature measurement. Radiation heat exchange will be dealt with in a later chapter.

3-4 FINNED HEATING SURFACES

As may be deduced from Eq. (3-5), the heat-transfer resistance of a plane wall is determined principally by the greatest single resistance. If this resistance is one of the convection resistances, the heat flow through the wall can be increased by putting fins on the surface where this large resistance occurs. Such finned surfaces are widely used, for example, in economizers in steam power plants, convectors for steam and hot-water heating systems, or electric transformers; for the cylinders of aircraft engines; etc.

The rectangular fin The following calculations are made for the plane finned surface as the simplest case. As long as the height of the fins on a tube is comparatively small with respect to the tube diameter, the formulas derived from the plane wall can also be used for the tube.

For fins of constant thickness with the height l, the formulas derived in Sec. 3-3 hold true. Using the symbols in Fig. 3-8, the cross-sectional area of the fin is bL and the circumference C is $2L$ when b is small compared with L. The value m, therefore, is derived by introducing the above values into the expression

$$m = \sqrt{\frac{hC}{kA}} = \sqrt{\frac{2h}{kb}} \tag{3-30}$$

The heat loss from such fins can be calculated from either Eq. (3-26) or (3-29).

It is of primary interest to recognize the conditions for which the finned surface has advantages over the unfinned surface. The answer to this question depends on whether the price, the weight, or the space needed is of the greatest importance. One must first find the circumstances under which the heat flow through a wall can be increased at all by fins. Obviously, fins are advantageous for the case where the heat flow through a fin increases with increased fin height. If the converse is true, it is advantageous to make the fins shorter until they are actually omitted. The limit of the conditions for which fins are advantageous is given by the expression

$$\frac{dQ_1}{dl} = 0 \tag{3-31}$$

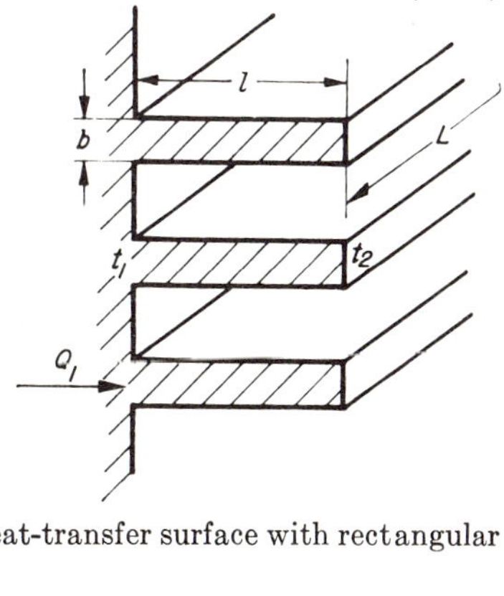

Fig. 3-8 Heat-transfer surface with rectangular fins.

Using Eq. (3-29) and considering the factors k, A, m, and ϑ_1 as constants, it suffices to differentiate only the fraction in Eq. (3-29). The resulting differentiation becomes zero when the numerator is zero or the denominator infinite. The last possibility results in a trivial statement only, for when $k = 0$, Eq. (3-31) is fulfilled. Consider then the numerator

$$\left(1 + \frac{h_2}{km} \tanh ml\right) \frac{m}{\cosh^2 ml} - \left(\frac{h_2}{km} + \tanh ml\right) \frac{h_2/k}{\cosh^2 ml} = 0$$

This equation can be simplified to the expression

$$m - \frac{h_2{}^2}{k^2 m} = 0$$

Introducing Eq. (3-30) gives, when $h = h_2$,

$$\frac{2k}{hb} = 1$$

or, rewriting,

$$\frac{1}{h} = \frac{b/2}{k} \tag{3-32}$$

The left-hand expression is the film heat-transfer resistance, and the right-hand term is the thermal-conduction resistance of a plane wall of thickness equal to one-half the fin thickness. When both resistances have the same magnitude, the limit is reached beyond which fins are useless. It should be borne in mind, however, that as fins become shorter, the heat flow becomes two-dimensional and therefore differs from the assumptions made in the derivation of Eq. (3-29). The heat-flow lines and isotherms in such an actual short fin are as shown in Fig. 3-9. The numerical value of Eq. (3-32) is influenced by this two-dimensional effect. One may be sure, however, that it is advantageous to use finned surfaces as soon as the

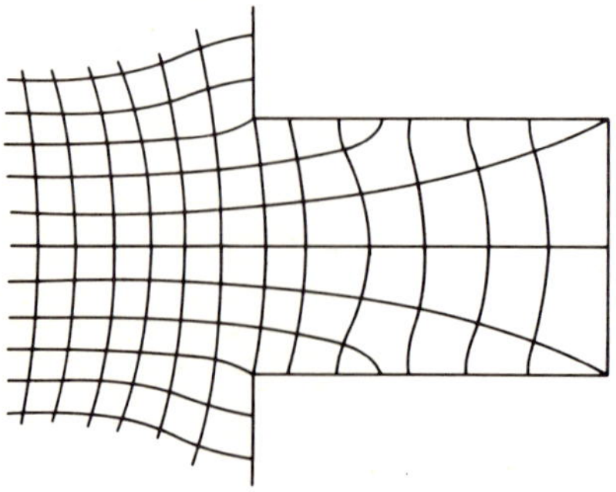

Fig. 3-9 Temperature distribution and heat-flow paths in a short rectangular fin.

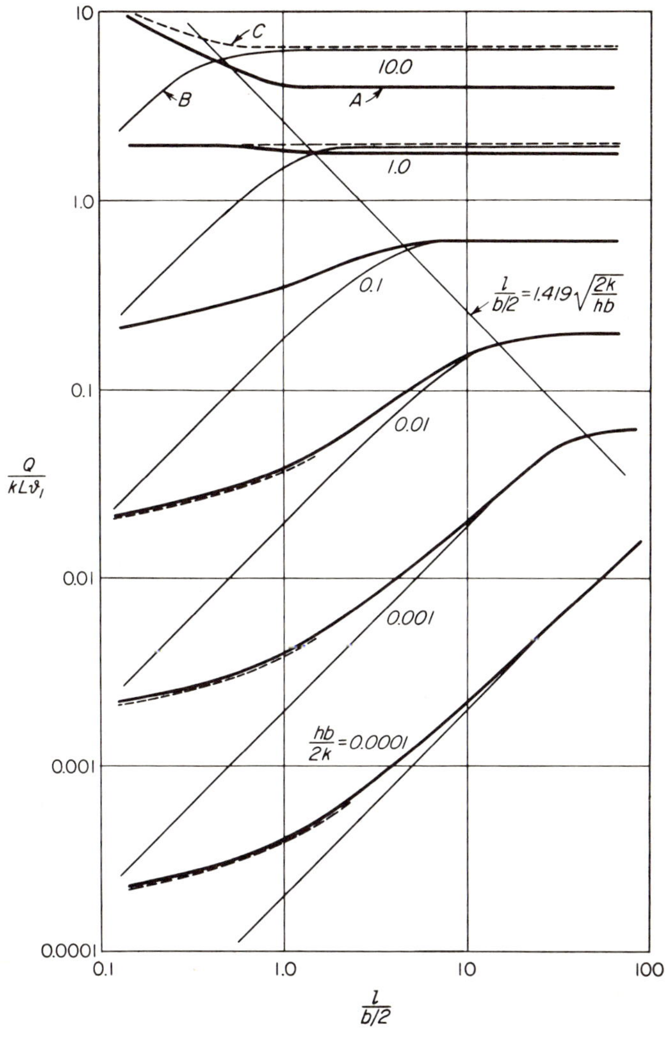

Fig. 3-10 Performance of short and long fins: (A), short fins; (B), long fins, insulated end; (C), long fin, heat loss from the end. [*Courtesy of Dr. D. L. Doughty.*]

condition

$$\frac{2k}{hb} > 5 \tag{3-33}$$

is fulfilled.

Figure 3-10 represents a comparison of the heat flow through the base of three fins and shows those fin lengths which can be treated as long fins with the subsequent use of the simpler one-dimensional heat-flow analysis.

Example 3-2 It is to be determined when iron fins of 0.30-cm thickness on a heating surface are advantageous. The thermal conductivity of iron can be found in Appendix Table B-1. The mean value of the thermal conductivity is taken as $k = 55$ W/m K. If the heat is transferred to air, the heat-transfer coefficient will fall in the range $10 < h < 100$ W/m^2 K. Using the higher value, we find

$$\frac{2k}{hb} = 365$$

For heat transfer to air (or other gases), fins are therefore advantageous. If the heat is to be transferred to water, the heat-transfer coefficients lie in the range $550 < h < 5{,}500$ W/m^2 K. If again the higher value is taken, the characteristic value becomes

$$\frac{2k}{hb} = 6.6$$

As this value is very low, the addition of fins to the surface will not materially increase the heat transfer. The effectiveness of fins for heat transfer to liquids can be increased by using thinner fins of a metal with higher thermal conductivity. Practically speaking the thickness of the fins cannot be reduced appreciably below 0.30 cm, and the conductivity of copper, the commercial material with the highest conductivity, is only six times that of iron. Therefore, fins provide no great improvement in heat transfer to liquids.

When fins are advantageous, the heat transfer is increased by placing them as near to each other as practicable. There is a limit, however, to the distance between two fins, for the heat-transfer coefficients decrease when the fluid boundary layers which occur on the surfaces of the fins influence each other. The distance between two fins must not be appreciably smaller, therefore, than twice the boundary-layer thicknesses. Boundary-layer-thickness calculations will be considered in another section of this book; however, as an order of magnitude, air flowing along a plate 30 cm long with a 15-m/s velocity creates a boundary layer which is approximately 0.25 cm thick. Low velocities, e.g., those occurring in a household convector, result in boundary layers approximately 1 cm thick.

The rectangular fin of minimum weight For the design of cooling devices on vehicles, especially aircraft, the problem of exchanging the greatest amount of heat with the least amount of weight in the heat exchanger is of paramount importance. This problem of minimum weight–maximum heat transfer–will be discussed for finned surfaces. The weight of one fin (Fig. 3-8) is

$$W = \rho Lbl = \rho LA_1 \tag{3-34}$$

where ρ is the density of the material for the fin and A_1 is the area of the fin cross section normal to L. The length L is fixed at a given dimension, whereas the two dimensions b and l are to be changed so as to give maximum heat flow for a given area A_1. In Eq. (3-26), for example, we insert

$m = \sqrt{2h/kb}$ and $A = bL$. (Note the difference between A and A_1: A is measured normal to the direction of heat flow, and A_1 is measured in the plane of the drawing in Fig. 3-8 to be perpendicular to dimension L.) If the height l of the fin is expressed in terms of the area A_1 ($l = A_1/b$), then

$$Q_1 = L\sqrt{2hkb}\,\vartheta_1 \tanh\left(\sqrt{\frac{2h}{kb}}\frac{A_1}{b}\right) \tag{3-35}$$

The expression in Eq. (3-35) becomes a maximum for $dQ_1/db = 0$; thus by differentiating and setting the result equal to zero, we get

$$\frac{1}{2\sqrt{b}}\tanh\left(\sqrt{\frac{2h}{kb}}\frac{A_1}{b}\right) - \sqrt{b}\sqrt{\frac{2h}{k}}\,A_1\frac{3}{2}b^{-\frac{5}{2}}\frac{1}{\cosh^2\left[\sqrt{2h/kb}\,(A_1/b)\right]} = 0 \tag{3-36}$$

Introducing the abbreviation

$$\sqrt{\frac{2h}{kb}}\frac{A_1}{b} = u$$

gives

$$\tanh u = \frac{3u}{\cosh^2 u} \tag{3-37}$$

This transcendental equation must be solved numerically or graphically, for example, by plotting both sides against u and thus determining the point of intersection of the two curves. Thus, the value $u = 1.419$ that satisfies the equation can be obtained. The maximum heat flow through a fin of given weight is obtained when the following equation is fulfilled:

$$\frac{l}{b/2} = 1.419\sqrt{\frac{2k}{hb}} \tag{3-38}$$

Here it can be seen that the ratio of height to half this fin thickness depends on the same characteristic value found in Eq. (3-32). The grouping $h(b/2)/k$ occurs so often in heat-conduction problems with convection boundary conditions that the *general form* of the group has been given a name, the *Biot modulus.* It is dimensionless and is similar to the *Nusselt number,* which will be encountered in the study of heat convection. However, there is an important difference: the thermal conductivity in the Biot modulus refers to the *conducting body,* while in the Nusselt number the thermal conductivity is that of the *convecting fluid.*

The temperature excess at the end of the optimum fin over the surrounding fluid is

$$\vartheta_2 = \frac{\vartheta_1}{\cosh u} = 0.457\vartheta_1 \tag{3-39}$$

Equation (3-39) provides a test as of whether the optimum height is obtained by measurement of ϑ_1 and ϑ_2.

When the heating surface has no fins, the area equivalent to the fin base transmits the heat flow $Q_1' = hbL\vartheta_1$. The ratio of the heat flow through the fins Q_1 to the heat flow Q_1' without the fins is, for the best fin according to Eq. (3-35),

$$\frac{Q_1}{Q_1'} = \sqrt{\frac{2k}{hb}} \tanh u = 0.889 \sqrt{\frac{2k}{hb}} \tag{3-40}$$

Equation (3-40) makes it possible to determine the heat-flow increase through the wall as a result of the addition of fins for the system considered.

Example 3-3 For an iron fin 0.30 cm thick the characteristic value $2k/hb$ was determined to be 365 for air and 6.6 for water. When these values are introduced into Eq. (3-38), the optimum ratio $l/(b/2)$ becomes 27 for the heat transfer to air and 3.64 for the heat transfer to water. An aluminum fin 0.10 cm thick, with a value of thermal conductivity $k = 208$ W/m K, has an optimum ratio $l/(b/2) = 91.5$ for air and 12.3 for water. It can be seen that the fin must be thicker when the ratio of the film resistance to the conduction resistance becomes smaller. Iron fins for heat transfer to water are of little use.

The steel fins of air-cooled cylinder barrels for reciprocating aeroengines are approximately 0.10 cm thick and 2.0 cm high with 0.40-cm spacing. The cylinder head is generally equipped with aluminum fins about 0.15 cm thick and 3.5 cm high with 0.50-cm spacing. The dimensions are very close to the optimum values, as may be calculated from Eq. (3-38). The radiators (really convectors) of liquid-cooled aeroengines also have very thin fins, approximately 0.01 to 0.02 cm thick and 0.50 to 1.0 cm high. Stationary heat-transfer equipment has fins which are generally thicker than the optimum value, as the saving of weight is not so important. Finned tubes of economizers for steam boilers have fins 0.40 to 0.60 cm thick and 1.50 to 2.50 cm high with 1.50-to 2.0-cm spacing; finned tubes for warm-water heating systems have fins 0.10 to 0.25 cm thick and 2.5 to 4.0 cm high with 1.0-to 2.50-cm spacing.

Fin effectiveness Equation (3-40) is one expression for *fin effectiveness*. Since the reference heat-flow system is somewhat arbitrary, two definitions for fin effectiveness need to be discussed: (1) effectiveness relative to the base area without the fin and (2) effectiveness relative to a similar fin of infinite thermal conductivity. It is clear that Eq. (3-40) belongs to type 1. As a means of illustration, both effectivenesses are shown below for the simple rectangular fin.

Infinitely long fin:

$$\eta = \frac{\sqrt{hCkA}\,\vartheta_1}{hA\vartheta_1} = \sqrt{\frac{kC}{hA}} \tag{3-41}$$

$$\varphi = \frac{\sqrt{hCkA}\,\vartheta_1}{hCl\vartheta_1} = \frac{1}{ml} \tag{3-42}$$

Finite fin:

$$\eta = \frac{\sqrt{hCkA}\,\vartheta_1 \tanh ml}{hA\vartheta_1} = \sqrt{\frac{kC}{hA}} \tanh ml \tag{3-43}$$

$$\varphi = \frac{\sqrt{hCkA}\,\vartheta_1 \tanh ml}{hCl\vartheta_1} = \frac{\tanh ml}{ml} \tag{3-44}$$

In Eqs. (3-41) to (3-44), η is the effectiveness relative to the base area, and φ is the effectiveness relative to a similar fin of infinite thermal conductivity.

Straight fin of triangular profile In determining the optimum fin, the question arises whether weight advantage can be gained by using a shape other than rectangular for the fin cross section that has been considered thus far. In the discussion that follows, a straight fin of triangular cross section will be considered. Such a fin is shown in Fig. 3-11. The mathematical treatment in this case is similar to that for the fin of rectangular cross section except that the area normal to the heat flow is a function of the distance along the fin, decreasing as the fin length increases. Thus, when written for constant thermal conductivity and heat-transfer coefficient, the differential equation becomes

$$\frac{d}{dx}\left[A(x)\frac{dt}{dx}\right] = \frac{hC}{k}(t - t_f) \tag{3-45}$$

The area $A(x)$ can be expressed directly in terms of the fin-length ratio x/l

$$A(x) = bL\frac{x}{l}$$

and for the case where $L \gg b$, the periphery becomes

$$C = 2L$$

If these values are substituted into Eq. (3-45), the following differential equation is the result:

$$\frac{d^2\vartheta}{dx^2} + \frac{1}{x}\frac{d\vartheta}{dx} - \beta\frac{\vartheta}{x} = 0 \tag{3-46}$$

where ϑ is the temperature excess over the surrounding fluid temperature as before and

$$\beta = \frac{2hl}{kb}$$

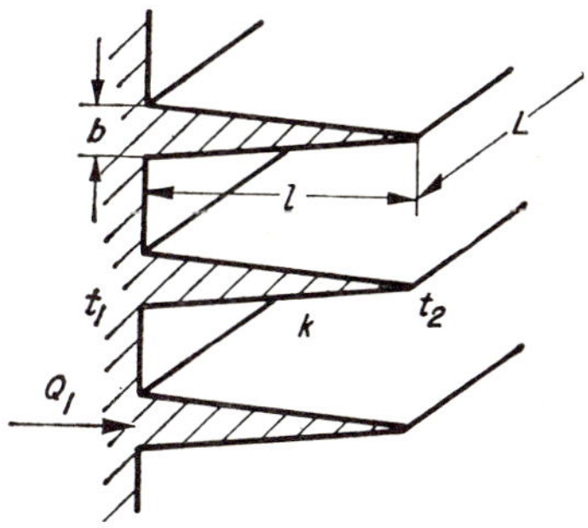

Fig. 3-11 Fins of triangular cross section.

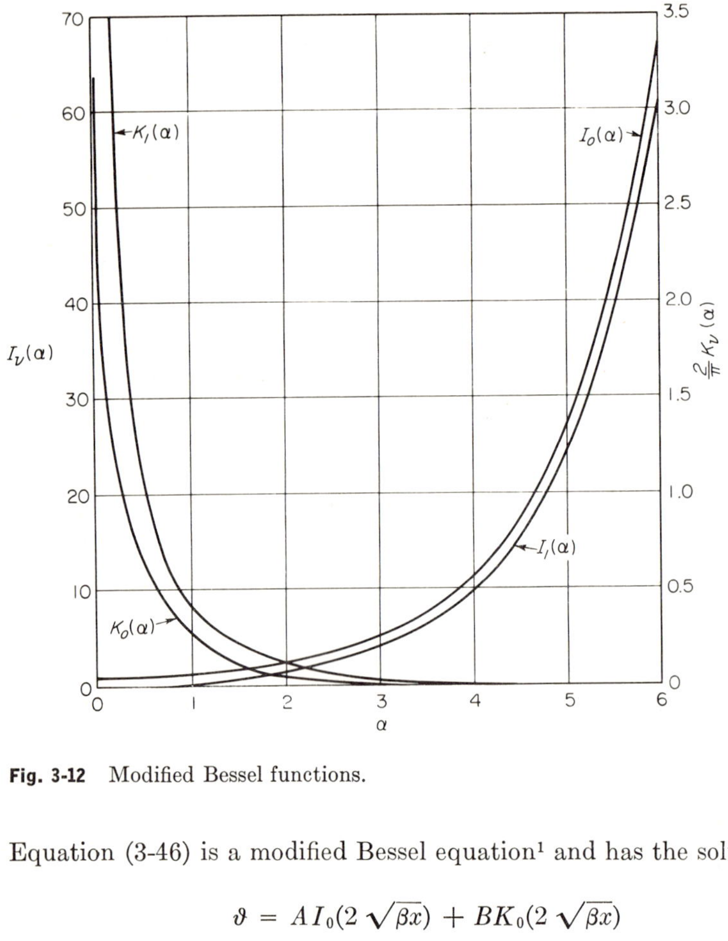

Fig. 3-12 Modified Bessel functions.

Equation (3-46) is a modified Bessel equation[1] and has the solution

$$\vartheta = AI_0(2\sqrt{\beta x}) + BK_0(2\sqrt{\beta x}) \tag{3-47}$$

The functions $I_0(\alpha)$ and $K_0(\alpha)$ are graphed in Fig. 3-12 and are given in tabular form in Table 3-2.

It can be seen that $K_0(\sqrt{\beta x})$ has the value infinity at the tip of the fin ($x = 0$), and as the temperature is physically not infinite at this location, the coefficient B for this term must be zero. The remaining boundary condition relates to the temperature at the base of the fin:

$$\vartheta = \vartheta_1 \qquad \text{for } x = l$$

Introduction of this boundary condition into Eq. (3-47) permits determi-

[1] See, for example, N. W. MacLachlan, "Bessel Functions for Engineers," Oxford University Press, London, 1934.

Table 3-2 Selected magnitudes of Bessel functions*

α	$I_0(\alpha)$	$I_1(\alpha)$	$\frac{2}{\pi} K_0(\alpha)$	$\frac{2}{\pi} K_1(\alpha)$
0.0	1.0000	0.0000		
0.2	1.01003	0.1005	1. 116	3. 040
0.4	1.04040	0.2040	0.7 095	1. 391
0.6	1.09205	0.3137	0.4 950	0.8 294
0.8	1.1665	0.4329	0.3 599	0.5 486
1.0	1.2661	0.5652	0.2 680	0.3 832
1.2	1.3937	0.7147	0.2 028	0.2 767
1.4	1.5534	0.8861	0.15 512	0.2 043
1.6	1.7500	1.0848	0.11 966	0.15 319
1.8	1.9896	1.3172	0.09 290	0.11 626
2.0	2.2796	1.5906	0.07 251	0.08 904
2.2	2.6291	1.9141	0.05 683	0.06 869
2.4	3.0493	2.2981	0.04 470	0.05 330
2.6	3.5533	2.7554	0.03 527	0.04 156
2.8	4.1573	3.3011	0.02 790	0.03 254
3.0	4.8808	3.9534	0.02 212	0.02 556
3.2	5.7472	4.7343	0.017 568	0.02 014
3.4	6.7848	5.6701	0.013 979	0.015 915
3.6	8.0277	6.7927	0.011 141	0.012 602
3.8	9.5169	8.1404	$0.0^{2}8$ 891†	$0.0^{2}9$ 999
4.0	11.3019	9.7595	$0.0^{2}7$ 105	$0.0^{2}7$ 947
4.2	13.4425	11.7056	$0.0^{2}5$ 684	$0.0^{2}6$ 327
4.4	16.0104	14.0462	$0.0^{2}4$ 551	$0.0^{2}5$ 044
4.6	19.0926	16.8626	$0.0^{2}3$ 648	$0.0^{2}4$ 027
4.8	22.7937	20.2528	$0.0^{2}2$ 927	$0.0^{2}3$ 218
5.0	27.2399	24.3356	$0.0^{2}2$ 350	$0.0^{2}2$ 575
5.2	32.5836	29.2543	$0.0^{2}1$ 888	$0.0^{2}2$ 062
5.4	39.0088	35.1821	$0.0^{2}15$ 181	$0.0^{2}16$ 531
5.6	46.7376	42.3283	$0.0^{2}12$ 214	$0.0^{2}13$ 262
5.8	56.0381	50.9462	$0.0^{3}9$ 832	$0.0^{2}10$ 648
6.0	67.2344	61.3419	$0.0^{3}7$ 920	$0.0^{3}8$ 556
6.2	80.72	73.89	$0.0^{3}6$ 382	$0.0^{3}6$ 879
6.4	96.98	89.03	$0.0^{3}5$ 146	$0.0^{3}5$ 534
6.6	116.54	107.30	$0.0^{3}4$ 151	$0.0^{3}4$ 455
6.8	140.14	129.38	$0.0^{3}3$ 350	$0.0^{3}3$ 588
7.0	168.6	156.04	$0.0^{3}2$ 704	$0.0^{3}2$ 891
7.2	202.9	188.3	$0.0^{3}2$ 184	$0.0^{3}2$ 331
7.4	244.3	227.2	$0.0^{3}17$ 646	$0.0^{3}1$ 880
7.6	294.3	274.2	$0.0^{3}14$ 262	$0.0^{3}15$ 172
7.8	354.7	331.1	$0.0^{3}11$ 530	$0.0^{3}12$ 248
8.0	427.6	399.9	$0.0^{4}9$ 325	$0.0^{4}9$ 891
8.2	515.6	483.0	$0.0^{4}7$ 543	$0.0^{4}7$ 991
8.4	621.9	583.7	$0.0^{4}6$ 104	$0.0^{4}6$ 458
8.6	750.5	705.4	$0.0^{4}4$ 941	$0.0^{4}5$ 220
8.8	905.8	852.7	$0.0^{4}4$ 000	$0.0^{4}4$ 221
9.0	1,093.6	1,030.9	$0.0^{4}3$ 239	$0.0^{4}3$ 415
9.2	1,320.7	1,246.7	$0.0^{4}2$ 624	$0.0^{4}2$ 763
9.4	1,595.3	1,507.9	$0.0^{4}2$ 126	$0.0^{4}2$ 236
9.6	1,927	1,824	$0.0^{4}17$ 226	$0.0^{4}1$ 810
9.8	2,329	2,207	$0.0^{4}13$ 962	$0.0^{4}14$ 658
10.0			$0.0^{4}11$ 319	$0.0^{4}11$ 872

* L. M. K. Boelter, V. H. Cherry, H. A. Johnson, and R. C. Martinelli, "Heat Transfer Notes." Copyright 1965.

† $0.0^{2}8$ 891 = 0.008891. McGraw-Hill Book Company, 1965. Used by permission.

nation of the remaining constant

$$A = \frac{\vartheta_1}{I_0(2\sqrt{\beta l})}$$

Thus the final equation for the temperature distribution becomes

$$\frac{\vartheta}{\vartheta_1} = \frac{I_0(2\sqrt{\beta x})}{I_0(2\sqrt{\beta l})} \tag{3-48}$$

The heat rate can be determined from the Fourier conduction law and the first derivative of Eq. (3-48), remembering that $dI_0(\alpha)/d\alpha = I_1(\alpha)$

$$Q_1 = L\vartheta_1\sqrt{2hkb}\,\frac{I_1(2\sqrt{\beta l})}{I_0(2\sqrt{\beta l})} \tag{3-49}$$

To optimize the fin of triangular cross section, let $2\sqrt{\beta l} = u$ in Eq. (3-49). A_1, the cross-sectional area, is to be held constant, allowing b and l to vary

$$A_1 = \frac{bl}{2}$$

$$b = \left(\frac{32A_1^2h}{u^2k}\right)^{\frac{1}{3}}$$

$$Q_1 = (16kh^2A_1)^{\frac{1}{3}}L\vartheta_1\,\frac{u^{-\frac{1}{3}}I_1(u)}{I_0(u)}$$

For Q_1 to be a maximum as u varies requires $dQ_1/du = 0$, leading to the relationship

$$\frac{4}{3}\frac{1}{u}\frac{I_1(u)}{I_0(u)} = 1 - \left[\frac{I_1(u)}{I_0(u)}\right]^2$$

which can be solved to yield the following relationship between l and b for maximum heat transfer:

$$\frac{l}{b/2} = 1.309\sqrt{\frac{2k}{hb}} \tag{3-50}$$

The temperature excess at the tip of the fin is

$$\vartheta_2 = 0.277\vartheta_1 \tag{3-51}$$

The ratio of the thickness of the triangular fin to the thickness of the rectangular fin with equal heat flow is 1.31; the ratio of the cross-sectional areas is 1:1.44. Therefore, the weight saved by using the triangular fin is 44 percent.

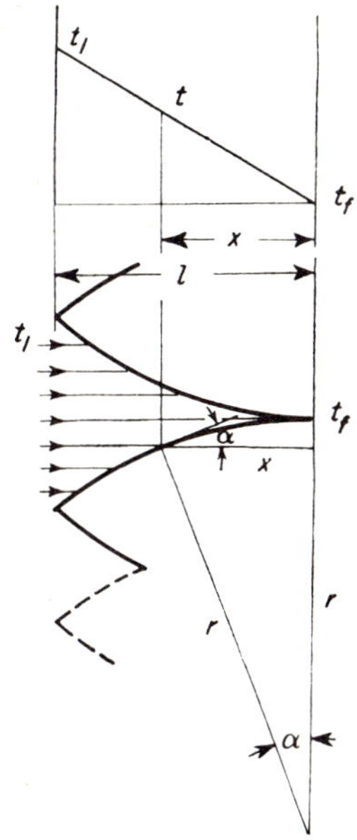

Fig. 3-13 Fin with smallest weight.

Fin of minimum weight[1] It is of further interest to determine the optimum shape of a fin having the smallest weight for a given heat flow. It is evident that in such a fin every part should be utilized to the same degree or that the specific rate of heat flow should be constant throughout the fin. The proof of this fact was presented by Schmidt.[2] The heat-flow lines in such a fin must have shapes as indicated in Fig. 3-13. The heat-flow lines are equally spaced and parallel to the fin axis. With a constant specific rate of heat flow the temperature decreases linearly along any flow line from the value t_1 at the root of the fin. The tip of the fin will approach the value of the fluid surrounding the fin only if the angle α goes to zero. For a finite value of α there will be a temperature discontinuity between the fin tip and the surrounding fluid. The temperatures within the fin are indicated in the upper part of Fig. 3-13. At a distance x from the fin tip the temperature is t. Because of the linearity of the temperature distribution, the difference between the temperature t and the fluid temperature t_f can be expressed as

$$t - t_f = \frac{x}{l}(t_1 - t_f)$$

[1] E. Schmidt, *Z. Ver. Deut. Ingr.*, **70**:885 (1926). Calculations were made for pin fins on a heating surface by R. Focke, *Forsch. Gebiete Ingenieurw.*, **13**:34–42 (1942); experimental investigation on different fin arrangements by R. H. Norris and W. A. Spofford, *Trans. ASME*, **64**:489–496 (1942).

[2] *Loc. cit.*

The constant specific rate of heat flow in the direction of the fin axis may be q W/m². Consider a surface element of the fin at a distance x. The element may be inclined toward the fin axis by the angle α. The specific rate of heat flow through this element is $q \sin \alpha$, and since the heat flowing through this surface element has to be transferred to the surrounding fluid, the equation

$$q \sin \alpha = h(t - t_f) = h \frac{x}{l} (t_1 - t_f)$$

is valid. This equation establishes the angle α of the surface element as a function of the distance x:

$$\sin \alpha = \frac{h\vartheta_1}{ql} x$$

The fraction on the right-hand side of the equation has a constant value. The contour of a fin defined in this way is a circle, for a circle is described by the equation $\sin \alpha = x/r$, which corresponds exactly to the above equation. The radius r of the circle has the value $ql/h\vartheta_1$. Weinig first pointed out the fact that the surface contours of fins of minimum weight are circles. It is not necessary to give the fin an infinitely thin edge as shown in Fig. 3-13. Any circular segment can be used to build up a contour for a fin of constant heat-flow distribution. Circular-arc fins having finite angles at the tip are shown by the dashed lines in Fig. 3-13. The difference in weight between a fin in the shape of a circular arc and one of a triangular shape, however, is very small. As the triangular shape is much easier to manufacture, it may, for practical purposes, be regarded as the best form.

Fin arrangement The cross-sectional area A_1 necessary for a given heat flow in the instance of a rectangular fin is derived by combining Eqs. (3-35) and (3-37) and solving for A_1:

$$A_1 = \left(\frac{Q_1}{\vartheta_1}\right)^3 \frac{1}{L^3} \frac{u}{\tanh^3 u} \frac{1}{4h^2k} = \frac{2.109}{4L^3h^2k} \left(\frac{Q_1}{\vartheta_1}\right)^3 \tag{3-52}$$

This equation shows that it is advantageous to make the fins as small as possible. To double the heat flow, the area of one fin must be increased eightfold, whereas it is sufficient to use two fins of the original size.

Equation (3-52) allows for the comparison of different materials for fins. The cross-sectional area A_1 is inversely proportional to the thermal conductivity k. The mass is proportional, therefore, to ρ/k. Table 3-3 lists the quotient of the density ρ to the thermal conductivity k for different materials. It can be seen that by using aluminum, for example, a weight savings of 50 percent can be realized as compared with copper. Iron fins have a tenfold weight and iron-alloy fins a fifty-fold weight. It is of no benefit, therefore, to make the fins on air-cooled

Table 3-3 Comparison of fin materials

Material	*Thermal conductivity k,* W/m K	*Density* ρ, kg/m^3	ρ/k, $\frac{\text{kg K}}{\text{W m}^2}$	$\frac{\rho/k}{(\rho/k)_{Al}}$
Copper	383	8970	2.34	1.94
Aluminum, pure	225	2722	1.21	1
Alloy	156	2670	1.71	1.41
Magnesium, pure	173	1761	1.02	0.84
Steel	55	7850	14.26	11.8
Stainless	14	7850	56.0	46.2

cylinders of aircraft engines of copper instead of aluminum, whereas the use of aluminum instead of iron or steel promises great advantages.

Circumferential fins Fins which are arranged around tubes are called circumferential fins and are quite important from an engineering point of view. Such a fin system is shown in Fig. 3-14. Here again the treatment is substantially the same as for the rectangular fin except that the area must be allowed to vary with the radius. The area normal to the heat-flux vector can be written

$$A = 2\pi rb$$

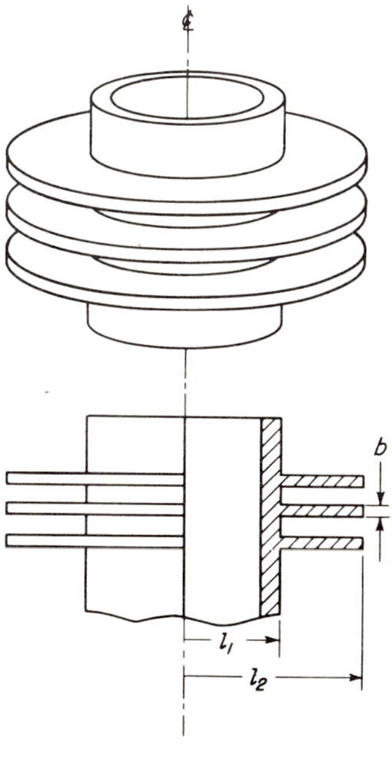

Fig. 3-14 Surface with circumferential fins.

and the periphery can be expressed as

$$C = 4\pi r$$

These values, when introduced into Eq. (3-45) with x replaced by the radius r, yield the differential equation

$$\frac{d^2\vartheta}{dr^2} + \frac{1}{r}\frac{d\vartheta}{dr} - \frac{2h}{kb}\vartheta = 0 \tag{3-53}$$

The appropriate boundary conditions for Eq. (3-53) may be

$$r = l_1 \qquad \vartheta = \vartheta_1$$

$$r = l_2 \qquad \frac{d\vartheta}{dr} = 0$$

The solution of Eq. (3-53) is obtained in the form of modified Bessel functions:

$$\frac{\vartheta}{\vartheta_1} = \frac{K_1(l_2\sqrt{\beta})I_0(r\sqrt{\beta}) + I_1(l_2\sqrt{\beta})K_0(r\sqrt{\beta})}{K_1(l_2\sqrt{\beta})I_0(l_1\sqrt{\beta}) + I_1(l_2\sqrt{\beta})K_0(l_1\sqrt{\beta})} \tag{3-54}$$

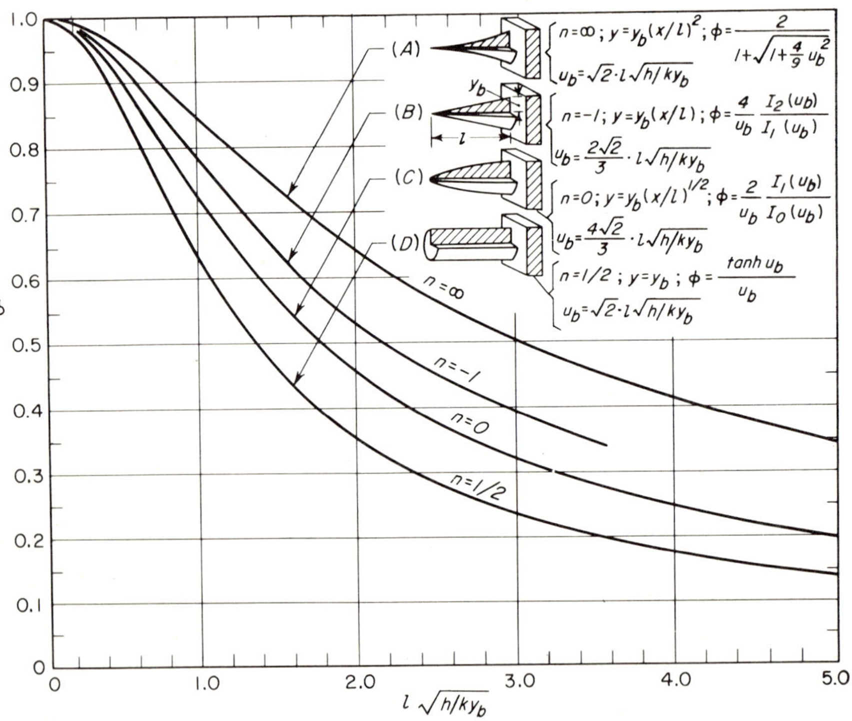

Fig. 3-15 Performance of pin fins. [*K. A. Gardner, Trans. ASME*, **67**:*621–631* (*1945*).]

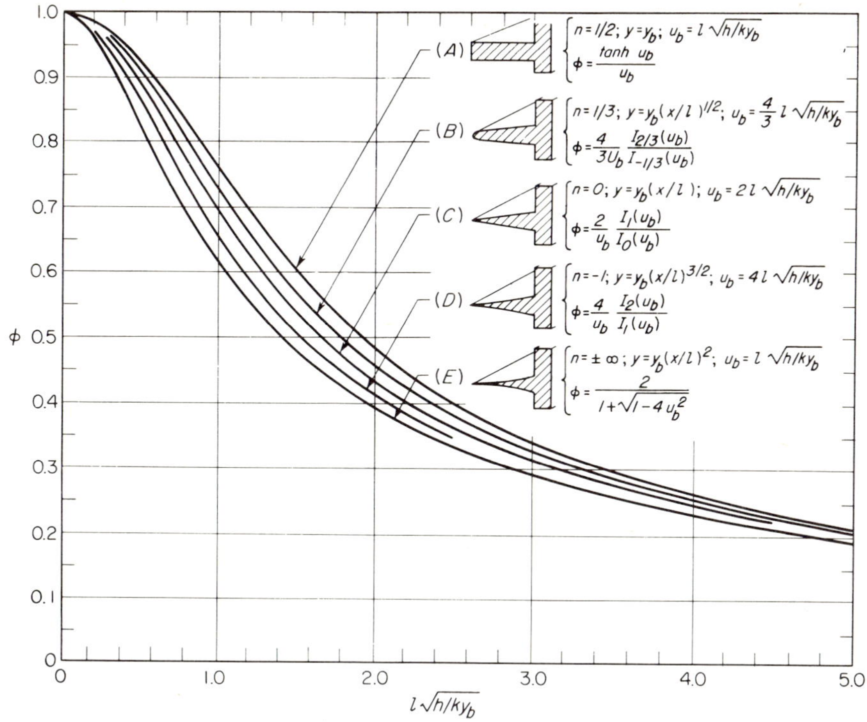

Fig. 3-16 Performance of straight fins. [*K. A. Gardner, Trans. ASME*, **67**:*621–631 (1945).*]

The heat flow into the base becomes

$$Q_1 = 2\pi l_1 bk \sqrt{\beta}\, \vartheta_1 \frac{I_1(l_2\sqrt{\beta})K_1(l_1\sqrt{\beta}) - K_1(l_2\sqrt{\beta})I_1(l_1\sqrt{\beta})}{I_1(l_2\sqrt{\beta})K_0(l_1\sqrt{\beta}) + K_1(l_2\sqrt{\beta})I_0(l_1\sqrt{\beta})} \qquad (3\text{-}55)$$

where $$\beta = \frac{2h}{kb}$$

Gardner[1] has generalized the relationships for extended-surface heat-exchanger systems to include not only the examples given in the preceding sections but also several other shapes. These results are useful in design and are presented in Figs. 3-15 to 3-18 in terms of the effectiveness based on a similar fin of infinite conductivity. Additional details of solutions for extended-surface heat-transfer devices are presented by Schneider.[2]

[1] K. A. Gardner, Efficiency of Extended Surface, *Trans. ASME*, **67**:621–631 (1945).

[2] P. J. Schneider, "Conduction Heat Transfer," Addison-Wesley, Reading, Mass., 1955.

Example 3-4 If it is desirable to increase the heat rejection from a cylinder barrel of an air-cooled engine by adding fins to enlarge the heat-transfer area, the usual type of fin is the circumferential fin shown in Fig. 3-14. The heat rejection from a finned barrel may be compared with the heat rejection of a barrel without fins. Equation (3-55) gives the heat loss of one fin of this type. Suppose the specifications require the fins to be 2.54 cm long and 0.228 cm thick and separated by 0.456 cm. The heat-transfer coefficient may be taken as $h = 284$ W/m² K due to forced flow over the fin. The thermal conductivity for steel from Appendix B is $k = 48$ W/m K. Thus if the cylinder wall is 393 K and the air temperature is 293 K,

$$l_1 = 7.62 \text{ cm} \qquad l_2 = 10.16 \text{ cm}$$

$$\vartheta_1 = 393 - 293 = 100 \text{ K}$$

$$\beta = \frac{2h}{kb} = \frac{2(284)(100)}{48(0.228)} = 5{,}200 \qquad \sqrt{\beta} = 72 \text{ m}^{-1}$$

$$l_1\sqrt{\beta} = \frac{7.62}{100}72 = 5.5 \qquad l_2\sqrt{\beta} = \frac{10.16}{100}72 = 7.3$$

From Table 3-2

$$I_1(7.3) = 207 \qquad K_1(7.3) = 0$$

$$I_1(5.5) = 38.5 \qquad K_1(5.5) = 0.0023$$

$$I_0(5.5) = 43.0 \qquad K_0(5.5) = 0.0021$$

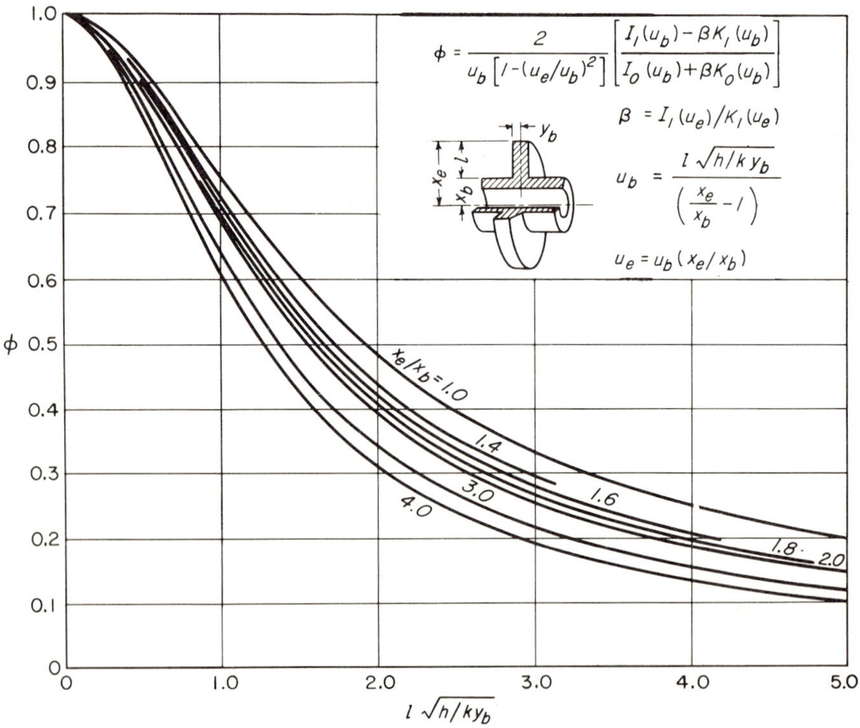

Fig. 3-17 Performance of circumferential fins of rectangular cross section. [*K. A. Gardner, Trans. ASME,* **67**:*621–631 (1945).*]

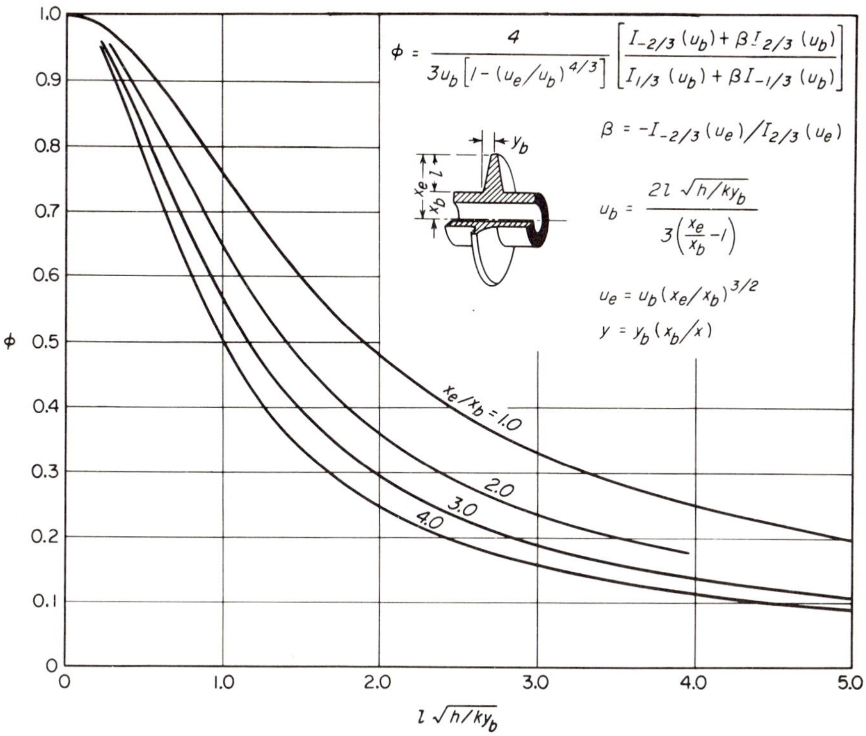

Fig. 3-18 Performance of circumferential fins of triangular cross section. [*K. A. Gardner, Trans. ASME,* **67**:*621–631 (1945).*]

Substituting the values above into Eq. (3-55)

$$Q_1 = 2\pi \frac{7.62}{100} \frac{0.228}{100} 48(72)(100) \frac{207(0.0023)}{207(0.0021)} = 414 \text{ W}$$

$$Q_B = 2\pi l_1 sh \, \Delta t = 2\pi \frac{7.62}{100} \frac{0.684}{100} 284(100) = 92.8 \text{ W}$$

$$\frac{Q_1}{Q_B} = 4.45$$

3-5 THE WALL WITH HEAT SOURCES

The previous considerations must be extended when heat is generated in any manner, i.e., when heat sources prevail in the heat-conducting body. An explicit case of heat sources in a conducting body is an electric conductor in which the dissipated electric energy is transformed into heat. The calculations of the temperatures which occur because of this heat generation are of specific interest in the design of electrical machines and apparatus. Other problem areas involving heat-source considerations lie in the chemical and nuclear fields. For example, refrigeration systems

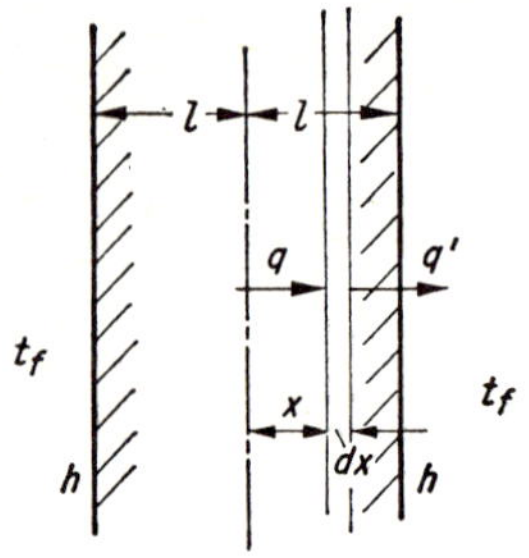

Fig. 3-19 The plane wall with heat sources.

must be designed to prevent production of inadmissibly high temperatures by the generation of heat during the setting of concrete. The flow losses in fluids are also transformed into heat. The temperatures caused by the dissipation of mechanical energy into heat in the case of high-speed gas flow (frictional heating) in aircraft currently is of the order of several hundreds of degrees and increases with the square of the Mach number. Considerable temperature increases occur from frictional heating in the oil films used for lubrication of fast-running bearings.

In this section, as an example, the treatment will be limited to solid bodies, and for simplicity the initial consideration will involve the plane or slab (Fig. 3-19). Uniformly distributed heat sources exist in the wall, and therefore a quantity of heat Q' per unit volume and time is generated. Hence Q' is independent of the space coordinate x. On each of the exposed surfaces the slab is bounded by a circulating fluid of temperature t_f. The heat-transfer coefficient for each surface is h. If the thermal conductivity is considered constant and the conditions are steady with time, Eq. (1-16) reduces to

$$\frac{d^2\vartheta}{dx^2} + \frac{Q'}{k} = 0 \tag{3-56}$$

where again ϑ is the excess temperature.

The boundary conditions for the determination of the two constants that will appear in the solution of Eq. (3-56) can be expressed as

$$\left(\frac{d\vartheta}{dx}\right)_{x=0} = 0$$

$$-k\left(\frac{\partial\vartheta}{\partial x}\right)_{x=l} = h\vartheta_{x=l}$$

The solution of Eq. (3-56) is

$$\vartheta = -\frac{Q'}{2k}x^2 + C_1x + C_2 \tag{3-57}$$

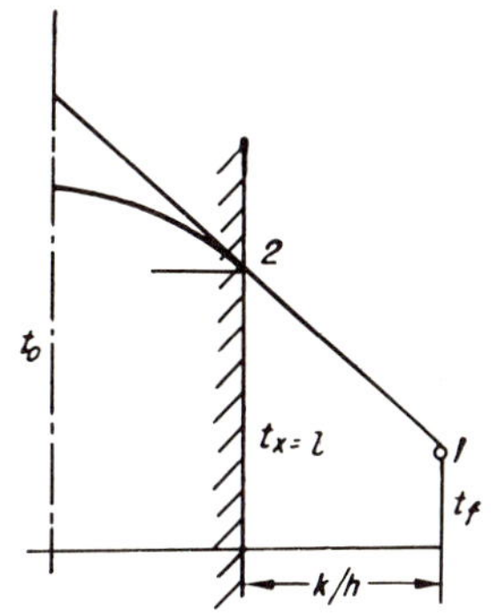

Fig. 3-20 Temperature distribution in the plane wall with heat sources.

By evaluating the constants in Eq. (3-57)

$$\vartheta = \frac{Q'}{2k}(l^2 - x^2) + \frac{Q'l}{h} \tag{3-58}$$

Thus, thermal traces in a plate with uniform heat generation are parabolas, as shown in Fig. 3-20. The maximum excess temperature is obtained from Eq. (3-58) when $x = 0$; thus

$$\vartheta_0 = \frac{Q'}{2k}l^2 + \frac{Q'l}{h} \tag{3-59}$$

The temperature gradient on the surface can be found easily by the method given in Fig. 3-20. If we plot the distance k/h from the surface, add the temperature t_f as ordinate, and connect point 1, which is obtained in this manner, with point 2 as defined by the surface temperature, the direction of the straight line connecting both points yields the temperature gradient at the wall. This result can be obtained directly by the solution of the second part of the boundary conditions for the gradient at the wall:

$$-\left(\frac{d\vartheta}{dx}\right)_{x=l} = \frac{\vartheta_{x=l}}{k/h} \tag{3-60}$$

Example 3-5 A cylindrical transformer coil made of insulated copper wire has an ID of 16.8 cm and an OD of 24.4 cm. The fraction $\phi = 0.6$ of the total cross section of the coil is copper, and the rest insulation. The current density in the conductors is $j = 200$ A/cm²; the resistivity of the copper is $\rho = 200 \times 10^{-6}$ Ω cm²/m. Hence the generation of heat per unit volume of space in the coil is

$$Q' = \Phi j^2 \rho = 0.6(200^2)(200 \times 10^{-6}) = 48{,}000 \text{ W/m}^3$$

The heat-transfer coefficient on both surfaces of the coil, which are cooled by air at 20.7 C, is $h = 22$ W/m² K; the effective thermal conductivity of the coil is $k = 0.350$ W/m K (mica, glue). If the coil in first approximation is considered as a plane wall with a thickness $2L = 3.80$ cm, the highest temperature prevailing in it can be approximated by Eq. (3-59)

$$\vartheta_0 = \frac{48{,}000(3.61)}{2(0.350)(100^2)} + \frac{48{,}000(1.9)}{22(100)} = 66.3 \text{ C}$$

Thus the temperature t_0 at the center of the spool is $66.3 + 20.7 = 87$ C. Accurate calculation of the thermal field in the coil considered as a hollow cylinder yields the value 88.3 C. Consequently, it can be seen that the calculation, as for a plane wall, yields useful results even on rather thick-walled coils.

3-6 THE BURIED CABLE

The use of heat sources (and sinks) makes it possible to analyze certain types of steady-state systems that are intractable by other means. Such a case is the buried cable. A cable is assumed to be buried in homogeneous surroundings such as firmly packed soil, and the cable is generating heat by some means, perhaps electrically (or in the case of a pipe carrying some fluid, perhaps chemically), such that the specific heat rate is constant. The medium surrounding the cable terminates in a level surface (Fig. 3-21) at a distance N above the centerline of the cable. The tem-

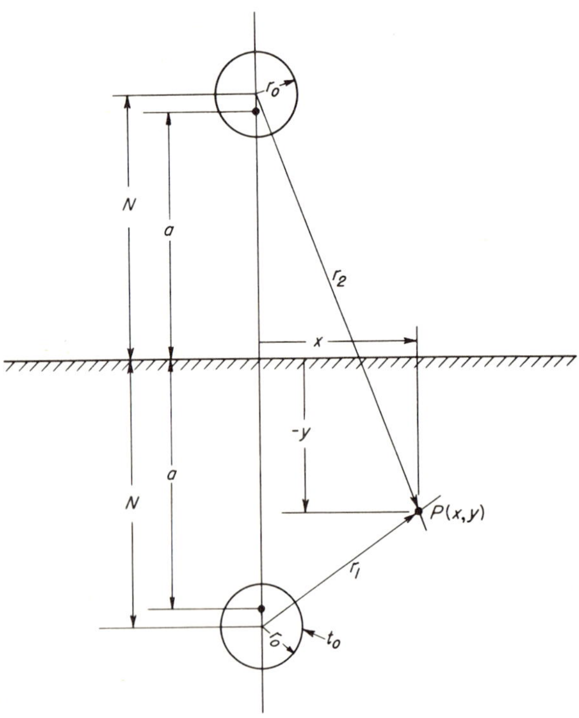

Fig. 3-21 The buried cable.

perature of the cable surface is taken to be uniformly t_0, and the surface temperature uniformly t_s. The excess temperature at any point $P(x,y)$ is then

$$\vartheta = t - t_s$$

and the temperature excess at the cable surface is

$$\vartheta_0 = t_0 - t_s$$

It is required to determine the temperature distribution around the cable within the soil.

To effect a solution, the cable is treated as a line heat source located a distance a below the surface causing heat to be conducted radially, the cable being assumed very long in the direction perpendicular to the plane of the paper. It is clear that if the soil extended infinitely in the x and y directions, the isotherms would be circles oriented concentrically about the cable center and there would be no linear constant-temperature surface as is required by the proposed problem. However, if we imagine that a mirror-image system exists, as shown in Fig. 3-21, except that the upper fictitious cable is taken not as a heat source but as a heat sink, at points equidistant from the source and sink the temperature will be just between the two. This system will yield an isotherm along $y = 0$.

Consider the cable. We know that the cable will have a heat flow

$$Q = \frac{\vartheta_0 - \vartheta_1}{(1/2\pi kL) \ln (r_1/r_0)} \tag{3-61}$$

and that due to the cable alone the temperature excess at point $P(x,y)$ will be

$$\vartheta_1 = \vartheta_0 - \frac{Q}{2\pi kL} \ln \frac{r_1}{r_0} \tag{3-62}$$

If the fictitious cable is to act as a heat sink, it must have the temperature $-\vartheta_0$ on its surface and must absorb the same amount of heat as that generated by the real cable. Thus its effect on the point $P(x,y)$ is

$$\vartheta_2 = -\vartheta_0 + \frac{Q}{2\pi kL} \ln \frac{r_2}{r_0} \tag{3-63}$$

Since the heat-conduction equation is linear, one solution can be superimposed upon another solution to yield a third solution. Therefore, Eqs. (3-62) and (3-63) can be added together to give a specification of the temperature excess at $P(x,y)$ as a result of both source and sink:

$$\vartheta = \frac{Q}{2\pi kL} \ln \frac{r_2}{r_1} \tag{3-64}$$

Now the radii r_1 and r_2 can be written in terms of x, y, and a with small error to read

$$r_1^2 = x^2 + (y + a)^2 \qquad \text{and} \qquad r_2^2 = x^2 + (y - a)^2$$

When these values are introduced into Eq. (3-64),

$$\vartheta = \frac{Q}{4\pi kL} \ln \frac{x^2 + (y - a)^2}{x^2 + (y + a)^2} \tag{3-65}$$

Equation (3-65) can likewise be written

$$e^{(4\pi kL/Q)\vartheta} = \frac{x^2 + (y - a)^2}{x^2 + (y + a)^2} \tag{3-66}$$

The equations for the isotherms can be obtained by holding the left-hand term a constant, which we can call C. Thus an equation in x and y parametric in C is obtained:

$$x^2 + \left(y - \frac{1 + C}{1 - C} a\right)^2 = \frac{4Ca^2}{(1 - C)^2} \tag{3-67}$$

Equation (3-67) represents a family of circles with origins at $x = 0$; $y = a(1 + C)/(1 - C)$, with radii $= 2a\sqrt{C}/(1 - C)$. The circle representing the $\vartheta = 0$ isotherm is the one in which $C = 1$ with a radius of infinity, therefore, a straight line. The center of this circle is located at negative infinity on the y axis.

The heat transfer from a cable such as that considered above can be obtained for any depth of submergence. It must be recognized that the actual line heat source will be higher than the centerline of the cable (the cable surface in reality is one of the isotherms) (Fig. 3-22) because

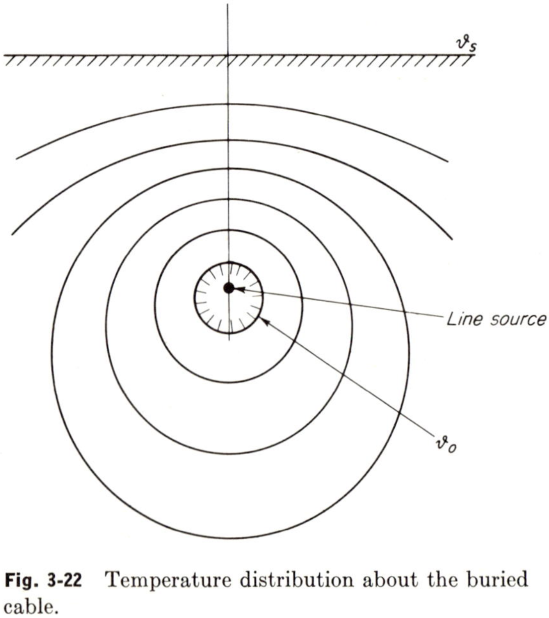

Fig. 3-22 Temperature distribution about the buried cable.

the value $[(1 + C)/(1 - C)] > 1$, since $C > 1$; thus the centers of the isotherm circles move downward as the radii increase. If D is the cable diameter and N is the depth of the cable centerline below the surface,

$$N = \frac{1 + C}{1 - C}\, a$$

$$\frac{D}{2} = \frac{2a\sqrt{C}}{1 - C} = \frac{2\sqrt{C}\,N}{1 + C}$$

The last equation can be written explicitly in C:

$$C^2 + \left(2 - 16\frac{N^2}{D^2}\right)C + 1 = 0 \tag{3-68}$$

The quadratic equation in C can be solved by the usual methods:

$$C = \left[8\left(\frac{N}{D}\right)^2 - 1\right] \pm 4\frac{N}{D}\sqrt{4\left(\frac{N}{D}\right)^2 - 1} = e^{(4\pi kL/Q)\vartheta} \tag{3-69}$$

The question of the correct sign for Eq. (3-69) must be determined from the physics of the system. Suppose N/D is very large; therefore

$$C = 8\left(\frac{N}{D}\right)^2 \pm 8\left(\frac{N}{D}\right)^2$$

If the negative sign is used, $C = 0$, which indicates that

$$e^{(4\pi kL/Q)\vartheta} = 0$$

and thus that ϑ is negative. This statement violates the initial assumption that the cable was a heat source. Thus, the positive sign must be retained. The final result is

$$C = \left[8\left(\frac{N}{D}\right)^2 - 1\right] + 4\frac{N}{D}\sqrt{4\left(\frac{N}{D}\right)^2 - 1} = e^{(4\pi kL/Q)\vartheta} \tag{3-70}$$

which can be solved for Q to yield

$$Q = \frac{2\pi kL\vartheta_0}{\ln\,[2(N/D) + \sqrt{4(N/D)^2 - 1}]} \tag{3-71}$$

where ϑ_0 is the temperature difference and the rest of the right-hand side of the equation makes up the thermal resistance

$$R = \frac{1}{2\pi kL}\ln\left[2\frac{N}{D} + \sqrt{4\left(\frac{N}{D}\right)^2 - 1}\right] \tag{3-72}$$

If $N/D \gg 1$, Eq. (3-71) reduces to

$$Q = \frac{2\pi kL\vartheta_0}{\ln\,[4(N/D)]} \tag{3-73}$$

3-7 THE CYLINDER WITH ECCENTRIC BORE

It is useful to note that Eq. (3-71) is also the solution for the heat flow through a circular cylinder whose bore is also circular but eccentric with the outer wall (Fig. 3-23). It is required, however, that the surfaces be maintained at constant but different temperatures, and that each surface represent an isotherm of the buried cable.

The heat flow Q across each isotherm in Fig. 3-23 is the same value as that generated from the heat source; thus an expression can be written for the heat flow across the isotherm ϑ_1

$$Q = \frac{\vartheta_1}{(1/2\pi kL) \ln [2(N_1/D_1) + \sqrt{4(N_1/D_1)^2 - 1}]} \tag{3-74}$$

and an expression for the same heat flow across the isotherm ϑ_2

$$Q = \frac{\vartheta_2}{(1/2\pi kL) \ln [2(N_2/D_2) + \sqrt{4(N_2/D_2)^2 - 1}]} \tag{3-75}$$

Combining Eqs. (3-74) and (3-75) for a temperature difference,

$$\vartheta_1 - \vartheta_2 = \frac{Q}{2\pi kL} \ln \frac{2(N_1/D_1) + \sqrt{4(N_1/D_1)^2 - 1}}{2(N_2/D_2) + \sqrt{4(N_2/D_2)^2 - 1}} \tag{3-76}$$

The thermal resistance for the cylinder with the eccentric bore is thus

$$R = \frac{1}{2\pi kL} \ln \frac{2(N_1/D_1) + \sqrt{4(N_1/D_1)^2 - 1}}{2(N_2/D_2) + \sqrt{4(N_2/D_2)^2 - 1}} \tag{3-77}$$

From the two equations preceding Eq. (3-68) and Fig. 3-23, the following

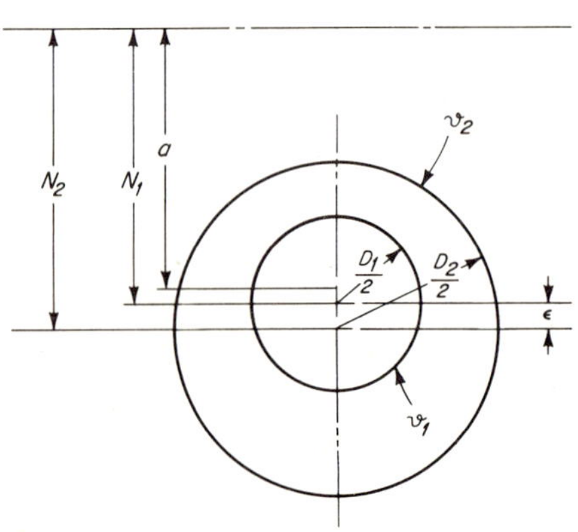

Fig. 3-23 The cylinder with eccentric bore.

relationships can be shown to be true:

$$N_1^2 - \left(\frac{D_1}{2}\right)^2 = N_2^2 - \left(\frac{D_2}{2}\right)^2 = a^2$$
$$N_2 - N_1 = \epsilon$$

where a is the depth of the heat source, and ϵ is the bore eccentricity of the cylinder. From these relationships

$$N_1 = \frac{(D_2/2)^2 - (D_1/2)^2 - \epsilon^2}{2\epsilon} \qquad N_2 = \frac{(D_2/2)^2 - (D_1/2)^2 + \epsilon^2}{2\epsilon}$$

Recognizing that Eq. (3-77) can be written in the form

$$R = \frac{1}{2\pi kL}\left(\cosh^{-1}\frac{2N_1}{D_1} - \cosh^{-1}\frac{2N_2}{D_2}\right) \tag{3-78}$$

and substituting for the values of N_1 and N_2, the expression for the thermal resistance becomes

$$R = \frac{1}{2\pi kL}\cosh^{-1}\frac{2D_1^2 + 2D_2^2 - \epsilon^2}{D_1D_2} \tag{3-79}$$

Thus the heat flow through a cylinder with an eccentric bore can be written as

$$Q = \frac{t_1 - t_2}{(1/2\pi kL)\cosh^{-1}[(2D_1^2 + 2D_2^2 - \epsilon^2)/D_1D_2]} \tag{3-80}$$

where t_1, t_2 = temperatures of the respective surfaces
D_1 = diameter of cylinder
D_2 = diameter of bore
ϵ = eccentricity of bore

3-8 THE BURIED ROD

Another physical system that can be analyzed by the use of sources and sinks is that of a rod of finite length inserted in a homogeneous semi-infinite body such that the rod is perpendicular to the free surface and one end is touching it. The rod, which is arranged as shown in Fig. 3-24, has a length L and a diameter D. The rod surface temperature is taken to be uniformly t_o. The temperature of the free surface is to be held at t_s. The medium into which the rod is inserted is homogeneous with constant thermophysical properties. The task is to find the thermal resistance.

The equation for heat transfer between two concentric spherical surfaces is given by Eq. (3-13) repeated here for convenience

$$Q = \frac{t_i - t_o}{(1/4\pi k)(1/r_i - 1/r_o)} \tag{3-81}$$

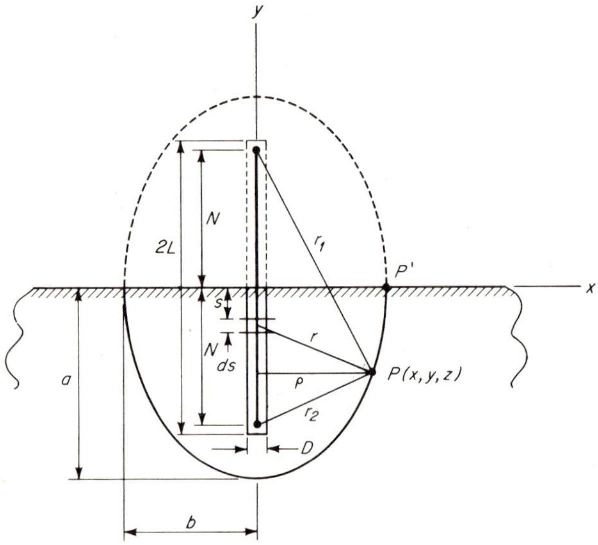

Fig. 3-24 The buried rod.

If $r_o \gg r_i$ and $t_o = 0$, Eq. (3-81) reduces to

$$t = \frac{Q}{4\pi kr} \tag{3-82}$$

the expression for the temperature field of a steady point heat source, where the subscript i is dropped so that t is identified at any $r > 0$.

The solution to the problem posed can be obtained by imagining the rod to be replaced by a line of point heat sources, N long for $x = 0$ and $y < 0$. This line of sources must have a mirror image of line heat sinks, N long for $x = 0$ and $y > 0$. By this arrangement one ensures that the plane $y = 0$ will be maintained at the prescribed zero temperature. The temperature distribution in the body due to all the sources and sinks can be obtained by summing the effects, for example, on point $P(x,y,z)$, noting that the temperature field is rotationally symmetrical about the y axis.

Consider a line heat source and a line heat sink as shown in Fig. 3-24. The strength of the source is Q', the heat rate per unit length, and the strength of the sink is thus $-Q'$. Using the same arguments as for the cylindrical case [Eqs. (3-61) to (3-64)], the contribution of the line heat source to the point $P(x,y,z)$ can be expressed as

$$t_1 = t_o - \int_0^{-N} \frac{Q'\,ds}{4\pi kr}$$

where t_o is the temperature of the source, or in this case the rod.

The contribution of the line heat sink to the same point $P(x,y,z)$ is

$$t_2 = -t_o + \int_0^{N} \frac{Q'\,ds}{2\pi kr}$$

where $-t_o$ is the temperature of the sink, or the negative of the rod temperature. Again, combination of the two contributions yields for the temperature of the point $P(x,y,z)$

$$t(x,y,z) = t_1 + t_2 = \int_0^{N} \frac{Q'\,ds}{4\pi kr} - \int_0^{-N} \frac{Q'\,ds}{4\pi kr}$$

Combining the two integrals gives

$$t = \frac{Q'}{4\pi k} \int_{-N}^{N} \frac{ds}{r} = \frac{Q'}{4\pi k} \int_{-N}^{N} \frac{ds}{\sqrt{(y-s)^2 + \rho^2}} \tag{3-83}$$

where the symbols are as shown in Fig. 3-24.

Integration of Eq. (3-83) yields

$$t = \frac{Q'}{4\pi k} \ln \frac{(y+N) + r_1}{(y-N) + r_2} \tag{3-84}$$

For constant values of t the isothermal surfaces are ellipsoids. If the major axis is $2a$ and the minor axis $2b$,

$$N = \sqrt{a^2 - b^2}$$

and

$$r_1 + r_2 = 2a$$

For any specific point such as P', $r_1 = r_2 = a$, and $y = 0$

$$t = \frac{Q'}{4\pi k} \ln \frac{a+N}{a-N} \tag{3-85}$$

For very large values of the ratio a/b the ellipsoid approaches the shape of the rod, and the value of N can be approximated,

$$N = \sqrt{a^2 - b^2} = a\sqrt{1 - \left(\frac{b}{a}\right)^2} \approx a\left[1 - \frac{1}{2}\left(\frac{b}{a}\right)^2 + \cdots\right]$$

which, when used in Eq. (3-85), for $a \gg b$, yields

$$t \approx \frac{Q'}{2\pi k} \ln \left(2\frac{a}{b}\right) \tag{3-86}$$

For the resulting slender ellipsoid, which may be taken to represent the rod, $a = L$ and $b = D/2$; thus where $Q' = Q/L$

$$t_o \approx \frac{Q}{2\pi kL} \ln \frac{4L}{D} \tag{3-87}$$

where the thermal resistance is

$$R = \frac{1}{2\pi kL} \ln \frac{4L}{D} \tag{3-88}$$

The heat-flow rate is

$$Q = \frac{t_o - t_s}{(1/2\pi kL) \ln (4L/D)} \tag{3-89}$$

where t_o is the surface temperature of the rod and t_s is the temperature of the free surface.

Table 3-4 presents the values of the thermal resistances of heat sources of several shapes submerged in a semi-infinite body of low thermal conductivity.

Table 3-4 Thermal resistances of certain heat sources submerged in a semi-infinite body of low thermal conductivity†

Description	*Thermal resistance R*
Sphere of diameter D with center at a distance N below surface	$R = \frac{1}{2\pi kD}\left(1 - \frac{D}{4N}\right)$
Horizontal cylinder of length L and diameter D with axis at distance N below surface	$R = \frac{1}{2\pi kL} \ln \left[2\frac{N}{D} + \sqrt{4\left(\frac{N}{D}\right)^2 - 1}\right]$ Eq. (3-72) $L \gg N$
Horizontal thin circular disk of diameter D well below the surface	$R \approx \frac{1}{4kD}$
Horizontal torus of mean diameter D_m and thickness d, with axis a distance N below the surface	$R = \frac{1}{2\pi^2 kD_m} \ln \frac{4N}{d}$ $N > d$ $D_m \gtrsim 20d$
Thin horizontal rectangle with larger and smaller sides D_1 and D_2 well below the surface; N is depth below surface	$R = \frac{1}{2\pi kD_1} \ln \frac{2\pi N}{D_2}$ $D_1 \gg D_2$ $N > 2D_2$

† A. L. London, unpublished data cited in W. H. McAdams, "Heat Transmission," 3d ed., McGraw-Hill, New York, 1954.

3-9 TWO-DIMENSIONAL STEADY HEAT CONDUCTION

The heat-conduction equation for an isotropic, homogeneous material in two dimensions in which the temperature distribution is invariant with time and in which there are no heat sources can be reduced to a statement of the Laplace equation:

$$\frac{\partial^2 t}{\partial x^2} + \frac{\partial^2 t}{\partial y^2} = 0 \tag{3-90}$$

This equation can be interpreted in consideration of the functions of a complex variable.[1] Consider Fig. 3-25, which represents the z plane, where $z = x + iy$. On this plane are drawn traces φ and ψ which are related to z by the expression

$$\varphi + i\psi = f(x + iy) = f(z)$$

φ and ψ are *conjugate functions of a complex variable.* Therefore the following relations hold:

$$\frac{\partial \varphi}{\partial x} + i\frac{\partial \psi}{\partial x} = f'(z) \qquad \frac{\partial \varphi}{\partial y} + i\frac{\partial \psi}{\partial y} = if'(z)$$

and from these relations we can write, by separating and equating the real and imaginary parts, the *Cauchy-Riemann relations*

$$\frac{\partial \varphi}{\partial y} = -\frac{\partial \psi}{\partial x} \qquad \frac{\partial \varphi}{\partial x} = \frac{\partial \psi}{\partial y} \tag{3-91}$$

The traces $\varphi = \text{const}$ and $\psi = \text{const}$ are thus orthogonal, i.e., perpendicular to each other at the points of intersection. Another property of conjugate functions is that each satisfies the Laplace equation. This property can be demonstrated by the partial differentiation of Eq. (3-91),

[1] See, for example, I. S. Sokolnikoff and Redheffer, 2d ed., *op. cit.*

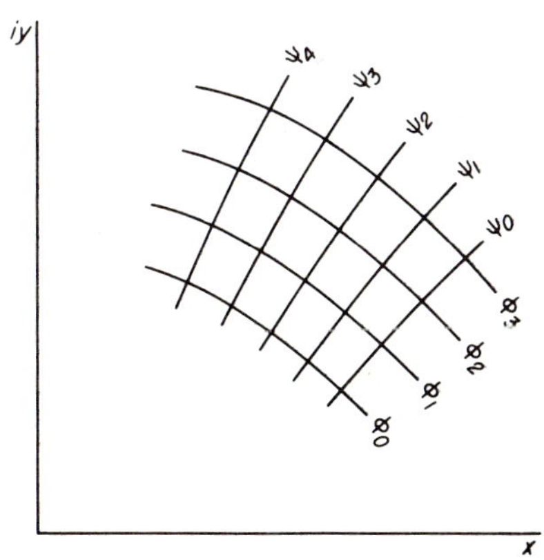

Fig. 3-25 Potential function and stream function in a complex domain.

first with respect to x and then with respect to y:

$$\frac{\partial^2 \varphi}{\partial y\, \partial x} = -\frac{\partial^2 \psi}{\partial x^2} \qquad \frac{\partial^2 \varphi}{\partial x\, \partial y} = \frac{\partial^2 \psi}{\partial y^2}$$

and

$$\frac{\partial^2 \varphi}{\partial y^2} = -\frac{\partial^2 \psi}{\partial x\, \partial y} \qquad \frac{\partial^2 \varphi}{\partial x^2} = \frac{\partial^2 \psi}{\partial x\, \partial y}$$

From these relations we obtain the two Laplace equations

$$\frac{\partial^2 \varphi}{\partial x^2} + \frac{\partial^2 \varphi}{\partial y^2} = 0 \qquad \frac{\partial^2 \psi}{\partial x^2} + \frac{\partial^2 \psi}{\partial y^2} = 0 \tag{3-92}$$

If we consider $\varphi = t$ the temperature, then ψ must be a path of constant heat flow and therefore related to Q.[1] Consider these values in the light of Fig. 3-26, which shows a coordinate system (s,n) so oriented that n is the normal to an isotherm and s is tangent to the isotherm at the same point. For the heat flux along n, we can write

$$-k\frac{\partial \varphi}{\partial n} = -k \cos\theta \frac{\partial \varphi}{\partial x} - k \sin\theta \frac{\partial \varphi}{\partial y}$$

[1] The conjugate functions in the Laplace equation are commonly called the *potential function* and the *stream function*. These correspond here to the *temperature* and the *heat flow*, respectively.

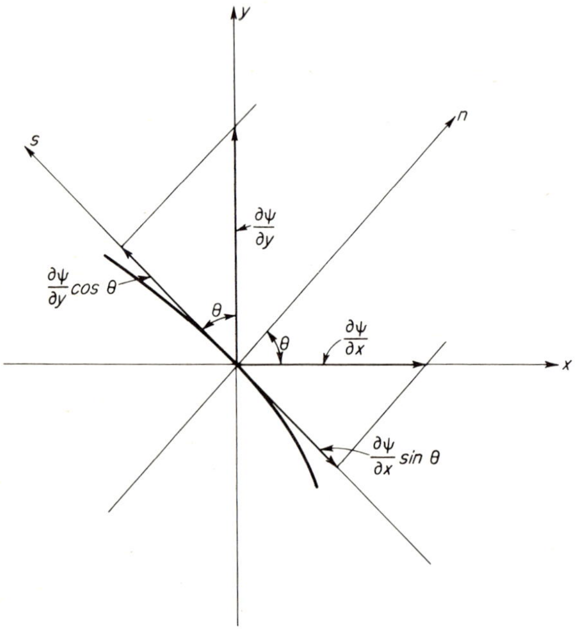

Fig. 3-26 Relations among components of the stream-function derivatives.

When the conditions in Eq. (3-91) are used, this becomes

$$-k\frac{\partial \varphi}{\partial n} = -k\cos\theta\frac{\partial \psi}{\partial y} + k\sin\theta\frac{\partial \psi}{\partial x} = -k\frac{\partial \psi}{\partial s} \tag{3-93}$$

[Equation (3-93) is shown graphically in Fig. 3-26.] Therefore,

$$\frac{\partial \varphi}{\partial n} = \frac{\partial \psi}{\partial s} \tag{3-94}$$

In particular, if the net of lines of constant temperature and lines related to the heat flow are taken in a fine mesh but still in finite increments, the following approximation of Eq. (3-94) can be employed:

$$\frac{\Delta\varphi}{\Delta n} = \frac{\Delta\psi}{\Delta s} \tag{3-95}$$

Equation (3-95) is exactly true when the rates of change of φ and ψ are constant, but regardless, it follows from Eq. (3-95) that the plane will be divided into *curvilinear squares*[1] if $\Delta\varphi = \Delta\psi$ and the increments are small enough. Further, the smaller the increments, the more nearly the curvilinear squares will approach geometrical squares. Equation (3-95) has an important application to a graphical technique to be employed in the next section.

The nature of ψ, the stream function, can be expressed if we calculate the heat flow through an elemental area (Fig. 3-27)

$$dQ_x = -kL\frac{\partial t}{\partial x}\,dy = -kL\frac{\partial \varphi}{\partial x}\,dy \tag{3-96}$$

[1] Curvilinear squares closely approximate true squares only in the limit where the number of squares approaches infinity. In finite form curvilinear squares may be considered for constructional purposes to have sides of lengths such that the average lengths of the sides opposite each other are equal. The internal angles are, of course, $\pi/2$ in each case.

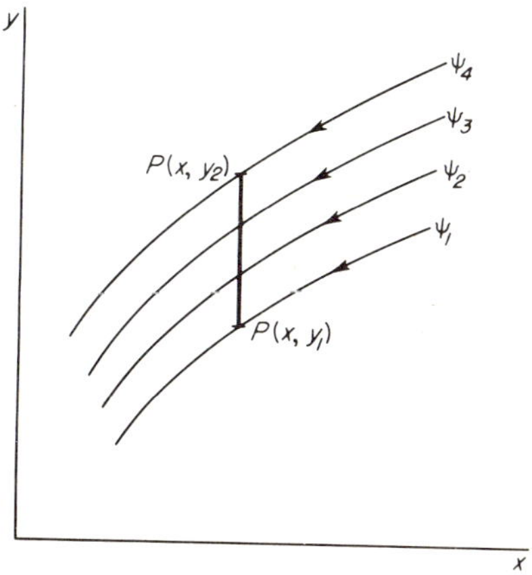

Fig. 3-27 The stream function.

where L is the dimension normal to the plane of the paper. Integrating Eq. (3-96) across a height $y_2 - y_1$, we obtain, again with the aid of Eq. (3-91),

$$Q_x = -kL \int_{y_1}^{y_2} \frac{\partial \varphi}{\partial x}\, dy = -kL \int_{y_1}^{y_2} \frac{\partial \psi}{\partial y}\, dy = -kL(\psi_2 - \psi_1)$$

Thus

$$\frac{Q_x}{kL} = -\Delta\psi \tag{3-97}$$

or the stream function ψ becomes, in form,

$$\psi = \frac{Q}{kL} \tag{3-98}$$

Thus the stream function ψ is the *total rate of heat flow per unit thermal conductivity per unit depth.* If ψ is constant, the rate of heat flow Q is constant and the surfaces represented by the lines ψ_1 and ψ_2 are those which bound a heat-flow tube in which Q is constant.

The flux plot The ideas discussed in the last few paragraphs can be applied to effect a solution of two-dimensional, steady-state heat-conduction problems by the technique known as *flux plotting.* In its crudest form, this method is essentially a freehand sketching of the flow lines (streamlines) and potential lines, carefully obeying the rules relating to orthogonality and curvilinear squares, to construct a *flow net* or *flux plot.* This method will be illustrated. Consider Fig. 3-28, which represents a parallelepiped the upper and lower surfaces of which are at constant temperatures t_1 and t_2, respectively, while the other surfaces are perfectly

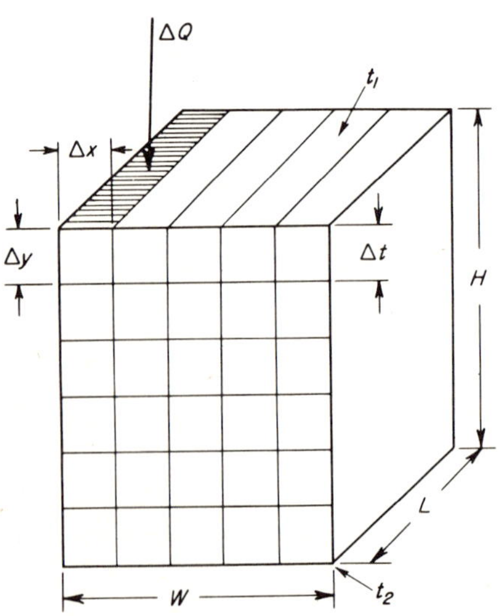

Fig. 3-28 Steady heat flow in a parallelepiped by flux plot.

insulated. The upper surface can be ruled off to make an arbitrary number of heat-flow lanes (stream tubes). Lines representing isotherms can be scribed at intervals Δy maintaining $\Delta y = \Delta x$ and the condition of perpendicularity of the isotherms and heat-flux lines at their intersections. The heat flow through any one of the flow tubes is

$$\Delta Q = -kL\,\Delta x\,\frac{\Delta t}{\Delta y} \tag{3-99}$$

Equation (3-99) can be rewritten to be analogous to Eq. (3-97) for the condition $\Delta y = \Delta x$:

$$-\frac{\Delta Q}{kL} = \Delta t \tag{3-100}$$

The construction can be continued until the flow field is completed. The result is N_C *flow tubes* and N_R *temperature increments* Δt. The total heat flow in the body is the sum of the incremental heat flows across the total temperature difference:

$$\frac{Q}{kL} = \frac{N_C\,\Delta x}{N_R\,\Delta y}(t_1 - t_2) \tag{3-101}$$

where $t_1 > t_2$. In the construction, $\Delta x = \Delta y$; thus

$$\frac{Q}{kL} = \frac{N_C}{N_R}(t_1 - t_2) \tag{3-102}$$

The ratio $N_C/N_R = \xi$ is the *shape factor*[1] of the body. The shape factor is purely geometrical and can be determined once and for all for any given system. In the body in Fig. 3-28 the shape factor can be calculated by counting the rows and columns:

$$\xi = \frac{N_C}{N_R} = \frac{5}{6} \tag{3-103}$$

It can easily be shown that if the heat-flow lines are interchanged with the isothermals (this corresponds to heat flow horizontally across the body rather than vertically), the shape factor ξ' for this alternate problem is reciprocally related to the shape factor of Eq. (3-103):

$$\xi' = \frac{1}{\xi}$$

The system considered was a very simple one and as such has a place only as an illustrative example. The method described here, how-

[1] The term *shape factor* occurs in several different usages in the study of heat transfer, each one having a different meaning which the reader will distinguish for himself. The property in common for all shape factors is that they are geometrical and thus can be determined once and for all for the bodies in question.

ever, is not restricted to simple systems and is often quite useful for rapid results where extreme accuracy is not paramount.

Example 3-6 A more complex system can be treated which represents heat flow in the wall of a square duct whose surface temperatures are constant. Since the duct is symmetrical, only one corner need be treated. The isotherms and heat-flow lines are sketched in, attention being paid to the requirements of the construction of the resulting curvilinear squares. The result after several adjustments is shown in Fig. 3-29.

The shape factor can be readily determined by noting that the heat-flow lanes are 20 in number and the temperature increments are 7 in number. The shape factor becomes

$$\xi = \frac{N_C}{N_R} = \frac{20}{7} = 2.85$$

and the heat transfer by conduction is

$$Q = 2.85kL(t_1 - t_2)$$

The heat flow can be evaluated on the basis of the material from which the duct wall is made, its length perpendicular to the plane of the page, and the temperature difference across its wall.[1]

The relaxation method Solution of the conduction equation in two (or three) dimensions with or without heat sources can be accomplished by means of a numerical method developed by Southwell[2] called the *relax-*

[1] There are a number of methods based on heat-conduction analogies with electric conduction, fluid flow, etc., which can be used to obtain a flux plot or for gross determination of the shape factor. The more useful of these systems are discussed in Schneider, *op. cit.*, chap. 13.

[2] R. V. Southwell, "Relaxation Methods in Engineering Science," Oxford University Press, New York, 1940; "Relaxation Methods in Theoretical Physics," Oxford University Press, New York, 1946; H. W. Emmons, The Numerical Solution of Heat-conduction Problems, *Trans. ASME*, **65**(6):607–612 (1943); G. M. Dusinberre, "Numerical Analysis of Heat Flow," McGraw-Hill, New York, 1949.

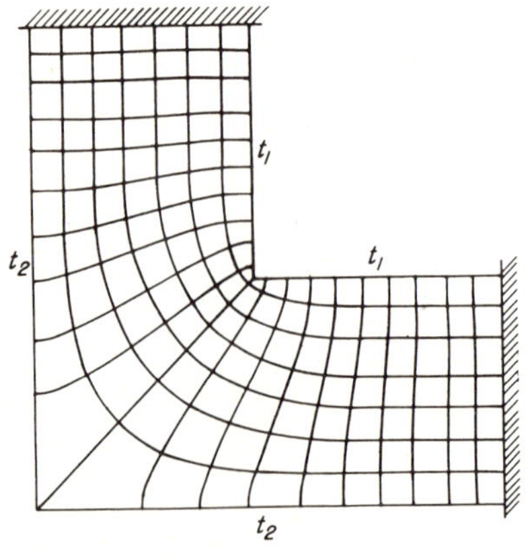

Fig. 3-29 Steady heat flow in an L-shaped corner by flux plot.

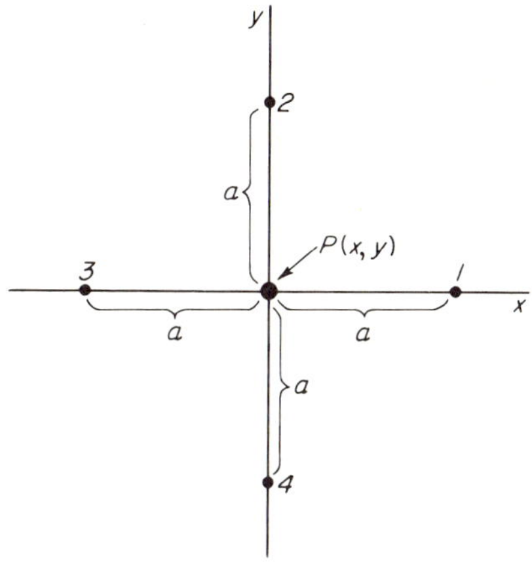

Fig. 3-30 Coordinate system for the relaxation method.

ation method. The method, of course, has general application to the solution of the Laplace-Poisson equation, but the discussion of it here will be based on the heat-conduction system. With the advent of high-speed computing machines the relaxation method has fallen into disuse; however, because of similarities to other methods a discussion here is felt to be worthwhile.

In steady state, with no internal heat sources, the Laplace equation is the differential equation whose solution with appropriate boundary conditions results in the temperature distribution in the body:

$$\frac{\partial^2 t}{\partial x^2} + \frac{\partial^2 t}{\partial y^2} = 0 \tag{3-104}$$

Consider a point in a body $P(x,y)$ through which is passed a rectangular coordinate system (Fig. 3-30). Other points in this body may be located at small distances away from $P(x,y)$, such as $P(x + a, y)$, $P(x, y + a)$, $P(x - a, y)$, $P(x, y - a)$. They are numbered 1, 2, 3, 4, respectively, for future convenience. If the temperature at point $P(x,y)$ is known, then by means of a Taylor series the temperatures at the surrounding points can be approximated to any desired degree of accuracy. For example, denoting the point $P(x,y)$ as 0,

$$t_1 = t_0 + \left(\frac{\partial t}{\partial x}\right)_0 a + \left(\frac{\partial^2 t}{\partial x^2}\right)_0 \frac{a^2}{2!} + \left(\frac{\partial^3 t}{\partial x^3}\right)_0 \frac{a^3}{3!} + \cdots \tag{3-105}$$

and
$$t_3 = t_0 - \left(\frac{\partial t}{\partial x}\right)_0 a + \left(\frac{\partial^2 t}{\partial x^2}\right)_0 \frac{a^2}{2!} - \left(\frac{\partial^3 t}{\partial x^3}\right)_0 \frac{a^3}{3!} + \cdots \tag{3-106}$$

If Eqs. (3-105) and (3-106) are added together, we obtain

$$t_1 + t_3 = 2t_0 + \left(\frac{\partial^2 t}{\partial x^2}\right)_0 a^2 + O(a^4) \tag{3-107}$$

where $O(a^4)$ indicates a remainder of terms of the order of a^4.

Equation (3-107) can be rewritten explicitly in terms of $(\partial^2 t/\partial x^2)_0$:

$$\left(\frac{\partial^2 t}{\partial x^2}\right)_0 = \frac{t_1 + t_3 - 2t_0}{a^2} + O(a^2)$$

Similar equations which result in a statement for $(\partial^2 t/\partial y^2)_0$ can be written:

$$\left(\frac{\partial^2 t}{\partial y^2}\right)_0 = \frac{t_2 + t_4 - 2t_0}{a^2} + O(a^2)$$

The sum of these second derivatives constitutes a solution of the Laplace equation accurate to order a^2 of the neglected terms:

$$\left(\frac{\partial^2 t}{\partial x^2}\right)_0 + \left(\frac{\partial^2 t}{\partial y^2}\right)_0 = \frac{t_1 + t_2 + t_3 + t_4 - 4t_0}{a^2} = 0 \tag{3-108}$$

Equation (3-108) results in an expression for the temperature t_0 in terms of the surrounding points at equilibrium:

$$t_0 = \frac{t_1 + t_2 + t_3 + t_4}{4} \tag{3-109}$$

The *objective of the method* is to establish a temperature field in the body such that each point satisfies Eq. (3-109). To do this *systematically*, a *residual* Q_0 is defined and written

$$Q_0 = t_1 + t_2 + t_3 + t_4 - 4t_0 \tag{3-110}$$

and the objective takes the form of making the residuals zero for each point in the body. Regarding Fig. 3-30 again in light of Eq. (3-110), it can be seen that a unit temperature change in any one of the neighboring points can change the residual Q_0 by ± 1, whereas a unit change in the temperature of the point itself changes the residual Q_0 by ∓ 4. This, then, is the key to the procedure. Figure 3-31 depicts the relaxation pattern to be used for convenience. The following simple example will suffice to clarify the process.

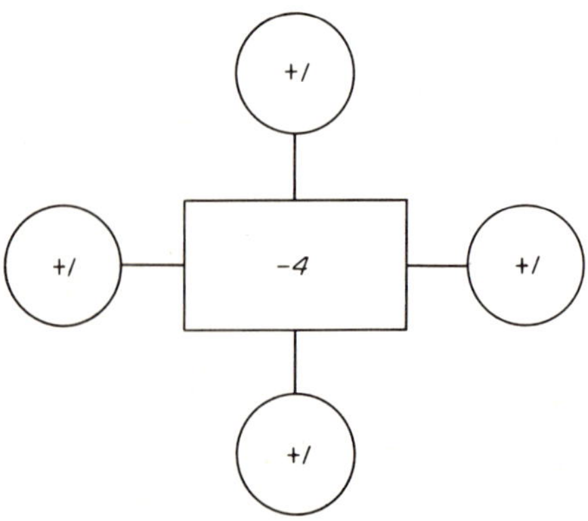

Fig. 3-31 Relaxation pattern for the Laplace equation.

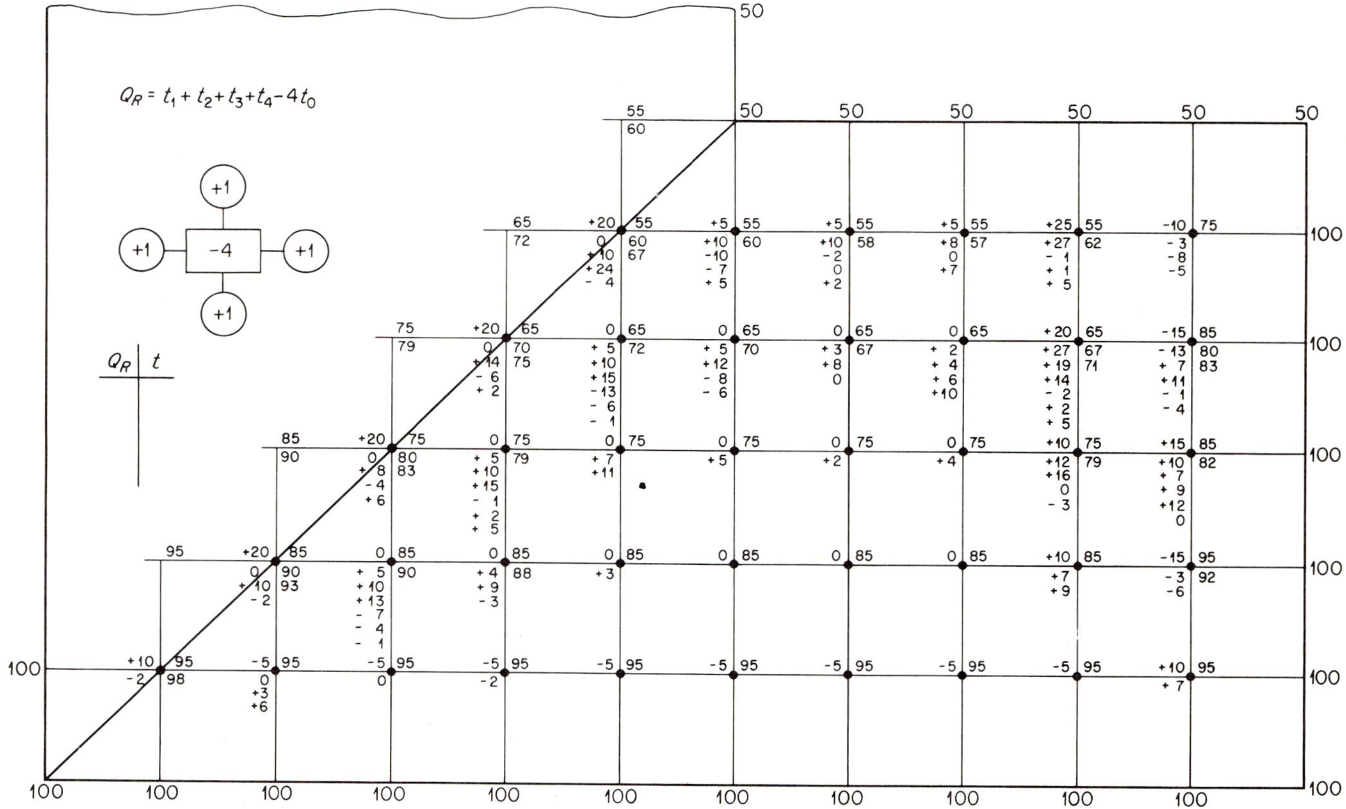

Fig. 3-32 Relaxation network for the L-shaped corner.

Example 3-7 As an example to demonstrate this method consider the L-shaped corner, shown in Fig. 3-32. The temperatures at the boundaries are fixed, as shown in larger-sized type. The initial temperature distribution is guessed at any values that might appear reasonable and compatible with the boundary conditions. These temperatures are shown to the right of the nodal points. The next step is to calculate the residuals for each of the nodal points by using the appropriate relaxation pattern, which in this case is that for the Laplace equation, since no heat sources are present. The residuals are shown to the left of the nodal points. Now attacking the largest residuals, proceed from point to point, changing the temperatures and adjusting the residuals. It is better to overshoot somewhat than to try to adjust the point exactly. As in any other trial-and-error process one improves with practice. In the figure the first few calculations have been made to illustrate the method. Only one-half of the corner has been computed because of the symmetry of the system.

If the body under consideration has distributed heat sources, the conduction equation is a form of the Laplace-Poisson equation

$$\frac{\partial^2 t}{\partial x^2} + \frac{\partial^2 t}{\partial y^2} = -\frac{Q'(x,y)}{k} = f(x,y) \tag{3-111}$$

Again, using the expressions for the partial derivatives,

$$\frac{t_1 + t_2 + t_3 + t_4 - 4t_0}{a^2} + O(a^2) = f(x,y) \tag{3-112}$$

The residual Q_0 in this case, neglecting terms of fourth or higher order, becomes

$$Q_0 = t_1 + t_2 + t_3 + t_4 - 4t_0 - a^2 f(x,y) \tag{3-113}$$

In the case of Eq. (3-113) the relaxation pattern must be changed to account for the heat-generation term $a^2f(x,y)$, as shown in Fig. 3-33. It is clear that the heat sources can be arbitrarily distributed throughout the body, the distribution being accommodated pointwise in the

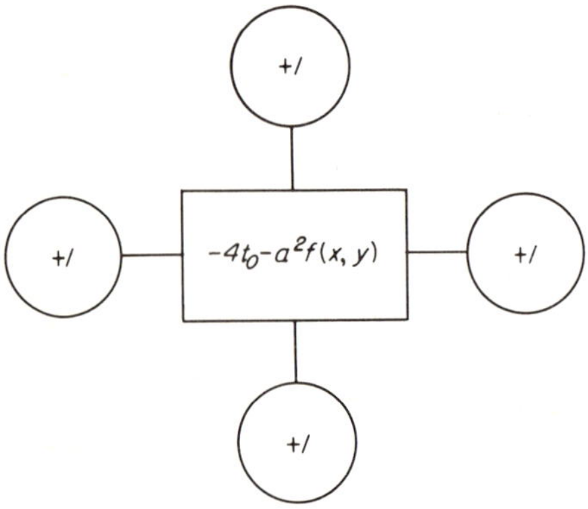

Fig. 3-33 Relaxation pattern for the Laplace-Poisson equation.

computation procedure. Thus the heat source is simply a modification of the effect of a change at the zero point.

Although the relaxation method can be applied to cylindrical coordinate systems by developing the relaxation pattern within the cylindrical system, the equations corresponding to Eqs. (3-110) and (3-113) are not easy to use. A change of variable, however, will reduce the cylindrical coordinate system to a rectangular coordinate system in the new variable. The steady heat-conduction equation in the two dimensions r and θ becomes

$$\frac{\partial^2 t}{\partial r^2} + \frac{1}{r}\frac{\partial t}{\partial r} + \frac{1}{r^2}\frac{\partial^2 t}{\partial \theta^2} = 0 \tag{3-114}$$

Let $\zeta = \ln r$; $\eta = \theta$; thus

$$\frac{\partial^2 t}{\partial \theta^2} = \frac{\partial^2 t}{\partial \eta^2}$$

$$\frac{\partial t}{\partial r} = \frac{1}{r}\frac{\partial t}{\partial \zeta}$$

$$\frac{\partial^2 t}{\partial r^2} = \frac{1}{r^2}\frac{\partial^2 t}{\partial \zeta^2} - \frac{1}{r^2}\frac{\partial t}{\partial \zeta}$$

When these values are substituted into Eq. (3-114), we get

$$\frac{1}{r^2}\left(\frac{\partial^2 t}{\partial \zeta^2} - \frac{\partial t}{\partial \zeta}\right) + \frac{1}{r^2}\frac{\partial t}{\partial \zeta} + \frac{1}{r^2}\frac{\partial^2 t}{\partial \eta^2} = 0$$

which becomes

$$\frac{\partial^2 t}{\partial \zeta^2} + \frac{\partial^2 t}{\partial \eta^2} = 0 \tag{3-115}$$

where now $t = t(\zeta,\eta)$ and the relaxation system of Eq. (3-110) applies. The new boundary conditions can be obtained from the relations $\zeta = \ln r$ and $\eta = \theta$. The results obtained can then be translated graphically from the $t(\zeta,\eta)$ system to the $t(r,\theta)$ system.

Consider Fig. 3-34*a*, showing a section of tube for which the temperature distribution is desired, and Fig. 3-34*b*, showing the tube section transformed into the $\zeta\eta$ plane by the transformation $\zeta = \ln r$; $\eta = \theta$. The boundary conditions are shown here as constant surface temperatures; however, if the surface temperatures varied with either r or θ, or both, the particular value at the boundary point would be associated with the same point transformed into the $\zeta\eta$ plane.

The rectangular relaxation net can be applied directly to the figure in the $\zeta\eta$ plane to establish the temperatures at each point. Once this has been accomplished, the temperature values for those points are associated with the transformed point in the $r\theta$ plane and the problem is complete.

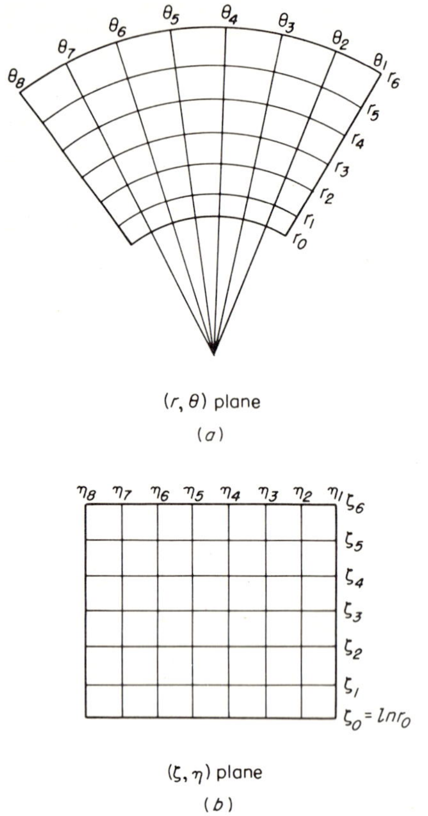

Fig. 3-34 The log transformation in cylindrical systems: (a) the tube in cylindrical coordinates; (b) the tube in transformed coordinates.

3-10 HEAT CONDUCTION IN A PERMEABLE SLAB

In certain applications such as turbine blades, rocket nozzles, and conduits transporting fluids at very high temperatures in which there exists no readily available material to withstand the extreme temperatures and cooling by usually considered convective means would not be sufficient, resort has been made to the use of porous material which is permeable to a fluid. In practice the cooling fluid is forced under pressure through the pores of the material. The material is maintained at a reasonably low temperature inasmuch as the body of the material is flooded with the coolant fluid.

The mechanics of the process of solid-fluid heat exchange under the specified conditions is fairly complex; however, if no phase change in the fluid takes place, and all heat conducted into the body is assumed to take place in the solid and not in the fluid, it has been demonstrated[1] that

[1] S. Weinbaum and H. L. Wheller, Jr., Heat Transfer in Sweat Cooled Metals, *J. Appl. Phys.* **20**(1):113–122 (1949).

temperatures in the solid and in the coolant fluid are essentially the same. This fact greatly simplifies the analysis.

The permeable slab Consider a slab of permeable material of thickness L, as shown in Fig. 3-35. The temperature of the hot face is assumed known, of value t_{s_2}. Heat flows into the slab in a negative x direction while the cooling fluid at a temperature initially at t_{f_1} flows at a mass velocity G in the positive x direction through the slab. Both heat and fluid flow are considered one-dimensional. The thermophysical properties of the fluid and the solid are invariant with temperature.

If the pores in the slab and the solids in the slab are thought to be parallel to each other, then for an area A normal to the x axis an area A_f may be assigned for flow of the cooling fluid and an area A_s assigned to the solid. The areas are related as

$$A = A_s + A_f$$

or per unit surface area

$$\frac{A_s}{A} + \frac{A_f}{A} = 1$$

Thus, for an incremental length δx the net heat flow into the solid by conduction is

$$\delta Q = kA_s \frac{d^2t_s}{dx^2}\,\delta x \tag{3-116}$$

The heat added to the fluid in its passage through the distance δx is

$$\delta Q = \rho c U_f A_f \frac{dt_s}{dx}\,\delta x = \rho c U_\infty A \frac{dt_s}{dx}\,\delta x$$

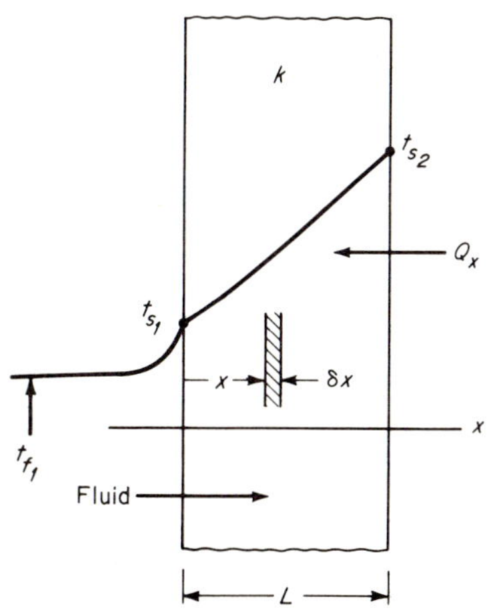

Fig. 3-35 The permeable slab.

or

$$\delta Q = Gc \frac{dt_s}{dx} A \, \delta x \tag{3-117}$$

where U_f = velocity of fluid in fluid passage
U_∞ = velocity of fluid before entering slab
ρ = mass density of fluid
c = specific heat of fluid
$G = \rho U_\infty$ = onflow mass velocity of fluid

If a steady-state condition is to be established, the heat conducted into the element δx must be equal to the heat absorbed by the fluid in the same element; therefore the statements of Eqs. (3-116) and (3-117) can be equated for an energy balance equation for the solid

$$\frac{d^2t_s}{dx^2} - \frac{Gc}{k(1 - A_f/A)} \frac{dt_s}{dx} = 0 \tag{3-118}$$

Changing the dependent variable to $\vartheta = t_s - t_{f_1}$ and letting

$$\frac{Gc}{k(1 - A_f/A)} = m$$

the differential equation becomes

$$\frac{d^2\vartheta}{dx^2} - m \frac{d\vartheta}{dx} = 0 \tag{3-119}$$

with boundary conditions

$$x = 0 \qquad \vartheta = \vartheta_1 = t_{s_1} - t_{f_1}$$
$$x = L \qquad \vartheta = \vartheta_2 = t_{s_2} - t_{f_1}$$

Heat flowing through the slab increases the temperature of the coolant fluid before the fluid enters the plate. An energy balance on this fluid yields the equation

$$\frac{d^2\vartheta'}{dx^2} - n \frac{d\vartheta'}{dx} = 0 \tag{3-120}$$

where $\quad n = \dfrac{Gc}{k_f} \quad$ and $\quad \vartheta' = t_f - t_{f_1}$

with boundary conditions

$$x = -\infty \qquad \vartheta' = 0$$
$$x = 0 \qquad k_f \frac{d\vartheta'}{dx} = k\left(1 - \frac{A_f}{A}\right) \frac{d\vartheta}{dx}$$

The condition at $x = 0$ states that the heat conducted through the slab is conducted into the fluid.

The solution of Eq. (3-119) subject to the stated boundary conditions is

$$\frac{\vartheta - \vartheta_1}{\vartheta_2 - \vartheta_1} = \frac{1 - e^{mx}}{1 - e^{mL}} \qquad 0 \le x \le L \tag{3-121}$$

while the solution of Eq. (3-120) subject to its boundary conditions is

$$\frac{\vartheta' - \vartheta'(-\infty)}{\vartheta_2 - \vartheta_1} = \frac{e^{nx}}{e^{mL} - 1} \tag{3-122}$$

The value ϑ_1 can be eliminated by combining Eqs. (3-121) and (3-122) to obtain a final expression for the temperature distribution through the slab

$$\frac{\vartheta}{\vartheta_2} = \exp\left[-\frac{GcL}{k(1 - A_f/A)}\left(1 - \frac{x}{L}\right)\right] \tag{3-123}$$

in the region $0 \leq x \leq L$.

The mean temperature in the slab is calculated from the expression

$$\bar{\vartheta} = \frac{1}{L}\int_0^L \vartheta \, dx$$

such that

$$\frac{\bar{\vartheta}}{\vartheta_2} = \frac{k(1 - A_f/A)}{GcL}\left\{1 - \exp\left[-\frac{GcL}{k(1 - A_f/A)}\right]\right\} \tag{3-124}$$

The temperature distribution and the mean temperature in the plate are shown graphically in Fig. 3-36.

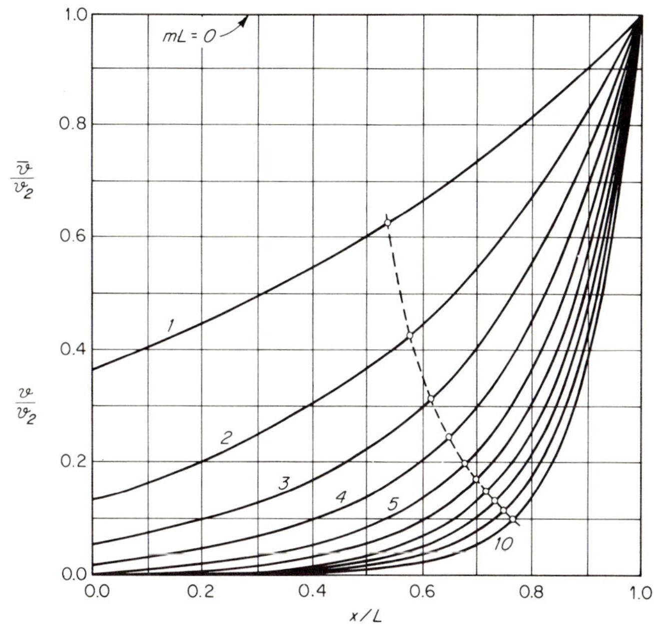

Fig. 3-36 Temperature distribution and mean temperature in a permeable slab. [*P. J. Schneider, "Conduction Heat Transfer," Addison-Wesley Publishing Company, Inc., Reading, Mass., 1957.*]

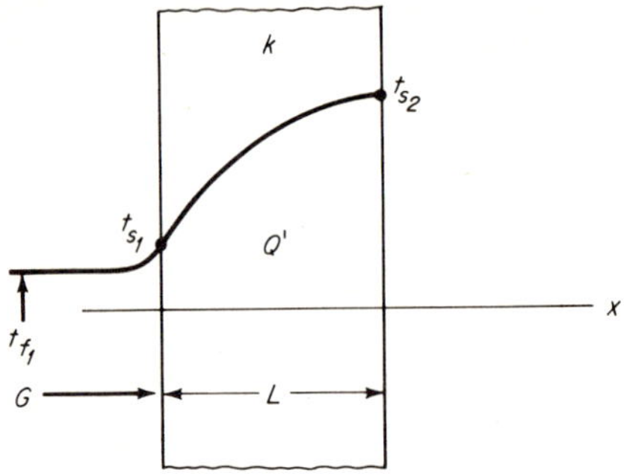

Fig. 3-37 Permeable slab with internal heat generation.

The permeable slab with heat generation[1] An interesting extension to the analysis of the permeable slab is the analysis of the permeable slab which is generating heat uniformly throughout its volume (Fig. 3-37). Equation (3-119) for the solid part of the slab must be changed by the addition of the heat-source term [see Eq. (1-12)] to read

$$\frac{d^2\vartheta}{dx^2} - m\frac{d\vartheta}{dx} + \frac{Q'}{k} = 0 \tag{3-125}$$

Equation (3-120) governing the fluid remains the same, or

$$\frac{d^2\vartheta'}{dx^2} - n\frac{d\vartheta'}{dx} = 0 \tag{3-126}$$

where ϑ, ϑ', m, and n are defined as before.

The solutions to the differential equations (3-125) and (3-126) are, respectively,

$$\vartheta = C_1 e^{mx} + \frac{Q'(1 - A_f/A)}{Gc}x + C_2 \tag{3-127}$$

and

$$\vartheta' = C_3 e^{nx} + C_4 \tag{3-128}$$

subject to the boundary conditions

$$\begin{aligned} x &= 0 & \vartheta &= \vartheta_1 \\ x &= L & \vartheta &= \vartheta_2 \\ x &= -\infty & \vartheta' &= 0 \\ x &= 0 & k_f\frac{d\vartheta'}{dx} &= k\left(1 - \frac{A_f}{A}\right)\frac{d\vartheta}{dx} \end{aligned}$$

It is worth mentioning here that the boundary condition at $x = L$ for the permeable slab with heat generation is a consequence of the heat generation and is not the result of any imposed heat flux at that surface, as in the case of the permeable slab with no heat generation.

[1] Leon Green, Jr., Gas Cooling of a Porous Heat Source, *J. Appl. Mech.*, **19**(2):173–178 (1952).

The total temperature rise in the fluid will be given by

$$\vartheta_2' = Q'\left(1 - \frac{A_f}{A}\right)\frac{L}{Gc}$$

and since the fluid temperature and the temperature of the solid at any location x are assumed to be equal in the region $0 \leq x \leq L$, $\vartheta_2' = \vartheta_2$.

Subject to the above boundary conditions, and eliminating ϑ_1 between the resulting equations, the equations for the temperatures in the solid and in the fluid are, respectively,

$$\frac{\vartheta}{\vartheta_2} = \frac{k(1 - A_f/A)}{GcL}\left\{1 - \exp\left[-\frac{GcL}{k(1 - A_f/A)}\left(1 - \frac{x}{L}\right)\right]\right\} + \frac{x}{L} \qquad 0 \leq x \leq L \quad (3\text{-}129a)$$

$$\frac{\vartheta'}{\vartheta_2} = \frac{k(1 - A_f/A)}{GcL}\left[1 - \exp\frac{GcL}{k(1 - A_f/A)}\right]\exp\left(\frac{GcL}{k_f}\frac{x}{L}\right) \qquad -\infty \leq x \leq 0 \quad (3\text{-}129b)$$

where it should be emphasized that $\vartheta_2 = \vartheta_2' = Q'(1 - A_f/A)L/Gc$. The results of Eqs. (3-129) are shown graphically in Fig. 3-38.

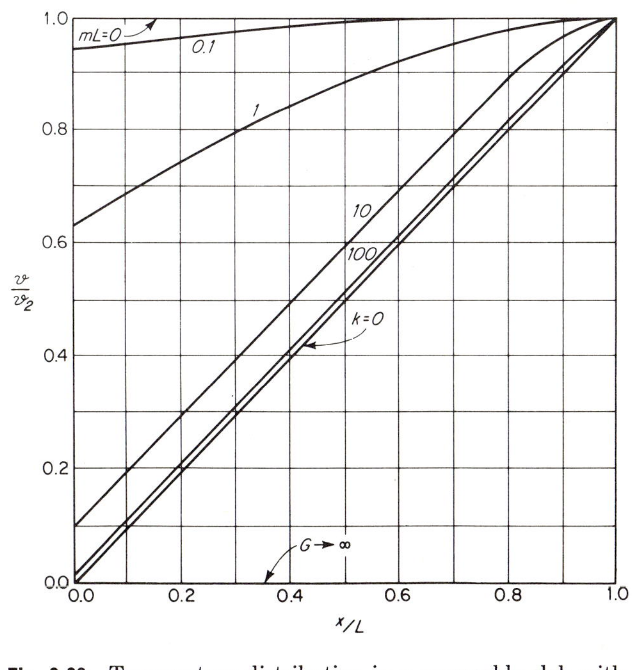

Fig. 3-38 Temperature distribution in a permeable slab with internal heat generation. [*Leon Green, Jr., J. Appl. Mech. ASME,* **19**:*2* (*1952*).]

From Eqs. (3-129) and Fig. 3-38 it may be noticed that uniformity of wall temperature, desirable if thermal stresses are to be alleviated, may be obtained with high thermal conductivity in the solid. Low thermal conductivity in the solid material leads to a linear distribution in temperature of the solid. If G becomes very large, the temperature of the solid approaches the coolant fluid temperature t_{f_1}.

The temperature difference across the slab ϑ_2 increases with decreasing fluid flow and reaches a maximum as $G \to 0$. From Eqs. (3-129) and calculating the limit as $G \to 0$ the maximum temperature difference can be shown to be

$$\vartheta_{2,\max} = \frac{Q'L^2}{2k} \tag{3-130}$$

Equation (3-130) may be compared with Eq. (3-59), where the similarity with the first term, the only appropriate one, is evident.

Similarly, as in the case of the nongenerating slab, the mean temperature of the permeable slab with heat generation can be calculated to be

$$\frac{\bar{\vartheta}}{\vartheta_2} = \left[\frac{k(1 - A_f/A)}{GcL}\right]^2 \left\{\frac{GcL}{k(1 - A_f/A)} + e^{-GcL/[k(1-A_f/A)]} - 1\right\} + \frac{1}{2} \tag{3-131}$$

3-11 PROBABILITY METHODS IN HEAT CONDUCTION

Concurrent with the development and extensive use of high-speed digital computers has arisen an interest in the application of the methods of probability for the solution of problems not having any underlying probabilistic aspects. One such example is the application of probability methods to the problems of heat conduction as presented by Haji-Sheikh and Sparrow.[1] The authors are able to treat both steady-state and transient problems in bodies of arbitrary shape, with and without internal heat sources; problems with nonlinear boundary conditions; and problems involving moving boundaries, i.e., changes of phase. Following the organization of this book, these methods could be developed either in this chapter or in subsequent chapters that are concerned with transient heat conduction and heat conduction with moving boundaries. However, since the probabilistic representation bears a relationship to the relaxation methods as applied to the steady-state heat-conduction problems, it seems better to treat the development here rather than elsewhere.

The Monte Carlo method as used is not a single procedure but an approach to solving problems resulting from mathematical models of physical phenomena in which the mathematical problem is made to cor-

[1] A. Haji-Sheikh and E. M. Sparrow, The Solution of Heat Conduction Problems by Probability Methods, *J. Heat Transfer*, **89**(3):121–131 (1967).

respond to a game of chance. Such a correspondence is not known for all problems, and the first task one faces in considering the application of Monte Carlo procedures to a specific problem is to discover or invent the corresponding game of chance.

All Monte Carlo procedures are, however, methods of solving mathematical problems *experimentally* and are characterized by the use of statistical sampling techniques to approximate the solution of the problem. The solution to the problem will be obtained by actually playing a game of chance, either by tossing coins, rolling dice, or some equivalent experimental technique. A practical, much more rapid but similar approach is to use the digital computer to play the game by drawing random numbers from a table or to generate random numbers in the computer, thus actually performing the experiment with the computer.

In order to demonstrate the concepts of a Monte Carlo procedure for the solution of problems in heat conduction, consider steady heat conduction without internal heat sources or sinks in two space dimensions (Fig. 3-39). In this problem the temperatures on the boundary are supposedly specified. The problem is to determine the temperature field within the body.

This problem is identical to the kinds of problems considered in Sec. 3-9 and of course can be represented by the equation

$$\frac{\partial^2 t}{\partial x^2} + \frac{\partial^2 t}{\partial y^2} = 0 \tag{3-132}$$

with the condition that the temperature is specified on the boundary.

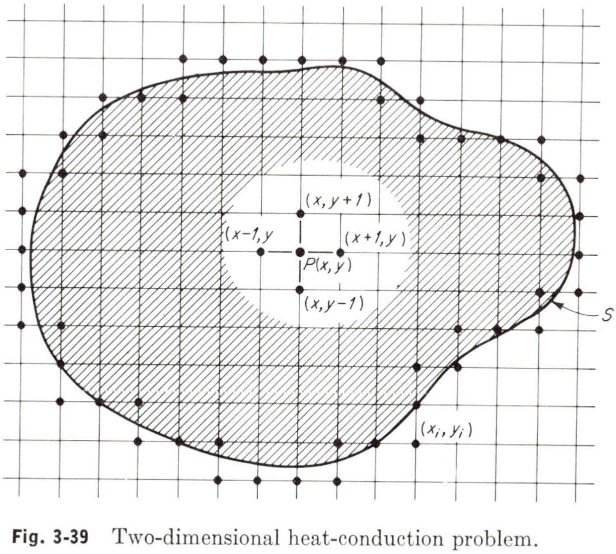

Fig. 3-39 Two-dimensional heat-conduction problem.

The body shape is defined by the boundary S. Superimposed on this shape is a mesh or grid, as shown. The boundary S is approximated by a set of points (x_i,y_i) in the mesh close to the boundary. The finer the mesh, the closer the approximation. A set of internal mesh points centered about the point (x,y) is shown.

It was established in Sec. 3-9, Eq. (3-109), that the temperature $t(x,y)$ at any mesh point (x,y) is expressible in terms of the temperatures of the surrounding mesh points, or

$$t(x,y) = pt(x + 1, y) + pt(x, y + 1) + pt(x - 1, y) + pt(x, y - 1) \tag{3-133}$$

where for a square mesh $p = \frac{1}{4}$.

Equation (3-133) may be given a probabilistic interpretation; i.e., if a random-walking particle is at the point (x,y), it has an equal probability of moving to any of the adjacent points, $(x + 1, y)$, $(x, y + 1)$, $(x - 1, y)$, $(x, y - 1)$, and that probability is $\frac{1}{4}$.

Suppose one lets a particle execute a random walk from point (x,y) and traces its history until it reaches a point on the boundary (x_i,y_i). The random walk executed by the particle could be determined by drawing chips from a box. If there are four chips numbered 1 to 4, chip number 1 indicates a move from point (x,y) to point $(x + 1, y)$, and so on with the others. More practical, as has been noted, is to use a computer. It will be found that the histories of a multiplicity of particles released from point (x,y) are different and terminate at different points on the boundary.

To begin the computation of the temperature at some internal point (x,y) a particle is set upon a random walk from that point and the walk continued until the particle reaches a mesh point on the boundary, where the walk is terminated and the boundary temperature there recorded. Let the temperature so recorded be $t_s(1)$. A second particle is set upon its walk and at the terminus another boundary temperature $t_s(2)$ is recorded. This procedure is repeated until the Nth particle has reached its terminus and the temperature $t_s(N)$ recorded. With these recorded results the temperature of the interior point $t(x,y)$ can be expressed as[1]

$$t(x,y) = \frac{1}{N} \sum_{j=1}^{N} t_s(j) \tag{3-134}$$

where N represents the number of random walks executed.

It is worth noting here that the Monte Carlo procedure permits the determination of the temperature of a single interior point, whereas other

[1] G. M. Brown, Monte Carlo Methods, in E. F. Beckenbach (ed.), "Modern Mathematics for the Engineer," McGraw-Hill, New York, 1956.

techniques, such as the relaxation method (Sec. 3-9), require the simultaneous determination of the temperatures of all interior points.

In Eq. (3-134) if $N \to \infty$, the temperatures then determined are identical with those obtained from an exact solution of the finite-difference equations for the problem. For N finite, the error involved from the finite sample can be estimated by using the central-limit theorem of probability.[1]

If the body under consideration has distributed heat sources in the interior, Eq. (3-111) applies, again with a specification of the boundary temperatures, and the finite-difference representation is that of Eq. (3-112) or

$$t(x,y) = pt(x+1, y) + pt(x, y+1) + pt(x-1, y) + pt(x, y+1) + \frac{a^2 Q'(x,y)}{4k} \tag{3-135}$$

where $Q'(x,y)$ = rate of heat generation per unit volume
k = thermal conductivity
a = mesh size

The value of the probability p is $\frac{1}{4}$, as before. In this case a particle on a random walk again has an equal probability of moving to any of the four adjacent points; however, at each new point there is a contribution from the heat-source term that must be scored for each move. Thus if m_j steps are required for the jth random walk originating at the interior point (x,y) to reach the boundary and $t_s(j)$ is the boundary temperature at the end of the walk, one tallies

$$t_s(j) + m_j \frac{Q_j'(x,y)a^2}{4k} \tag{3-136}$$

and after N such random walks the temperature of the internal point (x,y) is given by

$$t(x,y) = \frac{1}{N}\sum_{j=1}^{N} t_s(j) + \frac{a^2}{4k}\frac{1}{N}\sum_{j=1}^{N} Q_j'(x,y)m_j \tag{3-137}$$

The Monte Carlo procedures for solving more complicated heat-transfer problems are variations on the ideas described above. Application of the procedure to problems with nonlinear boundary conditions due to convection and radiation on the boundary, to problems involving change in phase and therefore moving boundaries, and to problems in transient heat conduction have been reported.[2]

[1] J. H. Curtiss, Sampling Method Applied to Differential and Difference Equations, *Proc. IBM Seminar Sci. Computation*, 1950.

[2] A. Haji-Sheikh, The Application of Monte Carlo Methods to Thermal Conduction Problems, Ph. D. dissertation, Department of Mechanical Engineering, University of Minnesota, 1965; Haji-Sheikh and Sparrow, *loc. cit.*

3-12 NUMERICAL METHODS IN STEADY HEAT CONDUCTION[1]

The relaxation method described earlier has been popular as a numerical procedure since 1940. The relaxation method is an excellent procedure for hand calculation or a desk calculator. Relaxation methods, however, require decisions during the computation process which are easily made by the human calculator but present some difficulty when incorporated in a digital-computer program. Briefly, the task of identifying the maximum residual is easily done by inspection in a hand calculation, but in a machine program the same process for N nodal points will require $N - 1$ subtractions to determine a single maximum residual, or approximately N^2 operations to reduce the residuals just once. Since each residual would in practice have to be reduced many times, the computation time would be excessive.

Furthermore, modern problems in conduction heat transfer are complicated because of complex geometries, nonlinear boundary conditions, and the need to consider the thermophysical properties as temperature-dependent. Analytical procedures leading to exact solutions are important but mostly are confined to the treatment of linear problems and the simplest of geometries. Even though exact solutions can be obtained, the results are often not in closed form and require numerical evaluation for specific parameters before use. It is therefore usually just as easy to resort directly to numerical methods to take advantage of the ready availability of modern, large, high-speed computers. Exact solutions, where they exist, provide standards by means of which approximate solutions and approximation methods can be compared for accuracy.

For some years analog methods and techniques were used for the solutions to problems in heat conduction that were intractable by formal analytical methods. During that period which roughly preceded the advent of large digital computers, much attention was given to thermal analyzers, machines of lumped electric RC circuits, for the solution of problems of transient heat conduction. However, such analog systems were never in great supply, nor were those which existed very versatile. Generally speaking, analog facilities have been superseded by the digital computers, which accomplish all that the analog system could and more.

The foundations for numerical solutions for practical problems in engineering are to be found in the books by Southwell,[2] Allen,[3] and

[1] From A. M. Clausing, Numerical Methods in Heat Transfer, in B. T. Chao (ed.), "Advanced Heat Transfer," pp. 157–216, University of Illinois Press, Urbana, 1969; reprinted with changes by permission of the author and publisher.

[2] *Op. cit.*

[3] D. N. G. Allen, "Relaxation Methods," McGraw-Hill, New York, 1954.

Dusinberre.[1] Development of methodology proceeded rapidly with the ready availability of large digital machines and easily assimilated computer languages.

In this section numerical methods in steady heat conduction will be discussed. The numerical procedures for unsteady heat conduction will be presented in Sec. 4-3. The methods described are suitable for use on a high-speed digital computer. The notation employed is appropriate for transcription to a computer language such as Fortran.

Governing difference equations The difference equations can be derived from an energy balance on a small, finite element fixed in space, in a manner quite similar to the procedures used in Chap. 1 except that here the finite size of the volume element is retained. As before, the shape of the element depends upon the coordinate system used; e.g., in the Cartesian coordinate system the element is a right parallelepiped.

The energy balance states

$$\begin{matrix}\text{Time rate of change} \\ \text{of internal energy in} \\ \text{interval } \Delta\tau\end{matrix} = \begin{matrix}\text{net heat conducted} \\ \text{into the element in} \\ \text{time interval } \Delta\tau\end{matrix} + \begin{matrix}\text{energy generated} \\ \text{within element in} \\ \text{time interval } \Delta\tau\end{matrix} \qquad (3\text{-}138)$$

In the steady-heat-conduction problems the left-hand side of the energy balance is zero.

Consider the applications of Eq. (3-138) to an interior element of a body in the Cartesian coordinate system. One such body is shown in Fig. 3-40. The energy balance yields

$$\rho c(t_{\tau+\Delta\tau} - t_\tau)\frac{\Delta V}{\Delta\tau} = \left(Q_x\Big|_{x-\frac{\Delta x}{2}} - Q_x\Big|_{x+\frac{\Delta x}{2}} + Q_y\Big|_{y-\frac{\Delta y}{2}} - Q_y\Big|_{y+\frac{\Delta y}{2}} + Q_z\Big|_{z-\frac{\Delta z}{2}} - Q_z\Big|_{z+\frac{\Delta z}{2}}\right) + Q'\,\Delta V \qquad (3\text{-}139)$$

where Q is again the rate of heat flow and Q' is the rate of heat generated per unit volume. In the finite-element system,

$$Q_n = \bar{k}A_n \frac{t(\text{adjacent element}) - t(\text{element})}{\text{distance between centers of elements}} \qquad (3\text{-}140)$$

where the thermal conductivity $\bar{k}$ is based on the average of the two nodal temperatures and A_n is the face area of the element normal to the

[1] *Op. cit.*

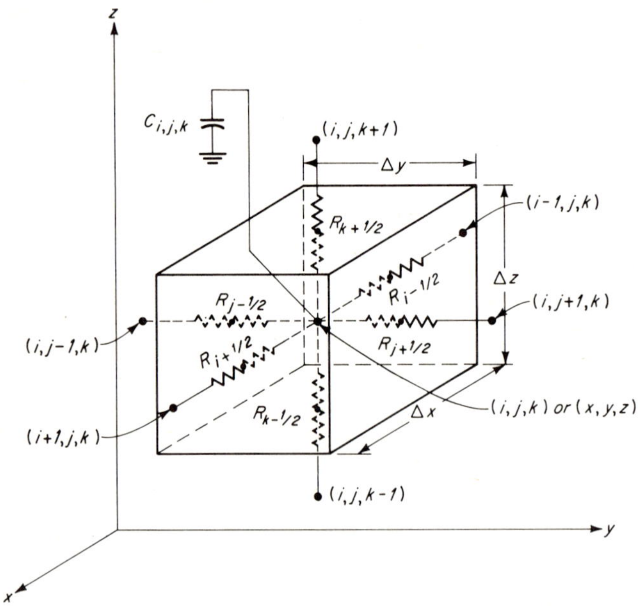

Fig. 3-40 Finite element: cartesian coordinates.

heat flow. Equation (3-140) is applicable to any orthogonal coordinate system and can be used for boundary nodes as well as interior nodes.

If the increments Δx, Δy, Δz, and $\Delta\tau$ are defined as

$$\begin{aligned} x &= i\,\Delta x \\ y &= j\,\Delta y \\ z &= k\,\Delta z \\ \tau &= n\,\Delta\tau \end{aligned}$$

where i, j, k and n are integers restricted as follows:

$$0 \le i \le \frac{x_1}{\Delta x} \qquad 0 \le j \le \frac{y_1}{\Delta y} \qquad 0 \le k \le \frac{z_1}{\Delta z} \qquad 0 \le n$$

and it is assumed that Δx, Δy, and Δz do not vary throughout the region considered ($0 \le x \le x_1$; $0 \le y \le y_1$; $0 \le z \le z_1$), the rate of heat flow into the face $x - \Delta x/2$ may be written, using Eq. (3-140), as

$$Q_x\bigg|_{x-\Delta x/2} = k_{i-\frac{1}{2}}(\Delta y\,\Delta z)\,\frac{t_{i-1,j,k} - t_{i,j,k}}{\Delta x} \tag{3-141}$$

or in terms of a thermal resistance

$$Q_x \Big|_{x-\Delta x/2} = \frac{t_{i-1,j,k} - t_{i,j,k}}{R_{i-\frac{1}{2}}} \tag{3-142}$$

where

$$R_{i-\frac{1}{2}} = \frac{\Delta x}{k_{i-\frac{1}{2}}(\Delta y\ \Delta z)}$$

and $k_{i-\frac{1}{2}}$ is the thermal conductivity evaluated at the mean temperature,

$$t = \frac{t_{i-1,j,k} + t_{i,j,k}}{2}$$

Similarly,

$$Q_x \Big|_{x+\Delta x/2} = \frac{t_{i,j,k} - t_{i+1,j,k}}{R_{i+\frac{1}{2}}} \tag{3-143}$$

If one proceeds in the same way for the other two coordinates, the results when substituted into Eq. (3-139) yield in the limit as $\Delta\tau \to 0$

$$\rho c\ \Delta V \frac{dt_{i,j,k}}{d\tau} = \frac{t_{i-1,j,k} - t_{i,j,k}}{R_{i-\frac{1}{2}}} + \frac{t_{i+1,j,k} - t_{i,j,k}}{R_{i+\frac{1}{2}}} + \frac{t_{i,j-1,k} - t_{i,j,k}}{R_{j-\frac{1}{2}}} + \frac{t_{i,j+1,k} - t_{i,j,k}}{R_{j+\frac{1}{2}}} + \frac{t_{i,j,k-1} - t_{i,j,k}}{R_{k-\frac{1}{2}}} + \frac{t_{i,j,k+1} - t_{i,j,k}}{R_{k+\frac{1}{2}}} + Q'\ \Delta V \tag{3-144}$$

Equation (3-144) with the employment of continuous time is the form of most equations in the set of simultaneous ordinary differential equations which must be solved. For steady-heat-conduction problems the left-hand side of the equation is zero. For unsteady-heat-conduction problems the time variable must be made discrete in order to employ digital computers. The discrete form of the unsteady-heat-conduction equation will be discussed in Chap. 4.

Equation (3-144) is applicable to the cylindrical and spherical coordinate systems as well as to the Cartesian coordinate system. The components appropriate to the three coordinate systems are shown in Table 3-5. The volume elements of the coordinate systems are shown in Figs. 3-41 and 3-42.

If the thermophysical properties are constant, Eq. (3-144) for each of the three major coordinate systems reduces to:

Cartesian:

$$\frac{1}{(\Delta x)^2}(t_{i-1,j,k} + t_{i+1,j,k} - 2t_{i,j,k}) + \frac{1}{(\Delta y)^2}(t_{i,j-1,k} + t_{i,j+1,k} - 2t_{i,j,k}) + \frac{1}{(\Delta z)^2}(t_{i,j,k-1} + t_{i,j,k+1} - 2t_{i,j,k}) + \frac{Q'}{k} = \frac{1}{\alpha}\frac{dt_{i,j,k}}{d\tau} \tag{3-145}$$

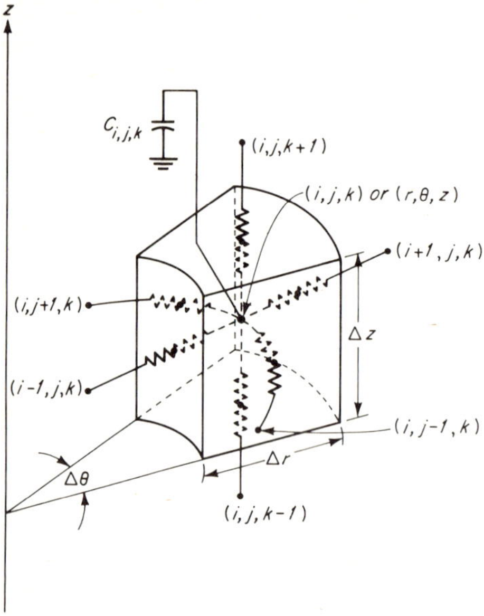

Fig. 3-41 Finite element: cylindrical coordinates.

Cylindrical:

$$\frac{1}{(\Delta x)^2}\left[t_{i-1,j,k}\left(1-\frac{\Delta r}{2r_i}\right)+t_{i+1,j,k}\left(1+\frac{\Delta r}{2r_i}\right)-2t_{i,j,k}\right]$$
$$+\frac{1}{(\Delta\varphi)^2 r_i^2}(t_{i,j-1,k}+t_{i,j+1,k}-2t_{i,j,k})$$
$$+\frac{1}{(\Delta z)^2}(t_{i,j,k-1}+t_{i,j,k+1}-2t_{i,j,k})+\frac{Q'}{k}=\frac{1}{\alpha}\frac{dt_{i,j,k}}{d\tau} \quad (3\text{-}146)$$

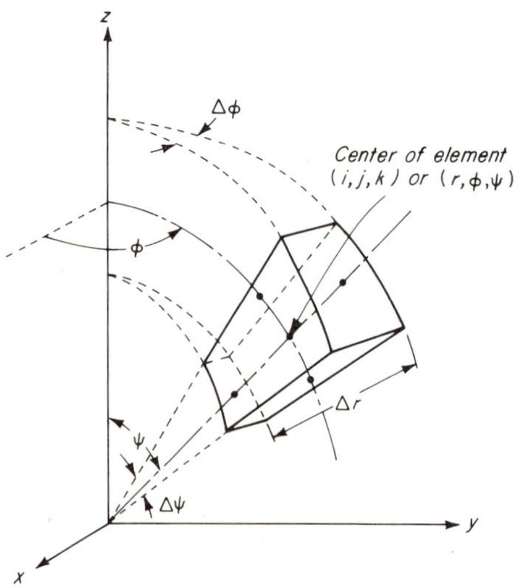

Fig. 3-42 Finite element: spherical coordinates.

Table 3-5 Components for Eq. (3-144) for three coordinate systems†

	Coordinate system		
	Cartesian	*Cylindrical*	*Spherical*
	$x = i\,\Delta x$ $y = j\,\Delta y$ $z = k\,\Delta z$	$r = i\,\Delta r$ $\theta = j\,\Delta\theta$ $z = k\,\Delta z$	$r = i\,\Delta r$ $\varphi = j\,\Delta\vartheta$ (*aximuth angle*) $\psi = k\,\Delta\psi$ (*polar angle*)
ΔV	$\Delta x\,\Delta y\,\Delta z$	$r\,\Delta r\,\Delta\theta\,\Delta z$	$r^2 \sin\psi\,\Delta r\,\Delta\varphi\,\Delta\psi$
$R_{i-\frac{1}{2}}$	$\dfrac{\Delta x}{\Delta y\,\Delta z\,k_{i-\frac{1}{2}}}$	$\dfrac{\Delta r}{(r_i - \Delta r/2)\,\Delta\theta\,\Delta z\,k_{i-\frac{1}{2}}}$	$\dfrac{\Delta r}{(r_i - \Delta r/2)^2 \sin\psi_k\,\Delta\varphi\,\Delta\psi\,k_{i-\frac{1}{2}}}$
$R_{i+\frac{1}{2}}$	$\dfrac{\Delta x}{\Delta y\,\Delta z\,k_{i+\frac{1}{2}}}$	$\dfrac{\Delta r}{(r_i + \Delta r/2)\,\Delta\theta\,\Delta z\,k_{i+\frac{1}{2}}}$	$\dfrac{\Delta r}{(r_i - \Delta r/2)^2 \sin\psi_k\,\Delta\varphi\,\Delta\psi\,k_{i+\frac{1}{2}}}$
$R_{j-\frac{1}{2}}$	$\dfrac{\Delta y}{\Delta x\,\Delta z\,k_{j-\frac{1}{2}}}$	$\dfrac{r_i\,\Delta\theta}{\Delta r\,\Delta z\,k_{j-\frac{1}{2}}}$	$\dfrac{\Delta\theta \sin\psi_k}{\Delta r\,\Delta\psi\,k_{j-\frac{1}{2}}}$
$R_{j+\frac{1}{2}}$	$\dfrac{\Delta y}{\Delta x\,\Delta z\,k_{j+\frac{1}{2}}}$	$\dfrac{r_i\,\Delta\theta}{\Delta r\,\Delta z\,k_{j+\frac{1}{2}}}$	$\dfrac{\Delta\theta \sin\psi_k}{\Delta r\,\Delta\psi\,k_{j+\frac{1}{2}}}$
$R_{k-\frac{1}{2}}$	$\dfrac{\Delta z}{\Delta x\,\Delta y\,k_{k-\frac{1}{2}}}$	$\dfrac{\Delta z}{\Delta r\,\Delta\theta\,r_i\,k_{k-\frac{1}{2}}}$	$\dfrac{\Delta\psi}{\sin(\psi_k - \Delta\psi/2)\,\Delta r\,\Delta\varphi\,k_{k-\frac{1}{2}}}$
$R_{k+\frac{1}{2}}$	$\dfrac{\Delta z}{\Delta x\,\Delta y\,k_{k+\frac{1}{2}}}$	$\dfrac{\Delta z}{\Delta r\,\Delta\theta\,r_i\,k_{k+\frac{1}{2}}}$	$\dfrac{\Delta\psi}{\sin(\psi_k - \Delta\psi/2)\,\Delta r\,\Delta\varphi\,k_{k+\frac{1}{2}}}$

† A. M. Clausing, Numerical Methods in Heat Transfer, in B. T. Chao (ed.), "Advanced Heat Transfer," University of Illinois Press, Urbana, 1969.

Spherical:

$$\frac{1}{(\Delta r)^2}\left[(t_{i-1,j,k} - t_{i,j,k})\left(1 - \frac{\Delta r}{2r_i}\right)^2 + (t_{i+1,j,k} - t_{i,j,k})\left(1 + \frac{\Delta r}{2r_i}\right)^2\right]$$
$$+ \frac{1}{r_i^2 \sin^2\psi_k(\Delta\varphi)^2}(t_{i,j-1,k} + t_{i,j+1,k} - 2t_{i,j,k})$$
$$+ \frac{1}{r_i^2 \sin\psi_k(\Delta\psi)^2}\left[(t_{i,j,k-1} - t_{i,j,k})\sin\left(\psi_k - \frac{\Delta\psi}{2}\right)\right.$$
$$\left. + (t_{i,j,k+1} - t_{i,j,k})\sin\left(\psi_k + \frac{\Delta\psi}{2}\right)\right] + \frac{Q'}{k} = \frac{1}{\alpha}\frac{dt_{i,j,k}}{d\tau} \qquad (3\text{-}147)$$

In each of Eqs. (3-145) to (3-147) the right-hand side would be zero for the case of steady heat conduction.

Boundary conditions in difference form Derivation of the boundary conditions in difference form must be accomplished such that the results are physically consistent with the difference form of the differential equation representing the interior of the body in question, e.g., Eq. (3-144). Physical consistency requires that each thermal resistance be uniquely defined. In this way it is possible to write an overall energy balance for the system which acts as a check on both the equations and the computer program and provides as well a criterion for the termination of the iterative process necessary for a solution.

Application of an energy balance to the finite elements of a schematic *RC* network is a certain method for obtaining a physically consistent set of difference equations. This method was used in the development of the equations for the interior of the body and will be used in the development of the difference equations for the boundary conditions.

The *RC* thermal network shown in Fig. 3-43 is physically consistent. Equation (3-145), modified for steady conditions and for no internal heat generation, applies for the case of constant thermophysical properties to the nine internal nodes. Equation (3-144) will be applicable, as well, if reduced for steady conditions and no heat sources but allowing the thermal conductivity to vary with temperature and position.

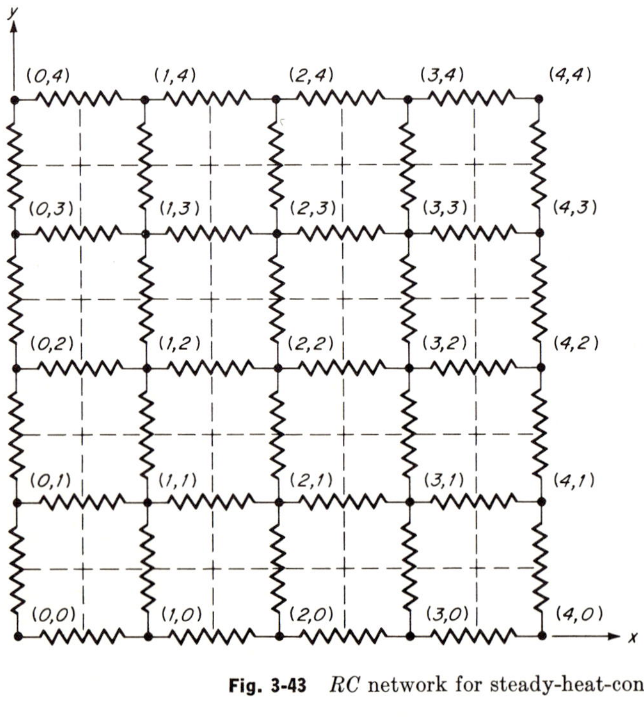

Fig. 3-43 *RC* network for steady-heat-conduction analysis.

The boundary condition of a *prescribed surface temperature* is the simplest condition with which to work. Consider the left-hand boundary of the solid depicted in Fig. 3-43 and any node $(0,j)$. From each such node there are three heat flows:

$$Q_{0,j+1\rightarrow 0,j} = +k\frac{\Delta A}{2}\frac{t_{0,j+1} - t_{0,j}}{\Delta y}$$

$$Q_{0,j-1\rightarrow 0,j} = +k\frac{\Delta A}{2}\frac{t_{0,j-1} - t_{0,j}}{\Delta y}$$

$$Q_{0,j\rightarrow 1,j} = -k\,\Delta A\,\frac{t_{0,j} - t_{1,j}}{\Delta x}$$

where $\Delta x = \Delta y$; $\Delta A = \Delta x\,\Delta y = \Delta y\,\Delta z$. The heat balance on the node yields for steady conditions

$$t_{0,j+1} + t_{0,j-1} + 2t_{1,j} - 4t_{0,j} = 0 \tag{3-148}$$

For the boundary condition of a *prescribed heat flux* at the surface along the left-hand boundary, for example,

$$-k\left(\frac{\partial t}{\partial x}\right)_{x=0} = q''(x,\tau)$$

which can be written in difference form for $q''(x)$ only as

$$\frac{2\Delta y\, q''}{k} + t_{0,j+1} + t_{0,j-1} + 2t_{1,j} - 4t_{0,j} = 0 \tag{3-149}$$

For a *convective boundary condition*

$$-k\left(\frac{\partial t}{\partial x}\right)_{x=0} = h(t_f - t_{0,j})$$

where h is the heat-transfer coefficient and t_f is the ambient-fluid temperature sweeping the surface. In difference form the relationship becomes

$$2\frac{h\,\Delta x}{k}(t_f - t_{0,j}) + t_{0,j+1} + t_{0,j-1} + 2t_{1,j} - 4t_{0,j} = 0 \tag{3-150}$$

or since $\mathrm{Bi} = h\,\Delta x/k$, the Biot number,

$$2\,\mathrm{Bi}\,t_f + t_{0,j+1} + t_{0,j-1} + 2t_{1,j} - (4 + 2\,\mathrm{Bi})t_{0,j} = 0 \tag{3-151}$$

For a *thermal-radiation boundary condition*

$$-k\left(\frac{\partial t}{\partial x}\right)_{x=0} = \sigma F_{\Delta A-S}(T^4 - T_S{}^4)$$

where σ is the Stefan-Boltzmann constant and $F_{\Delta A-S}$ is the *shape factor* between the element of interest and the surroundings. In difference form,

$$\frac{2\Delta x\,\sigma F_{\Delta A-S}}{k}(T_S{}^4 - T_{0,j}{}^4) + T_{0,j+1} + T_{0,j-1} + 2T_{1,j} - 4T_{0,j} = 0 \quad (3\text{-}152)$$

Similar difference equations and boundary conditions in difference form can be obtained for other geometries.

Numerical procedures, steady conditions The equations derived in the preceding paragraphs when applied to each nodal point of the conducting body are in general a set of simultaneous ordinary differential equations. Under steady conditions the derivatives with respect to time are zero at each node, and the equations become a set of simultaneous algebraic equations. In the case of constant thermophysical properties and linear boundary conditions the resulting equations are linear. Thus the solution of a linear partial differential equation has been reduced to the solution of a set of simultaneous linear algebraic equations.

These equations are of the form

$$[A][X] = [B] \quad (3\text{-}153)$$

where $[X]$ = unknown column vector, or matrix
$[B]$ = constant column matrix
$[A]$ = coefficient matrix

Equation (3-153) can be rewritten in expanded form as

$$\begin{aligned} a_{11}x_1 + a_{12}x_2 + a_{13}x_3 + \cdots + a_{1N}x_N &= b_1 \\ a_{21}x_1 + a_{22}x_2 + a_{23}x_3 + \cdots + a_{2N}x_N &= b_2 \\ \cdots\cdots\cdots\cdots\cdots\cdots\cdots\cdots\cdots \\ a_{N1}x_1 + a_{N2}x_2 + a_{N3}x_3 + \cdots + a_{NN}x_N &= b_N \end{aligned} \quad (3\text{-}154)$$

Many different procedures are available to effect a solution of the set of equations, some being more appropriate than others for application to problems of a specific nature. In general the methods for the solution of sets of simultaneous linear algebraic equations may be categorized into two general types: (1) direct methods which require a finite number of steps and (2) indirect methods which require, in theory, an infinite number of steps to reach a solution.

It is generally true that (1) direct methods are more suitable for general matrices for any nonsingular set whereas iterative methods are more successful with sparse matrices; (2) direct methods are more susceptible to inaccuracies as the result of round-off errors than indirect methods are; (3) indirect methods may not converge or may require a large number of operations for convergence; (4) indirect methods are generally easier to program and require less computer storage; and (5) indirect methods can often be extended to nonlinear algebraic equations whereas direct methods are limited to linear problems. Indirect methods are probably the more widely used in heat-conduction calculations, but there are instances where the direct methods are quite useful.

It is desirable to select methods which permit one to take advantage of the properties of the particular set of equations under consideration. The special characteristics of sets of equations are:

1. The number of simultaneous algebraic equations required for a meaningful solution of a problem is often quite large. For example, Clausing[1] notes that the two-dimensional mixed boundary-value problem requires 2,000 simultaneous equations. If the solution procedure required the storage of the complete coefficient matrix including the zero elements, 4 million storage locations would be required. This storage requirement far exceeds the high-speed storage space of most computers.
2. The coefficient matrix is sparse; i.e., many elements are zero. In general, a one-dimensional problem results in three nonzero elements in each row, a two-dimensional problem contains five nonzero elements, and a three-dimensional problem contains seven nonzero elements.
3. The coefficient matrix $[A]$ is strongly diagonal; i.e., the diagonal elements are large relative to the off-diagonal elements. This property permits the employment of iterative methods, which are, in general, limited to such systems. It is also a desirable property if an elimination procedure is used.

An excellent discussion of direct and iterative (indirect) methods is given by Clausing,[2] as well as a list of primary references to numerical methods and procedures.

[1] A. M. Clausing, Some Influences of Microscopic Constructions on the Thermal Contact Resistance, *Univ. Illinois Dept. Mech. Eng., Tech. Rept.* ME-TN-242-2, 1965.
[2] Numerical Methods in Heat Transfer, *loc. cit.*

4
Unsteady Heat Conduction

4-1 TRANSIENT HEAT CONDUCTION: ANALYTICAL METHODS

The previous discussion was related to heat-conduction systems where the temperature was invariant with time. It was assumed that sufficient time had elapsed to permit the transient condition, brought about by the steady application of heat to some boundary of the system, to die out. In this section the heating or cooling of the body as a function of time, the transient state, will be discussed.

Solids of infinite thermal conductivity These present the simplest cases. If the body has a large thermal conductivity and a corresponding low film heat-transfer coefficient, the heat flow to or from the body is controlled principally by the convection resistance and few temperature gradients or none exist in the body; i.e., the body is spacewise isothermal, the temperature varying only with time. Certain smaller bodies of finite thermal conductivity satisfy the above conditions. Consider the small billet shown in Fig. 4-1, which exists at some uniform temperature

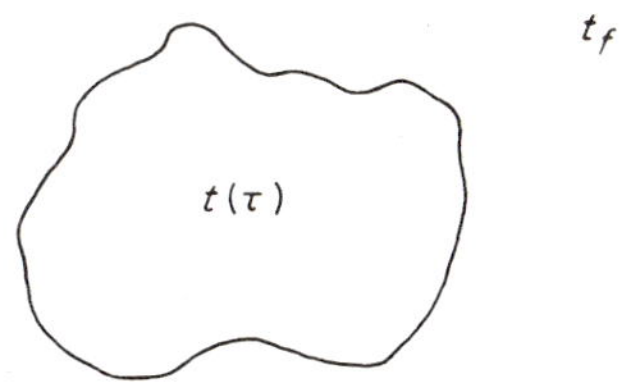

Fig. 4-1 The homogeneous billet.

when at time zero it is placed in a fluid of a different temperature. The problem is to determine the change in temperature with time as a function of the system characteristics. A heat balance on the billet, equating the time rate of change in heat capacity with the convective heat loss, yields

$$\rho c V \frac{dt}{d\tau} = -hA(t - t_f) \tag{4-1}$$

where V is the volume of the billet and A is the surface area. The appropriate initial condition is $t = t_0$ at $\tau = 0$. The surrounding fluid temperature t_f is a constant. Equation (4-1) can be rewritten in the form of a differential equation for the excess temperature where $\vartheta = t - t_f$

$$\frac{d\vartheta}{\vartheta} = -\frac{hA}{\rho c V}\, d\tau \tag{4-2}$$

The appropriate condition for the evaluation of the single constant appearing in the solution of Eq. (4-2) is the initial condition, the temperature of the billet at time zero

$$\tau = 0 \qquad \vartheta = \vartheta_0$$

With the use of this initial condition the solution for Eq. (4-2) becomes

$$\frac{\vartheta}{\vartheta_0} = e^{-(hA/\rho c V)\tau} \tag{4-3}$$

It is clear that a plot of Eq. (4-3) for $\ln (\vartheta/\vartheta_0)$ vs. τ will yield a family of straight lines with $hA/\rho c V$ as a parameter. Equation (4-3) can be rewritten to advantage as

$$\frac{\vartheta}{\vartheta_0} = e^{-(hL/k)(\alpha\tau/L^2)} \tag{4-4}$$

where hL/k = Biot modulus
$\alpha\tau/L^2$ = Fourier modulus
α = thermal diffusivity

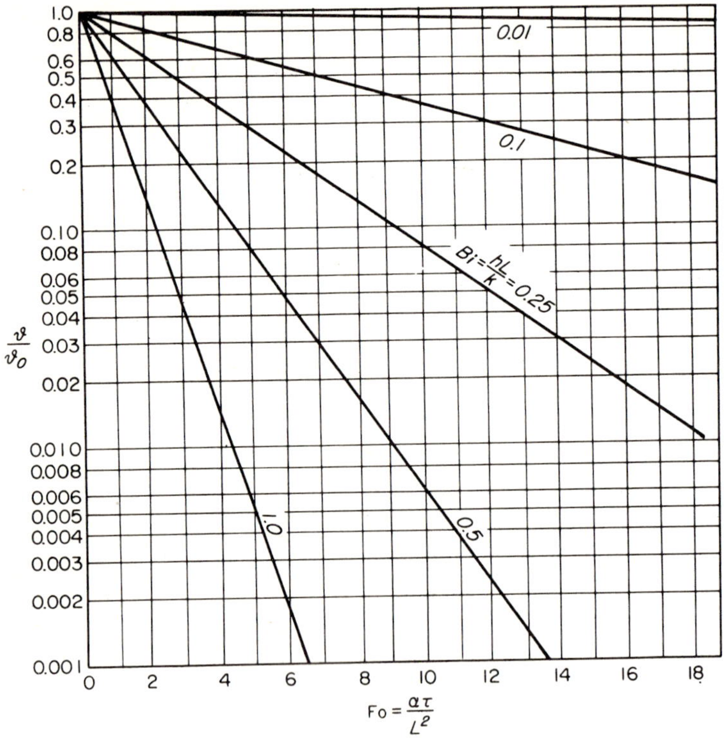

Fig. 4-2 Temperature-time history of the homogeneous billet.

The appropriate plot of Eq. (4-4) is ln (ϑ/ϑ_0) vs. $\alpha\tau/L^2$ with hL/k as a parameter (Fig. 4-2).

The value L in the Biot and Fourier modulus is recognized as a significant dimension of the system and is derived as the ratio of the volume of the body to the surface area. Thus for simple geometrical shapes the value of L can readily be obtained:

Sphere: $$L = \frac{\frac{4}{3}\pi r_0^3}{4\pi r_0^2} = \frac{r_0}{3}$$

Cylinder: $$L = \frac{\pi r_0^2 l}{2\pi r_0 l} = \frac{r_0}{2} \qquad l \gg r_0$$

Cube: $$L = \frac{l^3}{6l^2} = \frac{l}{6}$$

The Biot and Fourier moduli are dimensionless and have the great advantage that when used in the manner of Eq. (4-4), the temperature-time histories for all bodies of infinite thermal conductivity can be reduced to a single universal plot for all values of the convection boundary condition. The electrical equivalent of the transient heat-transfer system of the small billet is the discharge of an electric condenser in a circuit with a pure resistance. The electrical equation is

$$\frac{E}{E_0} = e^{-(\tau/RC)} \tag{4-5}$$

where E_0 = potential at time zero and $E = E(\tau)$
R = resistance
C = electric capacitance

This analogy led to the design of many computers for transient heat-conduction behavior based on electric lumped RC circuits.

Example 4-1 When it is desired to measure an unsteady temperature, it is very important to understand the response characteristics of the measuring device. The *time constant* τ^* is the pertinent quantity for first-order systems and may be identified from Eq. (4-3) as

$$\tau^* = \frac{\rho c V}{hA} = \frac{k}{\alpha h}\frac{V}{A}$$

In response to a step change in temperature, τ^* is the time required for the temperature change to reach e^{-1} of its final value. In response to a sinusoidal variation in temperature, $(\tau^*)^{-1}$ is the *break frequency*, above which attenuation is increasingly large.

Consider a cylindrical mercury bulb thermometer with a 0.3-cm diameter installed in an airstream. From Appendix B, for mercury

$$k = 8.6 \text{ W/m K}$$
$$\alpha = 0.0166 \text{ m}^2\text{/hr}$$

The thermal resistance of the glass wall is neglected, and the air-side heat-transfer coefficient is taken to be

$$h = 57 \text{ W/m}^2\text{ K}$$

Hence the Biot modulus is

$$\frac{hL}{k} = \frac{57(0.15)}{8.6(2)(100)} = 0.0049$$

and the time constant

$$\tau^* = \frac{8.6}{0.0166(57)}\frac{0.15}{2(100)}3{,}600 = 24.6 \text{ s}$$

The half time $\tau_{\frac{1}{2}}$, which is the time required for the change to reach half its final value, is related to τ^* by

$$\tau_{\frac{1}{2}} = \tau^* \ln 2 = 24.6(0.693) = 17.2 \text{ s}$$

The break frequency for the thermometer in the airstream is

$$\frac{1}{\tau_*} = 0.0406 \text{ rad/s} = 0.0065 \text{ Hz}$$

The magnitude of a sinusoidal temperature variation of 0.0065 Hz indicated by the thermometer will be only 71 percent of the magnitude of the sine wave in the airstream, and the attenuation will increase with increasing frequency.

Only for unsteady temperature changes, which are slower (with a sine-shaped temperature oscillation the period duration must be about tenfold), can one expect the thermometer to record the temperature trend accurately.[1]

Solids of infinite conductivity, boundary conditions, functions of the time Consider the billet of Fig. 4-1 but imagine that the fluid temperature begins at time zero to change linearly with time; i.e., the temperature of the fluid obeys the expression

$$t_f = B\tau \tag{4-6}$$

Equation (4-1) can be rewritten to consider the variation of the fluid temperature:

$$\frac{dt}{d\tau} + \frac{hA}{\rho cV} t = \frac{hAB}{\rho cV} \tau \tag{4-7}$$

The solution of Eq. (4-7) becomes

$$t = B\left(\tau - \frac{\rho cV}{hA}\right) + C_1 e^{-(hA\tau/\rho cV)} \tag{4-8}$$

The constant C_1 can be evaluated from the initial condition, which establishes the initial temperature at zero. The complete solution is therefore

$$t = B\tau - \frac{\rho cV}{hA} B(1 - e^{-(hA\tau/\rho cV)}) \tag{4-9}$$

which is demonstrated graphically in Fig. 4-3. It can be seen that the billet temperature always lags behind the fluid temperature. As soon as

[1] F. Lieneweg, *Wiss. Veroffentl. Siemens-Werken,* **16**:112–126 (1937); **17**:19–32 (1937); **19**:73–87 (1940); also a brief report on the subject in *Z. Ver. Deut. Ingr.,* **85**:272–273 (1941).

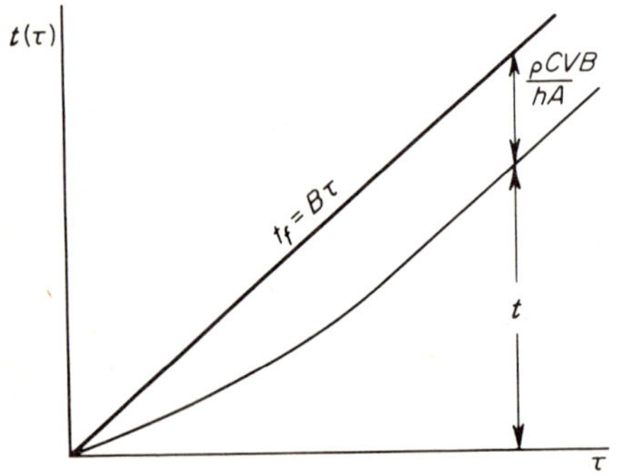

Fig. 4-3 Temperature-time history of a homogeneous billet subjected to a linearly varying fluid temperature.

the initial transient dies out, the lag becomes a constant. This can be seen from Eq. (4-9) for τ very large.

Example 4-2 If the thermometer used in Example 4-1 were used to monitor the temperature of an oven for an ordinary household stove, it would be of interest to calculate the lag of the thermometer while the oven is heating at a rate of 200 K/hr. Suppose the heat-transfer coefficient is $h = 10$ W/m² K; in this case,

$$\frac{\rho c V}{Ah} = \frac{r_0 k}{2\alpha h} = \frac{0.015(8.65)}{100(2)(0.01655)(10)} = 0.0392 \text{ hr}$$

$$\Delta t_{\text{lag}} = 0.0392(200) = 8.62 \text{ K}$$

The infinite flat plate Problems in transient heat conduction in systems with spacewise temperature distributions constitute the beginning of some rather formidable mathematical treatments. Fourier[1] occupied himself with solutions to this problem and developed the renowned method of the Fourier series.

For a slab of thickness l in the x direction and infinite in the y and z directions, insulated on the face $x = 0$, and losing heat by convection on the face $x = l$, the heat-conduction equation in the absence of heat sources reduces to

$$\frac{\partial \vartheta}{\partial \tau} = \alpha \frac{\partial^2 \vartheta}{\partial x^2} \tag{4-10}$$

where ϑ is the excess temperature anywhere in the body. The slab can be represented as shown in Fig. 4-4. Equation (4-10) can be solved for this set of boundary conditions by the method of separation of variables. Assume a solution of the form

$$\vartheta = F(\tau)G(x)$$

where, as indicated, F is a function of τ only and G is a function of x only:

$$\frac{\partial \vartheta}{\partial \tau} = G(x)F'(\tau)$$

$$\frac{\partial^2 \vartheta}{\partial x^2} = F(\tau)G''(x)$$

Substitution of these values in Eq. (4-10) gives

$$G(x)F'(\tau) = \alpha F(\tau)G''(x) \tag{4-11}$$

whereby the variables are separable:

$$\frac{1}{\alpha}\frac{F'}{F} = \frac{G''}{G} = \pm\lambda^2 \tag{4-12}$$

where λ is a constant.[2]

[1] Jean Baptiste Fourier, "Théorie analytique de la chaleur," Gauthier-Villars, Paris, 1822 (English translation by Alexander Freeman, Cambridge, 1878).

[2] λ is called on *eigenvalue* and the problem an *eigenvalue problem.*

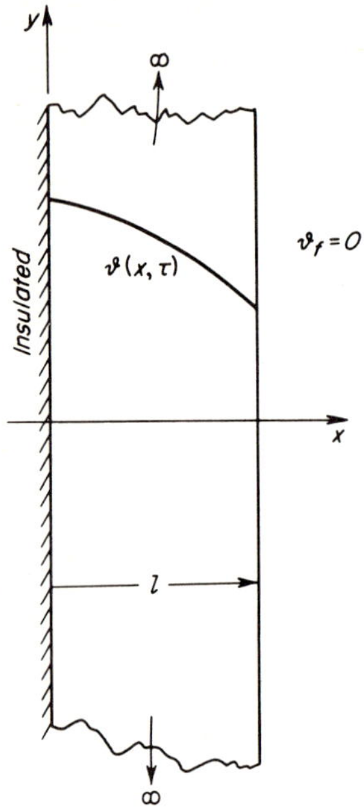

Fig. 4-4 Heat conduction in the infinite flat plate.

Equation (4-12) presents two differential equations for solution instead of the one in Eq. (4-10), but these are ordinary linear equations and so are amenable to solution. The two equations are

$$F' - (\pm)\,\lambda^2 \alpha F = 0 \tag{4-13}$$

$$G'' - (\pm)\,\lambda^2 G = 0 \tag{4-14}$$

Solution of Eq. (4-13) yields

$$F = C_1 e^{\pm\lambda^2 \alpha\tau}$$

and since we know that the excess temperature will decrease with time, we accept the negative sign. Equation (4-14) with the negative λ^2 can be solved to yield

$$G = C_2 e^{+i\lambda x} + C_3 e^{-i\lambda x}$$

The complete solution is, of course, $\vartheta = F(\tau)G(x)$; therefore

$$\vartheta = e^{-\lambda^2 \alpha\tau}\,(A \cos \lambda x + B \sin \lambda x) \tag{4-15}$$

The boundary and initial conditions are

$$\frac{\partial \vartheta}{\partial x} = 0 \qquad x = 0 \tag{4-16}$$

$$\frac{\partial \vartheta}{\partial x} = -\frac{h}{k}\vartheta_l \qquad x = l \tag{4-16a}$$

$$\vartheta = \vartheta_0 \qquad \tau = 0 \tag{4-16b}$$

There are really three constants to be determined, A, B, and λ, and there are appropriately three conditions adequate for their determination. Equation (4-16) requires the derivative $\partial\vartheta/\partial x = 0$ at $x = 0$.

$$\frac{\partial\vartheta}{\partial x} = e^{-\lambda^2\alpha\tau}(-A_1\lambda \sin \lambda x + B\lambda \cos \lambda x) \tag{4-17}$$

It is clear from Eq. (4-17) that $B \equiv 0$ in order that the temperature gradient $(\partial\vartheta/\partial x)_{x=0} = 0$. The solution now reads

$$\vartheta = Ae^{-\lambda^2\alpha\tau} \cos \lambda x \tag{4-18}$$

The second boundary condition [Eq. (4-16*a*)] can be used to obtain the λ:

$$\left(\frac{\partial\vartheta}{\partial x}\right)_{x=l} = -A_1 e^{-\lambda^2\alpha\tau}\lambda \sin \lambda l = -\frac{h}{k}\vartheta_l \tag{4-19}$$

From Eq. (4-18)

$$\vartheta_l = Ae^{-\lambda^2\alpha\tau} \cos \lambda l \tag{4-20}$$

Combining Eqs. (4-19) and (4-20), we obtain

$$\cot \lambda l = \frac{\lambda k}{h} \tag{4-21}$$

Equation (4-21) is a specification for λ, which can be obtained by plotting each side of the equation against λ, as in Fig. 4-5. From the intersec-

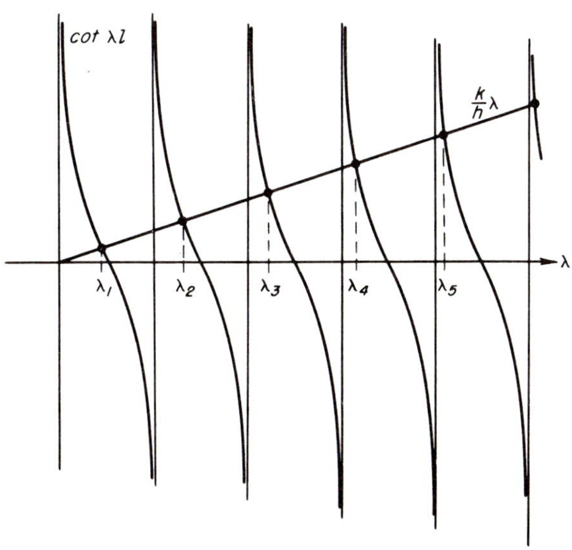

Fig. 4-5 Determination of the eigenvalues for the infinite-flat-plate problem.

tions of the two functions as many values of λ can be obtained as are necessary. The temperature distribution becomes

$$\vartheta = \sum_{n=1}^{\infty} A_n e^{-\lambda_n^2 \alpha\tau} \cos \lambda_n x \tag{4-22}$$

where a value of A_n for each λ_n must now be determined. The initial condition is utilized for this purpose. Assume that ϑ_0 can be expanded in an infinite series:

$$\vartheta_0 = A_1 \cos \lambda_1 x + A_2 \cos \lambda_2 x + A_3 \cos \lambda_3 x + \cdots \tag{4-23}$$

If Eq. (4-23) is multiplied through by $\cos \lambda_n x \, dx$ and integrated from $x = 0$ to $x = l$ (assuming the integration permissible), it can be shown that

$$\int_0^l A_m \cos \lambda_n x \cos \lambda_m x \, dx = 0$$

if $m \neq n$. The remaining values are

$$\int_0^l \vartheta_0 \cos \lambda_n x \, dx = \int_0^l A_n \cos^2 \lambda_n x \, dx$$

and these can be evaluated to yield the constants A_n. In general, then,

$$A_n = \frac{2\vartheta_0 \sin \lambda_n l}{\lambda_n l + \sin \lambda_n l \cos \lambda_n l} \tag{4-24}$$

Now for convenience let $\lambda_n = \delta_n / l$, so that the temperature distribution is, finally,

$$\vartheta = \sum_{n=1}^{\infty} e^{-\delta_n^2(\alpha\tau/l^2)} \frac{2\vartheta_0 \sin \delta_n \cos (\delta_n x/l)}{\delta_n + \sin \delta_n \cos \delta_n} \tag{4-25}$$

The heat loss for the semi-infinite slab is obtained from the Fourier conduction law [Eq. (1-2)]

$$dQ_i = -kA \left(\frac{\partial \vartheta}{\partial x}\right)_{x=l} d\tau$$

$$\left(\frac{\partial \vartheta}{\partial x}\right)_{x=l} = -\frac{2\vartheta_0}{l} \sum_{n=1}^{\infty} e^{-\delta_n^2(\alpha\tau/l^2)} \frac{\delta_n \sin^2 \delta_n}{\delta_n + \sin \delta_n \cos \delta_n} \tag{4-26}$$

When Eq. (4-26) is substituted into the Fourier conduction law,

$$\frac{Q_i}{A} = \frac{2k\vartheta_0}{l} \int_0^{\tau} \sum_{n=1}^{\infty} e^{-\delta_n^2(\alpha\tau/l^2)} \frac{\delta_n \sin^2 \delta_n}{\delta_n + \sin \delta_n \cos \delta_n} \, d\tau \tag{4-27}$$

The integration of Eq. (4-27) yields the heat flow per unit area:

$$\frac{Q_i}{A} = \frac{2kl}{\alpha} \vartheta_0 \sum_{n=1}^{\infty} \frac{\sin^2 \delta_n [1 - e^{-\delta_n^2(\alpha\tau/l^2)}]}{\delta_n^2 + \delta_n \sin \delta_n \cos \delta_n} \tag{4-28}$$

Equation (4-21) involving the eigenvalue λ_n can be written in terms of δ_n:

$$\frac{hl}{k} = \delta_n \tan \delta_n \tag{4-29}$$

The result shows that

$$\delta_n = \phi\left(\frac{hl}{k}\right)$$

The results of Eqs. (4-25), (4-28), and (4-29) can be presented graphically in the following functional forms:

$$\frac{\vartheta}{\vartheta_0} = f_1\left(\frac{hl}{k}, \frac{\alpha\tau}{l^2}, \frac{x}{l}\right)$$

$$\frac{\vartheta_c}{\vartheta_0} = f_2\left(\frac{hl}{k}, \frac{\alpha\tau}{l^2}, 0\right)$$

$$\frac{\vartheta_l}{\vartheta_0} = f_3\left(\frac{hl}{k}, \frac{\alpha\tau}{l^2}, 1\right)$$

$$\frac{Q_i}{Q_0} = f_4\left(\frac{hl}{k}, \frac{\alpha\tau}{l^2}\right)$$

The infinite circular cylinder A solution similar to that of the infinite flat plate has been made for the infinite circular cylinder. Consider the cylinder as shown in Fig. 4-6. If the cylinder is long in the z direction, the important variables are r, θ, and τ. An assumption of symmetry in θ reduces the problem to one of pure radial conduction in a transient state. The appropriate heat-conduction equation can be obtained from Eq. (1-17) by dropping the terms depending on θ and z and the heat-source terms in this case. Thus,

$$\frac{\partial\vartheta}{\partial\tau} = \alpha\left(\frac{\partial^2\vartheta}{\partial r^2} + \frac{1}{r}\frac{\partial\vartheta}{\partial r}\right) \tag{4-30}$$

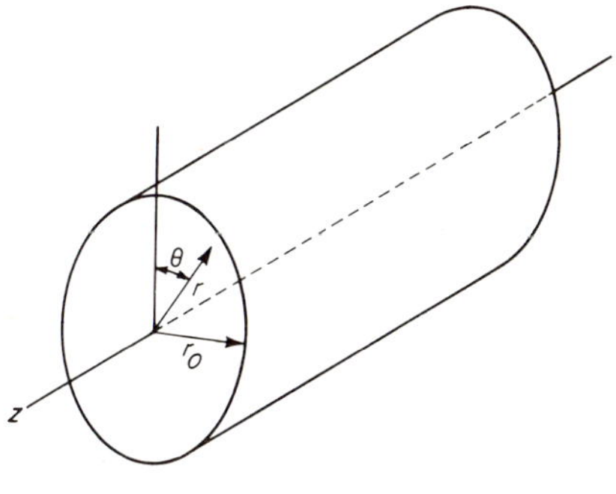

Fig. 4-6 The circular-cylinder coordinate system.

Assume for the moment that an initial temperature distribution exists in the cylinder and is a function of r only. This condition may be written

$$\vartheta = f(r) \qquad 0 \leq r \leq r_0 \qquad \tau \leqq 0 \tag{4-31}$$

If the cylinder is transferring heat by convection to a surrounding fluid at temperature t_f, the boundary condition on the surface, $r = r_0$, can be expressed

$$\left(\frac{\partial \vartheta}{\partial r}\right)_{r=r_0} = -\frac{h}{k}\vartheta(r_0,\tau) \qquad r = r_0 \qquad \tau > 0 \tag{4-32}$$

As in the case of the flat plate a solution can be obtained by the separation of variables. Assume such a solution as

$$\vartheta(r,\tau) = F(\tau)G(r) \tag{4-33}$$

Using Eq. (4-33) in Eq. (4-30), one obtains

$$F'(\tau)G(r) = \alpha\left[G''(r)F(\tau) + \frac{1}{r}G'(r)F(\tau)\right] \tag{4-34}$$

where, as before, the variables can be separated and the resulting expressions equated to a constant

$$\frac{G''}{G} + \frac{1}{r}\frac{G'}{G} = \frac{1}{\alpha}\frac{F'}{F} = \pm\lambda^2 \tag{4-35}$$

Equation (4-35) with the choice of the negative value of λ^2 to make the temperature finite at infinite values of τ becomes two equations, each to be solved,

$$F'(\tau) + \alpha\lambda^2 F(\tau) = 0 \tag{4-36}$$
$$r^2G''(r) + rG'(r) + \lambda^2 r^2 G(r) = 0 \tag{4-37}$$

The solution to Eq. (4-36) is

$$F(\tau) = C_1 e^{-\alpha\lambda^2\tau} \tag{4-38}$$

In Eq. (4-37) let $u = \lambda r$ such that the equation becomes

$$u^2\frac{d^2G}{du^2} + u\frac{dG}{du} + u^2G = 0$$

which can be identified as a *Bessel equation* with a solution

$$G(u) = C_2J_0(u) + C_3Y_0(u)$$

where the $J_0(u)$ and the $Y_0(u)$ are *Bessel functions of the first and second kind of zero order*. Selected magnitudes are shown in Fig. 4-7 and Tables

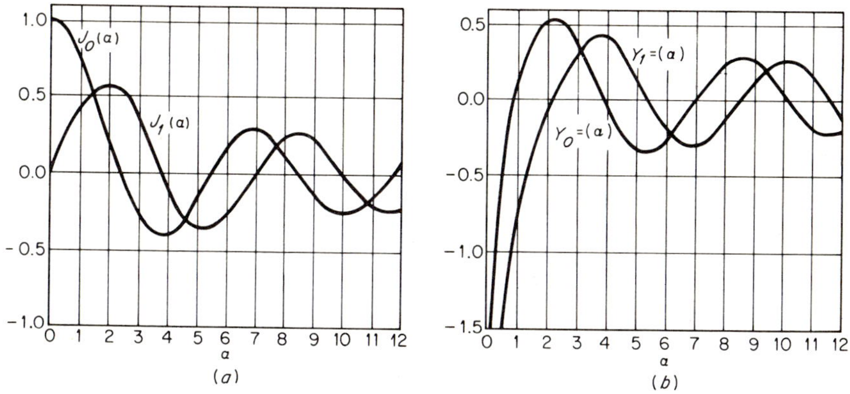

Fig. 4-7 Bessel functions of the first and second kind.

Table 4-1 Selected values of Bessel functions of the first kind

α	$J_0(\alpha)$	$J_1(\alpha)$	α	$J_0(\alpha)$	$J_1(\alpha)$	α	$J_0(\alpha)$	$J_1(\alpha)$
0.00	+1.000	0.000	5.00	−0.178	−0.328	10.00	−0.246	+0.435
0.20	+0.990	+0.099	5.20	−0.110	−0.343	10.20	−0.250	−0.007
0.40	+0.960	+0.196	5.40	−0.041	−0.345	10.40	−0.243	−0.056
0.60	+0.912	+0.287	5.60	+0.027	−0.334	10.60	−0.228	−0.101
0.80	+0.846	+0.369	5.80	+0.092	−0.311	10.80	−0.203	−0.142
1.00	+0.765	+0.440	6.00	+0.151	−0.277	11.00	−0.171	−0.177
1.20	+0.671	+0.498	6.20	+0.202	−0.233	11.20	−0.133	−0.204
1.40	+0.567	+0.542	6.40	+0.243	−0.182	11.40	−0.090	−0.223
1.60	+0.455	+0.570	6.60	+0.274	−0.125	11.60	−0.045	−0.232
1.80	+0.340	+0.582	6.80	+0.293	−0.065	11.80	+0.002	−0.232
2.00	+0.224	+0.577	7.00	+0.300	−0.005	12.00	+0.048	−0.223
2.20	+0.110	+0.556	7.20	+0.295	+0.054	12.20	+0.091	−0.206
2.40	+0.003	+0.520	7.40	+0.279	+0.110	12.40	+0.130	−0.181
2.60	−0.097	+0.471	7.60	+0.252	+0.159	12.60	+0.163	−0.149
2.80	−0.185	+0.410	7.80	+0.215	+0.201	12.80	+0.189	−0.111
3.00	−0.260	+0.339	8.00	+0.172	+0.235	13.00	+0.207	−0.070
3.20	−0.320	+0.261	8.20	+0.122	+0.258	13.20	+0.217	−0.027
3.40	−0.364	+0.179	8.40	+0.069	+0.271	13.40	+0.218	+0.016
3.60	−0.392	+0.096	8.60	+0.015	+0.273	13.60	+0.210	+0.059
3.80	−0.403	+0.013	8.80	−0.039	+0.264	13.80	+0.194	+0.098
4.00	−0.397	−0.066	9.00	−0.090	+0.245	14.00	+0.171	+0.133
4.20	−0.377	−0.139	9.20	−0.137	+0.217	14.20	+0.141	+0.163
4.40	−0.342	−0.203	9.40	−0.177	+0.182	14.40	+0.106	+0.185
4.60	−0.296	−0.257	9.60	−0.209	+0.140	14.60	+0.068	+0.200
4.80	−0.240	−0.299	9.80	−0.232	+0.093	14.80	+0.027	+0.206
5.00	−0.178	−0.328	10.00	−0.246	+0.435	15.00	−0.014	+0.205

Table 4-2 Selected values of Bessel functions of the second kind

α	$Y_0(\alpha)$	$Y_1(\alpha)$	α	$Y_0(\alpha)$	$Y_1(\alpha)$	α	$Y_0(\alpha)$	$Y_1(\alpha)$
0.00	$-\infty$	$-\infty$	5.00	−0.309	+0.148	10.00	+0.058	+0.249
0.20	−1.081	−3.324	5.20	−0.331	+0.079	10.20	+0.006	+0.250
0.40	−0.606	−1.781	5.40	−0.340	+0.010	10.40	−0.044	+0.242
0.60	−0.309	−1.260	5.60	−0.335	+0.057	10.60	−0.090	+0.224
0.80	−0.087	−0.978	5.80	−0.318	−0.119	10.80	−0.133	+0.197
1.00	+0.088	−0.781	6.00	−0.288	−0.175	11.00	−0.169	+0.164
1.20	+0.228	−0.621	6.20	−0.248	−0.222	11.20	−0.198	+0.124
1.40	+0.338	−0.479	6.40	−0.200	−0.260	11.40	−0.218	+0.081
1.60	+0.420	−0.348	6.60	−0.145	−0.286	11.60	−0.230	+0.035
1.80	+0.477	−0.224	6.80	−0.086	−0.300	11.80	−0.232	−0.012
2.00	+0.510	−0.107	7.00	−0.026	−0.303	12.00	−0.225	−0.057
2.20	+0.521	+0.002	7.20	+0.034	−0.293	12.20	−0.210	−0.099
2.40	+0.510	+0.101	7.40	+0.091	−0.273	12.40	−0.186	−0.137
2.60	+0.481	+0.188	7.60	+0.142	−0.243	12.60	−0.155	−0.169
2.80	+0.436	+0.264	7.80	+0.187	−0.204	12.80	−0.119	−0.194
3.00	+0.377	+0.325	8.00	+0.224	−0.158	13.00	−0.078	−0.210
3.20	+0.307	+0.371	8.20	+0.250	−0.107	13.20	−0.035	−0.218
3.40	+0.230	+0.401	8.40	+0.266	−0.054	13.40	+0.009	−0.218
3.60	+0.148	+0.415	8.60	+0.272	+0.001	13.60	+0.051	−0.208
3.80	+0.645	+0.414	8.80	+0.266	+0.054	13.80	+0.091	−0.191
4.00	−0.017	+0.380	9.00	+0.250	+0.104	14.00	+0.127	−0.167
4.20	−0.094	+0.368	9.20	+0.225	+0.149	14.20	+0.158	−0.136
4.40	−0.163	+0.326	9.40	+0.191	+0.187	14.40	+0.181	−0.100
4.60	−0.224	+0.274	9.60	+0.150	+0.217	14.60	+0.197	−0.061
4.80	−0.272	+0.214	9.80	+0.105	+0.238	14.80	+0.206	−0.020
5.00	−0.309	+0.148	10.00	+0.058	+0.249	15.00	+0.206	+0.021

4-1 and 4-2. In terms of the original variables since $u = \lambda r$,

$$G(\lambda r) = C_2 J_0(\lambda r) + C_3 Y_0(\lambda r) \tag{4-39}$$

Thus from Eqs. (4-38) and (4-39) the solution to Eq. (4-30) becomes, if one sums all possible such solutions,

$$\vartheta(r,\tau) = \sum_{n=0}^{\infty} e^{-\alpha\lambda_n^2\tau}[A_n J_0(\lambda_n r) + B_n Y_0(\lambda_n r)] \tag{4-40}$$

At the center of the cylinder, $r = 0$, the temperature must be finite. Since $Y_0(0) \to -\infty$, $B_n \equiv 0$. The remaining equation is then

$$\vartheta(r,\tau) = \sum_{n=0}^{\infty} A_n e^{-\alpha\lambda_n^2\tau} J_0(\lambda_n r) \tag{4-41}$$

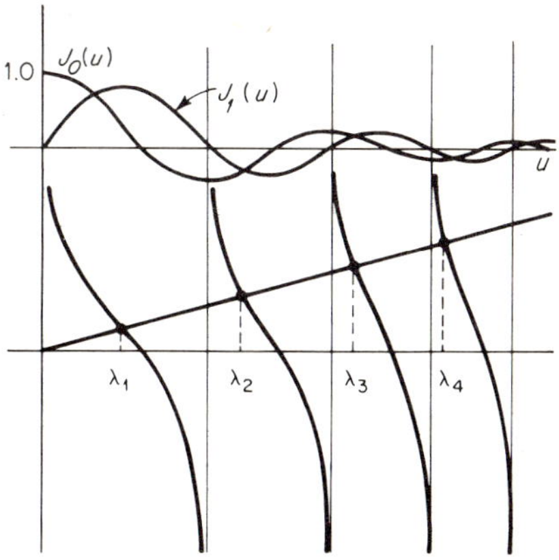

Fig. 4-8 Determination of the eigenvalues for the infinite-circular-cylinder problem.

The values of the λ_n and the constants A_n are to be determined. The λ_n may be evaluated by means of the boundary condition [Eq. (4-32)]

$$\left(\frac{\partial\vartheta}{\partial r}\right)_{r=r_0} = -\sum_{n=0}^{\infty} A_n e^{-\alpha\lambda_n^2\tau}\lambda_n J_1(\lambda_n r_0) = -\frac{h}{k}\vartheta(r_0,\tau)$$

$$= -\frac{h}{k}\sum_{n=0}^{\infty} A_n e^{-\alpha\lambda_n^2\tau} J_0(\lambda_n r_0)$$

or

$$\lambda_n r_0 \frac{k}{hr_0} = \frac{J_0(\lambda_n r_0)}{J_1(\lambda_n r_0)} \tag{4-42}$$

This is a transcendental equation for the values of λ_n. A graphical representation of the solution is shown in Fig. 4-8. While there exists no λ_0, the other λ_n are determinable. The initial condition must now be considered:

$$\vartheta(r,0) = \sum_{n=1}^{\infty} A_n J_0(\lambda_n r) = f(r)$$

Assume that $f(r)$ can be expanded in a series of Bessel functions,

$$f(r) = A_1 J_0(\lambda_1 r) + A_2 J_0(\lambda_2 r) + \cdots + A_n J_0(\lambda_n r) + \cdots \tag{4-43}$$

An important integral property of Bessel functions is indicated in the following equation:

$$\int_0^x x J_n(\alpha x) J_n(\beta x)\, dx = 0 \qquad \text{when } \alpha \neq \beta$$

but has a value

$$\int_0^x x J_n^2(\alpha x)\, dx \qquad \text{when } \alpha = \beta$$

Thus the A_n can be determined by multiplying Eq. (4-43) by a succession of $rJ_0(\lambda_n r)$, the general term of which is

$$\int_0^{r_0} rf(r)J_0(\lambda_n r)\,dr = \int_0^{r_0} A_n rJ_0^2(\lambda_n r)\,dr$$

whence

$$A_n = \frac{\int_0^{r_0} rf(r)J_0(\lambda_n r)\,dr}{\int_0^{r_0} rJ_0^2(\lambda_n r)\,dr} \tag{4-44}$$

Integration of Eq. (4-44) gives

$$A_n = \frac{2}{r_0^2[J_0^2(\lambda_n r_0) + J_1^2(\lambda_n r_0)]}\int_0^{r_0} rf(r)J_0(\lambda_n r)\,dr$$

and if, for example, $f(r) = \vartheta_0$, a constant, the A_n become, where $\lambda_n r_0 = \delta_n$,

$$A_n = \frac{2\vartheta_0}{\delta_n}\,\frac{J_1(\delta_n)}{J_0^2(\delta_n) + J_1^2(\delta_n)} \tag{4-45}$$

The final result for the temperature variation in the circular cylinder, initially with a constant temperature throughout, transferring heat to a fluid by combining Eqs. (4-41), (4-42), and (4-45) is

$$\frac{\vartheta}{\vartheta_0} = 2\sum_{n=1}^{\infty} \frac{(hr_0/k)J_0(\delta_n r/r_0)}{\delta_n^2 J_0(\delta_n) + (hr_0/k)^2 J_0(\delta_n)}\, e^{-\delta_n^2(\alpha\tau/r_0^2)} \tag{4-46}$$

Equation (4-46) shows that the temperature variation in the circular cylinder can be expressed as a function of three parameters,

$$\frac{\vartheta}{\vartheta_0} = f\left(\frac{\alpha\tau}{r_0^2}, \frac{hr_0}{k}, \frac{r}{r_0}\right) = f\left(\mathrm{Fo}, \mathrm{Bi}, \frac{r}{r_0}\right)$$

The heat loss from the cylinder is the heat that flows through the surface $r = r_0$, or, for the cylinder length L,

$$dQ_i = -2\pi r_0 Lk\left(\frac{\partial\vartheta}{\partial r}\right)_{r=r_0} d\tau \tag{4-47}$$

Noting that

$$\frac{dJ_0(u)}{dx} = -J_1(u)\,\frac{du}{dx}$$

the necessary derivative can be calculated from Eq. (4-46):

$$\left(\frac{\partial\vartheta}{\partial r}\right)_{r=r_0} = -\frac{2h\vartheta_0}{k}\sum_{n=1}^{\infty}\frac{\delta_n J_1(\delta_n)}{\delta_n^2 J_0(\delta_n) + (hr_0/k)^2 J_0(\delta_n)}\, e^{-\delta_n^2(\alpha\tau/r_0^2)}$$

which when substituted into Eq. (4-47) becomes

$$Q_i = \int_0^{\tau} 4\pi Lk\vartheta_0 \sum_{n=1}^{\infty} \frac{(hr_0/k)^2 e^{-\delta_n^2(\alpha\tau/r_0^2)}}{\delta_n^2 + (hr_0/k)^2}\, d\tau$$

This when integrated yields

$$Q_i = \pi r_0^2 L\rho c\vartheta_0 \sum_{n=1}^{\infty} \frac{4}{\delta_n^2} \frac{(hr_0/k)^2}{\delta_n^2 + (hr_0/k)^2} [1 - e^{-\delta_n^2(\alpha\tau/r_0^2)}] \tag{4-48}$$

The value $\pi r_0^2 L\rho c\vartheta_0 = Q_0$, the heat energy in the cylinder above the fluid temperature; thus

$$\frac{Q_i}{Q_0} = f\left(\frac{\alpha\tau}{r_0^2}, \frac{hr_0}{k}\right) = f(\text{Fo,Bi})$$

Thus, as in the case of the semi-infinite slab, the results for the infinite cylinder can be displayed as

$$\frac{\vartheta}{\vartheta_0} = g_1\left(\frac{hr_0}{k}, \frac{\alpha\tau}{r_0^2}, \frac{r}{r_0}\right)$$

$$\frac{\vartheta_c}{\vartheta_0} = g_2\left(\frac{hr_0}{k}, \frac{\alpha\tau}{r_0^2}, 0\right)$$

$$\frac{\vartheta_{r_0}}{\vartheta_0} = g_3\left(\frac{hr_0}{k}, \frac{\alpha\tau}{r_0^2}, 1\right)$$

$$\frac{Q_i}{Q_0} = g_4\left(\frac{hr_0}{k}, \frac{\alpha\tau}{r_0^2}\right)$$

Similar solutions have been made for the sphere, and all such results have been calculated and plotted in terms of the above parameters for rapid use by Gröber,[1] Gurney and Lurie,[2] Heisler,[3] and others. The results of the three solutions are shown graphically for a limited range of variables in Figs. 4-9 to 4-11.

Two- and three-dimensional solutions It is more often the case that transient heat-conduction problems concern finite bodies such as rectangular parallelepipeds or short cylinders. The solutions discussed so far cannot be used directly. However, a method due to Berger[4] and Newman[5] permits extension of the spacewise one-dimensional problems, for which solutions exist, to two and three space dimensions. The method for two dimensions, which follows, can easily be applied for three dimensions. Consider a long rectangular bar as shown in cross section in Fig.

[1] H. Gröber and S. Erk, "Die Grundgesetze der Wärmeübertragung," Springer-Verlag, Berlin, 1933.

[2] H. P. Gurney and J. Lurie, *Ind. Eng. Chem.*, **15**:1170–1172 (1923).

[3] M. P. Heisler, *Trans. ASME*, **69**:227–236 (1947).

[4] F. Berger, Über die Berechnung des Temperaturverlaufes in einem Rechtkant beim Abkühlen und Erwärmen, *Z. Angew. Math. Mech.*, **8**:479 (1928); also, *ibid.*, **11**:45 (1931).

[5] A. B. Newman, *AIChE*, **24**:44 (1930).

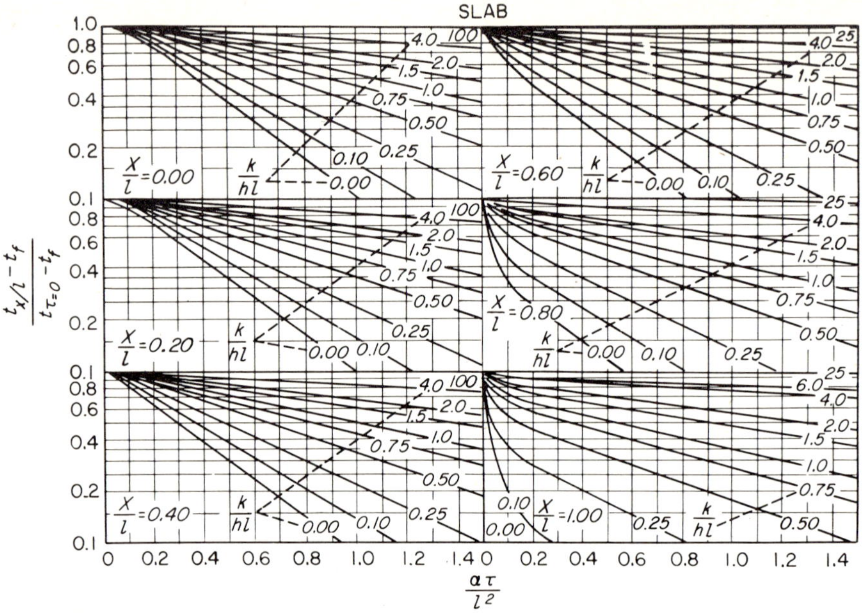

Fig. 4-9 Transient heat conduction in a slab. [*L. M. K. Boelter et al., "Heat Transfer Notes." Copyright 1965. McGraw-Hill Book Company. Used by permission.*]

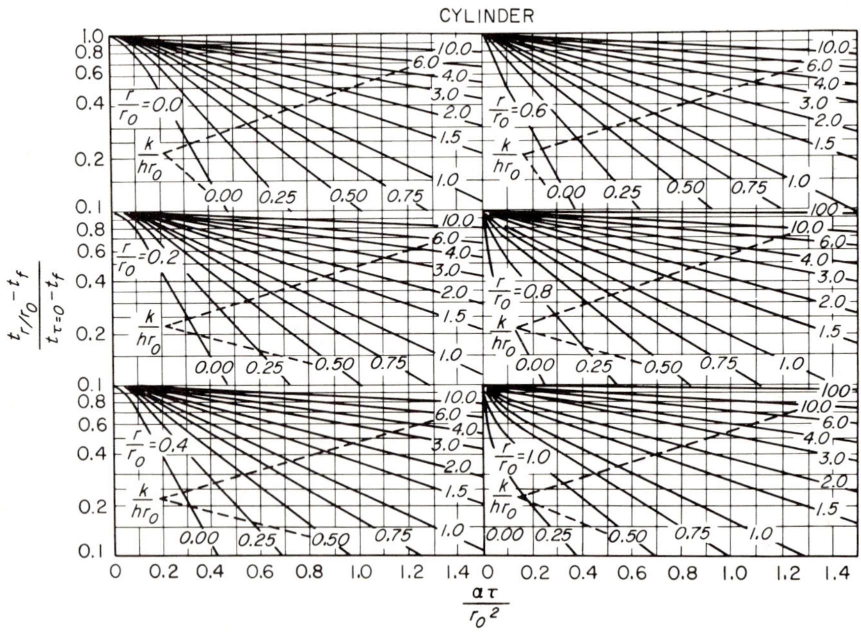

Fig. 4-10 Transient heat conduction in a cylinder. [*L. M. K. Boelter et al., "Heat Transfer Notes." Copyright 1965. McGraw-Hill Book Company. Used by permission.*]

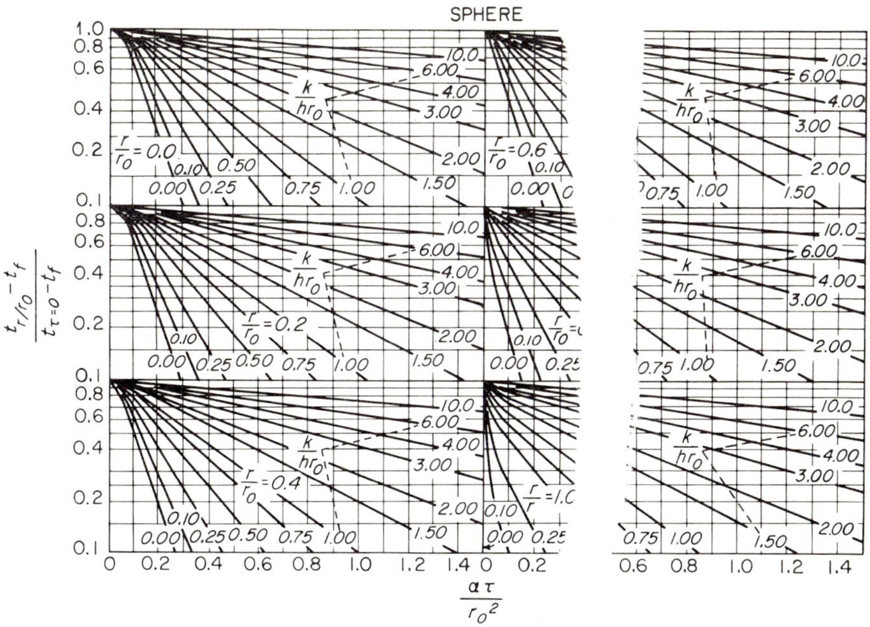

Fig. 4-11 Transient heat conduction in a sphere. [*L. M. K. Boelter et al., "Heat Transfer Notes." Copyright 1965. McGraw-Hill Book Company. Used by permission.*]

4-12. There is convection on the outer surfaces of the bar yielding heat-transfer coefficients h, which may be constant all around or equal on opposite sides. Only one-quarter section of the bar need be considered because of the symmetry of the system.

The differential equation representing the system is

$$\frac{\partial\vartheta}{\partial\tau} = \alpha\left(\frac{\partial^2\vartheta}{\partial x^2} + \frac{\partial^2\vartheta}{\partial y^2}\right) \tag{4-49}$$

The boundary conditions are

$$\frac{\partial\vartheta}{\partial x} = 0 \qquad x = 0$$

$$\frac{\partial\vartheta}{\partial y} = 0 \qquad y = 0$$

$$-k\left(\frac{\partial\vartheta}{\partial x}\right)_{x=A} = (h\vartheta)_A \qquad x = A$$

$$-k\left(\frac{\partial\vartheta}{\partial y}\right)_{y=B} = (h\vartheta)_B \qquad y = B$$

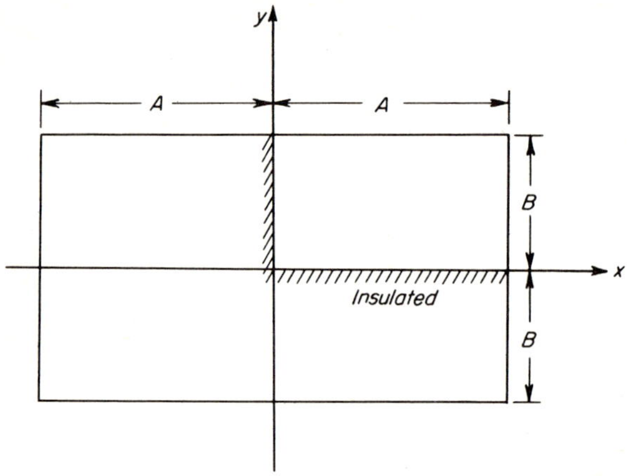

Fig. 4-12 Two-dimensional heat flow in an infinite rod.

Equation (4-49) and the boundary conditions are made dimensionless by letting

$$x' = \frac{x}{A} \qquad y' = \frac{y}{B} \qquad \tau'_x = \frac{\alpha\tau}{A^2} \qquad \tau'_y = \frac{\alpha\tau}{B^2} \qquad \vartheta' = \frac{\vartheta}{\vartheta_0}$$

and taking note of the fact that the dimensionless times are Fourier numbers. Substitution results in

$$\frac{\partial\vartheta'}{\partial\tau'_x} + \left(\frac{A}{B}\right)^2 \frac{\partial\vartheta'}{\partial\tau'_y} = \frac{\partial^2\vartheta'}{\partial x'^2} + \left(\frac{A}{B}\right)^2 \frac{\partial^2\vartheta'}{\partial y'^2} \tag{4-50}$$

$$\frac{\partial\vartheta'}{\partial x'} = 0 \qquad x' = 0$$

$$\frac{\partial\vartheta'}{\partial y'} = 0 \qquad y' = 0$$

$$-\frac{\partial\vartheta'}{\partial x'} = \frac{hA}{k}\vartheta' \qquad x' = 1$$

$$-\frac{\partial\vartheta'}{\partial y'} = \frac{hB}{k}\vartheta' \qquad y' = 1$$

A solution is assumed to be of the form $\vartheta' = \vartheta'_x\theta'_y$, where

$$\vartheta'_x = f_1\left(x', \frac{\alpha\tau}{A^2}, \frac{hA}{k}\right)$$

$$\vartheta'_y = f_2\left(y', \frac{\alpha\tau}{B^2}, \frac{hB}{k}\right)$$

Introduction of the assumed solution into Eq. (4-50) yields

$$\frac{1}{\vartheta'_x}\frac{\partial\vartheta'_x}{\partial\tau'_x} + \left(\frac{A}{B}\right)^2 \frac{1}{\vartheta'_y}\frac{\partial\vartheta'_y}{\partial\tau'_y} = \frac{1}{\vartheta'_x}\frac{\partial^2\vartheta'_x}{\partial x'^2} + \left(\frac{A}{B}\right)^2 \frac{1}{\vartheta'_y}\frac{\partial^2\vartheta'_y}{\partial y'^2} \tag{4-51}$$

Equation (4-51) is of the form

$$\phi_x(x,\tau) = \phi_y(y,\tau) = 0^1$$

Thus, we have two equations:

$$\frac{\partial \vartheta'_x}{\partial \tau'_x} = \frac{\partial^2 \vartheta'_x}{\partial x'^2} \qquad \frac{\partial \vartheta'_y}{\partial \tau'_y} = \frac{\partial^2 \vartheta'_y}{\partial y'^2} \tag{4-52}$$

Equations (4-52) are dimensionless forms of Eq. (4-10) which have solutions like Eqs. (4-25) and (4-28). Thus it has been shown that the product of the solution of two semi-infinite slab thicknesses $2A$ and $2B$ gives a solution for an infinite bar of cross section $2A$ by $2B$. It is easily seen that the parallelepiped of dimensions $2A$ by $2B$ by $2C$ has a solution which is the product of three semi-infinite slabs, dimensions $2A$, $2B$, and $2C$, respectively. Figure 4-13 shows thermal histories of bodies which were initially held at an excess temperature ϑ_0 and whose surfaces were suddenly reduced to $\vartheta = 0$.[2] Many other solutions are discussed in the books by Carslaw and Jaeger, Schneider, McAdams, Ozisik, and Arpaci.[3]

[1] The condition here requires that the two values equal a constant. However, either equation must apply in the absence of the other; thus the constant must be zero.

[2] Gröber and Erk, *op. cit.*

[3] H. S. Carslaw and J. C. Jaeger, "Conduction of Heat in Solids," 2d ed., Oxford University Press, New York, 1959; P. J. Schneider, "Conduction Heat Transfer," Addison-Wesley, Reading, Mass., 1955; W. H. McAdams, "Heat Transmission," 3d ed., McGraw-Hill, New York, 1954; M. N. Ozisik, "Boundary Value Problems of Heat Conduction," International Textbook, Scranton, Pa., 1968; V. Arpaci, "Conduction Heat Transfer," Addison-Wesley, Reading, Mass., 1966.

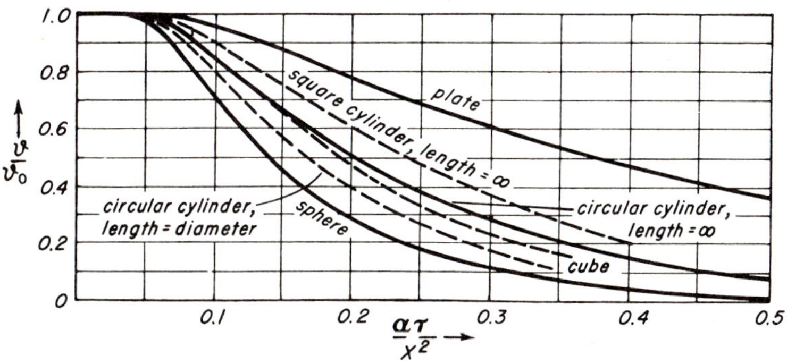

Fig. 4-13 Temperatures at the central axes of different shapes during the process of cooling; x is the thickness, side dimension, or diameter. [*From H. Gröber and S. Erk, "Die Grundgesetze der Wärmeübertragung," p. 58, fig. 28, Springer-Verlag, Berlin, 1933.*]

Example 4-3 In the heat treating of a steel billet shaped in the form of a rectangular parallelepiped measuring 0.6 by 0.6 by 1.2 m, it is required that all parts of the billet be heated to just about 643 K but that no part of the billet exceed 673 K. To do this, the billet, initially at 293 K, is placed in an oven which provides an atmosphere of inert gas at 673 K and a circulation designed to give a heat-transfer coefficient $h = 565$ W/m^2 K. How long must the billet remain in the oven for the center temperature to reach 643 K?

In order to use the existing solutions (which are applicable only for solids initially at a uniform temperature which at time zero are submerged in a fluid at zero temperature) we must state the billet problem in the same way. This is accomplished by inverting the boundary conditions and taking 673 K as the zero datum. This results in an equivalent problem statement that requires the billet to exist initially at $673 - 293 = 380$ K, submerged in a fluid of zero temperature at time zero, where now we wish to find the time it takes for the center temperature to reach $673 - 643 = 30$ K. From Appendix B, $k = 43$ W/m K, and the thermal diffusivity $\alpha = 0.053$ m^2/hr.

From the discussion of the product solution it is required that

$$\left(\frac{t}{t_0}\right)_A \left(\frac{t}{t_0}\right)_B \left(\frac{t}{t_0}\right)_C = \frac{t}{t_0} = \frac{30}{380} = 0.079$$

Since $A = B = 0.305$ m and $C = 0.61$ m, the expression above can become

$$\left(\frac{t}{t_0}\right)_A^2 \left(\frac{t}{t_0}\right)_C = 0.079 \qquad (a)$$

$$\frac{hA}{k} = \frac{hB}{k} = \frac{565(0.305)}{43} = 4.00 \qquad \text{and} \qquad \frac{hC}{k} = \frac{565(0.61)}{43} = 8.00$$

The solution to the problem becomes one of trial and error. We assume a time τ, calculate the two values $\alpha\tau/A^2$ and $\alpha\tau/C^2$, and use these values with the values hA/k and hC/k above to determine two values $(t/t_0)_C$ from the chart for $x/l = 0$ in Fig. 4-9. These values are substituted into expression (a) above. This process is repeated until expression (a) is satisfied. The time assumed for that case is the time required for the center temperature of the billet to reach 643 K. After several trials the final results are given below.

Assume that $\tau = 1.49$ hr:

$$\frac{\alpha\tau}{A^2} = \frac{0.053(1.53)}{0.305^2} = 0.871 \qquad \frac{k}{hA} = 0.25$$

$$\frac{\alpha\tau}{C^2} = \frac{0.053(1.53)}{0.61^2} = 0.218 \qquad \frac{k}{hC} = 0.125$$

From Fig. 4-9 for $x/l = 0$ (the center of each slab) we read for the above parameters

$$\left(\frac{t}{t_0}\right)_A = 0.314 \qquad \text{and} \qquad \left(\frac{t}{t_0}\right)_C = 0.802$$

Thus

$$0.314^2(0.802) = 0.079$$

Thus the time required for the center temperature of the billet to reach 643 K under the conditions specified in the problem is $\tau = 1.49$ hr approximately.

The infinite solid We now consider the case of the infinite solid, a body extending in all directions to infinity. We wish to consider the nature of the temperature distribution in the solid for $\tau > 0$ which would result from a temperature distribution $t = f(x)$ at $\tau = 0$. The equation to solve is

$$\frac{\partial t}{\partial \tau} = \alpha \frac{\partial^2 t}{\partial x^2} \tag{4-53}$$

Consider the expression

$$t = \frac{1}{\sqrt{\tau}} e^{-x^2/4\alpha\tau} \tag{4-54}$$

which can be differentiated with respect to τ once and with respect to x twice, substituted into Eq. (4-53), and shown to satisfy the equation exactly. The expression in Eq. (4-54) can be written in a slightly modified form which is still a solution of the differential equation:

$$t = \frac{f(\xi)}{2\sqrt{\pi\alpha\tau}} e^{-(x-\xi)^2/4\alpha\tau}\, d\xi \tag{4-54a}$$

where ξ is a parameter. This expression has the property of being zero everywhere at time zero except at $x = \xi$, where it is finite. At increasing values of τ the temperature distribution over the space coordinate x is as shown in Fig. 4-14. Equation (4-54a) can be interpreted physically as an instantaneous plane heat source at time zero located at $x = \xi$. The temperature at any location x then is the result of the source at ξ. The strength of the source is $f(\xi)$, and the quantity of heat liberated is $f(\xi)\rho c$ per unit area of the plane; that is, $f(\xi)$ is the temperature to which the volume (unit area x $\delta\xi$) would be raised by the heat liberated by the source.

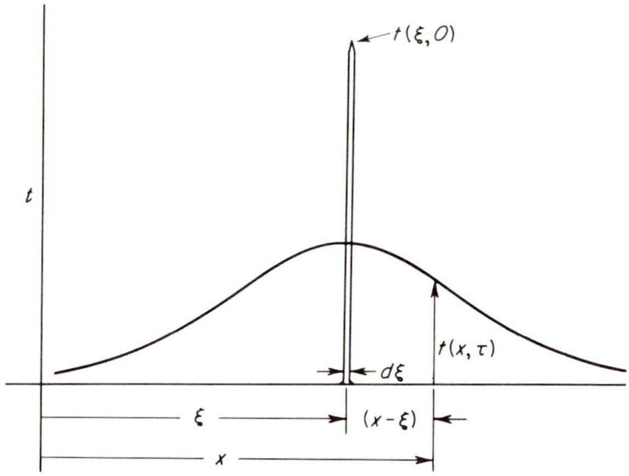

Fig. 4-14 The instantaneous heat source and ensuing temperature distribution in the infinite solid.

Since the conduction equation is linear, the sum of any number of particular solutions is also a solution; thus

$$t = \frac{1}{2\sqrt{\pi\alpha\tau}} \int_{-\infty}^{\infty} f(\xi) e^{-(x-\xi)^2/4\alpha\tau}\, d\xi \tag{4-55}$$

If a change of variable is made in Eq. (4-55), where

$$\xi = x + \beta\sqrt{4\alpha\tau}$$

then $$t(x,\tau) = \frac{1}{2\sqrt{\pi\alpha\tau}} \int_{-\infty}^{\infty} f(x + \beta\sqrt{4\alpha\tau})\, e^{-\beta^2} \sqrt{4\alpha\tau}\, d\beta$$

or $$t(x,\tau) = \frac{1}{\sqrt{\pi}} \int_{-\infty}^{\infty} f(x + \beta\sqrt{4\alpha\tau}) e^{-\beta^2}\, d\beta$$

If in the above expression $\tau = 0$, then $f(x + \beta\sqrt{4\alpha\tau}) = f(x)$; therefore,

$$t(x,0) = \frac{2}{\sqrt{\pi}} f(x) \int_0^{\infty} e^{-\beta^2}\, d\beta = f(x)$$

Thus the solution proposed in Eq. (4-54*a*) satisfies both the differential equation and the initial condition and therefore constitutes a solution to the problem.

The semi-infinite solid The semi-infinite solid may be a body bounded by a plane at $x = 0$ and extending to infinity in the positive x direction. Suppose the body to have a temperature $t = f(x)$ at time zero. The temperature at the surface is zero for $\tau \gtrless 0$. The temperature distribution for this body can be deduced from the solution for the infinite body [Eq. (4-55)]. If we imagine the solid to be continued in the negative x direction and the initial temperature distribution at $-\xi$ to be $-f(\xi)$, the initial temperature at ξ being $f(\xi)$, the plane $x = 0$ will remain at the prescribed zero temperature (Fig. 4-15). Thus from Eq. (4-55),

$$t = \frac{1}{2\sqrt{\pi\alpha\tau}} \left\{ \int_0^{\infty} f(\xi) e^{-(x-\xi)^2/4\alpha\tau}\, d\xi + \int_{-\infty}^{0} [-f(-\xi)] e^{-(x-\xi)^2/4\alpha\tau}\, d\xi \right\}$$

or, rewriting,

$$t = \frac{1}{2\sqrt{\pi\alpha\tau}} \int_0^{\infty} f(\xi) [e^{-(x-\xi)^2/4\alpha\tau} - e^{-(x+\xi)^2/4\alpha\tau}]\, d\xi \tag{4-56}$$

When the initial temperature t_0 is a constant, Eq. (4-56) can be simplified by substituting for the first term in the integral $\xi = x + 2\beta\sqrt{\alpha\tau}$ and for the second term $\xi = -x + 2\beta\sqrt{\alpha\tau}$. We thus obtain

$$t = \frac{t_0}{\sqrt{\pi}} \int_{-x/2\sqrt{\alpha\tau}}^{\infty} e^{-\beta^2}\, d\beta - \frac{t_0}{\sqrt{\pi}} \int_{x/2\sqrt{\alpha\tau}}^{\infty} e^{-\beta^2}\, d\beta$$

$$= \frac{t_0}{\sqrt{\pi}} \int_{-x/2\sqrt{\alpha\tau}}^{x/2\sqrt{\alpha\tau}} e^{-\beta^2}\, d\beta = \frac{2t_0}{\sqrt{\pi}} \int_0^{x/2\sqrt{\alpha\tau}} e^{-\beta^2}\, d\beta \tag{4-57}$$

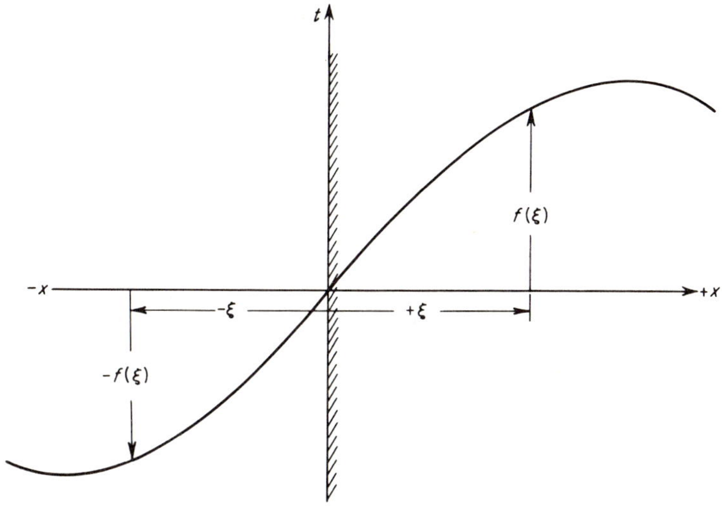

Fig. 4-15 Temperature-distribution model for the semi-infinite solid.

The definite integral in Eq. (4-57) has been tabulated, and a short tabulation is given in Table 4-3.

If we designate the definite integral in Eq. (4-57)

$$\operatorname{erf} z = \frac{2}{\sqrt{\pi}} \int_0^z e^{-\beta^2}\, d\beta$$

then the solution of the problem of the semi-infinite solid whose surface is kept at zero temperature and whose initial temperature is t_0 is given by

$$\frac{t}{t_0} = \operatorname{erf} \frac{x}{2\sqrt{\alpha\tau}} \tag{4-58}$$

The rate of heat loss from the face $x = 0$ becomes

$$\frac{Q}{A} = +k\left(\frac{\partial t}{\partial x}\right)_{x=0} = \frac{kt_0}{\sqrt{\pi\alpha\tau}} \tag{4-59}$$

Semi-infinite solid, transformation of coordinates Rather than build up a solution for the temperature distribution in a semi-infinite solid by superposition of instantaneous heat sources, it is possible to arrive at Eq. (4-58) and subsequently Eq. (4-59) by a transformation in the independent variables (x,τ).

The appropriate differential equation is Eq. (4-53), repeated here for convenience

$$\frac{\partial t}{\partial \tau} = \alpha \frac{\partial^2 t}{\partial x^2} \tag{4-60}$$

Table 4-3 The error function or probability integral†

$$\text{erf } z = \frac{2}{\sqrt{\pi}} \int_0^z e^{-\beta^2} d\beta$$

z	erf z	z	erf z	z	erf z	z	erf z
0.00	0.00000	0.66	0.64938	1.30	0.93401	1.94	0.99392
0.02	0.02256	0.68	0.66378	1.32	0.93807	1.96	0.99443
0.04	0.04511	0.70	0.67780	1.34	0.94191	1.98	0.99489
0.06	0.06762	0.72	0.69143	1.36	0.94556	2.00	0.99532
0.08	0.09008	0.74	0.70468	1.38	0.94902	2.05	0.99626
0.10	0.11246	0.76	0.71754	1.40	0.95229	2.10	0.99702
0.12	0.13476	0.78	0.73001	1.42	0.95538	2.15	0.99764
0.14	0.15695	0.80	0.74210	1.44	0.95830	2.20	0.99814
0.16	0.17901	0.82	0.75381	1.46	0.96105	2.25	0.99854
0.18	0.20094	0.84	0.76514	1.48	0.96365	2.30	0.99886
0.20	0.22270	0.86	0.77610	1.50	0.96611	2.35	0.9991107
0.22	0.24430	0.88	0.78669	1.52	0.96841	2.40	0.9993115
0.24	0.26570	0.90	0.79691	1.54	0.97059	2.50	0.9995930
0.26	0.28690	0.92	0.80677	1.56	0.97263	2.60	0.9997640
0.28	0.30788	0.94	0.81627	1.58	0.97455	2.70	0.9998657
0.30	0.32863	0.96	0.82542	1.60	0.97635	2.80	0.9999250
0.32	0.34913	0.98	0.83423	1.62	0.97804	2.90	0.9999589
0.34	0.36936	1.00	0.84270	1.64	0.97962	3.00	0.9999779
0.36	0.38933	1.02	0.85084	1.66	0.98110	3.10	0.9999884
0.38	0.40901	1.04	0.85865	1.68	0.98249	3.20	0.9999940
0.40	0.42839	1.06	0.86614	1.70	0.98379	3.30	0.9999969
0.42	0.44747	1.08	0.87333	1.72	0.98500	3.40	0.9999985
0.44	0.46623	1.10	0.88020	1.74	0.98613	3.50	0.99999925691
0.46	0.48466	1.12	0.88679	1.76	0.98719	3.60	0.99999964414
0.48	0.50275	1.14	0.89308	1.78	0.98817	3.70	0.99999983285
0.50	0.52050	1.16	0.89910	1.80	0.98909	3.80	0.99999992300
0.52	0.53790	1.18	0.90484	1.82	0.98994	3.90	0.99999996521
0.54	0.55494	1.20	0.91031	1.84	0.99074	4.00	0.99999998458
0.56	0.57162	1.22	0.91553	1.86	0.99147	4.20	0.99999999714
0.58	0.58792	1.24	0.92051	1.88	0.99216	4.40	0.99999999951
0.60	0.60386	1.26	0.92524	1.90	0.99279	4.60	0.99999999992
0.62	0.61941	1.28	0.92973	1.92	0.99338	4.80	0.99999999999
0.64	0.63459						1.00

† L. M. K. Boelter et al., "Heat Transfer Notes." Copyright 1965. McGraw-Hill Book Company. Used by permission.

with boundary and initial conditions

$$\begin{array}{lll} x = 0 & t = 0 & \tau \geq 0 \\ \tau = 0 & t = t_0 & x \geq 0 \end{array} \tag{4-61}$$

The problem posed therefore is a semi-infinite solid occupying the region $x \geq 0$, in which the temperature is initially uniform at a value t_0. At time zero the surface temperature is reduced to $t = 0$ and maintained there. The problem is to determine the temperature distribution in the solid as a function of x and τ, for values of $\tau > 0$.

The solution entails a search for a new independent variable $\eta(x,\tau)$ that will reduce the partial differential equation (4-60) to an ordinary differential equation solvable by quadratures.

Let the new independent variable be of the form

$$\eta = Cx\tau^m \tag{4-62}$$

where the constants C and m are to be determined. Therefore, the derivatives in Eq. (4-60) become

$$\frac{\partial t}{\partial \tau} = \frac{m\eta}{\tau}\frac{dt}{d\eta} \qquad \text{and} \qquad \frac{\partial^2 t}{\partial x^2} = C^2\tau^{2m}\frac{d^2t}{d\eta^2}$$

which when substituted into Eq. (4-60) yield the expression

$$m\eta\frac{dt}{d\eta} = \alpha C^2\tau^{2m+1}\frac{d^2t}{d\eta^2} \tag{4-63}$$

For Eq. (4-63) to be independent of τ, $2m + 1 = 0$ and $m = -\frac{1}{2}$. The equation becomes thus an ordinary differential equation of second order,

$$\frac{d^2t}{d\eta^2} + \frac{1}{2\alpha C^2}\eta\frac{dt}{d\eta} = 0 \tag{4-64}$$

Let $C = 1/\sqrt{4\alpha}$ such that $\eta = x/\sqrt{4\alpha\tau}$, and Eq. (4-64) becomes

$$\frac{d^2t}{d\eta^2} + 2\eta\frac{dt}{d\eta} = 0 \tag{4-65}$$

The solution of Eq. (4-65) becomes

$$t = C_1\int_0^\eta e^{-\eta^2}\,d\eta + C_2 \tag{4-66}$$

From the boundary condition $x = 0$; $t = 0$; $\eta = 0$; $C_2 = 0$, and from the initial condition $\tau = 0$; $\eta \to \infty$; $t = t_0$; thus

$$t_0 = C_1\int_0^\infty e^{-\eta^2}\,d\eta = C_1\frac{\sqrt{\pi}}{2}$$

from which relationship

$$C_1 = \frac{2t_0}{\sqrt{\pi}}$$

The final solution is thus

$$\frac{t}{t_0} = \frac{2}{\sqrt{\pi}} \int_0^{x/\sqrt{4\alpha\tau}} e^{-\eta^2}\, d\eta = \operatorname{erf} \frac{x}{\sqrt{4\alpha\tau}} \tag{4-67}$$

which is identical with Eq. (4-58).

Application to welding Equation (4-55) can be applied to the case where two sheets of heavy material are welded together by infusing molten metal between them. If it is assumed that there are no complications because of the phase change and also that the exposed surfaces of the sheets lose heat at a much slower rate than the heat flows by conduction into the metal, the problem can be idealized into an initial temperature distribution as shown in Fig. 4-16. This temperature distribution indicates that $t = 0$ for $x < l$ and $x > m$, $t = t_0$ for $l < x < m$. The solution can be written by writing Eq. (4-55) sectionally. The integral from l to m is the only segment that gives a contribution:

$$t = \frac{1}{2\sqrt{\pi\alpha\tau}} \int_l^m t_0 e^{-(x-\xi)^2/4\alpha\tau}\, d\xi \tag{4-68}$$

Making the change of variable as before,

$$t = \frac{t_0}{\sqrt{\pi}} \int_{(l-x)/2\sqrt{\alpha\tau}}^{(m-x)/2\sqrt{\alpha\tau}} e^{-\beta^2}\, d\beta$$

or rewriting in terms of the error function, we get

$$\frac{t}{t_0} = \frac{1}{2}\left(\operatorname{erf} \frac{m-x}{2\sqrt{\alpha\tau}} - \operatorname{erf} \frac{l-x}{2\sqrt{\alpha\tau}}\right) \tag{4-69}$$

Equation (4-69) is an adequate solution, but if $x = 0$ is equidistant from l and m, $b = (m - l)/2$, then Eq. (4-69) becomes, in terms of b,

$$\frac{t}{t_0} = \frac{1}{2}\left(\operatorname{erf} \frac{b-x}{2\sqrt{\alpha\tau}} + \operatorname{erf} \frac{b+x}{2\sqrt{\alpha\tau}}\right) \tag{4-70}$$

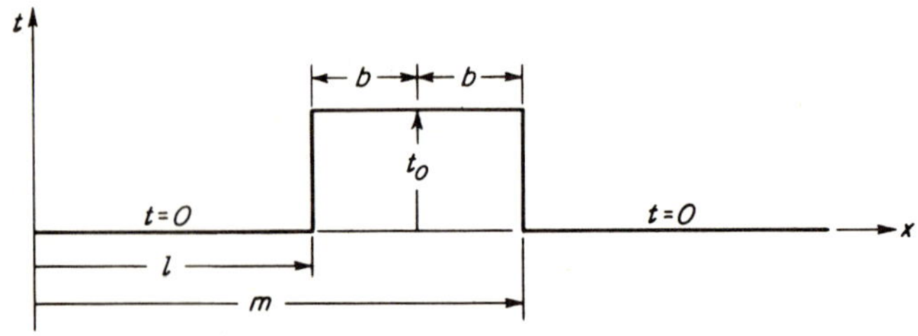

Fig. 4-16 Initial temperature distribution in infusion welding of two sheets.

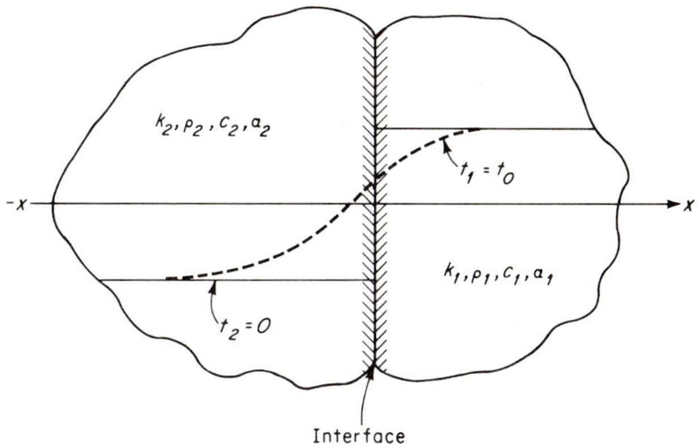

Fig. 4-17 Two semi-infinite bodies in contact.

Application to interfacial-temperature determination Equation (4-55) can also be used to determine the temperature distribution and the interfacial temperature of two semi-infinite solids of different materials initially at constant but different temperatures which are brought together at some time, $\tau = 0$ (Fig. 4-17).

The initial conditions are for no contact

$$\begin{aligned} \tau &= 0 \qquad x > 0 \qquad t_1 = t_0 \\ \tau &= 0 \qquad x < 0 \qquad t_2 = 0 \end{aligned} \tag{4-71}$$

The boundary conditions must reflect two facts: (1) the temperature of each body at the contact interface, in the absence of thermal contact resistance, will be the same, and (2) the heat flow from the body of higher temperature must be equal to the heat flow into the body of lower temperature. These conditions expressed mathematically are

$$x = 0 \qquad \tau > 0 \qquad \begin{aligned} t_1 &= t_2 \\ k_1 \frac{\partial t_1}{\partial x} &= k_2 \frac{\partial t_2}{\partial x} \end{aligned} \tag{4-72}$$

The temperature distribution in each body will necessarily be similar in form to Eq. (4-55); thus one may express the temperatures as

$$t_1 = A_1 + B_1 \operatorname{erf} \frac{x}{\sqrt{4\alpha\tau}} \qquad x > 0 \tag{4-73}$$

$$t_2 = A_2 + B_2 \operatorname{erf} \frac{|x|}{\sqrt{4\alpha\tau}} \qquad x < 0 \tag{4-74}$$

From the initial conditions [Eq. (4-71)] when applied to Eqs. (4-73) and (4-74), one can calculate

$$\begin{aligned} t_0 &= A_1 + B_1 \\ 0 &= A_2 + B_2 \end{aligned} \tag{4-75}$$

and for the boundary condition related to the temperature equality at the interface, $x = 0$,

$$A_1 = A_2 \tag{4-76}$$

The remaining boundary condition related to the heat flow at the interface, when applied to Eqs. (4-73) and (4-74), yields the equality,

$$\frac{B_1 k_1}{\sqrt{\alpha_1}} = -\frac{B_2 k_2}{\sqrt{\alpha_2}} \tag{4-77}$$

Solving for the four values A_1, A_2, B_1, and B_2 from the four equations (4-75), (4-76), and (4-77) and substituting the results in Eqs. (4-73) and (4-74), one obtains the temperature distributions in each body

$$t_1(x,\tau) = \frac{t_0}{1 + (k_2\sqrt{\alpha_1})/(k_1\sqrt{\alpha_2})}\left(1 + \frac{k_2\sqrt{\alpha_1}}{k_1\sqrt{\alpha_2}}\operatorname{erf}\frac{x}{\sqrt{4\alpha_1\tau}}\right) \tag{4-78}$$

$$t_2(x,\tau) = \frac{t_0}{1 + (k_2\sqrt{\alpha_1})/(k_1\sqrt{\alpha_2})}\left(1 - \operatorname{erf}\frac{|x|}{\sqrt{4\alpha_2\tau}}\right) \tag{4-79}$$

The interfacial temperature, where $t_1(0,\tau) = t_2(0,\tau) = t_i$, can be obtained from Eq. (4-78) or (4-79) for the condition $x = 0$:

$$\frac{t_i}{t_0} = \frac{k_1\sqrt{\alpha_2}}{k_1\sqrt{\alpha_2} + k_2\sqrt{\alpha_1}} \tag{4-80}$$

In the special case that the materials from which the two bodies are made are identical, the temperature ratio for the interface is $\frac{1}{2}$, as it should be.

The heat flow from the body at higher temperature to the body at lower temperature can be expressed as

$$Q = k_1\left(\frac{\partial t_1}{\partial x}\right)_{x=0}$$

or, with the use of Eq. (4-78),

$$Q = \frac{k_2\sqrt{\alpha_1}}{k_1\sqrt{\alpha_2} + k_2\sqrt{\alpha_1}}\,\frac{k_1 t_0}{\sqrt{\pi\alpha_1\tau}} \tag{4-81}$$

For the special case where $k_1 = k_2 = k$ and $\alpha_1 = \alpha_2 = \alpha$

$$Q = \frac{k t_0}{\sqrt{4\pi\alpha\tau}} \tag{4-82}$$

Semi-infinite solid with heat generation When heat is generated within a semi-infinite solid, certain solutions can be obtained from already known solutions, e.g., Eq. (4-56), by a transformation in the dependent variable. The following calculation will illustrate.

Consider a semi-infinite body in the half space $x > 0$, with uniformly distributed heat sources of strength Q' per unit volume. The body initially has a linear temperature distribution which can be described by the expression $t(x,0) = A + Bx$, where A and B are known constants. At time zero the free surface, $x = 0$, is brought to a zero temperature and held there. The problem is to determine the temperature distribution $t(x,\tau)$ for $x > 0$, $\tau > 0$.

If the thermophysical properties of the material are constant, the appropriate equation can be deduced from Eq. (1-16):

$$\frac{\partial t}{\partial \tau} - \alpha \frac{\partial^2 t}{\partial x^2} - \frac{Q'}{\rho c} = 0 \tag{4-83}$$

The initial condition is

$$t(x,0) = A + Bx \qquad x > 0 \qquad \tau = 0$$

and the boundary condition is

$$t(0,\tau) = 0 \qquad x = 0 \qquad \tau > 0$$

If, in Eq. (4-83) a change in the dependent variable is made to a new variable $u(x,\tau)$ such that

$$t = u - \frac{Q'}{2k} x^2$$

Eq. (4-83) becomes

$$\frac{\partial u}{\partial \tau} = \alpha \frac{\partial^2 u}{\partial x^2} \tag{4-84}$$

where now for the initial and boundary conditions

$$u = A + Bx + \frac{Q'}{2k} x^2 = f(x) \qquad x > 0 \qquad \tau = 0$$

$$u = 0 \qquad x = 0 \qquad \tau > 0$$

Equation (4-84) has a solution in Eq. (4-56). Substituting gives

$$u = \int_0^\infty \frac{A + B\xi + C\xi^2}{\sqrt{4\pi\alpha\tau}} [e^{-(x-\xi)^2/4\alpha\tau} - e^{-(x+\xi)^2/4\alpha\tau}]\, d\xi \tag{4-85}$$

where $C = Q'/2k$.

If $\xi = x + \sqrt{4\alpha\tau}\,\beta$ in the first integral and $\xi = -x + \sqrt{4\alpha\tau}\,\beta$ in the second integral, as before, the result is

$$u = \frac{1}{\sqrt{\pi}} \int_{-x/\sqrt{4\alpha\tau}}^{\infty} [A + B(x + \sqrt{4\alpha\tau}\,\beta) + C(x + \sqrt{4\alpha\tau}\,\beta)^2]\, e^{-\beta^2}\, d\beta$$
$$- \frac{1}{\sqrt{\pi}} \int_{x/\sqrt{4\alpha\tau}}^{\infty} [A + B(-x + \sqrt{4\alpha\tau}\,\beta) + C(-x + \sqrt{4\alpha\tau}\,\beta)^2] e^{-\beta}\, d\beta \tag{4-86}$$

Evaluation of the integral in Eq. (4-86) yields the expression

$$u = (A + 2\alpha\tau C + Cx^2) \operatorname{erf} \frac{x}{\sqrt{4\alpha\tau}} + 2Cx \sqrt{\frac{\alpha\tau}{\pi}}\, e^{-x^2/4\alpha\tau} + Bx \tag{4-87}$$

Remembering that $t = u - Cx^2$ and that $C = Q'/2k$, the temperature distribution in the body becomes

$$t(x,\tau) = \left[A + \frac{Q'\alpha}{k}\tau + \frac{Q'}{2k} x^2 \right] \operatorname{erf} \frac{x}{\sqrt{4\alpha\tau}}$$
$$+ \frac{Q'}{k} \sqrt{\frac{\alpha\tau}{\pi}}\, x e^{-x^2/4\alpha\tau} + Bx - \frac{Q'}{2k} x^2 \tag{4-88}$$

The heat flow at the surface $x = 0$ can be calculated from the Fourier heat-conduction equation

$$q = k \left(\frac{\partial t}{\partial x} \right)_{x=0}$$

which with the derivative calculated from Eq. (4-88) becomes

$$q = (A + B\sqrt{\pi\alpha\tau}) \frac{k}{\sqrt{\pi\alpha\tau}} + 2Q' \sqrt{\frac{\alpha\tau}{\pi}} \tag{4-89}$$

The semi-infinite solid with constant surface heat rate Quite often interest is directed to boundary conditions which specify the *heat rate* at a surface rather than the value or variation of the temperature of the surface. To this end it is convenient to modify Eq. (1-16) to show the dependent variable as the heat flux rather than the temperature. If in Eq. (1-16) the heat-source term is omitted, the remaining equation is multiplied by k and differentiated partially with respect to x, y, and z independently, the order of differentiation reversed in each case, and the three equations added vectorially, one obtains the expression

$$\frac{\partial q}{\partial \tau} = \alpha \left(\frac{\partial^2 q}{\partial x^2} + \frac{\partial^2 q}{\partial y^2} + \frac{\partial^2 q}{\partial z^2} \right) \tag{4-90}$$

where q is the heat flux as defined in Chap. 1.

Consider again a semi-infinite solid in the region $x > 0$, which has a constant rate of heat flow q_0 into the surface $x = 0$. The temperature distribution in the body is desired after the imposition of the surface heat flux.

Equation (4-90) as reduced for the required one-space-dimensional heat flow is

$$\frac{\partial q}{\partial \tau} = \alpha \frac{\partial^2 q}{\partial x^2} \tag{4-91}$$

$$\begin{array}{lll} q = 0 & x > 0 & \tau = 0 \\ q = q_0 & x = 0 & \tau > 0 \end{array}$$

The solution of Eq. (4-91) for the conditions shown is

$$q = q_0 \left(1 - \operatorname{erf} \frac{x}{\sqrt{4\alpha\tau}}\right) = q_0 \operatorname{erfc} \frac{x}{\sqrt{4\alpha\tau}} \tag{4-92}$$

where $\operatorname{erfc} z = 1 - \operatorname{erf} z$ is the *complementary error function.* The temperature distribution is obtained from

$$t(x,\tau) = -\int_x^\infty \frac{q(x,\tau)}{k}\,dx = -\frac{q_0}{k}\int_x^\infty \operatorname{erfc} \frac{x}{\sqrt{4\alpha\tau}}\,dx$$

or

$$t(x,\tau) = \frac{q_0}{k}\sqrt{4\alpha\tau}\,\operatorname{ierfc} \frac{x}{\sqrt{4\alpha\tau}}$$

where ierfc z is the *integral complementary error function.* Now

$$\operatorname{ierfc} z = \frac{1}{\sqrt{\pi}} e^{-z^2} - z \operatorname{erfc} z$$

Thus

$$t(x,\tau) = \frac{q_0}{k}\sqrt{\frac{4\alpha\tau}{\pi}}\left(e^{-x^2/4\alpha\tau} - x\sqrt{\frac{\pi}{4\alpha\tau}} \operatorname{erfc} \frac{x}{\sqrt{4\alpha\tau}}\right) \tag{4-93}$$

Equation (4-93) is the expression for the temperature distribution in the semi-infinite body for values of the time $\tau > 0$ as a result of a constant heat-flux input to the surface $x = 0$.

The semi-infinite solid with general boundary and initial conditions In general a solid may have an initial temperature distribution $t = f(x,y,z)$ and may subsequently have a surface-temperature variation $t_s = q(x,y,z,\tau)$. Problems posed in this way can be reduced to simpler problems by a change in the dependent variable t. Consider Eq. (1-16) omitting the term concerning the heat sources

$$\frac{\partial t}{\partial \tau} = \alpha \nabla^2 t \tag{4-94}$$

which governs the transient conduction of heat in three space dimensions. The boundary and initial condition for the general case above are

$$t = \begin{cases} f(x,y,z) & \text{inside the body for } \tau \leq 0 \\ g(x,y,z,\tau) & \text{on the surface of the body for } \tau > 0 \end{cases} \tag{4-95}$$

Assume $t = u(x,y,z,\tau) + v(x,y,z,\tau)$, which, when substituted into Eq. (4-94), yields two equations with different boundary and initial conditions:

$$\frac{\partial u}{\partial \tau} = \alpha \nabla^2 u \qquad \frac{\partial v}{\partial \tau} = \alpha \nabla^2 v \tag{4-96}$$

$$\begin{array}{llllll} u = f(x,y,z) & x > 0 & \tau = 0 & v = 0 & x > 0 & \tau = 0 \\ u = 0 & x = 0 & \tau > 0 & v = g(x,y,z,\tau) & \text{on surface} & \tau > 0 \end{array}$$

Thus the original problem [Eqs. (4-94) and (4-95)] has been reduced to the sum of the solutions of two much simpler problems.

If Eqs. (4-96) are considered in terms of the semi-infinite solid, i.e., one-dimensional heat flow in x only, the equation in $u(x,y,z,\tau)$ has a solution in Eq. (4-56), as soon as the initial temperature distribution $f(x,y,z)$ is known. The equation in $v(x,y,z,\tau)$ can be solved by use of the *Duhamel integral* or *superposition theorem.*[1]

If $t = F(x,y,z,\lambda,\tau)$ is the temperature at (x,y,z) at the time τ in a solid in which the initial temperature is zero while its surface temperature is $\varphi(x,y,z,\lambda)$, then the solution of the problem in which the initial temperature is zero and the surface temperature is $\varphi(x,y,z,\tau)$ is given by

$$t(x,y,z,\tau) = \int_0^\tau \frac{\partial}{\partial \tau} F(x,\, y,\, z,\, \lambda,\, \tau - \lambda)\, d\lambda \tag{4-97}$$

This equation has the following physical interpretation: λ represents the time at which a change occurs in the surface temperature, while τ is the time at which some point (x,y,z) in the interior of the body responds to that change in the surface temperature. As some delay occurs between a change in the surface temperature and a response indicated by a change in temperature at some interior point (x,y,z), the temperature at any point and at any $\tau > \lambda$ in the interior of the body must be a function not only of the time λ when the change in surface temperature occurred but also the time interval $\tau - \lambda$ from that occurrence until the response is felt at some interior point. Equation (4-97) expresses the temperature at some interior point (x,y,z) observed at time τ as a result of changes in the surface temperature at times represented by λ.

[1] The Duhamel integral has been widely discussed and used; see, for example, Carslaw and Jaeger, *op. cit.*, pp. 30–31, or T. von Kármán and M. A. Biot, "Mathematical Methods in Engineering," McGraw-Hill, New York, 1940.

If the function $F(x,y,z,\tau)$ is the response to a *unit step-function change with time only* in the temperature of the surface of the body, i.e., there is no spatial temperature variation on the body surface, Eq. (4-97) can be written in the form

$$t(x,y,z,\tau) = \int_0^\tau \varphi(\lambda) \frac{\partial}{\partial \tau} F(x, y, z, \tau - \lambda)\, d\lambda \tag{4-98}$$

This is the solution of the problem where the surface temperature has a variation $\varphi(\tau)$.

As an example of the application of the *Duhamel integral theorem* consider the problem of the semi-infinite solid initially at some uniform temperature t_i. At time zero the surface temperature is caused to vary as the square root of the time τ.

The governing equation and initial and boundary conditions are, where $\vartheta = t - t_i$, $\vartheta_0 = t_0 - t_i$,

$$\frac{\partial \vartheta}{\partial \tau} = \alpha \frac{\partial^2 \vartheta}{\partial x^2} \tag{4-99}$$

$$\begin{aligned} \frac{\vartheta}{\vartheta_0} &= 0 \qquad x > 0 \qquad \tau < 0 \\ \frac{\vartheta}{\vartheta_0} &= A\sqrt{\tau} \qquad x = 0 \qquad \tau > 0 \end{aligned} \tag{4-100}$$

To make use of Eq. (4-98) it is necessary to have a solution of Eq. (4-99) for a unit step-function change in the temperature at $x = 0$. That is, the initial and boundary conditions are

$$\begin{aligned} \frac{\vartheta}{\vartheta_0} &= 0 \qquad x > 0 \qquad \tau < 0 \\ \frac{\vartheta}{\vartheta_0} &= 1 \qquad x = 0 \qquad \tau \geq 0 \end{aligned}$$

Such a solution is, from Eq. (4-58),

$$\frac{\vartheta}{\vartheta_0} = \operatorname{erfc} \frac{x}{\sqrt{4\alpha\tau}} \tag{4-101}$$

In Eq. (4-98), therefore,

$$F(x, \tau - \lambda) = \operatorname{erfc} \frac{x}{\sqrt{4\alpha(\tau - \lambda)}}$$

and the derivative with respect to τ is

$$\frac{\partial F}{\partial \tau} = \frac{x e^{-x^2/4\alpha(\tau-\lambda)}}{\sqrt{4\pi\alpha(\tau - \lambda)^3}}$$

which can now be introduced into the integral of Eq. (4-98) to result in the expression

$$\frac{\vartheta(x,\tau)}{\vartheta_0} = \frac{x}{\sqrt{4\alpha\pi}} \int_0^\tau \frac{A\sqrt{\lambda}\, e^{-x^2/4\alpha(\tau-\lambda)}}{(\tau-\lambda)^{\frac{3}{2}}}\, d\lambda \tag{4-102}$$

In Eq. (4-102) let $\beta^2 = x^2/4\alpha(\tau - \lambda)$, and the equation becomes

$$\frac{\vartheta(x,\tau)}{\vartheta_0} = \frac{2A}{\sqrt{\pi}} \int_{x/\sqrt{4\alpha\tau}}^{\infty} \left[\tau - \frac{x^2}{4\alpha\beta^2}\right]^{\frac{1}{2}} e^{-\beta^2}\, d\beta$$

Solution of the last integral gives finally

$$\frac{\vartheta(x,\tau)}{\vartheta_0} = A\sqrt{\tau}\left(e^{-x\ /4\alpha\tau} - x\sqrt{\frac{\pi}{4\alpha\tau}}\ \text{erfc}\,\frac{x}{\sqrt{4\alpha\tau}}\right) \tag{4-103}$$

Comparison of Eq. (4-103) with Eq. (4-93) will show that

$$A\vartheta_0 = \frac{q_0}{k}\sqrt{\frac{4\alpha}{\pi}}$$

a constant, and that the equations are identical. Thus the temperature variation $A\vartheta_0\sqrt{\tau}$ on the surface corresponds to the case of constant heat-rate input at the surface. A calculation of the heat flux from Eq. (4-103) will give an identical value for the constant A as shown above; thus the heat flux is

$$q_0 = Ak\vartheta_0\sqrt{\frac{\pi}{4\alpha}} \tag{4-104}$$

Finite superposition The idea of superposition, central to the development of the Duhamel integral theorem, may be used with known heat-conduction solutions in a finite form to produce other solutions of interest. Two examples will illustrate the method.

The first is the case of a semi-infinite solid initially at zero temperature. During the period $0 \le \tau \le \tau_1$ the surface temperature is held at a constant temperature t_0. For $\tau > \tau_1$ the surface temperature is returned to zero and held there. The surface-temperature variation is shown by the heavier line in Fig. 4-18a.

In the period of time $0 < \tau < \tau_1$ the solution to the problem is known to be

$$\frac{\vartheta}{\vartheta_0} = \text{erfc}\,\frac{x}{\sqrt{4\alpha\tau}} \qquad 0 < \tau < \tau_1 \tag{4-105}$$

A solution for values of the time $\tau > \tau_1$ can be produced by subtracting a solution

$$\frac{\vartheta}{\vartheta_0} = \text{erfc}\,\frac{x}{\sqrt{4\alpha(\tau - \tau_1)}} \tag{4-106}$$

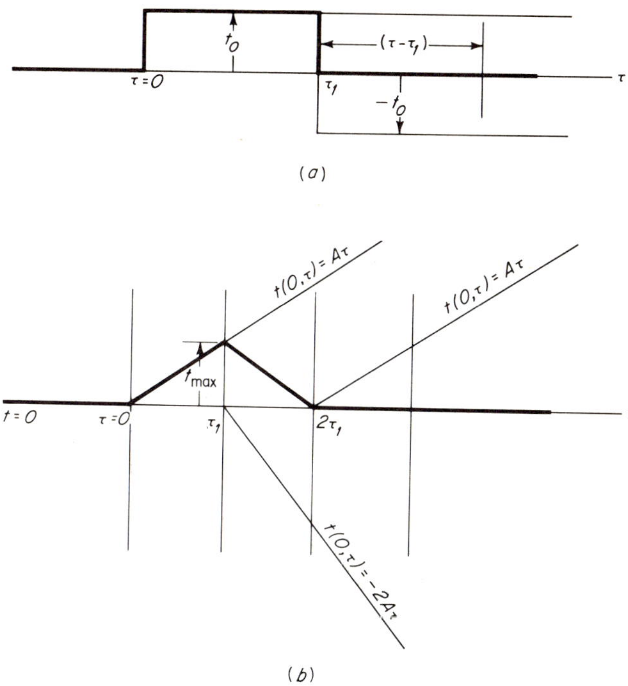

Fig. 4-18 Variations with time of surface temperature of a semi-infinite body.

from Eq. (4-105) to yield

$$\frac{\vartheta}{\vartheta_0} = \operatorname{erf} \frac{x}{\sqrt{4\alpha(\tau - \tau_1)}} - \operatorname{erf} \frac{x}{\sqrt{4\alpha\tau}} \qquad \tau > \tau_1 \tag{4-107}$$

Equations (4-105) and (4-107) together provide solutions for all values of the time greater than zero.

A second example of a variation in the surface temperature of a semi-infinite solid with time is shown by the heavier line in Fig. 4-18*b*. The solution can be built up from the solution for the linear variation of surface temperature with time, which can be calculated by means of the Duhamel integral in a manner similar to that leading to the solution given by Eq. (4-103). The solution for the linear variation of the surface temperature with time for a semi-infinite body is

$$\frac{\vartheta}{\vartheta_0} = A\tau \left[\left(1 + \frac{x^2}{2\alpha\tau}\right) \operatorname{erfc} \frac{x}{\sqrt{4\alpha\tau}} - \frac{x}{\sqrt{\pi\alpha\tau}}\, e^{-x^2/4\alpha\tau} \right] \tag{4-108}$$

Three such solutions, as shown graphically in Fig. 4-18*b*, can be combined to yield a solution for the surface-temperature variation shown. The equations to be added together are Eq. (4-108) and the following:

$$\frac{\vartheta}{\vartheta_0} = -2A\tau\left[\left(1 + \frac{x^2}{2\alpha(\tau - \tau_1)}\right)\operatorname{erfc}\frac{x}{\sqrt{4\alpha(\tau - \tau_1)}} - \frac{x}{\sqrt{\pi\alpha(\tau - \tau_1)}}\,e^{-x^2/4\alpha(\tau-\tau_1)}\right] \tag{4-109}$$

$$\frac{\vartheta}{\vartheta_0} = +A\tau\left[\left(1 + \frac{x^2}{2\alpha(\tau - 2\tau_1)}\right)\operatorname{erfc}\frac{x}{\sqrt{4\alpha(\tau - 2\tau_1)}} - \frac{x}{\sqrt{\pi\alpha(\tau - 2\tau_1)}}\,e^{-x^2/4\alpha(\tau-2\tau_1)}\right] \tag{4-110}$$

The sum of the three equations (4-108) to (4-110) forms a solution for values of $\tau > 2\tau_1$.

Other such solutions for different variations in the surface temperature can be synthesized from known solutions.

The Laplace transform method[1] The foregoing methods and examples were developed essentially from the classical work of Fourier. The methods become increasingly difficult to apply to complex problems. Alternative methods, based largely on the operational mathematical methods of Heaviside and particularly adaptable to transient-heat-conduction problems, are available. The advantages of the operational methods increase with the increasing complexity of the problem. One such method is the *Laplace transform.*

The Laplace transform method is of such power and interest that it is the subject matter of extensive treatises. In the interest of the brevity necessary in a work of the present scope, the method will be illustrated by two examples only. For the theorems and details of the method, reference should be made to standard works on the subject. The examples treated will be sufficient to show the power and neatness of the method.

The semi-infinite solid with surface-temperature variation with time Consider a problem treated earlier by the Duhamel integral theorem,[2]

[1] The Laplace transform method is fully discussed in Carslaw and Jaeger, *op. cit.;* H. S. Carslaw and J. C. Jaeger, "Operational Methods in Mathematical Physics," 2d ed., Cambridge University Press, London, 1931; and in R. V. Churchill, "Modern Operational Mathematics in Engineering," McGraw-Hill, New York, 1944.

[2] The Laplace transform theory includes the superposition theorem, the Duhamel integral theorem, as a part of its formalism. The theorem is sometimes called the *Faltung integral* theorem.

that of a semi-infinite solid without heat sources, initially held at a temperature t_i throughout. At time $\tau = 0$ the surface temperature is increased as the square root of the time. The results of the solution are shown by Eqs. (4-93), (4-103), and (4-104).

The Laplace transform in relation to the heat-conduction problem is to be applied to the temperature as a function of time and such spatial coordinates as are required. The use of the transform reduces the partial differential equation of heat conduction to an ordinary differential equation, together with transformed initial and boundary conditions, which must be solved. The solution so obtained must then be inverted, or transformed back into the original variable. The transform is usually fairly easy; however, the inversion process can be difficult. The operations are roughly analogous to the processes of differentiation and integration of functions. Tables of transforms and inverse transforms are available.[1]

The Laplace transform is stated as

$$L\{t(x,y,z,\tau)\} = \bar{t} = \int_0^\infty e^{-s\tau} t(x,y,z,\tau)\, d\tau \qquad \text{(4-111)}$$

Consider now the equation and initial and boundary conditions for the posed problem of the semi-infinite body whose surface temperature varies as the square root of the time. Where $\vartheta = t - t_i$, the excess temperature,

$$\begin{aligned} \frac{\partial \vartheta}{\partial \tau} &= \alpha \frac{\partial^2 \vartheta}{\partial x^2} \\ \vartheta &= A\sqrt{\tau} \qquad x = 0 \qquad \tau > 0 \\ \vartheta &= 0 \qquad x > 0 \qquad \tau < 0 \end{aligned} \qquad \text{(4-112)}$$

Application of Eq. (4-111) yields

$$L\left\{\frac{\partial \vartheta}{\partial \tau}\right\} = \int_0^\infty e^{-s\tau} \frac{\partial \vartheta}{\partial \tau}\, d\tau = [\vartheta e^{-s\tau}]_0^\infty + s\int_0^\infty e^{-s\tau}\vartheta\, d\tau = s\bar{\vartheta} - \vartheta(x,0)$$

$$L\left\{\frac{\partial^2 \vartheta}{\partial x^2}\right\} = \int_0^\infty e^{-s\tau} \frac{\partial^2 \vartheta}{\partial x^2}\, d\tau = \frac{\partial^2}{\partial x^2}\int_0^\infty e^{-s\tau}\vartheta\, d\tau = \frac{\partial^2 \bar{\vartheta}}{\partial x^2}$$

Thus Eq. (4-112) becomes in the transformed variable $\bar{\vartheta}(x,s)$

$$\frac{d^2\bar{\vartheta}}{dx^2} - \frac{s}{\alpha}\bar{\vartheta} = 0 \qquad \text{(4-113)}$$

[1] S. M. Selby (ed.), "Standard Mathematical Tables," 17th ed., Chemical Rubber Co., Cleveland, Ohio, 1969; M. Abramowitz and L. A. Stegun (eds.), Handbook of Mathematical Functions, *Natl. Bur. Std. U.S., Appl. Math.* 55, 1965.

Turning attention to the boundary conditions, we have

$$L\{A\sqrt{\tau}\} = \int_0^\infty e^{-s\tau} A\sqrt{\tau}\,d\tau = -\frac{A}{s}e^{-s\tau}\sqrt{\tau}\Big|_0^\infty + \frac{A}{s}\int_0^\infty \frac{e^{-s\tau}}{2\sqrt{\tau}}\,d\tau$$
$$= A\frac{\sqrt{\pi}}{2}s^{-\frac{3}{2}} \qquad (4\text{-}114)$$

The solution for the differential equation (4-113) is

$$\bar{\vartheta}(x,s) = C_1 \exp\left(-\sqrt{\frac{s}{\alpha}}\,x\right) + C_2 \exp\left(+\sqrt{\frac{s}{\alpha}}\,x\right) \qquad (4\text{-}115)$$

Note that $\vartheta(\infty,\tau) = 0$ and $\bar{\vartheta}(\infty,s) = 0$ as well; thus the constant $C_2 \equiv 0$. Application of the transformed boundary condition [Eq. (4-114)] yields

$$A\frac{\sqrt{\pi}}{2}s^{-\frac{3}{2}} = C_1 \exp\left(-\sqrt{\frac{s}{\alpha}}\,x\right)$$

which for $x = 0$, results in

$$C_1 = A\frac{\sqrt{\pi}}{2}s^{-\frac{3}{2}}$$

such that the solution of the Eq. (4-113) becomes

$$\bar{\vartheta}(x,s) = A\frac{\sqrt{\pi}}{2}s^{-\frac{3}{2}}\exp\left(-\sqrt{\frac{s}{\alpha}}\,x\right) = A\frac{\sqrt{\pi}}{2}\frac{1}{s}\left[s^{-\frac{1}{2}}\exp\left(-\sqrt{\frac{s}{\alpha}}\,x\right)\right] \qquad (4\text{-}116)$$

The *inverse transforms* are[1]

$$L^{-1}\left\{\frac{\exp(-\sqrt{s/\alpha}\,x)}{\sqrt{s}}\right\} = \frac{1}{\sqrt{\pi\tau}}e^{-x^2/4\alpha\tau}$$

$$L^{-1}\left\{\frac{1}{s}f(s)\right\} = \int_0^\tau F(\tau)\,d\tau$$

where $$F(\tau) = \frac{1}{\sqrt{\pi\tau}}e^{-x^2/4\alpha\tau}$$

Thus $$\vartheta(x,\tau) = \frac{A}{2}\int_0^\tau \frac{e^{-x^2/4\alpha\tau}}{\sqrt{\tau}}\,d\tau = \frac{Ax}{\sqrt{4\alpha}}\int_{x/\sqrt{4\alpha\tau}}^\infty \frac{e^{-\beta^2}}{\beta^2}\,d\beta \qquad (4\text{-}117)$$

Integration of Eq. (4-117) yields

$$\vartheta(x,\tau) = A\sqrt{\tau}\left(e^{-x^2/4\alpha\tau} - x\sqrt{\frac{\pi}{2\alpha\tau}}\operatorname{erfc}\frac{x}{\sqrt{4\alpha\tau}}\right) \qquad (4\text{-}118)$$

which is identical with Eqs. (4-93) and (4-103).

[1] A limited number of Laplace transforms are given in Table 4-4.

The sphere with constant initial temperature Consider a spherical body of radius r_0, with a constant initial temperature t_i, and no heat sources. At time $\tau > 0$ the surface temperature $r = r_0$ is maintained at a constant temperature $t_0 > t_i$.

Problems in radial heat conduction in spherical bodies can be reduced to problems of linear heat flow in a slab by the transformation $u = tr$, and while this transformation is not a necessary prelude to the use of the Laplace transform, it will be so used in this example to demonstrate further the use of both techniques.

The partial differential equation governing pure radial heat conduction in spheres is, from Eq. (1-18), where $\vartheta = t - t_i$,

$$\frac{\partial \vartheta}{\partial \tau} = \alpha \left(\frac{\partial^2 \vartheta}{\partial r^2} + \frac{2}{r} \frac{\partial \vartheta}{\partial r} \right) \tag{4-119}$$

The boundary condition and the initial conditions relevant to the posed problem are

$$\begin{array}{llll} t = t_i & \vartheta = 0 & 0 \le r \le r_0 & \tau < 0 \\ t = t_0 & \vartheta = \vartheta_0 = t_0 - t_i & r = r_0 & \tau > 0 \end{array}$$

In Eq. (4-119) let $u = \vartheta r$, to obtain

$$\frac{\partial u}{\partial \tau} = \alpha \frac{\partial^2 u}{\partial r^2} \tag{4-120}$$

$$\begin{array}{lll} u = 0 & 0 \le r \le r_0 & \tau < 0 \\ u = \vartheta_0 r_0 & r = r_0 & \tau > 0 \end{array}$$

The Laplace transforms of Eq. (4-120) and the boundary conditions are

$$\frac{d^2 \bar{u}}{dr^2} - \frac{s}{\alpha} \bar{u} = 0 \tag{4-121}$$

$$\begin{array}{ll} \bar{u} = \dfrac{\vartheta_0 r_0}{s} & r = r_0 \\ \bar{u} = 0 & r = 0 \end{array}$$

The solutions of Eq. (4-121) is

$$\bar{u} = C_1 \exp\left(-\sqrt{\frac{s}{\alpha}}\, r\right) + C_2 \exp\left(+\sqrt{\frac{s}{\alpha}}\, r\right)$$

and when the constants C_1 and C_2 are evaluated by means of the conditions expressed in Eq. (4-121), there results

$$\bar{u} = \frac{\vartheta_0 r_0}{s} \frac{\sinh\left(\sqrt{s/\alpha}\, r\right)}{\sinh\left(\sqrt{s/\alpha}\, r_0\right)} \tag{4-122}$$

Table 4-4 Laplace transform pairs

$L\{t(x,y,z,\tau)\} = \bar{t} = \int_0^\infty e^{-s\tau} t(x,y,z,\tau)\, d\tau$

$\bar{t}(s)$		$t(\tau)$
1. $\frac{1}{s}$		1
2. $s^{1/(n+1)}$	$n > -1$	$\frac{\tau^n}{\Gamma(n+1)}$
3. $\frac{1}{s+\alpha}$		$\exp(-\alpha\tau)$
4. $\frac{\omega}{s^2+\omega^2}$		$\sin \omega\tau$
5. $\frac{s}{s^2+\omega^2}$		$\cos \omega\tau$
6. $\exp\left(-\sqrt{\frac{s}{a}}\,x\right)$	$x > 0$, $a > 0$	$\frac{x}{2\sqrt{\pi a \tau^3}} \exp\left(\frac{-x^2}{4a\tau}\right)$
7. $\frac{1}{\sqrt{s/a}} \exp\left(-\sqrt{\frac{s}{a}}\,x\right)$	$x > 0$, $a > 0$	$\sqrt{\frac{a}{\pi\tau}} \exp\left(\frac{-x^2}{4a\tau}\right)$
8. $\frac{1}{s} \exp\left(-\sqrt{\frac{s}{a}}\,x\right)$	$x > 0$, $a > 0$	$\operatorname{erfc} \frac{x}{\sqrt{4a\tau}}$
9. $\frac{1}{s\sqrt{s/a}} \exp\left(-\sqrt{\frac{s}{a}}\,x\right)$	$x > 0$, $a > 0$	$2\sqrt{\frac{a\tau}{\pi}} \exp\left(\frac{-x^2}{4a\tau}\right) - x \operatorname{erfc} \frac{x}{\sqrt{4a\tau}}$
10. $\frac{1}{s^2} \exp\left(-\sqrt{\frac{s}{a}}\,x\right)$	$x > 0$, $a > 0$	$\left(\tau + \frac{x^2}{2a}\right) \operatorname{erfc} \frac{x}{\sqrt{4a\tau}} - x\sqrt{\frac{\tau}{\pi a}} \exp\left(\frac{-x^2}{4a\tau}\right)$
11. $\frac{1}{s^{1+n/2}} \exp\left(-\sqrt{\frac{s}{a}}\,x\right)$	$x > 0$, $a > 0$, $n = 0, 1, 2, \ldots$	$(4\tau)^{n/2} i^n \operatorname{erfc} \frac{x}{\sqrt{4a\tau}}$

12. $\dfrac{\exp(-\sqrt{s/a}\,x)}{\sqrt{s/a}+h}$	$x>0$ $a>0$	$\sqrt{\dfrac{a}{\pi\tau}}\exp\left(\dfrac{-x^2}{4a\tau}\right) - ha\exp(hx+ah^2\tau)\,\mathrm{erfc}\left(\dfrac{x}{\sqrt{4a\tau}}+h\sqrt{a\tau}\right)$
13. $\dfrac{1}{\sqrt{s/a}\,(\sqrt{s/a}+h)}\exp\left(-\sqrt{\dfrac{s}{a}}\,x\right)$	$x>0$ $a>0$	$a\exp(hx+ah^2\tau)\,\mathrm{erfc}\left(\dfrac{x}{\sqrt{4a\tau}}+h\sqrt{a\tau}\right)$
14. $\dfrac{1}{s(\sqrt{s/a}+h)}\exp\left(-\sqrt{\dfrac{s}{a}}\,x\right)$	$x>0$ $a>0$	$\dfrac{1}{h}\,\mathrm{erfc}\,\dfrac{x}{\sqrt{4a\tau}} - \dfrac{1}{h}\exp(hx+ah^2\tau)\,\mathrm{erfc}\left(\dfrac{x}{\sqrt{4a\tau}}+h\sqrt{a\tau}\right)$
15. $\dfrac{1}{s\sqrt{s/a}\,(\sqrt{s/a}+h)}\exp\left(-\sqrt{\dfrac{s}{a}}\,x\right)$	$x>0$ $a>0$	$\dfrac{2}{h}\sqrt{\dfrac{a\tau}{\pi}}\exp\left(\dfrac{-x^2}{4a\tau}\right) - \dfrac{1+hx}{h^2}\,\mathrm{erfc}\,\dfrac{x}{\sqrt{4a\tau}}$ $+\dfrac{1}{h^2}\exp(hx+ah^2\tau)\,\mathrm{erfc}\left(\dfrac{x}{\sqrt{4a\tau}}+h\sqrt{a\tau}\right)$
16. $\dfrac{1}{(s/a)^{(n+1)/2}(\sqrt{s/a}+h)}\exp\left(-\sqrt{\dfrac{s}{a}}\,x\right)$	$x>0$ $a>0$	$\dfrac{a}{(-h)^n}\exp(hx+ah^2\tau)\,\mathrm{erfc}\left(\dfrac{x}{\sqrt{4a\tau}}+h\sqrt{a\tau}\right)$ $-\dfrac{a}{(-h)^n}\sum_{r=0}^{n-1}(-2h\sqrt{a\tau})^r i^r\,\mathrm{erfc}\,\dfrac{x}{\sqrt{4a\tau}}$
17. $\dfrac{1}{s-\alpha}\exp\left(-\sqrt{\dfrac{s}{a}}\,x\right)$	$x>0$ $a>0$	$\dfrac{1}{2}\exp\alpha\tau\left[\exp\left(-x\sqrt{\dfrac{\alpha}{a}}\right)\mathrm{erfc}\left(\dfrac{x}{\sqrt{4a\tau}}-\sqrt{\alpha\tau}\right)\right.$ $\left.+\exp\left(x\sqrt{\dfrac{\alpha}{a}}\right)\mathrm{erfc}\left(\dfrac{x}{\sqrt{4a\tau}}+\sqrt{\alpha\tau}\right)\right]$
18. $K_0\left(\sqrt{\dfrac{s}{a}}\,x\right)$	$x>0$ $a>0$	$\dfrac{1}{2\tau}\exp\left(\dfrac{-x^2}{4a\tau}\right)$
19. $\dfrac{1}{s}\exp\dfrac{x}{s}$		$I_0(\sqrt{4x\tau})$
20. $s^{\nu/2}K_\nu(x\sqrt{s})$		$\dfrac{x^\nu}{(2\tau)^{\nu+1}}\exp\left(\dfrac{-x^2}{4\tau}\right)$

or, in terms of $\bar{\vartheta}$,

$$\bar{\vartheta} = \frac{\vartheta_0 r_0}{sr} \frac{\sinh(\sqrt{s/\alpha}\, r)}{\sinh(\sqrt{s/\alpha}\, r_0)} \tag{4-123}$$

The inverse transform of Eq. (4-123) is

$$\frac{\vartheta}{\vartheta_0} = \frac{r_0}{r} \sum_{n=1}^{\infty} \left[\operatorname{erf} \frac{(2n+1)r_0 + r}{\sqrt{4\alpha\tau}} - \operatorname{erf} \frac{(2n+1)r_0 - r}{\sqrt{4\alpha\tau}} \right] \tag{4-124}$$

where if $r \to 0$, the limit of Eq. (4-124) can be calculated to be

$$\frac{\vartheta(0,\tau)}{\vartheta_0} = \frac{2r_0}{\sqrt{\pi\alpha\tau}} \sum_{n=1}^{\infty} \exp\left[- \frac{(2n+1)^2 r_0^2}{4\alpha\tau} \right] \tag{4-125}$$

For $\tau = 0$ it can be determined from Eq. (4-125) that $\vartheta = \vartheta_0$, the correct initial condition. The heat flow to the sphere is the heat flow through the surface $r = r_0$, or

$$Q = +4\pi r_0^2 k \left(\frac{\partial\vartheta}{\partial r} \right)_{r=r_0}$$

From Eq. (4-124) the derivative is

$$\left(\frac{\partial\vartheta}{\partial r} \right)_{r=r_0} = \sum_{n=1}^{\infty} \left\{ \frac{\vartheta_0}{r_0} \left[\operatorname{erf} \frac{nr_0}{\sqrt{\alpha\tau}} - \operatorname{erf} \frac{(n+1)r_0}{\sqrt{\alpha\tau}} \right] + \frac{\vartheta_0}{\sqrt{\pi\alpha\tau}} \left(e^{[-(n+1)^2 r_0^2]/\alpha\tau} + e^{-n^2 r_0^2/\alpha\tau} \right) \right\} \tag{4-126}$$

Substituting Eq. (4-126) into the expression for the heat flow gives

$$\frac{Q}{Q_0} = \sum_{n=1}^{\infty} \left\{ \frac{\alpha}{r_0} \left[\operatorname{erf} \frac{nr_0}{\sqrt{\alpha\tau}} - \operatorname{erf} \frac{(n+1)r_0}{\sqrt{\alpha\tau}} \right] + \sqrt{\frac{\alpha}{\pi\tau}} \left(e^{[-(n+1)^2 r_0^2]/\alpha\tau} + e^{-n^2 r_0^2/\alpha\tau} \right) \right\} \tag{4-127}$$

where $Q_0 = 4\pi r_0^2 \rho c \vartheta_0$, the initial heat content of the sphere.

The results of Eq. (4-124) are shown graphically in Fig. 4-19.

Nonlinear heat conduction The foregoing discussion of conduction heat transfer by analytical methods considered the thermophysical properties, i.e., thermal conductivity, mass density, and the specific heat at constant pressure, to be independent of the temperature, an important result of which is that the heat-conduction equations are linear and therefore much more amenable in devising solutions.

Physically, the thermal conductivity particularly and the other two properties to a real but lesser degree exhibit variations with temperature.

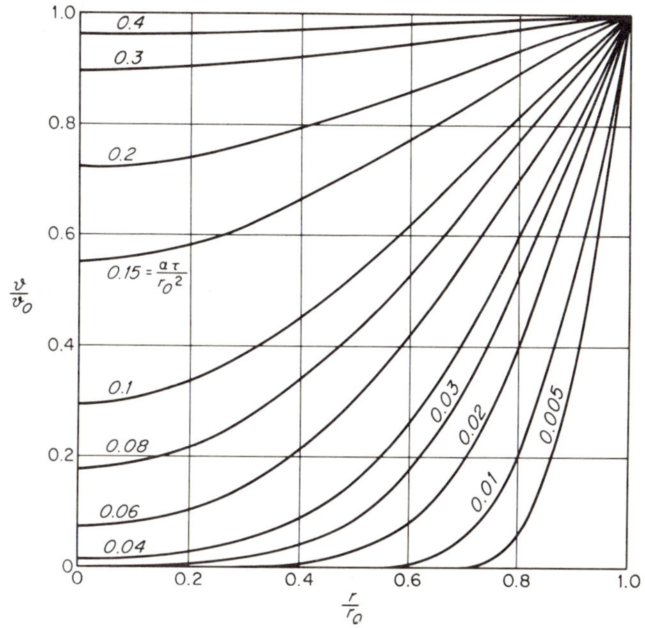

Fig. 4-19 Temperature-time history of a sphere of radius r_0 with constant initial temperature and constant but different surface temperature. [*H. S. Carslaw and J. C. Jaeger, "Conduction of Heat in Solids," 2d ed., Oxford University Press, New York, 1959.*]

In such instances the differential equations are nonlinear, and special techniques must be called into play to effect solutions. Fortunately, the assumption of constant thermophysical properties is an acceptable approximation for many situations; however, in other important instances such assumptions lead to intolerable errors in the results.

Where temperature changes are large, thermophysical property dependence upon temperature cannot be ignored. Where property changes are large, analytical solutions obtained from the nonlinear heat-conduction equation differ significantly from solutions obtained from the constant-property linear equations. The differences between the two results are most pronounced for small values of the time after the onset of a thermal change.

Unfortunately, and in spite of great interest in the probable results, not many analytical solutions for the nonlinear cases of heat conduction exist, and those which do are restricted to the relatively simple geometries of the semi-infinite and infinite solids.

While the nonlinear heat-conduction problems for geometries and conditions more complicated than those of the semi-infinite and infinite solids may be better accomplished by numerical methods on a high-speed computer, it is well to note here the several analytical techniques that have been exploited in the past.

The methods of approach to the solution of variable-property heat-conduction problems can be divided into five major categories:

1. Methods for linearization of the partial differential equations and conditions under which this can be accomplished[1]
2. Transformation of the independent variable, thus achieving a reduction of the nonlinear partial differential equation to a nonlinear ordinary differential equation[2]
3. Separation of variables[3,4]
4. Extension of the Neumann method[5]
5. Iteration method of Picard[6]

Basically, the analytical methods are directed to the solution of the heat-conduction equation of the form

$$\frac{\partial}{\partial x}\left[k(t)\,\frac{\partial t}{\partial x}\right] = C(t)\,\frac{\partial t}{\partial \tau}$$

where $C(t) = \rho(t)c(t)$. The search is for some simple functional relationship for $k(t)$ and $C(t)$ which not only will be representative of some material in a desired temperature range but which will also be such as to permit the solution of the equation, though not necessarily in closed form. Some of these methods are applied, discussed, and compared with experimental data, where such data exist, by Friedmann,[7] to whom the reader is referred for further study.

[1] I. A. Carnyi, The Methods of Linearization of Non-linear Equations of the Type of the Heat Conduction Equations, *Izv. Akad. Nauk SSSR*, **15**:829–838 (1951); M. L. Storm, Heat Conduction in Simple Metals, *J. Appl. Phys.*, **22**:940–951 (1951).

[2] J. Crank and M. E. Henry, Diffusion in Media with Variable Properties: Part I, *Trans. Faraday Soc.*, **45**:636–650 (1949); J. Crank, "The Mathematics of Diffusion," Clarendon Press, Oxford, 1956.

[3] J. K. L. MacDonald, "Non-linear Heat Conduction," Project-Squid Semiannual Report, New York University, New York, 1947.

[4] M. Sawada, On the General Solution of Basic Expressions for Heat Transmission, *J. Soc. Mech. Engr. Japan*, **35**:183 (1932).

[5] *Ibid.*

[6] M. R. Hopkins, Heat Conduction in a Medium Having Thermal Properties Depending upon the Temperature, *Proc. Roy. Soc. London, Ser. A*, **50**:703 (1938).

[7] N. E. Friedmann, Quasilinear Heat Flow, *Trans. ASME*, **80**:635–645 (1958).

4-2 TRANSIENT HEAT CONDUCTION: APPROXIMATE METHODS

As discussed in Chap. 1, certain approximate methods have great utility in the solution of problems in transient heat conduction, especially when the thermophysical properties vary greatly with temperature or when the boundary conditions include radiant heat flow or a change in phase. In all these cases either the heat-conduction equations or the boundary conditions become nonlinear, and the classical methods become extremely difficult or fail to provide a solution altogether.

THE INTEGRAL HEAT-CONDUCTION EQUATION

The integrated form of the heat-conduction equation as discussed in Sec. 1-4 leads to approximate solutions of the nonlinear heat-conduction problem by the reduction of the nonlinear boundary-value problem to an ordinary initial-value problem, the solution of which can be expressed frequently in a closed form. The integrated form of the heat-conduction equation can also be used to obtain approximate solutions to linear problems with spatially dependent thermal properties and problems where convection as well as conduction is involved. Many of these solutions have been examined and reported by Goodman[1] and others. The method will be illustrated by several examples.

The semi-infinite solid Consider the case of the semi-infinite solid with constant physical properties. The solid is held initially at a constant temperature t_i throughout. At time zero the surface temperature is reduced to another constant temperature t_0 and maintained there.

In terms of the excess temperature $\vartheta = t - t_0$ the governing equation and boundary conditions are

$$\frac{\partial \vartheta}{\partial \tau} = \alpha \frac{\partial^2 \vartheta}{\partial x^2} \tag{4-128}$$

where

$$\tau \leq 0 \qquad \vartheta(x,0) = t_i - t_0 = \vartheta_0$$
$$\tau > 0 \qquad \vartheta(0,\tau) = 0$$

The integrated form of Eq. (4-128) comparable to Eq. (1-31) becomes

$$\frac{d}{d\tau}\left[\int_0^\delta \vartheta(x,\tau)\,dx - \vartheta_\infty \delta(\tau)\right] = -\alpha \frac{\partial \vartheta}{\partial x}(0,\tau) \tag{4-129}$$

where $\delta(\tau)$ is the *penetration depth,* defined in Chap. 1.

[1] T. R. Goodman, Application of Integral Methods to Transient Nonlinear Heat Transfer, *Advan. Heat Transfer,* **1**:55–122 (1964).

Assume that $\vartheta(x,\tau)$ can be represented by a polynomial of the second degree, or

$$\vartheta(x,\tau) = A + Bx + Cx^2 \tag{4-130}$$

subject to the physical conditions

$$x = 0 \qquad \vartheta(0,\tau) = 0$$

$$x = \delta(\tau) \qquad \vartheta = \vartheta_0 \qquad \frac{\partial\vartheta}{\partial x} = 0$$

The resulting expression for the assumed temperature distribution is

$$\frac{\vartheta}{\vartheta_0} = 2\frac{x}{\delta} - \left(\frac{x}{\delta}\right)^2 \tag{4-131}$$

Using the temperature distribution in Eq. (4-131) with $\eta = x/\delta$ in Eq. (4-129), one obtains

$$\frac{d}{d\tau}\left[\int_0^1 \vartheta_0(2\eta - \eta^2)\delta\, d\eta - \vartheta_0\delta\right] = \frac{2\alpha\vartheta_0}{\delta} \tag{4-132}$$

Upon evaluation Eq. (4-132) becomes a first-order ordinary differential equation in $\delta(\tau)$:

$$\frac{d\delta^2}{d\tau} - 12\alpha = 0$$

the solution of which is, for the condition $\delta(0) = 0$,

$$\delta = \sqrt{12\alpha\tau} \tag{4-133}$$

The temperature distribution is obtained by substituting the above result for δ into Eq. (4-131):

$$\frac{\vartheta}{\vartheta_0} = \frac{2x}{\sqrt{12\alpha\tau}} - \frac{x^2}{12\alpha\tau} \tag{4-134}$$

The results of the exact solution for this problem are given by Eq. (4-58):

$$\frac{\vartheta}{\vartheta_0} = \operatorname{erf}\frac{x}{\sqrt{4\alpha\tau}}$$

In usual terms the heat flux at the surface $x = 0$ is obtained from the expression

$$q = k\left(\frac{\partial\vartheta}{\partial x}\right)_{x=0}$$

which, when used with the temperature distribution of Eq. (4-134), will result in the expression

$$q = \frac{k\vartheta_0}{\sqrt{3\alpha\tau}} \tag{4-135}$$

This may be compared with the exact result of Eq. (4-59)

$$q = \frac{k\vartheta_0}{\sqrt{\pi\alpha\tau}}$$

The comparison in this case shows a very small difference between the results of the approximate and the exact solutions.

The semi-infinite solid, surface temperature variable with time Consider now a semi-infinite solid, with constant physical properties, as in the preceding example, but where the surface heat flux can vary arbitrarily with time. Assume that the body is initially held at a constant temperature t_i; then at some time the surface $x = 0$ is subjected to a time-varying heat flux.

If $\vartheta = t - t_i$, the governing expression is given by Eq. (4-128). The boundary conditions are

$$\begin{array}{lll} x = 0 & \tau > 0 & q = F(\tau) \\ x > 0 & \tau \leq 0 & \vartheta = 0 \end{array}$$

Assume a quadratic form for the temperature profile as before,

$$\vartheta(x,\tau) = A + Bx + Cx^2 \tag{4-136}$$

Physically, the conditions to be met by the assumed profile are

$$\begin{array}{lll} x = 0 & & \dfrac{\partial\vartheta}{\partial x} = -\dfrac{F(\tau)}{k} \\ x = \delta & \vartheta(\delta,\tau) = 0 & \dfrac{\partial\vartheta}{\partial x} = 0 \end{array}$$

Using these relationships and Eq. (4-136), one obtains a temperature profile

$$\vartheta(x,\tau) = \frac{F(\tau)\delta(\tau)}{2k}\left[1 - 2\frac{x}{\delta} + \left(\frac{x}{\delta}\right)^2\right] \tag{4-137}$$

In this case, since $\vartheta(\delta,\tau) = 0$, the integrated form of the heat-conduction equation is

$$\frac{d}{d\tau}\int_0^\delta \vartheta(x,\tau)\,dx - = \alpha\frac{\partial\vartheta}{\partial x}(0,\tau) \tag{4-138}$$

If the results in Eq. (4-137) are introduced into Eq. (4-138) and the integral evaluated, there results a differential equation of the first order

$$\frac{d}{d\tau}\left[\frac{1}{6}F(\tau)\delta^2(\tau)\right] = \alpha F(\tau) \tag{4-139}$$

which can be integrated directly to yield

$$\delta(\tau) = \left[\frac{6\alpha}{F(\tau)} \int_0^\tau F(\tau)\, d\tau\right]^{\frac{1}{2}} \tag{4-140}$$

It remains only to specify the form of $F(\tau)$ to calculate $\delta(\tau)$, the penetration depth. If, for example, the heat flux into the surface $x = 0$ is a constant q_0, then

$$\delta(\tau) = \sqrt{6\alpha\tau} \tag{4-141}$$

and using Eq. (4-137), the temperature distribution is

$$\vartheta(x,\tau) = \frac{q_0}{k}\sqrt{\frac{3\alpha\tau}{2}}\left(1 - \frac{2x}{\sqrt{6\alpha\tau}} + \frac{x^2}{6\alpha\tau}\right) \tag{4-142}$$

or, if evaluated at $x = 0$, the surface of the solid,

$$\vartheta(0,\tau) = \frac{q_0}{k}\sqrt{\tfrac{3}{2}\alpha\tau} \tag{4-143}$$

The exact solution, given by Eq. (4-93),[1] is, for $x = 0$,

$$\vartheta(0,\tau) = \frac{q_0}{k}\sqrt{\frac{4\alpha\tau}{\pi}} \tag{4-144}$$

The difference between the approximate and exact solutions is 9 percent.

If instead of a quadratic expression a cubic expression is used to approximate the temperature profile in the body, an improved solution can be obtained. A cubic profile may be

$$\vartheta(x,\tau) = \frac{F(\tau)\delta(\tau)}{3k}\left[1 - \frac{x}{\delta(\tau)}\right]^3 \tag{4-145}$$

which when used in the integral heat-conduction equation results in an ordinary differential equation

$$\frac{d}{d\tau}\left[\frac{1}{12}F(\tau)\delta^2(\tau)\right] = \alpha F(\tau) \tag{4-146}$$

This, when solved for $\delta(\tau)$ for the specification $F(\tau) = q_0$, a constant, results in the following specifications for the penetration depth:

$$\delta(\tau) = \sqrt{12\alpha\tau} \tag{4-147}$$

Equation (4-147) gives a value 40 percent greater than the results calculated for the quadratic profile. The resulting temperature distribution is

$$\vartheta(x,\tau) = \frac{q_0}{k}\sqrt{\tfrac{4}{3}\alpha\tau}\left(1 - \frac{x}{\sqrt{12\alpha\tau}}\right)^3 \tag{4-148}$$

[1] See also Carslaw and Jaeger, "Conduction of Heat in Solids," p. 75.

or, at the surface $x = 0$,

$$\vartheta(0,\tau) = \frac{q_0}{k}\sqrt{\tfrac{4}{3}\alpha\tau} \tag{4-149}$$

which is within 2 percent of the exact solution for this problem as shown by Eq. (4-144).

The semi-infinite solid with convective heat loss on the surface One can make use of the results of the last problem to account for heat flux on the surface $x = 0$ because of convective heat losses to a fluid. Let the solid be held initially at a uniform temperature t_i. The free surface $x = 0$ is swept by a fluid at a constant temperature t_f. Let $\vartheta = t - t_i$ be the excess temperature. Then the boundary condition for the problem becomes

$$\left(\frac{\partial\vartheta}{\partial x}\right)_{x=0} = -\frac{F(\tau)}{k} = \frac{h}{k}[\vartheta(0,\tau) - \vartheta_f] \tag{4-150}$$

where $\vartheta_f = t_f - t_i$ is a constant. The heat-transfer coefficient h is as defined in Sec. 1-7. The temperature profile may be taken the same as in Eq. (4-145), which for the condition $x = 0$ results in an expression for $\delta(\tau)$

$$\delta(\tau) = \frac{3k\vartheta(0,\tau)}{F(\tau)} \tag{4-151}$$

Introducing $\delta(\tau)$ from Eq. (4-151) into Eq. (4-146) yields

$$\frac{d}{d\tau}\frac{\vartheta^2(0,\tau)}{F(\tau)/k} = \frac{4}{3}\alpha\frac{F(\tau)}{k} \tag{4-152}$$

If $F(\tau)/k = f(\vartheta(0,\tau))$, it is possible to separate the variables for an analytical solution. In the case of a convective boundary condition the heat flux at the surface does depend only upon $\vartheta(0,\tau) = \vartheta_0$.

The differential equation in Eq. (4-152) therefore becomes

$$\frac{1}{f(\vartheta_0)}\,d\,\frac{\vartheta_0^2}{f(\vartheta_0)} = \frac{4}{3}\alpha\,d\tau$$

which can be expanded to

$$\frac{2\vartheta_0 f(\vartheta_0) - \vartheta_0^2 f'(\vartheta_0)}{f^3(\vartheta_0)}\,d\vartheta_0 = \tfrac{4}{3}\alpha\,d\tau$$

and has a solution for the variation of the temperature of the surface at $x = 0$ as calculated by Goodman:[1]

$$\frac{4}{3}\left(\frac{h}{k}\right)^2\alpha\tau = \frac{1}{2}\left[\frac{1}{(1 - \vartheta_0/\vartheta_f)^2} - 1\right] + \ln\left(1 - \frac{\vartheta_0}{\vartheta_f}\right) \tag{4-153}$$

[1] *Loc. cit.*

The exact solution is shown in Carslaw and Jaeger:[1]

$$\frac{\vartheta(x,\tau)}{\vartheta_f} = \operatorname{erfc}\frac{x}{2\sqrt{\alpha\tau}} - \exp\left(\frac{hx}{k} + \frac{h^2\alpha\tau}{k}\right)\operatorname{erfc}\left(\frac{x}{2\sqrt{\alpha\tau}} + \frac{h}{k}\sqrt{\alpha\tau}\right) \tag{4-154}$$

The results of Eq. (4-154) are shown graphically in Fig. 4-20.

Comparison of Eq. (4-154) evaluated for $x = 0$ and Eq. (4-153) shows no discernible difference.

The semi-infinite solid with temperature-dependent thermophysical properties One of the advantages to be gained from the use of approximate methods in heat conduction is the *relative* ease with which the dependence of the thermophysical properties on temperature can be treated. The method can be demonstrated by the application of the integral form of the heat-conduction equation to the semi-infinite solid with temperature-dependent properties.

[1] Carslaw and Jaeger, "Conduction of Heat in Solids," p. 72.

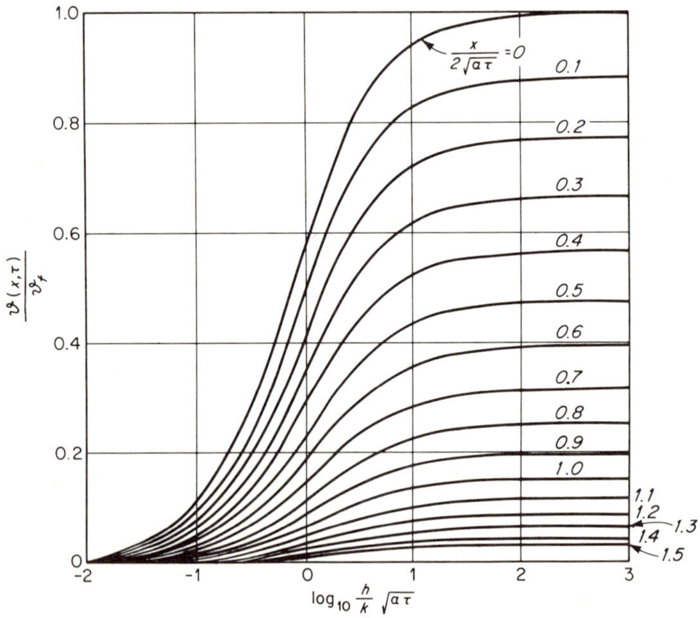

Fig. 4-20 Temperature distribution in the semi-infinite solid with heat loss by convection from the surface. [*H. S. Carslaw and J. C. Jaeger, "Conduction of Heat in Solids," 2d ed., Oxford University Press, New York, 1959.*]

Consider the semi-infinite solid whose thermal conductivity k, density ρ, and specific heat c are functions of the temperature $t(x,\tau)$. Assume further that the solid is held initially and uniformly at some temperature t_i. At the surface $x = 0$ one can specify either the heat flux or the temperature. Consider first the case in which the surface temperature is reduced to t_0 and maintained there. Specifications for the temperature distribution with time and the surface heat flux are desired from the calculations.

The appropriate equation in this case is given by Eq. (1-32), which expressed in terms of the excess temperature $\vartheta = t - t_i$ becomes

$$\rho c \frac{\partial \vartheta}{\partial \tau} = \frac{\partial}{\partial x}\left(k \frac{\partial \vartheta}{\partial x}\right) \tag{4-155}$$

Transformation in the dependent variable as suggested by Goodman[1]

$$v(x,\tau) = \int_0^{\vartheta} \rho c \, d\vartheta$$

will yield the equivalent of Eq. (1-33)

$$\frac{\partial v}{\partial \tau} = \frac{\partial}{\partial x}\left[\alpha(v) \frac{\partial v}{\partial x}\right] \tag{4-156}$$

When Eq. (4-156) is integrated with respect to x from $x = 0$ to $x = \delta(\tau)$, there results

$$\frac{d}{d\tau}\left[\int_0^{\delta} v \, dx - v_i \delta(\tau)\right] = -\alpha(v_0) \frac{\partial v}{\partial x}(0,\tau) \tag{4-157}$$

The initial and boundary conditions are

$$\begin{array}{lll} \tau \leq 0 & \vartheta(x,0) = 0 & v = 0 \\ \tau > 0 & \vartheta(0,\tau) = t_0 - t_i = \vartheta_0 & v = v_0 \end{array}$$

Assume in this case a cubic relationship for the temperature profile in the solid

$$\frac{v}{v_0} = \left(1 - \frac{x}{\delta}\right)^3 \tag{4-158}$$

Noting that $v_i = 0$, using Eq. (4-158) in Eq. (4-157), and performing the necessary operations gives for the penetration depth $\delta(\tau)$

$$\delta(\tau) = \frac{\sqrt{24}}{v_0}\left(\int_0^{\tau} \alpha_0 v_0^2 \, d\tau\right)^{\frac{1}{2}} \tag{4-159}$$

A knowledge of $v_0(\tau)$ is necessary for further evaluation. If, as in this case, a step change occurs in the surface temperature, then v_0 is a con-

[1] T. R. Goodman, The Heat Balance Integral: Further Consideration and Refinements, *J. Heat Transfer*, **83**:32 (1961).

stant, as is the thermal diffusivity at the surface α_0. In this instance the evaluation is simple:

$$\delta(\tau) = \sqrt{24\alpha_0\tau} \tag{4-160}$$

Equation (4-160) is quite like the case for constant properties except that now the thermal diffusivity α_0 must be evaluated for the temperature t_0. Suppose further that the density and specific heat vary but little with temperature such that most of the property variation occurs in the thermal conductivity. In such a case, for small ranges of temperatures α_0 can be expressed as a linear variation with the temperature

$$\alpha = \alpha_R\left(1 \pm a\frac{\vartheta}{\vartheta_0}\right) \tag{4-161}$$

where α_R is the thermal diffusivity at some reference temperature and a is the temperature coefficient for the thermal conductivity. In this instance v varies only with ϑ; thus from Eq. (4-158) the temperature distribution becomes

$$\frac{\vartheta}{\vartheta_0} = \left[1 - \frac{x}{\sqrt{24\alpha_R(1 \pm a)\tau}}\right]^3 \tag{4-162}$$

which is shown graphically in Fig. 4-21 in comparison with the exact solution of Yang.[1]

The heat flux calculated by means of Eq. (1-2) is

$$q = \vartheta_0 k_R \sqrt{\frac{3(1 \pm a)}{8\alpha_R\tau}} \tag{4-163}$$

For constant properties Eq. (4-163) readily reduces to

$$q = \vartheta_0 k \sqrt{\frac{3}{8\alpha\tau}}$$

which may be compared with the exact solution shown in Eq. (4-59). The difference is about 7 percent.

If the heat flux at the surface $x = 0$ is to be specified instead of the surface temperature, the required initial and boundary conditions are

$$x \geq 0 \qquad \tau \leq 0 \qquad \vartheta(x,0) = 0 \tag{4-164a}$$

$$x = 0 \qquad \tau > 0 \qquad k\frac{\partial\vartheta}{\partial x} = -F(\tau) \tag{4-164b}$$

[1] K. T. Yang, Transient Conduction in a Semi-infinite Solid with Variable Thermal Conductivity, *J. Appl. Mech.*, **25**:146 (1958).

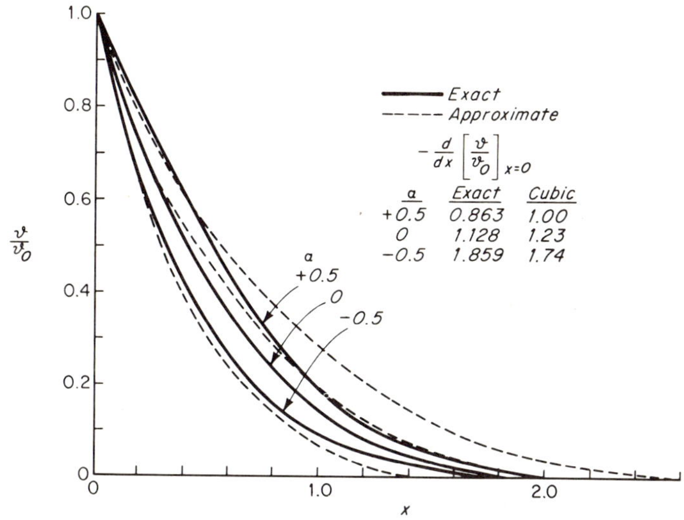

Fig. 4-21 Temperature profiles in a semi-infinite slab with linear dependence of thermal conductivity on temperature: comparison between exact and integral solution using cubic profile. [*T. R. Goodman, The Heat Balance Integral: Further Consideration and Refinements, J. Heat Transfer,* **83**:*32* (*1961*).]

such that the expression for the penetration depth becomes

$$\delta(\tau) = \left[\frac{12\alpha_0}{F(\tau)}\int_0^\tau F(\tau)\,d\tau\right]^{\frac{1}{2}} \tag{4-165}$$

If the heat flux at the surface $x = 0$ is constant, then $F(\tau) = q_0$ and the penetration depth becomes

$$\delta(\tau) = \sqrt{12\alpha_0\tau} \tag{4-166}$$

In Eq. (4-166) the thermal diffusivity α_0 is to be evaluated at the temperature of the surface $x = 0$; thus an evaluation of Eq. (4-166) requires a knowledge of the surface temperature, which is itself a function of time.

From Eqs. (4-158) and (4.164*b*) can be obtained the relationship

$$\left(\frac{\partial\vartheta}{\partial x}\right)_{x=0} = -\frac{3}{\rho c}\frac{v_0}{\delta} = -\frac{q_0}{k_0}$$

or

$$v_0 = \frac{q_0\delta}{3\alpha_0}$$

which can be combined with Eq. (4-166) to yield

$$v_0 = \frac{2}{\sqrt{3}}q_0\sqrt{\frac{\tau}{\alpha_0}} \tag{4-167}$$

If the density and specific heat can be assumed to vary only slightly with temperature, as before, the excess temperature at the surface ϑ_0 becomes

$$\sqrt{\alpha_0}\,\vartheta_0 = \frac{q_0}{\rho c}\sqrt{\tfrac{4}{3}\tau} \tag{4-168}$$

If, further, the thermal conductivity can be taken as a linear function of the temperature, such that for ρc constant the thermal diffusivity may be expressed as

$$\alpha = \alpha_R\left(1 \pm a\frac{\vartheta}{\vartheta_0}\right)$$

the excess temperature at the surface $x = 0$ becomes

$$\vartheta_0 = \frac{2}{\sqrt{3}}\frac{q_0}{k_R}\sqrt{\frac{\alpha_R\tau}{1 \pm a}} \tag{4-169}$$

Thus the temperature profile can be represented by

$$\vartheta(x,\tau) = \frac{2}{\sqrt{3}}\frac{q_0}{k_R}\sqrt{\frac{\alpha_R\tau}{1 \pm a}}\left[1 - \frac{x}{2\sqrt{3\alpha_R(1 \pm a)\tau}}\right]^3 \tag{4-170}$$

where again the subscript R indicates the reference value.

For constant properties Eq. (4-170) becomes identical with Eq. (4-148).

Semi-infinite solid with internal heat generation The integral form of the heat-conduction equation can be easily extended to the important physical case of a solid with internal heat sources. As an example consider a semi-infinite solid whose initial temperature is t_i. If $\vartheta = t - t_i$, the governing equation for constant properties is

$$\frac{\partial\vartheta}{\partial\tau} - \alpha\frac{\partial^2\vartheta}{\partial x^2} = \frac{Q'(\tau)}{\rho c} \tag{4-171}$$

with the imposed conditions

$$\begin{array}{lll} x \geq 0 & \tau < 0 & \vartheta = 0 \\ x = 0 & \tau \geq 0 & \vartheta = 0 \end{array}$$

Integrating Eq. (4-171) with respect to x from $x = 0$ to $x = \delta$, one obtains

$$\frac{d}{d\tau}\int_0^\delta \vartheta(x,\tau)\,dx - \vartheta(\delta,\tau)\frac{d\delta}{d\tau} + \alpha\frac{\partial\vartheta}{\partial x}(0,\tau) = \frac{Q'(\tau)}{\rho c}\delta \tag{4-172}$$

The second term on the left-hand side can be evaluated from Eq. (4-171) at $x > \delta$, where $\partial^2\vartheta/\partial x^2 = 0$ and where

$$\vartheta(\delta,\tau) = \int_0^\tau \frac{Q'(\tau)}{\rho c}\,d\tau \tag{4-173}$$

Thus the form of the integral heat-conduction equation becomes

$$\frac{d}{d\tau}\left[\int_0^\delta \vartheta(x,\tau)\,dx - \delta(\tau)\int_0^\tau \frac{Q'(\tau)}{\rho c}\,d\tau\right] = -\alpha\frac{\partial\vartheta}{\partial x}(0,\tau) \qquad \text{(4-174)}$$

Assumption of a cubic relationship for the temperature profile

$$\vartheta(x,\tau) = \frac{Q'(\tau)}{k}\,\alpha\tau\left[1 - \left(1 - \frac{x}{\delta}\right)^3\right] \qquad \text{(4-175)}$$

is seen by calculations to satisfy all imposed conditions. The use of Eq. (4-175) in Eq. (4-174) results finally in the equation for $\delta(\tau)$, where $Q''(\tau) = [Q'(\tau)/k]\tau$,

$$\frac{d}{d\tau}[Q''(\tau)\delta(\tau)] = \frac{12\alpha Q''(\tau)}{\delta(\tau)} \qquad \text{(4-176)}$$

which when solved for $\delta(\tau)$ with $Q'/k = \text{const}$,

$$\delta(\tau) = \sqrt{8\alpha\tau} \qquad \text{(4-177)}$$

and the temperature distribution becomes

$$\vartheta(x,\tau) = \frac{Q'}{k}\,\alpha\tau\left[1 - \left(1 - \frac{x}{\sqrt{8\alpha\tau}}\right)^3\right] \qquad \text{(4-178)}$$

The approximate solution in Eq. (4-178) may be compared with the exact solution as shown in Eq. (4-89), where correspondence of conditions is gained if the constants $A = B = 0$ in Eq. (4-89), to result in an exact expression for the heat flux

$$q = Q'\sqrt{\frac{4}{\pi}\alpha\tau} \qquad \text{(4-179)}$$

The heat flux from Eq. (4-178) becomes

$$q = Q'\sqrt{\tfrac{9}{8}\alpha\tau} \qquad \text{(4-180)}$$

The result in Eq. (4-180) underestimates the heat flux by 6 percent and thus is really quite acceptable.

Further applications of the integral heat-conduction equation will be made in Chap. 5 in heat-conduction problems with moving boundaries.

THE BIOT APPROXIMATE METHOD

The variational method for treating the heat-conduction equation permits the development of approximate solutions of some interest. The development of the necessary equations and a justification for the method were given in Chap. 1. The method will be illustrated by two examples: (1) the semi-infinite solid initially at a constant temperature upon which is imposed a constant heat flux and (2) a semi-infinite slab initially at a constant temperature after which one surface is established at a higher temperature and the other thermally insulated.

The semi-infinite solid with constant heat flux at the surface Consider again the semi-infinite solid initially at a constant temperature t_i. At some time zero there is imposed upon the surface $x = 0$ a constant heat flux q_0. The heat-conduction equation was shown in Chap. 1 to have an equivalent in Eq. (1-53), which is to be used in the solution of the posed problem. In this case only one of the generalized coordinates q_i need be used, and that one is selected to be the penetration depth δ.

If $\vartheta = t - t_i$, to proceed formally requires calculation of the values

$$E = \frac{1}{2}\int_V \rho c\vartheta^2\, dV \tag{4-181a}$$

$$D = \frac{1}{2}\int_V \frac{1}{k}|\dot{\mathbf{H}}|^2\, dV \tag{4-181b}$$

$$Q_i = \int_S \vartheta\mathbf{n}\cdot\frac{\partial\mathbf{H}}{\partial\delta}\, dS \tag{4-181c}$$

and use of these results in Eq. (1-53).

Assume a temperature profile of the form

$$\vartheta = \vartheta_0\left(1 - \frac{x}{\delta}\right)^2$$

where ϑ_0 is the excess temperature at the surface $x = 0$ and for the constant heat-flux condition must vary with the time. The heat flux into the body in an interval of time must equal the storage of thermal energy in the body during that interval of time; thus,

$$\int_0^\tau q_0\, d\tau = \rho c\vartheta_0\int_0^\delta\left(1 - \frac{x}{\delta}\right)^2 dx$$

from the solution of which is obtained a specification for the surface-temperature variation with time, or

$$\vartheta_0 = \frac{3q_0\tau}{\rho c\delta}$$

Thus the temperature distribution becomes

$$\vartheta(x,\tau) = \frac{3q_0\tau}{\rho c\delta}\left(1 - \frac{x}{\delta}\right)^2 \tag{4-182}$$

The conservation-of-energy expression is

$$-\rho c\vartheta = \frac{\partial\mathbf{H}}{\partial x}$$

from which $\mathbf{H}$ can be calculated with the use of Eq. (4-182) to be

$$\mathbf{H} = q_0\tau\left(1 - \frac{x}{\delta}\right)^3 \tag{4-183}$$

$$\dot{\mathbf{H}} = \frac{\partial \mathbf{H}}{\partial \tau} = q_0\left(1 - \frac{x}{\delta}\right)^3 + 3q_0\tau x\left(1 - \frac{x}{\delta}\right)^2 \frac{\dot{\delta}}{\delta^2} \tag{4-184}$$

where $\dot{\delta} = d\delta/d\tau$. Using Eq. (4-182) in (4-181a), E becomes

$$E = \frac{9q_0{}^2\tau^2}{10\rho c\delta}$$

and

$$\frac{\partial E}{\partial \delta} = -\frac{9q_0{}^2\tau^2}{10\rho c\delta^2} \tag{4-185}$$

Using Eq. (4-184) in Eq. (4-181b),

$$D = \frac{1}{2k}\left[\frac{q_0{}^2}{7}(\delta + \tau\dot{\delta}) + \tfrac{3}{35}\, q_0{}^2\tau^2\frac{\dot{\delta}^2}{\delta}\right]$$

and

$$\frac{\partial D}{\partial \dot{\delta}} = \frac{q_0{}^2\tau}{k}\left(\frac{3}{42} + \frac{3}{35}\frac{\dot{\delta}}{\delta}\tau\right) \tag{4-186}$$

The value $Q_i = 0$ in this case, and so substituting the results of Eqs. (4-185) and (4-186) into Eq. (1-53) gives

$$\tfrac{3}{42}\delta^2 + \tfrac{3}{35}\delta\dot{\delta}\tau = \tfrac{9}{10}\alpha\tau \tag{4-187}$$

which is to be solved for $\delta(\tau)$, or

$$\delta(\tau) = 2.81\sqrt{\alpha\tau} \tag{4-188}$$

Thus, the temperature distribution in the body becomes

$$\vartheta(x,\tau) = 1.065\,\frac{q_0}{k}\sqrt{\alpha\tau}\left(1 - \frac{x}{2.81\sqrt{\alpha\tau}}\right)^2 \tag{4-189}$$

and for $x = 0$ the temperature at the surface is

$$\vartheta(0,\tau) = 1.065\,\frac{q_0}{k}\sqrt{\alpha\tau} \tag{4-190}$$

Comparisons with the exact result Eq. (4-93) indicate an error of about 5.5 percent.[1]

The infinite slab As another example consider the infinite slab initially at a constant temperature t_i. At time zero the surface $x = 0$ is changed

[1] T. J. Lardner, Biot's Variational Principle in Heat Conduction, *AIAA J.*, **1**:1 (1963), presents a solution to this same problem when the heat flux at the surface cannot be satisfied identically by the selected temperature profile.

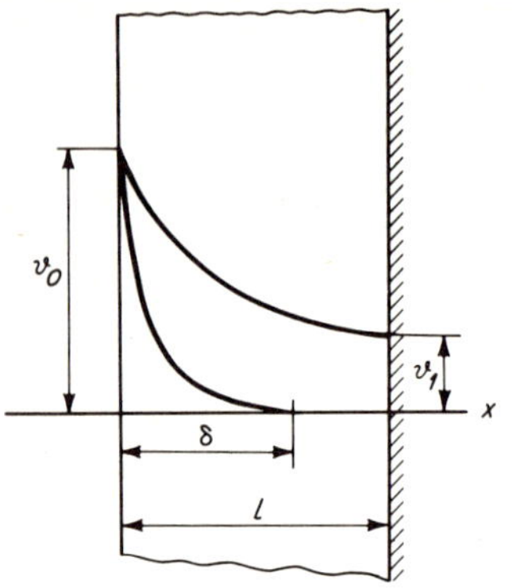

Fig. 4-22 Temperature profiles in the infinite slab.

to a temperature t_0 and maintained there while the surface $x = l$ is insulated. The thermophysical properties of the material are constant. Here the problem must be treated in two steps: (1) to cover the time interval necessary for the temperature at the face $x = l$ to change the *transit time* and (2) to cover the period for times greater than the transit time. The system and variables are shown schematically in Fig. 4-22. Assume a parabolic temperature profile where $\vartheta = t - t_i$ and $\vartheta_0 = t_0 - t_i$:

$$\vartheta(x,\tau) = \begin{cases} \vartheta_0\left(1 - \dfrac{x}{\delta}\right)^2 & \text{for } x < \delta \\ 0 & \text{for } x > \delta \end{cases} \tag{4-191}$$

From Eqs. (4-191) and (1-35)

$$\mathbf{H} = \tfrac{1}{3}\rho c\vartheta_0\delta\left(1 - \frac{x}{\delta}\right)^3$$

$$\frac{\partial \mathbf{H}}{\partial \tau} = \dot{\mathbf{H}} = \tfrac{1}{3}\rho c\vartheta_0\dot{\delta}\left(1 - \frac{x}{\delta}\right)^3 + \rho c\vartheta_0 x\left(1 - \frac{x}{\delta}\right)^2\frac{\dot{\delta}}{\delta}$$

$$E = \tfrac{1}{10}\rho c\vartheta_0{}^2\delta$$

$$D = \frac{13}{630}\frac{\rho^2c^2\vartheta_0{}^2}{k}\dot{\delta}^2\delta$$

$$Q_i = \tfrac{1}{3}\rho c\vartheta_0{}^2$$

from which can be calculated

$$\frac{\partial E}{\partial \delta} = \tfrac{1}{10}\rho c\vartheta_0{}^2$$

$$\frac{\partial D}{\partial \dot{\delta}} = \frac{13}{315}\frac{\rho^2c^2}{k}\vartheta_0{}^2\dot{\delta}\delta$$

The last three values when substituted into Eq. (1-53),

$$\frac{\partial E}{\partial \delta} + \frac{\partial D}{\partial \dot{\delta}} = Q_i$$

yield after clearing the equation

$$\frac{d\delta^2}{d\tau} = \frac{147}{13}\alpha$$

or

$$\delta(\tau) = 3.36\sqrt{\alpha\tau} \tag{4-192}$$

The *transit time* τ_l is defined as the time taken for ϑ at $x = l$ to change from zero; thus for $\delta = l$

$$\tau_l = \frac{13}{147}\frac{l^2}{\alpha} = 0.0885\frac{l^2}{\alpha}$$

The *transit-time Fourier number* is

$$\mathrm{Fo}_l = \frac{\alpha\tau_l}{l^2} = 0.0885$$

The temperature distribution for $\alpha\tau/l^2 < 0.0885$ is thus

$$\vartheta(x,\tau) = \vartheta_0\left(1 - \frac{x}{3.36\sqrt{\alpha\tau}}\right)^2 \tag{4-193}$$

To continue the calculation beyond $\tau = \tau_l$, a change in the temperature profile is needed to reflect the change of ϑ_l with time at the insulated surface $x = l$. Such a temperature profile may be

$$\vartheta(x,\tau) = [\vartheta_0 - \vartheta_l(\tau)]\left(1 - \frac{x}{l}\right)^2 + \vartheta_l(\tau) \tag{4-194}$$

In this case the generalized variable becomes $\vartheta_l(\tau)$ since the penetration depth $\delta(\tau)$ is no longer meaningful.

Again,

$$\mathbf{H} = \tfrac{1}{3}\rho cl(\vartheta_0 - \vartheta_l)\left(1 - \frac{x}{l}\right)^3 + \rho c\vartheta_l l\left(1 - \frac{x}{l}\right)$$

$$\dot{\mathbf{H}} = -\tfrac{1}{3}\rho cl\left(1 - \frac{x}{l}\right)^3\dot{\vartheta}_l + \rho cl\left(1 - \frac{x}{l}\right)\dot{\vartheta}_l$$

$$D = \frac{34}{315}\frac{\rho^2c^2l^3\dot{\vartheta}_l^2}{k}$$

$$\frac{\partial D}{\partial \dot{\vartheta}_l} = \frac{68}{315}\frac{\rho^2c^2l^3}{k}\dot{\vartheta}_l \tag{4-195}$$

$$E = \tfrac{1}{2}\rho c[\tfrac{1}{5}(\vartheta_0 - \vartheta_l)^2 l + \tfrac{2}{3}\vartheta_l(\vartheta_0 - \vartheta_l)l + \vartheta_l^2 l]$$

$$\frac{\partial E}{\partial \vartheta_l} = \frac{2}{15}\rho cl\vartheta_0 + \frac{8}{15}\rho cl\vartheta_l \tag{4-196}$$

$$Q_i = \tfrac{2}{3}\rho cl\vartheta_0 \tag{4-197}$$

Substituting Eqs. (4-195) to (4-197) in the variational statement [Eq. (1-53)], one obtains the differential equation

$$\dot{\vartheta}_l + \frac{42}{17}\frac{\alpha}{l^2}\vartheta_l = \frac{42}{17}\frac{\alpha}{l^2}\vartheta_0 \tag{4-198}$$

If the transit time $\tau_l = \frac{13}{147}(l^2/\alpha)$ is used in Eq. (4-198), one obtains

$$\dot{\vartheta}_l + \frac{1}{4.57\tau_l}\vartheta_l = \frac{1}{4.57\tau_l}\vartheta_0$$

which upon solution yields

$$\frac{\vartheta_l(\tau)}{\vartheta_0} = 1 - \exp\left[-0.219\left(\frac{\tau}{\tau_l} - 1\right)\right] \tag{4-199}$$

The temperature distribution for $\alpha\tau/l^2 > 0.0885$ becomes

$$\frac{\vartheta(x,\tau)}{\vartheta_0} = 1 - \left[1 - \left(1 - \frac{x}{l}\right)^2\right]\exp\left[-0.219\left(\frac{\tau}{\tau_l} - 1\right)\right] \tag{4-200}$$

The comparison of the approximate results in Eq. (4-200) with the exact solution is shown in Fig. 4-23. The results are satisfactory.

The heat flux at the surface $x = 0$ can readily be calculated to be

$$q = \frac{2k\vartheta_0}{l}\exp\left[-0.219\left(\frac{\tau}{\tau_l} - 1\right)\right] \tag{4-201}$$

THE VUJANOVIC AND STRAUSS METHOD[1]

Consider the case of the one-space-dimensional transient-heat-conduction problem, the semi-infinite body whose initial temperature is zero with an

[1] B. Vujanovic and A. M. Strauss, "Heat Transfer with Nonlinear Boundary Conditions via a Variational Principle," *AIAA J.*, **9**:(2), 327–330 (1971).

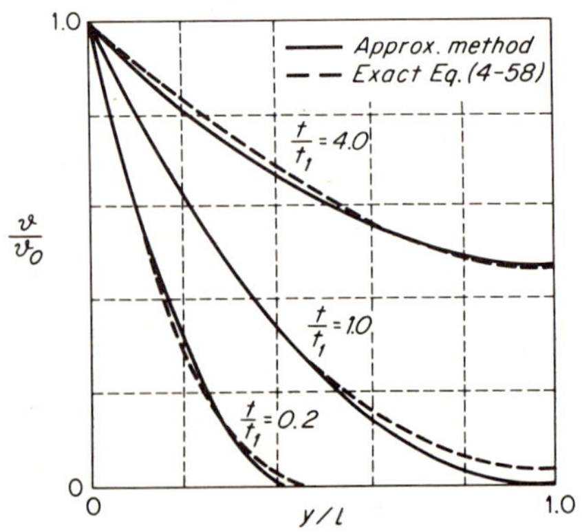

Fig. 4-23 Temperature distribution in the infinite slab; comparison of approximate with exact solutions [Eq. (4-58)]. [*M. A. Biot, New Methods in Heat Flow Analysis with Applications to Flight Structures, J. Aeron. Sci.*, **24**:*12* (*1957*).]

arbitrary heat-flux input at the surface $x = 0$. The boundary condition necessary for use in Eq. (1-65) can be obtained in the form required by Eq. (1-66) from a heat balance at the surface $x = 0$

$$k\left(\frac{\partial\vartheta}{\partial x}\right)_{x=0} + F(\vartheta_0,\tau) = 0 \tag{4-202}$$

where $F(\vartheta_0,\tau)$ = arbitrary heat flux at surface $x = 0$
$\vartheta_0(\tau)$ = excess temperature at surface
$\vartheta = \vartheta(x,\tau)$ = excess temperature over initial temperature

Assume a cubic profile for the temperature distribution

$$\vartheta(x,\tau) = \begin{cases} \vartheta_0\left(1 - \dfrac{x}{\delta}\right)^3 & \text{for } 0 \le x \le \delta \\ 0 & \text{for } x > \delta \end{cases} \tag{4-203}$$

where $\delta = \delta(\tau)$ the penetration distance. Both $\vartheta_0(\tau)$ and $\delta(\tau)$ are unknown functions of time to be determined.

Combining the action integral [Eq. (1-64)] with the lagrangian [Eq. (1-62)], one obtains

$$I = \rho c \int_{\tau_1}^{\tau_2} \int_0^{\delta} \left[\frac{\tau_0}{2}\left(\frac{\partial t}{\partial \tau}\right)^2 - \frac{\alpha}{2}\left(\frac{\partial t}{\partial x}\right)^2\right] e^{\tau/\tau_0}\, dx\, d\tau \tag{4-204}$$

In Eq. (4-204) the time interval $\tau_2 - \tau_1$ is arbitrarily chosen.

The temperature profile [Eq. (4-203)], when substituted into Eq. (4-204) and the integration with respect to x performed, yields

$$I = \rho c \int_{\tau_1}^{\tau_2} \left[\frac{\tau_0}{2}\left(\frac{1}{7}\,\delta\dot{\vartheta}_0^2 + \tfrac{1}{7}\vartheta_0\dot{\vartheta}_0\dot{\delta} + \frac{9}{105}\,\frac{\vartheta_0^2\dot{\delta}^2}{\delta}\right) - \tfrac{9}{10}\alpha\,\frac{\vartheta_0^2}{\delta}\right] e^{\tau/\tau_0}\, d\tau \tag{4-205}$$

or

$$I = \int_{\tau_1}^{\tau_2} L(\vartheta_0,\delta,\dot{\vartheta}_0,\dot{\delta},\tau)\, d\tau \tag{4-206}$$

If the action integral [Eq. (4-205)] possesses a stationary value relative to the penetration distance δ, which is an independent variable, the Euler-Lagrange equation

$$\frac{d}{d\tau}\frac{\partial L}{\partial \dot{\delta}} - \frac{\partial L}{\partial \delta} = 0 \tag{4-207}$$

must be satisfied. Substituting the lagrangian determined from the integration of Eq. (4-205) into Eq. (4-207) yields

$$\frac{\tau_0}{2}\frac{d}{d\tau}\left(\frac{\vartheta_0\dot{\vartheta}_0}{7} + \frac{18}{105}\,\frac{\vartheta_0^2\dot{\delta}}{\delta}\right) - \frac{\tau}{2}\left(\frac{\dot{\vartheta}_0^2}{7} - \frac{9}{105}\,\frac{\vartheta_0^2\dot{\delta}^2}{\delta^2}\right) + \frac{1}{2}\left(\frac{\vartheta_0\dot{\vartheta}_0}{7} + \frac{18}{105}\,\frac{\vartheta_0^2\dot{\delta}}{\delta}\right) - \frac{9}{10}\,\frac{\alpha\vartheta_0^2}{\delta^2} = 0 \tag{4-208}$$

For the classical solution let $\tau_0 = 0$ and obtain

$$\tfrac{1}{7}\vartheta_0\dot{\vartheta}_0 + \frac{6}{35}\frac{\vartheta_0{}^2\dot{\delta}}{\delta} = \frac{9}{5}\frac{\alpha\vartheta_0{}^2}{\delta^2} \tag{4-209}$$

From the boundary condition [Eq. (4-202)]

$$\delta = \frac{3k\vartheta_0}{F(\vartheta_0,\tau)} \tag{4-210}$$

which when combined with Eq. (4-209) gives the first-order differential equation

$$11\vartheta_0\dot{\vartheta}_0 - 6\frac{\vartheta_0{}^2}{F}\left(\dot{F}'\dot{\vartheta}_0 + \frac{\partial F}{\partial\tau}\right) = 7\frac{\alpha F^2}{k^2} \tag{4-211}$$

where $F' = \partial F/\partial\vartheta_0$. The dot as usual represents differentiation with respect to time. The initial condition for the problem is

$$\vartheta_0 = 0 \qquad \tau = 0$$

There are three cases in which Eq. (4-211) can be integrated analytically: (1) $F = \text{const}$; (2) $F = F(\tau)$; and (3) $F = F(\vartheta_0)$.

For the first case, $F/k = q_0/k$, and Eq. (4-211) becomes

$$11\vartheta_0\dot{\vartheta}_0 = 7\frac{\alpha q_0}{k} \tag{4-212}$$

or

$$\vartheta_0 = \frac{q_0}{k}\sqrt{\tfrac{14}{11}\alpha\tau} = 1.128\frac{q_0}{k}\sqrt{\alpha\tau} \tag{4-213}$$

Comparing Eq. (4-213) with Eqs. (4-149) and (4-144) obtained by use of the integral method and the exact solution

$$\vartheta_0 = \frac{q_0}{k}\sqrt{\tfrac{4}{3}\alpha\tau} = 1.15\frac{q_0}{k}\sqrt{\alpha\tau} \qquad \text{integral method}$$

$$\vartheta_0 = \frac{q_0}{k}\sqrt{\frac{4}{\pi}\alpha\tau} = 1.128\frac{q_0}{k}\sqrt{\alpha\tau} \qquad \text{exact}$$

shows that the method results in an improvement.

The value for $\delta(\tau)$ can be calculated by combining Eqs. (4-210) and (4-213) to obtain

$$\delta(\tau) = 3.384\sqrt{\alpha\tau} \tag{4-214}$$

and the temperature distributions can be calculated from Eqs. (4-203), (4-210), and (4-214) as

$$\vartheta(x,\tau) = 1.128\frac{q_0}{k}\sqrt{\alpha\tau}\left(1 - \frac{x}{3.384\sqrt{\alpha\tau}}\right)^3 \tag{4-215}$$

In the second case, that is, $F = F(\tau)$, Eq. (4-211) becomes

$$\frac{d}{d\tau}(\vartheta_0{}^2) - \frac{12}{11}(\vartheta_0)^2\frac{\dot{F}}{F} = \frac{14}{11}\frac{\alpha F^2}{k^2} \tag{4-216}$$

Assume $F = A\tau^n$, where $n = 0, 1, 2, \ldots$; thus the equation becomes

$$\frac{d(\vartheta_0)^2}{d\tau} - \frac{12}{11}\frac{n}{\tau}(\vartheta_0)^2 = \frac{14}{11}\frac{\alpha A^2\tau^{2n}}{k^2} \tag{4-217}$$

The solution to Eq. (4-217), making use of the fact that for $n = 0$ and $\tau = 0$, $\vartheta_0 = 0$, is

$$\vartheta_0 = \sqrt{\frac{\alpha A^2}{k^2}\left(\frac{14}{10n + 11}\right)}\,\tau^{n+\frac{1}{2}} \tag{4-218}$$

which for $n = 0$ reduces to the result shown in Eq. (4-213), where $q_0 = \sqrt{A}$. Equation (4-218) compares favorably with the results of the integral method of Goodman, but for $n > 0$ is larger than the exact solution by 20 to 25 percent.

In the third case, a semi-infinite body radiating thermally to an environment at a zero absolute temperature, $\partial F/\partial\tau$ in Eq. (4-211) becomes zero; that is, $F = F(\vartheta_0)$. Thus,

$$11\vartheta_0\dot{\vartheta}_0 - 6\frac{F'}{F}\vartheta_0{}^2\dot{\vartheta}_0 = 7\alpha\frac{F^2}{k^2} \tag{4-219}$$

Upon separation of variables this becomes

$$\int_0^{\vartheta_0}\frac{11F\vartheta_0 - 6F'\vartheta_0{}^2}{F^3}\,d\vartheta_0 = \int_0^{\tau}\frac{7\alpha\,d\tau}{k^2} \tag{4-220}$$

If F is taken as the radiant heat flux,

$$F(\vartheta_0) = \sigma(\vartheta_0 + T_0)^4$$

where σ is the Stefan-Boltzmann constant and T_0 is the reference to absolute temperature, the solution to Eq. (4-220) is

$$\frac{7\alpha\sigma^2T_0{}^6}{k^2}\tau = \frac{1}{\eta^8}\left(3 - \tfrac{37}{7}\eta + \tfrac{13}{6}\eta^2\right) + \tfrac{5}{42} \tag{4-221}$$

where
$$\eta = \frac{\vartheta_0}{T_0} + 1$$

Equation (4-221) is in agreement with the exact solution for this problem.

THE SCHMIDT GRAPHICAL METHOD

In many cases a quicker solution to the semi-infinite body and the slab can be obtained by the graphical approximation method proposed by

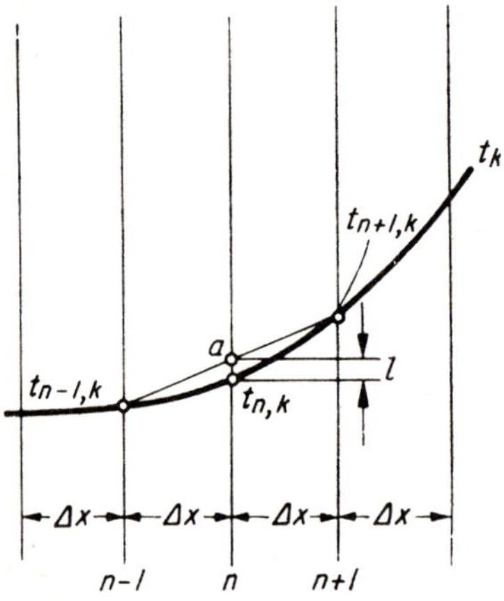

Fig. 4-24 Unsteady heat flow for an interior point in a plane wall determined by the Schmidt method.

Schmidt.[1] The heat-conduction equation (4-10) can be changed into an equation of finite differences by splitting the body depth (or thickness) into intervals Δx and the time into intervals $\Delta\tau$ and considering the temperature changes in these intervals (Fig. 4-24).[2] Equation (4-10) can be written as a difference equation:

$$\frac{\Delta_\tau t}{\Delta\tau} = \alpha\frac{\Delta_x{}^2 t}{(\Delta x)^2} \tag{4-222}$$

The subscripts indicate that the time τ or the location x is variable with the differential formation Δt. The length intervals and the time intervals can be numbered continuously. Thus the designation n (Fig. 4-24) refers to any arbitrary place in the plate, and the number k indicates any arbitrary instant in the time scale.

Hence, we can write the value $\Delta_\tau t$ in the form

$$\Delta_\tau t = t_{n,k+1} - t_{n,k}$$

and also the value

$$\Delta_x t = t_{n+1,k} - t_{n,k}$$

The expression $\Delta_x{}^2 t$ is the difference of two successive differences; hence we obtain

$$\Delta_x{}^2 t = (t_{n+1,k} - t_{n,k}) - (t_{n,k} - t_{n-1,k}) = t_{n+1,k} - 2t_{n,k} + t_{n-1,k}$$

[1] E. Schmidt, in "Beitrage zur technischen Mechanik und technischen Physik" (Föppl-Festschrift), Springer-Verlag, Berlin, 1924; also E. Schmidt, "Einfuhrung in die technische Thermodynamik," p. 262, Springer-Verlag, Berlin, 1936 (English translation by J. Kestin, Oxford University Press, New York).

[2] The graphical method of Schmidt is used in modern practice only for rapid, rough approximation; however, because of its relations to numerical methods considered later, a discussion here is worthwhile.

Therewith, the difference equation (4-222) becomes

$$t_{n,k+1} - t_{n,k} = \alpha \frac{\Delta\tau}{(\Delta x)^2} (t_{n+1,k} - 2t_{n,k} + t_{n-1,k}) \tag{4-223}$$

From Eq. (4-223) the temperature throughout the wall at the time $k + 1$ can be computed if the temperatures are known at the time k. By continued application of the equation, the development of the temperature field with time can be determined from a known initial temperature distribution. In lieu of the numerical calculations, the graphical solution illustrated in Fig. 4-24 can be used. The temperature field t_k, for the time k, is plotted in this figure. Now we connect the temperature points, which are two Δx intervals apart, by a straight line; e.g., in Fig. 4-24 the point $t_{n-1,k}$ on the vertical line $n - 1$ with the point $t_{n+1,k}$ on the ordinate $n + 1$. In this manner we arrive at the point of intersection a. The distance designated has the value

$$l = \frac{t_{n-1,k} + t_{n+1,k}}{2} - t_{n,k} = \tfrac{1}{2}(t_{n+1,k} - 2t_{n,k} + t_{n-1,k}) \tag{4-224}$$

Equation (4-224) agrees closely with the temperature difference defined by the right side of Eq. (4-223), the only difference being the occurrence of the coefficient $\frac{1}{2}$ in one case and the coefficient $\alpha\,\Delta\tau/\Delta x^2$ in the other. However, we still have at our disposal the time increment $\Delta\tau$, which can always be arranged with an arbitrarily chosen space increment Δx such that the condition $(\alpha\,\Delta\tau/\Delta x)^2 = \frac{1}{2}$ is fulfilled. By the procedure shown in Fig. 4-24 we have developed in point a a point in the temperature field for the time $k + 1$ as given by the expression

$$\Delta\tau = \frac{(\Delta x)^2}{2\alpha}$$

which expresses the time increment later than the time k. In the same manner by plotting straight lines, other points of the temperature field for the time trace $k + 1$ can be obtained, and thus the entire temperature field can be determined solely with a straightedge. In order to use this graphical solution, the temperature field in the wall must be known at some specific time. If, in addition, the change of the surface temperature with time is known, the temperature field in the wall can be plotted as shown in Fig. 4-25. Here the temperature t_0 for the time $\tau = 0$ and the temperatures of the surface (points 0, 1, 2, etc.) for successive time intervals are given.

Frequently, however, only the temperature t_f of the fluid adjacent to the wall and the heat-transfer coefficient on the surface of the wall are known. For such a case, the subdivision into layers is more suitably

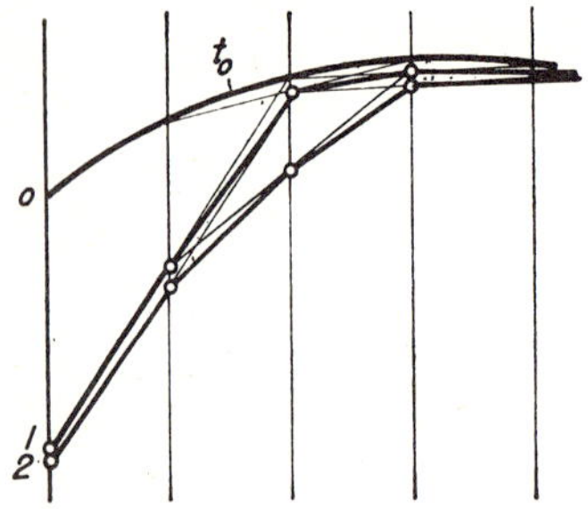

Fig. 4-25 Unsteady heat flow near the surface of a plane wall determined by the Schmidt method.

accomplished in the manner of Fig. 4-26. In this case the temperature gradient on the surface for each instant is defined by the surface condition

$$-k\frac{dt}{dx} = h(t - t_f)$$

Graphically, this condition means that the tangent to the temperature curve at the surface must pass through a reference point whose distance from the wall is k/h and whose ordinate is the fluid temperature t_f. This relationship has already been pointed out. It can be used here in the following manner: We extend the given temperature distributions 1, 2, 3, . . . at the beginning of the equalization process (Fig. 4-26) by the straight line (r,a) past the surface of the wall and start the process by connecting point 0 with 2, 1 with 3, etc., by straight lines. Thus a new temperature distribution 1′, 2′, 3′, . . . is obtained. This line can be extended again by connection with the guide point r, thus yielding the point 0′. The process is repeated. If during the course of time the sur-

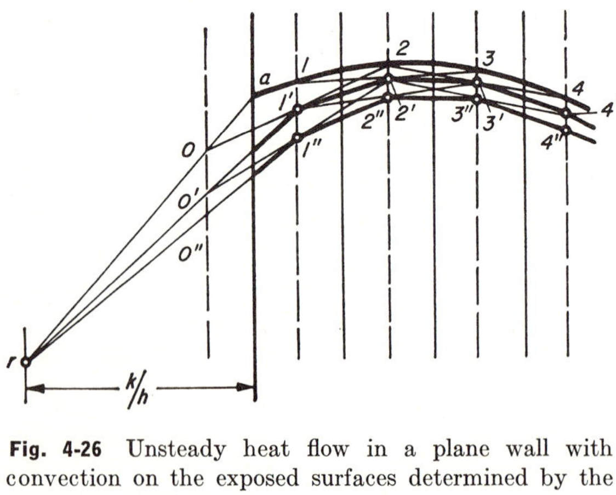

Fig. 4-26 Unsteady heat flow in a plane wall with convection on the exposed surfaces determined by the Schmidt method.

rounding temperature t_f or the heat-transfer coefficient h changes, such changes can easily be incorporated in the process by appropriate displacement of the guide point r in either the vertical or horizontal direction, respectively. This ease in handling the variable surface conditions is a marked advantage of the graphical procedure as compared with an analytical treatment of the differential equation, where a variable boundary condition leads to considerable mathematical difficulty. The Schmidt method has been extended by Nessi and Nissolle[1] to other body forms and to sectionally homogeneous systems.

Example 4-4 A living room with brick walls, which initially has an outside temperature of 0 C, is heated by some unit. It is to be determined how long it takes to establish a steady-state temperature distribution in the walls and in the room. The final temperature in the room is to be 20 C. The heat-transfer coefficient on the inside of the wall is $h_i = 6.8$ W/m² K, the heat-transfer coefficient on the outside $h_0 = 17.0$ W/m² K, the thermal diffusivity $\alpha = 0.00112$ m²/hr, and the thickness of the wall 0.4 m. The steady-state temperature in the wall is given by a straight line connecting the two points a and b, which are separated from the two wall surfaces by k/h_i and k/h_0, respectively, and have respective temperatures of 0 and 20 C. If we assume that the rate of heat transfer to the inside of the wall is constant, all temperature gradients in the wall during the heating process must have the same inclination to the inner surface. To start with the graphical method, the wall is divided into six layers, each with the thickness $\Delta x = 0.0665$ m. Then the time interval is given by

$$\Delta\tau = \frac{(\Delta x)^2}{2\alpha} = \frac{0.0665^2}{2(0.00112)} = 1.97 \text{ hr}$$

The construction of the temperature curves can be easily followed in Fig. 4-27. It can be seen that more than 80 hr, or 4 days, is needed before steady-state conditions are reached. After the time 9.85 hr, the length interval was made $2\Delta x$; accordingly, the time interval now becomes 4 (1.97) = 7.89 hr.

4-3 PERIODIC HEAT CONDUCTION

In many cases of practical interest, heat conduction occurs under conditions where the temperature (or heat-flux) boundary conditions change with time in a periodic manner. Examples of this phenomenon are evident in the cylinders of steam and internal-combustion engines, industrial processes where thermal cycling of the system is desirable from a controls point of view, and in many other instances. Such periodically transient systems can be treated by the Schmidt graphical method just described, but the method is apt to become graphically difficult. However, some analytical treatment is also possible, and by this means one

[1] A. Nessi and L. Nissolle, "Méthodes graphiques pour l'étude des installations de chauffrage et de réfrigération en régime discontinu," Dunod, Paris, 1929; see also M. Jakob, "Heat Transfer," pp. 380–398, Wiley, New York, 1949.

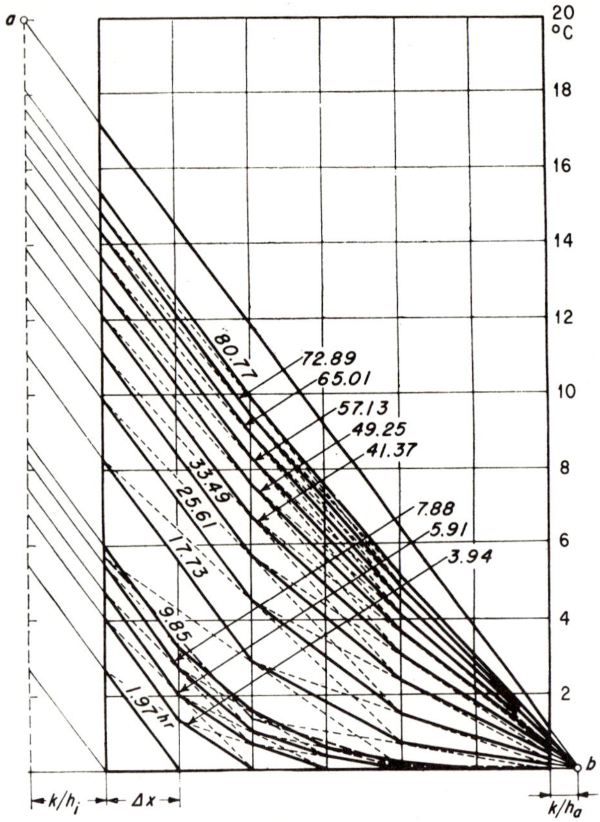

Fig. 4-27 Unsteady heat flow into a brick wall.

can examine the nature of periodic heat conduction to an extent not possible graphically.

The solid of infinite thermal conductivity, surroundings periodic with time Consider the same little billet discussed in Sec. 4-1 but now submerged in a fluid whose temperature is changing in a periodic manner. Any such periodic cycle can be represented by using a sufficient number of terms in a Fourier series;[1] therefore, we can write the following expression for the fluid-temperature variations:

$$t_f = \frac{a_0}{2} + \sum_{n=1}^{\infty} \left(a_n \cos \frac{2\pi n\tau}{\tau_0} + b_n \sin \frac{2\pi n\tau}{\tau_0} \right) \tag{4-225}$$

[1] R. V. Churchill, "Fourier Series and Boundary Value Problems," McGraw-Hill, New York, 1941.

where the a_n and b_n are Fourier coefficients and τ_0 is the period of the temperature oscillation. When Eq. (4-225) is introduced into Eq. (4-1),

$$\frac{dt}{d\tau} + \frac{hA}{\rho cV} t = \frac{hA}{\rho cV} \left[\frac{a_0}{2} + \sum_{n=1}^{\infty} \left(a_n \cos \frac{2\pi n\tau}{\tau_0} + b_n \sin \frac{2\pi n\tau}{\tau_0} \right) \right] \tag{4-226}$$

which can be integrated[1] to yield

$$t = e^{-hA\tau/\rho cV} \left\{ \int e^{hA/\rho cV} \frac{hA}{\rho cV} \left[\frac{a_0}{2} + \sum_{n=1}^{\infty} \left(a_n \cos \frac{2\pi n\tau}{\tau_0} + b_n \sin \frac{2\pi n\tau}{\tau_0} \right) \right] d\tau + C_1 \right\}$$

or when the operations are completed,[2]

$$t = \frac{a_0}{2} + \sum_{n=1}^{\infty} \frac{hA}{\rho cV} \frac{a_n \cos [(2\pi n\tau/\tau_0) - \delta] + b_n \sin [(2\pi n\tau/\tau_0) - \delta]}{\sqrt{(hA/\rho cV)^2 + (2\pi n/\tau_0)^2}} + C_1 e^{-hA\tau/\rho cV} \tag{4-227}$$

where

$$\delta = \tan^{-1} \frac{2\pi n}{\tau_0} \frac{\rho cV}{hA}$$

The constant C_1 can be determined from the specification of the temperature at time zero; however, at long times the transient term goes to zero, and the temperature oscillation becomes regular. The temperature-time history of the billet after the transient period can be expressed as

$$t = \frac{a_0}{2} + \sum_{n=1}^{\infty} \frac{a_n \cos [(2\pi n\tau/\tau_0) - \delta] + b_n \sin [(2\pi n\tau/\tau_0) - \delta]}{\sqrt{1 + \tan^2 \delta}} \tag{4-228}$$

The conclusions to be drawn from Eq. (4-228) are that the temperature of the billet always lags behind the fluid temperature by the phase angle δ and the amplitude of the temperature variation is diminished by the factor

$$\frac{1}{\sqrt{1 + \tan^2 \delta}}$$

Example 4-5 A resistance thermometer is used to measure the gas temperature in the cylinder of a machine running at 120 rpm. The output of the resistance thermometer is driving an oscilloscope. We want to estimate the probable error in the measurement. Assuming the temperature oscillation to be sinusoidal, we can use the results of Eq. (4-228) with $n = 1$. The resistance bulb is roughly cylindrical, made of platinum, and 0.050 cm in diameter. The thermal

[1] L. R. Ford, "Differential Equations," 2d ed., McGraw-Hill, New York, 1955.
[2] B. O. Pierce, "A Short Table of Integrals," 3d ed., Ginn, Boston, 1929.

conductivity of platinum is $k = 69.2$ W/m K, and the thermal diffusivity is $\alpha = 0.0871$ m²/hr. The heat-transfer coefficient in the cylinder may be 284 W/m² K. The significant dimension of the resistance bulb is $L = 0.050/4 = 0.0125$ cm. The machine makes $120(60) = 7{,}200$ cycles/hr; thus

$$\tan \delta = \frac{2\pi}{\tau_0}\frac{\rho c V}{hA} = \frac{2\pi}{\tau_0}\frac{\rho c}{k}\frac{Lk}{h} = \frac{2\pi(7{,}200)(0.0125)(69.2)}{0.0871(100)(284)} = 15.8$$

$$\tan^2 \delta = 250$$

$$\text{Amplitude ratio} = \frac{1}{\sqrt{1 + 250}} = 0.063$$

$$\delta = \tan^{-1} 15.8 = 86.3°$$

Therefore there is a 94 percent temperature-reading error and a lag in the reading of 86.3° or almost $\frac{1}{4}$ rev.

Semi-infinite solid, surface temperature periodic with time Consider again the semi-infinite solid extending from the surface at $x = 0$ to infinity, where the temperature of the exposed surface at $x = 0$ is varying periodically with time. The differential equation for the system is

$$\frac{\partial t}{\partial \tau} = \alpha \frac{\partial^2 t}{\partial x^2}$$

and the initial and boundary conditions to be satisfied are

$$\begin{array}{ll} \tau = 0 & t = 0 \\ x = 0 & t_0 = f(\tau) \\ x \rightarrow \infty & t_\infty \neq \infty \end{array}$$

A solution may be assumed to be of the form

$$t = F(\tau)G(x) \tag{4-229}$$

Since the temperature is to be periodic, it is necessary that the time τ and the space coordinate x both appear in the argument of some trigonometric function. This can be forced by making the exponent for the solution for $F(\tau)$ imaginary. Differentiating Eq. (4-229) and separating the variable, one obtains

$$\frac{F'(\tau)}{\alpha F(\tau)} = \frac{G''(x)}{G(x)} = \pm i\lambda^2 \tag{4-230}$$

where $i = \sqrt{-1}$.

As noted before in the treatment of product solutions, Eq. (4-230) is actually two equations:

$$F'(\tau) - (\pm i\lambda^2)\alpha F(\tau) = 0$$
$$G''(x) - (\pm i\lambda^2)G(x) = 0$$

from the solutions of which comes

$$t = C \exp\,(\pm i\lambda^2\alpha\tau \pm \lambda\sqrt{\pm ix}) \tag{4-231}$$

Equation (4-231) actually results in four particular solutions:[1]

$$\begin{aligned} t_1 &= C_1 \exp\,[-\sqrt{\tfrac{1}{2}}\,\lambda x + i(\lambda^2\alpha\tau - \sqrt{\tfrac{1}{2}}\,\lambda x)] \\ t_2 &= C_2 \exp\,[-\sqrt{\tfrac{1}{2}}\,\lambda x - i(\lambda^2\alpha\tau - \sqrt{\tfrac{1}{2}}\,\lambda x)] \\ t_3 &= C_3 \exp\,[\sqrt{\tfrac{1}{2}}\,\lambda x + i(\lambda^2\alpha\tau + \sqrt{\tfrac{1}{2}}\,\lambda x)] \\ t_4 &= C_4 \exp\,[\sqrt{\tfrac{1}{2}}\,\lambda x - i(\lambda^2\alpha\tau + \sqrt{\tfrac{1}{2}}\,\lambda x)] \end{aligned}$$

Two of these solutions are not physically admissible, since the temperature cannot be infinite for x infinite. These two (t_3 and t_4) are discarded, and the other two are added to obtain another particular solution:

$$t = \exp\,(-\sqrt{\tfrac{1}{2}}\,\lambda x)\{C_1 \exp\,[+i(\lambda^2\alpha\tau - \sqrt{\tfrac{1}{2}}\,\lambda x)] + C_2 \exp\,[-i(\lambda^2\alpha\tau - \sqrt{\tfrac{1}{2}}\,\lambda x)]\} \tag{4-232}$$

Equation (4-232) can be written in terms of trigonometric functions

$$t = \exp\,(-\sqrt{\tfrac{1}{2}}\,\lambda x)[A \cos\,(\lambda^2\alpha\tau - \sqrt{\tfrac{1}{2}}\,\lambda x) + B \sin\,(\lambda^2\alpha\tau - \sqrt{\tfrac{1}{2}}\,\lambda x)] \tag{4-233}$$

or in terms of a phase angle

$$t = C \exp\,(-\sqrt{\tfrac{1}{2}}\,\lambda x)\cos\,(\lambda^2\alpha\tau - \sqrt{\tfrac{1}{2}}\,\lambda x - \delta) \tag{4-234}$$

where $\delta = \tan^{-1}\dfrac{B}{A}$ and $C = \sqrt{A^2 + B^2}$

It is necessary to determine the constants A, B, and λ from the available boundary conditions. It is assumed that the function $t_0 = f(\tau)$ can be expressed by a Fourier series:

$$t_0 = \frac{a_0}{2} + \sum_{n=1}^{\infty}\left(a_n \cos\frac{2\pi n\tau}{\tau_0} + b_n \sin\frac{2\pi n\tau}{\tau_0}\right) \tag{4-235}$$

Equation (4-233) for $x = 0$ becomes

$$t_0 = A \cos \lambda^2\alpha\tau + B \sin \lambda^2\alpha\tau \tag{4-236}$$

Thus, comparing Eqs. (4-235) and (4-236), we see that

$$A = a_n \qquad B = b_n \qquad \lambda = \sqrt{\frac{2\pi n}{\alpha\tau_0}}$$

However, the constant term $a_0/2$, or the average temperature at $x = 0$, does not appear in Eq. (4-236). This term is actually the mean value of

[1] L. M. K. Boelter, V. H. Cherry, H. A. Johnson, and R. C. Martinelli, "Heat Transfer Notes," McGraw-Hill, New York, 1965, contains a detailed solution.

the oscillating temperature at $x = 0$ and is the result of the initial transient at time zero. We have developed such a step solution earlier which accounts for this variation:

$$t = \frac{a_0}{2}\left(1 - \operatorname{erf}\frac{x}{2\sqrt{\alpha\tau}}\right) \tag{4-237}$$

Equation (4-237) can be added to Eq. (4-234) for another particular solution:

$$t = \frac{a_0}{2}\left(1 - \operatorname{erf}\frac{x}{2\sqrt{\alpha\tau}}\right) + \sum_{n=1}^{\infty} \exp\left(-\sqrt{\frac{n\pi}{\alpha\tau_0}}\,x\right) \left[a_n \cos\left(\frac{2\pi n\tau}{\tau_0} - \sqrt{\frac{n\pi}{\alpha\tau_0}}\,x\right) + b_n \sin\left(\frac{2\pi n\tau}{\tau_0} - \sqrt{\frac{n\pi}{\alpha\tau_0}}\,x\right)\right] \tag{4-238}$$

That Eq. (4-238) satisfies the differential equation and its boundary conditions can be seen by evaluation at $x = 0$ and $x \to \infty$. Equation (4-238) is not satisfactory for the condition $\tau = 0$ and cannot be used in that region of small time.[1] However, for τ large the erf $(x/2\sqrt{\alpha\tau})$ goes to zero, and the resulting expression adequately represents the periodic temperature distribution for the case where the initial transient has diminished:

$$t = \frac{a_0}{2} + \sum_{n=1}^{\infty} \exp\left(-\sqrt{\frac{n\pi}{\alpha\tau_0}}\,x\right) C_n \cos\left(\frac{2\pi n\tau}{\tau_0} - \sqrt{\frac{n\pi}{\alpha\tau_0}}\,x - \delta_n\right) \tag{4-239}$$

where $\qquad C_n = \sqrt{a_n^2 + b_n^2} \qquad$ and $\qquad \delta_n = \tan^{-1}\dfrac{b_n}{a_n}$

For the simpler case where the surface temperature is a cosine function

$$t_0 = t_{0M} \cos\frac{2\pi n\tau}{\tau_0}$$

and only the temperature oscillations about the mean temperature are considered; i.e., this is equivalent to taking the temperature datum as $a_0/2$; t_{0M} is the maximum absolute value of the surface-temperature variation. Equation (4-239) reduces to

$$t = t_{0M} \exp\left(-\sqrt{\frac{n\pi}{\alpha\tau_0}}\,x\right) \cos\left(\frac{2\pi n\tau}{\tau_0} - \sqrt{\frac{n\pi}{\alpha\tau_0}}\,x\right) \tag{4-240}$$

The physical significance of the more complicated equations just written can be discussed in terms of the simpler Eq. (4-240). Consider Eq. (4-240) at a particular value of the space coordinate x by assuming x a

[1] The solution good for small τ is given in *ibid.*

constant. Since $\cos 2m\pi = +1$ when $m = 0, 1, 2, 3, \ldots$, it follows that t will have maxima when

$$\frac{2\pi n\tau}{\tau_0} - \sqrt{\frac{n\pi}{\alpha\tau_0}}\,x = 2m\pi$$

or

$$\tau_{t,\max} = m\frac{\tau_0}{n} + \frac{1}{2}\sqrt{\frac{\tau_0}{\alpha n\pi}}\,x$$

for any x.

For the surface $x = 0$, the temperature is

$$t_0 = t_{0M}\cos\frac{2\pi n\tau}{\tau_0}$$

and is a maximum when

$$\frac{2\pi n\tau}{\tau_0} = 2m\pi$$

or

$$\tau_{t_{0M}} = \frac{\tau_0}{n}m \qquad \text{at } x = 0$$

By comparing the values of time when the temperature maxima occur at a depth x and at the surface, it can be seen that the oscillations have the same period τ_0/n at each depth but the oscillations at depth x lag those on the surface by $\frac{1}{2}\sqrt{\tau_0/\alpha n\pi}\,x$. Furthermore the surface amplitude is diminished at the depth x by the factor

$$\exp\left(-\sqrt{\frac{n\pi}{\alpha\tau_0}}\,x\right)$$

These physical features are shown in Figs. 4-28 and 4-29 for values of $n = 1$.

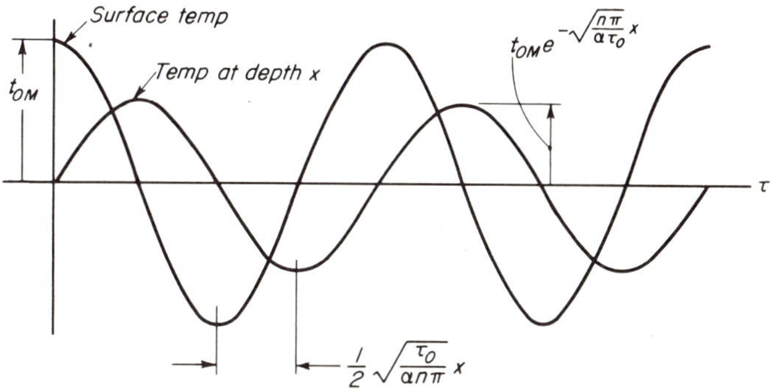

Fig. 4-28 Comparison of surface-temperature variation and temperature variation at depth x during periodic heat conduction in semi-infinite body.

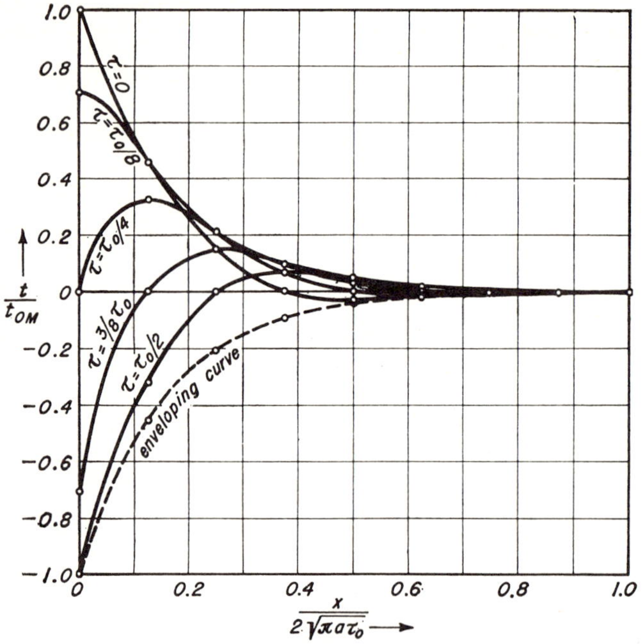

Fig. 4-29 Penetration of a temperature oscillation into an infinitely thick wall. *(From H. Gröber and S. Erk, "Die Grundgesetze der Wärmeübertragung," Springer-Verlag, Berlin, 1933.)*

If it is recalled that the cosine is an even function, that is,

$$\cos(-\beta) = \cos\beta$$

Eq. (4-240) can be written as

$$t = t_{0M} \exp\left(-\sqrt{\frac{n\pi}{\alpha\tau_0}}\,x\right)\cos\left(\sqrt{\frac{n\pi}{\alpha\tau_0}}\,x - \frac{2\pi n\tau}{\tau_0}\right) \tag{4-241}$$

The term $t_{0M}\cos(\sqrt{n\pi/\alpha\tau_0}\,x)$ represents a cosine wave of amplitude t_{0M} and wavelength x_0. The wavelength is equal to

$$x_0 = 2\sqrt{\frac{\pi\alpha\tau_0}{n}}$$

The term $t_{0M}\cos(\sqrt{n\pi/\alpha\tau_0}\,x - 2\pi n\tau/\tau_0)$ represents the same wave as before but moved in the positive x direction by the amount $2\pi n\tau/\tau_0$. The velocity of wave propagation thus becomes $2\sqrt{\pi\alpha n/\tau_0}$. The advancing wave is decreasing in amplitude with increasing depth by the factor

$\exp(-\sqrt{n\pi/\alpha\tau_0}\,x)$. These characteristics are shown in Fig. 4-30. Equation (4-241) also shows that the higher the frequency (the larger n), the less the penetration of the thermal wave; i.e., high-frequency thermal oscillations are rapidly damped out in comparison with the fundamental or lower harmonic oscillations.

Example 4-6 It is to be determined to what depth the daily and yearly temperature fluctuations penetrate the ground. From Appendix B one can obtain the thermal diffusivity of claylike soil, 0.00362 m²/hr, and the thermal diffusivity of sandstone, 0.00381 to 0.00455 m²/hr. Calculations are performed with the lower value 0.00362. We infer from Fig. 4-29 that the oscillations are unimportant when $x/(2\sqrt{\pi\alpha\tau_0}) = 0.8$. For the daily fluctuations, $\tau_0 = 24$ hr. Thus we obtain

$$x = 1.6\sqrt{\pi 0.00362(24)} = 0.838 \text{ m}$$

With the yearly fluctuation, the depth of penetration is then $\sqrt{365}$-fold, hence 15.9 m.

After the initial transient effect has dissipated, the heat in a periodic system must necessarily flow into and out of the body periodically, since the temperature gradient at the surface is sometimes positive and sometimes negative. The heat absorbed or rejected can be deduced from the Fourier conduction equation

$$dQ_i = -kA\left(\frac{\partial t}{\partial x}\right)_{x=0} d\tau \tag{4-242}$$

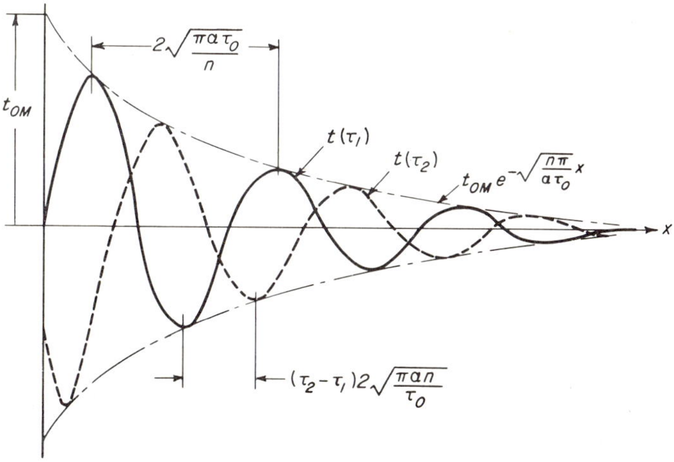

Fig. 4-30 Characteristics of a temperature oscillation penetrating an infinitely thick wall.

If Eq. (4-240) is differentiated, for example, with respect to x and substituted into Eq. (4-242), we obtain

$$dQ_i = -kAt_{0M}\sqrt{\frac{n\pi}{\alpha\tau_0}}\left(\sin\frac{2\pi n\tau}{\tau_0} - \cos\frac{2\pi n\tau}{\tau_0}\right)d\tau$$

or

$$\frac{dQ_i}{d\tau} = -\sqrt{2}\,kAt_{0M}\sqrt{\frac{n\pi}{\alpha\tau_0}}\sin\left(\frac{2\pi n\tau}{\tau_0} - \frac{\pi}{4}\right)$$

which, when integrated between limits 1 and 2, becomes

$$Q_i = Q_i\Big|_{\tau_1}^{\tau_2} = kAt_{0M}\sqrt{\frac{\tau_0}{2\pi n\alpha}}\cos\left(\frac{2\pi n\tau}{\tau_0} - \frac{\pi}{4}\right)\Big|_{\tau_1}^{\tau_2} \tag{4-243}$$

The relationships between the temperature maxima and minima and the heat-flow maxima and minima are shown in Table 4-5.

Using the values of time for Q_{max} and Q_{min} in Eq. (4-243), we can calculate the heat flowing into the solid during a half period:

$$Q_i\Big|_{\frac{1}{8}(\tau_0/n)}^{\frac{5}{8}(\tau_0/n)} = kAt_{0M}\sqrt{\frac{2\tau_0}{\pi n\alpha}} \tag{4-244}$$

These results are shown graphically in Fig. 4-31.

Semi-infinite solid, surrounding fluid temperature periodic with time If the fluid temperature adjacent to the exposed surface $x = 0$ varies

Table 4-5 Phase relationships in periodic heat flow†

$t_{x=0}$	τ
Max	0
0	$\frac{1}{4}\frac{\tau_0}{n}$
Min	$\frac{1}{2}\frac{\tau_0}{n}$
0	$\frac{3}{4}\frac{\tau_0}{n}$
Max	$\frac{\tau_0}{n}$

Q	τ	Q_i
0	$\frac{1}{8}\frac{\tau_0}{n}$	Max
Min	$\frac{3}{8}\frac{\tau_0}{n}$	0
0	$\frac{5}{8}\frac{\tau_0}{n}$	Min
Max	$\frac{7}{8}\frac{\tau_0}{n}$	0
0	$\frac{9}{8}\frac{\tau_0}{n}$	Max

† From L. M. K. Boelter et al., "Heat Transfer Notes." Copyright 1965. McGraw-Hill Book Company. Used by permission.

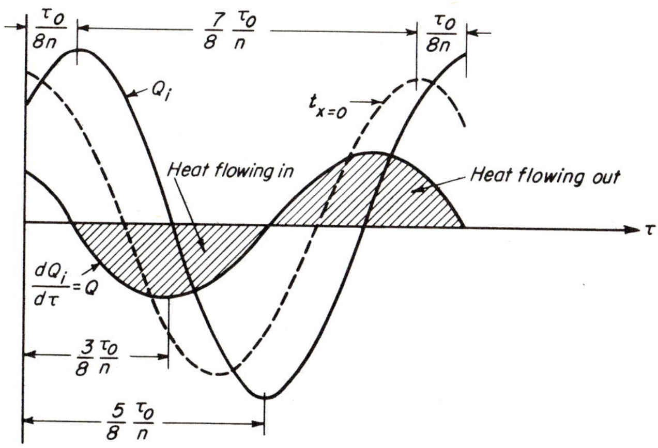

Fig. 4-31 Phase relationships in periodic heat flow. *(L. M. K. Boelter et al., "Heat Transfer Notes." Copyright 1965. McGraw-Hill Book Company. Used by permission.)*

periodically and yields a heat-transfer coefficient h, the solution follows in much the same way as in the preceding section except that now the boundary condition at $x = 0$ relates to the heat transfer by convection. The fluid temperature varies as

$$t_f = t_{fM} \cos \frac{2\pi n\tau}{\tau_0} \tag{4-245}$$

and the boundary condition imposed on the surface to account for the heat convection is

$$k\left(\frac{\partial t}{\partial x}\right)_{x=0} = -h(t - t_f) \tag{4-246}$$

A particular solution of the differential equation is given in Eq. (4-234), repeated here for convenience:

$$t = C \exp\left(-\sqrt{\tfrac{1}{2}}\,\lambda x\right) \cos\left(\lambda^2 \alpha\tau - \sqrt{\tfrac{1}{2}}\,\lambda x - \delta\right)$$

It is therefore necessary to determine the values of C, λ, and δ using Eqs. (4-245) and (4-246) to obtain a solution of the proposed problem. Upon evaluation of these required constants, the solution becomes

$$t = t_{fM} \frac{\exp\left(-\sqrt{\pi n/\alpha\tau_0}\,x\right)}{\sqrt{1 + 2\sqrt{\pi n k^2/\alpha\tau_0 h^2} + 2(\pi n k^2/\alpha\tau_0 h^2)}} \cos\left(\frac{2\pi n\tau}{\tau_0} - \sqrt{\frac{\pi n}{\alpha\tau_0}}\,x - \tan^{-1}\frac{1}{1 + \sqrt{\alpha\tau_0 h^2/{}_e n k^2}}\right) \tag{4-247}$$

An examination of Eq. (4-247) for the surface conditions ($x = 0$) will show that the surface temperature t_0 oscillates with the same frequency n/τ_0 as the surrounding fluid temperatures, but the amplitude of the surface-temperature oscillation is reduced by the factor

$$\frac{1}{\sqrt{1 + 2\sqrt{\pi n k^2/\alpha\tau_0 h^2} + 2(\pi n k^2/\alpha\tau_0 h^2)}} \tag{4-248}$$

From the above equations it can also be seen that for a large Biot modulus the surface temperature follows the fluid temperature more closely and for smaller Biot moduli the difference becomes greater. Also higher-frequency thermal oscillations are less effective in penetrating the body than are the lower-frequency oscillations.

Example 4-7 A case of periodic heat transfer is exhibited in the cylinder of a reciprocating internal-combustion engine. It is possible to calculate the depth of penetration of the temperature oscillations into the cylinder wall. Assume that the engine operates at 2,000 rpm. If the engine is the two-stroke type, it will produce 2,000 peak temperatures per minute, or the period of one oscillation is

$$\tau_0 = \frac{1}{60(2{,}000)} = \frac{1}{12 \times 10^4}$$

The thermal diffusivity of iron is $\alpha = 0.0595$ m²/hr. Thus again using the value $x/(2\sqrt{\pi\alpha\tau_0}) = 0.8$ from Fig. 4-29 as a measure of the practical depth of penetration, we can calculate x:

$$x = 1.6\sqrt{\frac{\pi(0.0595)}{12 \times 10^4}} = 1.99 \times 10^{-3}\text{ m}$$

Thus the temperature fluctuations penetrate only to a depth of 1.99 mm into the cylinder wall.

Because of the finite heat-transfer coefficient, the surface temperature of the cylinder wall will fluctuate with a much smaller amplitude than that of the gas in the cylinder. The amplitude reduction can be calculated with the aid of Eq. (4-248). Take $n = 1$ for the fundamental wave, use from the Appendix $k = 52$ W/m K, and estimate $h = 550$ W/m² K; then

$$\frac{\pi k^2}{\alpha\tau_0 h^2} = \frac{\pi(52^2)(12 \times 10^4)}{0.0595(550^2)} = 56{,}700$$

$$[1 + 2\sqrt{56{,}700} + 2(56{,}700)]^{-\frac{1}{2}} = 0.00297$$

If the gas fluctuates through a double amplitude of 2200 C, the cylinder-wall surface temperature has a double amplitude of

$$2200(0.00297) = 6.5\text{ C}$$

According to tests made by Meier[1] with an Otto engine at $n = 2{,}000$ rpm, the fluctuation of the surface temperature is approximately 11 C.

[1] A. Meier, *Forsch. Gebiete Ingenieurw.*, **10**:41–54 (1939).

4-4 NUMERICAL METHODS IN UNSTEADY HEAT CONDUCTION[1]

In Sec. 3-12 numerical methods for steady heat conduction were treated. In this section numerical methods for unsteady heat conduction will be discussed. It is well at the very beginning to point out some of the differences to be encountered between the steady- and unsteady-heat-conduction equations and methods.

In the first place, the differential equations governing steady heat conduction are *elliptic* equations. The conditions are everywhere independent of time. Specification of the dependent variable at the boundary everywhere is sufficient to determine the distribution of the dependent variable throughout the interior of the body. The boundary conditions are, of course, independent of the time.

By contrast, the differential equations appropriate to the problems of unsteady heat transfer, either transient or periodic, are *parabolic* equations. The boundary conditions are specified but may be functions of time. An initial spacewise distribution of the dependent variable is given. The solution yields the spacewise distribution of the dependent variable as a function of time.

In the numerical solution of steady-heat-conduction problems the distribution of the dependent variable is determined everywhere simultaneously, whereas in the case of unsteady heat conduction sequential solutions are constructed as a function of time, e.g., as shown in the development of the Schmidt graphical method.

The step-by-step construction of the solution to the unsteady-heat-conduction problem introduces the problem of *instabilities* in the process, a difficulty not experienced in the steady-heat-conduction solutions. If the *round-off* or numerical error increases as the solution continues, the procedure is *unstable*. If the *round-off errors* decrease, the procedure is *stable*.

Available in the technical literature are methods by which the stability of difference forms of the equations of unsteady heat conduction can be explored.[2] The method attributed to J. von Neumann assumes that the error at any specific time can be expanded in a harmonic series. The series is then analyzed to ascertain whether these initial errors grow or decay with time.

[1] From A. M. Clausing, Numerical Methods in Heat Transfer, in B. T. Chao (ed.), "Advanced Heat Transfer," pp. 157–216, University of Illinois Press, Urbana, 1969; reprinted with changes by permission of the author and publisher.

[2] G. G. O'Brien, M. A. Hyman, and S. Kaplan, A Study of the Numerical Solution of Partial Differential Equations, *J. Math. Phys.*, **29**:4, 223–251 (1951); R. D. Richtmyer, "Difference Methods of Initial Value Problems," Interscience, New York, 1957; D. Young, The Numerical Solution of Elliptic and Parabolic Differential Equations, in J. Todd (ed.), "Survey of Numerical Analysis," McGraw-Hill, New York, 1962.

Dusinberre[1] arrived at stability criteria from physical considerations in conjunction with the standard explicit method (forward difference in time). In the use of the method Dusinberre recognized that negative coefficients would be absurd from a physical point of view and recommended the consistent avoidance of negative coefficients. Clausing[2] notes that if the first and second laws of thermodynamics are not violated, the procedure will be stable.

Governing difference equations In the development of the difference equations in Sec. 3-12 the time rate of change of the internal energy was retained in differential form and then set to zero for steady considerations, the subject of that section. In the unsteady case it is necessary to put the temperature-time derivative in numerical form. One of the widely used methods of accomplishing this is the *forward-difference* or *standard explicit method.* This is precisely the procedure used in Sec. 4-2 with the Schmidt method.

The energy balance is

$$\Delta U_{n \to n+1} = \Delta\tau(Q_{\text{net},n} + Q'_n \, \Delta V) \tag{4-249}$$

where $\Delta U_{n \to n+1}$ represents the change in the internal energy during the increment of time, $\Delta\tau = \tau_{n+1} - \tau_n$. Again, $Q_{\text{net},n}$ is the net rate of heat flow by conduction from adjacent elements to the element in question in time $\Delta\tau$. Q'_n represents any constant, time-dependent or temperature-dependent element, such as (1) an internal heat source or sink, (2) the convective loss from the surface [see Eqs. (3-17) and (3-18)] for certain systems, or (3) a quantity required to satisfy certain boundary conditions when considering elements which are parts of the system boundary.

For the problem considered here, the difference equation becomes, in two space dimensions,

$$t_{i,j,n+1} = \frac{1}{M} [t_{i+1,j,n} + t_{i-1,j,n} + t_{i,j+1,n} + t_{i,j-1,n} + (M - 4)t_{i,j,n}] \tag{4-250}$$

where

$$M = \frac{(\Delta x)^2}{\alpha \, \Delta\tau} = \frac{(\Delta y)^2}{\alpha \, \Delta\tau}$$

which is the reciprocal of a difference *Fourier number.*

[1] G. M. Dusinberre, "Heat Transfer Calculations by Finite Differences," 2d ed., International Textbook, Scranton, Pa., 1961.

[2] *Op. cit.*

Equation (4-250) is one difference form of the differential equation for unsteady heat conduction,

$$\frac{1}{\alpha}\frac{\partial t}{\partial \tau} = \frac{\partial^2 t}{\partial x^2} + \frac{\partial^2 t}{\partial y^2} \tag{4-251}$$

where obviously the heat-source term has been dropped.

It can be seen from Eq. (4-250) that the temperature at the time τ_{n+1} is a weighted average of the known temperatures at the time τ_n. The new temperatures can be calculated in any order, or simultaneously if preferable.

It is necessary that M be assigned a value such that a violation of the second law is avoided. Consider any interior or boundary element which is exchanging heat with m surrounding elements whose temperatures will be denoted $t_{s,n}$, where the subscript s denotes the surrounding elements and takes on all values necessary. The thermal resistances connecting element 0 with adjacent elements of interest are denoted as R_{s-0}. For this case Eq. (4-249) yields

$$\frac{\rho c\,\Delta V}{\Delta \tau}(t_{0,n+1} - t_{0,n}) = \sum_{s=1}^{m} \frac{t_{s,n} - t_{0,n}}{R_{s-0}} \tag{4-252}$$

To preclude a possible violation of the second law of thermodynamics the time intervals must be small enough for the right side of Eq. (4-252) to retain the same sign when $t_{0,n}$ is replaced by $t_{0,n+1}$. Thus if

$$\sum_{s=1}^{m} \frac{t_{s,n} - t_{0,n}}{R_{s-0}} > 0 \tag{4-253}$$

then

$$\sum \frac{t_{s,n} - t_{0,n+1}}{R_{s-0}} \geq 0 \tag{4-254}$$

If Eq. (4-254) is substituted into Eq. (4-252) with the use of Eq. (4-250), the following inequality results:

$$\frac{\rho c\,\Delta V}{\Delta \tau} \geq \sum_{s=1}^{m} \frac{1}{R_{s-0}} \tag{4-255}$$

This is the criterion which must be satisfied if a potential violation of the second law of thermodynamics is to be avoided.

For the two-space-dimensional unsteady problem under consideration

$$\frac{\rho c\,\Delta V\,R_{s-0}}{\Delta \tau} = \frac{(\Delta x)^2}{\alpha\,\Delta \tau} = M \geq 4 \tag{4-256}$$

where M as noted earlier is the reciprocal of the *Fourier number*, or the *dimensionless time constant* for the system.

The boundary elements may result in a different criterion than that of the internal elements. Consider, for example, the network shown in Fig. 4-32. This network has been changed from that shown in Fig. 3-43 by a shift by one-half an element in both space coordinates, thus making the network more suitable for unsteady-heat-conduction considerations. In Fig. 3-43, if capacitances were added to each node to make the network usable for unsteady conditions, each node on the surface would have a capacitor. In the case of a boundary condition of a constant surface temperature that is different from the initial temperature, all the surface capacitors would be charged instantly. This makes Fig. 3-43 unattractive for unsteady calculations. The network shown in Fig. 4-32 removes this particular difficulty.

For Fig. 4-32, a convection boundary condition and an element on the left boundary,

$$M \geq 3 + \frac{2\,\text{Bi}}{2 + \text{Bi}} \tag{4-257}$$

For large Biot numbers $M \geq 5$, and the advantage of Fig. 4-32 is obvious. Other boundary conditions give rise to other values for M.

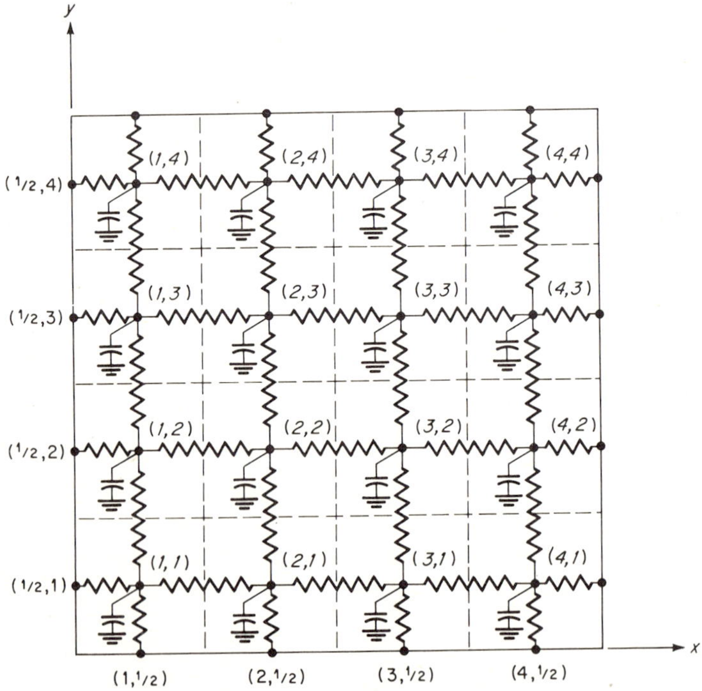

Fig. 4-32 *RC* network with zero-capacitance surface nodes.

Standard implicit method As has been pointed out in several places in the technical literature, the difficulties related to the problem of stability, i.e., the advertent or inadvertent violation of the second law of thermodynamics, may be largely eliminated by the use of the *backward-difference*, or *implicit, method* of formulation of the difference equations. In the implicit method the right-hand side of Eq. (4-249) is evaluated at the time $(n + 1)\,\Delta\tau$ rather than $n\,\Delta\tau$. Instead of Eq. (4-249) one obtains

$$\Delta U_{n\to n+1} = \Delta\tau(Q_{\text{net},n+1} + Q'_{n+1}\,\Delta V) \tag{4-258}$$

In difference form this becomes

$$t_{i,j,n} = -\frac{1}{M}[t_{i+1,j,n+1} + t_{i-1,j,n+1} + t_{i,j+1,n+1} + t_{i,j-1,n+1} - (4 + M)t_{i,j,n+1}] \tag{4-259}$$

which is an implicit equation; i.e., it is necessary in order to calculate the temperature distribution to solve a set of simultaneous, linear, algebraic equations for each time interval. However, in contrast with the explicit method the change in internal energy is no longer directly proportional to $\Delta\tau$. As $\Delta\tau$ becomes large, a large time step, $Q_{\text{net},n+1}$, becomes small; thus ΔU is bounded, and a violation of the second law is not possible. The formulation appears unconditionally stable, and formal stability analyses have confirmed this observation.

The major problem is thus the solution of a set of linear, simultaneous algebraic equations. For this purpose, one may choose from a variety of schemes such as a gaussian elimination, Gauss-Seidel iteration, or matrix-inversion methods. Several other difference formulations have been used in the past, of which the more important have been discussed by Clausing.[1]

[1] *Op. cit.*

5
Heat Conduction with Moving Boundaries

5-1 HEAT CONDUCTION IN MELTING AND SOLIDIFICATION

Many problems arise in engineering in which heat transfer is accompanied by a change of phase in the conducting medium or the composition changes because of some chemical reaction which is propagated through the medium. Such phenomena as these are generally accompanied by either the release or absorption of thermal energy in the active zone. The energy released or absorbed is commonly distributed or drawn from the surrounding system by the mechanism of heat conduction. Examples of such phenomena are the melting or solidification processes, chemical reactions such as combustion, the penetration of frost into the earth, and the aerodynamic phenomenon of ablation. The first published discussion of this class of problem was that of Stefan in 1891 in a study of the thickness of polar ice. More recently much attention has been directed toward the ablation of materials as a protective device in the design of atmospheric reentry vehicles.

The essential and common feature in these systems is that an interface exists separating two regions of different thermophysical properties, and this interface moves as some function of time. At the interface energy is liberated or absorbed. It is necessary for the solution of such problems to determine the manner and rate at which the interface will move. The problems so posed are nonlinear, so that specific solutions must be devised. The solutions involve considerable difficulties and despite their obvious practical value have not received their share of attention; however, a limited body of literature exists.[1]

There are few exact solutions available, the most notable being that of Neumann for the semi-infinite solid. Beyond those few exact solutions all problems have to be solved numerically by computers or by analogs. Numerical solutions are advantageous in that the variation in thermophysical properties, usually of some consequence in change of phase, can be taken into account with some ease.

In this section attention will be directed to some exact and some approximate methods for the solution of problems in solidification, melting, and ablation.

The solid phase of negligible heat capacity The simplest solution is that of the formation of a solid layer on a liquid maintained at the temperature of the phase change where the heat capacity in the solid is neglected. An example is the formation of ice on a body of water at 0 C. The arrangement is shown schematically in Fig. 5-1. The layer of the solid at some time τ has a thickness $x(\tau)$. The exposed surface is maintained at a temperature t_s which is lower than the temperature for the change of phase t_∞. The heat liberated at the interface must be conducted through the solid, there being of course no temperature gradient in the liquid in this approximation.

The situation can be adequately expressed by the following equation, in which the heat of fusion released is equated to that conducted through the solid layer per unit area:

$$\rho Q_L \frac{dx}{d\tau} = \frac{k(t_\infty - t_s)}{x} \tag{5-1}$$

[1] H. S. Carslaw and J. C. Jaeger, "Conduction of Heat in Solids," 2d ed., pp. 283–296, Oxford University Press, New York, 1959; H. Gröber, S. Erk, and V. Grigull, "Fundamentals of Heat Transfer," McGraw-Hill, New York, 1961; L. R. Ingersoll and J. Zobel, "Mathematical Theory of Heat Conduction," University of Wisconsin Press, Madison, 1954; J. H. Weiner, *Brit. J. Appl. Phys.*, **6**:361–363 (1955); H. G. Landau, Heat Conduction in a Melting Solid, *Quart. Appl. Math.*, **8**(1):81–94 (1951); T. R. Goodman, The Heat Balance Integral and Its Application to Problems Involving a Change in Phase, *Trans. ASME*, **8**(2):335–342 (1958); L. Roberts, On the Melting of a Semi-infinite Body of Ice Placed in a Hot Stream, *J. Fluid Mech.*, **4**(5):505–528 (1958); G. Horvay and J. G. Henzel, Freezing of Liquid Metal in a Mold, *General Electric Co. Res. Lab. Rept.* 57-RL-1844, December, 1957.

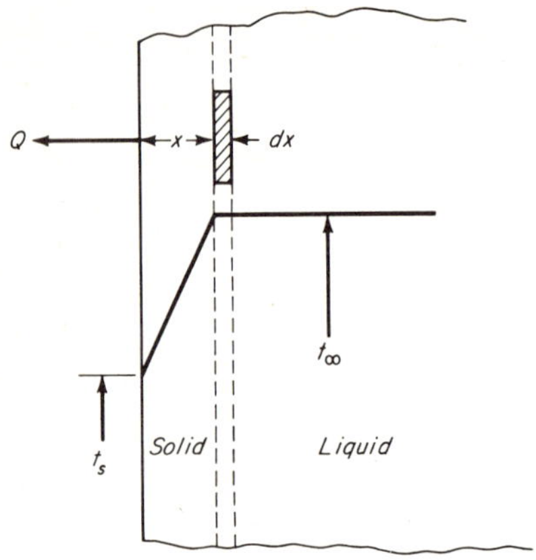

Fig. 5-1 Solidification process for negligible heat capacity.

The solution of Eq. (5-1) for the single condition $\tau = 0$, $x = 0$ gives the expression for the growth of the solid layer

$$x = \sqrt{\frac{2k(t_\infty - t_s)}{\rho Q_L}\,\tau} \tag{5-2}$$

where Q_L = latent heat of fusion
k = thermal conductivity of solid
ρ = mass density of solid
τ = time

Equation (5-2) shows that the solid layer grows like the square root of the time, a characteristic of all solutions involving phase changes and indeed reminiscent of the approximate solutions of the integral type in Chap. 4, where the penetration depth $\delta(\tau)$ had a like dependence on the time. Such a similarity leads one to believe that the integral method of solution for the heat-conduction equation would have great applicability to phase-change problems, which indeed it does, as will be seen. In spite of its simplicity, Eq. (5-2) gives an insight into the form more sophisticated solutions will take.

Melting and solidification in one dimension Solidification of a liquid will be considered here, with the understanding that the problem involving melting can be treated in a similar manner. In a problem such as this, we must consider the latent heat of fusion and at the same time account for the fact that such properties as the thermal conductivity, specific heats, and densities are different—sometimes markedly so—for the two phases. Consider Fig. 5-2 and suppose the substance fills the region $x > 0$ initially and is being solidified by the removal of heat at the exposed surface, which is maintained at a constant temperature T_1. At any time τ, the surface separating the liquid and solid phases is at $X(\tau)$. The bulk temperature of the liquid at a large distance away from the interface is T_2 and constant. Heat is conducted, therefore, from the liquid through the solid phase to the free surface. At the interface the system releases the latent heat of fusion as the liquid becomes a solid.

At some time τ, the region $x < X(\tau)$ consists of the solid phase with properties, k_1, α_1, ρ_1, c_1, and if t_1 is the temperature within this solid phase, it must satisfy the expression

$$\frac{\partial^2 t_1}{\partial x_1^2} - \frac{1}{\alpha_1}\frac{\partial t_1}{\partial \tau} = 0 \qquad t_1 = T_1 \text{ at } x_1 = 0 \tag{5-3}$$

The region $x > X(\tau)$ consists of the liquid phase with properties k_2, α_2, ρ_2, and c_2, and if t_2 is the temperature within this liquid phase (neglecting convection currents), it must satisfy the expression

$$\frac{\partial^2 t_2}{\partial x_2^2} - \frac{1}{\alpha_2}\frac{\partial t_2}{\partial \tau} = 0 \qquad t_2 = T_2 \text{ at } x \to \infty \tag{5-4}$$

In the solidification of water to ice, for example, there is an increase in volume (decrease in mass density), and this effect can be taken into account by noting that the ice surface will move away from the original

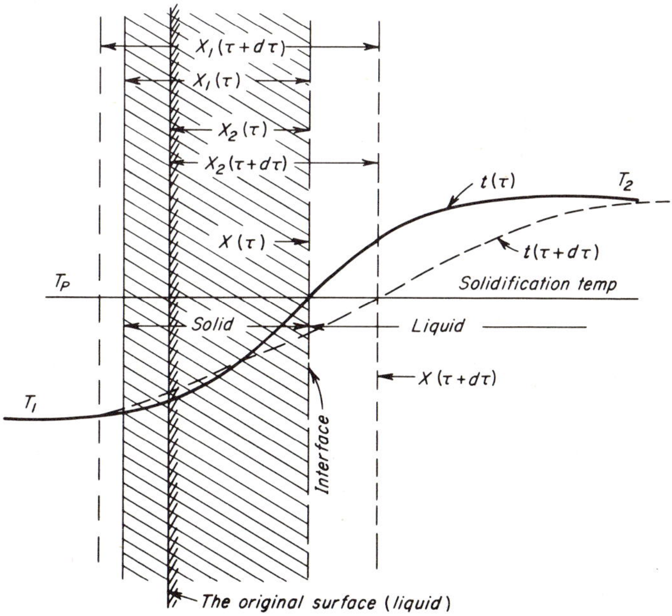

Fig. 5-2 Heat conduction during solidification of a liquid.

surface according to the density of each phase.[1] This movement can be expressed by the relationship

$$\frac{X_1}{X_2} = \frac{\rho_2}{\rho_1} = \beta \tag{5-5}$$

In addition it is required at the interface, which is at the temperature necessary for phase change, that $t_1 = t_2 = T_p$ at $x_1 = X_1(\tau)$, or $x_2 = X_2(\tau)$.

If Q_L is the latent heat of fusion of the material, then when the phase separation surface moves a distance dx, a quantity of heat

$$Q_L\rho_1 \frac{dX_1}{d\tau} = Q_L\rho_2 \frac{dX_2}{d\tau}$$

is liberated and must be conducted away. This requires that per unit area

$$k_1 \left(\frac{\partial t_1}{\partial x_1}\right)_{x_1=X_1} - k_2 \left(\frac{\partial t_2}{\partial x_2}\right)_{x_2=X_2} = Q_L\rho_1 \frac{dX_1}{d\tau} = Q_L\rho_2 \frac{dX_2}{d\tau} \tag{5-6}$$

From this point on the procedure follows the method due to Neumann.[2] Assume solutions of the form

$$\vartheta_1 = t_1 - T_P = (T_1 - T_P) + A \operatorname{erf} \frac{x_1}{2\sqrt{\alpha_1 \tau}} \tag{5-7}$$

$$\vartheta_2 = t_2 - T_P = (T_2 - T_P) + B \operatorname{erfc} \frac{x_2}{2\sqrt{\alpha_2 \tau}} \tag{5-8}$$

where A and B are constants, and thus Eqs. (5-7) and (5-8) satisfy Eqs. (5-3) and (5-4), respectively.[3] The condition relating to the interfacial temperature, that is, $t_1 = t_2 = T_P$ at $x_1 = X_1(\tau)$ and $x_2 = X_2(\tau)$, results in

$$T_P - T_1 = A \operatorname{erf} \frac{X_1}{2\sqrt{\alpha_1 \tau}} \tag{5-9}$$

$$T_P - T_2 = B \operatorname{erfc} \frac{X_2}{2\sqrt{\alpha_2 \tau}} \tag{5-10}$$

Now since Eqs. (5-9) and (5-10) must hold for all values of X_1 and X_2, these must be proportional to $\sqrt{\tau}$. Using Eq. (5-5), therefore, we have the relationships

$$\begin{aligned} X_1 &= K\beta \sqrt{\tau} \\ X_2 &= K \sqrt{\tau} \end{aligned} \tag{5-11}$$

[1] A. J. Rothmann, private communication, 1953.

[2] Carslaw and Jaeger, *op. cit.*, p. 283.

[3] erfc $z = 1 - \operatorname{erf} z$.

where K is a constant to be determined. When the results of Eqs. (5-7), (5-8), and (5-11) are used in Eq. (5-6),

$$\frac{Ak_1}{\sqrt{\pi\alpha_1}} e^{-K^2\beta^2/4\alpha_1} + \frac{Bk_2}{\sqrt{\pi\alpha_2}} e^{-K^2/4\alpha_2} = \frac{Q_L\rho_1 K\beta}{2} \tag{5-12}$$

and when Eqs. (5-9) and (5-10) are used in (5-12),

$$\frac{(T_P - T_1)k_1 e^{-K^2\beta^2/4\alpha_1}}{\sqrt{\pi\alpha_1}\,\mathrm{erf}\,(K\beta/2\sqrt{\alpha_1})} - \frac{(T_2 - T_P)k_2 e^{-K^2/4\alpha_2}}{\sqrt{\pi\alpha_2}\,\mathrm{erfc}\,(K/2\sqrt{\alpha_2})} = \frac{Q_L\rho_1 K\beta}{2} \tag{5-13}$$

Equation (5-13) can be solved numerically to give K in terms of T_1, T_2, and T_P and the thermal properties of the material. When K is known, A and B are found from Eqs. (5-9) to (5-11).

The temperature distribution in the solid becomes

$$\frac{t_1 - T_P}{T_1 - T_P} = 1 - \frac{\mathrm{erf}\,(x/2\sqrt{\alpha_1\tau})}{\mathrm{erf}\,(K\beta/2\sqrt{\alpha_1})} \tag{5-14}$$

and the temperature distribution in the liquid phase becomes

$$\frac{t_2 - T_P}{T_2 - T_P} = 1 - \frac{\mathrm{erfc}\,(x/2\sqrt{\alpha_2\tau})}{\mathrm{erfc}\,(K/2\sqrt{\alpha_2})} \tag{5-15}$$

From Eq. (5-11) it is clear that for $\tau \to 0$, $X_1 \to 0$ and $X_2 \to 0$, and from Eq. (5-8) it follows that for $x_2 > 0$ and $\tau \to 0$, the temperature $t_2 = T_2$; thus the initial condition is that the region $x > 0$ at $\tau = 0$ is all liquid at temperature T_2. Equation (5-13) has been solved for K for limited values of T_1 and T_2 around a solidification temperature of 0 C characteristic of water-ice systems for $\beta = 1$[1] and for $\beta = 1.09$,[2] the density ratio of water to ice. An adequate approximation for K in the case of the water-ice system is

$$K_0 = \sqrt{\frac{2(T_P - T_1)k_1}{Q_L\rho_1}} \qquad \beta = 1$$

$$K_\beta = \sqrt{\frac{2(T_P - T_1)k_1}{Q_L\rho_1\beta^2}} \qquad \beta > 1$$

Thus $K_0 = 1.09K_\beta$ in the water-ice case. It can be seen, then, that a slower freezing time is calculated if the change in density in the ice phase from that of the water phase is considered. The similarity of the last two equations to Eq. (5-2) is apparent.

For the special (and important) case where the liquid is initially at

[1] Carslaw and Jaeger, *op. cit.*, p. 286.

[2] Rothmann, *loc. cit.*

the temperature of the phase change T_P, Eq. (5-13) can be modified; that is $T_2 = T_P$, so that the second term vanishes, yielding

$$\frac{(T_P - T_1)k_1 e^{-K^2\beta^2/4\alpha_1}}{\sqrt{\pi\alpha_1}\,\text{erf}\,(K\beta/2\sqrt{\alpha_1})} = \frac{Q_L\rho_1 K\beta}{2} \tag{5-16}$$

The melting of a solid, approximate method As noted earlier, there is a marked similarity between the penetration depth as defined for the approximate method discussed in Chap. 4 and the location of the interface between phases in the problems on melting and solidification. The methods of Goodman[1] and Biot[2] are particularly adaptable to problems involving phase changes. In what follows, the integral formulation of the heat-conduction problem as used by Goodman will be used in the solution of the problem of the melting solid in one space dimension.

Assume a semi-infinite body (Fig. 5-3) extending from its free surface at $x = 0$ over the positive x dimension. Initially the temperature in the solid is t_∞, while at time $\tau \geq 0$ the surface temperature is raised to t_0 and held there. The problem will be simplified by assuming that the solid phase will be at the melting temperature; that is, $t_\infty = T_P$. Physically there is a temperature distribution in the solid; however, in many cases the effect is small, and the ultimate location of the melting line is

[1] *Loc. cit.*

[2] M. A. Biot, New Methods of Heat Flow Analysis with Applications to Flight Structures, *J. Aeron. Sci.*, **24**:12 (1957).

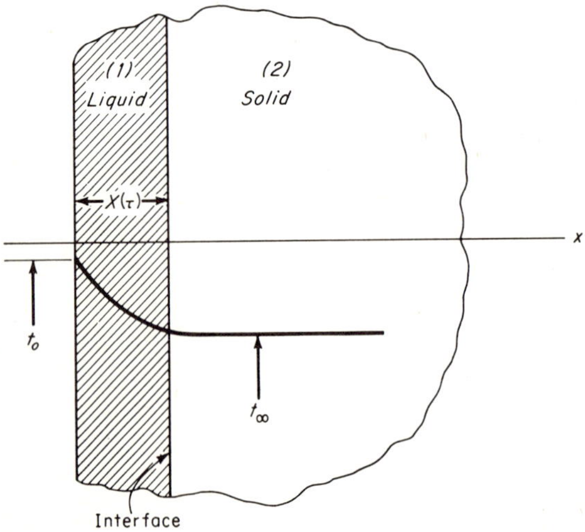

Fig. 5-3 The melting semi-infinite solid.

not seriously affected. Mathematically the decision to accept no temperature distribution in the solid eliminates an equation and greatly enhances the use of the approximate method. It can be seen then that the ensuing calculation pertains to the melt layer of liquid and that the penetration depth $\delta(\tau)$ is identical with the thickness of the liquid layer.

If $\vartheta = t - T_P$, where t is the temperature in the liquid melt, then two conditions are required:

$$\vartheta[X(\tau),\tau] = 0 \tag{5-17}$$

the condition at the interface, and

$$-\frac{\partial \vartheta}{\partial x} = \frac{\rho Q_L}{k} \frac{dX}{d\tau} \tag{5-18}$$

the heat balance across the interface. The appropriate boundary condition is

$$\vartheta_0 = (t_0 - T_P) \qquad x = 0 \qquad \tau > 0 \tag{5-19}$$

The integrated heat-conduction equation is, from Eq. (1-30), for the present case

$$\frac{d}{d\tau}\left[\int_0^X \vartheta(x,\tau)\,dx\right] = -\alpha\left[\frac{\rho Q_L}{k}\frac{dX}{d\tau} + \frac{\partial \vartheta}{\partial x}(0,\tau)\right] \tag{5-20}$$

Equations (5-17) to (5-19) and (5-3)[1] define the problem. The integral on the left of Eq. (5-20) is the total energy in the melt.

Once again let $\vartheta(x,\tau)$ be represented by a quadratic expression

$$\vartheta(x,\tau) = A(x - X) + B(x - X)^2 \tag{5-21}$$

which identically satisfies Eq. (5-17). Two other conditions are needed for the evaluation of A and B. These conditions are those shown by Eqs. (5-18) and (5-19), although a modification of Eq. (5-18) is necessary.

If Eq. (5-17) is differentiated with respect to τ,

$$\frac{\partial \vartheta}{\partial x}\frac{dX}{d\tau} + \frac{\partial \vartheta}{\partial \tau} = 0 \tag{5-22}$$

and the result solved for $dX/d\tau$ and substituted into Eq. (5-18) with the use of the relationship from Eq. (5-3), one obtains

$$\left(\frac{\partial \vartheta}{\partial x}\right)^2 = \frac{\alpha \rho Q_L}{k}\frac{\partial^2 \vartheta}{\partial x^2} \tag{5-23}$$

for the required condition. The nonlinearity of the problem is evident from Eq. (5-23).

[1] In this instance Eq. (5-3) represents the liquid layer rather than the solid layer as the problem here is one of melting.

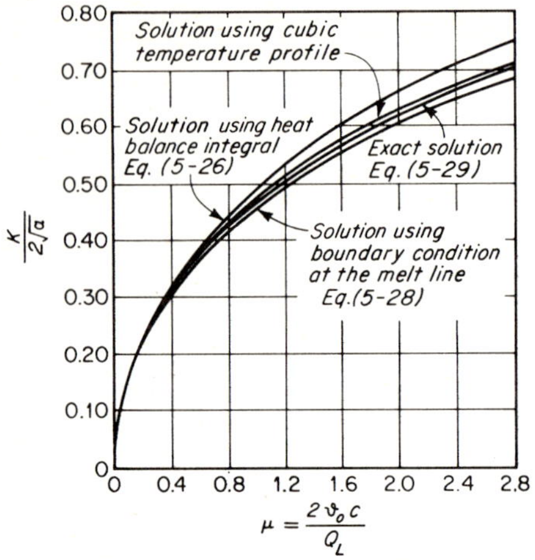

Fig. 5-4 Melting constant for fixed boundary temperature, semi-infinite solid. [*T. R. Goodman, Trans. ASME*, **8**(*2*):*505–528* (*1958*).]

From Eq. (5-21) and the conditions expressed by Eqs. (5-19) and (5-23), the values A and B are

$$A = \frac{Q_L}{cX}\left[1 \pm \left(1 + \frac{2c}{Q_L}\vartheta_0\right)^{\frac{1}{2}}\right]$$

$$B = \frac{A}{X} + \frac{\vartheta_0}{X^2}$$

where the choice of the negative sign in the expression for A is required as $A = (\partial\vartheta/\partial x)_{x=X} < 0$.

Use of the ensuing temperature distribution in the heat-balance integral [Eq. (5-20)] results in the differential equation

$$X\frac{dX}{d\tau} = \frac{6\alpha[1 - (1+\mu)^{\frac{1}{2}} + \mu]}{5 + \mu + (1+\mu)^{\frac{1}{2}}} \tag{5-24}$$

where $\mu = 2\vartheta_0 c/Q_L$, which when solved for $X(\tau)$ for the initial condition $X(0) = 0$ gives

$$X(\tau) = K\sqrt{\tau} \tag{5-25}$$

where

$$\frac{K}{2\sqrt{\alpha}} = \sqrt{3}\left(\frac{1 - \sqrt{1+\mu} + \mu}{5 + \sqrt{1+\mu} + \mu}\right)^{\frac{1}{2}} \tag{5-26}$$

Once $X(\tau)$ is known from Eqs. (5-25) and (5-26), the temperature distribution in the liquid melt can be determined from Eq. (5-21) and the values for A and B.

An alternate boundary condition exists as a possibility. If from the temperature distribution of Eq. (5-21), $\partial\vartheta/\partial x(X,\tau)$ is calculated and used in Eq. (5-18), a differential equation different from Eq. (5-24) is obtained

$$X \frac{dX}{d\tau} = \alpha \left(\sqrt{1+\mu} - 1\right)^{\frac{1}{2}} \tag{5-27}$$

where the constant K becomes

$$\frac{K}{2\sqrt{\alpha}} = \left(\frac{\sqrt{1+\mu} - 1}{2}\right)^{\frac{1}{2}} \tag{5-28}$$

Comparison with the exact solution given for the same case [Eq. (5-16)] for $\beta = 1$

$$\mu = 2\sqrt{\pi}\,\frac{K}{2\sqrt{\alpha}} \exp\left(\frac{K}{2\sqrt{\alpha}}\right)^2 \operatorname{erf}\frac{K}{2\sqrt{\alpha}} \tag{5-29}$$

shows that the boundary condition at the melt line is the better one. The solutions are shown in comparison graphically in Fig. 5-4, which also shows the similar result using the rather more complicated cubic expression for the temperature distribution.

Melting of a solid with a specified heat flux on the boundary In this case, as in the preceding one, the solid is assumed to be initially at the melting temperature T_P. There is therefore no temperature gradient in the solid (Fig. 5-5). In this case only Eq. (5-3) will apply, but again

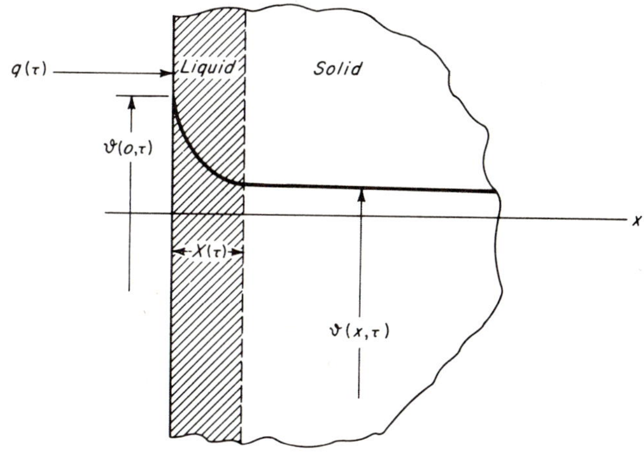

Fig. 5-5 Melting semi-infinite solid with arbitrary heat flux on the boundary.

for the liquid phase. The boundary conditions and the interfacial conditions are

$$k\frac{\partial\vartheta}{\partial x} = -q(\tau) \qquad x = 0 \qquad \tau > 0 \tag{5-30}$$

$$\vartheta(X,\tau) = 0 \qquad x = X(\tau) \tag{5-31}$$

$$\left(\frac{\partial\vartheta}{\partial x}\right)^2 = \frac{Q_L}{c}\left(\frac{\partial^2\vartheta}{\partial x^2}\right) \qquad x = X(\tau) \tag{5-32}$$

Equation (5-30) replaces the condition given by Eq. (5-19); otherwise the problem is the same as the preceding one.

The integral of the heat-conduction equation becomes

$$\frac{d}{d\tau}\left[\int_0^X \vartheta(x,\tau)\,dx + \frac{Q_L\alpha\rho}{k}X\right] = \frac{\alpha q(\tau)}{k} \tag{5-33}$$

where the symbols are as defined before and $q(\tau)$ is the heat flux to be specified at the surface $x = 0$. Equation (5-33) can be integrated for the condition $X(0) = 0$, that is, no melt thickness at time zero, to yield

$$\int_0^X \vartheta(x,\tau)\,dx + \frac{Q_L X}{c} = \frac{1}{\rho c}\int_0^\tau q(\tau)\,d\tau \tag{5-34}$$

A temperature distribution of a parabolic form may be assumed as before [Eq. (5-21)], which satisfies exactly the condition of Eq. (5-31). A and B are determined from application of the remaining two conditions [Eqs. (5-30) and (5-32)] such that

$$A = \tfrac{1}{2}[1 - (1 + 4\mu)^{\frac{1}{2}}]\frac{Q_L}{cX} \tag{5-35}$$

$$B = \frac{1}{8}\frac{Q_L}{cX^2}[1 - (1 + 4\mu)^{\frac{1}{2}}]^2 \tag{5-36}$$

where now $\mu = q(\tau)X(\tau)/\alpha\rho Q_L$. Using the temperature distribution so determined in Eq. (5-34), the result becomes

$$\frac{q(\tau)}{\alpha\rho^2 Q_L^2}\int_0^\tau q(\tau)\,d\tau = \frac{\mu}{6}[\mu + 5 + (1 + 4\mu)^{\frac{1}{2}}] \tag{5-37}$$

shown graphically in Fig. 5-6.

The exact solution for this problem was given by Evans, Isaacson, and MacDonald[1] in a Taylor series for μ as a function of τ for the case $q(\tau) = \text{const}$

$$\mu = \tau - \frac{\tau^2}{2!} + \frac{5\tau^3}{3!} - \frac{51\tau^4}{4!} + \frac{827\tau^5}{5!} - \cdots \tag{5-38}$$

[1] G. W. Evans, E. Isaacson, and J. K. L. MacDonald, Stefan-like Problems, *Quart. Appl. Math.*, **8**:312–319 (1950).

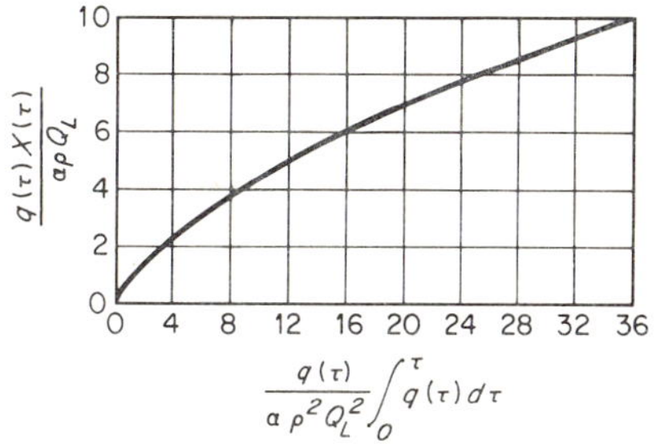

Fig. 5-6 Thickness of melting layer with time for a solid with an arbitrary heat flux at the surface. [*T. R. Goodman, Trans. ASME*, **80**:*335* (*1958*).]

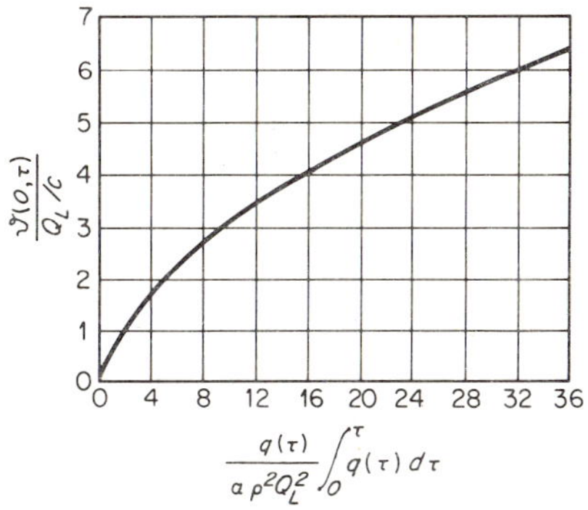

Fig. 5-7 Temperature-time history of the surface $x = 0$ for a semi-infinite melting solid with an arbitrary heat-flux input to the surface.

Equation (5-37) for $q(\tau)$ = const has been shown in a comparable series by Goodman to be

$$\mu = \tau - \frac{\tau^2}{2!} + \frac{5\tau^3}{3!} - \frac{51\tau^4}{4!} + \frac{795\tau^5}{5!} - \cdots \tag{5-39}$$

Comparison of the last two equations shows excellent agreement between the approximate and exact solutions. The approximate method in the present form is useful only for constant or monotonic functions of $q(\tau)$, failing for pulselike functions. The heat-balance integral can be used, however, for pulselike inputs of the heat flux by using a two-parameter method as demonstrated by Goodman.[1]

The temperature distribution in the liquid melt is given by Eqs. (5-21), (5-35), and (5-36) or

$$\frac{\vartheta(x,\tau)}{Q_L/c} = \frac{1}{2}[1 - (1 + 4\mu)^{\frac{1}{2}}]\left(\frac{x}{X} - 1\right) + \frac{1}{8}[1 - (1 + 4\mu)^{\frac{1}{2}}]^2\left(\frac{x}{X} - 1\right)^2 \tag{5-40}$$

The temperature-time history at the surface $x = 0$ is of interest and can be obtained from Eq. (5-40)

$$\frac{\vartheta(0,\tau)}{Q_L/c} = -\frac{1}{4} + \frac{1}{4}(1 + 4\mu)^{\frac{1}{2}} + \tfrac{1}{2}\mu \tag{5-41}$$

Equation (5-41) can be cross-plotted with Eq. (5-37) and is shown graphically in Fig. 5-7.

[1] T. R. Goodman, Integral Methods for Non-linear Heat Transfer, *Advan. Heat Transfer*, **1**:51–122 (1964); see particularly pp. 85–96.

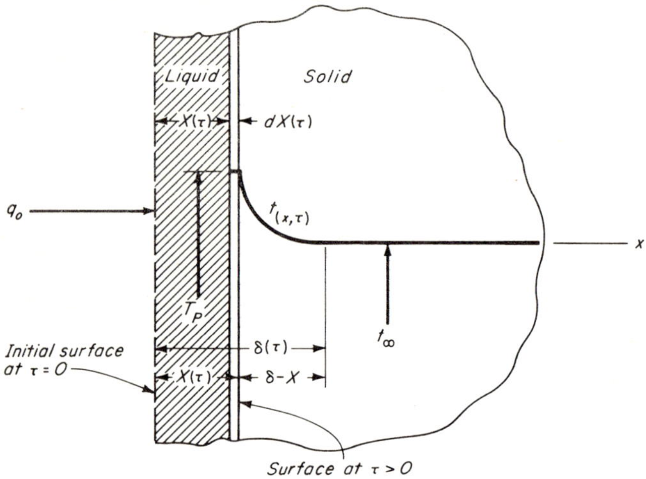

Fig. 5-8 Melting solid with complete removal of melt.

Melting of a solid with complete removal of melt (ablation) In the general problem of melting or solidification as treated earlier in this section there exists a temperature distribution in both the liquid and solid phases. A simplification already used is the assumption of a constant temperature equal to the phase-change temperature in the solid such that only the temperature distribution in the liquid need be considered.

Another simplification appropriate to problems of practical interest is to consider the melting liquid to be completely removed as it is formed, perhaps by aerodynamic forces, as in the process of *ablation*. In this case the surface recedes with time, the surface temperature remains constant at the phase-change temperature, and a temperature distribution exists only in the remaining solid (Fig. 5-8).

Assume that the semi-infinite solid is heated by the application of a constant heat flux q_0 to the surface $x = 0$. At time $\tau = 0$ the surface temperature has risen to the melting temperature, and the phase change ensues. The melted material is removed; thus the boundary of the solid and the melt line are identical and located at $x = X(\tau)$. The temperature distribution in the solid penetrates to a depth $\delta(\tau)$. The temperature for $x > \delta(\tau)$ is constant at a value t_∞.

Let $\vartheta(x,\tau) = t - t_\infty$ such that at the melt line

$$\vartheta(X,\tau) = T_P - t_\infty = \vartheta_P$$

the excess temperature for phase change. Assume further that the temperature distribution in the solid can be represented by a second-degree polynomial, or

$$\frac{\vartheta(x,\tau)}{\vartheta_P} = \left[1 - 2\frac{x - X}{\delta - X} + \left(\frac{x - X}{\delta - X}\right)^2\right] \tag{5-42}$$

such that

$$\vartheta(\delta,\tau) = 0 \qquad \text{and} \qquad \vartheta(X,\tau) = \vartheta_P$$

The heat balance at the melting face is

$$q_0 + k\frac{\partial\vartheta}{\partial x} = \rho Q_L \frac{dX}{\partial\tau} \tag{5-43}$$

for $\qquad x = X(\tau) \qquad \text{and} \qquad \tau > 0$

Using Eq. (1-28) and integrating with respect to x over the range X to δ gives

$$\frac{d}{d\tau}\left[\int_X^\delta \vartheta(x,\tau)\,dx - \vartheta(\delta,\tau)\delta(\tau) + \vartheta(X,\tau)X(\tau)\right] = \alpha\left[\frac{\partial\vartheta}{\partial x}(\delta,\tau) - \frac{\partial\vartheta}{\partial x}(X,\tau)\right] \tag{5-44}$$

From the known conditions $\vartheta(\delta,\tau) = 0$ and $\partial\vartheta/\partial x\ (\delta,\tau) = 0$. The value of $\partial\vartheta/\partial x\ (X,\tau)$ can be evaluated from Eq. (5-43) such that Eq. (5-44) becomes

$$\frac{d}{d\tau}\left[\int_X^\delta \vartheta(x,\tau)\,dx + \left(\vartheta_P + \frac{Q_L}{c}\right)X(\tau)\right] = \frac{q_0}{\rho c} \tag{5-45}$$

Evaluating the integral in Eq. (5-45) through the use of Eq. (5-42) yields the following differential equation:

$$\frac{d}{d\tau}\left[\frac{1}{3}(\delta - X)\right] + \left(1 + \frac{Q_L}{c\vartheta_P}\right)\frac{dX}{d\tau} = \frac{q_0}{\rho c\vartheta_P} \tag{5-46}$$

A second differential equation can be obtained from Eq. (5-43) if $\partial\vartheta/\partial x\ (X,\tau)$ is evaluated from Eq. (5-42),

$$\rho Q_L \frac{dX}{d\tau} = q_0 - \frac{2k\vartheta_P}{\delta - X} \tag{5-47}$$

Equations (5-46) and (5-47) are simultaneous equations, making the determination of $X(\tau)$ and $\delta - X$ possible. The initial condition for X is $X(0) = 0$. The initial condition for $\delta(\tau)$, the specification for δ when the surface reaches the melting temperature T_P, can be obtained from Eqs. (4-141) and (4-143), repeated here for convenience,

$$\delta(\tau) = \sqrt{6\alpha\tau}$$

which gives the value of $\delta(\tau)$ for a parabolic temperature profile, and

$$\vartheta(0,\tau) = \frac{q_0}{k}\sqrt{\tfrac{3}{2}\alpha\tau}$$

which gives the specification for the surface temperature as a result of a constant heat-flux input q_0.

Using the above equations, the value of $\delta(\tau_P)$ at the time the surface temperature first reaches ϑ_P may be calculated to be

$$\delta(\tau_P) = 2\frac{k\vartheta_P}{q_0} \tag{5-48}$$

which is the remaining required initial condition.[1]

Equations (5-46) and (5-47) have a steady-state solution if $dX/d\tau = A$, a constant. Thus from Eq. (5-47), $\delta - X$ must be constant; so from Eq. (5-46),

$$\frac{dX}{d\tau} = A = \frac{q_0}{\rho(Q_L + c\vartheta_P)} \tag{5-49}$$

This is exactly the same as the steady-state form of the exact solution given by Landau.[2]

Goodman[3] has solved Eqs. (5-46) and (5-47) by eliminating $dX/d\tau$ between them to obtain finally

$$\Omega = -\tfrac{1}{3}\left[\zeta - 2 + 2(1+\nu)\ln\frac{2(1+\nu)-\zeta}{2\nu}\right] \tag{5-50}$$

where $\qquad \Omega = \dfrac{q_0{}^2\tau}{\rho Q_L k\vartheta_P} \qquad \nu = \dfrac{Q_L}{c\vartheta_P} \qquad \zeta = \dfrac{q_0(\delta - X)}{\vartheta_P k}$

Equation (5-50) substituted into Eq. (5-47) yields

$$\lambda = -\tfrac{1}{3}\left[\zeta - 2 + 2\nu\ln\frac{2(1+\nu)-\zeta}{2\nu}\right] \tag{5-51}$$

where $\qquad \lambda = \dfrac{q_0 X}{k\vartheta_P}$

Equations (5-50) and (5-51) represent parametrically the variation of the melt line X with time τ. Goodman has shown the results graphically (Fig. 5-9).

Both Goodman and Biot have treated the melting solid for a variety of other boundary conditions of interest, such as the inclusion of vaporization and the boundary conditions resulting from aerodynamic heating or heat fluxes due to thermal radiation.

[1] Values more nearly in accord with the exact solution are shown elsewhere, e.g., Eqs. (4-188) and (4-190); however, Eq. (5-48) is used to be consistent with the assumption of a parabolic temperature profile.

[2] H. G. Landau, Heat Conduction in a Melting Solid, *Quart. Appl. Math.*, **8**(1):81–94 (1951), eq. 4-3.

[3] Goodman, Integral Methods for Non-linear Heat Transfer, *loc. cit.*

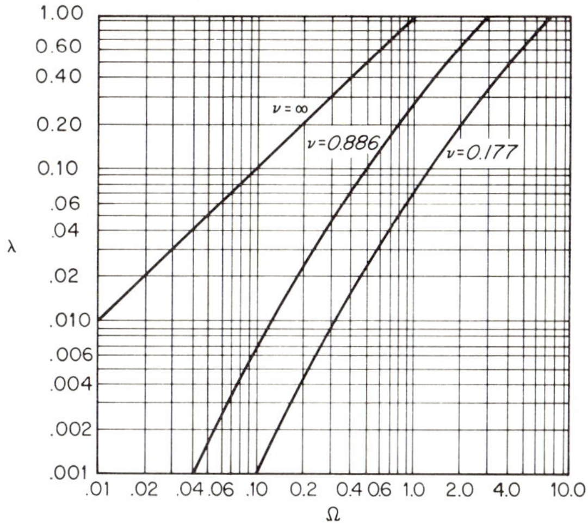

Fig. 5-9 Melt-line location as a function of time for a melting semi-infinite solid with complete removal of the melt. [*T. R. Goodman, Trans. ASME,* **80**:*335* (*1958*).]

5-2 MOVING HEAT SOURCES

Heat conduction in a body resulting from a moving heat source (or sink) is of great technical importance and has broad applications in the fields of arc welding, surface hardening, continuous casting or quenching, and cooling rotating systems with jets of coolant. Moving heat sources have been considered quite thoroughly by Rosenthal.[1] The analysis is based upon the notion that if a heat source is moved on a body of sufficiently large dimensions, a quasi-steady state occurs, which is to say that the system appears to be steady state from the standpoint of the observer located in and traveling with the source. For a three-dimensional system in the cartesian coordinate system, Eq. (1-16) is applicable if no heat sources are considered:

$$\frac{\partial t}{\partial \tau} = \alpha \left(\frac{\partial^2 t}{\partial x^2} + \frac{\partial^2 t}{\partial y^2} + \frac{\partial^2 t}{\partial z^2} \right)$$

If the properties of the material are considered to be constant, a quantity of heat at rate Q may be supplied by a point source moving along the x axis with a constant velocity U. Therefore, imagine an observer riding along with this heat source and having the material flow by him. If we supply to the observer a moving coordinate system of

[1] D. Rosenthal, The Theory of Moving Sources of Heat and Its Application to Metal Treatments, *Trans. ASME,* **68**:849–866 (1946).

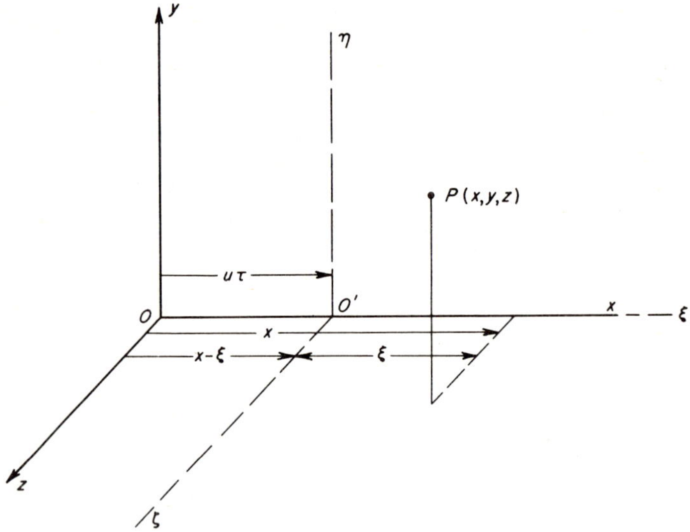

Fig. 5-10 Coordinate system for consideration of moving heat sources.

which he is the center, the system will appear in reference to the fixed system as shown in Fig. 5-10.

The point $P(x,y,z)$ in the fixed system becomes $P(\xi,\eta,\zeta)$ in the moving system, and since the translation in the x direction does not change the other coordinates $y = \eta$ and $z = \zeta$, then $P(x,y,z) = P(\xi,y,z)$.

It is necessary to make a transformation from the variables of the stationary coordinate system, where $t = f(x,y,z,\tau)$. In the new set of variables $\xi = x - U\tau$; thus $\partial\xi/\partial x = 1$, $\partial\xi/\partial\tau = -U$, and $\partial\tau'/\partial\tau = 1$. Thus the transformation proceeds:

$$\frac{\partial t}{\partial \tau} = \frac{\partial t}{\partial \xi}\frac{\partial \xi}{\partial \tau} + \frac{\partial t}{\partial \tau'}\frac{\partial \tau'}{\partial \tau} = -U\frac{\partial t}{\partial \xi} + \frac{\partial t}{\partial \tau'}$$

$$\frac{\partial^2 t}{\partial x^2} = \frac{\partial^2 t}{\partial \xi^2} \qquad \frac{\partial^2 t}{\partial y^2} = \frac{\partial^2 t}{\partial \eta^2} \qquad \frac{\partial^2 t}{\partial z^2} = \frac{\partial^2 t}{\partial \zeta^2}$$

If these substitutions are made in Eq. (1-16) neglecting the term $Q'/\rho c$, we obtain

$$\alpha\left(\frac{\partial^2 t}{\partial \xi^2} + \frac{\partial^2 t}{\partial y^2} + \frac{\partial^2 t}{\partial z^2}\right) = \frac{\partial t}{\partial \tau'} - U\frac{\partial t}{\partial \xi} \tag{5-52}$$

The value $\partial t/\partial\tau'$ is zero from the point of view of the observer at the source, and thus Eq. (5-52) becomes, in quasi-steady form,

$$\frac{\partial^2 t}{\partial \xi^2} + \frac{\partial^2 t}{\partial y^2} + \frac{\partial^2 t}{\partial z^2} = -\frac{U}{\alpha}\frac{\partial t}{\partial \xi} \tag{5-53}$$

The thin rod Consider a rod similar to the one in Sec. 3-3 which has constant cross-sectional area. The temperature distribution at any cross section is to be assumed constant; i.e., the physical condition allowing such an assumption is one where the resistance to heat loss from the rod surface is much greater than the internal resistance to heat flow in the rod. This ratio of resistances allows the equalization of the temperature at each location because of the low heat-transfer coefficient controlling the convective loss. In such a case there are no temperature gradients $\partial t/\partial y$ and $\partial t/\partial z$. Thus the appropriate differential equation for the problem becomes, in terms of the excess temperature $\vartheta = t - t_f$,

$$\frac{\partial \vartheta}{\partial \tau} = \alpha \frac{\partial^2 \vartheta}{\partial x^2} - \frac{Q'}{\rho c}$$

and if a surface-convection loss term is considered in the manner of Sec. 1-6, the above equation becomes

$$\frac{\partial \vartheta}{\partial \tau} = \alpha \left(\frac{\partial^2 \vartheta}{\partial x^2} - m^2 \vartheta \right) \tag{5-54}$$

where $m = \sqrt{hC/kA}$. Making the transformation to ξ in the quasi-stationary case, as above, Eq. (5-54) becomes

$$-\frac{U}{\alpha}\frac{\partial \vartheta}{\partial \xi} = \frac{\partial^2 \vartheta}{\partial \xi^2} - m^2 \vartheta \tag{5-55}$$

which has a solution of the form $\vartheta = e^{-(U/2\alpha)\xi} f(\xi)$, where $f(\xi)$ is to be determined. Thus, differentiating the expression for ϑ and substituting into Eq. (5-55), we obtain

$$f''(\xi) = \left[\left(\frac{U}{2\alpha} \right)^2 + m^2 \right] f(\xi) \tag{5-56}$$

Equation (5-56) has a solution

$$f(\xi) = Ae^{n\xi} + Be^{-n\xi}$$

where

$$n = \sqrt{\left(\frac{U}{2\alpha} \right)^2 + m^2} = \sqrt{\left(\frac{U}{2\alpha} \right) + \frac{hC}{kA}}$$

and since $\vartheta = e^{-(U/2\alpha)\xi} f(\xi)$, the expression for ϑ becomes

$$\vartheta = A \exp \left\{ + \left[\sqrt{\left(\frac{U}{2\alpha} \right)^2 + \frac{hC}{kA}} - \frac{U}{2\alpha} \right] \xi \right\} + B \exp \left\{ - \left[\sqrt{\left(\frac{U}{2\alpha} \right)^2 + \frac{hC}{kA}} + \frac{U}{2\alpha} \right] \xi \right\} \tag{5-57}$$

The appropriate boundary conditions to be applied to Eq. (5-57) are

$$\vartheta = 0 \qquad \xi \to \pm \infty$$

The solution takes two forms, depending on whether ξ is greater or less than zero. These are

$$\vartheta = A \exp\left\{+\left[\sqrt{\left(\frac{U}{2\alpha}\right)^2 + \frac{hC}{kA}} - \frac{U}{2\alpha}\right]\xi\right\} \qquad \xi < 0$$

$$\vartheta = B \exp\left\{-\left[\sqrt{\left(\frac{U}{2\alpha}\right)^2 + \frac{hC}{kA}} + \frac{U}{2\alpha}\right]\xi\right\} \qquad \xi > 0$$

At $\xi = 0$, the temperatures of both branches of the solution are equal and represent the maximum temperature. Thus $A = B = \vartheta_{\max}$.

$$\begin{aligned} \vartheta &= \vartheta_{\max} \exp\left\{+\left[\sqrt{\left(\frac{U}{2\alpha}\right)^2 + \frac{hC}{kA}} - \frac{U}{2\alpha}\right]\xi\right\} \qquad \xi < 0 \\ \vartheta &= \vartheta_{\max} \exp\left\{-\left[\sqrt{\left(\frac{U}{2\alpha}\right)^2 + \frac{hC}{kA}} + \frac{U}{2\alpha}\right]\xi\right\} \qquad \xi > 0 \end{aligned} \tag{5-58}$$

The heat flow in each direction from $\xi = 0$ can be obtained from Eqs. (5-58) by differentiating with respect to ξ and substituting into the expression

$$Q = -kA \frac{dt}{d\xi}$$

to yield

$$\begin{aligned} Q_1 &= kA\vartheta_{\max}\left[\sqrt{\left(\frac{U}{2\alpha}\right)^2 + \frac{hC}{kA}} - \frac{U}{2\alpha}\right] \qquad \xi < 0 \\ Q_2 &= kA\vartheta_{\max}\left[\sqrt{\left(\frac{U}{2\alpha}\right)^2 + \frac{hC}{kA}} + \frac{U}{2\alpha}\right] \qquad \xi > 0 \end{aligned} \tag{5-59}$$

Since the total output of the heat source must be equal to the sum of the positive and negative heat flow, this relationship permits calculation of the maximum temperature $\vartheta_{\max}$, which becomes, in terms of the total heat output of the source,

$$\vartheta_{\max} = \frac{Q_1 + Q_2}{2kA\sqrt{(U/2\alpha)^2 + hC/kA}} \tag{5-60}$$

The result of Eqs. (5-58) is given in Fig. 5-11, showing that more heat is conducted ahead of the source than is conducted away behind the source.

The results of Eqs. (5-58) to (5-60) can be modified to express the physical case of a thin rod insulated on the surface. For this case $h = 0$, and the equations readily reduce.

The point source If the heat flows in the three dimensions from a point source at 0′ moving at velocity U in the x direction (Fig. 5-10), the

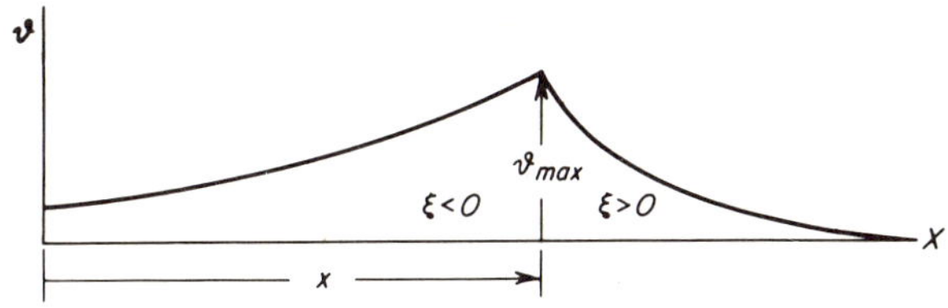

Fig. 5-11 Temperature distribution in a thin rod as a result of a heat source moving along the rod.

system is treated with more simplicity if the radius from $0'$ is used rather than the coordinates x, y, and z. Equation (5-53) has a solution of the form

$$\vartheta = e^{-(U/2\alpha)\xi} f(\xi,y,z)$$

which, when used in Eq. (5-53), results in the differential equation for the unknown function $f(\xi,y,z)$

$$\frac{\partial^2 f}{\partial \xi^2} + \frac{\partial^2 f}{\partial y^2} + \frac{\partial^2 f}{\partial z^2} - \left(\frac{U}{2\alpha}\right)^2 f = 0$$

This auxiliary equation can be written, for $r = \sqrt{\xi^2 + y^2 + z^2}$, in the form

$$\frac{d^2 f}{dr^2} + \frac{2}{r}\frac{df}{dr} - \left(\frac{U}{2\alpha}\right)^2 f = 0 \tag{5-61}$$

when the initial and surface temperatures are such that the isothermal surfaces are spheres, and thus the temperature depends on r alone in this quasi-steady condition. This requires boundary conditions on ϑ that will satisfy the notion of radial heat flow from the source. These may be

$$\begin{aligned} \vartheta &= 0 & r &\rightarrow \infty \\ -\frac{\partial \vartheta}{\partial r} &= \frac{q'}{4\pi k r^2} & r &= 0 \end{aligned} \tag{5-62}$$

where q' is the rate of heat flow from the source.

Solution of Eq. (5-61) can be effected by the transformation[1] $g = fr$ to yield

$$\frac{d^2 g}{dr^2} - \left(\frac{U}{2\alpha}\right)^2 g = 0$$

which has a standard solution

$$g = Ae^{+(U/2\alpha)r} + Be^{-(U/2\alpha)r}$$

[1] Carslaw and Jaeger, *op. cit.*, p. 198.

We want the solution for ϑ, which is

$$\vartheta = e^{-(U/2\alpha)\xi}\frac{g}{r}$$

Thus
$$\vartheta = \frac{1}{r}[Ae^{(U/2\alpha)(r-\xi)} + Be^{-(U/2\alpha)(r+\xi)}] \tag{5-63}$$

The boundary conditions proposed in Eqs. (5-62) are used in the solution of Eq. (5-63) to obtain finally

$$\vartheta = \frac{q'}{4\pi kr}e^{-(U/2\alpha)(r+\xi)} \tag{5-64}$$

Equation (5-64) gives the temperature distribution about a moving point source in an infinite medium. This result is approximately valid for a source moving on the surface of a semi-infinite medium if the exposed surface is losing heat at a rate which is negligible compared with the source strength. A physical application of this analysis is in the treatment of a welding electrode moving along the surface of a very thick plate where there are minimal losses. In the welding problem only the lower half plane need be considered as the body. Equation (5-64) must be modified to account for this to read

$$\vartheta = \frac{q''}{2\pi kr}e^{-(U/2\alpha)(r+\xi)} \tag{5-65}$$

where q'' is the heat release of the electrode.

The above methods apply to problems in arc welding, extrusion, quenching, annealing, and the passage of projectiles through a gun bore.

part II

Heat Transfer by Convection

II-1 INTRODUCTION

Conduction as it has been discussed in Part I is, in general, the sole agent which transports energy within a solid material. For fluids, however, this applies only under special conditions because fluids are usually in motion. Thus the moving fluid particles carry heat along in the form of internal energy, a mode of energy transport called *convection.* Heat transfer in a fluid is therefore usually caused by an interplay between convection and conduction. Motion in a fluid can be created by external influences, e.g., a blower or a pump. In this case one talks about *forced convection.* Even in the absence of such external influences, however, motion is frequently created in a fluid itself as a consequence of temperature differences. Buoyancy forces then create motion which is referred to as *free convection.* In a more general way, free convection is caused by motion created by any body force within the system in which the heat transfer occurs. Heat transfer under the combined action of convection and conduction will be discussed in Part II. This overall process is often referred to simply as convective heat transfer.

An analysis of convective heat transfer is more involved than that of heat transfer by conduction alone because the motion of the fluid must be studied simultaneously with the energy-transfer process. The laws of mechanics and thermodynamics have to be applied for this purpose. There are, in principle, two ways in which one can proceed. The fluid can be considered as consisting of particles (molecules, atoms, electrons), and the above-mentioned laws can be applied to them to discuss their motion and to obtain statistical information on the behavior of groups of such particles. Such a description gives an insight into the details of the energy-transfer processes; however, it is too cumbersome for most situations arising in engineering. A second approach applies the basic laws of mechanics and thermodynamics to portions of the fluid which contain a very large number of particles and is satisfied to obtain information on the parameters describing the movement of those portions. In this approach one forgets about the single particles and considers the fluid as a continuum. This approach is simpler and usually gives the answers required in engineering; on the other hand, the information is not so complete, and certain parameters, such as thermodynamic state and transport properties, have to be introduced empirically. This continuum approach will be used in Part II with the exception of Sec. 11-3, on rarefied gases, where it is no longer satisfactory. Parallel to the study of the heat-transfer process by continuum mechanics, molecular considerations can be used to obtain information on transport and thermodynamic properties.

The main portion of the resistance to heat transfer is usually concentrated in a thin layer immediately adjacent to the wall surface. Heat transfer is essentially a problem of the interplay of heat conduction and energy transport by the moving fluid within this layer. As soon as the heat has penetrated this layer, it is readily carried away by the core of the fluid. Hence, heat transfer is determined essentially by the thickness and the characteristics of this boundary layer; it in turn depends on all the parameters determining the flow along the surface.

For forced convection we shall later deduce, from the differential equations which describe the fluid-flow and the heat-flow processes, the fact that neither the velocity field nor the development of the boundary layer is influenced by heat transfer as long as the fluid property values which enter into the flow equations are independent of temperature. The development of the boundary layer and of the flow is then purely a problem in fluid mechanics. However, it can be seen from the property-value tables in Appendix B that all property values actually depend on temperature. In such a case an interrelation exists between the heat-flow and the fluid-flow processes. Such an interrelation makes understanding the heat-transfer process much more difficult, and in order to avoid this complication, a considerable portion of the discussion in this part will be devoted to an idealized fluid with properties which are independent of temperature. This procedure has an additional advantage: only for such an ideal fluid can relationships be developed which are universally applicable, whereas relations for fluids whose properties vary with temperature or pressure hold for a specific fluid or at best for a certain group of fluids only. The recognition of this fact and its utilization to develop universal relations were the most significant contributions of Nusselt (1916). Finally, many situations arise in applications in which the temperature differences within the flow field are sufficiently small to permit the variation of properties caused by temperature differences to be neglected. In these cases relationships developed for the ideal constant-property fluid accurately describe the actual heat-transfer process. For such a situation it makes no difference whether the fluid is a liquid or a gas, and the ideal-fluid relationships hold for both. In gases, the fluid properties, and especially the density, depend not only on temperature but also on pressure, and therefore the pressure variation must remain small if the density variation connected with it is to be considered small. This is the case as long as the flow velocities are small compared with the sound velocity. In airflow, for instance, it is found that constant-property relations describe the actual situation well up to velocities of approximately 100 m/s.

In engineering applications very large variations of the properties sometimes arise. For instance, the temperature dependence of the viscosity of oil is such that it influences the heat transfer in oil coolers considerably, even when the temperature differences are moderate. In nuclear reactors or in many high-temperature processes the temperature differences are so great that they cause large variations of the properties. In gases at high subsonic and especially supersonic velocities large pressure and temperature variations are inherently connected with the flow process. Heat transfer under such conditions will be discussed in later chapters of this part.

This text deals almost exclusively with heat transfer for steady-state processes. In an unsteady process (transient and periodic) the given relationships change as soon as the temperatures in the boundary layer lag noticeably owing to its heat capacity (approximately as in Fig. 4-19). However, this is the case for very rapid changes only, because the boundary-layer thicknesses are generally small. Heat transfer in the cylinders of diesel engines, for instance, can still be computed with steady-state heat-transfer coefficients,[1] and accelerations of missiles or aircraft of the order of 20 to 40 g's are necessary to cause a heat transfer to the skin of the craft which is markedly different from steady-state heat transfer.[2]

Fourier's law of heat conduction, discussed in Part I, expresses a linear relationship between heat flux and the driving force, a temperature gradient. In an analogous way, Newton introduced a linear relation between the heat flux q transferred to a solid surface per unit time and area by a fluid moving over it and a characteristic temperature difference Δt

$$q = h \, \Delta t$$

The proportionality factor h is called a heat-transfer coefficient or sometimes unit surface conductance. It was soon established that h is not a

[1] H. Pfriem, Unsteady Heat Transfer in Gases, Especially in Piston Engines, *VDI-Forschungsh.* 413, 1942.

[2] K. Stewartson, *Quart. J. Mech. Appl. Math.*, **4**(2):182–198 (1951); E. M. Sparrow, *Jet Propulsion*, **28**:403 (1958); E. M. Sparrow and J. L. Gregg, *NACA Tech. Note* 4311 (1958); R. Siegel, *Trans. ASME*, **80**:347 (1958).

constant but a function of many parameters describing the flow and heat-transfer process and that even the relation between heat flux and temperature difference is sometimes nonlinear. Nevertheless, the use of a heat-transfer coefficient is by now firmly established and will be continued in this text.

II-2 BOUNDARY LAYER AND TURBULENCE

Two concepts of fluid mechanics are of special importance in any convective-heat-transfer studies, namely, the concept of a boundary layer or shear layer and that of turbulence. They will be reviewed briefly in the present section before entering into a detailed discussion in the following chapters.

The movement of a fluid (liquid or gas) is determined by the forces to which the various fluid particles are subjected. A considerable number of such forces may occur in a specific problem: *body forces,* like gravity, electric or magnetic forces, centrifugal or Coriolis forces, and *surface forces,* which have to be introduced to describe the influence of the surrounding fluid on the fluid particle under consideration, like tensile forces and shear forces. Surface forces referred to a unit surface area are called *stresses.* Such stresses are also introduced in solid mechanics, where they are related to the local deformation of the medium. A fluid, on the other hand, is defined as a substance in which tensile and shear stresses are related to the time rate of the deformation. No such stresses exist in a fluid when it is at rest, with the exception of a tensile stress, the pressure. Among the remaining stresses, referred to as viscous stresses, the shear stresses are the important ones in our context.

An expression connecting the shear stress with the rate of deformation was introduced by Newton for a simple flow situation as illustrated in Fig. II-1, where it is assumed that all velocities are parallel to each other and that they change in a direction normal to the velocity vectors only. Such a flow is referred to as a *simple shear flow.* The shears of interest to us are those in any plane parallel to the velocity vectors like the plane 1-1 in Fig. II-1. The fluid above the plane exerts on the fluid below a shear τ in the flow direction, and the fluid below the plane exerts the same shear on the fluid above opposite to the flow direction. This shear stress (force per unit area) is given according to Newton by the equation

$$\tau = \mu \frac{du}{dy} \qquad \text{(II-1)}$$

For other types of flow the expressions connecting the velocities with the shear are more complicated and will be discussed later. Fluids in which

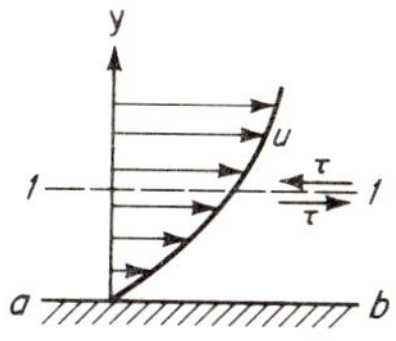

Fig. II-1 Simple shear flow of a fluid along a solid wall.

a linear relation, as in Eq. (II-1), exists between the shear stresses and the velocities are called newtonian fluids. Most fluids are newtonian, but some follow different laws. A few expressions which have been proposed to describe the shear stress for simple shear flow in such fluids are listed below:

$$\tau = m\left(\frac{du}{dy}\right)^n \qquad \text{Ostwald–de Waele fluid}$$

$$\tau = A \sinh^{-1}\left(\frac{1}{B}\frac{du}{dy}\right) \qquad \text{Eyring fluid}$$

$$\left.\begin{aligned} \tau &= \mu\frac{du}{dy} + \tau_0 && \text{if } \tau > \tau_0 \\ \frac{du}{dy} &= 0 && \text{if } \tau < \tau_0 \end{aligned}\right\} \qquad \text{Bingham plastic}$$

In these equations m, n, A, and B are constants. It may be observed that the fluid for which the shear is described by Bingham's law really cannot be classified as a fluid in the sense of the definition given above because the shear does not vanish at zero velocity. It constitutes a substance with a behavior between that of a fluid and a solid in the usual sense.

In this book, the discussion will be restricted to newtonian fluids. The transport property μ in Eq. (II-1) has the following dimension in the SI system

$$\frac{\text{N s}}{\text{m}^2} \qquad \text{or} \qquad \frac{\text{kg}}{\text{m s}}$$

The corresponding unit in the cgs system, g/cm s, is referred to as a *poise*.

A second property, called the *kinematic viscosity* ν, is also frequently used. It is the ratio of the viscosity μ to the fluid density ρ

$$\nu = \frac{\mu}{\rho}$$

Its dimension is m^2/s or cm^2/s in the two systems mentioned above. The unit cm^2/s is also referred to as a *stoke*. Appendix B contains tables and diagrams of both transport properties.

It is found that for most fluids the viscosity μ has values which are quite small. As a consequence, shear stresses according to Eq. (II-1) assume larger values only in regions where the velocity gradient du/dy is especially large; outside of these they can usually be neglected compared to the other forces. As a matter of fact, many aspects of the behavior of a fluid can often be described satisfactorily by neglecting the shear forces altogether. One then replaces the actual fluid by an idealized model, the *inviscid fluid*. In a real fluid, however, there always exist regions in a flow field in which the velocity gradients are large enough to make the influence of shear forces important, and these are the regions which are of special concern in heat-transfer calculations.

Figure II-2 indicates such regions. Figure II-2a is the sketch of the flow along a solid surface. It is found that fluid particles in immediate contact with the surface generally have a zero velocity if the wall is at rest. The velocity then increases rapidly in a thin layer of thickness δ called a *boundary layer*. Figure II-2b is a sketch of the situation in which two parallel streams with different velocities come together. A free boundary layer or shear layer is created in this instance with a thickness which increases in the downstream direction and in which velocity gradients again are large. Figure II-2c shows two boundary layers meeting at the downstream end of a thin plate and forming a region called a *wake*. A special situation exists in a channel with considerable length. Figure II-2d indicates that boundary layers form in such a channel along the walls and increase in thickness in the downstream direction until they meet in the center. Downstream from this point the channel is filled

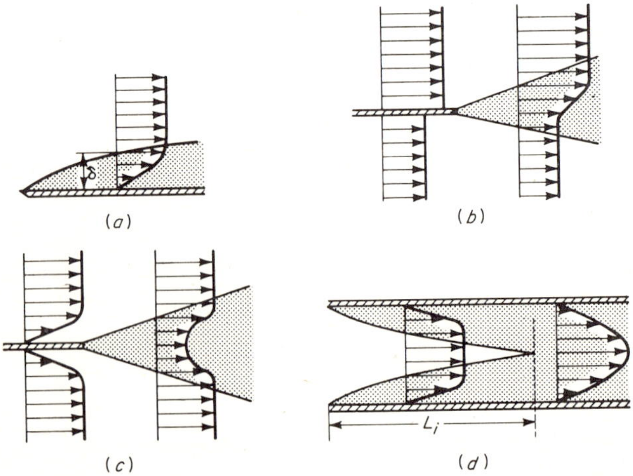

Fig. II-2 Boundary-layer flows.

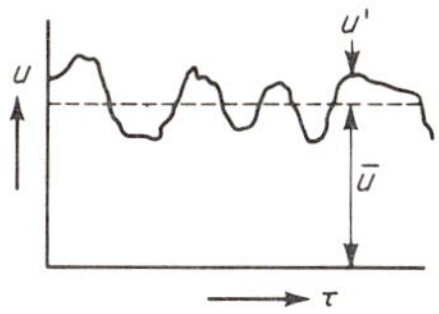

Fig. II-3 Timewise fluctuations of the velocity component u in turbulent flow.

with a fluid which must be considered as viscous in its full extent. The region upstream from the cross section where the boundary layers meet is called the *inlet region*, and its length is the *inlet length* L_i. The region downstream from the point where the boundary layers have met is referred to as the *developed-flow region*. The regions shown in Fig. II-2 are those with which we shall be especially concerned in our discussions of convective heat transfer.

In 1883, Osborne Reynolds demonstrated for the first time that two basically different forms of flow exist, namely, *laminar* and *turbulent flow*. In laminar flow, the individual streamlines run in an orderly fashion side by side, while in turbulent flow the streamlines are interwoven with each other in an irregular manner. The fluid particles execute fluctuating motions around some mean flow path, and vortices of different size and orientation occur. A quantitative description of turbulent flow is difficult because of its irregular character. One important parameter describing the intensity of the turbulent fluctuations goes back to a consideration already introduced by Reynolds. Imagine that one measures the velocity component u in turbulent flow at a fixed location as a function of time. The result will be a plot like the one presented in Fig. II-3. The velocity component u at any moment can then be written as

$$u = \bar{u} + u' \qquad \text{(II-2)}$$

where $\bar{u}$ is the time-averaged value of u and u' is the instantaneous value of the timewise fluctuation. Turbulent flow is called steady if $\bar{u}$ does not change with time when taken over a sufficiently long interval. But even for unsteady turbulent flow the time scale by which the mean velocity $\bar{u}$ changes is usually very much larger than the time scale for the turbulent fluctuations, and therefore the separation into an "instantaneous" mean and a fluctuating value can still be made. The time average $\overline{u'}$ of the fluctuating velocity is zero, according to our definition. Relations similar to the one in Eq. (II-2) exist for the velocity components v and w. Let us assume that $\bar{v} = \bar{w} = 0$, which means that the mean velocity is in the x direction. Then the turbulence intensity J is defined by the expression

$$J = \frac{\sqrt{\frac{1}{3}(\overline{u'^2} + \overline{v'^2} + \overline{w'^2})}}{\bar{u}} \qquad \text{(II-3)}$$

in which the term $\overline{u'^2}$ is the time mean of the square of the velocity fluctuation u'. Turbulence intensities encountered in boundary layers, duct flows, and free shear layers or behind screens are of the order of 5 to 10 percent. Normal wind tunnels may have 1 percent turbulence intensity, but by taking special care in the design it can be decreased to values of the order of 0.05 percent.

Other parameters are required to give a more complete description of the nature of the turbulent fluctuations and to describe their influence on heat transfer. For instance, the length scale of the turbulent fluctuations and of the vortices should surely be considered. A parameter which is an indication of the scale of turbulence can be obtained by hot-wire measurements, which determine the instantaneous velocity fluctuation u_1' and u_2' simultaneously but at a varying distance y. The parameter

$$R(y) = \frac{\overline{u_1' u_2'}}{\sqrt{\overline{u_1'^2}}\sqrt{\overline{u_2'^2}}} \tag{II-4}$$

assumes the value 1 for $y = 0$ and decreases toward 0 as y increases. The length L defined by

$$L = \int_0^\infty R\,dy \tag{II-5}$$

is then a measure of the average scale of the turbulent fluctuations and vortices.[1] Heat exchange by convection is promoted by the fluctuating motions in turbulent flow; consequently heat transfer in turbulent flow is generally much larger than in laminar flow.

Actually a connection exists between the shear layers discussed before and the turbulence in that such shear layers tend to break up into vortices and then to degenerate into the irregular statistical motions of turbulence. In this way they are the sources of turbulence in fluid streams. For example, the boundary layer which covers the surface of a flat plate exposed to a uniform flow (Fig. II-4) is found to increase in thickness in the downstream direction and to become turbulent when its thickness exceeds a certain value δ_c. Actually the point of transition to turbulence within the boundary layer is described by a critical value of a dimensionless parameter $u_s x_c/\nu$. Its importance in fluid mechanics was

[1] Of course similar length parameters can also be formed by varying the distance x or using different velocity components.

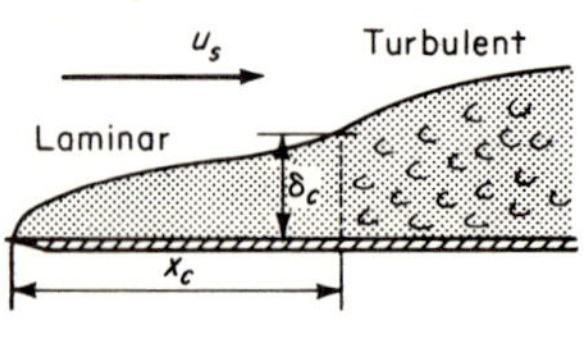

Fig. II-4 Laminar and turbulent boundary layer on a flat plate.

recognized by Reynolds, and it is therefore called the *Reynolds number*. The terms appearing in the Reynolds number are the free-stream velocity u_s outside the boundary layer, the distance x_c between the leading edge of the plate and the place where transition to turbulence occurs, and the kinematic viscosity $\nu = \mu/\rho$. Sometimes the boundary-layer thickness δ_c at the point of transition is used instead of length x_c as a characteristic length in the critical Reynolds number. Transition to turbulence occurs on the flat plate at a value of the critical Reynolds number based on x_c around 500,000. It is found, however, that this value depends on various characteristics of the flow, e.g., on the turbulence level in the approaching stream, on the surface roughness, and the leading-edge configuration of the flat plate.

The influence of the turbulence level in the free stream has been established by measurements at the National Bureau of Standards[1] in which the turbulence level was systematically varied (Fig. II-5). The intensity J of the free-stream turbulence is plotted on the ordinate, and the critical Reynolds number Re_c for transition, based on the length x_c, is on the abscissa. It can be seen that the transition Reynolds number increases with a decrease in the free-stream turbulence. This, however, holds only down to a certain turbulence level. Below the value of 0.08 percent, the transition Reynolds number no longer depends on the free-stream turbulence. This Reynolds number therefore constitutes an upper limit for laminar boundary-layer flow on the flat plate. At these small turbulence levels, it takes some time to develop the turbulence. For this reason one can define a lower and an upper transition Reynolds number, depending on whether the first indication of turbulence or its establishment is considered. The two curves in Fig. II-5 indicate these values. It is also found that the transition process generally fluctuates in time, so that the values in Fig. II-5 actually present time-averaged Reynolds numbers.

Through the last hundred years, many attempts have been made to predict the transition process analytically. These efforts have led to success insofar as analysis showed that shear flow becomes unstable under certain conditions. In such an analysis small sinusoidal fluctuations are superimposed on a steady shear flow, and one investigates under what conditions such fluctuations are decreased or increased in amplitude while they travel downstream. The results of such an analysis by Tollmien[2] for boundary-layer flow over a flat plate are presented in Fig. II-6. The abscissa presents the Reynolds number based on the boundary-layer dis-

[1] G. B. Schnbauer and H. K. Skramstad, *J. Res. Natl. Bur. Std.*, **38**:281 (1947).

[2] W. Tollmien, *Nachr. Wiss. Göttingen, Math. Physik. Kl.*, **1929**:21–44; Engl. trans. *NACA Tech. Memo* 609 (1931).

placement thickness[1] δ^* and the ordinate the ratio of boundary-layer displacement thickness to the wavelength λ of the superimposed fluctuation. The curve in this figure separates two regions: the outside region, where the flow is stable in the sense mentioned above, and the inside region, where the flow is unstable. Assuming that in reality small disturbances of any wavelength are always present, one defines that value Re_c of the Reynolds number at which at least one wavelength becomes unstable as the *critical* or *instability Reynolds number.* Tollmien found in this way an instability Reynolds number

$$\mathrm{Re}_c = \frac{u_s \delta_c^*}{\nu} = 420$$

[1] For its definition see page 286.

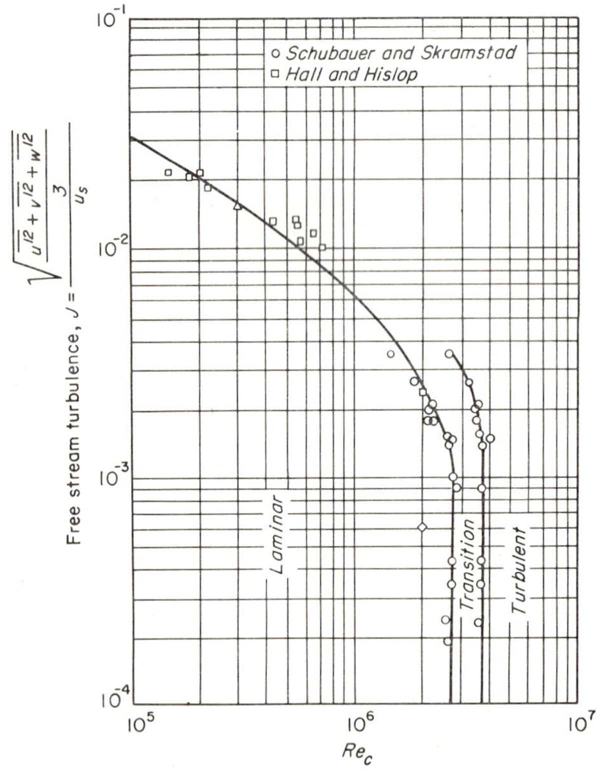

Fig. 11-5 Critical Reynolds number Re_c as affected by free-stream turbulence with intensity J. [*After G. B. Schubauer and H. K. Skramstad. J. Res. Natl. Bur. Std.,* **38**:*281 (1947) and A. A. Hall and G. S. Hislop, Aeron. Res. Comm. Rep. Mem., 1843, 1938.*]

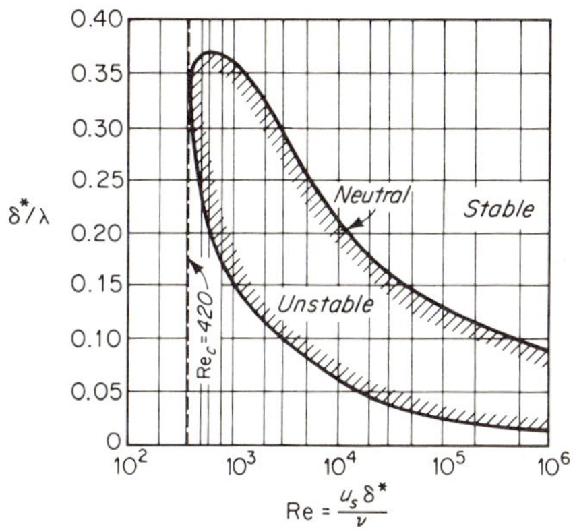

Fig. II-6 Stable and unstable regimes in boundary-layer flow over a flat plate. [*After W. Tollmen. Nacler. Wiss. Göttingen, Math. Physik. Kl., (1929) 21–44.*]

This corresponds to a value 1.1×10^5 of the instability Reynolds number based on distance x_c from the leading edge of the plate. Figure II-5 indicates that the corresponding experimental value for small free-stream turbulence is 2.5×10^6. The analysis, therefore, gives the right order of magnitude for the critical Reynolds number. The difference in the values may be caused by the fact that it takes some time for the velocity fluctuations to grow large enough to become measurable.

Critical Reynolds numbers have been determined analytically and experimentally for many other flow situations. They have been found, for instance, to depend strongly on the pressure variation existing along the surface of an object exposed to flow. A pressure drop in flow direction increases and a pressure rise decreases the critical Reynolds number. Free shear flows have considerably smaller critical Reynolds numbers than boundary layers. In duct flows, transition to turbulence can occur at different positions in the inlet region. The fully developed region is then either laminar or turbulent, depending on the value of a critical Reynolds number which, in this case, is based on some characteristic dimension of the duct cross section; for a round tube, for instance, it is the tube diameter. Under normal circumstances it is found that, for a Reynolds number below 2,300, the flow even in the developed region is normally laminar whereas for larger values of the Reynolds number it is usually turbulent. By carefully avoiding all disturbances, however, the critical Reynolds number can be increased considerably, and values as high as 500,000 have been reached. Knowledge of the transition Reynolds number is essential for heat-transfer calculations. The available space, however, does not permit further discussion of this aspect of fluid mechanics, and the reader is referred to the literature, e.g., the text by Schlichting.[1]

[1] H. Schlichting, "Boundary Layer Theory," 6th ed., McGraw-Hill, New York, 1968.

6
Derivation of the Conservation Equations

In the continuum approach, which is used in this text (exception in the section on rarefied-gas flow), the laws of mechanics and thermodynamics are applied to the medium in a conveniently selected control volume separated from the rest of the fluid by control surfaces. The lagrangian method considers this control volume as moving with the fluid, whereas in Euler's method, the control volume is considered fixed relative to a coordinate system with the fluid moving through it. Euler's method will be used in this section. If information on local quantities is required, a very small control volume should be selected. One must be careful, on the other hand, that the particle structure of the fluid does not create difficulties. This becomes evident if one considers, for example, the definition of the local density in the fluid. In principle, the density is obtained by considering a certain volume $\Delta\mathcal{V}$ in the fluid which may have a mass Δm. The density then is defined as

$$\bar{\rho} = \frac{\Delta m}{\Delta \mathcal{V}} \tag{6-1}$$

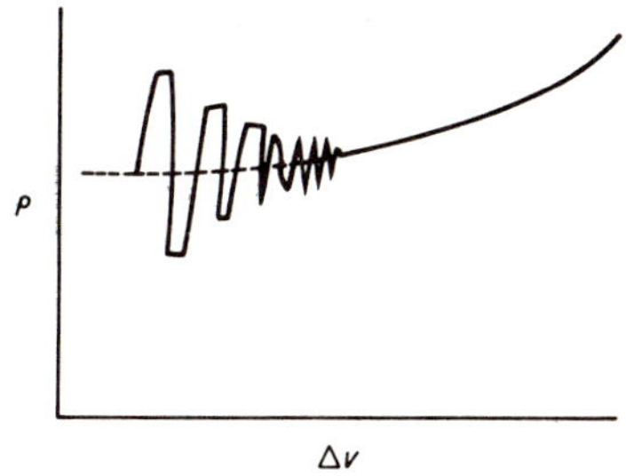

Fig. 6-1 Definition of a local density in a fluid.

If the density varies locally, a very small control volume must be selected; otherwise the value defined in Eq. (6-1) will be an average density. In general, this density will change as the size of the volume $\Delta\mathcal{V}$ is decreased, as indicated in Fig. 6-1. However, when the size of the control volume is so small that only a few molecules are contained in it, a further decrease will cause the density to change abruptly when a molecule is excluded. It will then vary irregularly with decreasing $\Delta\mathcal{V}$, as indicated in the left-hand part of Fig. 6-1. In the continuum model, it is assumed that the density asymptotically approaches a limiting value before this irregular fluctuation occurs. The limiting value is then called the *local density.* This obviously will impose a limit on the density gradient at the location under consideration beyond which the continuum approach ceases to be valid.

Similar considerations hold if one defines a velocity. Imagine for this purpose an area element ΔA located arbitrarily in the fluid and assume that one measures the mass flow Δm through it in a time increment $\Delta\tau$. The velocity component V_n in the direction of the surface normal is then defined by the equation

$$V_n = \frac{\Delta m}{\bar{\rho}\,\Delta A\,\Delta\tau} \tag{6-2}$$

In general, this gives only an average value of the velocity, averaged with respect to location and time, and one must again assume that the local variation of the velocity is such that a limiting value is closely approximated with decreasing size of ΔA and $\Delta\tau$, before the molecular structure causes difficulties. In the lagrangian method, where the control volume moves with the fluid, the velocity V through any part of the control surface must be zero. This, however, does not mean that the control volume always contains the same particles (because molecules in their irregular random movement will cross back and forth through the control surface) but only that the same mass of molecules moves out of the control volume through any ΔA as moves in. It is this crisscrossing of the molecules through the control surface that in gases is responsible

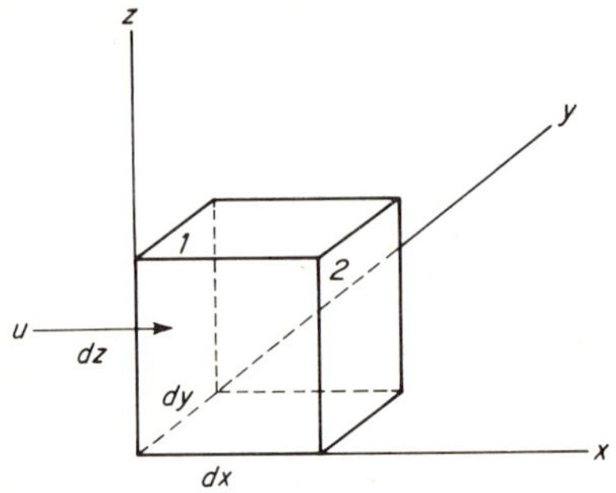

Fig. 6-2 Control volume in a system of cartesian coordinates.

for the fact that we have to introduce surface forces on the control volume in the continuum approach. In liquids, this effect is supplemented by forces between neighboring molecules.

6-1 CONTINUITY EQUATION

The first conservation law to be considered, the continuity equation, will be derived in cartesian coordinates because later we shall need the equations in this form. For this purpose we consider a system with the coordinates x, y, z located in our fluid and select a control volume with dimensions dx, dy, dz, as shown in Fig. 6-2, fixed with respect to this coordinate system. The fact that in all cases considered here no mass is created or destroyed can then be expressed in the way that the difference between the mass flow into the control volume and out of it must equal the increase in mass contained in the control volume. The mass entering per unit time through area 1 is $\rho u\, dy\, dz$. Considering the fact that in a continuum the mass flow can vary only steadily, the mass leaving the control volume through area 2 can be expressed

$$\rho u\, dy\, dz + \frac{\partial}{\partial x}(\rho u\, dy\, dz)\, dx$$

or, since the size of the area $dy\, dz$ does not change with x,

$$\rho u\, dy\, dx + \frac{\partial(\rho u)}{\partial x}\, dy\, dz\, dx$$

In this way the following scheme can be developed: flow leaving minus flow entering:

In x direction: $\quad \dfrac{\partial(\rho u)}{\partial x}\, dy\, dz\, dx$

In y direction: $\quad \dfrac{\partial(\rho v)}{\partial y}\, dx\, dz\, dy$

In z direction: $\quad \dfrac{\partial(\rho w)}{\partial z}\, dx\, dy\, dz$

The mass stored in the control volume per unit time is $(\partial/\partial\tau)(\rho\, dx\, dy\, dz)$ or $(\partial\rho/\partial\tau)\, dx\, dy\, dz$. The law of conservation of mass leads, therefore, to the equation

$$\frac{\partial\rho}{\partial\tau} + \frac{\partial(\rho u)}{\partial x} + \frac{\partial(\rho v)}{\partial y} + \frac{\partial(\rho w)}{\partial z} = 0 \tag{6-3}$$

The term $\partial\rho/\partial\tau$ indicates the change of density with time at a fixed location. The change of density in a volume element which moves along with the flow can be obtained from the total differential

$$\frac{d}{d\tau} = \frac{\partial}{\partial\tau} + \frac{\partial}{\partial x}\frac{dx}{d\tau} + \frac{\partial}{\partial y}\frac{dy}{d\tau} + \frac{\partial}{\partial z}\frac{dz}{d\tau}$$

For a volume element moving along with the flow

$$\frac{dx}{d\tau} = u \qquad \frac{dy}{d\tau} = v \qquad \frac{dz}{d\tau} = w$$

and

$$\frac{D}{d\tau} = \frac{\partial}{\partial\tau} + u\frac{\partial}{\partial x} + v\frac{\partial}{\partial y} + w\frac{\partial}{\partial z} \tag{6-4}$$

This operator is called the *substantial derivative.* The capital letter D is used in order to identify it clearly.

The continuity equation can now be written

$$\frac{D\rho}{d\tau} + \rho\left(\frac{\partial u}{\partial x} + \frac{\partial v}{\partial y} + \frac{\partial w}{\partial z}\right) = 0 \tag{6-5}$$

In vector notation, the two forms of the continuity equation read

$$\frac{\partial\rho}{\partial\tau} + \operatorname{div} \rho\mathbf{v} = 0 \tag{6-6}$$

$$\frac{D\rho}{d\tau} + \rho \operatorname{div} \mathbf{v} = 0 \tag{6-7}$$

where $\mathbf{v}$ indicates the velocity vector.

Another way to simplify the writing of the conservation equations, the tensor notation, is being used more and more. In this notation, the three coordinates of the cartesian system are written as x_1, x_2, x_3 or generally as x_i with i having values from 1 to 3. Analogously, the velocity components are v_1, v_2, v_3, and v_i stands for the three components of the velocity vector. Equation (6-3) is now

$$\frac{\partial\rho}{\partial\tau} + \frac{\partial(\rho v_i)}{\partial x_i} = 0 \tag{6-8}$$

The following summation rule of the tensor analysis has been used in writing this equation: When an index is repeated in a term, a summation over the values of this index must be carried out. Therefore $\partial(\rho v_i)/\partial x_i$ stands for

$$\sum_{i=1 \text{ to } 3} \frac{\partial(\rho v_i)}{\partial x_i}$$

In terms of vector operations, the second term denotes the scalar product of the vector ρv_i and the vector operator $\partial/\partial x_i$. Equation (6-5) thus becomes

$$\frac{D\rho}{d\tau} + \rho \frac{\partial v_i}{\partial x_i} = 0 \tag{6-9}$$

The simplification achieved in writing equations in mechanics in this form will become even more obvious with the other conservation laws.[1]

6-2 MOMENTUM EQUATION

The law of conservation of momentum can now be applied to our control volume. This law is essentially an extension of Newton's law stating that the sum of all forces acting on a mass must be equal to the mass multiplied by its acceleration. For a variable mass, this law is reformulated to state that the sum of all forces must equal the change per unit time of the momentum of the fluid mass involved. In our control volume, the fluid momentum can change either by the fact that the momentum flux out of the element is different from the momentum flux into the element or by a change of the momentum of the mass contained in the volume element. Forces acting on the fluid in the control volume can be body (volume) forces or surface forces. Since forces and momenta are vectors, we shall write the momentum law in the three components starting with the x component. We assume that a body force $\mathbf{g}$ per unit mass acts on the fluid within the control volume (gravitational, electric, magnetic, . . .). In order to describe the motion of our fluid particle, we introduce surface forces which describe the action of the surrounding fluid on that particle. A force per unit area is called a *stress*. Such a stress will act, for instance, on surface 1 of Fig. 6-3 in an arbitrary direction and can be separated into three components: one component will be normal to the surface 1, and two components will act within the surface in y and z directions. These stresses will be denoted by p with two

[1] R. Aris, "Vectors, Tensors, and the Basic Equations of Fluid Motion," Prentice-Hall, Englewood Cliffs, N.J., 1963; G. A. Hawkins, "Multilinear Analysis for Students in Engineering and Science," Wiley, New York, 1963.

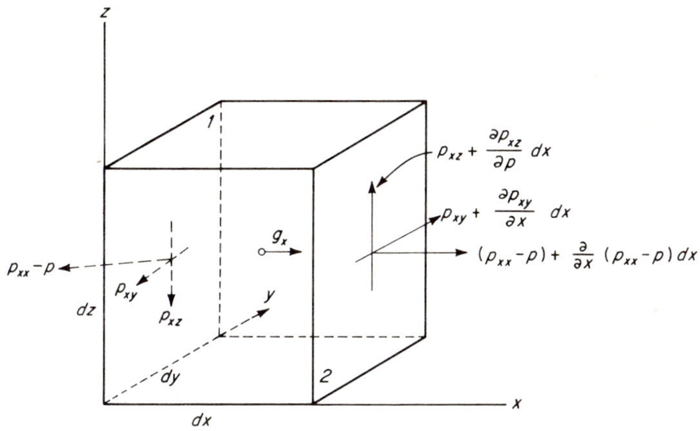

Fig. 6-3 Surface stresses and body forces acting on the fluid in a control volume.

indices added, the first index indicating the orientation of the area element by stating the direction of the surface normal and the second index indicating the direction of the stress component. In this way the stresses, as shown in Fig. 6-3, act on surfaces 1 and 2. Care must be exercised in establishing the direction in which the stress is considered positive. Conventionally, a stress which the fluid on the side of an increasing value of the coordinate exerts on the fluid on the side with decreasing value of the coordinate is counted positive when it acts in the direction in which the coordinate is counted positive. From this convention and from the fact that we must introduce the forces which the surrounding fluid exerts on the fluid in the control element, we establish the directions indicated by the arrows in Fig. 6-3 as positive stresses. p_{xy} and p_{xz} act within the plane of surface 1 and are called *shear stresses.* A further explanation is required in connection with the normal stress p_{xx}. It has been mentioned before that a fluid is defined as a substance in which shear stresses vanish when the velocity is reduced to zero. The normal stresses, however, do not vanish but degenerate to the fluid pressure, a stress characterized by the fact that it is normal to any surface of the fluid element and has the same value independent of its direction. We consider, therefore, the normal stress $p_{xx} - p$ to be composed of a viscous stress p_{xx}, which vanishes as the velocity decreases to zero, and of the fluid pressure p. The negative sign in the expression comes from the fact that the pressure is counted positive when it acts from the outside on the fluid element (compression stress) whereas the stress p_{xx} is, according to our convention, a tensile stress.

Now we can start to derive the various contributions to the change of x momentum of the mass in our control-volume element. The mass flow per unit time through control surface 1 is $\rho u\, dy\, dz$. The flux of x momentum is obtained when the mass flow is multiplied by the velocity component u in the x direction $\rho u^2\, dy\, dz$. The excess of x-momentum flux leaving surface 2 over that entering through surface 1 is

$$\frac{\partial}{\partial x}(\rho u^2\, dy\, dz)\, dx = \frac{\partial(\rho u^2)}{\partial x}\, dy\, dz\, dx$$

In the same way, the flux of x momentum through the other surfaces can be obtained by multiplying the corresponding mass fluxes with the velocity component u. This leads to the first column in the scheme in Table 6-1. The change of x momentum of the fluid within the control volume has been added in the last line of the table. The second column gives the surface forces in the first three lines and (in the last line) the volume force acting on the control volume. The equation describing the conservation of momentum is now obtained by balancing the forces against the momentum changes

$$\frac{\partial}{\partial x}(\rho u^2) + \frac{\partial}{\partial y}(\rho v u) + \frac{\partial}{\partial z}(\rho w u) + \frac{\partial}{\partial \tau}(\rho u) = \frac{\partial}{\partial x}(p_{xx} - p) + \frac{\partial p_{yx}}{\partial y} + \frac{\partial p_{zx}}{\partial z} + \rho g_x \tag{6-10}$$

Table 6-1

Mass flux in—	*Momentum flux in x direction (out minus in)*	*Net surface force in x direction*
x direction	$\frac{\partial}{\partial x}(\rho u^2)\, dy\, dz\, dx$	$\frac{\partial}{\partial x}(p_{xx} - p)\, dy\, dz\, dx$
y direction	$\frac{\partial}{\partial y}(\rho v u)\, dx\, dz\, dy$	$\frac{\partial p_{yx}}{\partial y}\, dx\, dz\, dy$
z direction	$\frac{\partial}{\partial z}(\rho w u)\, dx\, dy\, dz$	$\frac{\partial p_{zx}}{\partial z}\, dx\, dy\, dz$
	Momentum change	*Body force*
Volume element	$\frac{\partial}{\partial \tau}(\rho u)\, dx\, dy\, dz$	$\rho g_x\, dx\, dy\, dz$

By partial differentiation of the terms on the left-hand side we obtain

$$\rho u \frac{\partial u}{\partial x} + \rho v \frac{\partial u}{\partial y} + \rho w \frac{\partial u}{\partial z} + \rho \frac{\partial u}{\partial \tau} + u \frac{\partial(\rho u)}{\partial x} + u \frac{\partial(\rho v)}{\partial y} + u \frac{\partial(\rho w)}{\partial z} + u \frac{\partial \rho}{\partial \tau} = -\frac{\partial p}{\partial x} + \frac{\partial p_{xx}}{\partial x} + \frac{\partial p_{yx}}{\partial y} + \frac{\partial p_{zx}}{\partial z} + \rho g_x$$

and considering the conservation of mass [Eq. (6-3)],

$$\rho u \frac{\partial u}{\partial x} + \rho v \frac{\partial u}{\partial y} + \rho w \frac{\partial u}{\partial z} + \rho \frac{\partial u}{\partial \tau} = -\frac{\partial p}{\partial x} + \frac{\partial p_{xx}}{\partial x} + \frac{\partial p_{yx}}{\partial y} + \frac{\partial p_{zx}}{\partial z} + \rho g_x \tag{6-11}$$

This equation is simplified by introducing the substantial derivative $Du/d\tau$, which, according to Eq. (6-4), is

$$\frac{Du}{d\tau} = \frac{\partial u}{\partial \tau} + u \frac{\partial u}{\partial x} + v \frac{\partial u}{\partial y} + w \frac{\partial u}{\partial z}$$

The x-momentum equation can now be written as

$$\rho \frac{Du}{d\tau} = -\frac{\partial p}{\partial x} + \frac{\partial p_{xx}}{\partial x} + \frac{\partial p_{yx}}{\partial y} + \frac{\partial p_{zx}}{\partial z} + \rho g_x \tag{6-12}$$

Two equivalent equations exist for the y and z directions:

$$\rho \frac{Dv}{d\tau} = -\frac{\partial p}{\partial y} + \frac{\partial p_{xy}}{\partial x} + \frac{\partial p_{yy}}{\partial y} + \frac{\partial p_{zy}}{\partial z} + \rho g_y \tag{6-13}$$

$$\rho \frac{Dw}{d\tau} = -\frac{\partial p}{\partial z} + \frac{\partial p_{xz}}{\partial x} + \frac{\partial p_{yz}}{\partial y} + \frac{\partial p_{zz}}{\partial z} + \rho g_z \tag{6-14}$$

One equation is sufficient to express the law of conservation of momentum in tensor notation:

$$\rho \frac{Dv_i}{d\tau} = -\frac{\partial p}{\partial x_i} + \frac{\partial p_{ji}}{\partial x_j} + \rho g_i \tag{6-15}$$

The index i identifies the terms as vectors. $\partial p_{ji}/\partial x_j$ stands, as in Eqs. (6-8) and (6-9), for $\sum_j \frac{\partial p_{ji}}{\partial_{x_j}}$ according to the summation rule.

The equations derived up to now have a very general validity because they are based on the fundamental laws of mechanics. In order to make use of them, relationships must be introduced which connect the stresses with the deformation of the fluid particle. These relationships are empirical in nature and are occasionally referred to as *consti-*

tutive equations. It has been mentioned before that these relations are described by different laws for various kinds of fluids. In a newtonian fluid, a linear relation exists between stress and rate of deformation. Only such fluids will be considered in this text. For simple shear flow such a relation was given on page 246 as $\tau = \mu\, \partial u/\partial y$, where the shear stress p_{yx} was called τ. For the more general situation in which all three velocity components exist and depend on all three space coordinates, the relations are more complicated. They were established by G. G. Stokes in 1845, by a reasoning analogous to the derivation of the relations between stress and strain in a solid body.[1] From the laws of mechanics, it can be shown that

$$p_{xy} = p_{yx} \tag{6-16}$$

or, in general, that a change in the sequence of the indices on the stress terms does not change their magnitude.[2] The stresses occurring in the preceding equation are then given by the relations

$$\begin{aligned} p_{yx} &= \mu \left(\frac{\partial v}{\partial x} + \frac{\partial u}{\partial y} \right) \\ p_{xx} &= 2\mu \frac{\partial u}{\partial x} - \mu' \left(\frac{\partial u}{\partial x} + \frac{\partial v}{\partial y} + \frac{\partial w}{\partial \tau} \right) \end{aligned} \tag{6-17}$$

and corresponding ones for the other terms. The viscous stresses can again be expressed by a single equation in tensor notation, when the following unit tensor is introduced:

$$\delta_{ij} = \begin{cases} 1 & \text{for } i = j \\ 0 & \text{for } i \neq j \end{cases}$$

This tensor is called *Kronecker delta.* The equation for the viscous stresses is then

$$p_{ij} = -\mu' \frac{\partial v_k}{\partial x_k} \delta_{ij} + \mu \left(\frac{\partial v_i}{\partial x_j} + \frac{\partial v_j}{\partial x_i} \right) \tag{6-18}$$

The term $\partial v_k/\partial x_k$ again stands for a sum because the index k is repeated. The indices j identify the terms in the equation as components in the direction j. For $i = 1$ and $j = 1$, for instance, the above equation reads

$$p_{11} = -\mu' \left(\frac{\partial v_1}{\partial x_1} + \frac{\partial v_2}{\partial x_2} + \frac{\partial v_3}{\partial x_3} \right) + 2\mu \frac{\partial v_1}{\partial x_1}$$

[1] See, for instance, H. Schlichting, "Boundary Layer Theory," 6th ed., McGraw-Hill, New York, 1968.

[2] An exception to this is a situation in which the spin of the electrons around the nuclei is not randomly oriented in space.

which is identical with the second equation (6-17). For $i = 1, j = 2$, one obtains

$$p_{12} = \mu\left(\frac{\partial v_1}{\partial x_2} + \frac{\partial v_2}{\partial x_1}\right)$$

which is identical with the first equation (6-17).[1]

The viscosity μ' appearing in the second equation (6-17) has for monatomic molecules according to simple kinetic theory the value $\frac{2}{3}\mu$. On the other hand, different relations between μ' and μ are found for polyatomic molecules from the kinetic theory of gases[2] and from experimental studies of compression shocks. The bulk viscosity $\varkappa$ introduced through these studies is connected with the viscosity μ' by the relation $\varkappa = \mu' - \frac{2}{3}\mu$. The viscosity μ' will be found to drop out from the equations for any situation which is of interest in the context of the present text.

Introducing the relations between stresses and rates of deformation results in:

Continuity:
$$\frac{\partial \rho}{\partial \tau} + \frac{\partial(\rho u)}{\partial x} + \frac{\partial(\rho v)}{\partial y} + \frac{\partial(\rho w)}{\partial z} = 0 \tag{6-19}$$

Momentum:

$$\rho\frac{Du}{d\tau} = -\frac{\partial p}{\partial x} + 2\frac{\partial}{\partial x}\left(\mu\frac{\partial u}{\partial x}\right) + \frac{\partial}{\partial y}\left[\mu\left(\frac{\partial u}{\partial y} + \frac{\partial v}{\partial x}\right)\right] + \frac{\partial}{\partial z}\left[\mu\left(\frac{\partial u}{\partial z} + \frac{\partial w}{\partial x}\right)\right] - \frac{\partial}{\partial x}\left[\mu'\left(\frac{\partial u}{\partial x} + \frac{\partial v}{\partial y} + \frac{\partial w}{\partial z}\right)\right] + \rho g_x \tag{6-20a}$$

$$\rho\frac{Dv}{d\tau} = -\frac{\partial p}{\partial y} + 2\frac{\partial}{\partial y}\left(\mu\frac{\partial v}{\partial y}\right) + \frac{\partial}{\partial z}\left[\mu\left(\frac{\partial v}{\partial z} + \frac{\partial w}{\partial y}\right)\right] + \frac{\partial}{\partial x}\left[\mu\left(\frac{\partial v}{\partial x} + \frac{\partial u}{\partial y}\right)\right] - \frac{\partial}{\partial y}\left[\mu'\left(\frac{\partial u}{\partial x} + \frac{\partial v}{\partial y} + \frac{\partial w}{\partial z}\right)\right] + \rho g_y \tag{6-20b}$$

$$\rho\frac{Dw}{d\tau} = -\frac{\partial p}{\partial z} + 2\frac{\partial}{\partial z}\left(\mu\frac{\partial w}{\partial z}\right) + \frac{\partial}{\partial x}\left[\mu\left(\frac{\partial w}{\partial x} + \frac{\partial u}{\partial z}\right)\right] + \frac{\partial}{\partial y}\left[\mu\left(\frac{\partial w}{\partial y} + \frac{\partial v}{\partial z}\right)\right] - \frac{\partial}{\partial z}\left[\mu'\left(\frac{\partial u}{\partial x} + \frac{\partial v}{\partial y} + \frac{\partial w}{\partial z}\right)\right] + \rho g_z \tag{6-20c}$$

The continuity equation was added to complete the system, which was first derived by N. Navier (1827) and S. P. Poisson (1831) from a con-

[1] The pressure p was introduced in Eq. (6-10) to separate the viscous tensile stress from that part which remains when the velocity gradients decrease to zero. Accordingly the pressure p is described by the state equation which determines the state of a fluid at rest. In this way the pressure is connected with the other local state parameters.

[2] A. Busemann, Gasdynamik, in W. Wien and F. Harms (eds.), "Handbuch der Experimentalphysik," vol. IV, pt. 1, Akademische Verlagsgesellschaft, Leipzig, 1931.

sideration of intermolecular forces and by B. de Saint Venant (1843) and Stokes (1845) based on the assumption that the normal and shear stresses in a fluid are proportional to the deformation velocities. The equations are therefore referred to as *Navier-Stokes equations.*

In tensor notation the equations assume the simpler form

$$\frac{\partial \rho}{\partial \tau} + \frac{\partial(\rho v_i)}{\partial x_i} = 0 \tag{6-21}$$

$$\rho \frac{Dv_i}{d\tau} = -\frac{\partial p}{\partial x_i} + \frac{\partial}{\partial x_j}\left(\mu \frac{\partial v_i}{\partial x_j}\right) + \frac{\partial}{\partial x_j}\left(\mu \frac{\partial v_j}{\partial x_i}\right) - \frac{\partial}{\partial x_i}\left(\mu' \frac{\partial v_j}{\partial x_j}\right) + \rho g_i \tag{6-22}$$

Even in these equations, the number of unknowns is still greater than the number of equations, and additional relations are necessary for the properties ρ, μ, and μ'. The simplest assumption in this connection is the model of a constant-property fluid. For a constant-property medium, Eq. (6-20) reads after some rearrangement

$$\rho \frac{Du}{d\tau} = -\frac{\partial p}{\partial x} + \mu\left(\frac{\partial^2 u}{\partial x^2} + \frac{\partial^2 u}{\partial y^2} + \frac{\partial^2 u}{\partial z^2}\right) + (\mu - \mu')\left(\frac{\partial^2 u}{\partial x^2} + \frac{\partial^2 v}{\partial y\,\partial x} + \frac{\partial^2 w}{\partial z\,\partial x}\right) + \rho g_x$$

Inversion of the order of differentiation gives

$$\frac{\partial^2 u}{\partial x^2} + \frac{\partial^2 v}{\partial y\,\partial x} + \frac{\partial^2 w}{\partial z\,\partial x} = \frac{\partial}{\partial x}\left(\frac{\partial u}{\partial x} + \frac{\partial v}{\partial y} + \frac{\partial w}{\partial z}\right)$$

which is zero according to continuity.

The equations for a constant-property medium are, therefore, in tensor notation

Continuity: $$\frac{\partial v_i}{\partial x_i} = 0 \tag{6-23a}$$

Momentum: $$\rho \frac{Dv_i}{d\tau} = -\frac{\partial p}{\partial x_i} + \mu \frac{\partial}{\partial x_j}\frac{\partial v_i}{\partial x_j} + \rho g_i \tag{6-23b}$$

It may be observed that the viscosity μ' has disappeared from these equations. The equations contain the following unknown variable parameters: $v_i(u,v,w)$ and p. The number of equations is now sufficient to solve for them when proper boundary conditions are prescribed. This statement has an interesting and important consequence for a situation in which heat transfer occurs. The fact that the temperature occurs nowhere in the above equations means physically that the flow field is the same whether the temperature is uniform in the fluid or whether a temperature field is caused by heat transfer. This also means that all the information available in fluid mechanics for such a fluid can immediately be used as a basis for heat-transfer calculations. Heat transfer is superimposed on such a flow field but does not change it in any way.

The model of a constant-property fluid is a reasonably good approximation for a liquid as long as the temperature differences involved are moderate. It can also be used for gases, when additionally the pressure variations are small, which is the case at moderate velocities.

On the other hand, Eqs. (6-21) and (6-22) can be solved only in connection with the energy equation, when the temperature and pressure dependence of the transport properties must be accounted for. The energy equation will now be derived.

6-3 ENERGY EQUATION

In this section, the energy equation will be derived for a cartesian system of coordinates. The properties entering into this equation will be assumed as variable. The law of conservation of energy applied to the fluid contained in a control volume takes on the form of an energy balance expressing the fact that the sum of the energy fluxes entering the control volume through its boundaries must be equal to the increase of energy contained in the fluid filling the control volume. This energy balance can therefore be written in the form

$$d\dot{W} + d\dot{E}_c + d\dot{E}_M = d\dot{E}_i \tag{6-24}$$

$d\dot{W}$ indicates the energy transferred to the fluid in the control volume as work. The term *work* will be interpreted in this book as energy transfer caused by the forces acting on the fluid in the control volume. The term $d\dot{E}_c$ indicates energy transported with the fluid which crosses the boundaries into the control volume by convection. The term $d\dot{E}_M$ summarizes all other forms of energy transport across the boundaries caused by the molecular movements. Such an energy transport is, for instance, caused by heat conduction. There will, however, be other forms of energy transport specifically included in our considerations, for instance, those by diffusion. $d\dot{E}_i$ denotes the increase in the energy stored in the fluid within the control volume. The dot above the individual terms identifies this equation as a rate equation considering energy changes per unit time.

Equation (6-24) is an expression of the first law of thermodynamics. It will be noted that work added to the fluid is introduced as positive in this equation whereas in engineering thermodynamics work performed by the fluid on the surrounding is often assigned a positive sign. It is important that all modes of energy transport be included in the terms on the left-hand side of the above equation. It is less important which of the individual terms a specific mode is assigned to. This, however,

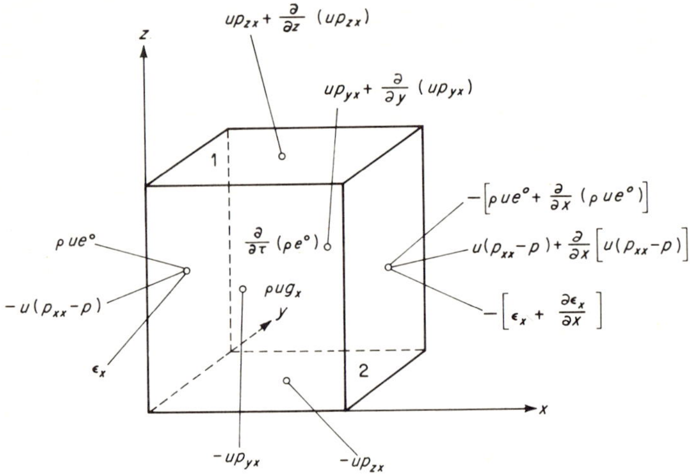

Fig. 6-4 Work performed by the surface stresses and body forces acting in the x direction on the fluid in a control volume.

becomes very important in formulating the second law of thermodynamics because one must then distinguish between reversible and irreversible modes of energy transfer. The second law will not be used in this book, and the reader should consult modern texts on thermodynamics in this connection.

Equation (6-24) will now be applied to the fluid in the volume element shown in Fig. 6-4. Work, according to our definition, is performed by the surface stresses and by the body forces. The work performed by the forces acting in the x direction on the volume element per unit area is indicated in the figure. The work is positive when the forces act in the flow direction and negative when they act opposite to the flow direction. The surface stresses in the x direction perform the following work per unit time on the fluid in the control volume:

$$\left[\frac{\partial}{\partial x}(up) - \frac{\partial}{\partial x}(up_{xx}) - \frac{\partial}{\partial y}(up_{yx}) - \frac{\partial}{\partial z}(up_{zx})\right] dx\, dy\, dz$$

Similar expressions hold for the work done by the stresses in the y and z directions. The body-force component q_x in the x direction performs the work

$$\rho u g_x\, dx\, dy\, dz$$

with analogous expressions for the other components.

Energy is contained in the fluid as internal energy or as kinetic energy. The sum of both of these quantities is called the *total energy*. Per unit mass, this total energy e° is expressed by

$$e^\circ = e + \frac{V^2}{2}$$

with V indicating the magnitude of the velocity vector.

The change per unit time of energy stored within the volume element is therefore

$$d\dot{E}_i = \frac{\partial}{\partial \tau} (\rho e^\circ)\, dx\, dy\, dz$$

The energy flux convected by the fluid through surface 1 of the control volume is

$$\rho u e^\circ\, dy\, dz$$

The net flux entering the volume element through surfaces 1 and 2 is therefore

$$-\frac{\partial}{\partial x} (\rho u e^\circ)\, dx\, dy\, dz$$

Analogous expressions hold for the fluxes in the y and z directions.

The components of the energy flux per unit area and time corresponding to the term $d\dot{E}_M$ can be denoted in tensor notation by ϵ_i. The corresponding energy flux into the volume element in the x direction is then

$$-\frac{\partial}{\partial x} \epsilon_x\, dx\, dy\, dz$$

Analogous expressions again hold for the fluxes in y and z direction.

All the terms developed above have now to be introduced into the energy-balance equation (6-24). This results in a very involved expression unless tensor notation is used. It is easily checked that in this notation the equation assumes the form

$$\frac{\partial}{\partial \tau} (\rho e^\circ) + \frac{\partial}{\partial x_i} (\rho v_i e^\circ) + \frac{\partial}{\partial x_i} (v_i p) + \frac{\partial \epsilon_i}{\partial x_i} = \frac{\partial}{\partial x_j} (v_i p_{ji}) + \rho v_i g_i \qquad \text{(6-25)}$$

keeping in mind that any term in which the same index appears twice stands for a sum. The term $(\partial/\partial x_j)\,(v_i p_{ji})$ therefore stands for nine terms since the two indices are repeated.

The action of body forces is sometimes introduced in a different way into the energy balance when these forces have a potential. In this case the total energy of the fluid is considered as being composed of internal energy, kinetic energy, and potential energy, and the work by body forces is omitted. That this is equivalent to the way the action of body forces was introduced here can be shown by a simple example in which it is assumed that body forces act in the x direction only, that the force field does not change with time, and that it has a potential

$$\varphi = -\int g\,dx$$

The change per unit time in potential energy stored per unit of the control volume is

$$\frac{\partial}{\partial \tau}(\rho\varphi) = \varphi \frac{\partial \rho}{\partial \tau}$$

The net flux of potential energy convected by the fluid into the control volume is

$$-\frac{\partial}{\partial x}(\rho u \varphi) = -\varphi \frac{\partial}{\partial x}(\rho u) - \rho u \frac{\partial \varphi}{\partial x} = -\varphi \frac{\partial}{\partial x}(\rho u) + \rho u g$$

Introducing these expressions into the proper terms and excluding the work term by body forces in Eq. (6-24) leads again to Eq. (6-25). The energy equation (6-25) can be simplified with the help of the continuity equation

$$\frac{\partial}{\partial x_i}\rho v_i + \frac{\partial \rho}{\partial \tau} = 0$$

and the equation defining the substantial derivative of the total internal energy

$$\frac{De^\circ}{d\tau} = \frac{\partial e^\circ}{\partial \tau} + v_i \frac{\partial e^\circ}{\partial x_i}$$

The energy balance takes the form

$$\rho \frac{De^\circ}{d\tau} + \frac{\partial}{\partial x_i}(v_i p) + \frac{\partial \epsilon_i}{\partial x_i} = \frac{\partial}{\partial x_j}(v_i p_{ji}) + \rho v_i g_i \qquad (6\text{-}26)$$

For steady-state situations, it is advantageous to replace the total internal energy e° by the total enthalpy i°. They are connected by the equation

$$e^\circ = i^\circ - \frac{p}{\rho}$$

Therefore

$$\rho \frac{De^\circ}{d\tau} = \rho \frac{Di^\circ}{d\tau} - \frac{Dp}{d\tau} + \frac{p}{\rho}\frac{D\rho}{d\tau} \qquad (6\text{-}27)$$

The substantial derivative of the density takes on the following form with consideration of the continuity equation:

$$\frac{D\rho}{d\tau} = -\rho \frac{\partial v_i}{\partial x_i}$$

With this, Eq. (6-27) becomes

$$\rho \frac{De^\circ}{d\tau} = \rho \frac{Di^\circ}{d\tau} - \frac{\partial p}{\partial \tau} - v_i \frac{\partial p}{\partial x_i} - p \frac{\partial v_i}{\partial x_i}$$

or

$$\rho \frac{De^\circ}{d\tau} = \rho \frac{Di^\circ}{d\tau} - \frac{\partial p}{\partial \tau} - \frac{\partial}{\partial x_i}(v_i p)$$

Introducing this last expression into the energy equation gives

$$\rho \frac{Di^\circ}{d\tau} - \frac{\partial p}{\partial \tau} + \frac{\partial \epsilon_i}{\partial x_i} = \frac{\partial}{\partial x_j}(v_i p_{ji}) + \rho v_i g_i \tag{6-28}$$

The advantage of this equation is that for a steady condition the partial derivative of the pressure toward the time is zero, so that the second term on the left-hand side drops out.

For some calculations it is also useful to have the energy equation in terms of the static enthalpy i instead of the total enthalpy i°. For this purpose, we multiply the momentum Eq. (6-15) by v_i and obtain

$$\rho v_i \frac{Dv_i}{d\tau} + v_i \frac{\partial p}{\partial x_i} = v_i \frac{\partial p_{ji}}{\partial x_j} + \rho v_i g_i$$

This actually is equivalent to a multiplication of the three-component equations indicated by Eq. (6-15) by the corresponding components of the velocity and to a summation of these three-component equations. The first term in this equation is equivalent to the expression

$$\rho \frac{D}{d\tau} \frac{V^2}{2}$$

in which V indicates the magnitude of the velocity vector. Subtracting the last equation from the energy equation (6-28) results in

$$\rho \frac{Di}{d\tau} - \frac{Dp}{d\tau} + \frac{\partial \epsilon_i}{\partial x_i} = p_{ji} \frac{\partial v_i}{\partial x_j} \tag{6-29}$$

The term $p_{ji}\, \partial v_i / \partial x_j$ is called the *dissipation function* Φ because it indicates the energy which is dissipated into heat. For a newtonian fluid it can be written with the use of Eq. (6-18) as

$$\Phi = p_{ji} \frac{\partial v_i}{\partial x_j} = \left[\mu \left(\frac{\partial v_i}{\partial x_j} + \frac{\partial v_j}{\partial x_i} \right) - \mu' \frac{\partial v_k}{\partial x_k} \delta_{ij} \right] \frac{\partial v_i}{\partial x_j} \tag{6-30}$$

A simple transformation of the energy equation (6-29) leads also to the form

$$\rho \frac{De}{d\tau} + p \frac{\partial v_i}{\partial x_i} + \frac{\partial \epsilon_i}{\partial x_i} = p_{ji} \frac{\partial v_i}{\partial x_j} \tag{6-31}$$

For a one-component fluid, the only mode of energy transfer which has to be considered in the term $d\dot{E}_M$ is heat transfer by conduction, which is described by *Fourier's law*. The heat flux per unit area and time according to this law is

$$\epsilon_i = -k \frac{\partial t}{\partial x_i} \tag{6-32}$$

and the heat flux entering the volume element in any of the three directions of the coordinate system is

$$-\frac{\partial \epsilon_i}{\partial x_i} = \frac{\partial}{\partial x_i}\left(k \frac{\partial t}{\partial x_i}\right)$$

This expression must be introduced into Eqs. (6-26) to (6-31) when dealing with a one-component fluid.

6-4 BOUNDARY-LAYER CONTINUITY AND MOMENTUM EQUATIONS

The conservation equations derived in the preceding sections describe the flow field and the thermal field in a viscous fluid. These equations, however, are so involved that a simplification is highly desirable in order to reduce the effort of obtaining solutions. Such simplifications can be made for very small and for very large Reynolds numbers. The simplification for large Reynolds numbers, the most important situation from the engineering point of view, was first accomplished by Ludwig Prandtl in 1904. He started from the experimentally observed fact that large velocity gradients normal to the streamlines occur only in regions close to the surface, and he concluded that it may be sufficient in an analysis of a flow field to consider the action of viscosity within these layers, which he called boundary layers, whereas the flow outside the boundary layers may be considered inviscid. He then proceeded to simplify the conservation equations by estimating the order of magnitude of the various terms in the conservation equations, and thus he derived the so-called *boundary-layer equations*. In the meantime scientists succeeded in deriving the boundary-layer equations through a more exact and formalized procedure.[1] Here we shall essentially follow Prandtl's derivation because of its simplicity. In doing so, we shall restrict ourself to two-dimensional flow of a one-component newtonian fluid and initially consider steady flow over a flat plate as indicated in Fig. 6-5. The x direction, identified by $i = 1$ in the following equations, will be measured parallel to the surface in flow direction and the y direction, identified by $i = 2$, normal to the surface. The velocity outside the boundary layer may vary with x in an arbitrary manner. The continuity and

[1] M. van Dyke, "Perturbation Methods in Fluid Mechanics," Academic, New York, 1964.

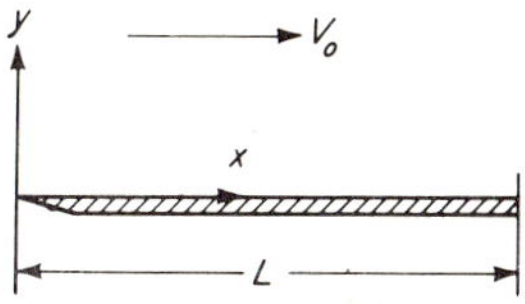

Fig. 6-5 Notation for flow over a flat plate.

momentum equations, according to the preceding section, are

$$\frac{\partial(\rho v_i)}{\partial x_i} = 0 \tag{6-33a}$$

$$\rho v_j \frac{\partial v_i}{\partial x_j} = -\frac{\partial p}{\partial x_i} + \frac{\partial p_{ji}}{\partial x_j} + \rho g_i \tag{6-33b}$$

The viscous stresses p_{ij} are described for a newtonian fluid by the relation

$$p_{ji} = p_{ij} = -\mu' \frac{\partial v_k}{\partial x_k}\delta_{ij} + \mu\left(\frac{\partial v_i}{\partial x_j} + \frac{\partial v_j}{\partial x_i}\right) \tag{6-34}$$

We now proceed to make these equations dimensionless by the use of a reference length L, which may, for instance, be taken as the length of the plate in flow direction, by the velocity V_0 at a reference point, and by the properties of the fluid ρ_0 and μ_0, again, at the selected reference point. The underlined dimensionless quantities are

$$\begin{aligned} &\underline{\rho} = \frac{\rho}{\rho_0} \qquad \underline{x}_i = \frac{x_i}{L} \qquad \underline{v}_i = \frac{v_i}{V_0} \qquad \underline{p} = \frac{p}{\rho_0 V_0^2} \\ &\underline{\mu} = \frac{\mu}{\mu_0} \qquad \underline{\mu}' = \frac{\mu'}{\mu_0} \qquad \underline{g}_i = \frac{g_i}{V_0^2/L} \end{aligned} \tag{6-35}$$

Replacing the dimensional quantities in the continuity equation (6-32) by the dimensionless quantities results in

$$\frac{\rho_0 V_0}{L}\frac{\partial(\underline{\rho}\underline{v}_i)}{\partial \underline{x}_i} = 0 \tag{6-36}$$

Since the expression $\rho_0 V_0/L$ cannot be zero, one obtains

$$\frac{\partial(\underline{\rho}\underline{v}_i)}{\partial \underline{x}_i} = 0 \tag{6-37}$$

The continuity equation assumes the same form in the dimensionless quantities as it does in the dimensional quantities. The momentum equation (6-33b) with Eq. (6-34) inserted assumes the following form when written in the dimensionless quantities:

$$\rho_0 \frac{V_0^2}{L}\underline{\rho}\underline{v}_j \frac{\partial \underline{v}_i}{\partial \underline{x}_j} = -\rho_0 \frac{V_0^2}{L}\frac{\partial \underline{p}}{\partial \underline{x}_i} + \mu_0 \frac{V_0}{L^2}\frac{\partial}{\partial \underline{x}_j}\left[-\underline{\mu}'\frac{\partial \underline{v}_k}{\partial \underline{x}_k}\delta_{ij} + \underline{\mu}\left(\frac{\partial \underline{v}_i}{\partial \underline{x}_j} + \frac{\partial \underline{v}_j}{\partial \underline{x}_i}\right)\right] + \rho_0 \frac{V_0^2}{L}\underline{\rho}\underline{g}_i$$

Dividing all terms of this equation by the factor $\rho_0 V_0^2/L$ results in the dimensionless expression

$$\underline{\rho}\underline{v}_j \frac{\partial \underline{v}_i}{\partial \underline{x}_j} = -\frac{\partial \underline{p}}{\partial \underline{x}_i} + \frac{1}{\mathrm{Re}_0}\frac{\partial}{\partial \underline{x}_j}\left[-\underline{\mu}'\frac{\partial \underline{v}_k}{\partial \underline{x}_k}\delta_{ij} + \underline{\mu}\left(\frac{\partial \underline{v}_i}{\partial \underline{x}_j}+\frac{\partial \underline{v}_i}{\partial \underline{x}_i}\right)\right] + \underline{\rho}\underline{g}_i \tag{6-38}$$

The factor preceding the viscous-stress term can be identified as the reciprocal of a Reynolds number

$$\mathrm{Re}_0 = \frac{\rho_0 V_0 L}{\mu_0} = \frac{V_0 L}{\nu_0} \tag{6-39}$$

based on the reference quantities, with ν_0 indicating the reference kinematic viscosity. One might conclude superficially that for large values of the Reynolds number the terms multiplied by the Reynolds number can be neglected in Eq. (6-38). This, however, would eliminate all viscous terms from the equation and would result in an equation describing the flow of an inviscid fluid. It is well known that solutions to the inviscid-flow equation cannot satisfy the experimentally established boundary condition of a velocity with a value zero at the surface of an immersed object. Prandtl therefore concluded that at least one of the dimensionless viscous terms must be large to compensate for the large value of the Reynolds number by which it is divided. This means that large velocity gradients must exist at least close to the surface of the plate and therefore leads again to the concept of a thin layer in which these gradients are concentrated.

In order to derive the boundary-layer equations, the continuity equation (6-37) is now written in cartesian coordinates

$$\begin{array}{ccccc} \dfrac{\partial(\underline{\rho}\underline{u})}{\partial \underline{x}} & + & \dfrac{\partial(\underline{\rho}\underline{v})}{\partial \underline{y}} & = & 0 \\ \dfrac{1\,1}{1} & & \dfrac{1\,\delta}{\delta} & & \end{array} \tag{6-40}$$

The reference density is now chosen so as to make the dimensionless density $\underline{\rho}$ of order 1. In the same way, the reference velocity V_0 is selected so that the dimensionless velocity $\underline{u}$ is of order 1 or less. The average of the velocity variation $\partial u/\partial x$ will then be of the same order of magnitude as V_0/L, or the dimensionless velocity variation $\partial \underline{u}/\partial \underline{x}$ will be of order 1. The expression $\partial(\underline{\rho u})/\partial \underline{x}$ will also be of order 1. The preceding observation is based on the assumption that the local velocity changes in a gradual way along the plate surface and that local sudden variations are excluded. It will be seen later that such sudden variations occur near the leading edge of a very thin flat plate in a stream of constant velocity and that therefore the flow in a small region close to the leading edge cannot be described by the boundary-layer equations. A similar situation occurs in supersonic flow when a shock traverses the boundary layer. Apart from this region, it can be concluded that the first term in Eq. (6-40) is of order 1. The equation then requires that the second term also be of order 1, which means that the dimensionless velocity gradient $\partial\underline{v}/\partial\underline{y}$ must be of order 1. This requires that the velocity $\underline{v}$ be of order $\underline{\delta}$ since its value at the wall surface is equal to zero. The order of magnitude of the various terms as it has been established by the foregoing considerations is written under Eq. (6-40). The main conclusion from the considerations so far is the fact that the velocity component v is of much smaller order of magnitude than the velocity component u for flow conditions which create a thin boundary layer.

The momentum equation for the x direction is now written down:

$$\underline{\varrho}\left(\underline{u}\frac{\partial \underline{u}}{\partial \underline{x}} + \underline{v}\frac{\partial \underline{u}}{\partial \underline{y}}\right) = -\frac{\partial \underline{p}}{\partial \underline{x}} + \frac{1}{\mathrm{Re}_0}\left[2\frac{\partial}{\partial \underline{x}}\left(\underline{\mu}\frac{\partial \underline{u}}{\partial \underline{x}}\right) + \frac{\partial}{\partial \underline{y}}\left(\underline{\mu}\frac{\partial \underline{u}}{\partial \underline{y}}\right)\right.$$

$$1 \quad \frac{1}{1} \qquad \underline{\delta} \quad \frac{1}{\underline{\delta}} \qquad\qquad \frac{1}{1} \qquad\qquad \frac{1}{1} \quad 1 \quad \frac{1}{1} \qquad \frac{1}{\underline{\delta}} \quad 1 \quad \frac{1}{\underline{\delta}}$$

$$\left. + \frac{\partial}{\partial \underline{y}}\left(\underline{\mu}\frac{\partial \underline{v}}{\partial \underline{x}}\right) - \frac{\partial}{\partial \underline{x}}\left(\underline{\mu}'\frac{\partial \underline{u}}{\partial \underline{x}}\right) - \frac{\partial}{\partial \underline{x}}\left(\underline{\mu}'\frac{\partial \underline{v}}{\partial \underline{y}}\right)\right] + \underline{\varrho}\underline{g}_x \qquad (6\text{-}41)$$

$$\frac{1}{\underline{\delta}} \quad 1 \quad \frac{\underline{\delta}}{1} \qquad \frac{1}{1} \quad 1 \quad \frac{1}{1} \qquad \frac{1}{1} \quad 1 \quad \frac{\underline{\delta}}{\underline{\delta}} \qquad 1$$

and the order of magnitude of the various terms is indicated. One of the terms containing the viscosity is seen to be of larger order of magnitude than the rest, namely,

$$\frac{\partial}{\partial \underline{y}}\left(\underline{\mu}\frac{\partial \underline{u}}{\partial \underline{y}}\right)$$

which is of order $1/\underline{\delta}^2$. The term is divided by the Reynolds number Re_0 in Eq. (6-41), and to make it commensurate with the remaining terms in the equation the Reynolds number must be large and of order $1/\underline{\delta}^2$.

Some additional consideration should be given to the dimensionless body-force term $\underline{\varrho}\underline{g}_x$. This term will be of order 1 provided $\underline{g}_x$ is of order 1 or g_x is of order $V_0{}^2/L$. An average value of the velocity V_0 occurring in engineering applications may be assumed as 100 m/s, and an average value of the dimension L may be taken as 1 m. The parameter V^2/L then assumes the value 10^4 m/s^2. The gravitational acceleration has the value 9.81 m/s^2. Our estimate therefore indicates that free-convection effects caused by gravitational acceleration can be neglected in most forced-flow problems.[1] The body-force term, however, will be retained in Eq. (6-41) to make this equation more general.

[1] The flow of a liquid in an open channel or the effect of surface waves on vessels are examples for situations in which the effect of gravitational forces cannot be neglected. Such flows are determined by the magnitude of the *Froude number* $\mathrm{Fr} = V_0/\sqrt{Lg}$ in addition to that of the Reynolds number.

The momentum equation in the y direction is

$$\underset{1}{\underline{\rho}}\left(\underline{u}\underset{\frac{\delta}{1}}{\frac{\partial \underline{v}}{\partial \underline{x}}} + \underset{\delta}{\underline{v}}\underset{\frac{\delta}{\delta}}{\frac{\partial \underline{v}}{\partial \underline{y}}}\right) = -\frac{\partial \underline{p}}{\partial \underline{y}} + \underset{\delta^2}{\frac{1}{\mathrm{Re}_0}}\left[2\underset{\frac{1}{\delta}}{\frac{\partial}{\partial \underline{y}}}\left(\underset{1}{\underline{\mu}}\underset{\frac{\delta}{\delta}}{\frac{\partial \underline{v}}{\partial \underline{y}}}\right) + \underset{\frac{1}{1}}{\frac{\partial}{\partial \underline{x}}}\left(\underset{1}{\underline{\mu}}\underset{\frac{\delta}{1}}{\frac{\partial \underline{v}}{\partial \underline{x}}}\right)\right.$$

$$\left. + \underset{\frac{1}{1}}{\frac{\partial}{\partial \underline{x}}}\left(\underset{1}{\underline{\mu}}\underset{\frac{1}{\delta}}{\frac{\partial \underline{u}}{\partial \underline{y}}}\right) - \underset{\frac{1}{\delta}}{\frac{\partial}{\partial \underline{y}}}\left(\underset{1}{\underline{\mu}'}\underset{\frac{\delta}{\delta}}{\frac{\partial \underline{v}}{\partial \underline{y}}}\right) - \underset{\frac{1}{\delta}}{\frac{\partial}{\partial \underline{y}}}\left(\underset{1}{\underline{\mu}'}\underset{\frac{1}{1}}{\frac{\partial \underline{u}}{\partial \underline{x}}}\right)\right] + \underset{1}{\underline{\rho}\underline{g}_y} \tag{6-42}$$

The order of magnitude of the various terms is again shown underneath, and it can be observed that all terms are of order $\underline{\delta}$ except the last, which may be of order 1. We conclude, therefore, that the pressure gradient $\partial \underline{p}/\partial \underline{y}$ is of this order of magnitude. The variation of the pressure through the boundary layer is then

$$\Delta \underline{p} = \int_0^{\delta} \frac{\partial \underline{p}}{\partial \underline{y}}\, d\underline{y}$$

It is therefore of order of magnitude $\underline{\delta}$. The variation of the pressure along the direction y through the boundary layer can therefore be neglected compared with quantities which are of order 1, and the pressure within the boundary layer can be considered to have a value prescribed by the inviscid flow outside the boundary layer.

For unsteady flow, an additional term appears in Eqs. (6-40) and (6-41) containing a derivative toward a nondimensional time $\underline{\tau}$. For this purpose a reference time must be selected which characterizes the rate with which the velocity V_0 changes with time. The continuity equation (6-37) will now have the form

$$\frac{L}{\tau_0 V_0}\frac{\partial \underline{\rho}}{\partial \underline{\tau}} + \frac{\partial(\underline{\rho}\underline{v}_i)}{\partial \underline{x}_i} = 0$$

It will depend on the magnitude of the expression $L/\tau_0 V_0$ whether the complete equation or the abbreviated form (6-37) is to be used. Returning to the example considered before, we select a velocity V_0 of 100 m/s and a reference length L of 1 m as a typical case. This results in a reference time $\tau_0 = \frac{1}{100}$ s, provided the parameter above is postulated to have the value 1. Only for a situation in which the velocity V_0 changes within a time of this order of magnitude does the flow have to be considered as unsteady. In many engineering applications, the velocity on the boundary of the problem changes at a considerably slower rate, and these processes can then be considered as quasi-steady in the sense that the steady-state boundary-layer equations derived before describe the problem at any instance with good accuracy. Rapid changes of velocity

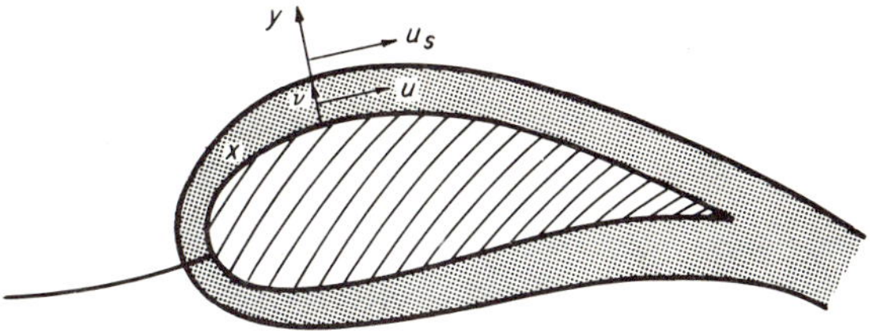

Fig. 6-6 Plane boundary-layer flow over a cylindrical body.

occur in turbulent flow, and the unsteady terms have to be retained in the corresponding equations.

The equations with the simplifications introduced above assume the following form when we return to the dimensional quantities:

$$\frac{\partial \rho}{\partial \tau} + \frac{\partial(\rho u)}{\partial x} + \frac{\partial(\rho v)}{\partial y} = 0 \tag{6-43}$$

$$\rho\left(\frac{\partial u}{\partial \tau} + u\frac{\partial u}{\partial x} + v\frac{\partial u}{\partial y}\right) = -\frac{\partial p}{\partial x} + \frac{\partial}{\partial y}\left(\mu\frac{\partial u}{\partial y}\right) + \rho g_x \tag{6-44}$$

$$p = p(x,\tau) \tag{6-45}$$

These are the boundary-layer continuity and momentum equations for unsteady flow. The two equations (6-43) and (6-44) with the appropriate boundary conditions are sufficient to obtain the two velocity components u and v since the pressure, according to Eq. (6-45), is prescribed by the inviscid flow outside the boundary layer. Experimentally it can be obtained by measurement of the static pressure along the plate surface. The thinness of the boundary layer permits application of these equations to boundary layers on curved surfaces if the radius of curvature of the surface is everywhere large compared to the boundary-layer thickness. The x axis is then measured along the surface in the flow direction, as indicated in Fig. 6-6. The y coordinate is normal to the surface.

6-5 BOUNDARY-LAYER ENERGY EQUATION

In order to derive the boundary-layer energy equation, we start from the energy equation for a one-component fluid in the form of Eq. (6-28). This equation will be repeated for convenience with the energy flux ϵ_i expressed by Fourier's equation (6-32):

$$\rho\frac{Di^\circ}{d\tau} - \frac{\partial p}{\partial \tau} = \frac{\partial}{\partial x_i}\left(k\frac{\partial t}{\partial x_i}\right) + \frac{\partial}{\partial x_j}(v_i p_{ji}) + \rho v_i g_i \tag{6-46}$$

The first step will again be to make the equation dimensionless. In addition to the dimensionless parameters introduced in the preceding section by the relations (6-35), the following dimensionless parameters will be used:

$$\underline{t} = \frac{t}{\Delta t_0} \qquad \underline{i}^\circ = \frac{i^\circ}{c_{p0}\,\Delta t_0} \qquad \underline{k} = \frac{k}{k_0} \tag{6-47}$$

One could use a reference temperature t_0 within the fluid to make the temperature and the total enthalpy within the boundary layer dimensionless, but it will be more convenient in later analyses to use the difference Δt_0 between a reference temperature upstream in the fluid and a temperature on the wall surface. c_{p0} denotes the specific heat at constant pressure and k_0 the thermal conductivity, both at the reference point. The energy equation (6-46) takes on the following form when it is written in the dimensionless parameters:

$$\frac{\rho_0 c_{p0}\,\Delta t_0\,V_0}{L}\,\rho\,\frac{D\underline{i}^\circ}{d\underline{\tau}} - \frac{\rho_0 V_0^3}{L}\frac{\partial \underline{p}}{\partial \underline{\tau}} = \frac{k_0\,\Delta t_0}{L^2}\frac{\partial}{\partial \underline{x}_i}\left(\underline{k}\,\frac{\partial \underline{t}}{\partial \underline{x}_i}\right) + \frac{\mu_0 V_0^2}{L^2}\frac{\partial}{\partial \underline{x}_j}(\underline{v}_i \underline{p}_{ji}) + \frac{\rho_0 V_0^3}{L}\,\underline{\rho}\,\underline{v}_i \underline{g}_i \tag{6-48}$$

The time τ has now been made dimensionless by the time L/V_0 and the viscous stress p_{ji} by $\mu_0 V_0/L$. The equation is made dimensionless by dividing it by the parameter $(\rho_0 c_{p0}\,\Delta t_0\,V_0)/L$, resulting in

$$\underline{\rho}\,\frac{D\underline{i}^\circ}{d\underline{\tau}} - \frac{V_0^2}{c_{p0}\,\Delta t_0}\frac{\partial \underline{p}}{\partial \underline{\tau}} = \frac{k_0}{\rho_0 c_{p0} V_0 L}\frac{\partial}{\partial \underline{x}_i}\left(\underline{k}\,\frac{\partial \underline{t}}{\partial \underline{x}_i}\right) + \frac{\mu_0 V_0}{\rho_0 c_{p0}\,\Delta t_0\,L}\frac{\partial}{\partial \underline{x}_j}(\underline{v}_i \underline{p}_{ji}) + \frac{V_0^2}{c_{p0}\,\Delta t_0}\,\underline{\rho}\,\underline{v}_i \underline{g}_i \tag{6-49}$$

The Reynolds number Re_0 is the only dimensionless factor based on reference quantities which appeared in the momentum equation (6-38). Some additional parameters of the same kind, however, occur in the energy equation (6-49), namely,

$$\mathrm{Ec}_0 = \frac{V_0^2}{c_{p0}\,\Delta t_0} \qquad \mathrm{Pr}_0 = \frac{\mu_0 c_{p0}}{k_0} = \frac{\nu_0}{\alpha_0} \tag{6-50}$$

α_0 denotes the reference thermal diffusivity ($\alpha = k/\rho c_p$). The first of these parameters is the ratio of the kinetic energy per unit mass at the reference velocity multiplied by 2 to the enthalpy difference of a fluid with constant c_p based on the temperature difference Δt_0. In the literature it is referred to as the Eckert number. The second parameter, which will frequently appear in heat-transfer analyses, is actually a dimensionless fluid property referred to as the *Prandtl number*. With these parameters, the energy equation takes on the form

$$\underline{\rho}\,\frac{D\underline{i}^\circ}{d\underline{\tau}} - \mathrm{Ec}_0\frac{\partial \underline{p}}{\partial \underline{\tau}} = \frac{1}{\mathrm{Re}_0\,\mathrm{Pr}_0}\frac{\partial}{\partial \underline{x}_i}\left(\underline{k}\,\frac{\partial \underline{t}}{\partial \underline{x}_i}\right) + \frac{\mathrm{Ec}_0}{\mathrm{Re}_0}\frac{\partial}{\partial \underline{x}_j}(\underline{v}_i \underline{p}_{ji}) + \mathrm{Ec}_0\,\underline{\rho}\,\underline{v}_i \underline{g}_i \tag{6-51}$$

The Eckert number assumes values of order 1 only at large velocities or small temperature differences. As an example, we shall calculate the magnitude of this parameter for air moving with 300 m/s velocity and a temperature difference Δt_0 of 100 C. The specific heat of air is approximately 1 kJ/kg C = 10^3 J/kg C (1 J = kg m^2/s^2)

$$\mathrm{Ec}_0 = \frac{9 \times 10^4}{10^3 \times 100} \frac{\mathrm{kg\ m^2\ C}}{\mathrm{J\ s^2\ C}} = 0.9$$

It can be seen that the parameter Ec_0 has values of order 1 in air when the velocity is of the order of the sound velocity and the temperature difference is of order 100 C. The situation is similar for other gases. For many engineering applications which operate with smaller velocities, the terms in Eq. (6-51) which are multiplied by the parameter Ec_0 can be neglected. In order to make the boundary-layer equations general, however, we shall retain these terms in Eq. (6-51). The Prandtl number Pr_0 of fluids varies in a wide range from a value of order 0.01 for liquid metals to a value of order 1,000 or more for viscous oils. Simplifications in the solution of boundary-layer problems are again possible for very small or very large Prandtl numbers. In the following derivations, we shall avoid such simplification in order to keep the boundary-layer equations general; however, we shall consider situations in which the Reynolds number Re_0 is very large. Therefore the terms in Eq. (6-51) which contain the Reynolds number in the denominator would vanish unless at least some of the terms which are multiplied by the inverse of the Reynolds number are very large. This indicates that not only the velocity field but also the temperature field is restricted to a thin layer of order of magnitude $\underline{\delta}$. The boundary-layer equation is then obtained from the energy equation (6-51) by estimating the order of magnitude of the individual terms, which is done in the following equation rewritten in the x and y coordinates:

$$\underline{\rho}\left(\frac{\partial \underline{i}^\circ}{\partial \underline{\tau}} + \underline{u}\frac{\partial \underline{i}^\circ}{\partial \underline{x}} + \underline{v}\frac{\partial \underline{i}^\circ}{\partial \underline{y}}\right) - \mathrm{Ec}_0 \frac{\partial \underline{p}}{\partial \underline{\tau}} = \frac{1}{\mathrm{Re}_0\,\mathrm{Pr}_0}\left[\frac{\partial}{\partial \underline{x}}\left(\underline{k}\frac{\partial \underline{t}}{\partial \underline{x}}\right) + \frac{\partial}{\partial \underline{y}}\left(\underline{k}\frac{\partial \underline{t}}{\partial \underline{y}}\right)\right]$$

$$1\ \ \frac{1}{1} \qquad 1\ \frac{1}{1} \qquad \delta\ \frac{1}{\delta} \qquad\qquad 1\ \frac{1}{1} \qquad\qquad \frac{\delta^2}{1} \qquad \frac{1}{1} \quad 1\ \frac{1}{1} \qquad \frac{1}{\delta} \quad 1\ \frac{1}{\delta}$$

$$+ \frac{\mathrm{Ec}_0}{\mathrm{Re}_0}\left[\frac{\partial}{\partial \underline{y}}\left(\underline{u}\,\underline{\mu}\frac{\partial \underline{u}}{\partial \underline{y}}\right) + \cdots\right] + \mathrm{Ec}_0\,(\underline{\rho}\,\underline{u}\,\underline{g}_x + \underline{\rho}\,\underline{v}\,\underline{g}_y) \tag{6-52}$$

$$\delta^2 \quad \frac{1}{\delta} \quad 1\ 1\ \frac{1}{\delta} \qquad\qquad\qquad 1(1\ 1\ 1\ +\ 1\ \delta\ 1)$$

The $\underline{y}$ coordinates are again of order $\underline{\delta}$: the $\underline{v}$ component of the velocity within the boundary layer must be assumed of order $\underline{\delta}$ by the argument presented in the preceding section. All the other quantities are assumed of order 1 to keep the equation as general as possible. It can easily be checked that the term $\partial/\partial y(\underline{u}\mu\partial\underline{u}/\partial\underline{y})$ has the largest order of magnitude among the terms describing work done by viscous stresses, and only this term has been retained in Eq. (6-52), whereas the presence of other terms is indicated by ellipses. It can also be observed that the dimensionless terms by which the parameter Ec_0 is multiplied are of order 1. These terms can therefore be neglected when Ec_0 is small compared to 1. Retaining only terms in the above equation which are of order 1 leads to the energy boundary-layer equation

$$\rho\left(\frac{\partial i^\circ}{\partial \tau} + u\frac{\partial i^\circ}{\partial x} + v\frac{\partial i^\circ}{\partial y}\right) - \frac{\partial p}{\partial \tau} = \frac{\partial}{\partial y}\left(k\frac{\partial t}{\partial y}\right) + \frac{\partial}{\partial y}\left(u\mu\frac{\partial u}{\partial y}\right) + \rho u g_x \qquad (6\text{-}53)$$

The equation is again written in the dimensional quantities. This equation corresponds to the general energy equation (6-28) written in total enthalpy for a one-component fluid. In some instances it is more convenient to use a boundary-layer energy equation in the static enthalpy corresponding to the general equation (6-29). The corresponding boundary-layer equation can be derived again by an order-of-magnitude argument from Eq. (6-29), or it can be obtained directly if the momentum boundary-layer equation (6-44) is multiplied by the velocity component u and subtracted from the energy equation (6-53). Multiplication of Eq. (6-44) by u results in

$$\rho\left(\frac{\partial}{\partial \tau}\left(\frac{u^2}{2}\right) + u\frac{\partial}{\partial x}\left(\frac{u^2}{2}\right) + v\frac{\partial}{\partial y}\left(\frac{u^2}{2}\right)\right) + u\frac{\partial p}{\partial x} = u\frac{\partial}{\partial y}\left(\mu\frac{\partial u}{\partial y}\right) + \rho u g_x$$

Subtracting this from Eq. (6-53) gives

$$\rho\left(\frac{\partial i}{\partial \tau} + u\frac{\partial i}{\partial x} + v\frac{\partial i}{\partial y}\right) - \frac{\partial p}{\partial \tau} - u\frac{\partial p}{\partial x} = \frac{\partial}{\partial y}\left(k\frac{\partial t}{\partial y}\right) + \mu\left(\frac{\partial u}{\partial y}\right)^2 \qquad (6\text{-}54)$$

The general relation between the total enthalpy i° and the static enthalpy i is

$$i^\circ = i + \frac{V^2}{2}$$

Within the boundary layer the v component of the velocity vector $\mathbf{v}$ is of a smaller order of magnitude than the u component, and the above relation simplifies to

$$i^\circ = i + \frac{u^2}{2}$$

This relation has been utilized in obtaining Eq. (6-54).

Equations (6-43) to (6-45) and (6-53) or (6-54) describe the velocity and temperature field in a boundary layer. In order to obtain a solution, one must have information about the fluid properties ρ, μ, i, and k. In many cases *local thermodynamic equilibrium* exists within the boundary layer, which means that the properties mentioned above are functions of the local state of the fluid expressible, for instance, by the local pressure p and the local temperature t. In a fluid consisting of several components which may react with each other chemically, one must also assume chemical equilibrium. Many calculations considering heat transfer in a dissociating or ionized gas, for instance, have been made on this basis. Boundary-layer solutions become much more involved when local equilibrium cannot be assumed because it is then necessary to consider the speed with which chemical reactions or the adjustment to thermodynamic equilibrium occur and these processes become interrelated with the flow and heat-transfer processes. Another complication arises when mass transfer is superimposed on the flow and heat-transfer processes. Such subjects will be taken up in Part IV.

The treatment of boundary-layer heat transfer, on the other hand, becomes especially simple when the fluid considered is one of constant properties. In this case, Eqs. (6-43) to (6-45) can be solved to obtain the velocity field in the boundary layer without referring to the boundary-layer energy equation. Physically, this means that the flow field around an object is the same regardless of whether the object is at a temperature equal, lower, or higher than that of the fluid. A solution of the energy equation (6-53) or (6-54) then results in a temperature field within the boundary layer which is only superimposed on the flow field without affecting it. This not only simplifies the analysis of boundary-layer problems considerably but also makes the results more general. For this reason, a considerable part of the discussion of convective heat

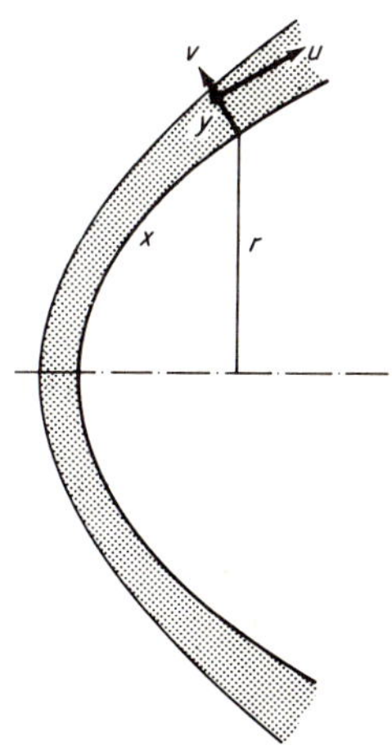

Fig. 6-7 Rotationally symmetric boundary-layer flow.

transfer will be based on the model of a constant-property fluid. For many applications, the influence of property variations is sufficiently small to permit results obtained for a constant-property fluid to serve as acceptable approximations to the actual situation.

The discussion in this chapter has been restricted to boundary layers on a plane surface. The equations continue to hold for plane boundary-layer flow over curved surfaces, as indicated in Fig. 6-6, but they become more involved for a general three-dimensional process and will not be discussed here. For a rotationally symmetric situation, however, the equations change only to a minor degree (Fig. 6-7). If r indicates the distance of a surface point under consideration from the axis of rotation, and if the x axis is oriented in downstream direction within the surface and the y direction is normal to it, the boundary-layer continuity equation assumes the form

$$r\frac{\partial \rho}{\partial \tau} + \frac{\partial (r\rho u)}{\partial x} + \frac{\partial (r\rho v)}{\partial y} = 0 \tag{6-55}$$

when the radius of curvature of the object is everywhere large compared to the boundary-layer thickness. The momentum and energy equations maintain their form as given in (6-44) and (6-53).

7
Forced Convection in Laminar Flow

In obtaining and discussing solutions to the conservation equations, in this chapter we consider the situation referred to as *forced convection;* this is a situation in which body forces are negligible. In reality, one sort of body forces, namely, gravitational forces, are always present as long as we consider processes in the environment of our planet earth. Their influence, however, is often quite small. An order-of-magnitude estimate, indicating when this effect can be neglected, was presented in the preceding chapter.

We start with consideration of a constant-property fluid. The results obtained in this way not only offer valuable information about the general trends in heat-transfer processes but result in relations which describe the flow and heat transfer quantitatively with good accuracy as long as temperature differences and velocities are moderate.

The discussion will also be restricted to two-dimensional flow and steady state.

7-1 THE LAMINAR-FLOW BOUNDARY-LAYER EQUATIONS

The boundary-layer equations (6-43) to (6-45) simplify for the situation mentioned above to

$$\frac{\partial u}{\partial x} + \frac{\partial v}{\partial y} = 0 \tag{7-1}$$

$$\rho\left(u\frac{\partial u}{\partial x} + v\frac{\partial u}{\partial y}\right) = -\frac{\partial p}{\partial y} + \mu\frac{\partial^2 u}{\partial y^2} \tag{7-2}$$

$$\frac{\partial p}{\partial y} = 0 \tag{7-3}$$

The following boundary conditions apply on a stationary solid surface:

$$y = 0 \qquad u = 0 \qquad v = 0 \tag{7-4}$$

$$y \to \infty \qquad u = u_s \tag{7-5}$$

The following sections will be concerned with solutions describing the flow field, since they will be required in a later study of heat transfer.

7-2 SIMILARITY SOLUTIONS FOR LAMINAR BOUNDARY-LAYER FLOW

Such solutions of the system of Eqs. (7-1) to (7-3) with the boundary conditions (7-4) and (7-5) depend on how the pressure p or the velocity u_s at the outer border of the boundary layer, respectively, changes along the surface. This stream velocity u_s is connected with the pressure variation along the surface by Bernoulli's equation

$$p + \rho\frac{u_s^2}{2} = \text{const} \tag{7-6}$$

Today solutions of the boundary-layer equations can be obtained for any prescribed pressure variation with the help of electronic computers, but they still require a considerable effort because the differential equations are nonlinear and depend on two variables x and y. For a basic study of boundary layers, it is then advantageous to inquire whether solutions exist for which the partial differential equations can be transformed into total ones. Prandtl[1] was the first to point out in 1904 that this is possible for longitudinal flow of a uniform stream over a flat plate, and in 1908 Blasius[2] obtained in this way the first boundary-layer solu-

[1] L. Prandtl, Über Flüssigkeitsbewegung bei sehr kleiner Reibung, *Proc. 3d Intern. Math. Koug. Heidelberg, 1904.*

[2] H. Blasius, Grenzschichten in Flüssigkeiten mit kleiner Reibung, *Z. Math. Phys.*, **56**:1 (1908).

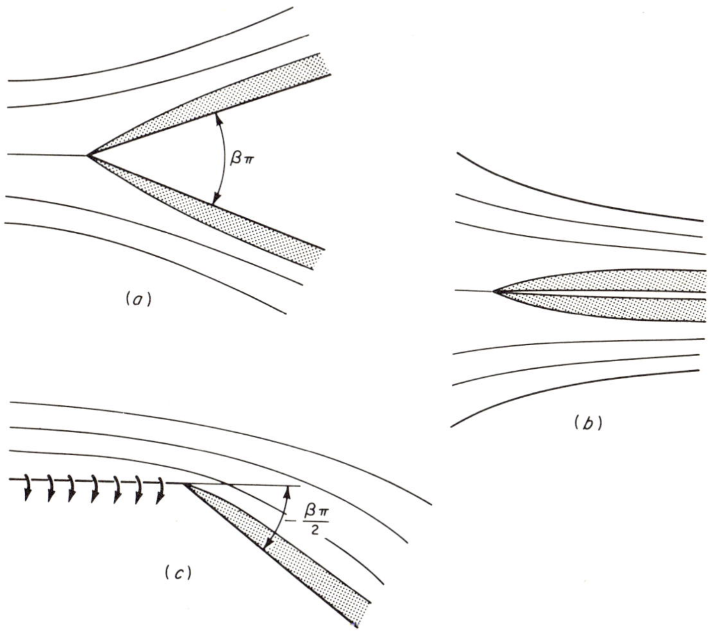

Fig. 7-1 Self-similar boundary layers.

tion. In 1930 Falkner and Skan[1] discovered a whole family of situations for which the same transformation is possible, and in 1939 Goldstein[2] investigated in detail under what conditions such a transformation can be carried out.

Thus it was established that the transformation is possible when the stream velocity u_s is proportional to some power of the coordinate x

$$u_s = Cx^m \tag{7-7}$$

Potential flow of a constant-property fluid over an unlimited wedge (Fig. 7-1a) creates along the surface a velocity variation of this sort, and for this reason this type of flow is referred to as *wedge-type flow*. The exponent m in Eq. (7-7) is connected with the opening angle $\beta\pi$ of the wedge by the relation

$$\beta = \frac{2m}{m+1} \tag{7-8}$$

[1] V. M. Falkner and S. W. Skan, Some Approximate Solutions of the Boundary Layer Equations, *Phil. Mag.*, **12**:865 (1931).

[2] S. Goldstein, A Note on Boundary Layer Equations, *Proc. Cambridge Phil. Soc.*, **35**:388 (1939).

A velocity variation corresponding to Eq. (7-7) can also be approximated experimentally in different ways, e.g., as shown in Fig. 7-1*b*. With the proper shape of the curved wall of the channel, a velocity distribution outside the boundary layer can be established along the flat plate which closely approximates the one prescribed by Eq. (7-7).

The continuity equation (7-1) can be eliminated from consideration by introduction of a stream function ψ satisfying the conditions

$$u = \frac{\partial \psi}{\partial y} \qquad v = -\frac{\partial \psi}{\partial x} \tag{7-9}$$

It is seen that introducing this stream function into the continuity equation satisfies it identically. The remaining equation (7-2) can then be transformed into a total one by introducing a new independent variable

$$\eta = \sqrt{\frac{m+1}{2}}\frac{y}{x}\sqrt{\mathrm{Re}_x} = \sqrt{\frac{m+1}{2}}\,y\sqrt{\frac{u_s}{\nu x}} = \sqrt{\frac{m+1}{2}}\sqrt{\frac{C}{\nu}}\,x^{(m-1)/2}y \tag{7-10}$$

and a new stream function which is a function of this variable η only:

$$f = \sqrt{\frac{m+1}{2}}\frac{\psi}{u_s x}\sqrt{\mathrm{Re}_x} = \sqrt{\frac{m+1}{2}}\frac{\psi}{\sqrt{\nu u_s x}} = \sqrt{\frac{m+1}{2}}\frac{1}{\sqrt{C\nu}}\,x^{-(m+1)/2}\psi \tag{7-11}$$

The velocity component u can now be written as

$$u = \frac{\partial \psi}{\partial y} = \sqrt{\frac{2C\nu}{m+1}}\,x^{(m+1)/2}\frac{df}{d\eta}\sqrt{\frac{m+1}{2}}\sqrt{\frac{C}{\nu}}\,x^{(m-1)/2} = Cx^m\frac{df}{d\eta} = u_s\frac{df}{d\eta} \tag{7-12}$$

The velocity component v is, by a similar calculation,

$$v = -\frac{\partial \psi}{\partial x} = -\sqrt{\frac{2C\nu}{m+1}}\,x^{(m-1)/2}\left(\frac{m+1}{2}f + \frac{m-1}{2}\frac{df}{d\eta}\eta\right) \tag{7-13}$$

In the same way the derivatives $\partial u/\partial x$, $\partial u/\partial y$, $\partial^2 u/\partial y^2$ can be expressed in f, η, and x. If one introduces these terms into Eq. (7-2) and expresses the term $\dfrac{1}{\rho}\dfrac{\partial p}{\partial x}$ from Bernoulli's equation

$$\frac{1}{\rho}\frac{\partial p}{\partial x} = -u_s\frac{\partial u_s}{\partial x} = -mC^2x^{2m-1}$$

one finds that the constant C and the variable x cancel out, and the following equation evolves:

$$\frac{d^3f}{d\eta^3} + f\frac{d^2f}{d\eta^2} - \beta\left[\left(\frac{df}{d\eta}\right)^2 - 1\right] = 0 \tag{7-14}$$

The only independent variable in this equation is η. The boundary conditions can also be expressed in the variables f and η. One obtains

$$\begin{aligned} \eta &= 0 \qquad \frac{df}{d\eta} = 0 \qquad f = 0 \\ \eta &= \infty \qquad \frac{df}{d\eta} = 1 \end{aligned} \tag{7-15}$$

In solving the total differential equation (7-14) with the boundary conditions (7-15), we also obtain a solution to the partial differential equations (7-1) and (7-2) with the boundary conditions (7-4) and (7-5). Numerical solutions have been obtained on an electronic computer by Hartree.[1] The calculated velocity profiles are presented in Fig. 7-2. The shape of these profiles depends solely on the parameter β and is independent of x. All velocity profiles along the surface are therefore similar in shape. Such boundary layers are called *self-similar* or *self-preserving*.

Of special interest are three values of the parameter β. $\beta - 0$ corresponds to $m = 0$ and, according to Eq. (7-7), to a constant velocity u_s.

[1] D. R. Hartree, On an Equation Occurring in Falkner and Skan's Approximate Treatment of the Equations of the Boundary Layer, *Proc. Cambridge Phil. Soc.*, **33**(II):223 (1937).

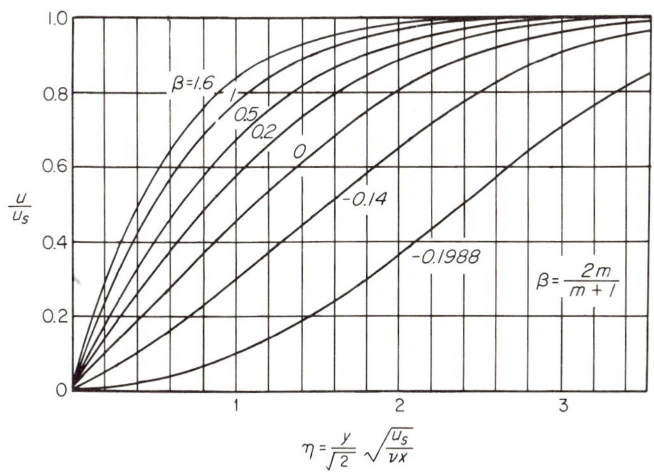

Fig. 7-2 Velocity profiles for laminar wedge-type flow. [*After D. R. Hartree, Proc. Cambridge Phil. Soc.,* **33**(*II*) *223* (*1937*).]

This is the situation usually referred to as flow over a flat plate. It may be recognized from the figure that, for such a condition, the velocity u increases at first almost linearly with the wall distance y and later turns asymptotically into the stream velocity u_s. Positive values of β correspond to a flow in which the velocity is zero for $x = 0$ and increases with increasing x. The value $\beta = 1$ ($m = 1$) describes, according to Fig. 7-1*a*, two-dimensional flow toward a flat plate which is oriented at a right angle to the upstream flow direction. Two-dimensional flow against a cylindrical object with a blunt leading edge creates conditions near the stagnation line which closely approximate those on such a flat plate. For this reason, this type of flow is referred to as two-dimensional stagnation flow. From Fig. 7-2 it can be recognized that the velocity profile for this type of flow is fuller than the velocity profile on a flat plate and that it starts out with a curvature at the plate surface.

For negative values of the parameter β, the velocity profiles take on an S shape. A negative β indicates a velocity u_s which starts from a value infinity at $x = 0$ and decreases in the direction of flow. In producing such a flow experimentally, one would need a wedge with a negative opening angle, which, of course, is impossible. However, an arrangement as shown in Fig. 7-1*c*, in which the boundary layer would be removed by suction on the horizontal wall, should approximate the required condition. Of specific interest is the profile with the value $\beta = -0.1988$, which has a zero velocity gradient at $y = 0$. It appears that no solutions to Eq. (7-14) can be found for larger negative values of β. However, one would expect that for such values the velocities u would at first be negative, would go through a value zero at some distance y, and would then asymptotically approach the value u_s. This would mean a backflow in upstream direction close to the surface, a condition which is found when flow separates from a surface. This will become clearer later on. The profile for $\beta = -0.1988$ is therefore called a *separation profile.*

Since all the velocity profiles approach the value u_s asymptotically, one must select some arbitrary definition for the boundary-layer thickness, e.g., by considering that value of y at the border of the boundary layer at which the velocity u differs from the velocity u_s by 1 percent. More appropriately, one uses the so-called *displacement thickness,* defined by the equation

$$\delta^* = \int_0^\infty \left(1 - \frac{u}{u_s}\right) dy \tag{7-16}$$

as a measure of the boundary-layer thickness or, alternately, the momentum thickness, defined by the equation

$$\delta_i = \int_0^\infty \frac{u}{u_s}\left(1 - \frac{u}{u_s}\right) dy \tag{7-17}$$

From Fig. 7-2 or Table 7-1 the boundary-layer thickness is obtained with the help of the equation

$$\delta = \sqrt{\frac{2}{m+1}}\sqrt{\frac{\nu x}{u_s}}\,\eta_\delta = \sqrt{\frac{2}{m+1}}\sqrt{\frac{\nu}{C}}\,x^{(1-m)/2}\eta_\delta \tag{7-18}$$

where η_δ indicates that value of η at which the velocity u approaches u_s to within 1 percent or satisfies Eq. (7-16) or (7-17). Equation (7-18) shows that for flat-plate flow ($m = 0$) the boundary layer starts out with a value zero at $x = 0$ and increases proportional to $\sqrt{x}$ in the downstream direction. For plane stagnation flow ($m = 1$) the boundary layer has the same thickness for all values of x. For the negative values of β in Fig. 7-2, the boundary layer increases in the downstream direction somewhat faster than at the flat plate.

The shear τ_w at the wall is of particular engineering interest. It is obtained from Newton's equation

$$\tau_w = \mu\left(\frac{\partial u}{\partial y}\right)_{y=0}$$

With the relation

$$\tau_w = \mu u_s\left(\frac{d^2f}{d\eta^2}\right)_0\left(\frac{\partial\eta}{\partial y}\right)_0 = \sqrt{\frac{m+1}{2}}\,\rho u_s\sqrt{\frac{\nu u_s}{x}}\left(\frac{d^2f}{d\eta^2}\right)_0 \tag{7-19}$$

the shear can be determined from the derivatives of the curves at $y = 0$ in Fig. 7-2. Values for $(d^2f/d\eta^2)_0$ are also listed in Table 7-1. The wall shear is conventionally expressed in dimensionless form by a friction factor c_f defined by the equation

$$\tau_w = c_f\rho\frac{u_s^2}{2}$$

Table 7-1 Parameters of the boundary layer for wedge-type flow†

Main flow parameter β	*Dimensionless displacement thickness* η^*	*Dimensionless momentum thickness* η_i	*Velocity gradient at the surface* $\left(\frac{d^2f}{d\eta}\right)_0$
1.6	0.5439	0.2503	1.5210
1.0	0.6479	0.2923	1.2326
0.5	0.8044	0.3502	0.9276
0.2	0.9839	0.4080	0.6866
0	1.2167	0.4695	0.4696
−0.14	1.5967	0.5388	0.2397
-0.198_8	2.3587	0.5852	0.0000

† E. Eckert, Die Berechnung des Wärmeübergangs in der laminaren Grenzschicht umströmter Körper, *VDI-Forschungsh.* 416, 1942; see Eq. (7-18).

With this and the preceding equation, the friction factor is

$$c_f = \sqrt{2(m+1)} \left(\frac{d^2f}{d\eta^2}\right)_0 \frac{1}{\sqrt{\text{Re}_x}}$$

So far, solutions to the boundary-layer equations have been discussed for flow over a solid surface. Continuity requires in this case that the velocity component v normal to the surface be zero at $y = 0$. For some applications, solutions of the equation are also of interest for which v_w has a finite value at the surface. This can be achieved, for instance, when the surface is fabricated of a porous material and when fluid is sucked away from the mainstream or ejected into it. This is done in various applications to influence the development of boundary layers, thus preventing flow separation for instance. Figure 7-1*c* shows such a porous wall with suction. A finite velocity at the surface is also often encountered in mass-transfer processes, evaporation of vapor from a liquid surface or condensation on it being examples of such processes.

The boundary-layer equations (7-1) to (7-3) remain unchanged when suction or ejection is applied, and only one of the boundary conditions (7-4) changes, insofar as now a finite value for v_w is prescribed. Similar solutions are obtainable only when v_w depends in a certain way on the distance x. This can be seen by writing Eq. (7-13) for $\eta = 0$

$$v_w = -\sqrt{\frac{(m+1)C\nu}{2}}\, f_w x^{(m-1)/2} \tag{7-20}$$

The stream function f is for similar solutions a function of the dimensionless coordinate η only. For $\eta = 0$, it is therefore a constant and the velocity v_w varies according to Eq. (7-20) in proportion to $x^{(m-1)/2}$. For the flat plate ($m = 0$) the blowing or suction velocity v_w must decrease inversely proportional to the square root of x; for plane stagnation flow ($m = 1$) it must be independent of x.

Solutions to the boundary-layer equations have been obtained numerically. Figure 7-3 presents, as an example, the velocity profiles for flow over a flat plate.[1] It can be recognized that suction decreases the boundary-layer thickness, whereas blowing increases it. Blowing also produces a velocity profile with an S shape. For a value $f_w = -0.438$, the slope of the velocity profile is zero at the surface ($\eta = 0$). No solutions which produce a velocity profile for finite η values can be obtained with blowing parameters f_w beyond this limit. This appears to indicate that for such values the boundary layer is lifted off the surface, producing a flow situation similar to separation.

[1] H. Schlichting and K. Bussmann, *Schriften Deut. Akad. Luftfahrtforschung,* **7B**(2) (1943).

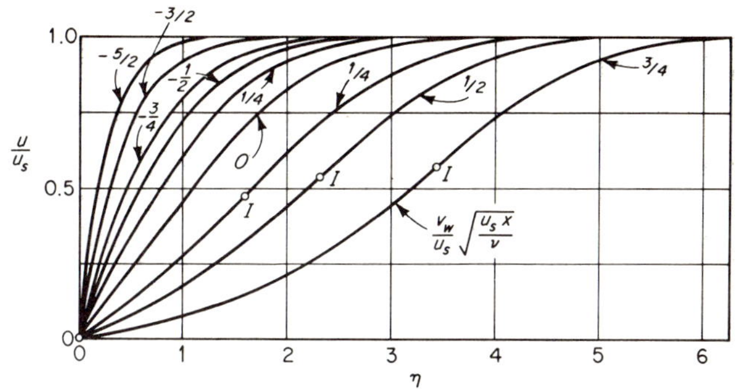

Fig. 7-3 Velocity profiles for laminar flow over a flat plate with fluid suction or injection. [*After H. Schlichting and K. Bussmann, Schriften Deut. Akad. Luftfalhrtforschung,* **7B**(*2*) (*1943*).]

A very simple asymptotic solution can be obtained for the flat plate when suction with a constant velocity $-v_w$ is applied. Equations (7-1) to (7-3) with $\partial p/\partial x = 0$ and with the boundary conditions

$$y = 0 \qquad u = 0 \qquad v = -v_w$$
$$y \to \infty \qquad u = u_s$$

describe this situation. A particular solution satisfying these boundary conditions is one for which the velocity u is independent of x. The continuity equation (7-1) then results in $v = -v_w$, and the momentum equation (7-2) reduces to

$$v_w \frac{du}{dy} + \nu \frac{d^2u}{dy^2} = 0$$

Integration of this equation and consideration of the boundary conditions results in the following expression for the velocity profile:

$$u = u_s(1 - e^{-v_w y/\nu})$$

This profile cannot describe conditions near the leading edge correctly because the boundary layer will obviously start with the thickness zero at $x = 0$. It describes the asymptotic profile toward which the velocity field converges with increasing x. This state, according to Iglisch,[1] is reached after an approximate length described by the relation

$$\frac{v_w}{u_s}\sqrt{\frac{u_s x}{\nu}} = 2$$

Another flow situation of general interest is rotationally symmetric flow directed normally toward a flat surface at right angles to the flow

[1] R. Iglisch, *Schriften deut. Akad. Luftfahrtforschung,* **8B**(1) (1944); *NACA Tech. Mem.* 1205, 1949.

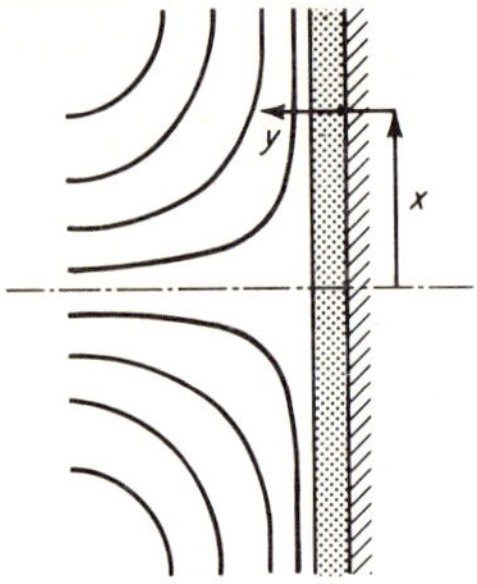

Fig. 7-4 Boundary-layer development in rotationally symmetric flow over a plate oriented normal to the upstream flow direction.

direction (Fig. 7-4). It is found that three-dimensional flow, creating a stagnation point on an object with a blunt nose, generates near the stagnation point flow conditions which closely correspond to those on such a flat surface. Therefore, this flow situation is referred to as *three-dimensional stagnation flow* or *stagnation-point flow*. The differential equations describing the boundary layer for rotationally symmetric flow are Eqs. (6-55), (7-2), and (7-3). For stagnation-point flow, the radius r is equal to x (Fig. 7-4), and the equations assume the form

$$\frac{\partial(xu)}{\partial x} + \frac{\partial(xv)}{\partial y} = 0 \tag{7-21}$$

$$u\frac{\partial u}{\partial x} + v\frac{\partial u}{\partial y} = -\frac{1}{\rho}\frac{dp}{dx} + \nu\frac{\partial^2 u}{\partial y^2} \tag{7-22}$$

with boundary conditions

$$\begin{aligned} &y = 0 \qquad u = 0 \qquad v = 0 \\ &y \to \infty \qquad u = u_s \end{aligned} \tag{7-23}$$

The velocity variation of the potential flow outside the boundary layer is found to be

$$u_s = Cx \tag{7-24}$$

The stream function now must be defined by the equations

$$u = \frac{1}{x}\frac{\partial\psi}{\partial y} \qquad v = -\frac{1}{x}\frac{\partial\psi}{\partial x} \tag{7-25}$$

in order to satisfy the continuity equation. Introduction of the independent variable

$$\eta = \sqrt{\frac{2C}{\nu}}\,y \tag{7-26}$$

and the dependent variable

$$f(\eta) = \sqrt{\frac{2}{C\nu}}\,\frac{\psi}{x^2} \tag{7-27}$$

again transforms the momentum equation of the boundary layer into the total differential equation

$$\frac{d^3f}{d\eta^3} + f\frac{d^2f}{d\eta^2} - \frac{1}{2}\left[\left(\frac{df}{d\eta}\right)^2 - 1\right] = 0 \tag{7-28}$$

The velocity components are given by

$$u = u_s \frac{df}{d\eta}$$
$$v = -\sqrt{2C\nu}\, f$$

It can be recognized that Eq. (7-28) is identical with Eq. (7-14) for two-dimensional wedge-type flow with the condition $\beta = \frac{1}{2}$ $(m = \frac{1}{3})$. This means that the velocity profile for this value of β in Fig. 7-2 also describes the velocity profile in the boundary layer of a three-dimensional stagnation flow with $v_w = 0$. The profile in the actual physical coordinates is obtained with the help of Eq. (7-26). This equivalence of the rotationally symmetric stagnation flow to a two-dimensional wedge-type flow is actually a special case of a more general correspondence principle discovered by Mangler.[1]

The *Mangler transformation* offers the possibility of determining the velocity field for a related laminar rotationally symmetrical boundary layer from a known velocity field within a steady, two-dimensional laminar boundary layer. The coordinates x and y of the rotationally symmetric boundary layer are related to the coordinates $\bar{x}$ and $\bar{y}$ for the two-dimensional flow by the following equations:

$$d\bar{x} = \frac{r^2}{L^2}\,dx \qquad d\bar{y} = \frac{r}{L}\,dy \tag{7-29}$$

or, integrated,

$$\bar{x} = \frac{1}{L^2}\int r^2\,dx \qquad \bar{y} = \frac{r}{L}\,y \tag{7-30}$$

r is the distance of the surface point of the rotationally symmetric object under consideration from the axis (Fig. 6-7), and L is an arbitrary constant with the dimension of a length. That the above relationships transform a rotationally symmetric boundary layer into a two-dimensional one can be checked by writing down the rotationally symmetric boundary-layer flow equations and introducing the above transformation into them. The continuity equation (6-55) reduces for a constant-property fluid and steady flow to

$$\frac{\partial(ru)}{\partial x} + \frac{\partial(rv)}{\partial y} = 0$$

[1] W. Mangler, *Z. Angew. Math. Mech.*, **28**:97 (1948).

It is satisfied identically by a stream function defined by

$$\frac{\partial \psi}{\partial y} = \frac{r}{L} u \qquad \frac{\partial \psi}{\partial x} = -\frac{r}{L} v$$

The momentum equation has for rotationally symmetric and plane flow the same form [Eq. (7-2)]. It reads, with the above stream function,

$$\frac{L}{r}\frac{\partial \psi}{\partial y}\frac{\partial}{\partial x}\left(\frac{L}{r}\frac{\partial \psi}{\partial y}\right) - \frac{L}{r}\frac{\partial \psi}{\partial x}\frac{\partial}{\partial y}\left(\frac{L}{r}\frac{\partial \psi}{\partial y}\right) = \frac{\partial}{\partial y}\left[\nu \frac{\partial}{\partial y}\left(\frac{L}{r}\frac{\partial \psi}{\partial y}\right)\right] - \frac{1}{\rho}\frac{\partial p}{\partial x}$$

With $\bar{x} = f(x,y)$, $\bar{y} = f(x,y)$, $r = f(x)$, the following differentiation rules apply:

$$\frac{\partial}{\partial x} = \frac{r^2}{L^2}\frac{\partial}{\partial \bar{x}} + \frac{y}{r}\frac{dr}{dx}\frac{\partial}{\partial \bar{y}} \qquad \frac{\partial}{\partial y} = \frac{r}{L}\frac{\partial}{\partial \bar{y}}$$

The above momentum equation transforms to

$$\frac{r^2}{L^2}\frac{\partial \psi}{\partial \bar{y}}\frac{\partial}{d\bar{x}}\frac{\partial \psi}{\partial \bar{y}} + \frac{\bar{y}}{L}\frac{dr}{dx}\frac{\partial \psi}{\partial \bar{y}}\frac{\partial^2 \psi}{\partial \bar{y}^2} - \frac{r^2}{L^2}\frac{\partial \psi}{\partial \bar{x}}\frac{\partial}{\partial \bar{y}}\frac{\partial \psi}{\partial \bar{y}} - \frac{\bar{y}}{L}\frac{dr}{d\bar{x}}\frac{\partial \psi}{\partial \bar{y}}\frac{\partial^2 \psi}{\partial \bar{y}^2}$$
$$= \frac{r^2}{L^2}\frac{\partial}{\partial \bar{y}}\left(\nu \frac{\partial}{\partial \bar{y}}\frac{\partial \psi}{\partial \bar{y}}\right) - \frac{r^2}{L^2}\frac{1}{\rho}\frac{\partial p}{\partial \bar{x}}$$

which reduces to

$$\frac{\partial \psi}{\partial \bar{y}}\frac{\partial}{\partial \bar{x}}\frac{\partial \psi}{\partial \bar{y}} - \frac{\partial \psi}{\partial \bar{x}}\frac{\partial}{\partial \bar{y}}\frac{\partial \psi}{\partial \bar{y}} = \nu \frac{\partial^3 \psi}{\partial \bar{y}^3} - \frac{1}{\rho}\frac{\partial p}{\partial \bar{x}}$$

This is the momentum equation for plane flow in the coordinates $\bar{x}$ and $\bar{y}$. The stream function ψ obtained by a solution to this equation therefore describes the stream function of a rotationally symmetric flow in the coordinates x and y simultaneously. The velocity $\bar{u}$ of the plane flow is

$$\bar{u} = \frac{\partial \psi}{\partial \bar{y}}$$

The velocity u of the rotationally symmetric flow is

$$u = \frac{L}{r}\frac{\partial \psi}{\partial y} = \frac{\partial \psi}{\partial \bar{y}} = \bar{u}$$

and therefore equal to the velocity $\bar{u}$. The same relation holds for the stream velocities

$$u_s = \bar{u}_s$$

This follows from the term $\frac{1}{\rho}\frac{\partial p}{\partial \bar{x}}$ in the momentum equation.

As an example of the use of the transformation equations, we shall consider rotationally symmetric stagnation flow, as it was discussed before.

For this flow

$$u_s = Cx \qquad r = x$$

Equation (7-30) results in $\bar{x} = x^3/3L^2$ or $x = \sqrt[3]{3L^2\bar{x}}$; therefore

$$u_s = C\sqrt[3]{3L^2\bar{x}}$$

The equivalent plane flow is accordingly one in which the stream velocity varies proportionally to the $\frac{1}{3}$ power of the distance $\bar{x}$. This result ($m = \frac{1}{3}$, $\beta = \frac{1}{2}$) was derived before, from Eq. (7-28). According to the transformation, the velocity at the location x, y in the rotationally symmetric field is equal to the velocity at the location $\bar{x}$, $\bar{y}$ of the plane field. For the same distance $x_0 = \bar{x}_0$, the reference length is $L = x_0/\sqrt{3}$, and the distance $\bar{y}_0 = \sqrt{3}\, y_0$. From this it follows that the thickness of the rotationally symmetric boundary layer at the distance x_0 is smaller by the factor $1/\sqrt{3}$ than that of the plane boundary layer at the same distance. The same ratio must hold for the local shear. The dimensionless wall distance η is for $m = \frac{1}{3}$, according to Eq. (7-10),

$$\eta = \sqrt{\frac{2C}{3\nu}}\,\bar{x}^{-\frac{1}{3}} \qquad \bar{y} = \sqrt{\frac{2C}{\nu}}\,\sqrt[3]{\frac{1}{3L^2}}\,y$$

which indicates that the boundary-layer thickness is constant along the surface for rotationally symmetric stagnation flow.

Wedge-type flows are characterized by the fact that velocity profiles at different locations x are similar in shape, which made it possible to reduce the partial to total differential equations. For body shapes for which the velocity u_s cannot be expressed by a power relation, this similarity does not exist, and an exact solution must start with the partial differential equations of the boundary layer. This is a tedious procedure even with the help of electronic computers, and many approximate procedures have been developed for this purpose. A group of them is built around the principle of *local similarity*, an expression for the fact that the boundary-layer development for a short distance downstream from a selected location x is mainly influenced by the flow parameters existing at this location (pressure gradient, boundary-layer thickness) and that the conditions upstream from the location x have only a small influence. The further away from x they are, the smaller they are. This principle will be applied here in a simple way which results only in a very approximate answer. It can, however, be used to describe the development of a boundary layer qualitatively. We shall consider a cylindrical object arranged with its axis normal to the upstream flow direction and with a cross section as indicated in Fig. 7-5*a*. Knowledge of the variation of the stream velocity u_s at the outer border of the boundary layer is required for the boundary-layer calculation and can be obtained from measurements of the pressure along the surface or from an analysis of the inviscid main flow. It is sketched in Fig. 7-5*b*. The

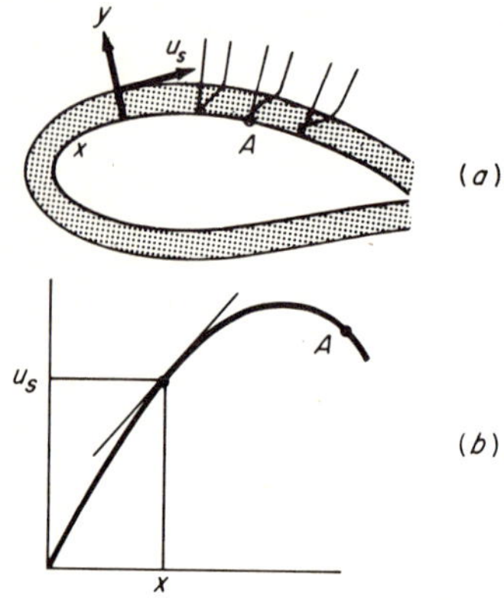

Fig. 7-5 Boundary-layer development for plane flow over a cylindrical body.

boundary-layer characteristics are to be determined at a location x. From the velocity gradient du_s/dx obtained from the tangent in Fig. 7-5*b*, one can derive an exponent m for wedge flow with the same gradient at the location x in the following way. The velocity variation for wedge-type flow is

$$u_s = Cx^m$$

The corresponding velocity gradient is

$$\frac{du_s}{dx} = Cmx^{m-1} = m\frac{u_s}{x}$$

From the last equation one obtains the exponent m

$$m = \frac{x\, du_s/dx}{u_s} \tag{7-31}$$

For our method, the following relation is postulated: the shape of the velocity profile at location x on the object of Fig. 7-5*a* is the same as for wedge-type flow with the value $\beta = 2m/(m+1)$ as calculated from Eq. (7-31), and the boundary-layer thickness can be obtained from Eq. (7-18). The gradient of the velocity profile at the surface will then result in the wall shear stress. In Fig. 7-5*b* it can be seen that β starts out with the value 1 and decreases in the direction of flow, reaching zero where the velocity u_s has a horizontal tangent and becoming negative further downstream. At a certain point (A) it will reach the value $\beta = -0.1988$; this point is interpreted as the point at which the flow separates from the surface of the object because velocity profiles for regions further downstream would exhibit a backward flow near the surface. The shortcoming of our procedure is obviously the fact that the length x which is not a parameter describing local conditions has been used. In an improvement of the method, this parameter is replaced by a characteristic for the local condition, e.g., by the local boundary-layer thickness δ. Various methods developed on the principle of local similarity are included in a paper[1] summarizing and comparing approximate methods of boundary-layer analyses. We shall return later to the principle of local similarity in heat-transfer calculations.

The availability of electronic digital computers makes it possible today to solve the complete nonsimilar boundary-layer equations by numerical integration for any particular problem. Methods for such an analysis are discussed in papers by Smith and Clutter,[2] by Flügge-Lotz and Baxter,[3] and Baxter and Flügge-Lotz.[4] An interesting approach has also been published by Sparrow, Quack, and Boerner.[5]

7-3 MISES TRANSFORMATION

A useful transformation of the laminar steady boundary-layer equations, due to Mises, will be discussed in this section. One can expect the second term on the left side of the boundary-layer equation, which describes the convection of x momentum in the y direction, to drop out if the independent variable y is replaced by ψ because no flow occurs across lines of constant stream function ψ. We shall see that this is actually the case. For this purpose the independent variables x and y will be changed to x and ψ. First we write the equations by which a transformation from the variables x and y to the variables ξ and η is accomplished:

$$\frac{\partial}{\partial x} = \frac{\partial}{\partial \xi}\frac{\partial \xi}{\partial x} + \frac{\partial}{\partial \eta}\frac{\partial \eta}{\partial x}$$
$$\frac{\partial}{\partial y} = \frac{\partial}{\partial \xi}\frac{\partial \xi}{\partial y} + \frac{\partial}{\partial \eta}\frac{\partial \eta}{\partial y} \tag{7-32}$$

Introducing x and ψ for ξ and η and utilizing the equations defining the stream function results, for plane flow, in

$$\frac{\partial}{\partial x} = \frac{\partial}{\partial x} - v\frac{\partial}{\partial \psi} \qquad \frac{\partial}{\partial y} = u\frac{\partial}{\partial \psi} \tag{7-33}$$

[1] D. B. Spalding and W. M. Pun, A Review of Methods for Predicting Heat-transfer Coefficients for Laminar Uniform-property Boundary Layer Flows, *Intern. J. Heat Mass Transfer*, **5**:239–249 (1962).

[2] A. M. O. Smith and D. W. Clutter, Solution of the Incompressible Laminar Boundary-layer Equations, *J. AIAA*, **1**:2062–2071 (1963).

[3] I. Flügge-Lotz and D. C. Baxter, Computation of the Compressible Laminar Boundary-layer Flow Including Displacement-thickness Interaction Using Finite-difference Methods, *Stanford Univ. Tech. Rept.* 131 (AFOSR2206), January, 1962.

[4] D. C. Baxter and I. Flügge-Lotz, *Z. Angew. Math. Phys.*, **9b**:81 (1958).

[5] E. M. Sparrow, H. Quack, and C. J. Boerner, *AIAA J.*, **8**:1936 (1970).

With these transformations, the boundary-layer momentum equation becomes

$$u\frac{\partial u}{\partial x} = -\frac{1}{\rho}\frac{dp}{dx} + \nu u \frac{\partial}{\partial \psi}\left(u\frac{\partial u}{\partial \psi}\right) \tag{7-34}$$

This equation can be further simplified when we replace the velocity u by the total pressure p°. The following equation defines the total pressure:

$$p^\circ = p + \rho\frac{u^2}{2}$$

with p indicating the static pressure which, according to boundary-layer theory, is independent of y or ψ for any fixed value of x. Differentiation of the preceding equation therefore results in

$$\frac{1}{\rho}\frac{\partial p^\circ}{\partial x} = \frac{1}{\rho}\frac{\partial p}{\partial x} + u\frac{\partial u}{\partial x}$$

$$\frac{1}{\rho}\frac{\partial p^\circ}{\partial \psi} = u\frac{\partial u}{\partial \psi}$$

The boundary-layer equation now assumes the form

$$\frac{\partial p^\circ}{\partial x} = \nu u \frac{\partial^2 p^\circ}{\partial \psi^2}$$

Finally the velocity component u in this equation can also be replaced by the total pressure, resulting in

$$\frac{\partial p^\circ}{\partial x} = \nu\sqrt{\frac{2(p^\circ - p)}{\rho}}\,\frac{\partial^2 p^\circ}{\partial \psi^2} \tag{7-35}$$

The unknown dependent variable in this equation is now the total pressure p°, whereas the static pressure p is prescribed to the boundary layer by the outside flow and the other quantities in this equation are constants. The boundary conditions are

$$\psi = 0 \qquad p^\circ = p$$

$$\psi \to \infty \qquad p^\circ = p + \rho\frac{{u_s}^2}{2} = \text{const} \tag{7-36}$$

The Mises transformation can also be applied to the boundary-layer energy equation, discussed in Sec. 7-11. Since Eq. (7-35) is of the same form as Fourier's equation

$$\frac{\partial t}{\partial \tau} = \alpha\frac{\partial^2 t}{\partial x^2}$$

describing unsteady one-dimensional heat conduction, computer programs available for the solution of the heat-conduction equation on electronic computers can be used to solve the boundary-layer equation.

It must be kept in mind, however, that the parameter α, which in the heat-conduction equation signifies the thermal diffusivity, is in the corresponding equation (7-35) in general a function of x and p° or, in terms of the heat-conduction equation, a function of temperature and time.

7-4 THE INTEGRATED BOUNDARY-LAYER MOMENTUM EQUATION

In situations where the pressure varies in an arbitrary way along a surface, the boundary-layer equations must be solved as partial differential equations, a task which requires a considerable effort even with an electronic computer. An approximate method is widely used which was proposed by Kármán[1] in 1921 and applied to the analysis of the boundary-layer development on a cylinder by Pohlhausen.[2] The method fulfills the conservation laws only in an integral manner for a control volume which extends through the whole boundary layer, as indicated in Fig. 7-6. The continuity and momentum equations for this control volume can either be obtained by an integration of Eqs. (7-1) and (7-2) or derived directly. We shall derive them for steady two-dimensional flow of a constant-property fluid (Fig. 7-6).

We use a system of coordinates whose x axis is parallel to the wall and whose y axis is normal to it. The velocities u have a direction parallel to the wall. Only small velocities v normal to the wall will occur. The control surface is formed by two planes 1-2 and 3-4, spaced apart from each other by the small distance dx; by a plane parallel to the wall at a distance l; and by the wall. The distance l is chosen greater than

[1] von Kármán, Über laminare und turbulente Reibung, *Z. Angew. Math. Mech.*, **1**:233 (1921).

[2] K. Pohlhausen, Zur näherungsweisen Integration der laminaren Reibungsschicht, *Z. Angew. Math. Mech.*, **1**:252 (1921).

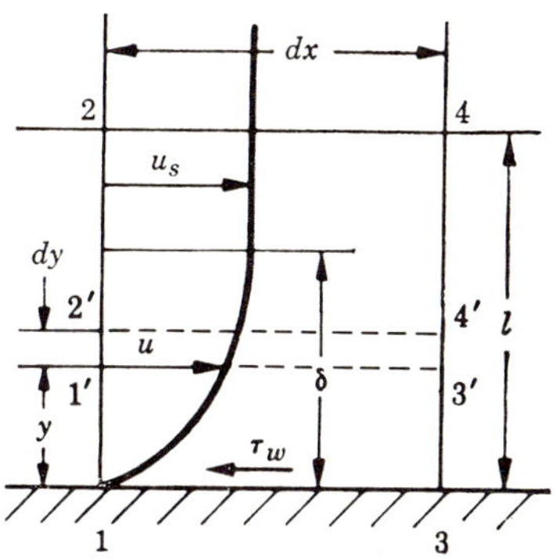

Fig. 7-6 Control volume for the derivation of the integrated boundary-layer equation.

the boundary-layer thickness δ. The velocity profile in the plane 1-2 has the sketched shape. At a distance y from the wall the velocity is u. The outside stream velocity u_s is reached at the distance δ from the wall and therefore within the length l. The control surfaces, which are parallel to the plane of the paper, are a unit distance apart. Through a strip with the height dy at a distance y the mass flow per unit time is $\rho u\, dy$. The mass entering the control volume through the plane 1-2 is obtained by integration and the excess of the mass flow through plane 3-4 over that entering through plane 1-2 by a differentiation with respect to dx. In this way the following expression is obtained

$$\rho\, dx \frac{d}{dx} \int_0^l u\, dy$$

Mass also leaves the control volume through plane 2-4. The velocity component in the y direction through this plane is v_s since the plane is located outside the boundary layer. The mass flow through this plane is therefore

$$\rho v_s\, dx$$

The continuity equation expressing the fact that the sum of all mass flows leaving the control volume must equal zero is

$$v_s + \frac{d}{dx} \int_0^l u\, dy = 0$$

The next step is to derive the momentum equation for the x direction. An expression for the mass entering the control volume through plane 1-2 has been derived above. Its momentum in the x direction is obtained by multiplying it by the velocity component u. The momentum flow therefore is $\rho u^2\, dy$. The momentum flow through the whole plane 1-2 is

$$\rho \int_0^l u^2\, dy$$

Progressing in the x direction, this momentum flow changes by

$$\rho \frac{d}{dx} \left(\int_0^l u^2\, dy \right) dx = \rho\, dx \frac{d}{dx} \int_0^l u^2\, dy$$

Through the solid wall 1-3 there is no flow.[1] There is, however, a mass flow through plane 2-4, and this flow carries x momentum. The total excess of x-momentum flow leaving the control volume is therefore

$$\rho v_s u_s\, dx + \rho\, dx \frac{d}{dx} \int_0^l u^2\, dy$$

[1] The corresponding analysis for fluid suction or injection into the boundary layer is found in Part IV.

Introducing the expression for the velocity component v_s from the mass conservation equation results in the expression

$$-\rho u_s\,dx\,\frac{d}{dx}\int_0^l u\,dy + \rho\,dx\,\frac{d}{dx}\int_0^l u^2\,dy = -\rho\,dx\,\frac{d}{dx}\int_0^l (u_s - u)u\,dy + \rho\,dx\,\frac{du_s}{dx}\int_0^l u\,dy$$

The following external forces act on the surface of the parallelepiped in the x direction: a shear stress τ_w along the wall 1-3 and the pressures p and $p + (dp/dx)\,dx$ on the planes 1-2 and 3-4, respectively. In the plane 2-4 no shear stresses occur, since the plane is beyond the boundary layer. By equating these forces with the increase of momentum,

$$\rho\,\frac{d}{dx}\int_0^l (u_s - u)u\,dy - \rho\,\frac{du_s}{dx}\int_0^l u\,dy = \tau_w + l\,\frac{dp}{dx} \qquad (7\text{-}37)$$

Boundary-layer theory shows that the pressure p varies in the y direction throughout the boundary layer by a negligibly small amount only. Bernoulli's equation (7-6) can therefore be used to bring the momentum equation (7-37) into a form which is more suitable for numerical calculations. According to Bernoulli's equation, the velocity u_s in the stream outside the boundary layer is related to pressure gradient in the following way:

$$\frac{dp}{dx} = -\rho u_s\,\frac{du_s}{dx} \qquad (7\text{-}38)$$

This equation can be used to express the last term on the right side of Eq. (7-37), keeping in mind that the stream velocity u_s does not depend on y:

$$l\,\frac{dp}{dx} = -\rho\,\frac{du_s}{dx}\int_0^l u_s\,dy$$

Substitution into Eq. (7-37) results in

$$\rho\,\frac{d}{dx}\int_0^\delta (u_s - u)u\,dy + \rho\,\frac{du_s}{dx}\int_0^\delta (u_s - u)\,dy = \tau_w \qquad (7\text{-}39)$$

The integration limits have now been changed, since in the range $\delta < y < l$ the term $u_s - u$ in the integrands is equal to zero.

The equation can further be written in a simpler form when the following abbreviations are used:

$$\delta^* = \int_0^\delta \left(1 - \frac{u}{u_s}\right) dy$$

$$\delta_i = \int_0^\delta \left(1 - \frac{u}{u_s}\right) \frac{u}{u_s}\,dy$$

The first term is called the *displacement thickness* of the boundary layer, since it indicates the extent to which the surface would have to be displaced in the direction toward the flow in order to create the same external flow field in a fluid which is free of viscosity. The second term δ_i is called the *momentum thickness* of the boundary layer, since it is connected with the flow of momentum or impulse through an area normal to the surface. The boundary-layer equation written with the two boundary-layer thicknesses is

$$\rho \frac{d}{dx}(u_s{}^2\delta_i) + \rho u_s \frac{du_s}{dx}\delta^* = \tau_w \tag{7-40}$$

It can also be written

$$u_s{}^2 \frac{d\delta_i}{dx} + (2\delta_i + \delta^*)u_s \frac{du_s}{dx} = \frac{\tau_w}{\rho} \tag{7-41}$$

The equation will now be used to obtain approximate solutions for boundary-layer flow around objects of arbitrary shape. In doing so, we shall again assume that the pressure distribution around this object has been obtained either by analysis of the inviscid flow field or by experiments. The method described in this section was suggested by Kármán,[1] worked out by Pohlhausen,[1] and later simplified by Holstein and Bohlen[2] and by Walz.[3] In this method a power law is used as an approximation of the boundary-layer velocity profile with a number of parameters which are determined from conditions which the actual velocity profile has to satisfy. We shall use the following power law

$$\frac{u}{u_s} = a + b\eta + c\eta^2 + d\eta^3 + e\eta^4 \tag{7-42}$$

with $\eta = y/\delta$. According to the five parameters a to e, one needs five conditions. We shall utilize the following ones:

$$y = 0 \qquad u = 0 \qquad \nu \frac{\partial^2 u}{\partial y^2} = -u_s \frac{du_s}{dx} \tag{7-43}$$

$$y \to \delta \qquad u = u_s \qquad \frac{\partial u}{\partial y} = 0 \qquad \frac{\partial^2 u}{\partial y^2} = 0 \tag{7-44}$$

The second condition in (7-43) is obtained from the boundary-layer equation when this equation is written for $y = 0$ and when the pressure gradient is expressed in the velocity u_s through Bernoulli's equation.

[1] See p. 297.

[2] H. Holstein and T. Bohlen, Ein einfaches Verfahren zur Berechnung laminarer Reibungsschichten, die dem Näherungsansatz von K. Pohlhausen genügen, *Lilienthal Ber.* 510, 5, 1940.

[3] A. Walz, Ein neuer Ansatz für das Geschwindigkeitsprofil der laminaren Reibungsschicht, *Lilienthal Ber.* 141, p. 8, 1941.

The last two conditions in (7-44) have the purpose of providing a smooth transition of the boundary-layer profile into the velocity u_s at the outer border of the boundary layer for $y = \delta$. Introduction of these conditions into the equation describing the velocity profile yields

$$\frac{u}{u_s} = (2\eta - 2\eta^3 + \eta^4) + \tfrac{1}{6}\lambda(\eta - 3\eta^2 + 3\eta^3 - \eta^4) \tag{7-45}$$

with the abbreviation

$$\lambda = \frac{\delta^2}{\nu}\frac{du_s}{dx} \tag{7-46}$$

The ratio of the displacement thickness δ^*, as defined in Eq. (7-16), and of the momentum thickness δ_i, as defined in Eq. (7-17), to the boundary-layer thickness δ can now be calculated. The result is

$$\frac{\delta^*}{\delta} = \frac{3}{10} - \frac{\lambda}{120} \tag{7-47}$$

$$\frac{\delta_i}{\delta} = \frac{37}{315} - \frac{\lambda}{945} - \frac{\lambda^2}{9{,}072} \tag{7-48}$$

In the same way, a parameter describing the wall shear τ_w is obtained from Newton's equation

$$\frac{\tau_w \delta}{\mu u_s} = 2 + \frac{\lambda}{6} \tag{7-49}$$

With the substitutions of Eqs. (7-47) to (7-49) into the boundary-layer equation (7-41), the only remaining unknown is δ, and the equation can be solved for it by numerical integration. The solution, however, is very much simplified when Eq. (7-41) is changed to the form

$$u_s \frac{d}{dx}\left(\frac{\delta_i^2}{\nu}\right) = 2\frac{\tau_w \delta_i}{\mu u_s} - 2\left(2 + \frac{\delta^*}{\delta_i}\right)\frac{\delta_i^2}{\nu}\frac{du_s}{dx} \tag{7-50}$$

by multiplying all terms by the expression

$$\frac{\delta_i}{\nu u_s}$$

It is now convenient to introduce a shape parameter

$$\varkappa = \frac{\delta_i^2}{\nu}\frac{du_s}{dx} \tag{7-51}$$

which, through the relation

$$\varkappa = \frac{\delta_i^2}{\nu}\frac{du_s}{dx} = \left(\frac{\delta_i}{\delta}\right)^2 \lambda = f(\lambda) \tag{7-52}$$

can be expressed as a function of the parameter λ appearing in Eqs. (7-45) to (7-49). The terms appearing on the right-hand side of the boundary-layer equation (7-50) can now all be expressed as functions of this shape parameter, so that this equation reduces to

$$u_s \frac{d}{dx}\left(\frac{\delta_i^2}{\nu}\right) = F(\varkappa)$$

Walz has found that this function F can be expressed with good approximation by a linear equation

$$F(\varkappa) = 0.470 - 6\varkappa \tag{7-53}$$

so that the boundary-layer equation finally reads

$$\frac{d}{dx}\left(\frac{u_s\delta_i^2}{\nu}\right) = 0.470 - 5\frac{u_s\delta_i^2}{\nu}\frac{1}{u_s}\frac{du_s}{dx} \tag{7-54}$$

which can easily be integrated to

$$\frac{u_s\delta_i^2}{\nu} = \frac{0.470}{u_s^5}\int_0^x u_s^5\,dx \tag{7-55}$$

The solution of this equation then describes the momentum thickness δ_i of the boundary layer as a function of x. The shape of the velocity profiles at the various locations x is determined by the shape parameter $\varkappa$. The value $\varkappa = 0.0783$ holds for the stagnation point of a blunt body, a value zero for that distance x where the velocity u_s reaches its maximum value and the value -0.1567 (according to $\lambda = -12$) holds for the point of separation. Velocity profiles described by Eq. (7-45) agree in shape quite well with those in Fig. 7-2.

As an example of the use of the method discussed in this section, the laminar boundary layer can be calculated for flow over a flat plate. In this case $\lambda = \varkappa = 0$, $u_s = $ const, and Eq. (7-55) results in

$$\frac{u_s\delta_i^2}{\nu} = 0.470x$$

or

$$\frac{\delta_i}{x} = \frac{0.686}{\sqrt{\mathrm{Re}_x}}$$

The mean shear stress along the plate surface from $x = 0$ to $x = x$ is

$$\bar{\tau}_w = \frac{\delta_i \rho u_s^2}{x}$$

and a mean friction factor defined by the equation $\bar{\tau}_w = \bar{c}_f \rho u_s^2/2$ is

$$\bar{c}_f = \frac{1.372}{\sqrt{\mathrm{Re}_x}}$$

The velocity profile is, according to Eq. (7-45),

$$\frac{u}{u_s} = 2\frac{y}{\delta} - 2\left(\frac{y}{\delta}\right)^3 + \left(\frac{y}{\delta}\right)^4$$

where the relation between δ and δ_i is, according to Eq. (7-48), $\delta = \frac{315}{37}\delta_i$.

These results may be compared with an analysis for the flat plate,[1] which is based on a third-degree polynomial. The exact solution of the boundary-layer equations in Sec. 7-2 results in the value

$$\bar{c}_f = \frac{1.327}{\sqrt{\mathrm{Re}_x}}$$

7-5 THE LAMINAR BOUNDARY-LAYER ENERGY EQUATION

With the knowledge of the flow field in laminar boundary layers of a constant-property fluid, we can now turn to a study of the temperature field and the heat transfer coupled with it.

The restrictions, which we impose, are steady, two-dimensional conditions with regard to flow and temperature fields, a constant-property fluid, absence of body forces, and velocities sufficiently small to make the dimensionless Eckert number of an order of magnitude small compared to 1. The difference between total enthalpy i° and static enthalpy i becomes negligible, and both parameters can be written as c_pt. There is actually no difference between the specific heat c_p at constant pressure and the specific heat c_v at constant volume for a fluid with constant density because no work by expansion occurs; however, c_p will be used in the following discussion because relations for a constant-property fluid approximate actual conditions in a real fluid somewhat better when the specific heat appearing in these equations is interpreted as c_p.

Equation (6-53) now simplifies to

$$\rho c_p\left(u\frac{\partial t}{\partial x} + v\frac{\partial t}{\partial y}\right) = k\frac{\partial^2 t}{\partial y^2} \tag{7-56}$$

The last two terms on the right-hand side of Eq. (6-53) vanish because of vanishing Eckert number and body-force parameter.

Boundary conditions must be prescribed along the outer edge of the boundary layer and along the wall surface. The temperature in a constant-property fluid is constant in the stream outside the boundary layer, since the effect of heat conduction there is negligible and since the constant density forbids temperature changes by expansion of the fluid. Along the wall surface, a wide variety of boundary conditions can exist. The wall temperature, for instance, may vary along the surface in some

[1] E. R. G. Eckert, "Introduction to Heat and Mass Transfer," McGraw-Hill, New York, 1963.

prescribed way, or the heat flux into the fluid may determine the temperature derivative normal to the wall at the surface.

The boundary-layer energy equation resembles the momentum equation of the boundary layer, but two important differences exist. In the energy equation (7-56), the values u and v must be considered as known parameters, determined from a solution of the flow equations. Correspondingly, the boundary-layer energy equation is a linear equation in the temperature t, which, from a mathematical standpoint, makes the problem of obtaining solutions to this equation considerably simpler because the law of superposition holds. This means that as soon as a number of solutions of this equation are known, new solutions can readily be obtained by addition or subtraction of any of the known solutions. Another difference in the two equations is connected with the fact that a term corresponding to the pressure gradient is not contained in the energy equation. From this it can be suspected (which will later be confirmed) that the influence of pressure variations along the surface is smaller on heat transfer than on flow parameters like the drag.

7-6 HEAT TRANSFER ON A FLAT PLATE WITH CONSTANT TEMPERATURE

A special solution, to be considered first, concerns flow with constant velocity u_s and with a temperature t_s over a solid surface with constant temperature t_w. The momentum equation and the energy equation describing the flow boundary layer and the thermal boundary layer and the appropriate boundary conditions are

$$u\frac{\partial u}{\partial x} + v\frac{\partial u}{\partial y} = \nu\frac{\partial^2 u}{\partial y^2} \qquad \begin{matrix} y = 0 & u = 0 & v = 0 \\ y \to \infty & u = u_s & \end{matrix}$$

$$u\frac{\partial t}{\partial x} + v\frac{\partial t}{\partial y} = \alpha\frac{\partial^2 t}{\partial y^2} \qquad \begin{matrix} y = 0 & t = t_w \\ y \to \infty & t = t_s \end{matrix}$$

The properties have been combined in these equations into the terms

$$\nu = \frac{\mu}{\rho} \qquad \alpha = \frac{k}{\rho c_p}$$

It is evident that the two differential equations are very similar. This becomes even more striking when the variables in the equations are made dimensionless with the stream velocity u_s and a characteristic length L, for instance with the length of the plate, and when the temperatures are measured with the wall surface temperature as reference point and made dimensionless by the temperature difference $t_s - t_w$. The new variables are

$$u^* = \frac{u}{u_s} \qquad v^* = \frac{v}{u_s} \qquad x^* = \frac{x}{L} \qquad y^* = \frac{y}{L} \qquad \theta^* = \frac{t - t_w}{t_s - t_w}$$

The equations and boundary conditions now assume the form

$$u^* \frac{\partial u^*}{\partial x^*} + v^* \frac{\partial u^*}{\partial y^*} = \frac{\nu}{u_s L} \frac{\partial^2 u^*}{\partial y^{*2}} \qquad \begin{array}{lll} y^* = 0 & u^* = 0 & v^* = 0 \\ y^* \to \infty & u^* = 1 & \end{array}$$

$$u^* \frac{\partial \theta^*}{\partial x^*} + v^* \frac{\partial \theta^*}{\partial y^*} = \frac{\alpha}{u_s L} \frac{\partial^2 \theta^*}{\partial y^{*2}} \qquad \begin{array}{ll} y^* = 0 & \theta^* = 0 \\ y^* \to \infty & \theta^* = 1 \end{array}$$

The reciprocals of the Reynolds number u_sL/ν and of the Peclet number u_sL/α appear in these equations. It is evident that the equations and their boundary conditions become identical in the dimensionless parameters, when $\nu = \alpha$ or

$$\mathrm{Pr} = \frac{\nu}{\alpha} = 1$$

This means that the dimensionless temperature profile θ^* is then identical to the dimensionless velocity profile u^* and is therefore presented by the velocity profile for $\beta = 0$ in Fig. 7-2.

With this information, a simple relation between the heat-transfer coefficient and the friction coefficient on the plate can be determined. The heat-transfer coefficient h is defined by the following equation, in which q_w indicates the heat flux from the wall into the fluid per unit area and time:

$$q_w = -h(t_s - t_w) = -k \left(\frac{\partial t}{\partial y}\right)_w$$

The right-hand term expresses the fact that heat is transferred from the solid wall to the fluid by conduction only. This assumes $v_w = 0$ and excludes a situation with mass flow through the surface (which will be treated in Part IV). Introducing the dimensionless quantities, the heat-transfer coefficient becomes

$$h = \frac{k}{L} \left(\frac{\partial \theta^*}{\partial y^*}\right)_w \tag{7-57}$$

The friction factor c_f is connected with the shear stress τ_w at the wall surface by the relation

$$\tau_w = c_f \rho \frac{u_s^2}{2}$$

On the other hand the shear at a solid surface is given by Newton's equation $\tau_w = \mu(\partial u/\partial y)_w$. With the dimensionless parameters, the friction factor is therefore

$$c_f = 2 \frac{\mu}{\rho u_s L} \left(\frac{\partial u^*}{\partial y^*}\right)_w \tag{7-58}$$

Heat transfer is often described by a dimensionless parameter called *Stanton number*

$$\mathrm{St} = \frac{h}{\rho c_p u_s} \tag{7-59}$$

From Eqs. (7-57) and (7-58), the relation $(\partial\theta^*/\partial y^*)_w = (\partial u^*/\partial y^*)_w$, and $\text{Pr} = \mu c_p/k = 1$, one obtains

$$\text{St} = \frac{c_f}{2} \tag{7-60}$$

which, for a fluid with a Prandtl number equal to 1, allows a simple calculation of the heat-transfer coefficient from the friction factor in Sec. 7-2. The relation is useful because gases at low velocities and moderate temperature differences can be approximated as constant-property fluids and because the Prandtl number for gases has values which do not differ greatly from 1.

7-7 HEAT TRANSFER IN WEDGE-TYPE FLOW

In Sec. 7-2 we found that velocity profiles in a laminar boundary layer at various stations along a surface are similar to each other when the free-stream velocity outside the boundary layer is prescribed by the equation

$$u_s = Cx^m$$

Such a velocity was found to exist, for example, in the flow of a constant-property fluid over a wedge when the opening angle of the wedge $\beta\pi$ is connected with the exponent m in the above equation by the relation $\beta = 2m/(m + 1)$. Mathematically, the similarity condition implied that the boundary-layer momentum equation could be transformed into a total differential equation by introduction of the following independent and dependent variables:

$$\eta = \sqrt{\frac{m+1}{2}}\, y \sqrt{\frac{u_s}{\nu x}} \qquad f(\eta) = \sqrt{\frac{m+1}{2}}\, \frac{\psi}{\sqrt{\nu u_s x}} \tag{7-61}$$

We shall now discuss under what condition similarity also exists in the thermal boundary layer which covers the surface when the temperature in the fluid mainstream is different from the surface temperature of the wedge. In a constant-property fluid, no mechanical energy can be converted into internal energy, and as a consequence the temperature t_s at the outer border of the boundary layer is locally constant even though the velocity varies. The wall-surface temperature t_w, on the other hand, may well vary with the distance x, which suggests checking whether similarity exists between the temperature profiles at various stations x, provided the following equation describes the difference between wall temperature and stream temperature:

$$t_w - t_s = C_t x^\gamma \tag{7-62}$$

In order to prove this, the differential equation describing the temperature field in the thermal boundary layer

$$u\frac{\partial t}{\partial x} + v\frac{\partial t}{\partial y} = \alpha\frac{\partial^2 t}{\partial y^2}$$

will be transformed by introduction of the variables η and f and of the dimensionless temperature parameter

$$\theta = \frac{t - t_s}{t_w - t_s} \tag{7-63}$$

assuming that f and θ are functions of the one independent variable η only. Introduction of these new variables into the boundary-layer energy equation is accomplished in the same way as for the momentum equation in Sec. 7-2. The result is

$$\frac{d^2\theta}{d\eta^2} + \mathrm{Pr}\, f\frac{d\theta}{d\eta} - (2 - \beta)\,\mathrm{Pr}\,\gamma\frac{df}{d\eta}\theta = 0 \tag{7-64}$$

with the boundary conditions

$$\begin{aligned} \eta &= 0 \qquad & \theta &= 1 \\ \eta &\to \infty \qquad & \theta &= 0 \end{aligned} \tag{7-65}$$

It is observed that the original variable x no longer appears in Eq. (7-64) and that the boundary conditions can also be expressed in the new variable alone. From this we conclude that the transformation of the boundary-layer energy equation into the total differential equation (7-64) was successful. Solutions to this equation must again be obtained by numerical means. The temperature profiles described by this equation depend on three parameters: the Prandtl number Pr, the parameter m or β determining the velocity variation along the surface, and the parameter γ determining the temperature variation along the surface.

In Figs. 7-7 to 7-9, showing such temperature profiles, one of the above parameters has been varied in each figure. Figure 7-7 shows the influence of the Prandtl number on the temperature profiles for laminar boundary-layer flow over a flat plate ($\beta = 0$) with constant wall temperature ($\gamma = 0$). It can be recognized that the shape of the temperature profile, and especially the boundary-layer thickness, varies very strongly with Prandtl number. According to the discussion in the preceding section, the velocity profile is identical with the temperature profile in this figure for Pr = 1. For Prandtl numbers larger than 1, the thermal boundary layer is thinner than the flow boundary layer. The reverse applies for Prandtl numbers smaller than 1. Liquid metals have Prandtl numbers of order 0.01. For such fluids the thermal boundary layer is much thicker than the velocity boundary layer.

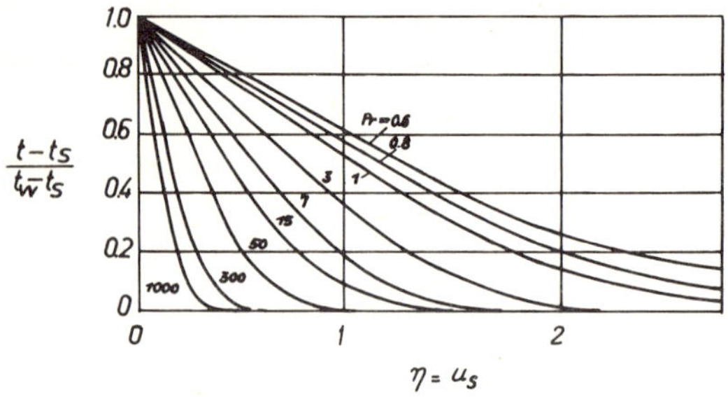

Fig. 7-7 Temperature profiles for laminar boundary-layer flow over a flat plate with constant wall temperature. [*After E. Pohlhausen Z. Angew. Math. Mech.*, **1**:*115* (*1921*).]

Figure 7-8 demonstrates the influence of the parameter β on the temperature profiles. The figure again holds for a constant wall-surface temperature ($\gamma = 0$). Two groups of profiles are presented: one for Pr = 0.7 and the other for Pr = 10. The figure indicates that the influence of the parameter β on the temperature profile is not too strong and at any rate much weaker than the influence of the Prandtl number. Figure 7-9 presents temperature profiles for various values of the parameter γ, that is, for varying wall temperature. Considered is laminar boundary-layer flow over a flat plate ($\beta = 0$) of a fluid with a

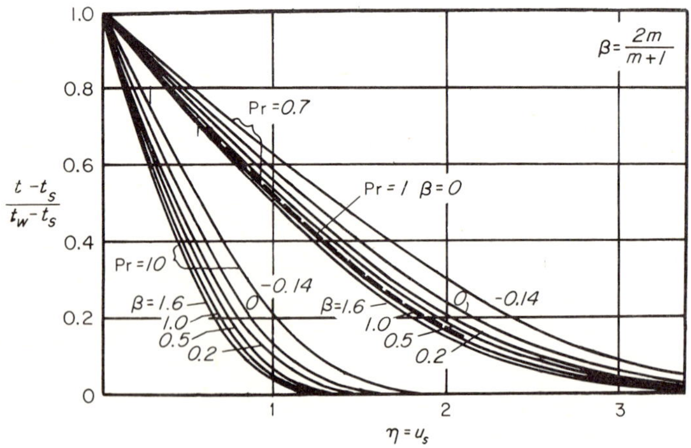

Fig. 7-8 Temperature profiles for laminar wedge-type flow with constant surface temperature. [*After E. Eckert, VDI-Forschungsh. 416, 1942.*]

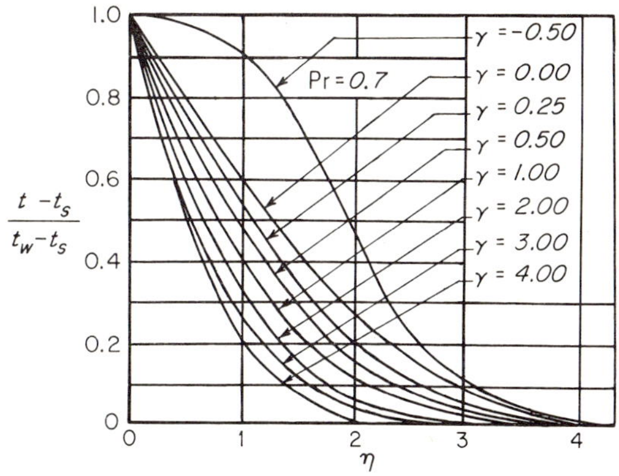

Fig. 7-9 Temperature profiles for laminar boundary-layer flow over a flat plate with variable wall temperature. [*After S. Levy, J. Aeron. Sci.* **19**:*341* (*1952*).]

Prandtl number Pr = 0.7. The parameter γ obviously has quite a strong influence on the shape of the temperature profiles, especially when it assumes negative values. From the figure one concludes that, for a value $\gamma = -0.5$, the temperature gradient at the surface, and therefore the heat flux from the wall into the fluid, is zero even though a finite temperature difference between wall surface and mainstream exists. This is physically explained in the following way: $\gamma = -0.5$ means that the difference between wall and stream temperature decreases in the downstream direction. In a situation where the wall temperature is higher than the stream temperature, the wall temperature will therefore also decrease in the downstream direction. If we focus our attention at a certain distance x, the fluid particles which arrive at that location close to the wall will have come from a region where the wall temperature was higher and will therefore have convected their enthalpy along and will protect the wall surface from the cooler fluid portions in the outer region of the boundary layer. This actually resembles a condition found in film cooling, which will be discussed later. For values of γ beyond -0.5, the temperature profiles reach a maximum at some distance from the wall surface, and heat flows into the wall even when the wall temperature is higher than the stream temperature.

The gradients of the temperature profiles in these figures at $\eta = 0$ determine the heat flux q_w from the wall into the fluid and the heat-transfer coefficient defined by

$$q_w = h(t_w - t_s) \tag{7-66}$$

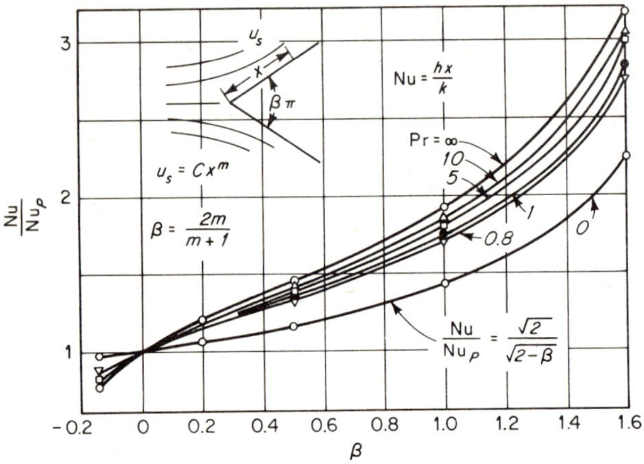

Fig. 7-10 Ratio of Nusselt number for laminar wedge-type flow to Nusselt number for flat-plate flow at constant surface temperature.

At the surface, a heat flux is caused by conduction only. Therefore, Fourier's equation gives

$$q_w = -k\left(\frac{\partial t}{\partial y}\right)_w = -k(t_w - t_s)\left(\frac{d\theta}{d\eta}\right)_0 \sqrt{\frac{m+1}{2}}\sqrt{\frac{u_s}{\nu x}}$$

and

$$h = -k\sqrt{\frac{m+1}{2}}\sqrt{\frac{u_s}{\nu x}}\left(\frac{d\theta}{d\eta}\right)_0$$

$(d\theta/d\eta_0)$ can be obtained from Figs. 7-7 to 7-9. The heat-transfer coefficient may assume negative values. This happens for $\gamma < -0.5$ in Fig. 7-9. For $\gamma < -0.5$, h has the value zero. Such heat-transfer coefficients are usually presented in dimensionless form as Nusselt numbers

$$\mathrm{Nu} = \frac{hx}{k}$$

The Nusselt number for similar wedge-type flow is therefore given by

$$\mathrm{Nu} = -\sqrt{\frac{m+1}{2}}\left(\frac{d\theta}{d\eta}\right)_0 \sqrt{\mathrm{Re}_x} \tag{7-67}$$

with $\mathrm{Re}_x = u_s x/\nu$.

Figures 7-10 to 7-12 demonstrate the influence of the individual parameters on Nusselt numbers. Figure 7-10 presents the ratio of the Nusselt number Nu for laminar boundary-layer flow over a wedge to the Nusselt number Nu_P for laminar boundary-layer flow over a flat plate. The wall temperature is specified constant. Figure 7-11 presents the

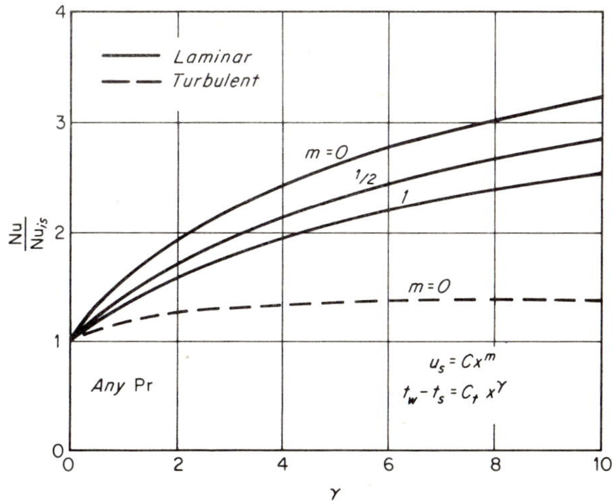

Fig. 7-11 Ratio of Nusselt number for a surface with locally varying temperature to the Nusselt number for an isothermal surface. Wedge-type flow with laminar and turbulent boundary layer.

ratio of the Nusselt number Nu for various values of the parameter γ describing the wall-temperature variation to the Nusselt number Nu_{is} of an isothermal surface; m appears as a parameter and has the same value for both Nusselt numbers. Figure 7-12 shows the influence of the Prandtl number on the Nusselt number. The diagram holds for flow over a flat plate ($\beta = 0$) with constant wall temperature ($\gamma = 0$). It

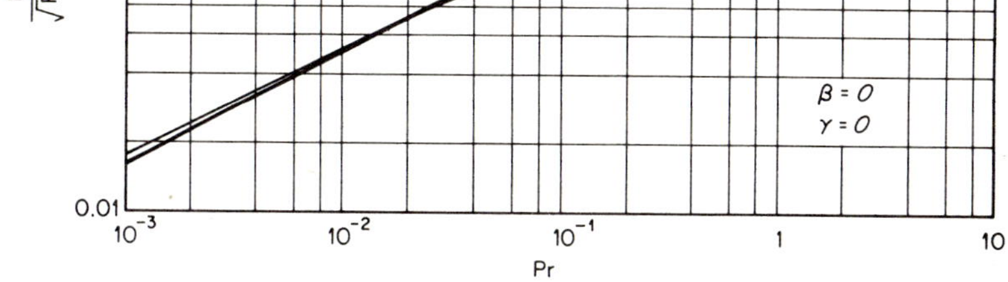

Fig. 7-12 Nusselt number for laminar boundary-layer flow of fluid with various Prandtl numbers over a flat plate.

can be observed that the rate of increase in the Nusselt number with growing Prandtl number changes somewhat. The Nusselt number increases in proportion to the square root of the Prandtl number for Prandtl numbers around 0.01 and in proportion to the cube root of the Prandtl number for Prandtl numbers around 100. The rate of increase can actually be calculated in a simple way for the two limiting conditions $\mathrm{Pr} \to 0$ and $\mathrm{Pr} \to \infty$. Equation (7-64) simplifies for constant wall temperature to

$$\frac{d^2\theta}{d\eta^2} + \mathrm{Pr}\, f \frac{d\theta}{d\eta} = 0$$

This equation can be integrated in a straightforward way.[1] The solution which satisfies the boundary conditions is

$$1 - \theta = \frac{\int_0^\eta \exp\left(-\int_0^\eta \mathrm{Pr}\, f\, d\eta\right) d\eta}{\int_0^\infty \exp\left(-\int_0^\eta \mathrm{Pr}\, f\, d\eta\right) d\eta} \tag{7-68}$$

The gradient $(d\theta/d\eta)_0$ which enters into the Eq. (7-67) for the Nusselt number is

$$\left(\frac{d\theta}{d\eta}\right)_0 = \frac{-1}{\int_\eta^\infty \exp\left(-\int_0^\eta \mathrm{Pr}\, f\, d\eta\right) d\eta} \tag{7-69}$$

For decreasing Prandtl number, the thermal boundary layer becomes larger and larger compared to the velocity boundary layer, so that for very small Prandtl numbers almost the entire temperature field is in a region where the velocity has the value u_s. An approximation which becomes better as the Prandtl number decreases can therefore be obtained by assuming

$$\frac{u}{u_s} = \frac{df}{d\eta} = 1 \qquad f = \eta$$

Introduction of this expression into Eq. (7-69) gives

$$\left(\frac{d\theta}{d\eta}\right)_0 = \frac{1}{\int_0^\infty \exp\left(-\frac{\mathrm{Pr}\,\eta^2}{2}\right) d\eta} = \sqrt{\frac{2\,\mathrm{Pr}}{\pi}}$$

For large Prandtl numbers,s on the other hand, the thermal boundary layer is very thin compared to the velocity boundary layer, and the velocities in the region of the thermal boundary layer can be well approximated by the expression

$$\frac{u}{u_s} = C\eta \qquad f = C\frac{\eta^2}{2}$$

[1] See for instance Eckert, *op. cit.*, p. 118.

which, when introduced into Eq. (7-69), then leads to

$$\left(\frac{d\theta}{d\eta}\right)_0 = \frac{1}{\int_0^\infty \exp\left(-\Pr C \frac{\eta^3}{6}\right) d\eta} = K \sqrt[3]{\Pr}$$

with K denoting a constant.

With these limiting values in mind, one finds that the Nusselt number for flow over a flat plate ($\beta = 0$) with constant wall temperature ($\gamma = 0$) can be well approximated by the following equations:

$$\mathrm{Nu}_x = 0.332 \sqrt{\mathrm{Re}_x} \sqrt[3]{\Pr} \qquad \text{for } \Pr > 0.6 \tag{7-70}$$

$$\mathrm{Nu}_x = 0.500 \sqrt{\mathrm{Re}_x \Pr} \qquad \text{for } 0.005 < \Pr < 0.05 \tag{7-71}$$

These equations are compared with exact solutions [Eq. (7-69)] in Fig. 7-12. For other β and γ values the Nusselt number can then be obtained from Figs. 7-10 and 7-11.

That the differential equation describing the thermal boundary layer is linear makes it possible to extend the solutions obtained for a wall-temperature variation according to Eq. (7-62) to a situation in which the wall temperature varies arbitrarily along x.[1] Such a temperature variation can be expressed by the series

$$t_w - t_s = a_0 + a_1 x + a_2 x^2 + \cdots = \sum_{i=0}^{\infty} a_i x^i \tag{7-72}$$

The temperature field for the boundary condition (7-72) can be obtained from the temperature fields with boundary conditions (7-62) by linear superposition. The heat flux at the wall surface is then described by

$$q_w = -k \sum \left(\frac{\partial t_i}{\partial y}\right)_0$$

With Eqs. (7-61), (7-63), and (7-67) one obtains

$$\begin{aligned} q_w &= \frac{k\,\mathrm{Nu}_0}{x}\left(a_0 + a_1 x \frac{\mathrm{Nu}_1}{\mathrm{Nu}_0} + a_2 x^2 \frac{\mathrm{Nu}_2}{\mathrm{Nu}_0} + \cdots\right) \\ &= \frac{k\,\mathrm{Nu}_0}{x} \sum_{i=0}^{\infty} a_i x^i \frac{\mathrm{Nu}_i}{\mathrm{Nu}_0} \end{aligned} \tag{7-73}$$

in which Nu_0 is the Nusselt number for a constant wall temperature and Nu_i is the Nusselt number for a wall-temperature variation according to Eq. (7-62) with $\gamma = i$. The ratio of Nusselt numbers appearing in the

[1] D. R. Chapman and M. W. Rubesin, Temperature and Velocity-Profiles in the Compressible Laminar Boundary Layer with Arbitrary Distribution of Surface Temperature, *J. Aeron. Sci.*, **16**:547 (1949).

above equation can be taken from Fig. 7-11. In practical application, the method is restricted, insofar as some wall-temperature variations cannot be sufficiently well approximated by the series in Eq. (7-72) with the finite number of terms for which the Nu_i are known.

The discussion of the velocity boundary layer in Sec. 7-2 included the situation in which a finite velocity v exists at the wall surface. The thermal boundary layer and the heat transfer for this case will be treated in Sec. 11-3.

7-8 APPROXIMATE SOLUTIONS OF THE BOUNDARY-LAYER ENERGY EQUATION

In situations with arbitrary variation of the wall-surface temperature, approximate answers can be obtained by various methods. One of them was discussed in the preceding section, and two more will be considered in the present section.

Integrated boundary-layer energy equation The first procedure is based on the integrated energy equation of the boundary layer. For its derivation, we assume that the actual boundary-layer temperature profiles can be approximated by profiles which merge into the temperature t_s of the approaching fluid at a finite distance δ_t, which may be called the *thermal-boundary-layer thickness.* Its value will generally be different from the flow-boundary-layer thickness δ, defined in Sec. 7-2. The integrated energy equation will now be obtained from an energy balance on a volume element at the surface, built up by two planes 1-2 and 3-4 in Fig. 7-13, separated from each other by the length dx, by a plane 2-4 parallel to the wall at the distance l, and by the wall 1-3. The length l is assumed greater than both boundary-layer thicknesses δ and δ_t. The calculation is again confined to steady two-dimensional problems, to a constant-property fluid, and to velocities which are sufficiently small to allow temperature increases caused in the boundary layer by internal friction to be neglected. In Fig. 7-13 the volume element has unit length normal to the plane of the

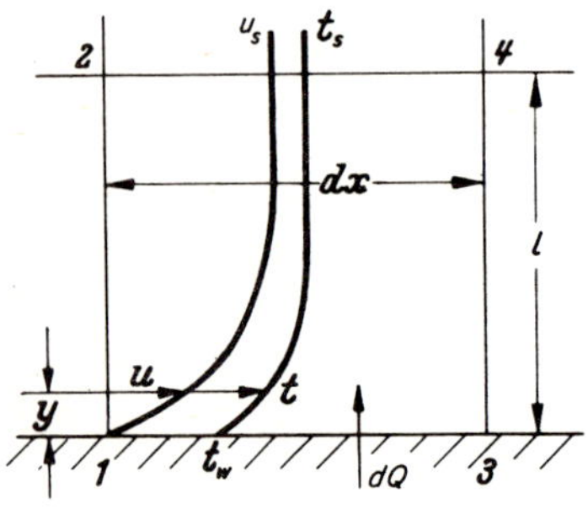

Fig. 7-13 Control volume for the derivation of the integrated thermal-boundary-layer equation.

paper. Through the plane 1-2 a quantity of heat, which is determined by the following integral, is transported per unit time:

$$\rho c_p \int_0^l tu\,dy$$

where c_p is the specific heat at constant pressure per unit mass. This heat flow changes with the length dx by

$$\rho c_p\,dx \frac{d}{dx} \int_0^l tu\,dy$$

The heat flow leaving the volume element through plane 3-4 is greater by this amount than the heat flow entering the element through plane 1-2. Through the plane 2-4 there flows the fluid mass $\rho\,dx \frac{d}{dx} \int_0^l u\,dy$, as was derived in Sec. 7-4. This transports an amount of heat given by

$$\rho c_p t_s\,dx \frac{d}{dx} \int_0^l u\,dy$$

into the volume element. Through the plane 1-3 a certain amount of heat flows from the volume element to the wall or in the opposite direction. Since the fluid particles adjacent to the wall have the velocity zero, this heat must be transferred by conduction through this fluid layer. Therefore Fourier's equation is applicable. With the nomenclature used here the heat flow per unit time is

$$dQ = -k\,dx \left(\frac{\partial t}{\partial y}\right)_w$$

where k is the thermal conductivity of the fluid. The sum of the heat carried into the volume element by convection and by conduction must be zero for steady state:

$$\rho c_p \frac{d}{dx} \int_0^l t_s u\,dy - \rho c_p \frac{d}{dx} \int_0^l tu\,dy - k \left(\frac{\partial t}{\partial y}\right)_w = 0 \qquad (7\text{-}74)$$

Introducing the thermal diffusivity of the fluid $\alpha = k/\rho c_p$ gives

$$\frac{d}{dx} \int_0^l (t_s - t)u\,dy = \alpha \left(\frac{\partial t}{\partial y}\right)_w \qquad (7\text{-}75)$$

This is the heat-flow equation of the boundary layer by which the heat transfer can be calculated. It holds for both turbulent and laminar flow.

Plane plate with stepwise-varying wall temperature A solution to the integrated boundary-layer equation will be obtained for the situation sketched in Fig. 7-14. Fluid with a temperature t_s flows over a flat plate creating a laminar boundary layer with thickness δ. The temperature

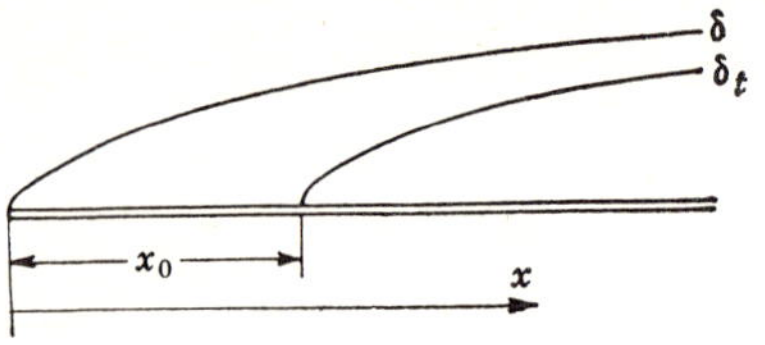

Fig. 7-14 Flow-boundary-layer thickness δ and thermal-boundary-layer thickness δ_t for laminar flow over a flat plate with a stepwise temperature variation.

of the plate surface varies along the length x in a stepwise fashion. Over the distance x_0 from the leading edge, the surface temperature is assumed equal to the temperature in the approaching fluid stream; whereas for values $x > x_0$ the surface temperature t_w is assumed different from the fluid temperature but with a locally uniform value. A thermal boundary layer will therefore start at the distance x_0 from the leading edge of the plate, its thickness δ_t increasing in the downstream direction as sketched in Fig. 7-14. The thickness of this boundary layer will now be calculated using the heat-flow equation (7-75). For this purpose a suitable statement must be made about the shape of the temperature profile in the boundary layer.

The closer the assumed profile agrees with the real one, the better the result of our calculation will be. Therefore, an expression with a number of free functions is chosen. These are determined such that the assumed profile satisfies conditions which are known to hold for the real one:

$$y = 0 \qquad t = t_w$$

$$y = \infty \qquad t = t_s \qquad \frac{\partial t}{\partial y} = 0$$

Equation (7-56), written for $y = 0$, results in

$$\frac{\partial^2 t}{\partial y^2} = 0$$

According to these four conditions, a polynomial with four terms will be used.

$$t = \alpha + by + cy^2 + dy^3$$

If, in addition, the temperature difference $\theta = t - t_w$ is introduced, and if the functions are determined from the above conditions, the temperature profile becomes

$$\theta' = \frac{\theta}{\theta_s} = \frac{3}{2}\frac{y}{\delta_t} - \frac{1}{2}\left(\frac{y}{\delta_t}\right)^3 \qquad \text{(7-76}a\text{)}$$

with $\qquad \theta = t - t_w \qquad \theta_s = t_s - t_w$

The velocity profile which is also needed for this calculation has to fulfill the following conditions:

$$y = 0 \qquad u = 0 \qquad \frac{d^2u}{dy^2} = 0$$

$$y \rightarrow \infty \qquad u = u_s \qquad \frac{du}{ds} = 0$$

Correspondingly, an expression with four terms is used to describe the velocity within the boundary layer

$$u = a' + b'y + c'y^2 + d'y^3 \tag{7-76b}$$

With the boundary conditions stated above, this equation takes on the form

$$\frac{u}{u_s} = \frac{3}{2}\frac{y}{\delta} - \frac{1}{2}\left(\frac{y}{\delta}\right)^3$$

It can be observed that the two expressions for the temperature and for the velocity profile are similar and thus satisfy the condition found to exist in a fluid with a Prandtl number equal to 1 (Sec. 7-6). Now the integral in the heat-flow equation can be calculated:

$$\int_0^l (t_s - t)u\,dy = \int_0^l (\theta_s - \theta)u\,dy$$

$$= \theta_s u_s \int_0^{\delta_t} \left[1 - \frac{3}{2}\frac{y}{\delta_t} + \frac{1}{2}\left(\frac{y}{\delta_t}\right)^3\right]\left[\frac{3}{2}\frac{y}{\delta} - \frac{1}{2}\left(\frac{y}{\delta}\right)^3\right] dy$$

It may be presumed that the thermal boundary layer is thinner than the hydrodynamic boundary layer. Then the integral need be extended only to $y = \delta_t$, since for $y > \delta_t$, $\theta = \theta_s$ and the integrand is therefore zero. By introducing the ratio $\zeta = \delta_t/\delta$ of the thermal to the hydrodynamic boundary layer and evaluating the integral, there follows

$$\int_0^l (\theta_s - \theta)u\,dy = \theta_s u_s \delta(\tfrac{3}{20}\zeta^2 - \tfrac{3}{280}\zeta^4)$$

As we assumed δ_t to be smaller than δ and therefore $\zeta < 1$, the second term in the right-hand expression is small compared with the first and can be neglected. According to Fig. (7-14), the ratio ζ is a function of x. The integral above must be introduced into Eq. (7-75). For a solution of the resulting equation, the thickness of the velocity boundary layer must be calculated first using Eq. (7-39), with the result

$$\delta = 4.64\sqrt{\frac{\nu x}{u_s}}$$

Equation (7-75) then gives the following expression for the ratio of the two boundary-layer thicknesses:[1]

$$\zeta = \frac{1}{1.026\sqrt[3]{\mathrm{Pr}}}\sqrt{1 - \left(\frac{x_0}{x}\right)^{\frac{3}{4}}} \tag{7-77}$$

If the plate is heated over the whole length ($x_0 = 0$),

$$\zeta = \frac{1}{1.026\sqrt[3]{\mathrm{Pr}}} \tag{7-78}$$

Viscous oils have a Prandtl number $\mathrm{Pr} = 1{,}000$ or more. For these fluids the thickness of the thermal boundary layer is only one-tenth the thickness of the hydrodynamic boundary layer. Gases have Prandtl numbers smaller than 1. For these, ζ is greater than 1, and therefore the assumption $\zeta < 1$ in the preceding calculation does not hold true. However, since the smallest value for gases is $\mathrm{Pr} = 0.6$ and therefore $\zeta = 1.16$, the error introduced is quite small. The only materials with a very small Prandtl number are liquid or molten metals, for which the above results are not valid.

The heat flow from the plate per unit area is given by

$$q_w = -k\left(\frac{\partial \vartheta}{\partial y}\right)_w$$

On the other hand a heat-transfer coefficient h can be introduced:

$$q_w = h(t_w - t_s) = -h\theta_s$$

Equating these two expressions gives

$$h = \frac{k}{\theta_s}\left(\frac{\partial \vartheta}{\partial y}\right)_w = \frac{3}{2}\frac{k}{\delta_t} = \frac{3}{2}\frac{k}{\zeta\delta} \tag{7-79}$$

The heat-transfer coefficient is therefore inversely proportional to the thermal-boundary-layer thickness. With the ratio ζ and the value of δ

$$h = 0.332k\,\frac{\sqrt[3]{\mathrm{Pr}}}{\sqrt[3]{1 - (x_0/x)^{\frac{3}{4}}}}\sqrt{\frac{u_s}{\nu x}} \tag{7-80}$$

and for the plate heated over its entire length

$$h = 0.332k\sqrt[3]{\mathrm{Pr}}\sqrt{\frac{u_s}{\nu x}} \tag{7-81}$$

As shown in Fig. 7-15, the heat-transfer coefficient begins with an infinite value at the starting point of the heated section of the plate and decreases

[1] For a detailed calculation see Eckert, *op. cit.*, p. 110.

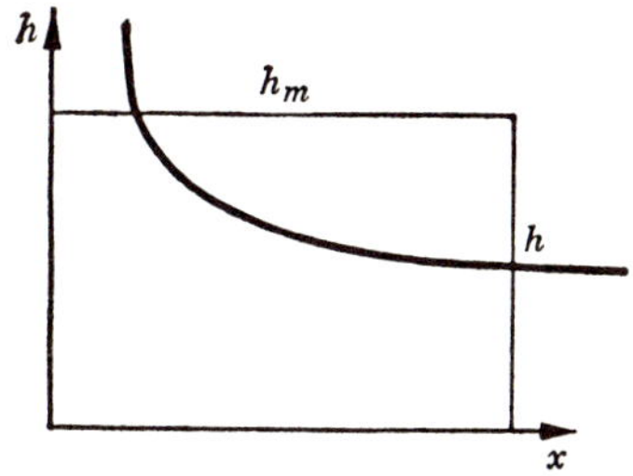

Fig. 7-15 Local and average heat-transfer coefficient on a flat plate with uniform temperature and laminar boundary layer.

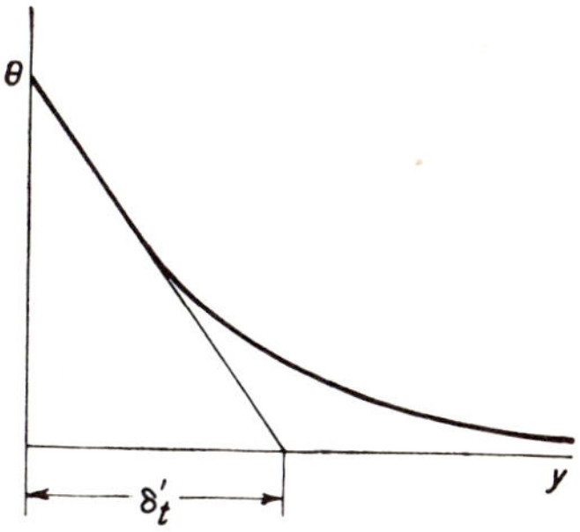

Fig. 7-16 Definition of the equivalent film thickness δ_t'.

with increasing x. It is advantageous to write Eq. (7-80) in dimensionless form:

$$\frac{hx}{k} = 0.332 \sqrt[3]{\mathrm{Pr}} \sqrt{\frac{u_s x}{\nu}} \frac{1}{\sqrt[3]{1 - (x_0/x)^{\frac{3}{4}}}}$$

On the right there appears again the Reynolds number calculated with the free-stream velocity u_s and the length x. The dimensionless expression on the left side is called the *Nusselt number* of Nusselt modulus Nu. Nu_x is written to indicate that the Nusselt number is calculated with the length x

$$\mathrm{Nu}_x = \frac{hk}{k} \tag{7-82}$$

With this notation the above equation can be written

$$\mathrm{Nu}_x = 0.332 \sqrt[3]{\mathrm{Pr}} \sqrt[3]{\mathrm{Re}_x} \frac{1}{\sqrt[3]{1 - (x_0/x)^{\frac{3}{4}}}} \tag{7-83}$$

and for the plate heated over its entire length,

$$\mathrm{Nu}_x = 0.332 \sqrt[3]{\mathrm{Pr}} \sqrt{\mathrm{Re}_x} \tag{7-84}$$

The Nusselt number can be expressed as the ratio of two lengths. For this purpose we introduce an equivalent film thickness δ_t', which is the length of the subtangent to the temperature profile within the boundary layer at the wall. Figure 7-16 shows this in detail. For the temperature profile represented by the cubic parabola, the relationship $\delta_t' = \frac{2}{3}\delta_t$ holds. The film heat-transfer coefficient is according to Eq. (7-79),

$$h = \frac{k}{\delta_t'}$$

and therefore the Nusselt number

$$\mathrm{Nu}_x = \frac{hk}{k} = \frac{x}{\delta_t'}$$

is the ratio of the length x to the equivalent film thickness δ_t'. Since the the boundary-layer thickness is always small compared with the length x, the Nusselt number will be large. The thermal-boundary-layer thickness decreases with increasing Reynolds and Prandtl numbers. Both values therefore increase the Nusselt number.

For engineering calculations not the local but the average value of the heat-transfer coefficient is often important. For the plate heated over its entire length this value is

$$\bar{h} = \frac{1}{x}\int_0^x h\,dx = \frac{C}{x}\int_0^x \frac{dx}{\sqrt{x}} = \frac{C}{x}\,2\sqrt{x} = 2\frac{C}{\sqrt{x}} = 2h \qquad (7\text{-}85)$$

In C are combined all values which do not depend on x. The average value of the heat-transfer coefficient is therefore twice the local value at the end of the plate. The heat transfer from a plate heated over its entire length was also calculated by an exact solution of the differential equations of the boundary layer. This calculation resulted in the expression in (7-84) with a numerical value 0.332. Such close agreement of the results with our approximate procedure is, of course, coincidental.

If the property values in the equation depend on temperature, they must be introduced at a mean temperature. From calculations which solved the laminar boundary-layer equations for a fluid with variable properties, it can be shown that Eq. (7-84) gives the right answer for air when the property values are introduced at a reference temperature determined by the equation

$$t^* - t_s = 0.50(t_w - t_s)\dagger \qquad (7\text{-}86)$$

This reference temperature can be used for gases. For liquids our knowledge of the reference temperature is less certain. Experiments have verified Eq. (7-84) well for gases.

Equations (7-83) to (7-84) have been obtained on the assumption that the ratio ζ of the thermal- to the flow-boundary-layer thickness is smaller than 1. It has been found that this is true for fluids with a Prandtl number larger than 1. An extension of the calculation procedure presented in this section to a ratio ζ larger than 1 is quite straightforward. One has only to carry out the integration in two steps, since the velocity u is given by Eq. (7-76*b*) for $0 < y < \delta$ and by $u = u_s$ for $\delta < y < \delta_t$.

† E. R. G. Eckert, *Trans. ASME*, **56**:1273–1283 (1956).

The following equation results for the plate heated to a constant temperature on its entire length:

$$\mathrm{Nu}_x = \frac{\sqrt{\mathrm{Re\ Pr}}}{1.55\sqrt{\mathrm{Pr}} + 3.09\sqrt{0.372 - 0.15\ \mathrm{Pr}}} \tag{7-87}$$

This relation is useful for calculating heat transfer in liquid metals which have Prandtl numbers between 0.005 and 0.05. In this range, the denominator is only a weak function of Pr, so that the Nusselt number depends essentially on the product Re Pr, which is also referred to as the *Peclet number*. An exact solution of the laminar-boundary-layer equations results in a relation having, in place of the denominator in the above equation, a weak function of Pr which varies by ± 5 percent around the value 1.98 for the above Prandtl number range.[1] It will be found that the agreement of the above equation with this result is very good.

Plane plate with arbitrarily varying wall temperature In the preceding section the case was investigated in which a plate exposed to steady flow has a temperature t_s up to $x = x_0$ and a temperature t_w downstream from x_0. The wall temperature therefore varies along the plate in a stepwise fashion.

In engineering applications other wall-temperature variations are often of interest. For certain wall-temperature variations the superposition method discussed on page 313 can be used. Such cases can also be calculated with the method used in the previous paragraph. A variation of the surface temperature along x has two effects on the temperature boundary layer. It influences the shape of the temperature profile and the boundary-layer thickness. The first effect becomes evident when Eq. (7-56) without the dissipation term is differentiated with respect to y and then written for $y = 0$. The result of this calculation is

$$\alpha \left(\frac{\partial^3 t}{\partial y^3}\right)_{y=0} = \left(\frac{\partial u}{\partial y}\right)_{y=0} \frac{\partial t_w}{\partial x}$$

This equation expresses a relation between the third derivative of the temperature profile and the wall-temperature gradient. It can be used again to determine a constant in the assumed expression by which the temperature profile is approximated. In addition, the temperature difference θ is now a function of x and has to be kept within the differential operator d/dx in the equation on page 317.

Another calculation method makes use of the fact that the energy equation (7-56) is a linear differential equation and develops a solution

[1] E. M. Sparrow and J. L. Gregg, *J. Aeron. Sci.*, **24**:852 (1957).

of the temperature field for an arbitrary wall-temperature variation by superposition of a large number of stepwise variations. This method, usually referred to as *Duhamel's theorem* (1833), is used extensively in electric-circuit analysis. It was applied for the first time to the present problem by M. W. Rubesin in a thesis in 1945. It proved very convenient and has the advantage that it can be applied not only for a flat plate with constant velocity u_s but also for bodies of different shapes and for turbulent- as well as for laminar-flow conditions.

Equation (7-56) is a linear differential equation describing the temperature field t within the boundary layer. Let us assume that we have a number of particular solutions t_i of this equation for specific boundary conditions. Then it can be easily seen by substitution into Eq. (7-56) that

$$t = \sum_1^n C_i t_i$$

is also a solution. The constants C_i can be used to adjust this new solution to the required boundary conditions. The heat flow at the surface is

$$q = -k\left(\frac{\partial t}{\partial y}\right)_w = -k\sum_1^n C_i\left(\frac{\partial t_i}{\partial y}\right)_w$$

We shall now assume that each particular solution corresponds to a condition in which the wall temperature is equal to the stream temperature up to a certain location and then suddenly changes by an amount Δt_{wi}. For each particular solution a film heat-transfer coefficient can be defined according to the equation

$$-k\left(\frac{\partial t_i}{\partial y}\right)_w = h_i\,\Delta t_{wi}$$

with Δt_{wi} denoting the jump in the wall temperature or, in other words, the difference between stream and wall temperature for the particular situation. Then for the temperature field t,

$$q = \sum_1^n C_i h_i\,\Delta t_{wi}$$

From the wall-temperature variation it is immediately seen that in this case all constants C_i have the value 1.

The method outlined above will now be used to calculate heat transfer to a plate with arbitrarily varying wall temperature. When the wall temperature varies in a stepwise fashion with steps $\Delta t_{w1}, \Delta t_{w2}, \ldots, \Delta t_{wi}, \ldots$ occurring at locations $\xi_1, \xi_2, \ldots, \xi_i \ldots$ (see Fig. 7-17),

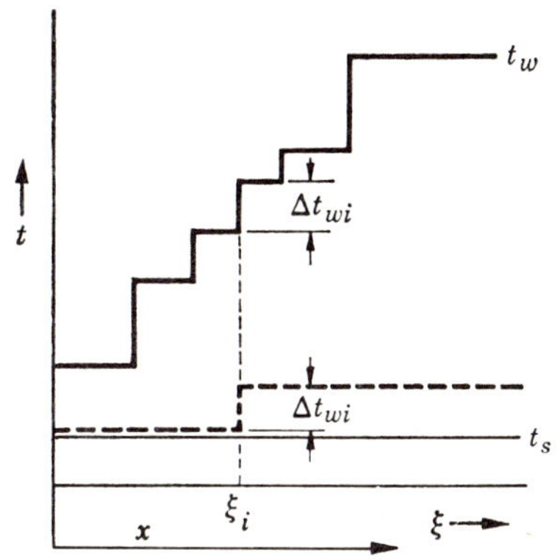

Fig. 7-17 Stepwise variation of wall temperature.

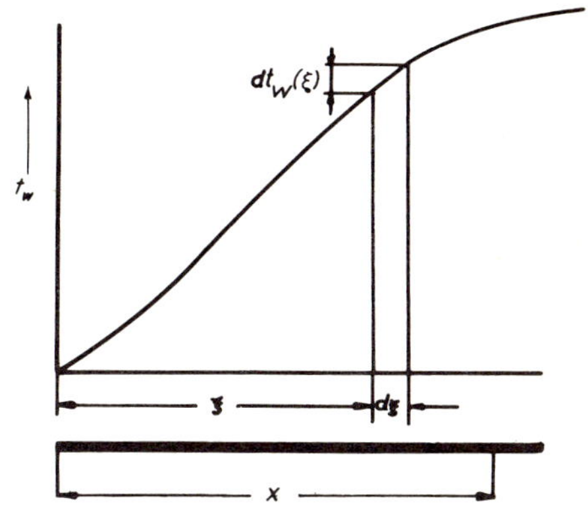

Fig. 7-18 Continuous variation of wall temperature.

the heat flow leaving the wall at location x is, according to the last equation above,

$$q = \sum_{1}^{n} h(x,\xi_i)\ \Delta t_{wi} \tag{7-88}$$

where h denotes the heat-transfer coefficient which describes the heat flow at location x when only one step Δt_{wi} in the wall temperature occurs at location ξ_i. This heat-transfer coefficient for laminar-flow conditions is given by Eq. (7-83) when x_0 is replaced by ξ_i. An additional temperature step $\Delta t_{w0} = t_{w0} - t_s$ at $\xi = 0$ must be added; t_s indicates the fluid temperature outside the boundary layer, where the plate has no unheated starting section.

A continuously varying wall temperature can be interpreted as a succession of infinitesimally small temperature steps dt_w occurring at infinitesimally closely spaced locations $d\xi$. From this consideration the heat flow at the location x for a wall temperature which varies as indicated in Fig. 7-18 can be obtained by replacing the series in the above equation by an integral. This results in the following relation:

$$q = \int_0^x h(x,\xi)\ dt_w\ (\xi)$$

In order to express this integral in the independent variable ξ, the equation is transformed into

$$q = \int_0^x h(x,\xi)\ \frac{dt_w(\xi)}{d\xi}\ d\xi \tag{7-89}$$

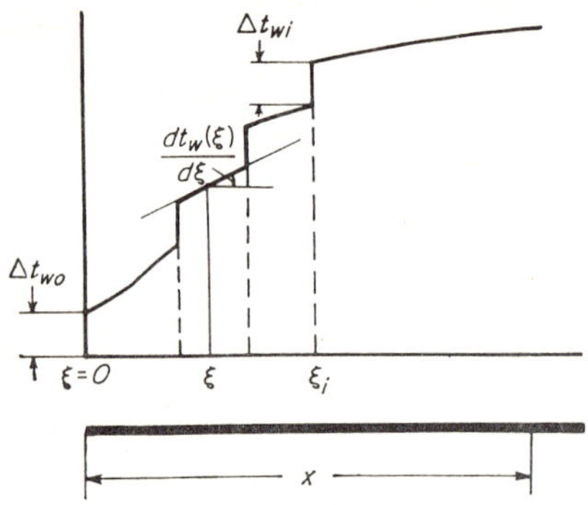

Fig. 7-19 Stepwise and continuous variation of wall temperature.

If finite steps occur in the wall temperature simultaneously with the continuous variation as shown in Fig. 7-19, the following equation has to be used:

$$q = \int_0^x h(x,\xi) \frac{dt_w(\xi)}{d\xi} \, d\xi + \sum_0^n h(x,\xi_i) \, \Delta t_w \, (\xi_i) \tag{7-90}$$

It must be kept in mind that a finite temperature difference between wall and stream temperature at the leading edge of the plate (for $x = 0$) must also be considered as a finite step Δt_w and must be taken care of by a term in the series in Eq. (7-88) or (7-90). By a summation and integration in the above equations the heat flow along a flat plate can be calculated for any prescribed wall-temperature variation. In some engineering applications the heat flow rather than the wall temperature is prescribed at the surface of the plate. Tribus and Klein have considered this problem in a paper in which they improved and generalized the method.[1] The method can be used for a turbulent boundary-layer flow also when the turbulent heat-transfer coefficients for a plate with stepwise-varying wall temperature are inserted for $h(x,\xi)$.

For engineering calculations the evaluation of the integral in Eq. (7-88) or (7-90) is somewhat tedious. For this reason an approximate procedure has been developed,[2] which will be briefly outlined. The way in which the temperature t_w varies along the surface of a flat plate with length L in the flow direction is prescribed. The task is to find the local heat flux $q(x)$ from the surface into a fluid stream of temperature t_s at an arbitrary location x and the total heat flux $Q(x)$ from a surface of length x in the flow direction and of unit width. The length L is subdivided

[1] H. Tribus and I. Klein, Forced Convection from Nonisothermal Surfaces, "Heat Transfer" (a symposium), University of Michigan, Engineering Research Institute, 1953.

[2] E. R. G. Eckert, J. P. Hartnett, and R. Birkebak, *J. Aeron. Sci.*, **24**:549–550 (1957).

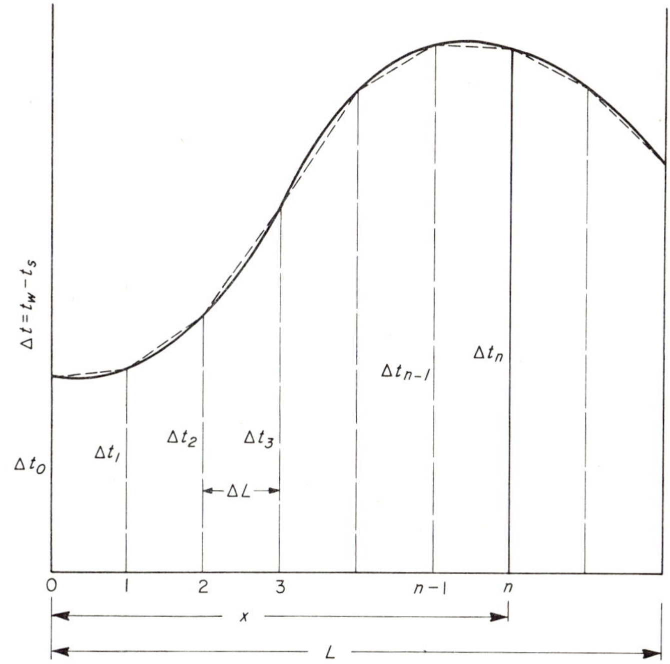

Fig. 7-20 An approximation to continuously varying wall temperature by straight-line segments.

into an arbitrary number of equal steps of length ΔL. Figure 7-20 shows a plot of the local temperature difference $\Delta t = t_w - t_s$. The temperature differences Δt_0, Δt_1, Δt_2 occur at locations 0, 1, 2. The following equation then gives the local or total heat flux for laminar- or turbulent-flow conditions at the location n when the proper constants are inserted from Table 7-2:

$$q(x) \text{ or } \frac{Q(x)}{x} = h_0(x) \left\{ \Delta t_0 + a(\Delta t_n - \Delta t_0) + b\,\frac{\Delta L}{x}\,[(2n-1)\,\Delta t_n - \Delta t_0 - 2(\Delta t_1 + \Delta t_2 + \Delta t_3 + \cdots + \Delta t_{n-1})] \right\} \tag{7-91}$$

Table 7-2

	Local heat flow		Total heat flow	
	a	*b*	*a*	*b*
Laminar	0.995	0.446	0.969	−0.432
Turbulent	0.991	0.117	0.982	−0.478

$h_0(x)$ is the local or average heat-transfer coefficient, respectively, for a plate with constant wall temperature. The equation gives the heat flux for any location x on the surface which coincides with one of the stations (1, 2, 3, . . .). n indicates the number of the station coinciding with location x.

For the development of Eq. (7-91), the actual difference between wall temperature and stream temperature was replaced by a broken line (Fig. 7-20). The integral appearing in Eq. (7-89) was solved for a linear wall-temperature variation, and the result was approximated by a second-order equation. From these elements the heat flux for the broken line could be derived. For turbulent flow the equation in "Heat and Mass Transfer"[1] which describes the Nusselt number for a stepwise temperature variation was used in the integration of Eq. (7-89). In a comparison with available exact solutions Eq. (7-91) was found to be in error by only a few percent.

The equation $q_c(x) = h_0(x)\,\Delta t_n$ describes the local heat flux at location n when the temperature difference Δt is constant above the surface and equal to Δt_n. Since the constant a in Eq. (7-91) is almost equal to 1, the difference between the local heat flux $q(x)$ and $q_c(x)$ depends on the value for the constant b. It is evident from Table 7-2 that the difference between q and q_c is much larger for laminar than for turbulent flow. In other words, the upstream temperature history makes itself felt to a much larger degree in laminar than in turbulent flow. This statement holds also for flow through tubes and ducts. For turbulent flow it is necessary to take the previous history into account only when the temperature upstream from the position n varies very rapidly.

Local similarity The concept of local similarity as introduced in Sec. 7-2 can also be applied to calculate the heat transfer on objects of arbitrary shape. In its simplest form, it uses Eq. (7-31) to calculate the shape parameter m for a location x. Equation (7-8) then determines the alternate shape parameter β, and Fig. 7-10, together with Eq. (7-70) or (7-71), determines a Nusselt number which can be used to calculate an approximate heat-transfer coefficient at the desired location x. For constant wall temperature it is found that the answer obtained in this way is often sufficient as a first approximation for engineering calculations. Figure 7-21 shows, as an example, Nusselt numbers obtained in this way for a cylinder with circular cross section and compares them with the results of experiments.[2] The basic shortcoming of this simple procedure is again the fact that the distance x from the stagnation point used is not a local parameter. Numerous methods have been devised to improve this fea-

[1] E. R. G. Eckert and R. M. Drake, "Heat and Mass Transfer," p. 217, McGraw-Hill, New York, 1959.

[2] E. Eckert and O. Drewitz, *Luftfahrtforschung*, **19**:189 (1942).

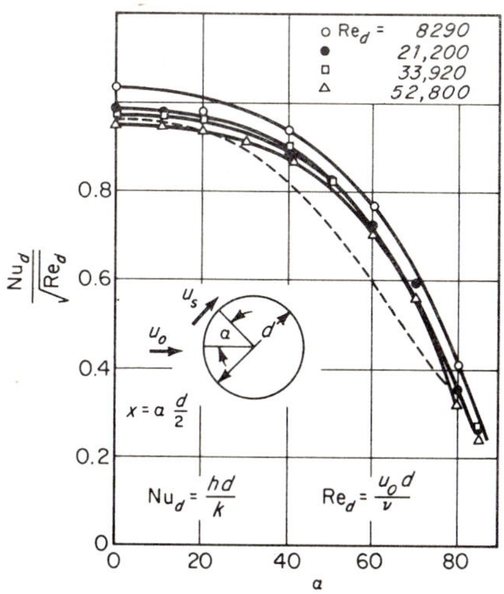

Fig. 7-21 Measured and calculated (*dashed line*) local heat-transfer parameter for flow over a cylinder with circular cross section. [*Calculation is based on local-similarity measurements by E. Schmidt and K. Wenner, Forsch. Gebiete Ingenieurw.*, **12**:*65* (*1941*).]

ture. A review of approximate methods for predicting heat-transfer coefficients for laminar, constant-property boundary-layer flows has been published by Spalding and Pun.[1]

7-9 LAMINAR DUCT FLOW

The characteristics of the flow in a duct have already been briefly discussed, and it has been pointed out that the velocity profiles develop gradually downstream from the inlet in the entrance region and in the developed region reach a shape which no longer changes in the downstream direction. For fluids with small viscosity or, more accurately, large Reynolds number, it is to be expected that the length of the entrance region will be fairly extensive and that the simplifications which led from the Navier-Stokes equations to the boundary-layer equations can be applied for duct flow as well. We shall therefore assume that the pressure is uniform over any cross section of the duct and that only certain of the shear stresses have values large enough to be considered.

The conservation equations describing the flow in the inlet region may now be derived with the restriction to steady flow of a constant-property fluid, with absence of body forces, in a duct with circular cross section. The flow under this condition will be rotationally symmetric. The choice of a cylindrical coordinate system is appropriate, to which the conservation equations developed in Chap. 6 can be transformed. Here, however, the equations will be rederived for the specific situation.

A ring-shaped control volume with radius r, thickness dr, and height

[1] *Loc. cit.*

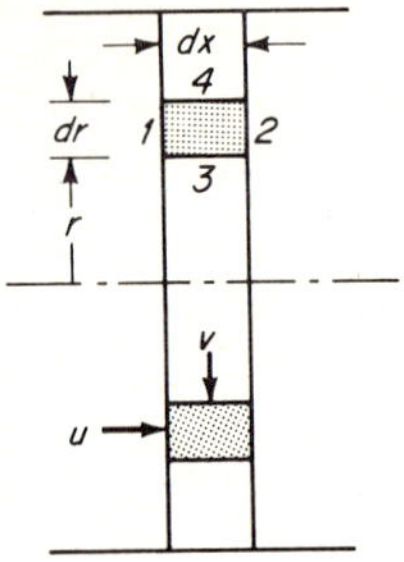

Fig. 7-22 Ring-shaped control volume for derivation of the flow equations for laminar tube flow.

dx is selected as shown in Fig. 7-22. The mass flow per unit time through the control surface 1 is

$$\rho u 2\pi r\, dr \tag{7-92}$$

The mass flow through control surface 2 differs from that through control surface 1 by

$$\rho \frac{\partial}{\partial x}(u) 2\pi r\, dr\, dx$$

The mass flow through the cylindrical control surface 3 is

$$\rho v 2\pi r\, dx \tag{7-93}$$

and the difference between the mass flow leaving the control volume through surface 4 and that entering through surface 3 is

$$\rho \frac{\partial}{\partial r}(rv) 2\pi\, dr\, dx$$

For a constant-property fluid, no mass can be stored in the control volume, and the mass flow leaving the control volume must equal the mass flow entering it. This leads to the continuity equation

$$\frac{\partial u}{\partial x} + \frac{1}{r}\frac{\partial (rv)}{\partial r} = 0 \tag{7-94}$$

The momentum equation in the direction x can now be derived. The momentum flow entering through surface 1 is obtained by multiplying Eq. (7-92) with the x component of the velocity

$$\rho u^2 2\pi r\, dr$$

The difference in the momentum flow leaving through 2 and entering through 1 is

$$\rho \frac{\partial u^2}{\partial x} 2\pi r\, dr\, dx$$

The flow of x momentum entering through surface 3 is again obtained by multiplying the mass flow with the velocity component u

$$\rho u v 2\pi r \, dx$$

The difference between x momentum leaving through 4 and entering through 3 is

$$\rho \frac{\partial (ruv)}{\partial r} 2\pi \, dr \, dx$$

The following forces act on the control surfaces in the x direction: a net pressure force with the value

$$-\frac{dp}{dx} 2\pi r \, dr \, dx$$

a shear force on surface 3 with the value

$$\mu \frac{\partial u}{\partial r} 2\pi r \, dx$$

and, therefore, a net shear force on the control volume

$$\mu \frac{\partial}{\partial r}\left(r \frac{\partial u}{\partial r}\right) 2\pi \, dr \, dx$$

Equating the net momentum fluxes to the net forces results in

$$\frac{\partial u^2}{\partial x} + \frac{1}{r}\frac{\partial (ruv)}{\partial r} = -\frac{1}{\rho}\frac{dp}{dx} + \frac{\nu}{r}\frac{\partial}{\partial r}\left(r\frac{\partial u}{\partial r}\right) \tag{7-95}$$

The left-hand side of the equation can be simplified. Partial differentiation gives

$$u\frac{\partial u}{\partial x} + u\frac{\partial u}{\partial x} + v\frac{\partial u}{\partial r} + \frac{1}{r} u \frac{\partial (rv)}{\partial r}$$

Two terms drop out through consideration of the continuity equation, and the momentum equation takes the form

$$u\frac{\partial u}{\partial x} + v\frac{\partial u}{\partial r} = -\frac{1}{\rho}\frac{dp}{dx} + \frac{\nu}{r}\frac{\partial}{\partial r}\left(r\frac{\partial u}{\partial r}\right) \tag{7-96}$$

with the boundary conditions

$$\begin{aligned} r &= R \qquad & u &= 0 \qquad & v &= 0 \\ r &= 0 & \frac{\partial u}{\partial r} &= 0 & v &= 0 \\ x &= 0 & u &= u_0 \end{aligned} \tag{7-97}$$

The last boundary condition is an approximation to the actual condition at the inlet of a tube, but it is satisfactory for describing the velocity field except for the immediate neighborhood of the inlet. The momentum equation in the radial direction has been replaced by the statement that the pressure does not depend on the radial coordinate. We now have two equations available for three unknowns, namely, u, v, p. On the other hand, the number of boundary conditions exceeds the required number. We can obtain an additional equation by multiplying Eq. (7-95) by r and integrating it over r from $r = 0$ to $r = R$. The second term on the left-hand side of this equation becomes zero after integration and the following relation evolves:

$$\frac{\partial}{\partial x}\int_0^R u^2 r\,dr = -\frac{1}{\rho}\frac{dp}{dx}\frac{R^2}{2} + \nu R\left.\frac{\partial u}{\partial r}\right|_{r=R} \tag{7-98}$$

The three equations (7-94), (7-96), and (7-98), which now describe the flow in the inlet section, are nonlinear partial differential equations, and no exact solutions have yet been obtained. Approximate solutions, however, have been calculated by assuming the shapes of the velocity profiles, or by a perturbation method, or by linearization of the equations. Some results obtained in this way will be discussed after the flow in the developed region has been investigated.

Flow in this region is characterized by the fact that the velocities do not depend on x. From the continuity equation and the boundary conditions it follows that the velocity component v is zero in this regime, and the momentum equation (7-96) simplifies to

$$\frac{1}{\rho}\frac{dp}{dx} = \frac{\nu}{r}\frac{d}{dr}\left(r\frac{du}{dr}\right)$$

It will be recognized that the left-hand side of this equation is a function of x only and the right-hand side a function of r only. This condition is fulfilled when both sides are equal to a constant. This means that the pressure drop dp/dx is constant in this region. The velocity profile is then obtained by two integrations of the equation

$$\frac{\nu}{r}\frac{d}{dr}\left(r\frac{du}{dr}\right) = C$$

With the boundary conditions, this results in the well-known expression for the velocity profile

$$u = u_c\left[1 - \left(\frac{r}{R}\right)^2\right] \tag{7-99}$$

The mean velocity $\bar{u}$ is connected with the centerline velocity u_c by the relation

$$u = \frac{u_c}{2} \tag{7-100}$$

The pressure drop is obtained from Eq. (7-98), which reduces to

$$\frac{dp}{dx}\frac{R}{2} = \mu \left.\frac{du}{dr}\right|_{r=R}$$

Introduction of the velocity gradient at the wall results in the following relation for the friction factor:

$$f = \frac{64}{\bar{u}D/\nu} = \frac{64}{\mathrm{Re}_D} \tag{7-101}$$

when it is defined by the relation

$$\Delta p = f\frac{x}{D}\rho\frac{\bar{u}^2}{2} \tag{7-102}$$

The type of flow just investigated is referred to as *Hagen-Poiseuille* flow. The friction factor for the developed regime in ducts with other cross sections is described by equations which differ from Eq. (7-101) only by changes in the numerical constant. Usually the hydraulic diameter $D_h = 4A/C$ is used in Eqs. (7-101) and (7-102) for noncircular geometries. In this case the constant for a channel composed of two parallel plates has the value 96. For a rectangular duct with various side ratios a/b, for an annular duct with various radius ratios r_i/r_o, and for a triangular duct, the values of the constant are listed in Table 7-3.

The velocity profile described in Eq. (7-99) is a parabola, as shown in Fig. 7-23. The figure also exhibits velocity profiles in the entrance

Table 7-3 Values for constant C in equation $f = \frac{C}{\mathrm{Re}}$

Rectangular duct

a/b	1	5	10	∞
C	57	77	84	96

Annular duct

r_i/r_o	1	4	7	10
C	96	94	91	89.5

Isosceles triangle

$C = 53$

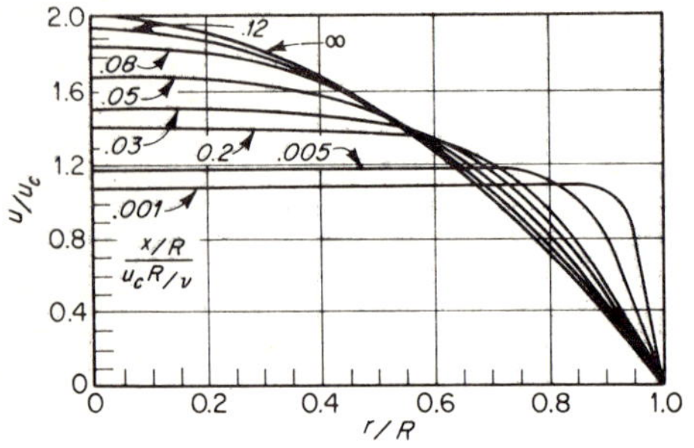

Fig. 7-23 Laminar velocity profiles in the inlet section of a tube. [*From E. Sparrow, S. H. Lin, and T. Lundgren, Phys. Fluids,* **7**:*338 (1964).*]

section of a tube.[1] The pressure drop in the inlet section of a duct with circular cross section can be obtained from

$$p_0 - p = \left(f\frac{x}{D} + K\right)\rho\frac{\bar{u}^2}{2} \tag{7-103}$$

and from the value of the parameter K in Fig. 7-24. In Eq. (7-103) p_0 is the total pressure in the inlet cross section and p is the static pressure at a distance x from the inlet. Usually in engineering applications one must deal with fairly large Reynolds numbers. At a Reynolds number of 1,000, for instance, the length of the inlet section is approximately $50D$, according to Fig. 7-24. When the Reynolds number drops to 100

[1] As calculated in E. M. Sparrow, S. H. Lin, and T. S. Lundgren, *Phys. Fluids*, **7**:338 (1964).

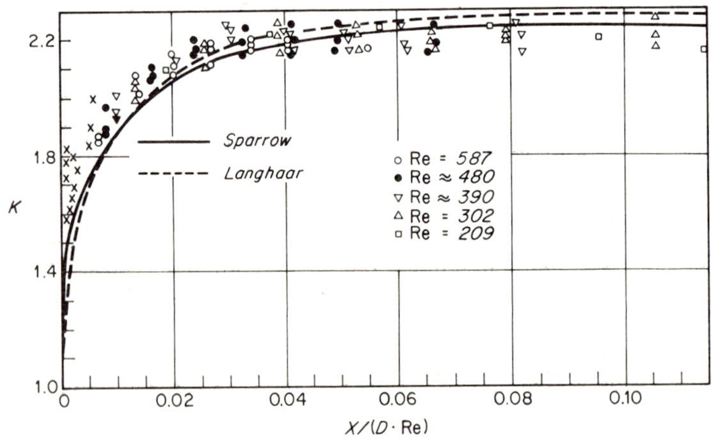

Fig. 7-24 Coefficient K for the calculation of the pressure drop caused by laminar flow through the inlet section of a tube. [*S. T. McComas and E. R. G. Eckert, T. Appl. Mech.,* **32**:*765 (1965).*]

or less, the entrance length will be fairly short. This will cause the actual flow field in the inlet region to be different from the one analyzed in this section for the following reasons: the velocity component v will be of the same order of magnitude as the velocity component u, and the assumption that the pressure is independent of r is no longer justified. The velocity distribution in the inlet cross section will also depend on the shape of the inlet and on the Reynolds number and will exhibit deviations from the assumed uniform velocity. This has only a minor effect in most of the inlet section as long as it is large compared to the tube diameter. It will, however, become important when the entrance length is short.

7-10 HEAT TRANSFER IN LAMINAR DUCT FLOW

In this section, heat transfer between steady laminar flow and the walls of a duct with circular cross section will be discussed. Forced convection, i.e., absence of body forces, and a constant-property fluid will be assumed. Under this condition, the flow field is not influenced by heat transfer, and the information in Sec. 7-9 applies. According to that section, the velocity profiles develop in the inlet region to the parabolic profile, which then remains unchanged in the fully developed region. A large variety of boundary conditions can exist with respect to the temperature field, some of which have found an extensive treatment. An early study by Graetz in 1883 considered a tube with an unheated entrance length and a wall temperature different from that of the entering fluid temperature, but otherwise constant, beginning at some location in the developed-flow region. Nusselt,[1] unaware of the solution by Graetz, solved the problem once more in considerable detail. This analysis will now be discussed.

The velocity field in the developed region is described by a parabolic velocity profile

$$u = 2\bar{u}\left(1 - \frac{r^2}{R^2}\right) \tag{7-104}$$

which does not change in the downstream direction. The temperature field is obtained from an energy balance on a ring-shaped control volume, as sketched in Fig. 7-25. The energy flow entering the control volume in the axial direction is $2\pi\rho c_p utr\,dr$. This makes the difference between the energy convected out of and into the volume element in the axial direction $2\pi\rho c_p ur\,\partial t/\partial x\,dr\,dx$. There is no convective energy transport in the radial direction because the radial velocity component is zero. Energy is transported into the control volume additionally by conduction.

[1] W. Nusselt, *Z. Ver. Deut. Inst.* **54**:1154 (1910).

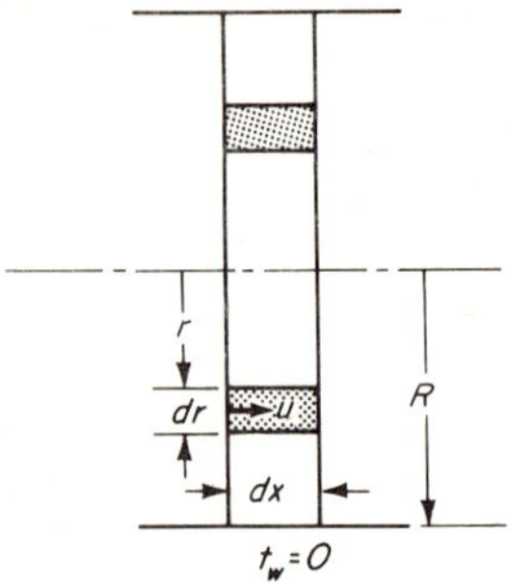

Fig. 7-25 Ring-shaped control volume for the derivation of the energy equation for laminar flow through a tube.

The radial heat flux into the volume element is $-2\pi kr\,\partial t/\partial r\,dx$, and the surplus of heat conducted out over that conducted in is

$$-2\pi k\,\frac{\partial}{\partial r}\left(r\,\frac{\partial t}{\partial r}\right)dr\,dx$$

The temperature gradients in the axial direction are usually much smaller than the temperature variation in radial direction. Correspondingly, the axial conduction will be neglected, and it will be checked later through the results of the analysis when this is permissible. It is additionally assumed that the velocities are small enough to make the temperature increase by internal friction negligible. The energy balance then reads

$$\frac{1}{r}\,k\,\frac{\partial}{\partial r}\left(r\,\frac{\partial t}{\partial r}\right) = \rho c_p u\,\frac{\partial t}{\partial x} \tag{7-105}$$

Dimensionless coordinates $r' = r/R$ and $x' = x/R$ can now be introduced. The energy balance now is

$$k\,\frac{1}{r'}\,\frac{\partial}{\partial r'}\left(r'\,\frac{\partial t}{\partial r'}\right) = \rho c_p u R\,\frac{\partial t}{\partial x'}$$

Introducing Eq. (7-104) for the velocity profile with some rearrangement results in

$$\frac{1}{r'(1-r'^2)}\,\frac{\partial}{\partial r'}\left(r'\,\frac{\partial t}{\partial r'}\right) = \frac{2\bar{u}R}{\alpha}\,\frac{\partial t}{\partial x'} \tag{7-106}$$

This is the differential equation which must be solved for the boundary conditions

$$\begin{aligned} x' &= 0 & t &= t_0 = \text{const} \\ r' &= 0 & \frac{\partial t}{\partial r'} &= 0 \\ r' &= 1 & t &= 0 \end{aligned} \tag{7-107}$$

The last condition indicates that we count temperatures from the wall temperature. We shall try to obtain a solution of this partial differential equation with a function for the temperature t which is the product of a function of x' and one of r'

$$t = \Phi(x')\psi(r')$$

Introducting this expression into the energy equation (7-106) and dividing both sides by the product $\Phi(x')\psi(r')$ results in

$$\frac{1}{r'(1-r'^2)}\frac{1}{\psi}\frac{d}{dr'}\left(r'\frac{d\psi}{dr'}\right) = \frac{2\bar{u}R}{\alpha}\frac{1}{\Phi}\frac{d\Phi}{dx'}$$

an equation which contains only a function of r' on the left-hand side and a function of x' on the right-hand side. This then can be satisfied only when both sides of the equation are equal to a constant. Denoting this constant by $-\beta_i^2$, the equation can now be separated into two total differential equations

$$\frac{1}{\Phi}\frac{d\Phi}{dx'} = -\beta_i^2\frac{\alpha}{2\bar{u}R} \tag{7-108}$$

$$\frac{1}{r'}\frac{d}{dr'}\left(r'\frac{d\psi}{dr'}\right) + \beta_i^2(1-r'^2)\psi = 0 \tag{7-109}$$

Integration of Eq. (7-108) results in

$$\Phi = C_i \exp\left(-\beta_i^2\frac{\alpha}{2\bar{u}R}x'\right)$$

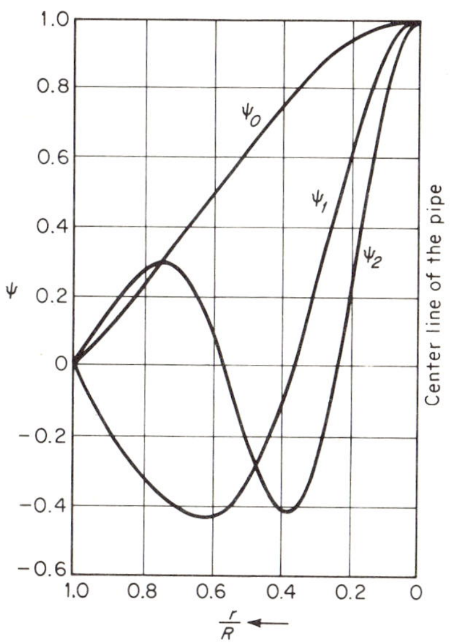

Fig. 7-26 Functions ψ for the calculation of the temperature field in the thermally developing region of laminar flow through a tube. [*After W. Nusselt, Z. Ver. Deut. Ingr.*, **54**:*1154* (*1910*).]

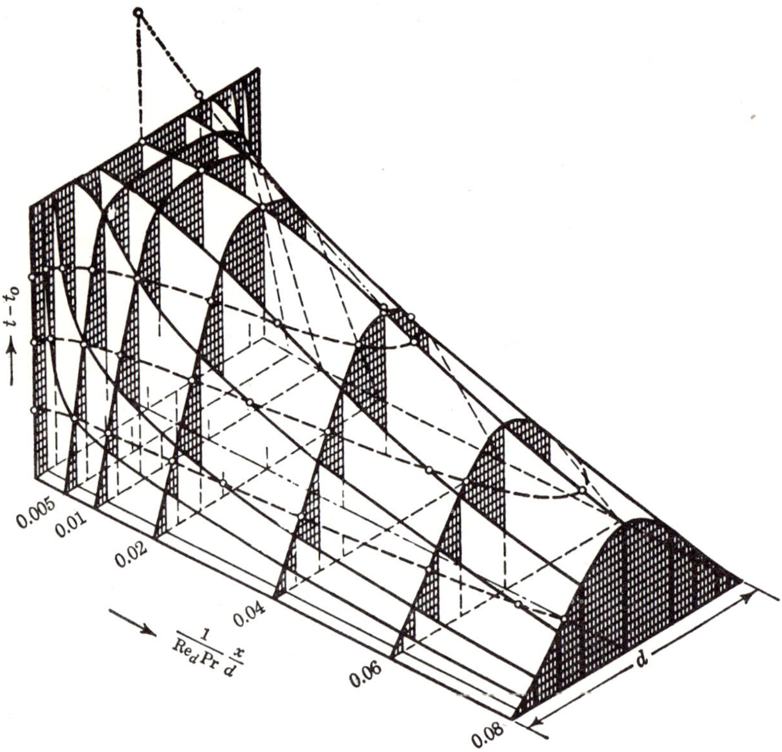

Fig. 7-27 Temperature profile in laminar flow in the entrance region of a tube with constant wall temperature and developed flow condition. [*From L. Prandtl, "Stromungslehre," Vieweg, Brunswick, 1949.*]

This expression describes the variation of the temperature field in the x direction, and the physical situation obviously requires the exponent to be negative. It was for this reason that a negative constant $-\beta_i^2$ was introduced in Eqs. (7-108) and (7-109).

The equation for the function ψ was integrated numerically by Nusselt, starting the integration at $r' = 0$ and prescribing at this location $d\psi/dr' = 0$ and $\psi = 1$. The last condition is admissible, as will be seen shortly. It is found that the boundary condition $\psi = 0$ at $r' = 1$ can be fulfilled by an infinite series of values β_i^2. The first three functions are shown in Fig. 7-26. The corresponding values of β are $\beta_n = 2.705$, $\beta_1 = 6.667$, and $\beta_2 = 10.67$. The general solution for the temperature field which fulfills the radial boundary conditions is therefore

$$t = \sum C_i \exp\left(-\beta_i^2 \frac{\alpha}{2\bar{u}R} \frac{x}{R}\right) \psi_i \frac{r}{R} \tag{7-110}$$

The constants C_i can now be used to fulfill the boundary condition at $x' = 0$ through the equation

$$t_0 = \sum C_i \psi_i \frac{r}{R}$$

In this way, the temperature field, as shown in Fig. 7-27, was determined. It is easily checked that the parameter $\frac{\alpha}{2\bar{u}R}\frac{x}{R}$ in Eq. (7-110) converts to twice the value of the expression $\frac{1}{\mathrm{Re}_d\,\mathrm{Pr}}\frac{x}{d}$ used in the figure, where d indicates the tube diameter.

The larger the value of β, the more rapidly the e function in Eq. (7-110) decreases. Downstream some distance from the inlet, therefore, the term in this equation with the lowest β value will dominate; the shape of the temperature profile is then the one for the lowest β value in Fig. 7-26, and it does not change at greater distances downstream. This, then, can be called the thermally developed region. Figure 7-27 implies that this region starts at a value $\frac{1}{\mathrm{Re}_d\,\mathrm{Pr}}\frac{x_d}{d} = 0.05$.

A question remains to be checked, namely, whether we were correct in neglecting axial heat conduction. At the location $x = 0$, Eq. (7-110) implies that the fluid temperature has a finite gradient for $x > 0$ but has the gradient zero for $x < 0$. Such a discontinuity cannot occur physically and appears in the analysis as a consequence of the neglect of heat conduction in the x direction. One can expect this discrepancy to be restricted to a small portion of the developing region when $x_d/d \gg 1$. For a Prandtl number equal or larger than 1 and a Reynolds number of 1,000, the distance ratio $x_d/d > 50$, and the results of the preceding analysis can be accepted as describing the actual situation over most of the thermally developing region. For a liquid metal with a Prandtl number 0.01 and a Reynolds number 1,000, the distance ratio x_d/d is equal to 0.5, and axial conduction surely cannot be neglected.

A heat-transfer coefficient is conventionally defined for duct flow with the equation

$$q_w = h(t_B - t_w) \qquad (7\text{-}111)$$

in which q_w is the heat flux per unit wall area and time and t_B is the fluid bulk temperature defined by the equation

$$t_B = \frac{\int \rho c_p u t \, dA}{\int \rho c_p u \, dA} \qquad (7\text{-}112)$$

with the integrations extended over the cross section of the duct under consideration. The bulk temperature, therefore, is the temperature obtained when the fluid flowing through this cross section is mixed.

For a constant-property fluid flowing through a tube with circular cross section, Eq. (7-112) reduces to

$$t_B = \frac{\int_0^R utr\, dr}{\int_0^R ur\, dr} \tag{7-113}$$

For the tube with constant wall temperature this bulk temperature can be calculated with Eq. (7-110). Relation (7-111) then determines the heat-transfer coefficient h and the Nusselt number

$$\mathrm{Nu}_d = \frac{hd}{k} \tag{7-114}$$

as a dimensionless heat-transfer parameter.

This Nusselt number, presented as curve a in Fig. 7-28, approaches with increasing x an asymptotic value 3.65.

The flow of oils through tubes in heat exchangers often occurs under conditions for which the parameter $\frac{1}{\mathrm{Re}_d\ \mathrm{Pr}}\frac{x}{d}$ has small values because of the large Prandtl numbers involved. For these conditions, the temperature field penetrates only a short distance from the wall into the fluid,

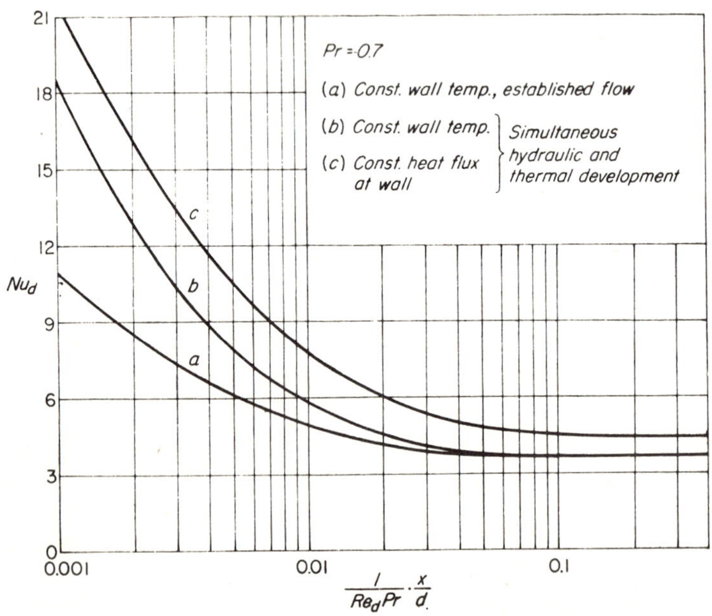

Fig. 7-28 Local Nusselt number for the entrance region of a tube. [*After W. Nusselt; and W. M. Kays, Trans. ASME,* **77**:*1265* (*1965*).]

whereas the core of the fluid remains at the temperature with which it entered the tube. The boundary-layer equation (7-56) can be used for an analysis of this situation because the effect of transverse curvature can also be neglected as long as the thickness of this thermal layer is small compared to the tube radius. The velocity component v is zero for developed flow. The profile of the velocity u within this layer can be approximated by a linear increase with wall distance, the rate of increase being determined from the parabolic Poiseuille profile. A calculation of this kind performed by J. Lévêque in 1928 resulted in the equation

$$\frac{hd}{k} = 1.16 \text{ Gz}^{\frac{1}{3}}$$

The dimensionless parameter Gz, called the Graetz number, is defined as

$$\text{Gz} = \frac{c_p \dot{m}}{kx} = \frac{\pi}{4} \text{Re}_d \text{ Pr} \frac{d}{x}$$

This equation is found to be a good approximation for Graetz numbers larger than 1,500.

Another frequently encountered boundary condition is a constant heat flux q_w starting at a cross section where the velocity field is already developed. Equation (7-105) must now be solved with the boundary conditions

$$\begin{aligned} x &= 0 \qquad & t &= t_0 = \text{const} \\ r &= 0 \qquad & \frac{\partial t}{\partial r} &= 0 \\ r &= R \qquad & k\frac{\partial t}{\partial r} &= q_w = \text{const} \end{aligned}$$

An energy balance of the fluid flowing through the tube immediately results in the statement that for constant fluid properties the fluid bulk temperature increases linearly in the flow direction.

For thermally developed flow this must also hold for the temperature at any distance r from the tube axis. Therefore, a constant heat rate results in

$$\frac{\partial t}{\partial x} = C$$

The energy equation thus reduces in this case to a total differential equation in r which can be solved by simple integration and results in the following expression for the temperature profile:

$$\theta = t - t_w = \frac{2\rho c_p \bar{u} R^2}{k} \frac{\partial t}{\partial x} \left[\frac{1}{4}\left(\frac{r}{R}\right)^2 - \frac{1}{16}\left(\frac{r}{R}\right)^4 - \frac{3}{16}\right]$$

The temperature profile is the same at any station x. The heat-transfer coefficient is determined by the temperature gradient at the wall. For the Nusselt number based on the local difference between wall temperature and fluid bulk temperature, a short calculation gives

$$\mathrm{Nu}_d = 4.36 \tag{7-115}$$

The more involved calculation[1] for the thermally developing region results in a local Nusselt number shown by curve c in Fig. 7-28 with an asymptotic value 4.36. The figure contains a third curve b showing the Nusselt number for laminar flow through a tube with constant wall temperature t_w starting at the entrance cross section so that the velocity and the temperature fields develop simultaneously.[2]

An energy balance on a control volume extending over the full cross section of the tube gives the result that the bulk temperature of a constant-property fluid rises linearly from the entrance to the exit when the heat flux at the wall is constant. This is an advantage for setting up experiments with the purpose of measuring local heat-transfer coefficients because the local bulk temperature is known by a measurement of the entrance temperature and the exit bulk temperature. The wall-surface temperature also rises linearly but only in the thermally developed region.

Laminar heat transfer as investigated above is often disturbed by free-convection effects. This will be discussed later.

In Fig. 7-29 the average Nusselt number resulting from the calculations of Graetz and Nusselt is compared with formulas derived by Kraussold[3] and by Sieder and Tate[4] from their experiments. Notwithstanding the fact that the experiments were carried out principally with oil, the agreement with the calculated results is fairly reasonable. Some results of experiments with gases by Nusselt[5] are inserted and shown by points. Experiments by Boehm[6] gave similar results. The logarithmic mean temperature difference (LMTD) is to be used in connection with the Nusselt numbers in Figs. 7-28 and 7-29. Hausen[7] developed a formula for the average Nusselt number which represents the results of Graetz and Nusselt very well:

$$\overline{\mathrm{Nu}}_d = 3.65 + \frac{0.0668(d/x)\ \mathrm{Re}_d\ \mathrm{Pr}}{1 + 0.04[(d/x)\ \mathrm{Re}_d\ \mathrm{Pr}]^{\frac{2}{3}}} \tag{7-116}$$

For greater temperature differences and oils the influence of the variable viscosity can be taken into account by multiplying the right-hand

[1] R. Siegel, E. M. Sparrow, and T. M. Hallman, *Appl. Sci. Res.*, **7A**:386 (1958).

[2] W. M. Kays, *Trans. ASME*, **77**:1265 (1955).

[3] H. Kraussold, *VDI-Forschungsh.* 351, 1931.

[4] E. N. Sieder and G. E. Tate, *Ind. Eng. Chem.*, **28**:1429–1435 (1936).

[5] W. Nusselt, Habilitationsschrift, Technische Hochschule, Dresden, 1909.

[6] J. Boehm, *Wärme*, **66**:143–152 (1943).

[7] H. Hausen, *VDIZ. Beih. Verfahrenstech.*, **4**:91–98 (1943).

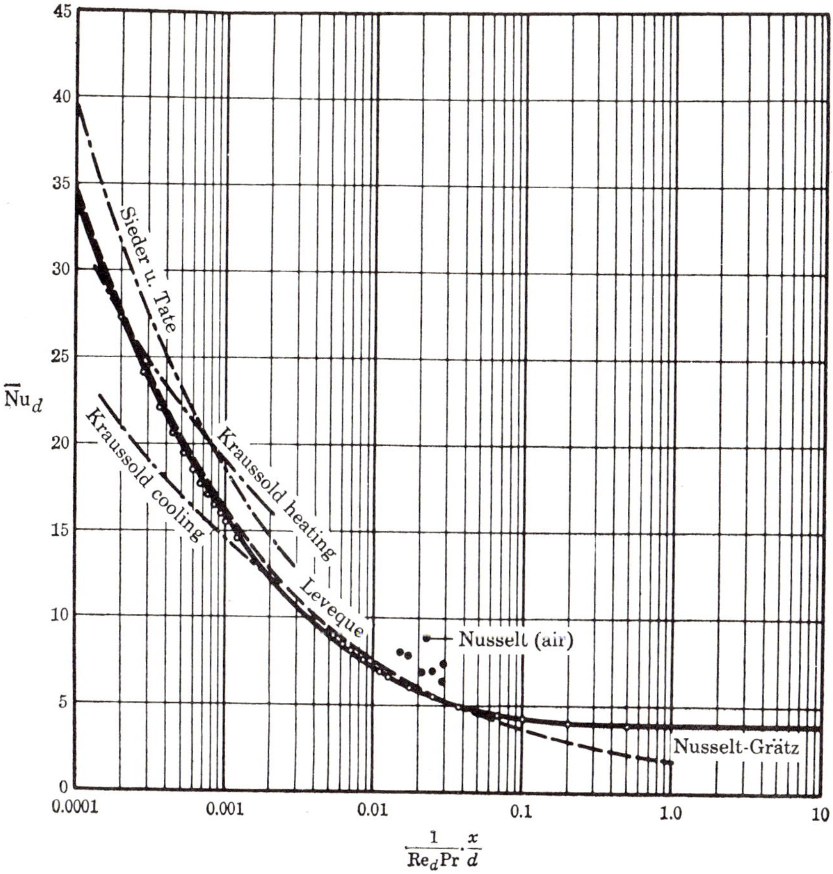

Fig. 7-29 Comparison of measured and calculated average Nusselt numbers in the entrance region of a tube with constant wall temperature and developed flow. [*From H. Hausen, VDI Z. Beih, Vertahrenstech.,* **34**:*91* (*1943*).]

side of Eq. (7-116) by the ratio $(\mu_B/\mu_w)^{0.14}$, where μ_B is the viscosity at the fluid bulk temperature and μ_w the viscosity at wall temperature. The properties in the dimensionless parameters of the equation are then inserted at the bulk temperature. For gases it is recommended that Eq. (7-116) be used without the addition of the viscosity ratio and that the properties be introduced at the reference temperature as given by Eq. (7-86).

In the literature, information on heat transfer related to laminar flow through tubes in the inlet region is often presented as a function of a dimensionless parameter called the Graetz number which is the reciprocal of the parameter used as the abscissa in Figs. 7-28 and 7-29. A summarizing paper considers various temperature conditions along the tube wall.[1]

Some peculiar features are exhibited by heat transfer connected with laminar flow through ducts with noncircular cross sections. This situation was studied analytically[2] for hydrodynamically and thermally developed flow through a duct whose cross section has the shape of the sector of a circle and for the condition that the heat flux from the duct wall into the fluid is constant in the direction of the duct axis. It was found that the local heat-transfer coefficient changes considerably around the duct periphery, approaching the value zero in the corners, and that the average heat-transfer coefficient depends very much on the boundary conditions. Two boundary conditions around the circumference of the duct were investigated: a wall temperature which is constant around the periphery and a locally constant heat flux. The Nusselt number averaged around the circumference for a constant wall temperature was found to be 7 times the value for a constant heat flux when the opening angle of the sector was 20°. For a 60° opening angle, the ratio of the two Nusselt numbers is 2.5. The heat-transfer coefficient in the Nusselt numbers was defined as the average heat flux q_w at the wall divided by the difference between the fluid bulk temperature and the average wall temperature (averaged around the duct periphery). The above result indicates that in using published information on heat-transfer coefficients, one must establish the pertinent boundary conditions carefully.

7-11 SEPARATED LAMINAR SHEAR LAYERS

Velocity field Viscous forces are important in boundary layers, in duct flow, and in separated shear layers, the last of which will now be discussed briefly. On a surface with a step, as indicated in Fig. 7-30, flow coming from the left will separate at the corner of the step. In a fluid with small viscosity, it will develop a shear layer with boundaries as shown in the

[1] R. H. Norris and D. D. Streid, *Trans. ASME*, **62**:525–533 (1940).

[2] E. R. G. Eckert, T. F. Irvine, Jr., and J. T. Yen, *Trans. ASME*, **80**:1433 (1958).

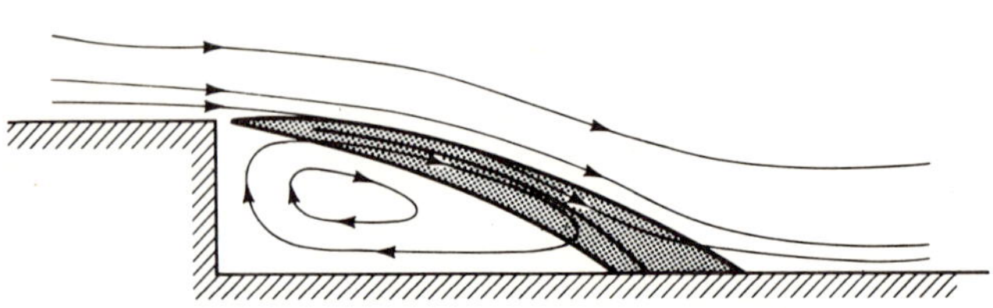

Fig. 7-30 Free-shear layer and recirculation zone for flow over a wall with downstream-facing step.

figure. Between this layer and the surface downstream from the step, one finds a *separated flow region,* within which the fluid rotates as indicated in the figure. The shear layer reattaches to the surface behind the separated region. The figure somewhat simplifies the actual situation. In reality a boundary layer already exists on the surface upstream from the step, and boundary layers will build up along the surface contour in the separated region. Separated shear layers are quite unstable and soon transform into turbulent flow. This transition occurs for a constant-property fluid at a Reynolds number of approximately 5×10^4.† In supersonic and hypersonic flow, free shear layers have been found to be much more stable, and transition Reynolds numbers as high as 10^6 have been observed. In this section, the laminar shear layer will be discussed.

The boundary-layer equations developed in Secs. 6-4 and 6-5 are valid for free shear layers as well as for attached boundary layers. Only the boundary conditions change. In analyses of such free shear layers, the following boundary conditions have been applied:

$$\begin{aligned} y &= \infty & u &= u_s \\ y &= 0 & v &= 0 \\ y &= -\infty & u &= 0 \end{aligned}$$

The condition at $y = 0$ means that the abscissa coincides with a streamline of the shear layer. The boundary condition at $y = -\infty$ is not quite in agreement with reality because the rotating fluid in the separated region will generate some velocity at this point. However, these velocities are usually small and therefore neglected. If the shear layer is considered to start with a thickness zero at the step, and if the velocity u_s outside this layer can be expressed as some power of the distance x measured from the point of separation, the boundary-layer equations can again be transformed into total differential equations.

A solution for such a shear layer is contained in a paper by Chapman[1] for the condition that the pressure, and therefore the velocity u_s outside the shear layer, is independent of x. This solution will be discussed in some detail. Chapman started with the boundary-layer equation as obtained by the Mises transformation. Equation (7-34) reduces for constant pressure to

$$\frac{\partial u}{\partial x} = \nu \frac{\partial}{\partial \psi}\left(u \frac{\partial u}{\partial \psi}\right) \tag{7-117}$$

† A detailed study of laminar separated shear layers and transition to turbulence is reported in R. G. Goldstein, V. L. Erikson, R. M. Olson and E. R. G. Eckert, "Laminar Separation—Reattachment and Transition of Flow over a Downstream Facing Step," *J. Basic Eng.*, **92**:732 (1970).

[1] D. R. Chapman, *NACA Tech. Rept.* 958, 1950.

The corresponding boundary conditions are

$$\begin{aligned} \psi &= -\infty \qquad & u &= 0 \\ \psi &= \infty \qquad & u &= u_s \end{aligned} \tag{7-118}$$

The above equation can be transformed into a total differential equation by the substitutions

$$\zeta = \frac{\psi}{\sqrt{\nu u_s x}} \qquad u^* = \frac{u}{u_s} \tag{7-119}$$

The shear-layer equation now reads

$$-\frac{\zeta}{2}\frac{du^*}{d\zeta} = \frac{d}{d\zeta}\left(u^*\frac{du^*}{d\zeta}\right) \tag{7-120}$$

with the boundary conditions

$$\begin{aligned} \zeta &= -\infty \qquad & u^* &= 0 \\ \zeta &= \infty \qquad & u^* &= 1 \end{aligned}$$

Equation (7-120) was transformed by Chapman into an integral equation. For this purpose u^* is temporarily considered as a known function of ζ. With introduction of a new variable for $du^*/d\zeta$, the equation transforms into a first-order total differential equation which can easily be integrated. A second integration then gives the result

$$u^* = C\int_0^\zeta \frac{F}{u^*}\,d\zeta + u_0^* \tag{7-121}$$

with

$$F = \exp\left(-\int_0^\zeta \frac{\zeta}{2u^*}\,d\zeta\right) \tag{7-122}$$

u_0^* denotes the dimensionless velocity at the location $\zeta = 0$. The constant C finally is obtained from the boundary conditions as

$$C = \frac{1 - u_0^*}{\displaystyle\int_0^\infty \frac{F}{u^*}\,d\zeta} \tag{7-123}$$

Equation (7-121) is used for an iteration procedure, in which a guess at the dimensionless velocity profile u^* is inserted into the right-hand side of the equation. The u^* resulting from the integration prescribed by Eq. (7-121) is reinserted as a second approximation, and so forth, until convergence is achieved. Figure 7-31 presents the dimensionless velocity u^* obtained in this way as a function of ζ. To obtain the velocity profile as a function of the coordinate y, one integrates the relation $u = \partial\psi/\partial x$ for $x = \text{const}$. This results in

$$y = \int_0^\psi \frac{1}{u}\,d\psi = \sqrt{\frac{\nu x}{u_s}}\int_0^\zeta \frac{1}{u^*}\,d\zeta \tag{7-124}$$

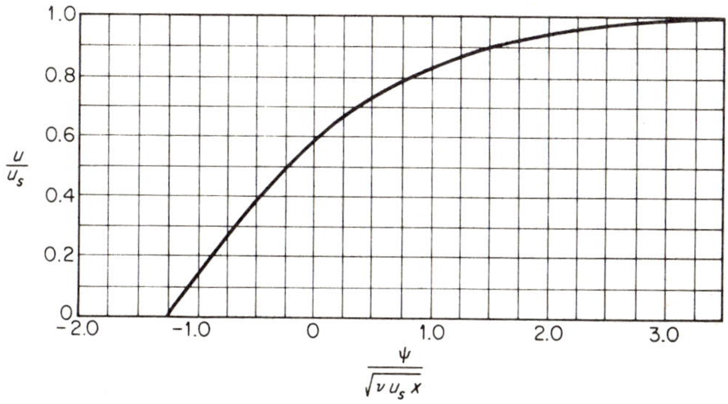

Fig. 7-31 Velocities in the laminar free shear layer plotted over the dimensionless stream function. [*After D. R. Chapman.*]

The distance y is counted from the streamline $\psi = 0$. From Fig. 7-32, which presents this velocity profile, it can be seen that the streamline $y = 0$ starts at the origin ($x = 0$, $y = 0$). This streamline therefore separates the main flow arriving from the left in Fig. 7-30 from the recirculating flow. The above solution can be used as an approximation for separated-flow regions like the one sketched in Fig. 7-30 as long as the boundary-layer thickness of the flow arriving at the step is small compared to the thickness of the shear layer. Similar shear layers are encountered in other flow situations, e.g., behind a cylinder exposed to flow normal to its axis.

Heat transfer The temperature field in the laminar shear layer is described by the boundary-layer energy equation as discussed in Sec. 7-6. The Mises transformation can be applied to this energy equation. The same procedure as in the previous paragraph leads to the result

$$\frac{\partial t}{\partial x} = \alpha \frac{\partial}{\partial \psi}\left(u \frac{\partial t}{\partial \psi}\right) \tag{7-125}$$

with the boundary condition

$$x = \infty \qquad t = t_s \qquad x = -\infty \qquad t = t_0 \tag{7-126}$$

For a fluid with a Prandtl number equal to 1, the thermal diffusivity α is equal to the kinetic viscosity ν. Equation (7-125) is therefore identical with Eq. (7-117), and the temperature profile becomes identical to the velocity profile u^* if it is interpreted as

$$\frac{t - t_0}{t_s - t_0}$$

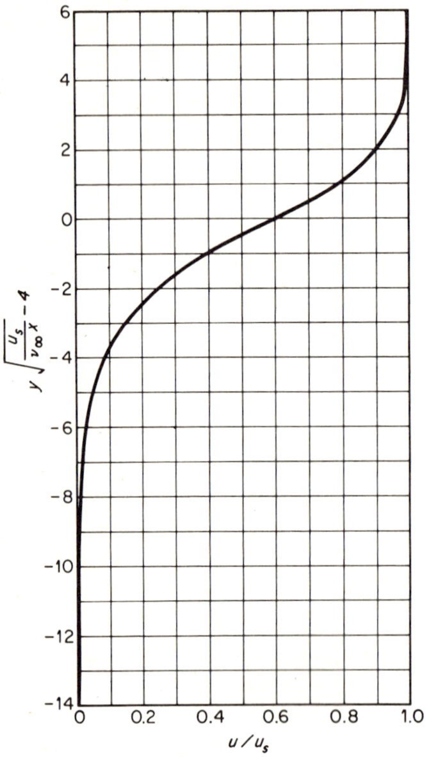

Fig. 7-32 Dimensionless velocity profile in the laminar shear layer for constant pressure in the adjoining mainstream. [*After D. R. Chapman.*]

The heat transferred from the fluid above the shear layer to the fluid below it in Fig. 7-30 has to be transported by conduction through the separating streamline (or, more correctly, stream surface). It is therefore

$$q = -k\left(\frac{\partial t}{\partial y}\right)_{y=0} = -k(t_s - t_0)\left(\frac{du^*}{d\xi}\right)_{\xi=0}\frac{\partial \xi}{\partial y}$$

From Eq. (7-124) one obtains

$$\frac{\partial y}{\partial \xi} = \frac{1}{u^*}\sqrt{\frac{\nu x}{u_s}}$$

The heat-transfer coefficient finally becomes

$$h = \frac{q}{t_s - t_0} = -k\left(u^*\frac{du^*}{d\xi}\right)_{\xi=0}\sqrt{\frac{u_s}{\nu x}}$$

and the Nusselt number as a dimensionless expression for this heat-transfer coefficient is

$$\mathrm{Nu}_x = \frac{hx}{k} = \left(u^*\frac{du^*}{d\xi}\right)_{\xi=0}\sqrt{\mathrm{Re}_x} = 0.199\sqrt{\mathrm{Re}_x} \qquad (7\text{-}127)$$

Table 7-4

Pr	0.1	0.25	0.5	0.72	1	1.5	2	10
$\dfrac{\mathrm{Nu}_x}{\sqrt{\mathrm{Re}_x}}$	0.0416	0.0842	0.132	0.165	0.199	0.250	0.293	0.675

The numerical value in this equation has been obtained from the velocity profile in Fig. 7-32. The corresponding values for fluids with other Prandtl numbers were obtained by Chapman[1] by a numerical solution of Eq. (7-125). The values are listed in Table 7-4.

In the separated flow sketched in Fig. 7-30, the heat flux then has to be transported from the separated fluid region to the wall surface. A boundary layer between the separated region and the wall constitutes a second thermal resistance. For laminar flow this resistance appears to be small compared with the thermal resistance of the shear layer because experimental results agree well with the results of the analysis presented in this section.

Energy is actually transferred through the separating streamline not only by conduction but also as mechanical work produced by the shear stresses acting at this streamline. These shear stresses perform work since the fluid moves along the streamline with a finite velocity. This energy flux is small compared to the heat transfer by conduction as long as the velocity u_s or, more exactly, the Eckert number, is small. For high-velocity flow, however, it leads to the situation where the temperature in the separated region is larger than the temperature in the mainstream provided no heat is removed through the solid walls. A similar situation occurs also in the normal boundary layer and will be discussed in Sec. 10-1. A relation will also be given there by which the temperature in the separated region can be calculated.

A review of the present state of our knowledge about flow and heat transfer in separated and reattaching flow is presented in a paper by Chilcott,[2] and an extensive study of laminar heat transfer in flow downstream from a stepdown contour is given in by Aung.[3]

7-12 COMPUTER ANALYSIS OF LAMINAR HEAT TRANSFER

The availability of high-speed digital computers with large storage capacity today makes it possible to obtain solutions to the conservation equa-

[1] D. R. Chapman, *NACA. Tech. Note* 3792, 1956.

[2] R. E. Chilcott, *Intern. J. Heat Mass Transfer*, **10**:783–797 (1967).

[3] W. Aung, Heat Transfer in the Separated Region beyond a Rearward-facing Step, Ph.D. thesis, University of Minnesota, 1969.

tions describing flow and heat transfer for many situations of engineering interest. This includes boundary-layer-type flow as well as flow with recirculation, steady flow as well as unsteady flow, and can also account for the effect of variable properties. For this purpose the conservation equations are written in finite-difference form. As a consequence, parameters describing the velocity and temperature fields are determined for an array of discrete points only, which are located at the intersections of two systems of lines. The simplest grid is composed of two groups of straight lines intersecting normally, corresponding to a cartesian coordinate system. In some cases it is more convenient to use some other system of intersecting lines, e.g., a general orthogonal coordinate system. The independent variables are usually the space coordinates. One or the other of them, however, may be replaced by the stream function or the vorticity. It is beyond the scope of this book to discuss the computation procedures in detail, and the reader is referred to the literature. A group at the Imperial College in London headed by Spalding is making a concerted effort to develop such computation schemes, and two books describe their findings in detail.[1]

In writing the conservation equations, it is necessary to approximate the expressions for the various flux terms by the use of finite differences instead of the differentials occurring in the differential equations. This can be done in various ways, and only ease of computation or speed of convergence determines which form is preferable. A form which proved useful will be discussed, taking as an example the energy equation describing a two-dimensional temperature field in a fluid with constant properties. The effect of viscous dissipation is assumed negligible. A grid of straight normally intersecting lines may be used, and the energy balance may be considered for the volume element bounded by surfaces represented by the dashed lines in Fig. 7-33. The purpose of the energy balance is to determine the temperature at point 0 provided the tem-

[1] S. V. Patankar and D. B. Spalding, "Heat and Mass Transfer in Boundary Layers," Morgan-Grampian, London, 1967; A. D. Gosman, W. M. Pun, A. K. Runchal, D. B. Spalding, and M. W. Wolfshtein, "Heat and Mass Transfer in Recirculating Flows," Academic, London, 1969.

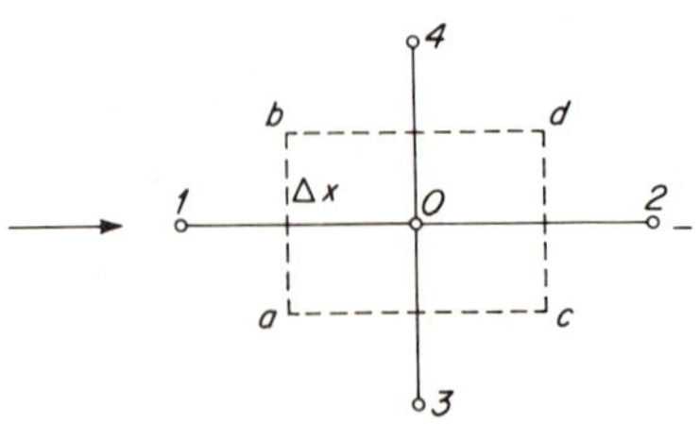

Fig. 7-33 Element for finite-difference formulation of energy equation.

peratures at the surrounding points 1, 2, 3, 4 are known. Energy fluxes will occur by conduction and by convection. The heat conduction can be treated as described in Part I. The heat flux per unit area through the surface *ab* can be described by the equation

$$q_{1-0} = k\frac{t_1 - t_0}{\Delta x} \tag{7-128}$$

Analogous expressions describe the conductive heat fluxes through the other surfaces into the volume element. Enthalpy fluxes occurring by convection will also be needed for the energy balance, and an expression must be found to describe, for instance, the enthalpy flux through the surface *ab*. One might decide to describe it with the velocity and the temperature which exists at the boundary of the volume element, approximated by the arithmetic mean of the temperatures t_1 and t_0. It was found, however, that a numerical computation converges better when the enthalpy flux is expressed with parameters existing at the point located upstream of the boundary. With the flow direction indicated in Figure 7-33 this is the point 1. The convective enthalpy flux per unit area of the surface *ab* is correspondingly written as

$$q'_{1-0} = \rho c_p u_1 t_1 \tag{7-129}$$

The enthalpy flux leaving the volume element through surface *cd* per unit area is correspondingly expressed in the parameters at point 0

$$q'_{0-2} = \rho c_p u_0 t_0 \tag{7-130}$$

point 2 and point 0, respectively, would, on the other hand, be used to express the convective fluxes provided the flow occurs from right to left. This way of introducing convective fluxes is referred to in the literature as the *upwind* or *donor-cell* method. It appears to have been introduced into transport equations written in finite-difference form by Runchal and Wolfshtein.[1]

In the example discussed here, the energy equation is obtained by equating the sum of all energy fluxes entering the control volume in Fig. 7-33 with the energy stored in it. The energy equation can also be obtained from the appropriate differential equation. In this case it is well to see that the energy balance on the finite control volume is strictly fulfilled.

Computation of the solution of the equations differs for various types of flow. For boundary-layer flow, the computation can start at

[1] A. K. Runchal and M. W. Wolfshtein, A Finite Difference Procedure for the Integration of the Navier-Stokes Equations, *Imperial College Mech. Eng. Dept. Rep.* SF/TN/1, 1964.

the upstream end of the flow field and move along in a straightforward way in the flow direction because heat conduction is neglected in the flow direction and because the convective terms refer to parameters on the upstream side only. Spalding's group refers to such flows as *parabolic flows*. The calculation becomes more involved in situations where recirculation occurs, as in regions of flow separation, or where heat conduction opposite to the flow direction must be considered, as in low Reynolds number or liquid-metal flows. In this case an iteration scheme must be devised which starts with an assumed field in the whole region of interest. Flows of this type are classified as *elliptic flows*.

Many flow and heat-transfer situations have already been investigated analytically in this way, and it is to be expected that in the future information required for engineering studies will more and more be obtained by analysis, whereas experiments will be relegated to occasional checks.

7-13 LAMINAR HEAT-TRANSFER RELATIONS

The heat-transfer relation obtained in the previous sections will be summarized below for convenience.

Boundary-layer flow Velocity profiles in laminar boundary layers of a constant-property fluid are self-similar when the free-stream velocity is described by an equation

$$u_s = Cx^m \tag{7-131}$$

The parameter

$$\beta = \frac{2m}{m+1} \tag{7-132}$$

is frequently used in the literature instead of the exponent m. The temperature profiles in such a boundary layer are self-similar when the wall-surface temperature varies according to the equation

$$t_w - t_s = C_t x^\gamma \tag{7-133}$$

The Nusselt number describing the heat transfer in these situations is given by the following functional relationship:

$$\frac{\mathrm{Nu}_x}{\sqrt{\mathrm{Re}_x}} = f(\mathrm{Pr},\beta,\gamma) \tag{7-134}$$

For flow over a flat plate ($\beta = 0$) with a constant wall temperature ($\gamma = 0$), the Nusselt numbers are presented in Fig. 7-12. They are also well approximated by the relations

$$\mathrm{Nu}_x = 0.332\sqrt{\mathrm{Re}_x}\,\sqrt[3]{\mathrm{Pr}} \qquad \text{for Pr} > 0.6 \tag{7-135}$$

$$\mathrm{Nu}_x = 0.500\sqrt{\mathrm{Re}_x\,\mathrm{Pr}} \qquad \text{for } 0.005 < \text{Pr} < 0.05 \tag{7-136}$$

The influence of a variation of the free-stream velocity ($\beta \neq 0$) can be obtained for a constant wall temperature ($\gamma = 0$) from Fig. 7-10, which presents the ratio of the Nusselt number for variable free-stream velocity to the Nusselt number on a flat plate. Figure 7-11 finally makes it possible to determine the influence of a locally varying wall temperature. Again the ratio of the Nusselt number for various values of the parameter γ to the Nusselt number for a constant wall temperature is presented. A solution of the integrated boundary-layer equation leads to the relation

$$\mathrm{Nu}_x = 0.332\sqrt{\mathrm{Re}_x}\,\sqrt[3]{\mathrm{Pr}}\,\frac{1}{\sqrt[3]{1-(x_0/x)^{\frac{3}{4}}}} \qquad \text{for Pr} > 0.6 \tag{7-137}$$

for a step change in the surface temperature of a flat plate.

Tube flow The following relations hold for heat transfer in the developed region of laminar tube flow:

For t_w = const: $$\mathrm{Nu}_d = 3.65 \tag{7-138}$$

For q_w = const: $$\mathrm{Nu}_d = 4.36 \tag{7-139}$$

Local heat-transfer coefficients in the entrance region can be obtained from Fig. 7-28. Close to the entrance the following equation describes the average heat transfer over the length x:

$$\mathrm{Nu}_d = 1.62\left(\frac{\pi}{4}\,\mathrm{Re}_d\,\mathrm{Pr}\,\frac{d}{x}\right)^{\frac{1}{3}} \tag{7-140}$$

The equation holds for constant wall temperature and for the condition

$$\frac{\pi}{4}\,\mathrm{Re}_d\,\mathrm{Pr}\,\frac{d}{x} > 10 \tag{7-141}$$

It is especially useful for viscous fluids (oils) which have a long thermal-development length. The following equation approximates conditions in the development length as well as in developed flow. It describes the Nusselt number averaged over the length x:

$$\mathrm{Nu}_d = 3.65 + \frac{0.0668(d/x)\,\mathrm{Re}_d\,\mathrm{Pr}}{1 + 0.04[(d/x)\,\mathrm{Re}_d\,\mathrm{Pr}]^{\frac{2}{3}}} \tag{7-142}$$

Laminar shear layer Heat transfer in a laminar shear layer which starts at $x = 0$ with 0 thickness is described by the equation

$$\mathrm{Nu}_x = C\sqrt{\mathrm{Re}_x} \tag{7-143}$$

Values for the constant C can be obtained from Table 7-3. The relation holds for constant free-stream velocity outside the shear layer.

For situations which lead to nonsimilar boundary layers or for flows through ducts with involved geometry or boundary conditions, numerical calculations as outlined in Sec. 7-12 are probably the best methods for obtaining heat-transfer information. This is also true for separated-flow regions, where it is often difficult to separate the shear layers from the recirculation zones.

8

Forced Convection in Turbulent Flow

8-1 MOMENTUM EQUATION AND VELOCITY PROFILE IN TURBULENT SHEAR LAYERS

In principle, the unsteady conservation equations in Secs. 6-1 and 6-2 should also describe turbulent flow. Their solution would result in the actual path which the fluid particles follow, consisting of the irregular random motions in addition to the mean motion. Practically, however, even with the most advanced electronic computer, it is not possible to obtain anything approaching such a solution. One is therefore satisfied in turbulent flow to get information on time-averaged quantities. In 1895 Reynolds described a procedure by which one can derive the conservation equations expressed in time-averaged quantities. We shall discuss this procedure for a constant-property fluid. For this purpose the instantaneous-flow parameters are written as the sum of the time-averaged values (indicated by a bar) and the instantaneous deviations from these values (indicated by a prime). Using again tensor notation, we can write

$$\begin{aligned} v_i &= \bar{v}_i + v_i' \\ p &= \bar{p} + p' \end{aligned} \tag{8-1}$$

It follows from the definition of the fluctuating quantities that their time-averaged values $\overline{v_i'}$, $\overline{p'}$ are equal to zero. This does not hold, however, for the squares or products of such values

$$\overline{v_i'v_j'} \neq 0$$

In addition, the following rule applies: $\overline{\partial v_i/\partial x_j} = \partial \bar{v}_i/\partial x_j$. The continuity equation, when Eq. (8-1) is introduced, reads

$$\frac{\partial \overline{v_i}}{\partial x_i} + \frac{\partial v_i'}{\partial x_i} = 0$$

Now all terms in this equation can be averaged over the time, with the result

$$\frac{\partial \bar{v}_i}{\partial x_i} + \frac{\partial \overline{v_i'}}{\partial x_i} = 0$$

According to the definition, the terms $\partial \overline{v_i'}/\partial x_i$ are equal to zero, and the continuity equation becomes

$$\frac{\partial \overline{v_i}}{\partial x_i} = 0 \tag{8-2}$$

It can be seen that this equation has the same form for a turbulent flow as for a laminar one. Actually the same relation must also be satisfied for the instantaneous fluctuating quantities, a fact which follows immediately from the above equations.

The same procedure can now be applied to the momentum equation. Introduction of Eq. (8-1) gives for a constant-property fluid

$$\rho\left(\frac{\partial \bar{v}_i}{\partial \tau} + \frac{\partial v_i'}{\partial \tau} + \bar{v}_j \frac{\partial \bar{v}_i}{\partial x_j} + \bar{v}_j \frac{\partial v_i'}{\partial x_j} + v_j' \frac{\partial \bar{v}_i}{\partial x_j} + v_j' \frac{\partial v_i'}{\partial x_j}\right)$$
$$= -\frac{\partial \bar{p}}{\partial x_i} - \frac{\partial p'}{\partial x_i} + \mu\left(\frac{\partial^2 \bar{v}_i}{\partial x_j^2} + \frac{\partial^2 v_i'}{\partial x_j^2}\right) \qquad \text{(8-3)}\dagger$$

One must remember that ∂x_j^2 is equivalent to $\partial x_j\, \partial x_j$, so that the two last terms stand for two sums because the index j is repeated. The term $v_j'\, \partial v_i'/\partial x_j$ can be transformed by starting with

$$\frac{\partial (v_i'v_j')}{\partial x_j} = v_j' \frac{\partial v_i'}{\partial x_j} + v_i' \frac{\partial v_j'}{\partial x_j}$$

The last term on the right-hand side is zero because $\partial v_j'/\partial x_j = 0$. Therefore

$$v_j' \frac{\partial v_i'}{\partial x_j} = \frac{\partial (v_i'v_j')}{\partial x_j}$$

† The variation of $\bar{v}_i$ with time is postulated as being slow enough for a meaningful separation into a fluctuating quantity v_i' and a mean value $\overline{v_i}$ to be possible.

In the next step, all terms in Eq. (8-3) are averaged over the time. The terms

$$\overline{\frac{\partial v_i'}{\partial \tau}} \qquad \bar{v}_j \overline{\frac{\partial v_i'}{\partial x_j}} \qquad \overline{v_j'} \frac{\partial \bar{v}_i}{\partial x} \qquad \overline{\frac{\partial^2 v_i'}{\partial x_j{}^2}} \qquad \overline{\frac{\partial p'}{\partial x_i}}$$

become equal to zero, and the momentum equation assumes the following final form:

$$\rho\left(\frac{\partial \bar{v}_i}{\partial \tau} + \bar{v}_j \frac{\partial \bar{v}_i}{\partial x_j}\right) = -\frac{\partial \bar{p}}{\partial x_i} + \mu \frac{\partial^2 \bar{v}_i}{\partial x_j{}^2} - \rho \frac{\partial (\overline{v_i' v_j'})}{\partial x_j} \tag{8-4}$$

It can be recognized that the momentum equation for turbulent flow differs from the corresponding equation for laminar flow by the terms on the right-hand side containing the fluctuating velocity components. The reason for these terms is that momentum flux depends essentially on the square or product of velocity components and that the average of such a quantity is different from the square or product of the average quantities. This fact is also responsible for the surface stresses appearing in the laminar momentum equation. In that case, the irregular motions of the individual particles (molecules, atoms, or electrons) cause the surface stresses. In turbulent flow, the irregular motion is not only performed by individual particles but also by large groups of them. The terms containing the fluctuating quantities in the momentum equations can therefore be interpreted as resulting from turbulent stresses:

$$p_{ij,t} = \rho \overline{v_i' v_j'}$$

Conventional instruments for the measurement of velocity and pressure, like pitot tubes, cannot detect the fluctuating quantities because their response time is too slow. Such quantities, however, can be measured directly with special instruments, e.g., a hot wire. With proper electronic circuitry, one can also measure the turbulent stresses. As an example of such measurements, Fig. 8-1 presents the time averages $\sqrt{\overline{u_1'^2}}$, $\sqrt{\overline{v_2'^2}}$, $\sqrt{\overline{w_3'^2}}$, divided by the mean stream velocity u_s in a turbulent boundary layer on a flat plate. It can be observed that all are approximately of the same magnitude and of order 4 to 8 percent of the free-stream velocity.

Little analysis has been performed on the complete set of conservation equations, but they can again be transformed to boundary-layer equations by the procedure used for laminar flow. The fluctuating velocity components v_i' are in a boundary layer smaller by an order of magnitude than the mean values $\bar{v}_i$. The fluctuations, however, are rapid in time and in space, so that the derivatives $\partial v_i'/\partial x_j$ cannot be considered as small; establishing the order of magnitude of the various terms in Eq. (8-4) then leads to the following system of the continuity

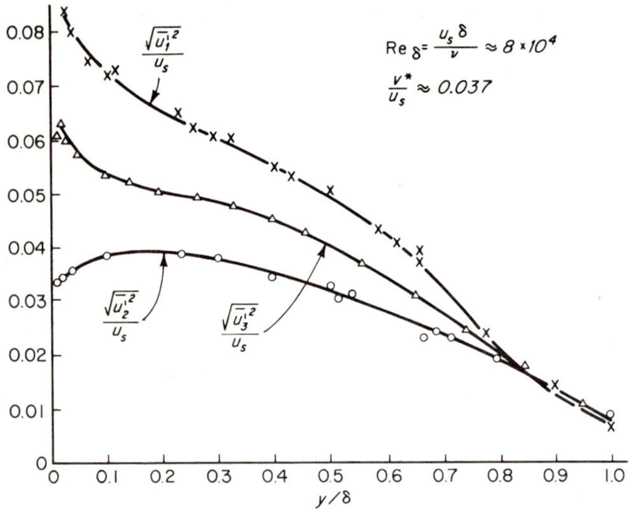

Fig. 8-1 Root-mean-square fluctuations of the velocity components in a boundary layer along a smooth wall with constant pressure. [*From P. S. Klebanoff, NACA Tech. Note 3178, 1954.*]

equation and the two momentum equations for two-dimensional flow of a fluid with constant properties:

$$\frac{\partial \bar{u}}{\partial x} + \frac{\partial \bar{v}}{\partial y} = 0$$
$$\rho \left(\frac{\partial \bar{u}}{\partial \tau} + \bar{u} \frac{\partial \bar{u}}{\partial x} + \bar{v} \frac{\partial \bar{u}}{\partial y} + \frac{\partial \overline{u'^2}}{\partial x} + \frac{\partial \overline{u'v'}}{\partial y} \right) = - \frac{\partial \bar{p}}{\partial x} + \mu \frac{\partial^2 \bar{u}}{\partial y^2} \tag{8-5}$$
$$\rho \frac{\partial \overline{v'^2}}{\partial y} = - \frac{\partial \bar{p}}{\partial y}$$

The last equation can be integrated to give $\bar{p} + \rho\overline{v'^2} = p_s$ when the turbulent fluctuations are zero or negligible outside the boundary layer. Introducing the last equation into the momentum equation for the x direction in the system (8-5) results in the equation

$$\rho \left(\frac{\partial \bar{u}}{\partial \tau} + \bar{u} \frac{\partial \bar{u}}{\partial x} + \bar{v} \frac{\partial \bar{u}}{\partial y} \right) = - \frac{\partial p_s}{\partial x} + \mu \frac{\partial^2 \bar{u}}{\partial y^2} - \rho \frac{\partial \overline{u'v'}}{\partial y} \tag{8-6}$$

A term $-\rho\, \partial/\partial x\, (\overline{u'^2} - \overline{v'^2})$ has been neglected in the last equation because it was found that in normal boundary layers it is of smaller order of magnitude than the other terms in that equation. Equation (8-6) assumes that the time variation of the velocity u is such that a meaningful separation into a time-averaged and fluctuating contribution is possible.

Since the turbulent-boundary-layer equations contain too many unknowns to be solved as such, the basic problem in the analysis of turbulent flow is to find relations which connect the turbulent stresses with the mean flow parameters. For boundary layers, this means finding a relation between the turbulent shear

$$\tau_t = -\rho \overline{u'v'} \tag{8-7}$$

and the time mean values of velocity and pressure. During the past 50 years great effort has been devoted to finding such a relation and many expressions have been proposed, either empirically or based on various models assumed to describe the turbulent process of transfer of momentum. This task, however, still is not solved today in the sense that no relation is available which holds universally for all flow situations. On the other hand, relations have been developed which allow the analytic calculation of flow and heat-transfer parameters for certain classes of boundary-layer flows; however, one must be careful in applying them to flow situations which are widely different from those for which agreement with experiments has been established. The following paragraphs will discuss the relations which have proved most useful for heat-transfer analyses.

The first person to propose such a relation was Boussinesq (1877). He wrote in analogy to Newton's law

$$\tau_t = \rho \epsilon_m \frac{\partial \bar{u}}{\partial y} \tag{8-8}$$

The term ϵ_m in this equation is called *turbulent diffusivity of momentum.* This relation, which is frequently used, has its shortcomings. Newton's equation after which the relation is patterned is justified in connecting the laminar shear with the local derivative of the velocity because the mean-free-path length of the molecules is in general very small compared to the boundary-layer thickness. As a consequence, the local shear is influenced by the velocity variation very close to the place where the shear occurs. In turbulent flow, on the other hand, one finds that in the turbulent mixing process fluid particles are transported over distances which are of the same order of magnitude as the boundary-layer thickness and one must therefore expect the turbulent shear to be influenced by the flow situation in a regime which cannot be considered as small. Actually it has been found experimentally that in certain flow situations a finite shear occurs at a location where the transverse velocity variation is zero and vice versa. This has been found, for instance, in a wall jet and in the flow through a parallel-wall channel with a rough and a smooth wall. Equation (8-8) requires that both vanish at the same location. For normal boundary layers, however, this inconsistency does not appear to cause any significant difficulty. It has also been found experimentally that the turbulent diffusivity ϵ_m depends strongly on the specific flow situation, and no major progress was made in the analysis of turbulent flow until Prandtl[1] proposed an expression for the turbulent diffusivity in 1925. He argued that, according to experience, wall friction, which is closely connected with the shear stress in the boundary layer, increases in turbulent flow approximately in proportion to the square of the velocity. Therefore he introduced a velocity gradient in the expression for the turbulent diffusivity and wrote

$$\epsilon_m = l^2 \left| \frac{\partial \bar{u}}{\partial y} \right| \tag{8-9}$$

The velocity gradient was multiplied by the square of a length l in order to make the equation dimensionally correct. This length l Prandtl called the *mixing length*, and he interpreted it as that length which fluid particles travel in the turbulent mixing process before they lose their identity. For boundary-layer flow he assumed that near a solid surface the mixing length is proportional to the wall distance, and from experimental results be determined $l = 0.4y$.

Kármán[2] obtained another expression for the turbulent diffusivity by reasoning that this parameter should depend at a certain location only on the velocity field in its surrounding. This field can obviously be expressed by the velocity gradient and by higher derivatives: $\partial u/\partial y$, $\partial^2\bar{u}/\partial y^2$, ... Usually it is found that the influence of such derivatives becomes smaller the higher their order. Therefore Kármán inquired which combination of the first two derivatives gave a dimensionally correct expression for the turbulent diffusivity. In this way is found the expression

$$\epsilon_m = \varkappa^2 \frac{(\partial \bar{u}/\partial y)^3}{(\partial^2 \bar{u}/\partial y^2)^2} \tag{8-10}$$

where $\varkappa$ is a dimensionless constant. This expression is more general than Prandtl's expression and has therefore been especially useful in heat-transfer calculations. For a constant-property fluid, however, it leads to the same expression for the velocity field, as will be shown in the following discussion.

Equation (8-10) will now be used to derive an equation describing

[1] L. Prandtl, *Z. Angew. Math. Mech.*, **5**:136 (1925).

[2] T. von Kármán, *Proc. 3d Intern. Congr. Appl. Mech., Stockholm*, 1930, pt. I, p. 85; *NACA Memo.* 611, 1931.

the turbulent velocities close to a smooth plate with constant pressure along its surface and for flow which is steady in the values $\bar{v}_i$. In this region, the streamlines are essentially parallel to the wall, and the flow has the character of a simple steady shear flow with constant pressure for which the variation of $\bar{u}$ in the x direction and, according to the continuity equation, the velocity component $\bar{v}$ are equal to zero. The boundary-layer momentum equation (8-6) shows that in this model the shear τ (composed of laminar shear $\tau_l = \mu\, \partial\bar{u}/\partial y$ and turbulent shear

$$\tau_t = \rho\epsilon_m \frac{\partial \bar{u}}{\partial y}$$

is independent of the wall distance and equal to the wall shear τ_w. The boundary-layer equation therefore simplifies to

$$\tau = \tau_l + \tau_t = \tau_w = \mu \frac{d\bar{u}}{dy} + \rho\epsilon_m \frac{d\bar{u}}{dy} \tag{8-11}$$

Neglecting the laminar contribution to the shear and introducing Eq. (8-10), one obtains the following equation (the bars over the velocity will now be dropped, and u will be interpreted as the time-averaged velocity):

$$\tau_w = \rho\varkappa^2 \frac{(du/dy)^4}{(d^2u/dy^2)^2} \tag{8-12}$$

In order to integrate this equation, we replace du/dy by a new variable x, take the square root, separate variables, and obtain after integration

$$\frac{1}{x} = \pm \frac{\varkappa}{\sqrt{\tau_w/\rho}}\, y + C$$

Introduction of du/dy and setting $C = 0$ then leads to the expression

$$\tau_w = \rho\varkappa^2 y^2 \left(\frac{du}{dy}\right)^2$$

for the wall shear, which can be seen to be identical to Prandtl's expression. Separation of the variables and an additional integration results in

$$\ln y + C' = \frac{\varkappa}{\sqrt{\tau_w/\rho}}\, u$$

or

$$\frac{u}{\sqrt{\tau_w/\rho}} = \frac{1}{\varkappa} \ln y + C''$$

The term $\sqrt{\tau_w/\rho}$ has the dimension of a velocity, is called *shear velocity*, and denoted by v^*. Accordingly, the left-hand side of the equation is dimensionless. In order to make the individual terms on the right-hand side also dimensionless, the integration constant is changed, and part of it combined with the first term. In this way, one obtains

$$u^+ = \frac{1}{\varkappa} \ln y^+ + C \tag{8-13}$$

in which $$u^+ = \frac{u}{\sqrt{\tau_w/\rho}} \qquad y^+ = y\,\frac{\sqrt{\tau_w/\rho}}{\nu} \tag{8-14}$$

Nikuradse obtained the following values for the constants in Eq. (8-13):

$$\varkappa = 0.4 \qquad C = 5.5$$

It was found experimentally that Eq. (8-13) describes measured velocity profiles in a certain portion of the flow field not only in boundary layers but also in tube flow in a wider range than one might expect from the derivation.

Figure 8-2 is a semilogarithmic diagram of the results of measured turbulent velocity profiles in the neighborhood of the wall of a tube with fully developed turbulent flow. The profiles are plotted in the dimensionless crossed variables. Equation (8-13), which appears as a straight line in the diagram, represents the measured values for dimensionless wall distances larger than 30. Close to the wall, however, the measured values deviate from this line and merge at vanishing wall distance with a

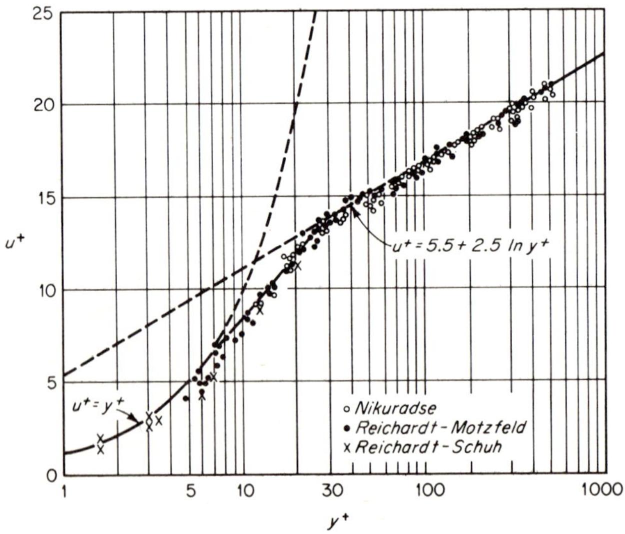

Fig. 8-2 Universal velocity profile in a tube with fully developed turbulent flow.

curve represented by the relation $u^+ = y^+$. In dimensional parameters, this equation is $\tau_w = \mu\, du/dy$. The deviation of the measured values from Eq. (8-13) is explained by the fact that for wall distances smaller than $y^+ = 30$ the contribution of the laminar shear cannot be neglected. Many expressions have been proposed which describe the actual variation of u^+ at smaller wall distances. An expression developed by Deissler[1] has proved quite successful for heat-transfer analyses. According to him, the wall shear in this layer is described by Eq. (8-11), which, in the dimensionless crossed variables, takes on the form

$$1 = \left(1 + \frac{\epsilon_m}{\nu}\right)\frac{du^+}{dy^+}$$

The turbulent diffusivity ϵ_m in this equation is given by the expression

$$\frac{\epsilon_m}{\nu} = n^2u^+y^+(1 - e^{-n^2u^+y^+}) \tag{8-15}$$

The term in parentheses becomes important only for heat-transfer calculations with fluids of large Prandtl number. Deissler found that the numerical value 0.125 of the dimensionless constant n gives the best agreement with measurements, and he used the value $y^+ = 26$ as the outer limit of this layer. He then determined the constant C in Eq. (8-13), which he used to describe the velocity profile in the turbulent shear layer, in such a way that the velocity profile is continuous at $y^+ = 26$. This results in $u^+ = 2.78 \ln y^+ + 3.8$. A similar equation

$$\frac{\epsilon_m}{\nu} = K^2y^{+2}(1 - e^{-y^+/A^+})^2 \left|\frac{\partial u^+}{\partial y^+}\right|$$

was proposed by van Driest[2] with the constants $K = 0.4$ and $A^+ = 26$ determined from experiments by addition of a damping factor to Prandtl's missing-length expression.

The velocity profile presented in Fig. 8-2 and described by the preceding equations was found to have quite a universal character. Not only does it hold for boundary layers exposed to a pressure rise or pressure drop in flow direction but also up to a certain distance from the wall in tube flow. The layer for which this relation applies is called the *wall layer*, and the relation in Fig. 8-2 is referred to as the *universal law of the wall*, with the argument that the fluid in this regime develops essentially under the influence of the presence of the wall. This layer extends to two-tenths or three-tenths of the boundary-layer thickness.

It was also found that Eq. (8-13) describes the velocity profile in the fully turbulent wall region of boundary-layer and tube flow with a rough

[1] R. G. Deissler, *Trans. ASME*, **73**:101 (1951); **76**:73 (1954).

[2] E. R. van Driest, *J. Aeron. Sci.*, **23**:1007 (1956).

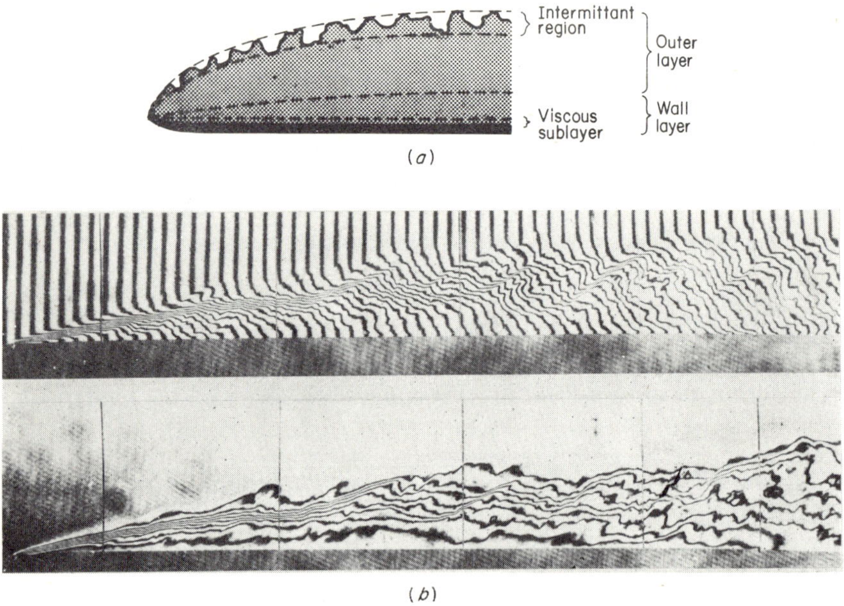

Fig. 8-3 (*a*) Regions within a turbulent boundary layer. (*b*) Interferograms of a turbulent boundary layer of a flat plate with carbon dioxide injection through the porous wall. [*Courtesy of V. P. Motulevich.*]

wall if it is used to express the difference between the velocity u and the velocity u_0 at a reference distance y_0. This relation is referred to as the *velocity-defect law.*

An instantaneous shadowgraph of a turbulent boundary layer presents a picture as sketched in Fig. 8-3*a*. One observes a sharp boundary between the turbulent flow, which appears grainy in the shadowgraph, and the flow in the mainstream. The irregular three-dimensional protuberances of this boundary move downstream with time. A hot wire placed in this region and fixed relative to the wall exhibits an intermittent signal with regions of turbulent fluctuations and regions of undisturbed steady velocity. This *intermittent layer* constitutes the outer portion of the *outer layer,* which extends through 70 to 80 percent of the whole boundary layer.

Remarkable interferograms of a turbulent boundary layer have been obtained by Motulevich[1] with a Zehnder-Mach interferometer using a

[1] V. P. Motulevich, Gas Flow with Porous Bodies under the Conditions of Intensive Injection, *Intern. Seminar Heat Mass Transfer Flows with Separated Regions and Measurement Techniques, Hercegnovi, Yugoslavia,* Sept. 1–13, 1969.

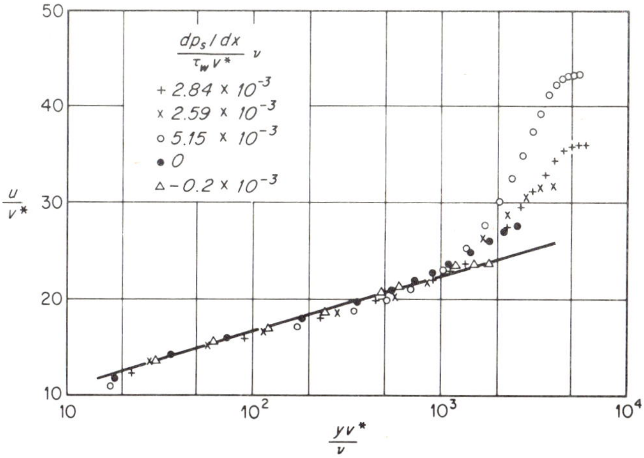

Fig. 8-4 Velocity profile in a turbulent boundary layer with positive and negative pressure gradient. [*From L. Tillmann, Ing. Archiv,* **17**:*288* (*1949*).]

ruby laser with an exposure time of 3×10^{-8} s as light source (Fig. 8-3*b*). The boundary layer shown develops in an airstream along a flat plate when carbon dioxide is ejected through the plate wall. The two photographs are made with two different adjustments of the interferometer.

The results of measurements of the time-averaged velocity including this outer layer are presented in Fig. 8-4. It can be observed that the profile shape in this layer (beyond $y^+ = 10^3$) is strongly affected by a pressure variation along the wall. The deviations are especially pronounced when the pressure increases in the flow direction. It has been observed that this outer layer resembles the flow in the wake behind a streamlined object for which Prandtl proposed a turbulent diffusivity which is constant in the direction normal to the main flow. Using this similarity, Clauser proposed for the outer layer of a boundary layer the equation

$$\epsilon_m = 0.0188 u_s \delta^* \tag{8-16}$$

where u_s denotes the velocity in the mainstream outside the boundary layer and δ^* the displacement thickness of the boundary layer. Coles,[1] who studied this region closely, developed relations describing the velocity field and the turbulent diffusivity in this region in more detail. He refers to them as the *law of the wake*. A general representation of the flow in

[1] D. Coles, *J. Fluid Mech.*, **1**:191 (1956).

the outer region is made difficult by the fact that the flow there has "a long memory" and responds to influences far upstream whereas the wall region is determined much more by local conditions.

In fully developed turbulent tube flow no intermittency exists, and the deviations from the universal law of the wall are less pronounced than in a boundary layer, so that the universal velocity profile has been used to describe the velocity field up to the tube axis. Actually, the shear in developed tube flow is described by the equation $\tau = \tau_w(1 - y/R)$, with R indicating the tube radius, and this should be considered in the integration of Eq. (8-11). The effect on wall shear and heat transfer, however, was found to be small.[1]

The mixing-length concept has been used extensively and successfully for a long time. It gave satisfactory answers to many engineering problems, and it can be expected to maintain its usefulness in the future as well. It has, however, the disadvantage that the local variation of the mixing length must be known in the analyses. It has been found that this relation differs for various flow situations, and thus only a limited extrapolation from experimentally investigated situations is possible.

The availability of computers with high speed and large memory has sparked development of a new approach, in which the required turbulence parameters are described by differential equations similar in form to the conservation equations. The development of this analysis is still in flux; however, the results are very promising.

One of these equations is a relation for the time-averaged turbulent energy $k = \overline{v_i'^2}/2$, an equation proposed by Prandtl[2] in 1945 but not used then because of our inability to solve the resulting system of differential equations. The equation for the turbulent energy is derived by logical reasoning and dimensional analysis. It considers the turbulent energy within a volume element at a fixed location and states that this energy is changed by convection, by turbulent mixing, by generation, and by destruction of turbulence energy within the volume element. The individual terms for the equation will now be developed for two-dimensional turbulent-boundary-layer type of flow of a constant-property fluid following essentially Prandtl's ideas.

A term describing the generation of turbulence was derived by Reynolds. It takes the following form in time-averaged quantities:

$$e_1 = \frac{1}{\rho}\tau_t\frac{\partial \bar{u}}{\partial y}$$

[1] R. G. Deissler, Turbulent Heat Transfer and Friction in Smooth Passages, *High Speed Aerodyn. Jet Propulsion*, **5**:288 (1959).

[2] L. Prandtl, *Nachr. Akad. Wiss. Göttingen, Math. Phys. Klasse*, **1945**:6–19.

where e_1 indicates energy generated per unit mass and time. The turbulent shear τ_t can be described by Boussinesq's equation

$$\tau_t = \rho \epsilon_m \frac{\partial \bar{u}}{\partial y}$$

The turbulent diffusivity ϵ_m has been expressed by Prandtl from dimensional reasoning in the form

$$\epsilon_m = C_m \sqrt{k}\, L \tag{8-17}$$

With this, the generation of turbulence becomes

$$\epsilon_1 = C_m \sqrt{k}\, L \left(\frac{\partial \bar{u}}{\partial y}\right)^2 \tag{8-18}$$

The destruction of turbulence is considered as being caused by the resistance of the turbulence lumps in the fluctuating velocity field. With L expressing the scale of the turbulence (the size of the turbulence lumps) the resistance W can be written as

$$W = \rho C_D L^2 u'^2$$

The work caused by this resistance per unit time and mass is then

$$e_2 = C_D \frac{u'^3}{L} = C_D \frac{k^{\frac{3}{2}}}{L} \tag{8-19}$$

It is of interest to consider a situation in which the generation and destruction of turbulence balance out whereas convection and diffusion are negligible. From Eqs. (8-18) and (8-19) one obtains

$$C_m \sqrt{k}\, L \left(\frac{\partial \bar{u}}{\partial y}\right)^2 = C_D \frac{k^{\frac{3}{2}}}{L}$$

Eliminating k from this and Eq. (8-17) results in

$$\epsilon_m = \frac{C_m^{\frac{3}{2}}}{C_D^{\frac{1}{2}}} L^2 \frac{\partial \bar{u}}{\partial y} \tag{8-20}$$

It can be observed that the equation is identical in form to the mixing-length expression (8-9) of Prandtl. With the numerical values in Table 8-1 it is found that the turbulence scale L is approximately $\frac{1}{2}l$.

It was found experimentally that such a balance between generation and destruction occurs at certain locations in a flow field. In general, however, convection and diffusion of turbulent energy also have to be considered. The loss of turbulent energy in the control volume by diffusion can be expressed in the form

$$e_3 = \frac{\partial}{\partial y}\left(\epsilon_m \frac{\partial k}{\partial y}\right) = \frac{\partial}{\partial y}\left(C_m \sqrt{k}\, L \frac{\partial k}{\partial y}\right) \tag{8-21}$$

which is analogous to the equations describing laminar heat conduction or mass diffusion. With this equation and Eqs. (8-18) and (8-19) the turbulent-energy equation takes the form

$$\frac{Dk}{d\tau} = C_m \frac{\partial}{\partial y}\left(\sqrt{k}\, L \frac{\partial k}{\partial y}\right) + C_m \sqrt{k}\, L \left(\frac{\partial \bar{u}}{\partial y}\right)^2 - C_D \frac{k^{\frac{3}{2}}}{L} \tag{8-22}$$

in which the left-hand term expresses the convective transport of turbulent energy from the volume element by the mean flow.

Equations (8-17) and (8-22) are now available for the analysis in addition to the time-averaged continuity and Navier-Stokes equations (8-5). It is found that this system still contains more unknowns than equations. Various investigators[1] therefore introduced an empirical relation for the turbulence scale L or made use of the experimental result that in some flow situations the turbulent shear was found to be proportion to k. Relying on similar reasoning, Kovasznay and Nee[2] developed an equation for the effective viscosity (the sum of the laminar and turbulent viscosities) instead of the turbulent kinetic energy.

Another possibility being investigated is to introduce one more differential equation for the turbulence scale L or for some parameter describing this scale. In 1942 Kolmogorov[3] actually proposed such an equation for the "frequency" $\sqrt{k}/L$. In 1951 Rotta[4] developed an equation for the product kL, which will be discussed here because it has found considerable use, especially by Spalding and his associates. As used by this group, it has the following form:

$$\frac{D(kL)}{d\tau} = \frac{\partial}{\partial y}\left\{\frac{\epsilon_m}{\sigma_{kL}}\left[\frac{\partial (kL)}{\partial y} + C_{kL} L \frac{\partial k}{\partial y}\right]\right\} + C_B \epsilon_m L \left(\frac{\partial \bar{u}}{\partial y}\right)^2 - C_s k^{\frac{3}{2}} - C_w \left(\frac{L}{y}\right)^q k^{\frac{3}{2}} \tag{8-23}$$

Again the left-hand term describes the convection of the parameter kL; the first term on the right side expresses the transport of the parameter kL by turbulent mixing. The remaining terms describe the following process, which has been well established. Turbulence is created by flow instability on a relatively large scale, is gradually broken up into smaller-scale turbulence, and is finally transformed by the action of viscosity into heat. The second term on the right-hand side of Eq. (8-23) expresses the decrease in the scale of turbulence by the action of the turbulent shear stresses which break up the eddies. The third term on the right-

[1] P. Bradshaw, D. H. Ferris, and N. P. Atwell, *J. Fluid Mech.*, **28**:593 (1967).

[2] L. S. G. Kovasznay and V. W. Nee, *Bull. Amer. Phys. Soc.*, (2)**10**:683 (1965).

[3] A. N. Kolmogorov, *Izv. Akad. Nauk SSSR*, ser. fix., **1942**:(1–2):56–58.

[4] J. C. Rotta, *Z. Physik*, **129**:547–551 (1951); **131**:517–577 (1951).

hand side is introduced to describe the effect of the viscosity in transforming turbulence energy into heat, and the fourth term to express the damping effect of a solid surface on the neighboring flow field.

Equations (8-17), (8-22), and (8-23) together with the time-averaged conservation equations

$$\frac{\partial \bar{u}}{\partial x} + \frac{\partial \bar{v}}{\partial y} = 0$$

$$\rho \frac{D\bar{u}}{d\tau} = -\frac{\partial \bar{p}}{\partial x} + \rho \frac{\partial}{\partial y}\left(\epsilon_m \frac{\partial \bar{u}}{\partial y}\right) \tag{8-24}$$

now constitute a closed system. It contains a large number of constants: C_m, C_D, C_s, C_B, C_{kL}, C_w, σ_k, σ_{kL}. Rodi and Spalding[1] obtained numerical values through a best fit of solutions to the system of equations with experimental results for free shear layers and jets. In the same way Ng and Spalding[2] determined the constants for boundary-layer flow, which are given in Table 8-1. The exponent q was found to be 4. For the round jet, which does not quite fit into this scheme, the values $C_D = 0.55$, $C_s = 0.0397$ were found to apply.

Equations (8-22) to (8-24) assume that the effect of viscosity is small. This means that they can be applied only to situations in which the ratio ϵ_m/ν is larger than, say, 30. Close to a solid wall other relations must be used. Spalding's group, for instance, takes the logarithmic velocity profile and matches it with the results of the calculation discussed above. The solution procedure for the differential equations is the same as for laminar flow (Sec. 7-12).

The turbulence parameter kL is not the only one that can be used. Various combinations k^mL^n have actually been considered. Kolmogorov used $\sqrt{k}/L$; Spalding is experimenting with k/L^2 and $k^{\frac{3}{2}}/L$. Other turbulence parameters have been used by Harlow and associates.[3] Spalding

[1] W. Rodi and D. B. Spalding, *Wärme- Stoffübertragung*, **3**:35 (1970).

[2] K. H. Ng and D. B. Spalding, *Imperial College Mech. Eng. Dept.* BL/TN/A/25, 1970.

[3] F. H. Harlow and P. I. Nakayama, *Phys. Fluids*, **10**:2323 (1967); also a number of Los Alamos reports.

Table 8-1 Constants in Eqs. (8-17), (8-22), and (8-23)

	C_m	C_D	C_s	C_B	C_{kL}	C_w	σ_k	σ_{kL}
Shear layers, jets	1	0.09	0.057	1	−0.7	0	1	0.3
Boundary layers	1	0.1	0.055	0.84	−0.4	2.2	2	−0.4

also mentions the possibility that constants in Table 8-1 are actually functions of some dimensionless turbulence parameter or that relations for all correlations $\overline{u_i'u_j'}$ of the fluctuating velocity components should be introduced. It has already been mentioned that in certain cases it is not possible to describe the turbulent mixing process by equations containing gradients because the large-scale fluctuations, particularly, travel over considerable distances. In summary, however, this new development, made possible by the availability of electronic computers, can be expected to make a decided improvement in our ability to predict turbulent flow and heat exchange.

The balance of turbulent kinetic energy $v_i'^2/2$ in boundary-layer and tube flow has also been studied experimentally in considerable detail. In Fig. 8-5, which presents a compilation of results by Townsend, it can be observed that turbulent production and dissipation greatly increase as the wall is approached. Maximum production actually occurs in a region beyond the one for which results are shown in Fig. 8-5, namely, at a dimensionless distance $y^+ \sim 12$. In the region very close to the wall, convection and diffusion of energy again become important.

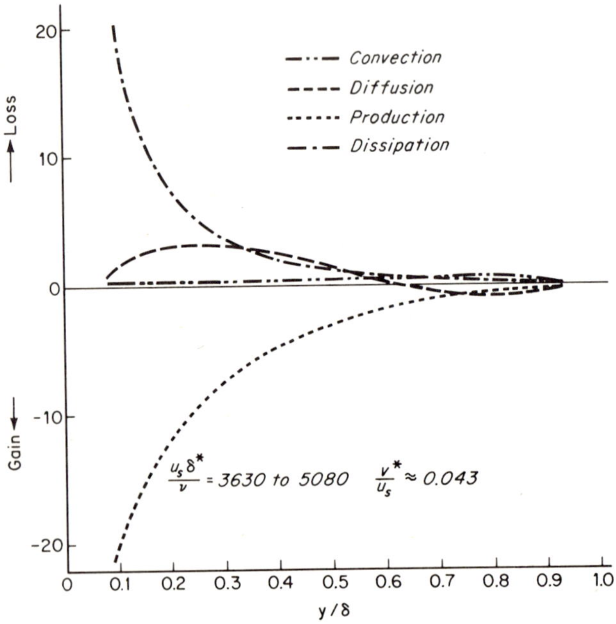

Fig. 8-5 Contributions to the turbulent energy in a boundary layer. [*Data from A. A. Townsend, "The Structure of Turbulent Shear Flow," p. 234, Cambridge University Press, London, 1956.*]

Space does not permit discussion of the results of recent turbulence research in more detail.[1]

We now demonstrate how the universal profile can be used to calculate the pressure drop for turbulent flow through a pipe with diameter D and with smooth walls in the developed flow region, with the stipulation that it describes the velocity field up to the tube axis. A Reynolds number is customarily defined for developed pipe flow by the relation

$$\mathrm{Re}_D = \frac{\bar{u}D}{\nu}$$

where $\bar{u}$ now indicates the velocity averaged over the cross section. Introducing the dimensionless parameters (indicated by a superscript plus)

$$\bar{u}^+ = \frac{\bar{u}}{\sqrt{\tau_w/\rho}} \qquad R^+ = \frac{R\sqrt{\tau_w/\rho}}{\nu}$$

with R the tube radius, one obtains

$$\mathrm{Re}_D = 2\bar{u}^+R^+ \tag{8-25}$$

The average velocity $\bar{u}$ is defined by

$$\bar{u} = \frac{2}{R^2}\int_0^R ur\,dr = \frac{2}{R^2}\int_0^R u(R-y)\,dy$$

or

$$\bar{u}^+ = \frac{2}{R^{+2}}\int_0^{R^+} u^+(R^+ - y^+)\,dy^+ \tag{8-26}$$

The universal velocity profile in Fig. 8-2 or Eqs. (8-13) and (8-15) can now be used to evaluate the right-hand side of the equation for any assumed value of R^+. The friction factor f is defined by the relation

$$\Delta p = f\frac{L}{D}\rho\frac{\bar{u}^2}{2} \tag{8-27}$$

Expressing the pressure drop Δp over the length L by the wall shear with the momentum equation gives

$$\Delta p\,\frac{D^2\pi}{4} = \tau_w D\pi L$$

or

$$\tau_w = \tfrac{1}{8}f\rho\bar{u}^2 \tag{8-28}$$

Introducing the dimensionless parameters results in

$$f = \frac{8}{\bar{u}^{+2}} \tag{8-29}$$

[1] Excellent summaries are presented in J. O. Hinze, "Turbulence," McGraw-Hill, New York, 1959, and in J. C. Rotta, Turbulent Boundary Layers in Incompressible Flow, *Progr. Aeron. Sci.*, **2**:1–219 (1962).

The friction factor is therefore obtained from Eqs. (8-29) and (8-26) as function of R^+. Equation (8-25) determines Re_D as function of R^+, and elimination of R^+ finally results in the friction factor as a function of the Reynolds number. Prandtl found that the friction factor obtained in this way can be well approximated by the equation

$$\frac{1}{\sqrt{f}} = 2.0 \log \left(\mathrm{Re}_D \sqrt{f}\right) - 0.8 \tag{8-30}$$

A simpler equation which approximates experimental data well up to a Reynolds number of 10^5 is Blasius' relation

$$f = \frac{0.316}{(\mathrm{Re}_D)^{\frac{1}{4}}} \tag{8-31}$$

It can be obtained by an analysis[1] similar to the one presented above when the velocity profile is approximated by the equation $u = Cy^{\frac{1}{7}}$. For laminar tube flow the equation

$$f = \frac{64}{\mathrm{Re}_D} \tag{8-32}$$

was found in Sec. 7-9.

Figure 8-6 compares friction factors calculated with these equations with experimental results.

For rough tubes similar equations have been developed in which the degree of roughness enters as a new parameter.[2] An inherent difficulty in investigations of the influence of surface roughness is the fact that no satisfactory geometric description of a rough surface by a limited number of parameters has yet been found. It is generally assumed that the most important parameter is the ratio of the average height of the roughness elements to the tube diameter. In his extensive experiments on friction in tubes with rough surfaces, Nikuradse[3] produced a defined roughness pattern by gluing sand of fairly uniform size to the tube surface to form a cover which was made as dense as possible. Friction factors determined in this way are plotted in Fig. 8-6 with the Reynolds number on the abscissa and with R/k_s as parameter (k_s is the average roughness height; R is the tube radius). It can be seen that in laminar flow and in turbulent flow with small Reynolds numbers, the roughness has no influence on friction. The tube wall in this range is said to be *hydraulically smooth*. This fact can be visualized as resulting from a situation in which the roughness elements are completely embedded in

[1] See E. R. G. Eckert, "Introduction to Heat and Mass Transfer," p. 91, McGraw-Hill, New York, 1963.

[2] H. Schlichting, "Boundary Layer Theory," 6th ed., McGraw-Hill, New York, 1968.

[3] J. Nikuradse, *NACA Tech. Mem.*, 1292, 1950.

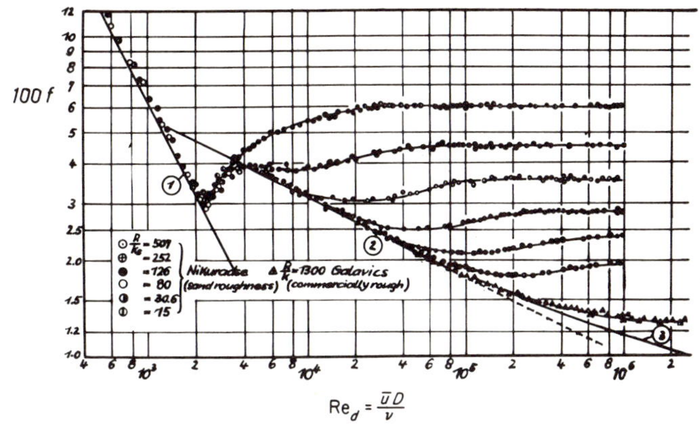

Fig. 8-6 Friction factors for turbulent flow through tubes with smooth and rough walls: (1) Eq. (7-101); (2) Eq. (8-31), (3) Eq. (8-30). [*From H. Schlichting, "Boundary Layer Theory," 4th ed., Copyright 1968. McGraw-Hill Book Company. Used by permission.*]

the viscous sublayer. For sufficiently large Reynolds numbers, the friction-factor curves become horizontal, indicating that the pressure drop in this range increases in proportion to the square of the mean velocity. In this range, local separation on the roughness elements constitutes the main contribution to the pressure drop. Investigations of other roughness geometries have revealed that the friction curves often differ in shape from those in Fig. 8-6. Measurements on commercially rough tubes, which have been summarized by Moody,[1] indicate that the friction factor on such surfaces decreases monotonically toward a constant value with increasing Reynolds number and that the dip apparent in the sand roughness curves in Fig. 8-6 is missing. Such a curve is shown in Fig. 8-6 as the line representing the measurements by Galavics.

The wall shear exerted by a turbulent boundary layer on a flat plate with a smooth surface at constant pressure or the friction factor c_f expressing this shear according to the equation

$$\tau_w = c_f \rho \frac{u_s^2}{2} \tag{8-33}$$

can be obtained by a calculation similar to the one for tube flow starting with the integrated momentum equation

$$\tau_w = \rho \frac{d}{dx} \int_0^\delta u(u_s - u)\, dy \tag{8-34}$$

[1] L. F. Moody, *Trans. ASME.*, **66**:671 (1944).

if one is again satisfied to accept the universal velocity profile as a sufficient approximation for the complete boundary layer. The details of such a calculation can, for instance, be found in a paper by Sparrow, Eckert, and Minkowycz.[1] Analytical expressions which approximate the friction factor for a turbulent boundary layer on a smooth wall with zero pressure gradient have been developed by various investigators. The following ones may be mentioned.

The equation by Schultz-Grunow[2]

$$c_f = \frac{0.370}{(\log \mathrm{Re}_x)^{2.584}} \tag{8-35}$$

for the local friction factor and the relation by Prandtl and Schlichting[3]

$$c_{fm} = \frac{0.455}{(\log \mathrm{Re}_x)^{2.58}} \tag{8-36a}$$

and by Kármán and Schoenherr[4]

$$\frac{1}{\sqrt{c_{fm}}} = 4.13 \log (\mathrm{Re}_x \, c_{fm}) \tag{8-36b}$$

for the average friction factor. Both neglect the laminar portion of the boundary layer upstream from the point of transition to turbulence and are therefore useful only for sufficiently high Reynolds numbers. The following relation includes the contribution of the laminar boundary layer assuming that the transition to turbulence occurs at a critical Reynolds number 5×10^5 but is restricted to moderate Reynolds numbers (below 10^7)

$$c_{fm} = \frac{0.074}{(\mathrm{Re}_x)^{1/5}} - \frac{1{,}700}{\mathrm{Re}_x} \tag{8-37}$$

This relation was obtained by Prandtl with an expression which makes the velocity in the turbulent shear layer proportional to the one-seventh power of the wall distance.[5]

Figure 8-4 demonstrates that the universal velocity profile cannot be used to describe the velocity field in the outer part of the boundary layer when the pressure varies along the surface. The integrated momentum equation (8-34) for this purpose must also be changed to include the

[1] E. M. Sparrow, E. R. G. Eckert, and W. J. Minkowycz, *J. Appl. Mech.*, **30**:37–43 (1963).

[2] F. Schultz-Grunow, *Luftfahrtforschung*, **17**:239 (1940).

[3] See Schlichting, *op. cit.*, p. 602.

[4] K. E. Schoenherr, *Trans. Soc. Naval Arch. Marine Engr.*, **40**:279 (1932).

[5] For a description of this analysis, the reader is referred by Eckert, *op. cit.*, p. 91.

effect of a pressure gradient. It can be written in the form

$$\rho u_s^2 \frac{d\delta_i}{dx} + \rho u_s \delta_i \left(2 + \frac{\delta^*}{\delta_i}\right) \frac{du_s}{dx} = \tau_w$$

with δ_i indicating the momentum thickness and δ^* the displacement thickness of the boundary layer.[1] Relations derived from experiments for the ratio $H = \delta^*/\delta_i$ and for the shear stress τ_w are needed to solve this equation, and such expressions have been proposed by a number of investigators.[2] It is found, however, that with this information, one can only interpolate between the situations for which the experimental information was obtained. With the availability of electronic computers, it is more profitable to turn to the time-averaged boundary-layer equations (8-5) and solve them. The turbulent stresses appearing in the equation must be expressed either by relations for the turbulent diffusivity, e.g., Eqs. (8-9), (8-10), (8-15), (8-16), or through the turbulent energy [Eq. (8-17)]. Intensive research may well produce more accurate and general expressions in the near future. A detailed description of the procedure for solving the parabolic differential equations (8-5) has been presented by Spalding and Patankar.[3] Their book also contains comparisons between calculated and measured flow parameters and the computer program in Fortran language.

8-2 ENERGY EQUATION AND HEAT TRANSFER IN TURBULENT SHEAR FLOW

To calculate heat transfer in turbulent flow, the energy equation will be derived for the time-averaged temperature field. The consideration leading to this equation is the same as that previously applied to the momentum equation.

The general energy equation for unsteady flow as derived in Sec. 6-3 should again hold for turbulent flow as well as for laminar flow and should describe the temperature of the fluid particles as they move through their tortuous paths. Equation (6-28) is repeated here for convenience:

$$\rho \frac{Di^\circ}{d\tau} - \frac{\partial p}{\partial \tau} = \frac{\partial}{\partial x_i}\left(k \frac{\partial t}{\partial x_i}\right) + \frac{\partial}{\partial x_j}(v_i p_{ji}) - \rho v_i g_i$$

In the present discussion we restrict ourself to low-velocity flow of constant-property fluids with negligible body forces. Accordingly, the tem-

[1] For its derivation see, for instance, *ibid.*

[2] For example, A. E. von Doenhoff and N. Tetervin, *NACA Rept.* 772, 1943.

[3] D. B. Spalding and S. V. Patankar, "Heat and Mass Transfer in Boundary Layers," Morgan-Grampian, London, 1967.

perature increase caused by the action of viscous forces is negligible, and the second term on the right-hand side can be dropped. It was also found that for low velocities, i.e., small Eckert number Ec, the second left-hand term can be neglected. Absence of body forces means that the last term on the right-hand side is equal to zero. At small velocities there is also practically no difference between the total enthalpy and the static enthalpy, and the latter can be written as $i = c_p t$. The energy equation then reads

$$\rho c_p \frac{Dt}{d\tau} = \frac{\partial}{\partial x_i}\left(k \frac{\partial t}{\partial x_i}\right)$$

According to the procedure initiated by Reynolds, the temperature is now written as the sum of a time-averaged value $\bar{t}$ and a fluctuating part t'. The same procedure is used for the velocity components and the pressure. Introducing these values into the energy equation and time-averaging each term results in the relation

$$\rho c_p \left(\frac{\partial \bar{t}}{\partial \tau} + \bar{v}_i \frac{\partial \bar{t}}{\partial x_i}\right) = k \frac{\partial^2 \bar{t}}{\partial x_i^2} - \rho c_p \frac{\partial}{\partial x_i} \overline{t'v_i'} \tag{8-38}$$

For a two-dimensional situation with boundary-layer character in which the extent of the velocity and temperature field in one direction is small compared to the other direction, certain terms in the above equation can again be neglected. In this way one obtains the equation

$$\rho c_p \left(\frac{\partial \bar{t}}{\partial \tau} + \bar{u} \frac{\partial \bar{t}}{\partial x} + \bar{v} \frac{\partial \bar{t}}{\partial y}\right) = k \frac{\partial^2 \bar{t}}{\partial y^2} - \rho c_p \frac{\partial}{\partial y} \overline{t'v'} \tag{8-39}$$

The first term on the right-hand side describes the heat flux into the volume element by conduction, and the second term the heat flux caused by the turbulent fluctuations. The total heat flux through a unit area normal to y, considering both contributions, is then

$$q = -k \frac{\partial \bar{t}}{\partial y} + \rho c_p \overline{t'v'}$$

Equation (8-39) can again be integrated only when more information for the term $\overline{t'v'}$ is added. Following Boussinesq, one introduces a turbulent diffusivity ϵ_H for heat by the equation

$$q = -k \frac{d\bar{t}}{dy} - \rho c_p \epsilon_H \frac{d\bar{t}}{dy} \tag{8-40}$$

It was found that calculations which assume that ϵ_H is proportional to the turbulent diffusivity ϵ_m of momentum lead to answers which for various flow situations agree with experimental results. The ratio of both diffusivities is referred to in the literature as the *turbulent Prandtl number*

$$\mathrm{Pr}_t = \frac{\epsilon_m}{\epsilon_H} \tag{8-41}$$

Reynolds suggested that the same mechanism of turbulent exchange causes transfer of momentum and of heat. This theorem, called *Reynolds' analogy,* leads to the statement $\epsilon_H = \epsilon_m$ or $\mathrm{Pr}_t = 1$.† In many analyses of heat transfer, the assumption that the turbulent Prandtl number is 1 was quite successful. In some cases the results can be improved when a value different from 1 is introduced for Pr_t. Recent measurements appear to indicate that actually the turbulent Prandtl number is not a constant and that a corresponding statement has to be considered as an approximation, which in various boundary-layer situations has led to good agreement with measured wall heat fluxes, especially when the value of the turbulent Prandtl number existing close to the wall is introduced into the analysis. This point will be discussed in greater detail later. In the wall region of the turbulent boundary layer, the actual flow situation was simplified to *simple shear flow,* characterized by $\partial u/\partial x = 0$. From the continuity equation, it then followed that $v = 0$; as a consequence, the last convective term on the left-hand side of Eq. (8-39) is zero. With the assumption that for a wall with constant temperature the convective term $u\,\partial t/\partial x$ can be neglected, the boundary-layer equation simplifies for steady flow over a wall with constant temperature to

$$q_w = -k\frac{d\bar{t}}{dy} + \rho c_p \epsilon_H \frac{d\bar{t}}{dy} \tag{8-42}$$

in which q_w indicates the heat flux at the wall surface.

An integration of Eq. (8-42) leads to a *universal temperature pvofile,* which, however, is a function of the Prandtl number Pr and the turbulent Prandtl number Pr_t, as will now be demonstrated for turbulent flow through a tube in the thermally developed regime. Introduction of a dimensionless temperature

$$T^+ = \frac{(t_w - t)\rho c_p \sqrt{\tau_w/\rho}}{q_w} \tag{8-43}$$

as well as of the previously used dimensionless velocity and wall distance

$$u^+ = \frac{u}{\sqrt{\tau_w/\rho}} \qquad y^+ = \frac{(\sqrt{\tau_w/\rho})\,y}{\nu}$$

† See, for instance, E. R. G. Eckert and R. M. Drake, "Heat and Mass Transfer," McGraw-Hill, New York, 1959.

transforms Eq. (8-42) into

$$\left(\frac{1}{\mathrm{Pr}} + \frac{\epsilon_m}{\nu\,\mathrm{Pr}_t}\right)\frac{dT^+}{dy^+} = 1 \tag{8-44}$$

Integration leads to an expression for the dimensionless temperature profile

$$T^+ = \int_0^{y^+} \frac{dy^+}{1/\mathrm{Pr} + \epsilon_m/(\nu\,\mathrm{Pr}_t)} \tag{8-45}$$

This integration can be carried out when relations for the dimensionless turbulent diffusivity ϵ_m/ν and values for Pr and Pr_t are introduced, e.g., Eqs. (8-13) and (8-15). The results of such an integration for $\mathrm{Pr}_t = 1$ are presented in Fig. 8-7. The Prandtl number of the fluid appears as parameter. It can easily be checked that for a Prandtl number and a turbulent Prandtl number equal to 1 the universal temperature profile is identical to the universal velocity profile. Those portions of the profiles which are represented by straight lines in Fig. 8-7 indicate the region in which laminar conduction can be neglected compared to the turbulent energy transport. The figure indicates that this region extends closer to the wall as the Prandtl number becomes larger, a fact which is immediately evident from Eq. (8-45). The Prandtl number has also the effect of making the profiles fuller, which indicates that more of the resistance to heat flux is concentrated in the viscous sublayer close to the wall as

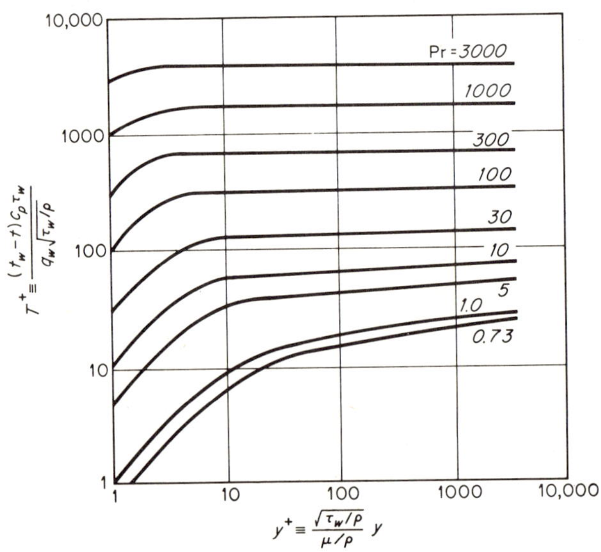

Fig. 8-7 Universal temperature profiles. [*From R. G. Deissler, High Speed Aerodyn. Jet Propulsion,* **5**:*288* (*1959*).]

the Prandtl number increases. The main interest from an engineering point of view is in the heat-transfer coefficient, defined by

$$q_w = h(t_w - t_B) \tag{8-46}$$

with t_B denoting the fluid bulk temperature. In a dimensionless form, the heat-transfer coefficient is represented by a Nusselt number

$$\mathrm{Nu} = \frac{hD}{k} = \frac{2q_w R}{k(t_w - t_B)}$$

The universal temperature profile will now be used to calculate the heat transfer to the wall of a tube in the thermally developed region. The temperature of the tube wall is specified constant, and the universal profiles will be accepted as a sufficiently accurate description of the temperature field up to the tube axis. With a dimensionless tube radius $R^+ = (R\sqrt{\tau_w/\rho})/\nu$ and with the dimensionless bulk temperature defined as

$$T_B{}^+ = \frac{(t_w - t_B)\rho c_p \sqrt{\tau_w/\rho}}{q_w}$$

the Nusselt number can be written as

$$\mathrm{Nu} = \frac{2R^+ \,\mathrm{Pr}}{T_B{}^+} \tag{8-47}$$

The bulk temperature is defined by the equation

$$t_w - t_B = \frac{\int_0^R ru(t_w - t)\,dr}{\int_0^R ru\,dr} = \frac{2\int_0^R u(t_w - t)(R - y)\,dy}{R^2\bar{u}}$$

Introducing the dimensionless values into this equation results in

$$T_B{}^+ = \frac{2}{R^{+2}\bar{u}^+}\int_0^{R^+} u^+T^+(R^+ - y^+)\,dy^+ \tag{8-48}$$

The value u^+ is described by the universal velocity profile and the value T^+ by the universal temperature profiles, both as functions of the dimensionless wall distance y^+. The integration in Eq. (8-48) can therefore be carried out for any assumed value R^+. The value $\bar{u}^+$ has been calculated before. Therefore the dimensionless bulk temperature $T_B{}^+$ is obtained in this way, usually by numerical integration. The assumed value R^+ determines a Reynolds number, according to the equation

$$\mathrm{Re} = 2R^+\bar{u}^+$$

In this way, then, the Nusselt number is found from Eq. (8-47) as a function of Reynolds number for any value of the Prandtl number. Figure 8-8*a* shows the results of such a calculation performed by Deissler.

Actually the shear and the heat flux are not constant but vary in thermally developed pipe flow downstream from the inlet region so that the shear is proportional to the distance from the tube axis, whereas the heat flux increases somewhat at first with increasing wall distance and then drops approximately linearly to the value zero at the axis. A better approximation to the actual situation is therefore found in the statement that the ratio of heat flux to shear is constant. This leads, in the region where the turbulent transport dominates, to the statement that the local temperature derivative in the r direction is proportional to the corresponding velocity derivative. The same statement is obtained from the preceding equations when both heat flux and shear are assumed constant. Deissler has also carried out calculations which take the actual variation of the fluxes into account, with the result that the influence on the Nusselt number was found to be very small. He has also shown[1] that, for very large Prandtl numbers, the integration of Eq. (8-48) can be carried out analytically with the result

$$\mathrm{Nu} = \frac{2n\sqrt{f}}{\pi}\,\mathrm{Re}\,\mathrm{Pr}^{\frac{1}{4}} \tag{8-49}$$

The constant n in this equation has the value 0.109; f indicates the friction factor. The equation shows that for large Prandtl numbers the Nusselt number varies proportional to the $\frac{1}{4}$ power of the Prandtl number. Figure 8-8b presents the variation of the Nusselt number with Prandtl number for a Reynolds number equal to 10,000 and compares it with experimental information.

Recently, considerable research has been done to determine the heat-transfer characteristics of tubes and other passages with rough surfaces because it is recognized that savings in the size of heat exchangers can be obtained by using passages with rough walls. The investigations are concerned with sand-roughened walls as well as with walls having regular discrete roughness elements like grooves or ridges. Bruck[2] has analyzed these investigations, and the following discussion, based on his findings, will be restricted to tubes with sand roughness for which the analysis was mainly based on experimental investigations by Nunner[3] and by Dipprey and Sabersky.[4]

[1] R. G. Deissler, *NACA Tech. Rept.* 1210, 1955.

[2] E. Bruck, *Wärme-Stoffübertragung*, **2**:87 (1969).

[3] W. Nunner, Wärmeübergang und Druckabfall in rauhen Rohren, *VDI-Forschungsh.* 455, 1956.

[4] D. F. Dipprey and R. H. Sabersky, Heat and Momentum Transfer in Smooth and Rough Tubes at Various Prandtl Numbers, *Intern. J. Heat Mass Transfer*, **6**:329 (1963).

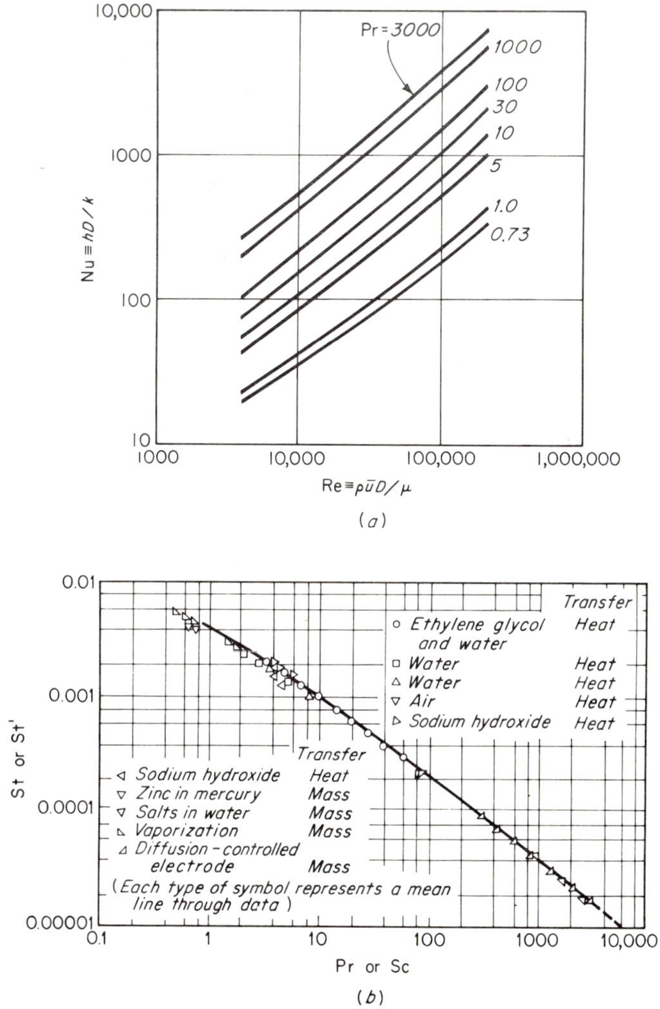

Fig. 8-8(*a*) Nusselt number in fully developed turbulent pipe flow. [*From R. G. Deissler, High Speed Aerodyn. Jet Propulsion,* **5**:*288* (*1959*).] (*b*) Stanton numbers as a function of Prandtl number or Schmidt number, respectively, for heat and mass transfer in fully developed turbulent pipe flow. [*From R. G. Deissler, High Speed Aerodyn. Jet Propulsion,* **5**:*288* (*1959*).] The Stanton number St′ for mass transfer as a function of the Schmidt number Sc is equal to the Stanton number St for heat transfer as a function of the Prandtl number Pr according to Part IV.

In the discussion of flow over a rough surface it was pointed out that the ratio of the height of the roughness element k to the viscous sublayer δ_l determines whether the flow can be considered hydraulically smooth or whether the roughness increases the wall shear. With the assumption that the velocity profile close to the wall is described by a universal relation, the following equation can be written for the viscous-sublayer thickness δ_l

$$\frac{\delta_l v^*}{\nu} = \frac{\delta_l \sqrt{\tau_w/\rho}}{\nu} = C$$

in which C denotes a constant. The ratio of the roughness height to the sublayer thickness then assumes the form

$$\frac{k}{\delta_l} = C \,\mathrm{Re}_D \frac{k}{D} \sqrt{\frac{f}{8}}$$

and the roughness parameter

$$k^+ = \mathrm{Re}_d \frac{k}{D} \sqrt{\frac{f}{2}}$$

has been found useful for the correlation of data for rough tubes.

The Nusselt number Nu and the friction factor f for the sand-rough tube can then be related to the Nusselt number Nu_0 and the friction factor f_0 of a tube with smooth wall by an effectiveness parameter

$$\eta = \frac{\mathrm{Nu}/\mathrm{Nu}_0}{f/f_0} = \frac{\mathrm{Nu}/f}{\mathrm{Nu}_0/f_0}$$

Bruck's analysis indicates that this effectiveness parameter η can be expressed as a function of the roughness parameter k^+ and the Prandtl number Pr of the fluid. Figure 8-9, which presents this relationship, can be used to calculate the Nusselt number Nu for the flow through a tube with a sand-rough wall, utilizing relations for the Nusselt number Nu_0 and the friction factor f_0 for the smooth-walled tube and obtaining the friction factor f for the rough tube from Fig. 8-6. According to Bruck, Fig. 8-9 can also be used to calculate the Nusselt number for a tube with discrete regular roughness elements provided that the friction factor f

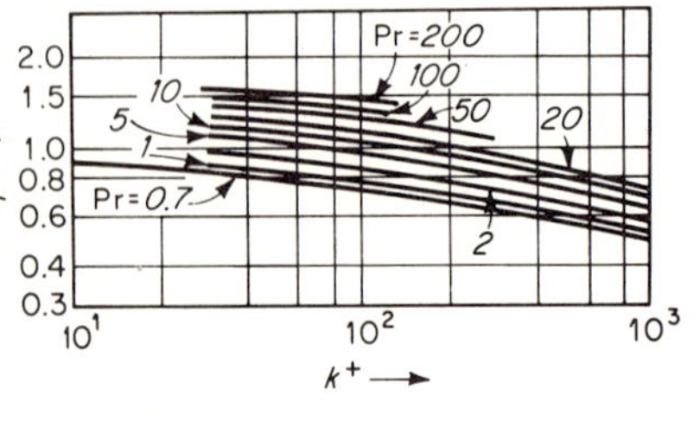

Fig. 8-9 Effectiveness parameter η for turbulent heat transfer in tubes with sand roughness as a function of the roughness parameter k^+. [*From E. Bruck, Wärme- Stoffübertragung,* **2**:87 (1969).]

for such a tube is determined by measurement at the proper value of the roughness parameter. Recent measurements, however, have cast doubt on this statement.[1]

Heat transfer to the wall of a rough tube is influenced by the contact resistance between the roughness elements and the wall. Figure 8-9 holds for a situation in which this resistance is zero.

Calculations of heat transfer through passages of arbitrary shape are usually based on the hydraulic diameter of the passage, postulating that relations for a tube with circular cross section can be used for ducts with arbitrary cross section when the tube diameter in these relations is replaced by the hydraulic diameter $d_h = 4A/C$, where A is the cross-sectional area and C the circumference. This procedure is generally satisfactory as long as the shape of the duct cross section is not extreme in the sense that it contains sharp corners, that heat transfer occurs over the major portion of the circumference, that the flow is turbulent, and that all the walls are smooth or rough. Special considerations have to be introduced when only part of the tube walls are rough.[2]

Heat-transfer analysis in a boundary layer on a smooth flat plate with constant temperature can be made in a way like that for tube flow if one is satisfied to accept the universal velocity and temperature profiles as approximations through the whole boundary layer. Equation (7-75), which holds for turbulent as well as for laminar flow, must be used for this purpose. An interesting advance which allows for a local variation in the wall temperature was made by Spalding.[3] The analysis is still based on the universal velocity profile and on the assumption that the shear stress can be considered as constant in a direction normal to the surface. It starts with the Mises form of the energy equation which for laminar flow was given as Eq. (7-125). For turbulent flow it has the form

$$\frac{\partial t}{\partial x} = \frac{\partial}{\partial \psi}\left[(\alpha + \epsilon_H)\, u \,\frac{\partial t}{\partial \psi}\right]$$

The equation is transformed to dimensionless variables. In addition to those given in the preceding pages, Spalding introduced

$$x^+ = \int_0^x \frac{1}{\nu}\sqrt{\frac{\tau_w}{\rho}}\, dx$$

together with an effective Prandtl number defined as

$$\mathrm{Pr}_e = \frac{\nu + \epsilon_m}{\alpha + \epsilon_H}$$

[1] R. L. Webb, E. R. G. Eckert, and R. T. Goldstein, *Intern. J. Heat Mass Transfer*, **14**:601 (1971).

[2] W. B. Hall, *J. Mech. Eng. Science*, **4**:287 (1962).

[3] D. B. Spalding, *Intern. J. Heat Mass Transfer*, **7**:761 (1964).

The energy equation takes the form

$$\frac{\partial t}{\partial x^+} = \frac{1}{1 + \epsilon_m/\nu} \frac{1}{u^+} \frac{\partial}{\partial u^+} \left(\frac{1}{\mathrm{Pr}_e} \frac{\partial t}{\partial u^+} \right)$$

The Stanton number as a dimensionless parameter of the local heat-transfer coefficient can be expressed in the form

$$\mathrm{St} = \frac{h}{\rho c_p u_s} = \frac{1}{\mathrm{Pr}} \sqrt{\frac{c_f}{2}}\, \mathrm{Sp}$$

The Spalding parameter Sp has been tabulated for a step change in the wall temperature

$$t_w = \begin{cases} t_s & \text{for } x < 0 \\ \text{const} & \text{for } x > 0 \end{cases}$$

and for a step change in wall heat flux

$$q_w = \begin{cases} 0 & \text{for } x < 0 \\ \text{const} & \text{for } x > 0 \end{cases}$$

Figure 8-10 presents this parameter as function of the dimensionless distance x^+. The solid lines refer to the step change in wall temperature and the dashed lines to the step change in heat flux. The analysis presents an exact solution of the time-averaged boundary-layer energy equa-

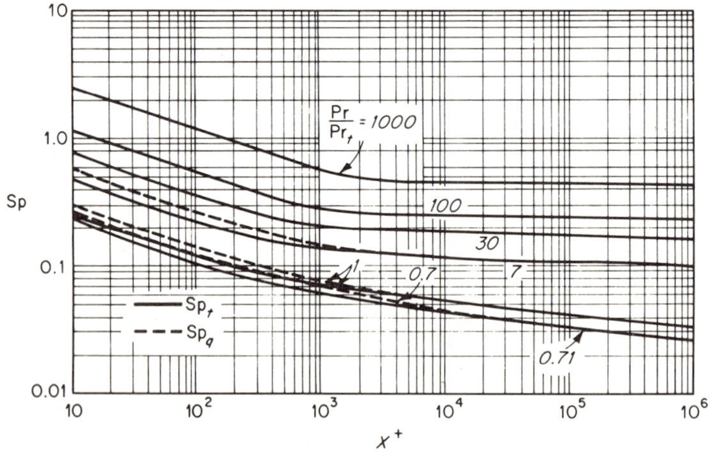

Fig. 8-10 Spalding number for heat transfer in turbulent tube flow with locally varying temperature or wall heat flux. [*From G. O. Gardner and J. Kestin, Intern. J. Heat Mass Transfer,* **6**:*289* (*1963*) *and A. G. Smith and V. L. Shale, ibid.,* **5**:*1179* (*1962*).]

tion and is useful for investigating the effect of a strong, local variation in wall temperature. For normal conditions, it is found that simpler methods[1] give results which are only slightly different. A disadvantage of the present method is that it requires a knowledge of the local friction factor c_f. The difficulties in obtaining good information on this parameter have already been discussed.

The availability of electronic computers again suggests a solution of the partial differential equation in the time-averaged variables, using for instance, Eq. (8-39). The turbulent enthalpy transport described by the last right-hand term must again be expressed through empirical relations. A constant turbulent Prandtl number Pr_t has been specified for this purpose almost universally. The calculation procedure described by Spalding and Patankar will again be found useful for such a calculation.

For engineering purposes, analytical correlations are more convenient than results of analyses or experiments presented in diagrams like Fig. 8-8. Many such correlations have been developed and are found in handbooks and in the scientific literature. As an example, two relations may be listed.

The heat transfer for thermally developed turbulent flow through a tube with circular cross section and smooth walls with constant heat flux at the tube wall is given by

$$\mathrm{St} = \frac{\mathrm{Nu}_d}{\mathrm{Re}_d\,\mathrm{Pr}} = \frac{f/2}{1.07 + 12.7\sqrt{f/2}\,(\mathrm{Pr}^{\frac{2}{3}} - 1)}$$

The friction factor f must be introduced from Eq. (8-30). Webb[2] found that this equation, developed by Petukhov and Popov, agrees with recent experiments and analysis within ± 2 percent for $0.7 < \mathrm{Pr} < 50$. Converted to an equation describing mass transfer through the heat- and mass-transfer analogy, it agrees also with experiments in the range $800 < \mathrm{Sc} < 10{,}000$ (Sc denotes the Schmidt number; see Part IV).

Heat transfer for a turbulent boundary layer on a flat plate is given by[3]

$$\mathrm{St} = \frac{\mathrm{Nu}_x}{\mathrm{Re}_x\,\mathrm{Pr}} = \frac{0.0297\,\mathrm{Re}_x^{-\frac{1}{5}}}{1 + 1.48\,\mathrm{Re}^{-\frac{1}{10}}\,\mathrm{Pr}^{-\frac{1}{6}}\,(\mathrm{Pr} - 1)}$$

Variable properties of gases can be accounted for to a first approximation by introducing the properties into the dimensionless parameters in the above equation at a reference temperature equal to the arithmetic mean

[1] Eckert and Drake, *op. cit.*, p. 178.

[2] R. L. Webb, "A Critical Evaluation of Analytical Solutions and Reynolds Analogy Equations for Turbulent Heat and Mass Transfer in Smooth Tubes." To be published in *Wärme-Stoffübertragung.*

[3] Eckert, *op. cit.*, p. 215, where it is derived from Reynolds analogy.

between the wall temperature and the fluid bulk temperature or the temperature outside the boundary layer, respectively. The customary procedure for liquids is to introduce the properties at bulk or stream temperature and to multiply the right-hand side of the above equation by $(\mu_B/\mu_w)^{0.14}$.

It has been mentioned that almost all analyses have been based on the assumption that the turbulent Prandtl number is a constant. The value of the constant, however, had to be selected according to the specific situation. For heat transfer in turbulent tube flow and in boundary layers attached to a solid surface, a value of 1 led to satisfactory agreement with measured results, whereas an analysis of the recovery temperature on a solid wall for the same flow situation had to be based on a somewhat lower value, approximately 0.8, to obtain agreement with experimental information. For free shear layers, a value of 0.5 was derived from comparison with experimental results. The above values were obtained in normal fluids excluding liquid metals.

Attempts have also been made to measure local values of the turbulent Prandtl number. Such measurements have mainly been performed in fully developed turbulent tube flow. Results for air and mercury are presented in Figs. 8-11 and 8-12. Experimental difficulties were obviously so great that no satisfactory agreement between the results obtained by various investigators could be obtained. There appears, however, to be a systematic difference between the values obtained in air and in mercury. The values for air in Fig. 8-11 are all below 1; the values for mercury are above 1. There also appears to be a trend of decreasing turbulent Prandtl number for mercury with increasing Reynolds number in the central part of the tube. This fact is puzzling because one would expect a turbulent Prandtl number; which is independent of the laminar Prandtl number provided the turbulent transport of heat is

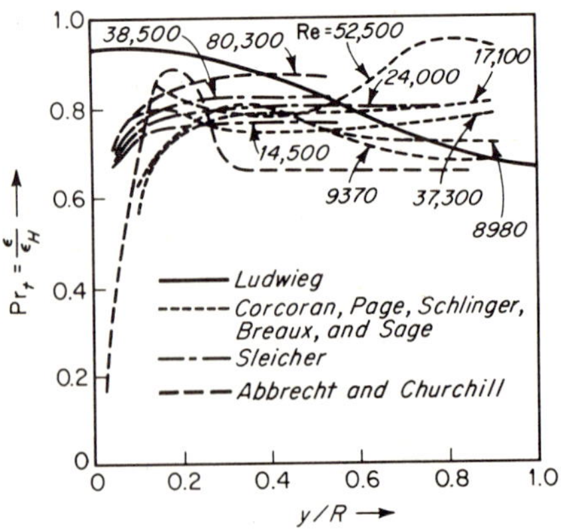

Fig. 8-11 Turbulent Prandtl number for heat transfer in flow of air through a tube. [*From E. R. G. Eckert, in D. M. McDowell and J. D. Jackson (eds.), Manchester University Press, 1970.*]

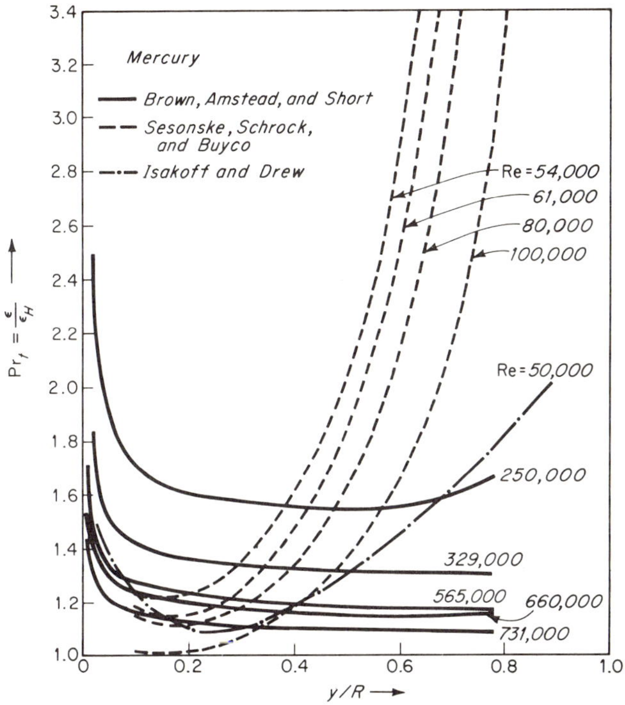

Fig. 8-12 Turbulent Prandtl number for heat transfer in flow of mercury through a tube. [*From E. R. G. Eckert, in D. M. McDowell and J. D. Jackson (eds.), "Osborne Reynolds and Engineering Science Today," Manchester University Press, 1970.*]

caused solely by the mixing of fluid particles, which is a flow process and should therefore be independent of the laminar Prandtl number for a constant-property fluid. The measured trend of the turbulent Prandtl number may indicate that the laminar and the turbulent contributions to the heat flux cannot simply be added up to obtain the total heat flux as was done in Eq. (8-40). Interactions between processes may occur and may have to be considered at least for a low Prandtl number fluid. In this connection one should also keep in mind that the description of the turbulent heat transport in Eq. (8-40) assumes it to be a local phenomenon whereas flow visualization and hot-wire measurements indicate that it occurs partially over distances which are of the same order of magnitude as the boundary-layer thickness. The considerations in this paragraph lead to the conclusion that considerable research remains to be done before an accurate and general theory of turbulent transport of momentum and heat is available.

8-3 HEAT TRANSFER IN FREE TURBULENT SHEAR LAYERS

The free shear layers connected with flow separation change to turbulence at sufficiently high Reynolds numbers. In this section, the shear layer behind a step will be considered, as shown in Fig. 7-30. The Reynolds number may be so high that the shear layer in its full extent can be considered turbulent. The analysis of a free shear layer is actually simpler than that for a boundary layer because it is not bounded by a solid surface in the neighborhood of which the laminar transport processes become important. Turbulent exchange therefore dominates throughout, and the momentum equation (8-6) simplifies for steady flow and constant pressure to

$$u \frac{\partial u}{\partial x} + v \frac{\partial u}{\partial y} = \epsilon_m \frac{\partial^2 u}{\partial y^2} \tag{8-50}$$

The turbulent diffusivity in free-shear-layer analysis is often expressed as

$$\epsilon_m = \frac{x u_s}{4\sigma^2} \tag{8-51}$$

x is the distance from the origin of the shear layer in the flow direction (Fig. 7-30), u_s is the velocity in the mainstream outside the shear layer, and σ is a constant which must be determined from experiments. The velocity on the other side of the shear layer is again assumed zero. Equation (8-51) states that the diffusivity ϵ_m is constant across the shear layer at any distance x. It is known today that this does not correspond to reality; however, the simple expression (8-51) is found to lead to fairly good agreement with the results of measurements. The momentum equation (8-50) can again be transformed by introduction of a new independent variable $\xi = \sigma y/x$ and a dependent variable F defined by the equation

$$F(\xi) = \frac{2\psi}{x u_s} \tag{8-52}$$

assumed to be a function of ξ only. In this equation ψ is the stream function which satisfies the continuity equation. The following total differential equation is obtained

$$F''' + 2\sigma^2 F F'' = 0 \tag{8-53}$$

A prime indicates differentiation with respect to ξ. The x component of the velocity is

$$u = \frac{\partial \psi}{\partial y} = \sigma \frac{u_s}{2} F' \tag{8-54}$$

The boundary conditions are therefore

$$x = +\infty \qquad F' = \frac{2}{\sigma}$$

$$x = -\infty \qquad F' = 0$$

Görtler[1] found that the solution to this equation can be approximated very satisfactorily by an error function according to the equation

$$u = \frac{u_s}{2}\left(1 + \frac{2}{\sqrt{\pi}} \int_0^\xi e^{-z^2}\, dz\right) \tag{8-55}$$

The constant σ in Eq. (8-51) was found by a comparison of the velocity profile given by Eq. (8-55) with measured ones. The value $\sigma = 13.5$ fitted the data best.

The energy equation again becomes identical with the momentum equation when the turbulent Prandtl number $\mathrm{Pr}_t = \epsilon_m/\epsilon_H$ is assumed equal to 1. The boundary conditions become identical when u/u_s is replaced by $(t - t_0)/(t_s - t_0)$. The velocity and temperature fields are therefore identical in these variables. The heat q transferred per unit area and time from the main flow through the shear layer to the separated region must be equal to the heat flux q transported by turbulent exchange through the separating streamline. Therefore

$$q = -\rho c_p \epsilon_m \left(\frac{\partial t}{\partial y}\right)_{y=0} = \rho c_p \epsilon_m (t_s - t_0) \left[\frac{d(u/u_s)}{d\xi}\right]_0 \frac{\sigma}{x}$$

From Eq. (8-55) one obtains

$$\left[\frac{d(u/u_s)}{d\xi}\right]_0 = \frac{1}{\sqrt{\pi}}$$

The heat-transfer coefficient is

$$h = \frac{q}{t_s - t_0} \frac{\rho c_p u_s}{4\sigma \sqrt{\pi}}$$

In dimensionless form it is most conveniently expressed as a Stanton number. With the value $\sigma = 13.5$, the Stanton number becomes

$$\mathrm{St} = \frac{h}{\rho c_p u_s} = 0.0105 \tag{8-56}$$

At high flow velocities, one must again consider that energy is transported by the action of turbulent shear stresses through the separating streamline.

The heat flux to the solid wall in the separated region of Fig. 7-30 must overcome a second thermal resistance posed by a boundary layer

[1] H. Görtler, *Z. Angew. Math. Mech.*, **22**:244 (1942).

at the wall surface. For Reynolds numbers which make the free shear layer turbulent, it was found by experiments performed in an airflow that almost all the thermal resistance is concentrated in the laminar sublayer at the wall surface. This is in agreement with what one would expect, since a turbulent shear layer does not have a region with vanishing turbulent transport (viscous sublayer). The temperature in the separated region was found to be essentially constant and quite close to the temperature in the free stream, approximately 90 percent of the temperature drop from the wall surface to the free stream occurring in the viscous sublayer at the wall. Heat-transfer coefficients in the separated region were found to be somewhat larger than on a surface without step, provided the shear layer was turbulent, with a maximum occurring at the reattachment point of the shear layer.[1]

Usually a boundary layer arrives at the corner in Fig. 7-30 from which the free shear layer starts. In many cases its thickness cannot be neglected. There may also be a pressure variation imposed by the mainstream on the shear layer. The analysis then has to deal again with partial differential equations, and the method of Spalding and Patankar can usefully be employed for such a calculation.

[1] R. A. Seban, *J. Heat Transfer*, **86**:259 (1964).

9

Dimensional Analysis; Forced Convection in Separated Flow

9-1 DIMENSIONAL ANALYSIS OF HEAT TRANSFER

Information on the various types of heat transfer which have been discussed up to this point could be obtained by analytical methods. For other types, however, our basic understanding of the underlying processes is still too poor to permit a calculation. This is especially true for situations where flow separation is involved. A very common and important example is the tube with circular cross section in a stream flowing normal to its axis. It has been shown in the preceding chapters that heat transfer to such a cylinder can be calculated along the upstream portion over which a boundary layer exists. The downstream part of the surface, however, borders on the separated-flow region, which at high Reynolds numbers is filled with irregular vortices, and great difficulties are encountered in the attempt to calculate convective heat transfer for this portion. The alternative possibility is to rely on experiments and to generalize the results by dimensional analysis. This method postulates that all physical processes can be expressed as a relationship among dimensionless parameters and specifies how to find those parameters. The application of dimensional analysis to heat transfer enabled Nusselt[1] in a fundamental paper published in 1915 to coordinate systematically for the first time earlier experimental results and to plan new experiments. That year can therefore be considered as marking the birth of the science of heat transfer. Today there are several methods for determining the dimensionless parameters on which a certain physical process depends. The most important and difficult part in such an analysis is not to find the form of the parameters but to find the number which fully describes the process under consideration. According to the author's experience, the most effective and reliable way of obtaining this answer is the method which starts from the differential equations and makes them dimensionless. This method will therefore be applied here. The objection may be raised that in some cases the equations governing the process may not be known. However, it will be seen later that only an approximate knowledge of the equations is required, and if this knowledge is missing, no satisfactory dimensional analysis can be carried out by any other method either.

In all the flow processes discussed in Chaps. 7 to 9 it was found that the flow parameters of interest (boundary-layer thickness, friction factor), when presented in a dimensionless form, were functions of the Reynolds number. In the same way, the heat-transfer parameters like the Nusselt or Stanton numbers and thermal-boundary-layer thickness, presented in dimensionless form, turned out to be functions of the dimensionless Reynolds and Prandtl numbers. We shall now try to obtain this information directly from the differential equations without solving them.

Let us first look at the flow process. Since we want our results to include separated-flow regions, we must use the complete conservation equations to describe this process. We shall restrict our considerations to steady flow of a constant-property fluid. In Sec. 6-2 these equations were given in cartesian coordinates. The momentum equation in the absence of body forces and the continuity equation will be repeated here:

$$\rho v_j \frac{\partial v_i}{\partial x_j} = -\frac{\partial p}{\partial x_i} + \mu \frac{\partial^2 v_i}{\partial x_j{}^2}$$

$$\frac{\partial v_i}{\partial x_i} = 0$$

These equations are four equations for the four unknowns v_i, p. In addition to the differential equations, boundary conditions must be given describing the specific flow situation. To fix our ideas, we shall be concerned with flow of a fluid normal to the axis of a cylinder of circular

[1] W. Nusselt, *Gesundh. Ingr.*, **38**:477–490 (1915).

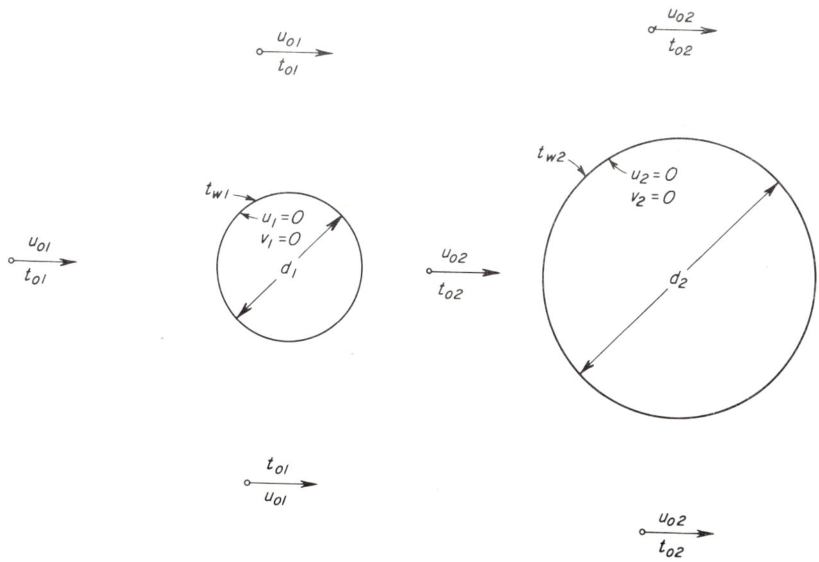

Fig. 9-1 Dimensional analysis on circular cylinders.

cross section and infinite length. Figure 9-1 is a sketch of this situation and of the necessary boundary conditions. The diameter of the cylinder is d, and the cylinder is assumed at rest. Accordingly, the velocity on its surface must be zero. The velocity upstream and on both sides at a sufficient distance from the cylinder is u_0 and is assumed constant along the stream boundaries. The pressure is also constant on the outer boundary of the flow field. Its absolute value is immaterial, since only the pressure differential appears in the above equation.

As a first step in the dimensional analysis, we shall make the differential equations and the boundary conditions dimensionless by dividing all parameters which have the dimension length by the prescribed length d and all parameters with the dimension velocity by the prescribed velocity u_0. The pressure can be made dimensionless with the term ρu_0^2 based on the constant density and the prescribed velocity u_0 on the boundary. Indicating dimensionless quantities by a prime, we write

$$x_i' = \frac{x_i}{d} \qquad v_i' = \frac{v_i}{u_0} \qquad p' = \frac{p}{\rho u_0^2} \tag{9-1}$$

Introducing the dimensionless quantities into the momentum equation results in

$$\rho \frac{u_0^2}{d} \left(v_j' \frac{\partial v_i'}{\partial x_j'} \right) = -\rho \frac{u_0^2}{d} \frac{\partial p'}{\partial x_i'} + \mu \frac{u_0}{d^2} \left(\frac{\partial^2 v_i'}{\partial x_j'^2} \right)$$

This equation is still not dimensionless but can be made so by dividing it by $\rho u_0^2/d$:

$$v'_j \frac{\partial v'_i}{\partial x'_j} = -\frac{\partial p'}{\partial x'_i} + \frac{1}{\text{Re}} \frac{\partial^2 v'_i}{\partial x'^2_j} \tag{9-2}$$

The dimensionless continuity equation is

$$\frac{\partial v'_i}{\partial x'_i} = 0 \tag{9-3}$$

and the boundary conditions are:

In the freestream, upstream of the cylinder:

$$v'_i = \frac{u_0}{u_0} = 1$$

At the cylinder surface: $\qquad v'_i = 0 \qquad$ (9-4)

The notation Re_d has been introduced for the dimensionless group $\rho u_0 d/\mu$. Now it is obvious that the solution of the differential equations will present the dimensionless dependent variables v'_i, p' as functions of the independent variables x'_i and of all constant parameters contained in the differential equations and in the boundary conditions. In our case the only constant parameter is Re. Therefore, the solution must have the form

$$v'_i = f_v(x'_i, \text{Re}) \tag{9-5}$$

$$p' = f_p(x'_i) \tag{9-6}$$

The significance of this result may need some further discussion. Consider for this purpose two cylinders with diameters d_1 and d_2 exposed to two fluid streams with upstream velocities u_{01} and u_{02}, respectively (Fig. 9-1). The above equations indicate that the dimensionless velocity components and the dimensionless pressures are for both cases the same functions of the dimensionless coordinates and of the Reynolds number. When the Reynolds number has the same value for both cases, then at similarly located points (with the same dimensionless coordinates) the dimensionless velocities and pressures will have the same values in the field around cylinders 1 and 2. Physical quantities which become identical in a dimensionless presentation—such as dimensionless values v'_i, p' plotted vs. the dimensionless location x'_i—are called *physically similar*, and we can now state: The velocity fields and pressure fields for flow around cylinders are similar for all situations regardless of cylinder diameter and upstream velocity provided the Reynolds number has the same value for all cases considered.

When we want to generalize our results to flow around or through other body geometries, it is immediately evident that physical similarity in the sense defined above can exist only for geometrically similar bodies. In addition, other conditions which have to be fulfilled to produce physical similarity may come from the velocities at the boundaries of the problem. When, for instance, the cylinder rotates with a circumferential veloity v_c at its surface, the solution of the differential equations must fulfill the condition that the velocity vector **V** is, at the cylinder surface, identical in amount and direction with the velocity v_c. In dimensionless form this boundary condition prescribes that at the cylinder surface $\mathbf{V}' = \mathbf{V}_c/u_0$, and the parameter will appear in the functional relations (9-5) and (9-6).

Therefore, it can be stated generally that the velocity and pressure fields for steady flow of fluids with constant properties are similar when geometrical similarity of the boundaries of the fields exists, when the velocities along the boundaries are similar, and when the Reynolds number has a constant value. The condition of similar velocities on the boundaries sometimes imposes restrictions. It implies, for instance, that velocity variations in time, when they occur, are similar. Such fluctuations occur in turbulent flow, and it is a well-known fact that flow parameters often depend on the degree of turbulence in the stream.

Let us now assume that the cylinder in Fig. 9-1 is kept at a temperature t_w whereas the upstream fluid has a temperature t_0. Accordingly, a temperature field will establish itself in the fluid around the cylinder which can, in principle, be calculated from the energy equation (6-29). For steady state and constant properties this equation reads

$$\rho c_p\left(v_i\,\frac{\partial t}{\partial x_i}\right) - \left(v_i\,\frac{\partial p}{\partial x_i}\right) = k\left(\frac{\partial^2 t}{\partial x_i^{\,2}}\right) + \mu\,\frac{\partial v_i}{\partial x_j}\left(\frac{\partial v_i}{\partial x_j} + \frac{\partial v_j}{\partial x_i}\right) \tag{9-7}$$

The last term on the right-hand side indicates the dissipation function. It is sufficient for dimensional analysis to consider the first term in this function, since the other terms are dimensionally identical with the first. It is immediately evident that the problem under consideration is independent of the temperature level, since only temperature differentials appear in the above equation. We can therefore measure the temperatures from an arbitrary reference value, and we shall choose the temperature t_0. Introducing $\vartheta = t - t_0$, building a dimensionless excess temperature $\vartheta' = \vartheta/\vartheta_w = (t - t_0)/(t_w - t_0)$, and introducing this parameter, together with the others defined before, into the energy equation results in

$$\frac{\rho c_p u_0 \vartheta_w}{d}\left(v_i'\,\frac{\partial \vartheta'}{\partial x_i'}\right) - \frac{\rho u_0^{\,3}}{d}\left(v_i'\,\frac{\partial p'}{\partial x_i'}\right) = \frac{k\vartheta_w}{d^2}\,\frac{\partial^2 \vartheta'}{\partial x_i'^2} + \mu\,\frac{u_0^{\,2}}{d}\,\frac{\partial v_i'}{\partial x_j'}\left(\frac{\partial v_i'}{\partial x_j'} + \frac{\partial v_j'}{\partial x_i'}\right) \tag{9-8}$$

The equation is made dimensionless by multiplying it by $d/\rho c_p u_0 \vartheta_w$:

$$v_i' \frac{\partial \vartheta'}{\partial x_i'} - \text{Ec}\left(v_i' \frac{\partial p'}{\partial x_i'}\right) = \frac{1}{\text{Re Pr}} \frac{\partial^2 \vartheta'}{\partial x_i'^2} + \frac{\text{Ec}}{\text{Re}} \frac{\partial v_i'}{\partial x_j'} \left(\frac{\partial v_i'}{\partial x_j'} + \frac{\partial v_j'}{\partial x_i'}\right) \tag{9-9}$$

The dimensionless parameter $k/\rho c_p u_0 d$ appearing in the first right-hand term can be identified as the reciprocal of the product Re Pr, and the parameter $\mu u_0/\rho c_p \vartheta_w d$ is the quotient of a parameter $u_0^2/c_p \vartheta_w$ (to which Schlichting[1] gave the name Eckert number) and of the Reynolds number. The boundary conditions are:

In the free stream, upstream from the cylinder: $\vartheta' = 0$
At the cylinder surface: $\vartheta' = 1$

In order to obtain a solution, the velocity components must be introduced into the energy equation from Eq. (9-5). The solution, if it could be obtained, would be of the form

$$\vartheta' = f\left(x_i', \text{Re}, \text{Re Pr}, \text{Ec}, \frac{\text{Ec}}{\text{Re}}\right)$$

A function of the parameters Re, Ec, Re Pr, and Ec/Re must also be expressible as a function of Re, Pr, and Ec. Therefore,

$$\vartheta' = f(x_i', \text{Re}, \text{Pr}, \text{Ec}) \tag{9-10}$$

As discussed in Sec. 6-5, for the velocities that usually occur in engineering heat transfer, the terms preceded by the parameter Ec can be neglected. Correspondingly, the dimensionless temperature field for low-velocity flow is

$$\vartheta' = f(x_i', \text{Re}, \text{Pr}) \tag{9-11}$$

With the generalizations made before, the situation is equivalent to the following statement.

The temperature fields around geometrically similar objects are similar when the temperatures and velocities around the boundaries are similar and when the Reynolds and Prandtl numbers as well as the parameter Ec have a constant value.

For engineering purposes, the heat exchange between the fluid and the cylinder wall is a very important quantity. It can be calculated as soon as the heat-transfer coefficient is known, which is defined by the equation

$$h(t_w - t_0) = -k\left(\frac{\partial t}{\partial n}\right)_w$$

[1] H. Schlichting, "Boundary Layer Theory," p. 250, McGraw-Hill, New York, 1955.

where n indicates the direction normal to the wall surface. Introducing the dimensionless temperature ϑ', one obtains

$$h\vartheta_w = \frac{k\vartheta_w}{d}\left(\frac{\partial\vartheta'}{\partial n'}\right)_w$$

or

$$\mathrm{Nu} = \frac{hd}{k} = \left(\frac{\partial\vartheta'}{\partial n'}\right)_w$$

The Nusselt number is therefore nothing more than the dimensionless temperature gradient at the surface. By differentiating Eq. (9-11) and by introducing the result into the above expression one obtains for low-velocity flow

$$\mathrm{Nu} = f(x_i',\mathrm{Re},\mathrm{Pr}) \tag{9-12}$$

where x_i' now indicate the coordinates of an arbitrary surface point and where Nu_d is the local Nusselt number at that point. The average Nusselt number of the whole surface is given by

$$\overline{\mathrm{Nu}} = f(\mathrm{Re},\mathrm{Pr}) \tag{9-13}$$

Summarizing the results obtained so far, it can be stated that in steady flow of a fluid with constant properties around geometrically similar objects or through geometrically similar ducts, all dimensionless flow parameters, e.g., friction factors, are functions of location and Reynolds number provided the velocities on the boundaries are similar. The dimensionless heat-transfer parameters like the Nusselt or Stanton numbers are functions of location, of the Reynolds number, of the Prandtl number, and, for high velocities, of the Ec parameter provided the velocities and the temperatures along the boundaries are similar.

Now we turn to a discussion of dimensional analysis for a fluid with variable properties. It will be seen later that a general treatment for such a fluid is not possible. We shall therefore start by considering flow of a gas for which the properties vary in the following way:

$$\rho = \frac{p}{RT} \qquad \mu = C_\mu T^\alpha \qquad k = C_k T^\alpha \qquad c_p = \text{const} \qquad \mathrm{Pr} = \text{const} \tag{9-14}$$

R denotes the gas constant and T the absolute temperature. It is easily verified that the exponent in the expressions for viscosity and heat conductivity must be the same when Pr and c_p are constants. Considering again the flow around a circular cylinder, we now must deal not only with velocity, pressure, and temperature fields but also with density, viscosity, and heat-conductivity fields. The latter can again be expressed conveniently in a dimensionless form as

$$\rho' = \frac{\rho}{\rho_0} \qquad \mu' = \frac{\mu}{\mu_0} \qquad k' = \frac{k}{k_0} \tag{9-15}$$

with the subscript 0 indicating the conditions at a sufficient distance from the cylinder. These fields can easily be determined from the following relations when the temperature and pressure fields are known:

$$\rho' = \frac{p'}{T'} \qquad \mu' = T'^{\alpha} \qquad k' = T'^{\alpha} \tag{9-16}$$

For a further analysis, the momentum, continuity, and energy equations for a fluid with variable properties must be known. A knowledge of the dimensional structure, however, is sufficient, and the equations are therefore abbreviated (this time in x, y, z coordinates)

$$\rho u \frac{\partial u}{\partial x} + \cdots = -\frac{\partial p}{\partial x} + \left[\frac{\partial}{\partial x}\left(\mu \frac{\partial u}{\partial x}\right) + \cdots\right] \tag{9-17}$$

$$\frac{\partial(\rho u)}{\partial x} + \cdots = 0 \tag{9-18}$$

$$\rho c_p u \frac{\partial T}{\partial x} + \cdots = \left[u \frac{\partial p}{\partial x} + \cdots\right] + \left[\frac{\partial}{\partial x}\left(k \frac{\partial T}{\partial x}\right) \cdots\right] + \left[2\mu \left(\frac{\partial u}{\partial x}\right)^2 + \cdots\right] \tag{9-19}$$

It can be seen that the equations are quite similar to those for a constant-property fluid.

The equation can be made dimensionless by introduction of the primed values:

$$\frac{p'}{T'}\left(u' \frac{\partial u'}{\partial x'} + \cdots\right) = -\frac{\partial p'}{\partial x'} + \frac{1}{\text{Re}}\left[\frac{\partial}{\partial x'}\left(T'^{\alpha} \frac{\partial u'}{\partial x'}\right) + \cdots\right] \tag{9-20}$$

$$\frac{\partial}{\partial x'}\left(\frac{p'}{T'} u'\right) + \cdots = 0 \tag{9-21}$$

$$\frac{p'}{T'}\left(u' \frac{\partial T'}{\partial x'} + \cdots\right) = \text{Ec}\left(u' \frac{\partial p'}{\partial x'}\right) + \frac{1}{\text{Re Pr}}\left[\frac{\partial}{\partial x'}\left(T'^{\alpha} \frac{\partial T'}{\partial x'}\right) + \cdots\right] + \frac{\text{Ec}}{\text{Re}}\left[2T'^{\alpha}\left(\frac{\partial u'}{\partial x'}\right)^2 + \cdots\right] \tag{9-22}$$

The boundary conditions must prescribe velocity, pressure, and temperature on the boundaries of our problem, but the temperatures cannot be expressed as measured from an arbitrary reference condition since the absolute temperature appears in the previous equations. In the dimensionless parameters the boundary conditions are:

In the free stream, upstream from the cylinder:

$$u' = 1 \qquad T' = 1 \qquad p' = 1$$

At the cylinder surface:

$$u' = v' = w' = 0 \qquad T' = \frac{T_w}{T_0} \tag{9-23}$$

(No pressure can be arbitrarily prescribed at the cylinder surface.) The solution of the system of equations with the boundary conditions will be

$$u' = f_u\left(x',y',z',\mathrm{Re},\mathrm{Pr},\mathrm{Ec},\alpha,\frac{T_w}{T_0}\right) \tag{9-24}$$

$$T' = f_T\left(x',y',z',\mathrm{Re},\mathrm{Pr},\mathrm{Ec},\alpha,\frac{T_w}{T_0}\right) \tag{9-25}$$

The parameter Ec in this case is proportional to the square of the Mach number upstream, since the velocity of sound for a gas with constant c_p is given by $a = \sqrt{\gamma RT} = \sqrt{(\gamma - 1)c_pT}$.

$$\mathrm{Ec} = \frac{u_0^2}{c_p(T_w - T_0)} = (\gamma - 1)\frac{u_0^2}{a_0^2}\frac{T_0}{T_w - T_0} = (\gamma - 1)(\mathrm{Ma})^2\frac{1}{T_w/T_0 - 1}$$

The average heat transfer to the cylinder is then described by the equation

$$\overline{\mathrm{Nu}} = f\left(\mathrm{Re},\mathrm{Pr},\mathrm{Ma},\frac{T_w}{T_0},\alpha,\gamma\right) \tag{9-26}$$

The number of parameters in the function f is already quite large, and the number of situations with physical similarity is correspondingly reduced. Actually the equations used above to describe the properties of a gas are only approximations to the actual situation. The more accurately an equation describes the actual variation of the properties, the more constants it will contain and the more parameters will appear in the dimensionless equations describing flow and heat transfer. This restricts the similarity to such a degree that in a practical sense it soon fails to exist. Only by describing the properties of a class of fluids in a dimensionless form by the same equations can similarity be maintained within this class, and common dimensionless equations be expected which describe flow and heat transfer.

9-2 FLOW SEPARATION

It has been mentioned before that flow separates from a step in the surface contour as shown in Fig. 7-30. It forms a shear layer between the main flow and the separated-flow region. At small Reynolds numbers, this shear layer and the flow in the separated region are laminar; however, they change to turbulence at a Reynolds number around 5×10^4 when the characteristic length in this number is based on the downstream distance from the step. Beyond this Reynolds number the fluid in the shear layer becomes unstable and exhibits waves, which increase in amplitude in the downstream direction, roll up into vortices, and finally break down into the statistically irregular turbulence. The flow in the separated region itself will then be turbulent or exhibit a fluctuating, unsteady character caused by the fact that the vortices existing in this region are periodically carried away with the main flow. The point of reattachment is approximately five step heights away from the step. A boundary layer will build up between the separated flow and the bounding surfaces.[1] Since these features of separated flow make an analytical approach rather difficult, our information rests mainly on experiments.

The extent and characteristics of the separated-flow region or wake in supersonic flow are strongly influenced by expansion fans and reattachment shocks arising on steps in the surface contour. On the other hand, the characteristics of boundary layers in supersonic flow are quite similar to those of subsonic boundary layers. As a consequence, much of the information obtained in subsonic boundary-layer flow can be transferred to supersonic flow. The main difference is caused by the fact that the heat generated by viscous dissipation cannot be neglected in the energy equation and that of necessity large temperature differences exist, so that the variation of the properties must be taken into account. Both these effects will be studied in the following sections. On the other hand the interaction of expansion fans and shock waves with shear layers and boundary layers, which was mentioned above, makes heat transfer in separated regions of supersonic flow quite different from heat transfer in separated regions of subsonic flow.

Flow separation also occurs on the downstream side of blunt objects. As examples, the flow over a cylinder and a sphere will be discussed. These geometries occur frequently in engineering applications and have been extensively investigated in a large Reynolds number range. The influence of Reynolds number on the flow will therefore be discussed at this time. Qualitatively, the flow field is influenced by the Reynolds number in the same way for other body shapes.

The discussion will at first be concerned with a cylinder of circular cross section and of infinite length exposed to a flow normal to its axis. At Reynolds numbers below 1, no separation occurs, and in moving around the cylinder the fluid closes in at the downstream stagnation point. The shape of the streamlines, however, is different from those in an inviscid fluid. The effect of viscosity causes the streamlines to move further apart on the downstream side than on the upstream side of the cylinder. With increasing Reynolds number, the flow starts to separate, creating on the downstream side of the cylinder a separation bubble in which the fluid circulates in a very regular and steady manner. The size of the separation bubble increases with increasing Reynolds number. Analyses of this flow situation have been performed by a numerical solution of the complete Navier-Stokes equations. At a Reynolds number

[1] R. A. Seban, *J. Heat Transfer*, **86**:259 (1964).

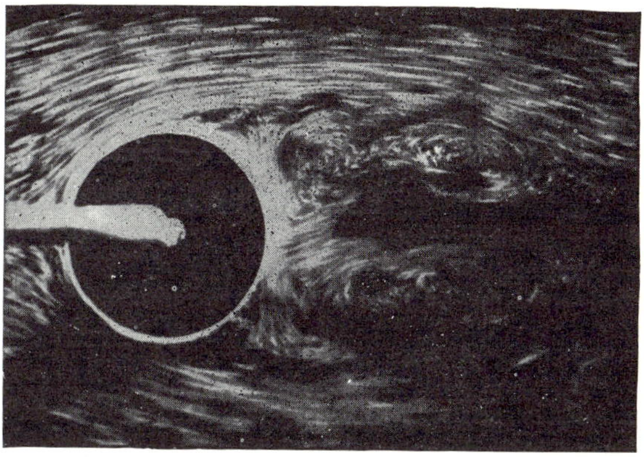

Fig. 9-2 Flow around a cylinder. [*Photograph by H. L. Rubach, Mitt. Forschungsarb. 185, 1916.*]

of about 100, the vortices in the separation bubble start to separate alternately from the cylinder surface and to move downstream, creating in this way the well-known Kármán vortex street. Figure 9-2 is an instantaneous exposure photograph of such a flow situation. The effect of viscosity extends far into the flow field at Reynolds numbers below 1. The affected region decreases in thickness gradually as the Reynolds number increases and at a Reynolds number of approximately 1,000 becomes sufficiently small for the boundary-layer simplification to be applied. The reason for the flow separation in the higher Reynolds number region can be understood by considering the pressure distribution around the cylinder surface. The dashed-dotted line in Fig. 9-3 presents the pressure distribution generated by an inviscid fluid which does not separate from the cylinder surface. The pressure decreases at first from the stagnation point ($\alpha = 0$) and reaches its minimum value at $\alpha = 90°$. On the rearward side it increases again and reaches, at $\alpha = 180°$, the same value as at the forward stagnation point. The same pressure distribution occurs in a fluid with friction during the early period when the flow is started from rest. The viscosity, however, causes the formation of a boundary layer which increases in thickness with time. The fluid particles outside the boundary layer are able to move against the pressure increase on the downstream side of the cylinder by changing their kinetic energy into pressure energy. The fluid particles in the boundary layer, however, have lost part of their energy by internal friction. They can therefore move only a certain distance into the region of increasing pressure before their kinetic energy is consumed. Then they reverse their flow direction. In this way the flow separates a short time after the start from rest, and this separation persists when a steady state has been reached.

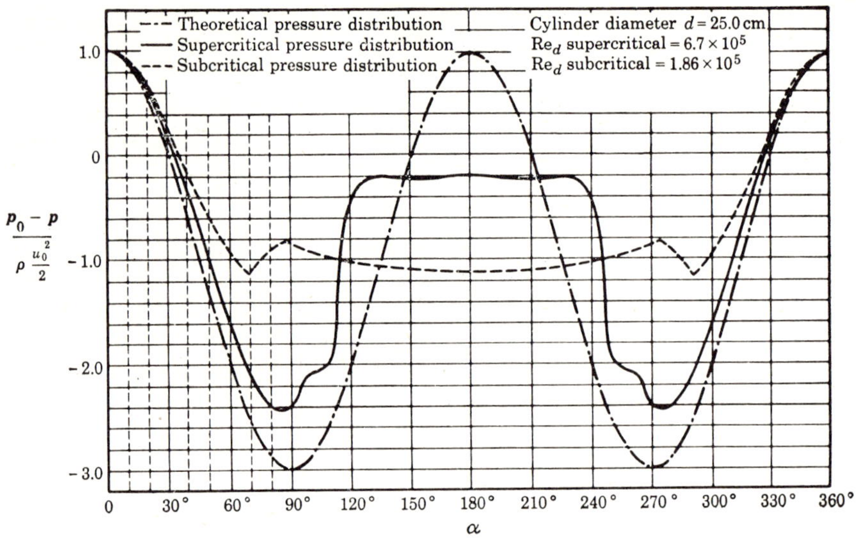

Fig. 9-3 Pressure distribution around the circumference of a circular cylinder: p = local pressure, p_0 = upstream pressure, u_0 = upstream velocity, ρ = density. [*From L. Flachsbart, in W. Wien and F. Harms (eds.), "Handbuch Experimental Physik," vol. 4, part II, p. 316, Akademische Verlagsgesellschaft, Leipzig, 1932.*]

Because of this separation, the pressure distribution along the back side of the cylinder is also changed. How far the fluid particles penetrate the region of increasing pressure depends on their kinetic energy. This is on an average greater in the turbulent boundary layer than in the laminar one. As a consequence, the turbulent boundary layer separates at approximately $\alpha = 110°$ and the laminar boundary layer at $\alpha = 82°$. These two flow situations are referred to as *subcritical* and *supercritical*. The corresponding pressure distributions are shown in Fig. 9-3. The pressures on the rear side are smaller than on the front. This gives a force in the flow direction called *form resistance*. To it is added a frictional resistance caused by the shear stresses along the cylinder walls. Both parts are included in the drag coefficient f_c defined by the equation

$$D = f_c L d \rho \frac{u_0^2}{2} \tag{9-27}$$

where d = diameter
L = length
D = drag of cylinder
ρ = density
u_0 = free-stream velocity
Ld = cross section normal to air flow

Figure 9-4 shows the drag coefficient f_c for various Reynolds numbers. The Reynolds number is calculated with the free-stream velocity u_0 and the diameter d. At very small free-stream velocities no separation occurs, and the drag is caused predominantly by shear stresses. From $Re_d \approx 100$ vortices separate alternately from the left and right sides (Kármán's vortex street) and are carried away by the flow in a regular pattern. The frequency n with which they peel off is described by the *Strouhal number* nd/u_0, which for Reynolds numbers greater than 200 has a value 0.21. The sudden decrease in drag at $Re_d \approx 4 \times 10^5$ is caused by the fact that the boundary layer has become turbulent ahead of the separation point. At larger Reynolds numbers, the vortex formation is irregular and breaks up into small vortices in the wake. At low Reynolds numbers, the drag of the cylinder is mostly frictional resistance and at high Reynolds numbers (say, larger than 1,000) mostly form resistance. Each of these flow regimes influences the heat transfer in a distinct way.

The development of a laminar or turbulent boundary layer along the front portion of the cylinder can be calculated by the methods presented in the preceding chapters. An analysis of the wake formation downstream from the cylinder is more involved since it requires a solution of the complete unsteady Navier-Stokes equations. Today's electronic computers, however, make it possible to obtain numerical solutions to such problems. Figure 9-5, an example of results which can be obtained in this way, considers vortex formation behind a bar with rectangular cross section which is indicated on the left-hand rim of the figure. The lines are streamlines. They demonstrate the vortex formation very

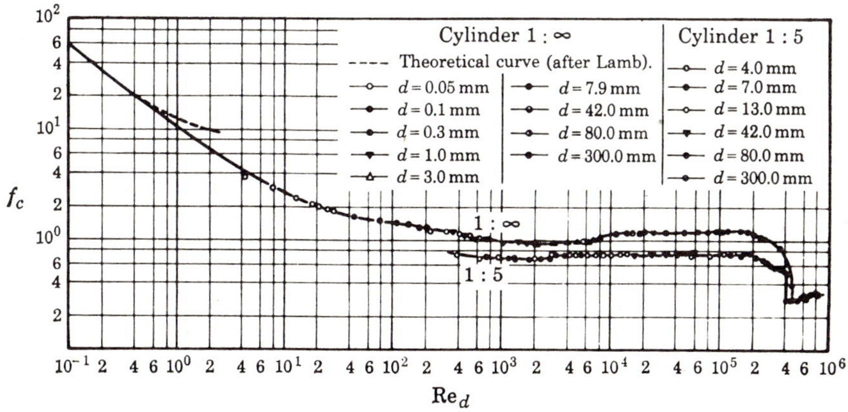

Fig. 9-4 Drag coefficient of a circular cylinder in a flow normal to its axis. [*From C. Wieselsberger, Ergeb. Aerodyn. Versuchsanstalt Göttingen,* **2**:*24* (*1932*).]

clearly. The flow is limited by two plain walls indicated by the horizontal lines. Slip was allowed to exit along the walls of the bar and the flow channel. The vortex was therefore created at the corners of the bar.

Rotationally symmetric flow around any immersed object exhibits essentially the same characteristic behavior as flow around cylindrical objects. Flow around a sphere, for instance, again shows a separation on the sides of the sphere and a wake region filled with vortices in the back. This separation occurs at larger Reynolds numbers, whereas for very low Reynolds numbers (below 1) laminar flow closes around the rear portion of the sphere. The drag connected with this type of flow is described by Stokes's law:

$$D = 3\pi d\mu u_0 \tag{9-28}$$

The drag coefficient of a sphere for a large Reynolds number range is indicated in Fig. 9-6. This coefficient is defined in the same way as that for a cylinder. Again a characteristic drop is observed at a Reynolds number around 3×10^5. It was found that the Reynolds number at which the drop in drag occurs for a sphere with smooth surface depends on the turbulence in the free stream because the turbulence level determines whether the boundary layer ahead of the separation point is laminar or turbulent. This connection between the turbulence level in the free stream and the critical Reynolds number indicated by the drop in the drag of a sphere was interpreted correctly by Prandtl. It makes a sphere useful for measuring turbulence in an airstream, for instance, in wind tunnels. Today this method has been almost universally replaced

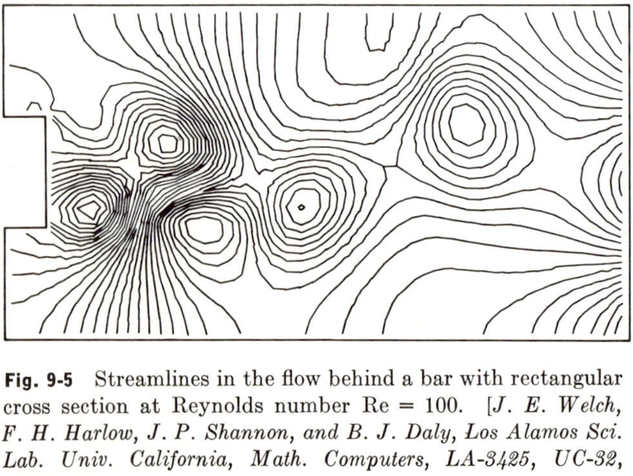

Fig. 9-5 Streamlines in the flow behind a bar with rectangular cross section at Reynolds number Re = 100. [*J. E. Welch, F. H. Harlow, J. P. Shannon, and B. J. Daly, Los Alamos Sci. Lab. Univ. California, Math. Computers, LA-3425, UC-32, TID-4500, 1966.*]

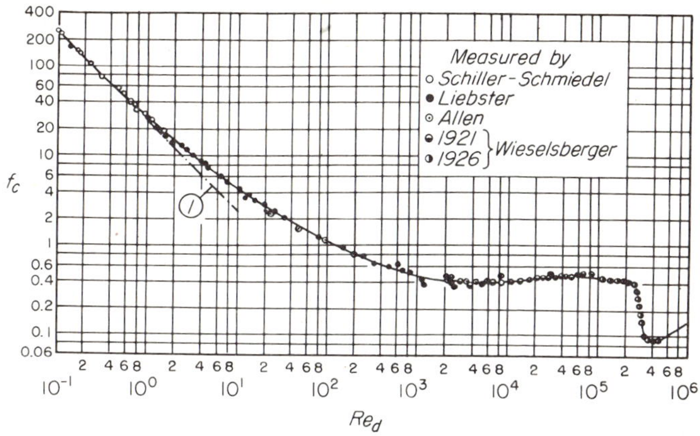

Fig. 9-6 Drag coefficient of a sphere: (1) Eq. (9-28). [*From H. Schlichting, "Boundary Layer Theory," 6th ed. Copyright 1968. McGraw-Hill Book Company. Used by permission.*]

by a direct measurement of the turbulent velocity fluctuations (u',v',w') with the hot-wire anemometer.

The development of a boundary layer around the front portion of a rotationally symmetric object can be handled, with small modifications, by the methods described for two-dimensional boundary layers in the previous paragraph. Very useful, also, is a transformation, pointed out by Mangler, which is discussed on page 291.

9-3 CYLINDERS AND TUBE BUNDLES IN CROSSFLOW

The flow around a cylinder, as discussed in Sec. 9-2, influences the temperature field in the fluid created when the cylinder surface is kept at a temperature different from the value in the approaching stream. Figure 9-7 compares the temperature field as obtained by an interference photograph with the temperature field obtained by a solution of the Navier-Stokes and the energy equations, both at a low Reynolds number. The lines in both figures are lines of constant temperature. The agreement is quite satisfactory. The asymmetry in the wake behind the cylinder in Fig. 9-7*b* is caused by free-convection effects which have not been considered in the analysis. Figure 9-8 shows the temperature field at a Reynolds number of 1,260. It can be observed that the thickness of the thermal boundary layer is considerably less than in the preceding figure. The point of separation has moved upstream.

The distribution of the local Nusselt numbers at low, intermediate,

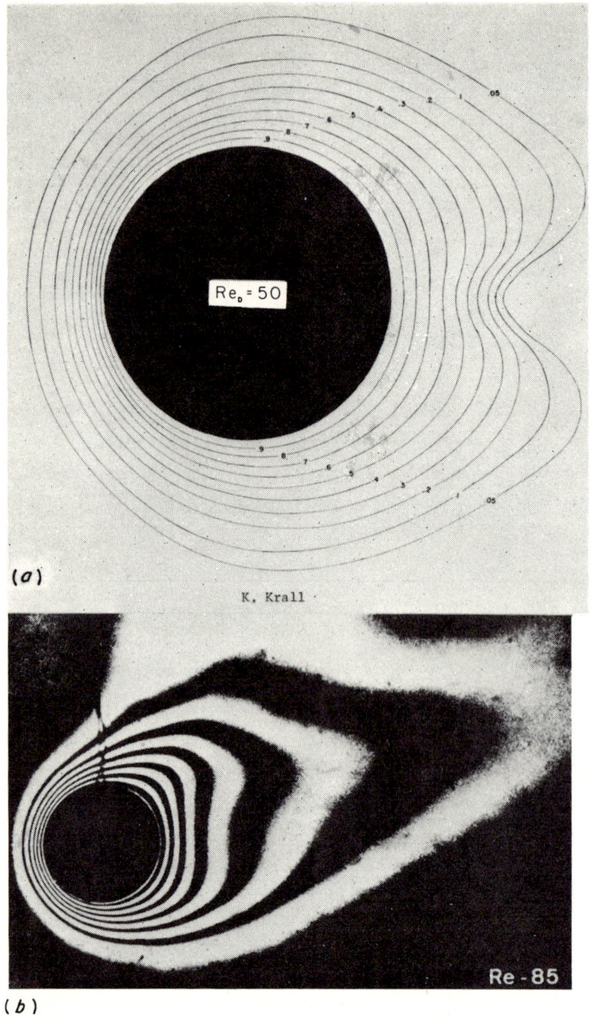

Fig. 9-7 Isotherms around a cylinder cooled by an airstream normal to its axis. (*a*) Result of analysis at Re = 90. (*K. M. Krall, Local Heat Transfer around a Transverse Circular Cylinder in Slip Flow, University of Minnesota, Ph.D. Thesis, 1969.*) (*b*) Interferogram at Re = 85. [*E. R. Eckert and E. Soehngen, Trans. ASME,* **74**:*343* (*1952*).]

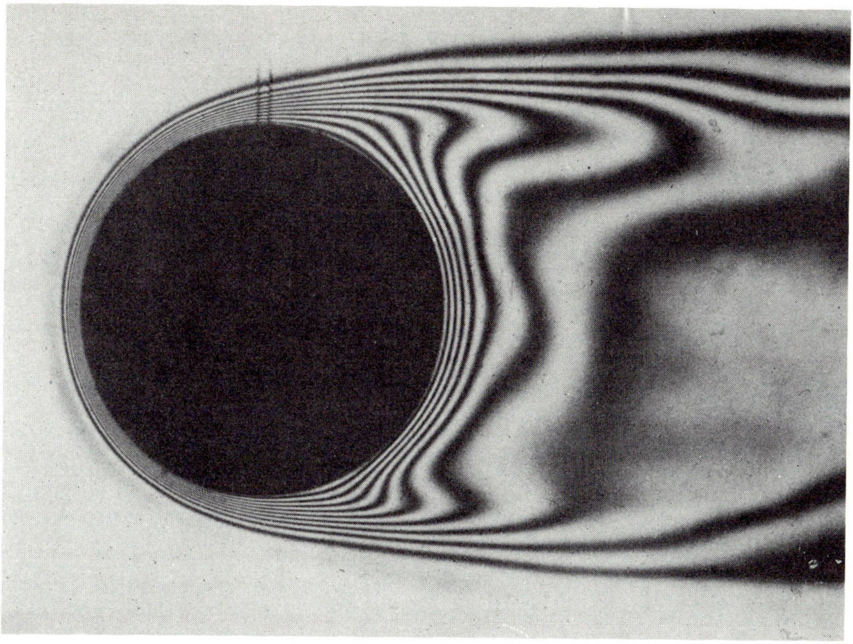

Fig. 9-8 Isotherms around a cylinder cooled by an airstream normal to its axis as revealed by an interference photograph. Re = 1,260. [*Photograph by E. R. G. Eckert and E. Soehngen, Trans. ASME,* **74**:*313* (*1952*).]

and high Reynolds numbers is presented in Fig. 9-9*a* to *c*. The local heat transfer is very small on the downstream part of the cylinder at the smallest Reynolds number. Beyond a Reynolds number of around 50,000 the heat transfer actually becomes larger on the downstream half than on the front half of the cylinder surface. The local distribution changes characteristically when the boundary layer becomes turbulent, as shown by the figure for Re_d = 426,000. The figure also indicates that the point of separation has now shifted to a value beyond 90°.

For engineering calculations, the main interest is in the total heat flow from or to a tube, or, what is equivalent, in the heat-transfer coefficient averaged over the circumference. Hilpert[1] was the first to report the results of accurate measurements for airflow in a wide range of Reynolds numbers. Figure 9-10 presents the results. The Nusselt and Reynolds numbers are calculated with the tube diameter as reference length and with the free-stream velocity as reference velocity. Many subsequent investigations have essentially verified Hilpert's results except in the range of Reynolds numbers below 600, where recently measured Nusselt numbers are approximately 30 percent lower than those measured

[1] R. Hilpert, *Forsch. Gebiete Ingenieurw.*, **4**:215 (1933).

by Hilpert. Zhukauskas et al.,[1] who made an extensive survey of available information for heat transfer to tubes in crossflow, recommend the following equations:

At $\mathrm{Re}_f = 1$ to 1×10^3:†

$$\mathrm{Nu} = (0.43 + 0.50\,\mathrm{Re}^{0.5})\,\mathrm{Pr}^{0.38}\left(\frac{\mathrm{Pr}_f}{\mathrm{Pr}_w}\right)^{0.25} \tag{9-29}$$

At $\mathrm{Re} = 1 \times 10^3$ to 2×10^5:

$$\mathrm{Nu} = 0.25\,\mathrm{Re}^{0.6}\,\mathrm{Pr}^{0.38}\left(\frac{\mathrm{Pr}_f}{\mathrm{Pr}_w}\right)^{0.25} \tag{9-30}$$

The ratio $\mathrm{Pr}_f/\mathrm{Pr}_w$ appearing in these equations expresses the influence of property variations on heat transfer. If this ratio of Prandtl numbers deviates markedly from a value of 1, properties appearing in Nu, Re, and Pr have to be introduced at the fluid bulk temperature. The method

[1] A. A. Zhukauskas, V. Makarevicius, and A. Shlanciauskas, "Heat Transfer in Banks of Tubes in Crossflow of Fluid," Mintis, Vilnius, Lithuania, 1968.

† The term 0.43 was added by the author of this chapter to extend the range to small Reynolds numbers according to E. R. G. Eckert and E. Soehngen, *Trans. ASME*, **74**:343 (1952).

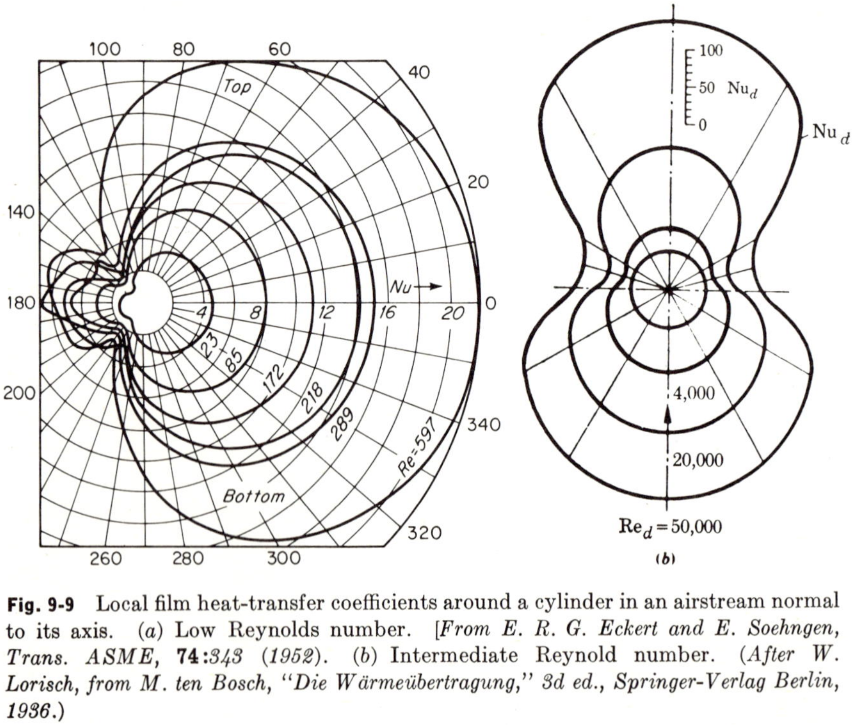

Fig. 9-9 Local film heat-transfer coefficients around a cylinder in an airstream normal to its axis. (*a*) Low Reynolds number. [*From E. R. G. Eckert and E. Soehngen, Trans. ASME,* **74**:*343* (*1952*). (*b*) Intermediate Reynold number. (*After W. Lorisch, from M. ten Bosch, "Die Wärmeübertragung," 3d ed., Springer-Verlag Berlin, 1936.*)

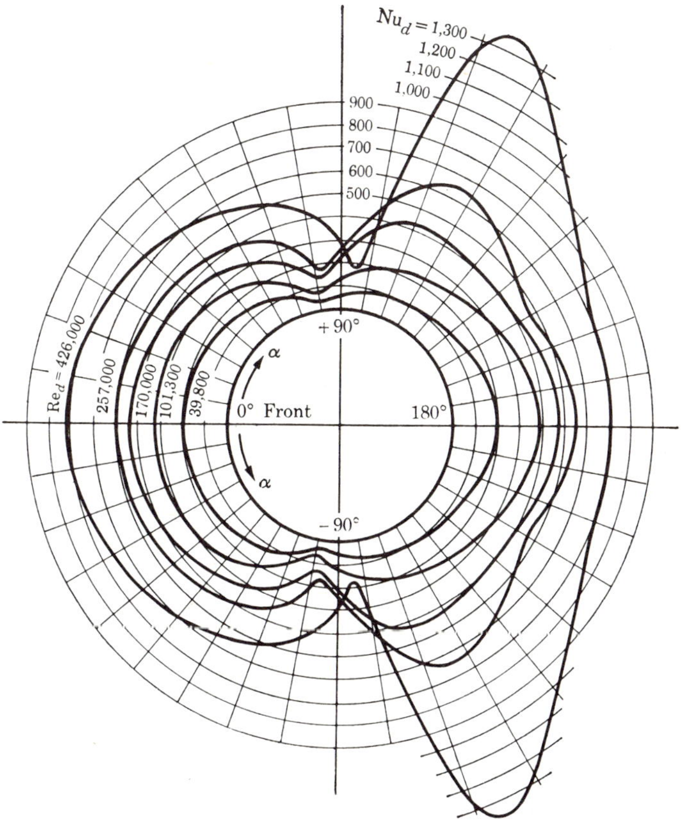

(c)

Fig. 9-9 (c) High Reynolds number. [*From E. Schmidt and K. Wenner, Forsch. Gebiete Ingenieurw.*, **12**:*65* (*1941*).]

of handling a variation of properties as proposed by the above equations is probably best suited for liquids. For gases another method is suggested by Douglas and Churchill,[1] derived from experiments at gas temperatures between 15 and 1000 C and wall temperatures between 20 and 1100 C. On this basis it is recommended that Eqs. (9-29) and (9-30) be used for gases without the Prandtl number ratio and that the properties be introduced at a reference temperature equal to the arithmetic mean of the fluid bulk temperature and the wall temperature.

A high turbulence level in the approaching stream increases not

[1] W. Y. M. Douglas and S. W. Churchill, Recorrelation of Data for Convective Heat Transfer between Gases and Single Cylinders with Large Temperature Differences, *AICHE Heat Transfer Symp., Natl. Meeting, Louisville, Ky., 1955,* Preprint 16.

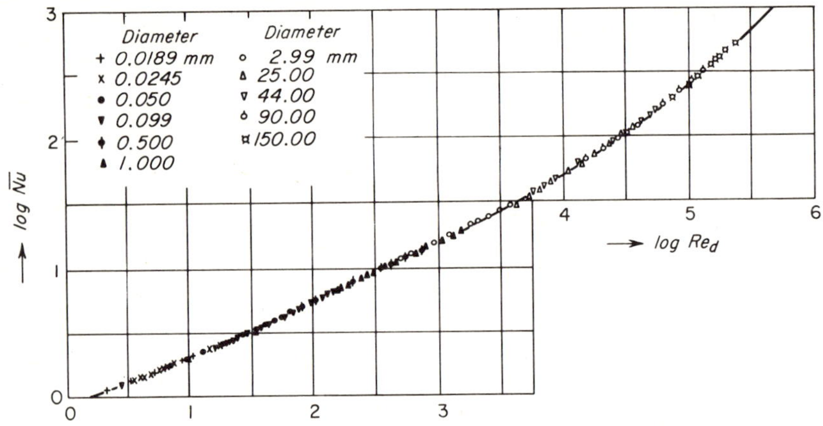

Fig. 9-10 Average film heat-transfer coefficients on a cylinder in an airstream normal to its axis. [*From R. Hilpert, Forsch. Gebiete Ingenieurw.*, **12**:*65* (*1941*).]

only the average heat-transfer coefficient but also the local heat transfer on the upstream part of the cylinder circumference which is covered by a laminar boundary layer.[1] Increases in Nu up to 60 percent have been measured for turbulence levels up to 7 percent.

Heat transfer between the surface of tube bundles and a fluid flowing normal to the tube axis depends on the geometric arrangement of the tubes. Experimental results are mainly available for two arrangements, i.e., tubes either aligned or staggered in flow direction (Fig. 9-11). On the basis of a review of the available experimental results, Zhukauskas et al.[2] recommend the relations in Fig. 9-12*a* for aligned tubes and the relations in Fig. 9-12*b* for staggered tubes. The figures themselves pre-

[1] W. H. Giedt, *J. Aeron. Sci.*, **18**:725 (1951); E. W. Comings, I. T. Clapp, and T. F. Taylor, *Ind. Eng. Chem.*, *40:* 1076; (1948) J. Kestin and P. F. Maeder, *NACA Tech. Note* 4018, 1954; *Advan. Heat Transfer*, **3**:1 (1966).

[2] *Op. cit.*

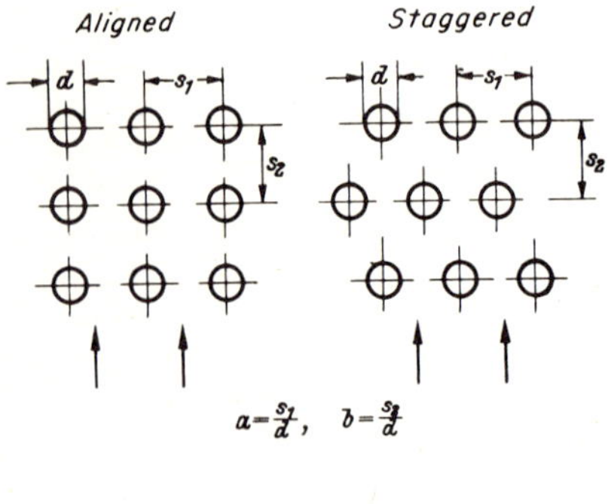

Fig. 9-11 Bundles with aligned and staggered tubes.

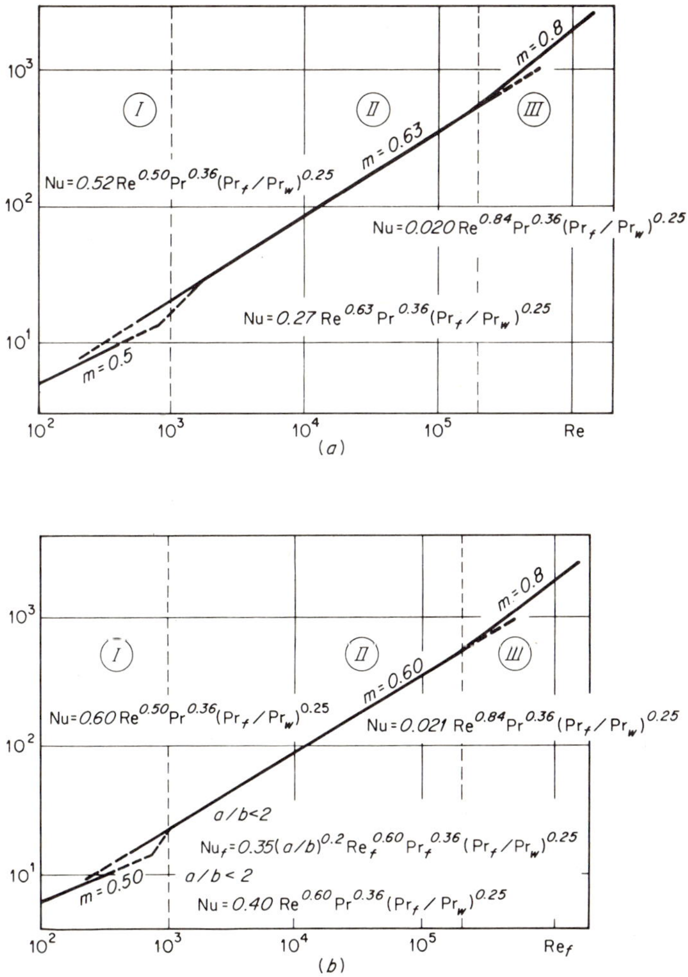

Fig. 9-12 Recommended relation for Nusselt numbers for (*a*) aligned and (*b*) staggered tube banks. [*From A. A. Zhukauskas, "Heat Transfer in Banks of Tubes," Mintis, Vilnius, Lithuania, 1968.*]

sent mean values of the Nusselt number for a fluid with a Prandtl number equal to 1 and a ratio $\mathrm{Pr}_f/\mathrm{Pr}_w$ close to 1. The equations apply for the condition in the interior of a tube bank. Near the entrance the heat-transfer coefficients assume different values. The first row has approximately the same Nusselt number as a single tube, whereas for the first four rows the average heat-transfer coefficient is smaller by approximately 10 percent than further downstream in the interior.

The pressure loss Δp which a fluid flowing through a tube experiences is conventionally expressed by

$$\Delta p = nf\rho \frac{u^2}{2} \tag{9-31}$$

where n = number of tube rows in flow direction
ρ = density of fluid
u = average velocity through area between tubes

The drag coefficient f is plotted for aligned and for staggered tube banks in Fig. 9-13*a* and *b*. Figure 9-13*a* presents the friction factor for aligned tubes in a square arrangement. A correction factor u for other tube arrangements can be taken from the insert figure, and the friction factor for such an arrangement is obtained by multiplying the friction factor for the square arrangement by this correction factor, u. Figure 9-13*b* similarly contains the friction factor for a tube arrangement in the form of isosceles triangles. The friction factor for other configurations of staggered tubes is again obtained by multiplying the friction factor read from the large diagram by the correction factor u obtained from the insert. The friction factors in Fig. 9-13*a* and *b* again apply to the interior of a tube bank. Banks with a smaller number of rows have larger friction factors. For example, a bank with four rows has a value greater by 8 percent. For rough surfaces, heat transfer and drag are increased. Increases up to 20 percent have been measured for very rough surfaces.

Much experimental information has recently been collected on compact heat-exchanger surfaces with large heat-transfer areas per unit volume, which they attain by using passages with small dimensions and by liberal application of fins and other extended surfaces. The available information can be found in the book by Kays and London,[1] which also contains valuable information on friction and heat transfer connected with flow through single tubes.

Less information exists on tube bundles with flow in the direction of the tube axis. Tentatively, they can be calculated for turbulent flow with the equation describing heat transfer in a single tube. The hydraulic diameter which should be introduced instead of the tube diameter d is

$$d_h = d\left(\frac{4ab}{\pi} - 1\right)$$

For heat exchangers in stationary plants the magnitude of the heating surface is most important, since the weight and price depend primarily upon it. This raises the question of which tube configuration

[1] W. M. Kays and A. L. London, "Compact Heat Exchangers," 2d ed., McGraw-Hill, New York, 1964.

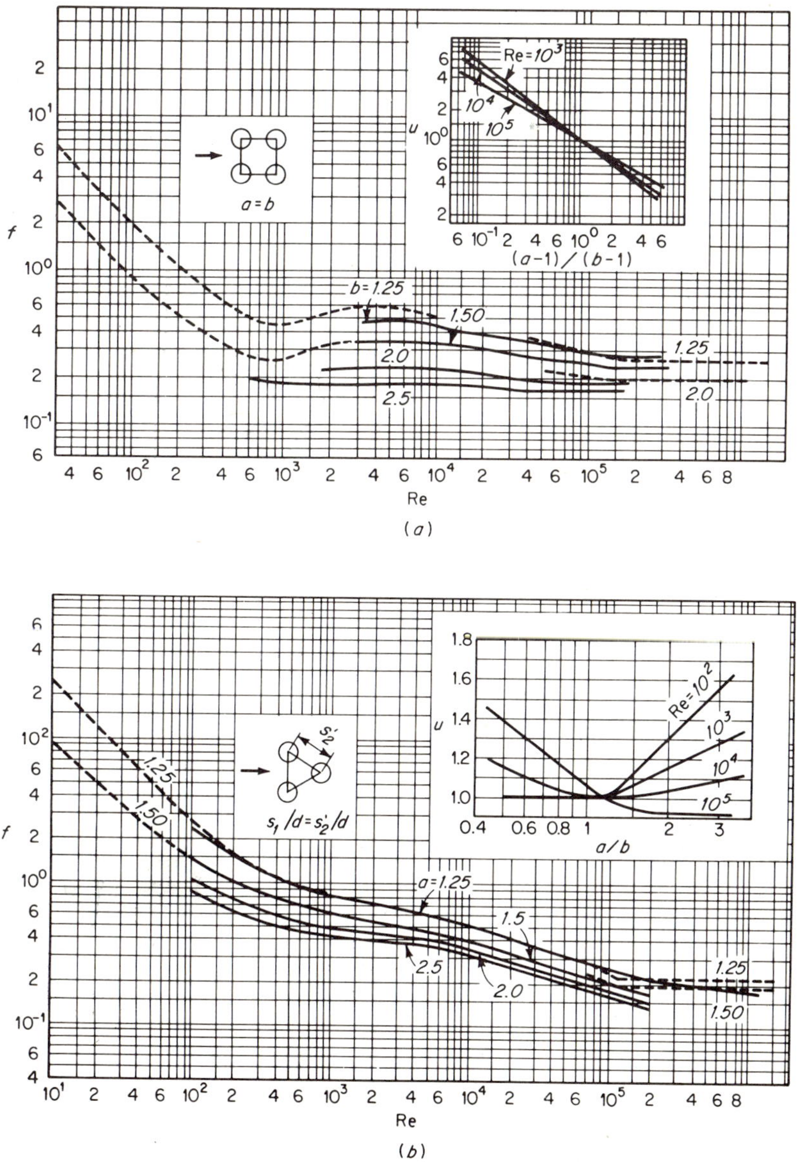

Fig. 9-13 Friction factor f for tube banks: (a) square arrangement (large diagram) and correction factor u for different geometrical arrangement of aligned tube banks; (b) arrangement of isosceles triangles (large diagram) and correction factor for different arrangement of staggered tube banks. [*From A. A. Zhukauskas, "Heat Transfer in Banks of Tubes," Mintis, Vilnius, Lithuania, 1968.*]

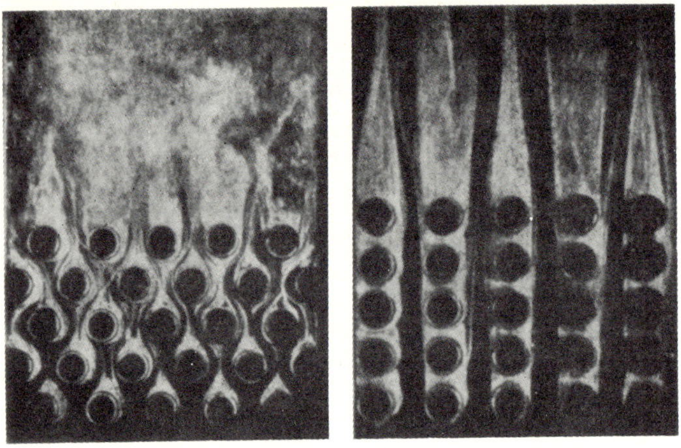

(*a*) (*b*)

Fig. 9-14 Visualization of the flow through banks of (*a*) staggered and (*b*) aligned tubes. [*According to H. Thoma; photographs by W. Lorisch: Mitt. Forschungsarb. 322, 1929.*]

transfers a certain heat flow with a given pressure loss and with the smallest heating surface. In this problem each configuration demands a different velocity, depending on the heat-transfer coefficient. A calculation[1] shows that tubes in crossflow are better than in flow parallel to the axis. The latter have a pressure loss 3 to 15 times greater when the heating surface area is the same. By spacing a given number of tubes closer together in the rows, thus decreasing the number of tube rows required, a given quantity of heat can be transferred with a reduction in the overall pressure loss through the tube bundle. Therefore, it is advantageous to make the distance between the tubes in such a heat exchanger as small as manufacturing and maintenance allow. At small Reynolds numbers, staggered tubes are somewhat better; at large Reynolds numbers, aligned tubes.

Figure 9-14*a* and *b* shows photographs of the flow through two tube banks made according to a proposal of Thoma.[2] The porous surfaces of the tubes were wetted with hydrochloric acid, and ammonia vapor was mixed with the air. Where this vapor mixes with the acid vapor, a white fog of ammonium hydrochloride arises. In this way the boundary layers and the wakes can be seen in the photographs. These are also the regions where the airstream is cooled or heated by the tubes.

[1] E. Eckert, *Forsch. Gebiete Ingenieurw.*, **16**:133 (1949–1950).

[2] H. Thoma, "Hochleistungskessel," Springer-Verlag, Berlin, 1921; also W. Lorisch, *Mitt. Forschungsarb.*, **1929**:322.

9-4 SPHERES AND PACKED BEDS

Heat transfer at the surface of spheres is determined by the flow conditions discussed in Sec. 9-2. A laminar boundary layer covers the upstream portion of the surface and separates from the side of the sphere, creating an irregularly fluctuating flow condition along the downstream portion. On the side of the sphere the boundary layer may become turbulent at large Reynolds numbers. This will influence the location of the flow separation. Near the forward stagnation point of a blunt object the stream velocity varies with the distance from the stagnation point according to the equation

$$u = Cx \tag{9-32}$$

For potential flow around a sphere with radius r, the constant C is $3u_0/2r$ (u_0 is the free-stream velocity). The heat-transfer coefficient in this region can be calculated from the solution of the boundary-layer equations for rotationally symmetric flow. The result is an equation which holds for any blunt object

$$\mathrm{Nu}_x = f(\mathrm{Pr})\sqrt{\mathrm{Re}_x} \tag{9-33}$$

with $\mathrm{Nu}_x = hx/k$ and $\mathrm{Re}_x = ux/\nu$ (x distance from stagnation point measured along surface, u = local velocity).[1] The function of Pr can be approximated for $0.5 < \mathrm{Pr} < 5$ by $f(\mathrm{Pr}) = 0.763\ \mathrm{Pr}^{0.4}$. Near a stagnation point, according to Eqs. (9-32) and (9-33), the heat transfer is independent of location. For larger distances from the stagnation point, an exact solution of the differential equations describing flow and heat transfer in the boundary layer requires a solution of partial differential equations. The average heat-transfer coefficient for a sphere must be obtained from experiments because of the difficulty of performing an adequate analysis in the separated-flow region on the downstream part of the sphere. Extensive and exact measurements have been performed by Raithby[2] with the following results. In an airstream with low turbulence level (0.05 to 0.2 percent), the results can be expressed by the equation

$$\mathrm{Nu}_d = 2 + 0.210\ \mathrm{Re}^{0.606} \tag{9-34}$$

The average deviation of the measured values from this equation was 0.8 percent in the Reynolds number range 3.6×10^3 to 5.2×10^4. The sphere was supported by a rod from the downstream side. An increase in the free-stream turbulence intensity to 5 percent caused an increase

[1] M. Sibulkin, *J. Aeron. Sci.*, **19**:570–71 (1952).

[2] G. D. Raithby and E. R. G. Eckert, The Effect of Turbulence Parameters, and Support Position, on the Heat Transfer from Spheres, *Intern. J. Heat Mass Transfer*, **11**(8):1233–1252 (1968).

in the Nusselt number by 7.5 and 17.5 percent for Reynolds numbers 3.6×10^3 and 5.2×10^4, respectively. The ratio of the diameter of the sphere to the scale of turbulence in the free stream was varied from near zero to 12. This caused an increase in the Nusselt number to 5 percent. Assuming a dependence of the Nusselt number on the $\frac{1}{3}$ power of the Prandtl number, one can generalize the above equation to the form

$$\mathrm{Nu} = 2 + 0.236\,\mathrm{Re}^{0.606}\,\mathrm{Pr}^{\frac{1}{3}} \tag{9-35}$$

For $\mathrm{Re} \rightarrow 0$, this equation goes toward $\mathrm{Nu}_d = 2$, which is the value for heat transfer by conduction in a spherical shell.

The flow and heat transfer through packed beds are of considerable importance in engineering applications. Storage-type heat exchangers, air-conditioning equipment, and devices in the chemical industry use such arrangements. Flow through sintered and porous materials belongs in the same category. One basic difficulty for a description of pressure drop and heat transfer in such beds is the large variety and often the poor definition of the geometries which are involved. In this paragraph only one well-defined arrangement will be discussed, namely, a bed consisting of spherical particles with uniform size.[1] It was found that the relationships describing conditions in such a bed can be utilized with sufficient accuracy for beds built of more irregularly shaped particles as long as their shape does not deviate too much from that of a sphere. In this case it is recommended to use, instead of the particle diameter, the expression

$$d_p = \frac{6}{S_p} \tag{9-36}$$

in which S_p denotes the surface area per unit volume. An important parameter in the description of a packed bed is the porosity ϵ, defined as the ratio of the volume of the voids to the overall volume. When a fluid is directed upward through a packed bed, it is found that at a certain velocity the fluid begins to carry the particles so that they start moving apart. The bed is now in a *fluidized state*. In such a state the average porosity is considerably greater than under the condition of low-flow velocities. Fluidization often does not produce a uniform density. For instance, when the particles are small, the air tends to create diverse channels. In beds consisting of particles with nonuniform size, a separation of the particles often takes place, with the smaller particles gradually locating themselves near the upper end of the layer. The particles in a fluidized bed are in continuous movement, and a vigorous mixing of the particles occurs. Relationships for pressure drop and heat transfer

[1] Measurements on wire screen matrices have been reported by J. E. Coppage and A. L. London, *Chem. Eng. Progr.*, **52**:57–63 (1956).

in a fluidized bed depend on the particular state it is in and are therefore extremely difficult to describe. Much research has still to be done in this connection.

Ergun[1] made a survey of the investigations on the pressure drop Δp in a nonfluidized packed bed of height L consisting of spherical particles with diameter d_p and arrived at the relation

$$\frac{\Delta p}{L} \frac{d_p}{\rho V_0^2} \frac{\epsilon^3}{1 - \epsilon} = 150 \frac{1 - \epsilon}{\mathrm{Re}_p} + 1.75 \tag{9-37}$$

from experiments which covered a range of the porosity ϵ between 0.40 and 0.65. In the above equation, ρ is the density of the fluid and V_0 is the velocity which the fluid would have if no particles were present (the approach velocity). The Reynolds number is based on this approach velocity V_0 and the particle diameter

$$\mathrm{Re}_p = \frac{V_0 d_p}{\nu}$$

Equation (9-37) can be transformed into the form

$$\Delta p = a\mu V_0 + b\rho V_0^2$$

in which the parameters a and b depend only on the geometry of the bed. This equation is often interpreted as indicating that the total resistance consists of a viscous part (first term) and a form resistance (second term). Figure 9-15 indicates the correlation obtained by this relationship.

The arrangement of the particles and the porosity of the bed are usually different near the side walls from in the interior of the bed. Accordingly, the above relationship holds only as long as the diameter of the whole bed is larger than at least 10 times the particle diameter. For smaller beds the resistance is found to be larger because of friction of the fluid at the wall.

In the fluidized state the pressure drop must be equal to the weight of the bed; therefore, the beginning of fluidization can be calculated in the following way:

$$\frac{\Delta p}{L} = g\rho_p$$

The particle density ρ_p is defined as the density of the bed in the packed state before fluidization occurs. Equation (9-37) for the pressure drop can then be used to calculate the approach velocity V_0 for beginning fluidization. The expansion of the bed by fluidization ranges between 20 and 50 percent.

Heat transfer in randomly packed beds of spheres is quite difficult

[1] Sabri Ergun, *Chem. Eng. Progr.*, **48**:227 (1952).

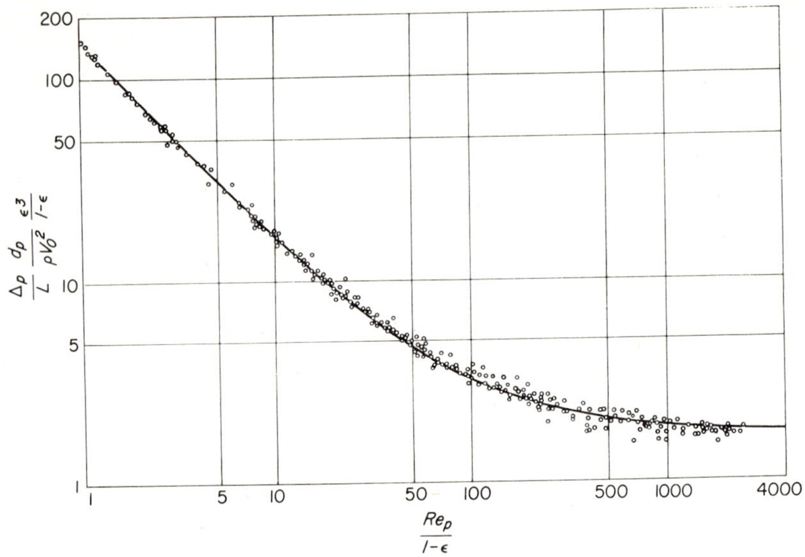

Fig. 9-15 Correlation for pressure drop in packed beds. [*Sabri Ergun, Chem. Eng. Progr.*, **48**:*227* (*1952*).]

to measure and therefore not very well known. A relationship which appears to describe heat-transfer conditions reasonably well has been obtained by Denton[1]

$$\mathrm{Nu}_p = 0.80(\mathrm{Re}_p)^{0.7}(\mathrm{Pr})^{\frac{1}{3}}$$

This relationship holds in a range of Reynolds numbers between approximately 500 and 50,000 and for a porosity $\epsilon = 0.370$. The parameters in the equation are based on sphere diameter. A considerable spread of data obtained by different investigators must be attributed to unknown effects and especially to complicated geometry. The situation is still more uncertain in a bed in the fluidized state.

In applications another type of heat transfer is often important, namely, the heat exchanged from the fluid and bed to the walls of the vessel containing the bed. This type will not be treated here because of space limitation.

[1] W. H. Denton, *Proc. Gen. Discussion Heat Transfer ASME-IME, London, 1951*, p. 370. The Prandtl number relation was incorporated by the author of this chapter.

10
Heat Transfer at High Velocities

10-1 BOUNDARY-LAYER FLOW OF A CONSTANT-PROPERTY FLUID

The energy equation for two-dimensional steady boundary-layer flow written in the total enthalpy i° can be obtained readily from Eq. (6-53)

$$\rho\left(u\frac{\partial i^\circ}{\partial x} + v\frac{\partial i^\circ}{\partial y}\right) = \frac{\partial}{\partial y}\left(k\frac{\partial t}{\partial y}\right) + \frac{\partial}{\partial y}\left(\mu u\frac{\partial u}{\partial y}\right) + \rho u g_x \qquad (10\text{-}1)$$

At high velocities, body forces caused by the gravitational acceleration are negligibly small. The situation may differ if body forces are created by electric or magnetic fields. In the present section, such body forces will be neglected ($g_x = 0$). It is now convenient to change the temperature t, appearing in the first term on the right-hand side of the equation, to total enthalpy. This can be done in a boundary layer, since the pressure is practically constant in the y direction. Therefore, the first law of thermodynamics $dq = di - 1/\rho\, dp$ states

$$\left(\frac{\partial q}{\partial y}\right)_p = c_p\left(\frac{\partial t}{\partial y}\right)_p = \left(\frac{\partial i}{\partial y}\right)_p$$

The total enthalpy i° is connected with the static enthalpy i by the equation

$$i^\circ = i + \frac{u^2}{2}$$

Therefore, introducing this expression and the Prandtl number into the energy equation transforms it to

$$\rho\left(u\frac{\partial i^\circ}{\partial x} + v\frac{\partial i^\circ}{\partial y}\right) = \frac{\partial}{\partial y}\left(\frac{\mu}{\mathrm{Pr}}\frac{\partial i^\circ}{\partial y}\right) + \frac{\partial}{\partial y}\left[\left(1 - \frac{1}{\mathrm{Pr}}\right)\mu\frac{\partial}{\partial y}\left(\frac{u^2}{2}\right)\right] \tag{10-2}$$

This equation is valid for a fluid with variable properties and will be discussed later as such. In the present section, the model of a constant-property fluid will be used. For such a model, $i = c_p t$, and the energy equation is

$$\rho c_p\left(u\frac{\partial t^\circ}{\partial x} + v\frac{\partial t^\circ}{\partial x}\right) = k\frac{\partial^2 t^\circ}{\partial y^2} + \left(1 - \frac{1}{\mathrm{Pr}}\right)\mu\frac{\partial^2}{\partial y^2}\left(\frac{u^2}{2}\right) \tag{10-3}$$

A solution to this equation (and with it information on heat transfer) can be obtained in a very simple manner when a gas with a Prandtl number $\mathrm{Pr} = 1$ is considered. In this case, the last term on the right-hand side of the equation is zero, and the remaining part is identically equal to Eq. (7-56), describing the temperature field in a low-velocity fluid. The solutions there can immediately be applied for the present situation, the only change being that the temperature t must now be interpreted as total temperature t°. This holds also for turbulent boundary layers, as can easily be established by adding to the equations in Sec. 8-2 the viscous-dissipation terms.

Heat transfer at high velocities of a constant-property fluid, with a Prandtl number different from 1, will now be discussed for the special case of a fluid moving over a surface with constant temperature t_w and having a constant velocity u_s outside the boundary layer (flat plate). The similarity condition of the velocity profiles and the temperature profiles at various stations along the plate, which was found to exist in a low-velocity flow, can be suspected to persist when the effect of viscous dissipation becomes important at large flow velocities. This suggests the introduction of the dimensionless variables

$$\eta = y\sqrt{\frac{u_s}{2\nu x}} \qquad f = \frac{\psi}{\sqrt{2\nu u_s x}}$$

and results in the equation

$$\frac{d^2 t^\circ}{d\eta^2} + \mathrm{Pr}\, f\frac{dt^\circ}{d\eta} + \frac{u_s^2}{2c_p}(\mathrm{Pr} - 1)\frac{d^2}{d\eta^2}\left(\frac{df}{d\eta}\right)^2 = 0 \tag{10-4}$$

The dimensionless stream function f must be regarded as a known function of η obtained by a solution of the continuity and momentum equations (see Sec. 7-2). Equation (10-4) is a second-order linear inhomogeneous differential equation. The solution of such an equation can be constructed from a particular solution of the complete differential equation and a general solution of the corresponding homogeneous equation. We shall first discuss the solution to the complete equation with the particular boundary condition

$$\eta = 0 \qquad q_w = 0 \qquad \text{or} \qquad \frac{dt^\circ}{d\eta} = 0$$
$$\eta = \infty \qquad t^\circ = t_s^\circ$$

The equation will be made dimensionless by introduction of the dimensionless parameter

$$\theta_a^\circ = \frac{t^\circ - t_s^\circ}{u_s^2/2c_p}$$

It then assumes the form

$$\theta_a^{\circ\prime\prime} + \Pr f\theta_a^{\circ\prime} + (\Pr - 1)(f'^2)'' = 0 \tag{10-5}$$

where differentiation with respect to η is indicated by a prime, with the boundary conditions

$$\eta = 0 \qquad \theta_a^{\circ\prime} = 0$$
$$\eta = \infty \qquad \theta_a^\circ = 0$$

The solution of this problem can be obtained by the method of variation of coefficients.[1] The result is

$$\theta_a^\circ = \Pr \int_0^\infty \varphi\, d\eta - \Pr \int_0^\eta \varphi\, d\eta$$
$$\varphi = \exp\left(-\Pr \int_0^\eta f\, d\eta\right)\left[\int_0^\eta f''^2 \exp\left(\Pr \int_0^\eta f\, d\eta\right) d\eta\right]$$

The resulting temperature profile is presented in Fig. 10-1 as dashed line *a* for a Prandtl number smaller than 1. For a Prandtl number equal to 1, the temperature parameter θ_a° and therefore the temperature t° is constant throughout the boundary layer, as was concluded in the foregoing discussion. For a fluid with a Prandtl number smaller than 1, the temperature t_w at the wall surface is smaller than the total temperature t_s° outside the boundary layer, as shown in Fig. 10-1. It then increases with increasing wall distance η and after overshooting asymptotically approaches the temperature t_s°. The reverse situation exists for fluids

[1] It was originally obtained by Pohlhausen not for Eq. (10-5) but for the corresponding one written in static temperatures.

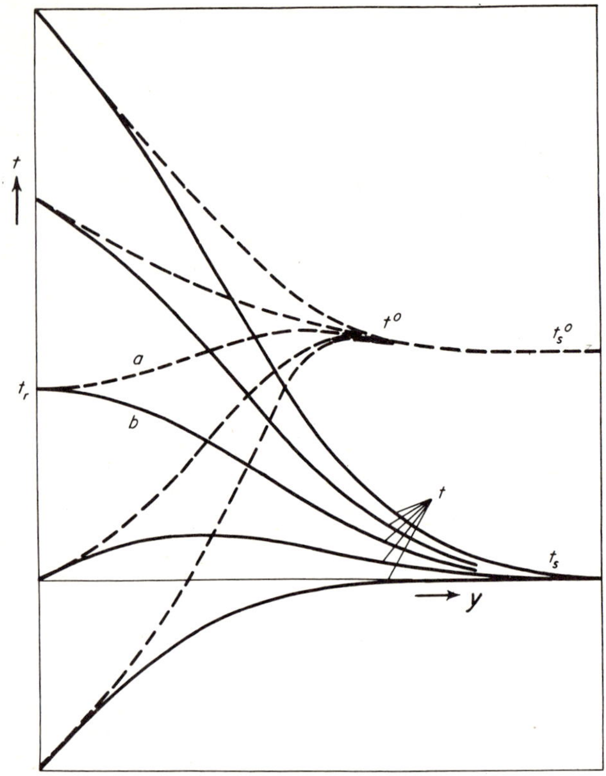

Fig. 10-1 Profiles of the total temperature t^0 and the static temperature t through a laminar boundary layer. a and b are for the condition $q_w = 0$ (adiabatic wall).

with a Prandtl number larger than 1. Also plotted as the full curve b is the parameter

$$\theta_a = \frac{t - t_s}{u_s^2/2c_p}$$

which describes the static temperature t within the boundary layer for a Prandtl number smaller than 1. The relation

$$\theta_a^\circ = \frac{t^\circ - t_s^\circ}{u_s^2/2c_p} = \frac{t + u^2/2c_p - t_s - u_s^2/2c_p}{u_s^2/2c_p} = \theta_a + \left(\frac{u}{u_s}\right)^2 - 1$$

connects the two temperature parameters. The curve representing either parameter has a zero slope at the wall surface (at $\eta = 0$). This agrees with the boundary condition $q_w = 0$.

Of special interest is the temperature which the wall surface assumes when there is no convective heat transfer. This temperature is called *recovery temperature* t_r and is described by a dimensionless parameter called the *recovery factor* r

$$r = \frac{t_r - t_s}{u_s^2/2c_p} = \frac{t_r - t_s}{t_s^\circ - t_s} \tag{10-6}$$

It has been found that in the Prandtl number range between 0.6 and 15 the recovery factor can be closely approximated by the simple relation

$$r = \sqrt{\Pr} \tag{10-7}$$

The homogeneous part of Eq. (10-3) is

$$\rho c_p \left(u \frac{\partial t^\circ}{\partial x} + v \frac{\partial t^\circ}{\partial y} \right) = k \frac{\partial^2 t^\circ}{\partial y^2} \tag{10-8}$$

It again has the same form as the energy equation for low flow velocities, so that the solution to this equation has already been obtained in Sec. 7-7, where it was given in the form of a dimensionless temperature parameter

$$\theta^\circ = \frac{t - t_s}{t_w - t_s}$$

The fact that the differential equation (10 3) is linear makes it possible to obtain new solutions by linear combination of known ones. The general solution to the complete differential equation (10-3) can therefore be written in the form

$$t^\circ = C_1 \theta^\circ (t_w - t_s) + C_2 + \frac{u_s^2}{2c_p} \theta_a^\circ$$

The solutions θ obtained in Sec. 7-7 are again interpreted as total temperature θ°. If one prescribes the following boundary conditions

$$\eta = 0 \qquad t^\circ = t_w$$
$$\eta = \infty \qquad t^\circ = t_s^\circ$$

one can calculate the constants C_1 and C_2. The solution to the complete differential equation takes the form

$$t^\circ = \theta^\circ \left[(t_w - t_s^\circ) - (r - 1) \frac{u_s^2}{2c_p} \right] + \frac{u_s^2}{2c_p} \theta_a^\circ + t_s^\circ$$

Introduction of the recovery temperature t_r simplifies this equation to

$$t^\circ = \theta^\circ (t_w - t_r) + \frac{u_s^2}{2c_p} \theta_a^\circ + t_s^\circ$$

Figure 10-1 presents the temperature t° as dashed lines and t as full lines for heat flux to and from the surface.

The heat flux at the surface into the fluid can now be calculated

$$q_w = -k\left(\frac{\partial t}{\partial y}\right)_w = -k\left(\frac{\partial t^\circ}{\partial y}\right)_w = k(t_w - t_r)\left(\frac{d\theta}{d\eta}\right)_0\sqrt{\frac{u_s}{2\nu x}}$$

With the expression for the dimensionless temperature gradient $(d\theta/dy)_0$ in Sec. 7-7, it takes the form

$$q_w = 0.322\sqrt[3]{\mathrm{Pr}}\,k\sqrt{\frac{u_s}{\nu x}}\,(t_w - t_r) \tag{10-9}$$

for $\mathrm{Pr} > 0.6$. This equation suggests defining a heat-transfer coefficient with the expression

$$q_w = h(t_w - t_r) \tag{10-10}$$

The Nusselt number, as a dimensionless expression for this heat-transfer coefficient, then becomes

$$\mathrm{Nu} = 0.332\sqrt[3]{\mathrm{Pr}}\sqrt{\mathrm{Re}_x} \tag{10-11}$$

which is exactly the same expression as that obtained in Sec. 7-7 for low-velocity flow.

A very simple method for the calculation of heat transfer has thus evolved. One redefines a heat-transfer coefficient with the difference between the actual surface temperature t_w and the temperature t_r which the surface assumes when exposed to the same fluid stream but without convective heat transfer ($q_w = 0$). The heat-transfer coefficient then has the same value as in low-velocity flow. This follows from the linear character of the energy equation for a constant-property fluid and is therefore exact for laminar and turbulent flow of such a fluid. It will be found that it presents a good approximate method for a variable-property fluid as well. The only new information required is knowledge of the temperature-recovery factor as defined by Eq. (10-6). For laminar boundary-layer flow over a flat plate, its value is given by Eq. (10-7). A pressure gradient was found to have only a small influence on the recovery factor, and even for a laminar shear layer as discussed in Sec. 7-11, Eq. (10-7) is a good approximation of the actual recovery factor for Prandtl numbers between 0.25 and 10. For turbulent boundary layers, the following relation was found experimentally

$$r = \sqrt[3]{\mathrm{Pr}} \tag{10-12}$$

10-2 BOUNDARY-LAYER FLOW OF A VARIABLE-PROPERTY FLUID

So far, solutions to the conservation equations have been obtained for a constant-property fluid. In this section, as an example of a variable-property fluid, heat transfer will be discussed for a gas which flows in a laminar boundary layer in steady two-dimensional flow over a flat plate with u_s = const. The plate may have a constant surface temperature. Heat transfer in gases has recently been extensively studied because very large temperature differences are connected with many engineering applications utilizing gases. Often the velocities are also quite high, so that the effect of viscous dissipation will have to be included in our consideration. The effect of body forces, however, will be neglected. The three conservation equations for a steady two-dimensional laminar boundary layer on a flat plate are

$$\frac{\partial(\rho u)}{\partial x} + \frac{\partial(\rho v)}{\partial y} = 0 \tag{10-13}$$

$$\rho u \frac{\partial u}{\partial x} + \rho v \frac{\partial u}{\partial y} = \frac{\partial}{\partial y}\left(\mu \frac{\partial u}{\partial y}\right) \tag{10-14}$$

$$\rho u \frac{\partial i^\circ}{x} + \rho v \frac{\partial i^\circ}{\partial y} = \frac{\partial}{\partial y}\left(\frac{\mu}{\mathrm{Pr}} \frac{\partial i^\circ}{\partial y}\right) + \frac{\partial}{\partial y}\left[\left(1 - \frac{1}{\mathrm{Pr}}\right) \mu \frac{\partial}{\partial y}\left(\frac{u^2}{2}\right)\right] \tag{10-15}$$

with the boundary conditions

$$\begin{array}{llll} y = 0 & u = 0 & v = 0 & i^\circ = i_w \\ y \rightarrow \infty & u = u_s & & i^\circ = i_s^\circ \end{array}$$

In these equations, the properties are assumed to vary, and therefore relations for them are required. The conservation equations developed in Chap. 6 and consequently Eqs. (10-13) to (10-15) are based on the assumption that local thermodynamic equilibrium is closely approximated within the boundary layer. This means that the properties are everywhere functions of the state parameters at that point.

Therefore relations of the form

$$\rho = f(p,i) \qquad \mu = f(p,i) \qquad \mathrm{Pr} = f(p,i) \tag{10-16}$$

will be available for the fluid under consideration. The static pressure and enthalpy appear in these equations because thermodynamic equilibrium is postulated for a reference frame which moves along with the fluid. For a constant-property fluid, the first two of the above equations could be solved independently of the energy equation (10-15). For a variable-property fluid, however, all three equations are interrelated through the properties and must be solved simultaneously. Physically this means that the velocity field is influenced by the imposed heat-transfer condition. The continuity equation (10-13) can again be eliminated through introduction of a stream function ψ defined by

$$\rho u = \frac{\partial \psi}{\partial y} \qquad \rho v = -\frac{\partial \psi}{\partial x}$$

which satisfies that equation identically. For a constant-property fluid it was found that the two remaining equations can be transformed into total differential equations by introduction of an independent variable η, which is essentially equal to $y/\sqrt{x}$, and of a dimensionless stream function which is postulated to be a function of η only. Several investigators showed independently that a similar transformation also works for a variable-property fluid. Equations (10-13) to (10-15) are thus transformed into total differential equations which are quite similar in form to the corresponding constant-property equations. This transformation is usually referred to as the *Dorotnitzyn-Stewartson transformation*

$$\eta = \frac{\rho_s u_s}{2\mu_s x} \int_0^y \frac{\rho}{\rho_s}\, dy \qquad f(\eta) = \frac{\psi}{\sqrt{2\rho_s \mu_s u_s x}}$$

In order to transform the momentum equation, one now calculates

$$u = \frac{1}{\rho}\frac{\partial \psi}{\partial y} = u_s \frac{df}{d\eta} = u_s f'$$

A prime is used here and in the following relations to indicate differentiation with respect to η. Additional differentiations give

$$\frac{\partial u}{\partial x} = -\frac{1}{2}\frac{u_s}{x}\eta f'' + u_s \sqrt{\frac{\rho_s u_s}{2\mu_s x}}\, f'' \frac{\partial}{\partial x}\int_0^y \frac{\rho}{\rho_s}\, dy$$

and

$$\frac{\partial u}{\partial y} = u_s \frac{\rho}{\rho_s}\frac{\rho_s u_s}{2\mu_s x} f''$$

Further differentiation results in $\partial/\partial y\ (\mu\ \partial u/\partial y)$, and a similar differentiation gives v. Introducing all these terms into the momentum equation and using the abbreviation

$$C = \frac{\rho\mu}{\rho_s \mu_s}$$

results in

$$(Cf'')' + ff'' = 0 \tag{10-17}$$

The same transformation can be applied to the energy equation, with the result

$$\left(\frac{C}{\mathrm{Pr}} i^{\circ\prime}\right)' + f i^{\circ\prime} + \frac{u_s^2}{2}\left[C\left(1 - \frac{1}{\mathrm{Pr}}\right)(f'^2)'\right]' = 0 \tag{10-18}$$

The boundary conditions to both equations become

$$\begin{array}{llll} \eta = 0 & f = 0 & f' = 0 & i^\circ = i_w \\ \eta \to \infty & & f' = 1 & i^\circ = i_s^\circ \end{array}$$

It will be observed that Eq. (10-17) becomes identical with the momentum equation for an incompressible fluid when C is a constant having a numeri-

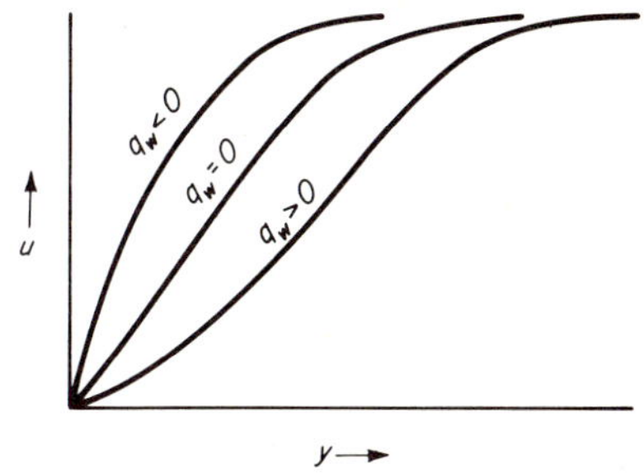

Fig. 10-2 Velocity profiles in the laminar boundary layer of a fluid with variable properties.

cal value of 1. The condition $\rho\mu$ = const is found to be a reasonably good approximation for various gases as long as the temperature is not too high. For air, as an example, the density is inversely proportional to the temperature, and the viscosity μ varies at moderate temperature as $T^{0.8}$. With increasing temperatures, the exponent drops toward a value 0.5. In this section, we shall be satisfied with the approximation $\rho\mu$ = const. A more general treatment will be considered in the following section. If, in addition, the Prandtl number can be considered constant, which again is well fulfilled for gases, then Eq. (10-18) and the corresponding boundary conditions also assume a form identical to the energy equation (10-4) of an incompressible fluid, with the boundary conditions on page 421, when $c_p t^\circ$ is replaced by i°. This, then, means that the solutions to the corresponding incompressible equations are also solutions to Eqs. (10-17) and (10-18) and that the dimensionless stream function f or velocity ratio u/u_s and the total enthalpy i° are the same functions respectively of the dimensionless wall distance η in both cases. The velocity and enthalpy profiles as a function of the actual wall distance y for a fixed location x, however, will be different. The incremental wall distance dy for a corresponding step in η can be obtained from the equation

$$dy = \sqrt{\frac{2\mu_s x}{\rho_s u_s}}\,\frac{\rho_s}{\rho}\,d\eta$$

This means that the η coordinate must be distorted, depending on the density variation within the boundary layer to obtain the coordinate y. Examples for the velocity profile are shown in Fig. 10-2. The same distortion must be applied to enthalpy profiles. The enthalpies at the wall surface for zero heat flux are then the recovery enthalpies, and it follows that the recovery enthalpy i_r or the enthalpy recovery factor r_i defined by the equation

$$r_i = \frac{i_r - i_s}{u_s^2/2} \tag{10-19}$$

is not influenced by the distortion. The recovery factor is therefore again given by the equation

$$r_i = \sqrt{\mathrm{Pr}} \tag{10-20}$$

Consequently Eq. (10-7) also gives the recovery factor for a fluid following the relation $\rho\mu$ = const.

No difference exists between the temperature and the enthalpy recovery factor as long as the specific heat is considered constant. Recovery factors calculated in this way are in good agreement with measured values. For laminar airflow at moderate temperatures the recovery factor is 0.84. In turbulent boundary-layer flow of air over a flat plate a value of 0.88 was measured. In the transition region between laminar and turbulent boundary layers the recovery factor rises from the value 0.84 to a peak and then decreases to the turbulent value 0.88 (Fig. 10-3). Interesting are the results of measurements on a cylinder in a subsonic airflow normal to its axis.[1] The cylinder was manufactured of rubber in order to eliminate temperature equalization by conduction in the solid material. The measured surface temperatures were used to calculate local recovery factors using Eq. (10-6) and introducing the

[1] E. Eckert and W. Weise, *Forsch. Gebiete Ingenieurw.* **13**:246–254 (1942).

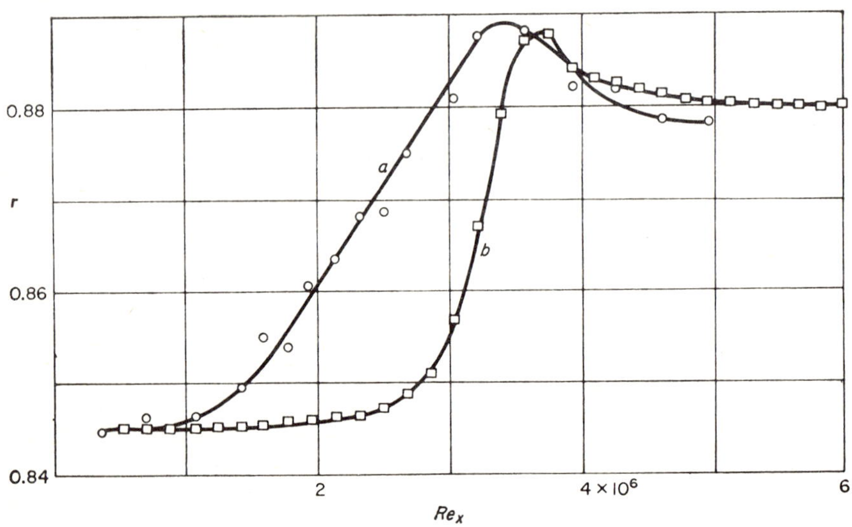

Fig. 10-3 Laminar, transitional, and turbulent temperature recovery factors r for air (measured on a cone at Ma = 3.12): (a) high stream turbulence; (b) low stream turbulence (~1 percent). [*From J. C. Evvard, H. Tucker, and W. C. Burgess, Inst. Aeron. Sci. Preprint 438, 1954.*]

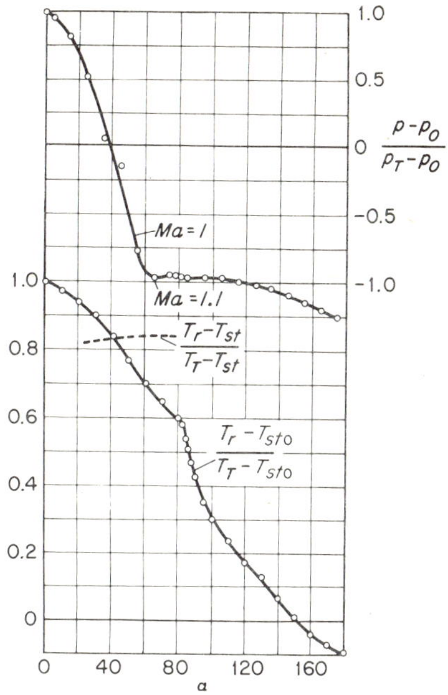

Fig. 10-4 Pressure distribution and temperature recovery factor r for high-velocity subsonic air flow normal to a circular cylinder. α is the angular distance from the stagnation line; T_T, p_T are total temperature and pressure upstream, T_{st0}, p_0 are static temperature and pressure upstream; p, T_{st} are static pressure and temperature outside the boundary layer; T_r is the recovery temperature. [*From E. Eckert and W. Weise, Forsch. Gebiete Ingenieurw.*, **13**:*246* (*1943*).]

upstream static temperature and the upstream velocity. An example of the results is shown in Fig. 10-4. It is seen that the recovery value at the stagnation point is 1, which means that the surface temperature is equal to the total temperature in the stream. The low recovery values along the downstream part of the surface are remarkable. Over some range the recovery factor is even negative, which means that the surface temperature is lower than the upstream static temperature. Ryan[1] confirmed this observation. No low temperatures in separated regions were observed in supersonic flows.

Another type of flow which leads to regions with low total temperature is vortex flow. Ranque utilized this effect to separate a gas stream into a portion with high energy content (high total temperature) and another one with low energy content. Figure 10-5 shows such a vortex tube as investigated by Hilsch, sometimes called a *Hilsch tube*. It consists of a tube into which air (or another gas) is blown tangentially through one or more nozzles a. In this way a strong vortex flow is created in the tube. Part of the flow leaves the tube through the orifice b; the rest leaves at the far end of the tube c. A valve at this location serves to adjust the strength of the two airstreams. It is found that the

[1] L. F. Ryan, *Mitt. Inst. Aerodynamik, Eidgen. Tech. Hochschule, Zurich*, no. 18, 1951, pp. 7–50.

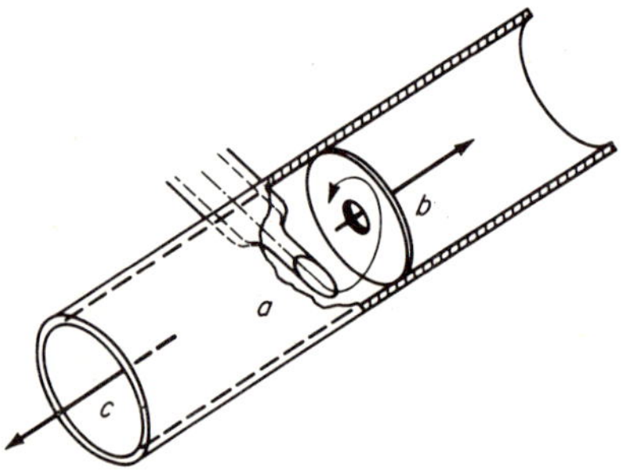

Fig. 10-5 Sketch of a vortex tube (Hilsch tube).

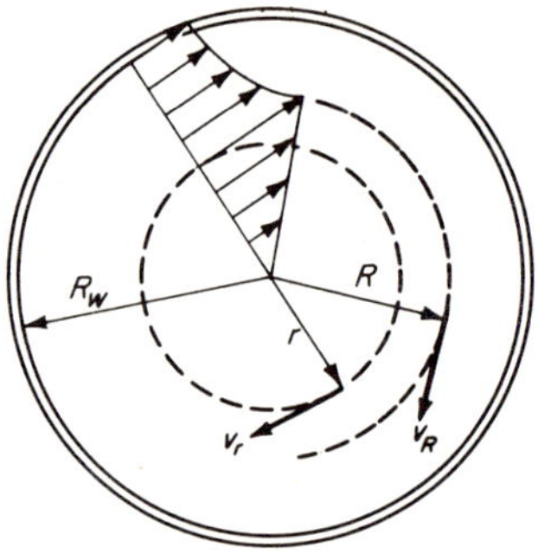

Fig. 10-6 Circumferential velocity in rotating flow through a tube.

total temperature in the air leaving the tube through the orifice is considerably lower (as much as 60 C at 1.3 atm inlet pressure) than the temperature of the air upstream from the inlet nozzle. The temperature of the air leaving the tube at c is correspondingly higher. Sprenger[1] found that various flow types exist which lead to strong energy separation effects, and Vonnegut[2] developed a thermometer which measures the static temperature in a high-velocity gas stream. It consists essentially of a vortex tube with suitable opening and a thermometer arranged in the low-energy air at the tube axis.

An understanding of the essential processes leading to the energy separation in the vortex tube can be obtained by some simple considerations of a fluid rotating in a tube of radius R_w as sketched in Fig. 10-6. A small axial movement (normal to the plane of the drawing) has little influence on the rotational flow but allows it to be steady. Experiments[3] show that the flow field can be subdivided into an outer ring, in which the circumferential velocity v_r is roughly proportional to $1/r$, and an inner area, in which the fluid approaches what is termed *solid-body rotation*, with the velocity described by

$$v_r = Cr \tag{10-21}$$

The radius of this inner region is R. The measurements also demonstrated that the energy separation is essentially confined to this inner region. We shall first consider the flow in this region to be laminar. No shear stresses and therefore no viscous dissipation exist in a solid

[1] H. Sprenger, *Z. Angew. Math. Physik,* **2**:293 (1951); *Mitt. Inst. Aerodynamik, Eidgen. Tech. Hochschule, Zurich.*, no. 21, 1954.

[2] B. Vonnegut, *Rev. Sci. Instr.*, **1950**:21–22, 136.

[3] J. P. Hartnett and E. R. G. Eckert, *Trans. ASME,* **79**:751 (1957).

rotation flow, and energy is transported solely by conduction, so that the energy equation is

$$\frac{d}{dr}\left(r\frac{dT}{dr}\right) = 0$$

with T indicating the static gas temperature. The solution satisfying the boundary conditions $T = T_R$ at $r = R$ and $dT/dr = 0$ at $r = 0$ is $T = T_R$ for any r. The static temperature is accordingly constant throughout the inner region. The total temperature is then

$$T_r^\circ = T_R + \frac{v_r^2}{2c_p} \tag{10-22}$$

if the specific heat c_p can be considered constant. The difference between the total temperature T_0° at the center of the tube and the value T_R° at the radius R is

$$T_R^\circ - T_0^\circ = \frac{v_R^2}{2c_p} \tag{10-23}$$

A similar energy separation also occurs in a high-speed boundary layer, even when it borders on an adiabatic wall. This can be seen in Fig. 10-1. The total temperature in a gas (for instance with a value $\Pr = 0.7$) at $q_w = 0$ is lower near the wall and higher in the outer layer than in the free stream outside the boundary layer. The recovery factor r is a measure of the magnitude of the separation effect. For a laminar boundary layer it is given by Eq. (10-7), and the difference between the total temperature in the free stream and at the wall surface is, accordingly,

$$T_s^\circ - T_r^\circ = (1 - \sqrt{\Pr})\frac{u_s^2}{2c_p} \tag{10-24}$$

Gases have Prandtl numbers not too far from 1, and the separation effect in a boundary layer is therefore considerably smaller than in a rotating flow. This is demonstrated by a comparison of Eqs. (10-24) and (10-23).

An additional process is active in a rotating turbulent flow, as was pointed out by Kassner and Knörnschild.[1] It is analogous to a process occurring in the atmosphere and being connected with its stability. The air in the atmosphere is stable when a fluid particle moved to a higher level has a density which is larger than the density in the new surrounding, because it will then have the tendency to move back to its original position. The atmosphere is in neutral equilibrium when the fluid particle finds its density to be just equal to the density in the new surrounding. It is now argued that intense turbulent mixing leads to this situation and that the movement of the particles is so fast that they have no time to exchange

[1] R. Kassner and E. Knörnschild, *Wright-Patterson Air Force Base*, Tech. Rept. F-TR-2198-ND, 1948.

heat with the surroundings, so that their temperature change in moving from one level to the other is adiabatic and follows the relation

$$\frac{T}{T_0} = \left(\frac{p}{p_0}\right)^{(\varkappa-1)/\varkappa} \tag{10-25}$$

with p and p_0 indicating the pressures at the two levels and $\varkappa$ denoting the ratio of specific heats at constant pressure and volume. Measurements in the atmosphere have verified these conclusions.

In a vortex tube, a radial pressure gradient is established by centrifugal forces according to the equation $dp/dr = \rho v_r^2/r$. Therefore, strong turbulence should again lead to a temperature variation as given by Eq. (10-25). The temperatures in this equation must be interpreted as static temperatures, since the density of the fluid particles determines the stability. Separation of variables in the above equation results in

$$\frac{dp}{\rho} = \frac{v_r^2}{r}\,dr$$

which can easily be integrated when Eqs. (10-21) and (10-25) as well as $p = \rho RT$ are utilized. Introducing total temperatures finally results in

$$T_r^\circ - T_0^\circ = \frac{v_r^2}{c_p}$$

The difference in the total temperatures at $r = R$ and at the center of the tube is

$$T_R^\circ - T_0^\circ = \frac{v_R^2}{c_p} \tag{10-26}$$

A comparison of Eqs. (10-23) and (10-26) indicates that this temperature difference, in turbulent rotating flow, can be up to twice as large as in laminar rotating flow. The temperature differences calculated above agree in order of magnitude with measured values.[1] A more detailed analysis considering the same processes has been carried out by Deissler and Perlmutter.[2]

For temperature measurements in high-velocity gas streams, two instruments proved suitable: the cylinder thermometer (Fig. 10-7) and the diffuser thermometer (Fig. 10-8). The cylinder thermometer measures the surface temperature on a cylinder in longitudinal flow. When the boundary-layer thickness is small compared with the cylinder diameter, the r values for the flat plate (Fig. 10-3) can be used and the static temperature calculated from Eq. (10-6) when the velocity is known. The diffuser thermometer (Fig. 10-8) measures the total temperature. The gas enters the diffuser through the opening b and is decelerated in tube a. The temperature of the gas is measured by the thermocouple d.

[1] Hartnett and Eckert, *loc. cit.*

[2] R. G. Deissler and M. Perlmutter, *Heat Transfer Fluid Mech. Inst.* 1958, pp. 40–53.

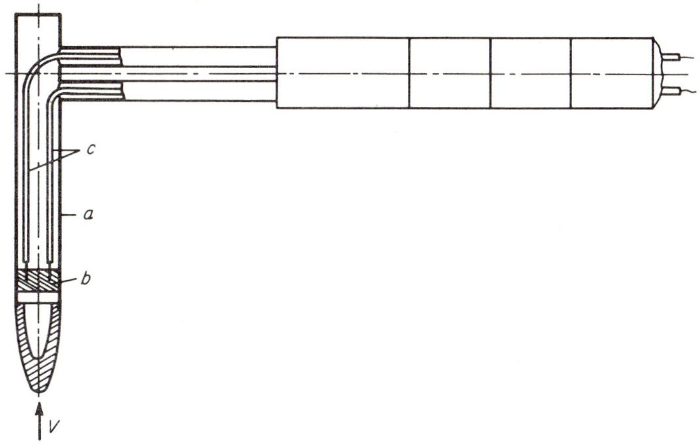

Fig. 10-7 Cylinder thermometer: (*a*) thin-walled tube. (*b*) copper bar; (*c*) thermocouples. [*From E. Eckert, Z. Ver. Deut. Ingr.*, **84**:*813* (*1940*).]

Afterward the gas escapes through the small holes *c*. If the holes were not present, the gas in the interior would cool off and would indicate an erroneous temperature. Both thermometers have their field of application. The advantage of the diffuser thermometer is that it immediately measures the total temperature. On the other hand, the cylinder thermometer can be made in very small dimensions and follows changes in gas temperature very rapidly.

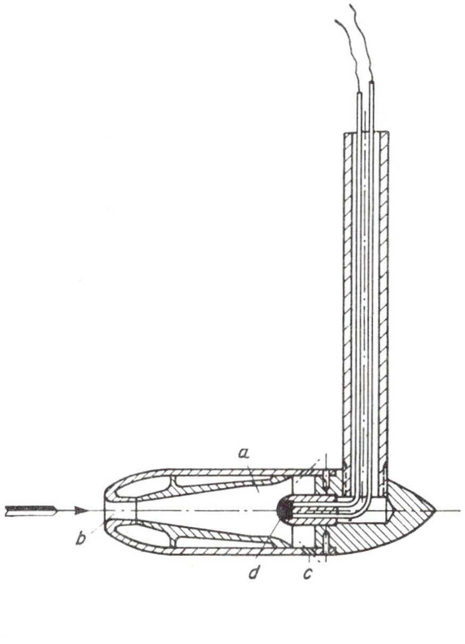

Fig. 10-8 Diffuser thermometer. [*From A. Franz, Jahrb. 1938 Deut. Luftfahrt-Forsch.*, **2**:*215*.]

We now consider heat transfer in high-velocity flow. For the situation that the enthalpy i_w at the wall of the flat plate considered in the first part of this section is prescribed, the heat flux at the surface can be obtained from the relation (see page 422)

$$q_w = -k_w \left(\frac{\partial t}{\partial y}\right)_w = -\frac{k_w}{c_{pw}} \left(\frac{\partial i^\circ}{\partial y}\right)_w = \frac{k_w}{c_{pw}} \sqrt{\frac{\rho_s u_s}{2\mu_s x} \frac{\rho_w}{\rho_s}}\, (i^{\circ\prime})_0$$

The derivative of the total enthalpy i° with respect to η can easily be obtained from the derivative of the temperature t° with respect to η in Sec. 10-1. From Eqs. (7-67) and (7-70) it follows that

$$(i^{\circ\prime})_0 = \left(\frac{di^\circ}{d\eta}\right)_0 = 0.332\sqrt{2}\,\mathrm{Pr}^{\frac{1}{3}}\,(i_w - i_r)$$

Therefore the heat flux is

$$q_w = \sqrt{\frac{\rho_s \mu_s \mu_s}{x}} \frac{1}{\mathrm{Pr}^{\frac{2}{3}}}\, 0.332(i_w - i_r)$$

If one defines a heat-transfer coefficient h_i based on enthalpies by the equation

$$q_w = h_i(i_w - i_r) \tag{10-27}$$

one obtains

$$h_i = 0.332 \sqrt{\frac{\rho_s \mu_s \mu_s}{x}} \frac{1}{\mathrm{Pr}^{\frac{2}{3}}} \tag{10-28}$$

We can now define a Stanton number with the heat-transfer coefficient h_i

$$\mathrm{St}_i = \frac{h_i}{\rho_s u_s} = \frac{0.332}{\sqrt{\mathrm{Re}_s}\,\sqrt[3]{\mathrm{Pr}^2}} \tag{10-29}$$

which agrees exactly with the Stanton number derived from Eq. (7-70) with the relation St = Nu/(Re Pr). The Stanton number and recovery factor of a gas with properties following the relations $\rho\mu$ = const and Pr = const are therefore described by the constant-property relations when they are defined with enthalpies rather than with temperatures. It is easily checked that the heat-transfer coefficient calculated from the equation above does not change its value regardless of the temperature at which the properties are introduced.

It has been found[1] that the equation describes the heat-transfer coefficient h_i with good approximation for a fluid with a different temperature dependence of the properties when they are introduced at a reference enthalpy i^* described by the equation

$$i^* = i_s + 0.5(i_w - i_s) + 0.22(i_r - i_s) \tag{10-30}$$

[1] E. R. G. Eckert, *Trans. ASME*, **78**:1273 (1956).

This has been demonstrated for air[1] and has also been verified for nitrogen and oxygen.[2] The heat-transfer coefficient h_i based on an enthalpy difference is connected with the heat-transfer coefficient h based on a temperature difference by the simple equation $h_i = h/c_p$ as long as the specific heat is constant. For a gas with a temperature-dependent c_p, this simple relation no longer exists, and the enthalpy recovery factor r_i as defined by Eq. (10-19) is different from the temperature recovery factor r defined by Eq. (10-6). It was also found that Eqs. (10-20) and (10-29) with properties at the reference enthalpy Eq. (10-30) are good approximations even for dissociating or ionizing gases, where the specific heat undergoes strong variation with temperature.[3] When the Prandtl number also varies with temperature, it is best to introduce it into Eq. (10-20) at the same reference enthalpy [Eq. (10-30)]. It appears that the same procedure leads also to a good approximation for a turbulent boundary layer when the equations in Sec. 8-2 together with Eqs. (10-12) and (10-30) are used.

10-3 VARIABLE PROPERTIES IN PLANE AND ROTATIONALLY SYMMETRIC BOUNDARY LAYERS

Recent aerospace developments demanded solutions to the so-called reentry problem. A spacecraft reentering the atmosphere on its return to the earth is surrounded by a layer of air with an extremely high temperature. This generates very high heat fluxes into the skin of the craft, and analytical methods had to be developed to predict their magnitude. Especially high heat-transfer coefficients occur at and near the stagnation point of such bodies, which generally have a blunt shape. The temperatures surrounding the skin are so high that dissociation and even ionization of the air may occur. The present section will discuss the basic features of analyses of this problem. Included is also the treatment of an approximate method by Lees[4] for calculating heat transfer over an extended portion of the spacecraft to the extent to which it is covered by a laminar boundary layer, with some alterations introduced by Eckert and Tewfik.[5]

The boundary-layer equations as developed in Secs. 6-4 and 6-5 can be written in such a way that they apply to plane as well as to rotationally symmetric flow

$$\frac{\partial}{\partial s}(\rho u r_0^n) + \frac{\partial}{\partial y}(\rho v r_0^n) = 0 \tag{10-31}$$

[1] E. R. G. Eckert, *Trans. ASME*, **78**:1273 (1956).

[2] H. A. Simon, C. S. Liu, and J. P. Hartnett, *Intern. J. Heat Mass Transfer*, **10**:406–409 (1967).

[3] E. R. G. Eckert and O. E. Tewfik, *J. Aerospace Sci.*, **27**:464 (1960).

[4] L. Lees, *Jet Propulsion*, **26**:259 (1956).

[5] Eckert and Tewfik, *loc. cit.*

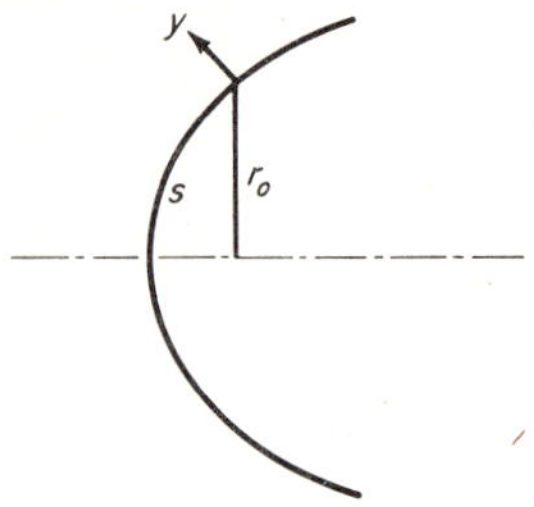

Fig. 10-9 Coordinates in rotationally symmetric flow.

$$\rho\left(u\frac{\partial u}{\partial s}+v\frac{\partial u}{\partial y}\right)=\frac{\partial}{\partial y}\left(\mu\frac{\partial u}{\partial y}\right)-\frac{dp_s}{ds} \tag{10-32}$$

$$\rho\left(u\frac{\partial i^\circ}{\partial s}+v\frac{\partial i^\circ}{\partial y}\right)=\frac{\partial}{\partial y}\left(\frac{\mu}{\Pr}\frac{\partial i^\circ}{\partial y}\right)+\frac{\partial}{\partial y}\left[\left(1-\frac{1}{\Pr}\right)\mu\frac{u^2/2}{\partial y}\right] \tag{10-33}$$

The symbol s indicates the distance from the stagnation point measured along the surface, y is the distance from the surface measured normal to it, and r_0 denotes the distance of the surface point under consideration from the axis of rotation (Fig. 10-9). The subscript s refers to parameters measured at the outer edge of the boundary layer and the subscript w to parameters measured at the wall surface. The momentum equation now contains the term dp_s/ds, indicating the pressure variation along the surface. The boundary conditions are

$$\begin{array}{llll} y=0 & u=0 & v=0 & i^\circ=i_w \\ y\to\infty & u=u_s & & i^\circ=i_s^\circ \end{array} \tag{10-34}$$

The parameter n is allowed to assume the value 0 or 1. For $n = 0$, the equations describe plane flow and for $n = 1$, rotationally symmetric flow. Equations of the form (10-16) are again required to solve the system of Eqs. (10-31) to (10-33) and are available provided local thermodynamic equilibrium exists in the boundary layer. Such relations for the properties exist even in a multicomponent gas mixture with chemical reactions occurring as long as chemical equilibrium is also established locally and as long as no mass transfer to the surface occurs. Equations (10-31) to (10-33), for instance, describe conditions in a dissociating or ionizing gas with local thermodynamic and chemical equilibrium.

The enthalpy i_w of the fluid at the wall surface will be assumed constant. The equations are again transformed by introducing a stream function defined by

$$\rho u r_0{}^n=\frac{\partial\psi}{\partial y} \qquad \rho v r_0{}^n=\frac{\partial\psi}{\partial s} \tag{10-35}$$

the dimensionless coordinates

$$\eta=\frac{\rho_s u_s r_0{}^n}{\sqrt{2\bar{s}}}\int_0^y\frac{\rho}{\rho_s}\,dy \tag{10-36}$$

with $\bar{s}$ defined by

$$\bar{s} = \int_0^s \rho_s \mu_s u_s r_0^{2n}\, ds \tag{10-37}$$

and

$$f(\eta) = \frac{\psi}{\sqrt{2\bar{s}}} \qquad g(\eta) = \frac{i^\circ}{i_s^\circ - i_w} \tag{10-38}$$

These are obtained by combining the Dorotnitzyn-Stewartson transformation and the Mangler transformation, discussed before. The dimensionless stream function f and the enthalpy parameter g are assumed to be functions of one variable η only. In addition, the abbreviations

$$C = \frac{\rho\mu}{\rho_s \mu_s} \qquad \beta = \frac{2\bar{s}}{u_s}\frac{du_s}{d\bar{s}} \tag{10-39}$$

will be used. The transformed differential equations have the form

$$(Cf'')' + ff'' + \beta\left(\frac{\rho_s}{\rho} - f'^2\right) = 0 \tag{10-40}$$

$$\left(\frac{C}{\Pr} g'\right) + fg' + \frac{u_s^2}{2(i_s^\circ - i_w)}\left[2C\left(1 - \frac{1}{\Pr}\right)f'f''\right]' = 0 \tag{10-41}$$

with the boundary conditions

$$\eta = 0 \qquad f' = 0 \qquad f = 0 \qquad g = \frac{i_w}{i_s^\circ - i_w} = g_w \tag{10-42}$$

$$\eta = \infty \qquad f' = 1 \qquad g = \frac{i_s^\circ}{i_s^\circ - i_w} = g_s \tag{10-43}$$

A prime again indicates differentiation with respect to η. As in the previous analyses,

$$u = u_s f'$$

Equations (10-40) and (10-41) are total differential equations under certain restrictions only, namely, when

$$\beta\left(\frac{\rho_s}{\rho} - f'^2\right),\ C,\ \Pr,\ \frac{u_s^2}{2(i_s^\circ - i_w)} \tag{10-44}$$

are constants or functions of η. The first term vanishes, and the others are functions of η for flow with u_s = const, because the static enthalpy i_s in the stream on which the properties depend is then a constant.

In the neighborhood of a stagnation point, u_s is small, so that the third term in Eq. (10-41) can be neglected. In addition, $i_s \approx i_s^\circ$, and the rest of the terms in (10-41) are again functions of η.

In both cases, the Eqs. (10-40) and (10-41) are total differential equations. Numerical solutions obtained on electronic computers have been reported in the recent literature including dissociation and ionization processes within the boundary layer.[1]

[1] For references see E. R. G. Eckert and E. Pfender, Plasma Heat Transfer, *Advan. Heat Transfer*, **4**:229–316 (1967).

Lees[1] used the same system of equations to develop an approximate method for calculating heat transfer as it occurs on the surface of space vehicles during the reentry through the atmosphere. The high supersonic velocities during this process cause the air between the shock wave and the surface to assume very high temperatures, whereas the surface of the spacecraft must be maintained at relatively low temperatures. Therefore, the enthalpy ratio i_w/i_s° in this application is small. The method is based on the approximation of local similarity as discussed in Sec. 7-2. The terms (10-44) can then be approximately considered as functions of η only. The system of Eqs. (10-40) and (10-41) with the boundary conditions as listed before was studied by Cohen and Reshotko[2] for a gas satisfying the conditions $C = 1$, $\Pr = 1$. With a somewhat differently defined β parameter

$$\bar{\beta} = \frac{2\bar{s}}{\mathrm{Ma}_s}\frac{d(\mathrm{Ma}_s)}{d\bar{s}} \tag{10-45}$$

and an enthalpy ratio $G = i^\circ/i_s^\circ$, the momentum equation (10-40) transforms to

$$f''' + ff'' + \bar{\beta}(G - f'^2) = 0 \tag{10-46}$$

$\bar{\beta}$ is a constant provided the Mach number Ma_s follows the relation $\mathrm{Ma}_s = C\bar{s}^m$. Such a flow can therefore be considered as a wedge-type flow. The energy Eq. (10-41) has now the form

$$G'' + \Pr fG' = 0 \tag{10-47}$$

From the solutions of Eqs. (10-46) and (10-47), the heat flux at the wall can be obtained in the following way:

$$q_w = -k_w\left(\frac{\partial t}{\partial y}\right)_w = -\frac{k_w}{c_{pw}}\left(\frac{\partial i^\circ}{\partial y}\right)_w$$

which, utilizing the relation $\Pr = \mu c_p/k$, can be written as

$$q_w = \frac{1}{\Pr_w}(g')_w\frac{u_s r_0{}^n \rho_w \mu_w}{\sqrt{2\bar{s}}}(i_w - i_s^\circ) \tag{10-48}$$

when the relation $g = G/(1 - G_w)$ is used. The dimensionless enthalpy gradient g'_w for the condition $\beta =$ const, $c_p =$ const, $\Pr = 1$, and $C = 1$, is presented in Fig. 10-10. The enthalpy ratio i_w/i_s° appears in the figure as a parameter. The figure shows that the dependence of the gradient g'_w on i_w/i_s° and on β is comparatively weak, especially for a highly cooled

[1] Lees, *loc. cit.*

[2] C. B. Cohen and E. Reshotko, *NACA Tech. Rept.* 1293, 1956.

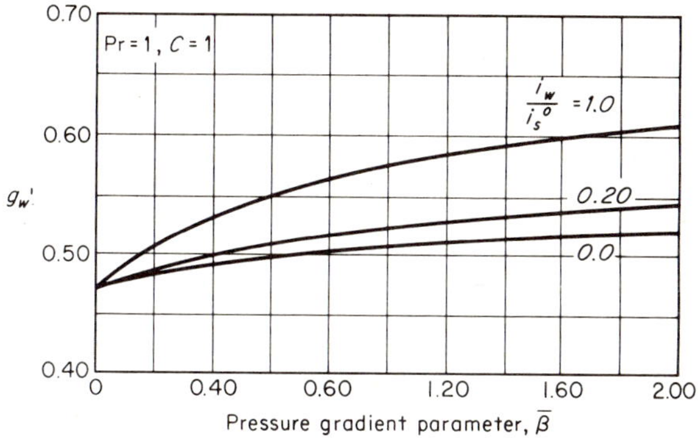

Fig. 10-10 Heat-transfer parameter as a function of pressure-gradient parameter and ratio of wall to stream enthalpy for laminar wedge-type flow of a variable-property fluid. [*After Cohen and Reshotko.*]

wall $i_w/i_s^\circ \ll 1$. Lees therefore proposed to use a mean value $(g_w)' = 0.5$. He also indicated that Eq. (10-48) is a good approximation for gases with a product $\rho\mu$ which is not constant when $\rho_w\mu_w$ is replaced by $\rho_s\mu_s$. For gases with a Prandtl number different from 1, it was found in the various analyses of this text that the Nusselt number varies approximately proportionally to $\mathrm{Pr}^{\frac{1}{3}}$. To account for this, the term Pr_w in the above equation should be replaced by $\mathrm{Pr}_w^{\frac{2}{3}}$. According to Eckert and Tewfik, Eq. (10-48) can also be used, in an even wider range of parameters, with good approximation for gases with property variations different from $\rho\mu$ = const when all the properties appearing in that equation are introduced at a reference enthalpy i^* defined by Eq. (10-30) and when the total enthalpy i_s° is replaced by the recovery enthalpy i_r. With these changes, the equation for the heat flux becomes

$$q_w = \frac{0.35}{\mathrm{Pr}^{*\frac{2}{3}}} \frac{\mu^*\rho^* u_s r_0{}^n}{\sqrt{\bar{s}^*}} (i_w - i_r) \tag{10-49}$$

The asterisk indicates that the property must be introduced at the reference enthalpy i^*. The equation again suggests defining a heat-transfer coefficient by Eq. (10-27), which leads to a Stanton number

$$\mathrm{St}^* = \frac{h_i}{\rho^* u_s} = \frac{0.35\mu^* r_0{}^n}{\sqrt{\bar{s}^*}} \tag{10-50}$$

The parameter $\bar{s}^*$ appearing in this equation must be obtained from the relation

$$\bar{s}^* = \int_0^s \rho^*\mu^* u_s r_0{}^{2n}\, ds \tag{10-51}$$

Equation (10-50) can now be used to calculate Stanton numbers for some important special flow situations.

Plane plate For this situation, u_s, ρ, μ, are independent of $s = x$, and $n = 0$. Equation (10-51) transforms to

$$s^* = \rho^* \mu^* u_s x$$

Introducing this into Eq. (10-50) gives for the Stanton number

$$\mathrm{St}^* = \frac{0.35}{\mathrm{Pr}^{*\frac{2}{3}}} \sqrt{\frac{\mu^*}{\rho^* u_s x}} = \frac{0.35}{\mathrm{Pr}^{*\frac{2}{3}} \sqrt{\mathrm{Re}_x^*}} \tag{10-52}$$

Plane stagnation flow For this flow, $r_0 = x$, $u_s = Cx$, and the properties ρ and μ in the neighborhood of the stagnation line can be considered as constant. n is again equal to 0. Equation (10-51) becomes with $s = x$

$$s^* = \rho^* \mu^* C \int_0^x x\,dx = \tfrac{1}{2} \rho^* \mu^* u_s x$$

and the Stanton number is

$$\mathrm{St}^* = \frac{0.5}{\mathrm{Pr}^{*\frac{2}{3}} \sqrt{\mathrm{Re}_x^*}} \tag{10-53}$$

Rotationally symmetric stagnation flow The stream velocity is $u_s = Cx$, $r_0 = x$; the properties ρ and μ are again considered constant; n has now the value 1. This gives

$$s^* = \rho^* \mu^* C \int_0^x x^3\,dx = \tfrac{1}{4} \rho^* \mu^* u_s x^3$$

The Stanton number is

$$\mathrm{St}^* = \frac{0.7}{\mathrm{Pr}^{*\frac{2}{3}} \sqrt{\mathrm{Re}_x^*}} \tag{10-54}$$

For flow around objects with arbitrary shape, β will vary along x. In this case Lees proposes to use the concept of local similarity by calculating the local heat flux for each location along the surface, using Eq. (10-50) and introducing the parameters in this equation as they exist at the point under consideration. This method has been used extensively for engineering calculations and has been found to approximate actual conditions well even for situations in which dissociation occurs within the boundary layer as a consequence of high temperatures. This will be discussed in greater detail in the chapter on mass transfer.

11
Special Heat-transfer Problems

11-1 HEAT TRANSFER IN LIQUID METALS

This section deals with heat transfer in liquid metals because these fluids have some characteristic features which are not present in others. The heat-conductivity values for liquid metals are much larger than those for other fluids, and correspondingly the Prandtl numbers are very small, covering a range of approximately 0.005 to 0.03. As a consequence of the high heat conductivity, the dominating part in the combined conduction and convection process is often heat conduction.

For laminar flows through ducts, the solutions obtained in Sec. 7-10 must be modified because heat conduction in an axial direction cannot be neglected, especially in the regions near the ends of the heated section. The temperature field at the entrance of the heated section extends into the upstream unheated region, and in the same way the temperature field at the exit of the heated section extends some distance downstream. This has been investigated analytically by Hennecke.[1]

[1] D. K. Hennecke, *Wärme- Stoffübertragung*, **1**:177–184 (1968).

For a constant wall temperature, heat conduction in the axial direction may even measurably affect the developed thermal region. In the case of constant heat flux, all temperatures increase or decrease linearly in the axial direction, and axial conduction creates a heat flux which is constant along the duct in the thermally developed regime. This, then, does not influence the temperature field in any cross section in this region; however, it has led in the past to errors in the determination of heat-transfer coefficients from experimental results. Conventionally, heat-transfer coefficients in duct flow are defined with the local difference of wall temperature and fluid bulk temperature. Since the bulk temperature is difficult to measure within the heated duct length, one measures the uniform temperature with which the fluid enters the heated duct section and provides a mixing device at the exit of the heated duct section, so that a temperature-sensing device placed in the fluid stream after the mixing measures the exit bulk temperature. For a fluid with constant specific heat and negligible axial conduction, it can be stated that the fluid bulk temperature rises linearly from the entrance to the exit cross section, as is shown by the dashed line t_B' in Fig. 11-1*a*. A measurement of the wall temperature t_w (as shown by the full line in the same figure) is then sufficient to determine the difference between wall temperature and bulk temperature in any cross section. This procedure, however, ceases to apply when axial conduction in the fluid becomes important. The fluid bulk temperature then changes to values t_B indicated by the full line in the same figure. A correction Δt has to be applied to the bulk temperatures t_B' determined as described above. This correction is constant in

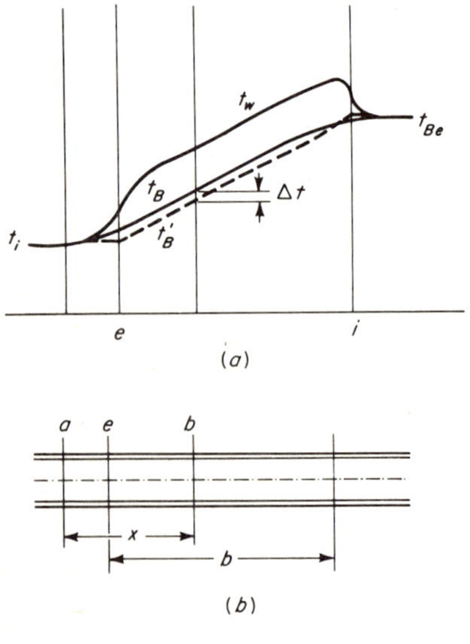

Fig. 11-1 Variation of wall temperature t_w and bulk temperature t_B in flow of a liquid metal through a tube with uniform heat flux q_w: e = entrance of heated duct section, i = exit of heated duct section, L = length of heated section.

the central part of the duct and can be calculated in this region by a simple energy balance on the control volume shown in Fig. 11-1*b*. The increase in bulk temperature in the fluid between the cross sections *aa* and *bb* is given by

$$\rho\bar{u}c_pA(t_B - t_i) = Q + kA\frac{dt_B}{dx} \tag{11-1}$$

where $\bar{u}$ = average fluid velocity
A = duct cross section
Q = heat addition to fluid up to cross section *bb*
t_i denotes the fluid temperature some distance upstream of the entrance to the heated duct at a location where the fluid temperature is still constant over the cross section. The axial temperature gradient dt_B/dx can be expressed by

$$\frac{dt_B}{dx} = \frac{t_{Be} - t_i}{L}$$

where t_{Be} denotes the bulk exit temperature and L the heated duct length. Equation (11-1) becomes

$$\rho\bar{u}c_pA(t_B - t_i) = Q + kA\frac{t_{Be} - t_i}{L} \tag{11-2}$$

Without axial conduction the bulk temperature t'_B is described by

$$\rho\bar{u}c_pA(t'_B - t_i) = Q \tag{11-3}$$

Subtraction of Eq. (11-3) from (11-2) leads to the desired equation for the temperature correction Δt

$$\Delta t = \frac{k}{\rho\bar{u}c_p}\frac{t_{Be} - t_i}{L} \tag{11-4}$$

This correction becomes important for large thermal-conductivity values k (liquid metals) and for small velocities $\bar{u}$. The necessity of applying this correction in experimental determinations of heat-transfer coefficients was pointed out by Trefethen.[1]

Turbulent duct flow of liquid metals can be calculated in a simple way as long as the heat transport by turbulent mixing can be considered small compared with the conductive transport. This is the case at moderate Reynolds numbers beyond the critical value. To a first approximation, the turbulent-velocity profile can be replaced by one of constant

[1] L. Trefethen, *Trans. ASME*, **78**:1207 (1956). Of course, it may also be necessary to include heat conduction in the duct wall.

velocity. Such a flow is usually referred to as *slug flow*. The calculations discussed in Sec. 7-10 can easily be modified for this simplified velocity distribution. One thus finds that the Nusselt number for constant heat flow into a tube is

$$\mathrm{Nu}_d = 8 \tag{11-5}$$

and for constant wall temperature

$$\mathrm{Nu}_d = 5.8 \tag{11-6}$$

Actually the equation describing heat transfer for slug flow has the same form as the equation for unsteady heat conduction in a solid body. Heat transfer in the entrance section of a tube can therefore be easily obtained from solutions to the unsteady-heat-conduction process into a rod with circular cross section[1] when axial conduction can be neglected.

Measurements with liquid metals have indicated that for turbulent flow through circular tubes, the heat exchange caused by turbulent mixing becomes appreciable and cannot be neglected when the parameter $\mathrm{Re}_d\,\mathrm{Pr}$, called the *Peclet number*, exceeds a value of approximately 50. Much used in this region is the equation

$$\mathrm{Nu}_d = 7 + 0.025(\mathrm{Re}_d\,\mathrm{Pr})^{0.8} \tag{11-7}$$

developed by Lyons and usually referred to as the *Lyons-Martinelli equation*. The equation describes heat transfer for thermally developed conditions, turbulent flow through a tube, and constant heat flux at the wall.

The constant in this equation is somewhat smaller than the value 8 in Eq. (11-5), taking into account that the turbulent velocity profile is not quite constant over the cross section.

According to the preceding discussion, the Nusselt number is expected to vary with the Peclet number $\mathrm{Re}_d\,\mathrm{Pr}$ as sketched in Fig. 11-2 for constant heat flux into the fluid. Up to the critical Reynolds number, at which transition to turbulent flow occurs, the Nusselt number has the value 4.36 (Sec. 7-10). The transition Reynolds number is assumed in Fig. 11-2 to occur at a Peclet number of 40. Beyond this value, the Nusselt number rises to a value of 7, as expected for turbulent flow and moderate Peclet numbers and finally follows Eq. (11-7). This behavior was suggested by Dwyer.[2] Measurements by various investigators in the

[1] Heat transfer for turbulent flow through ducts with various cross sections is calculated this way by J. P. Hartnett and T. F. Irvine, *AIChE J.*, **3**:313–317 (1957).

[2] See O. E. Dwyer, in Atomic Energy Commission, "Liquid Metals Handbook," 2d ed., Department of the Navy, Washington, D.C., 1952.

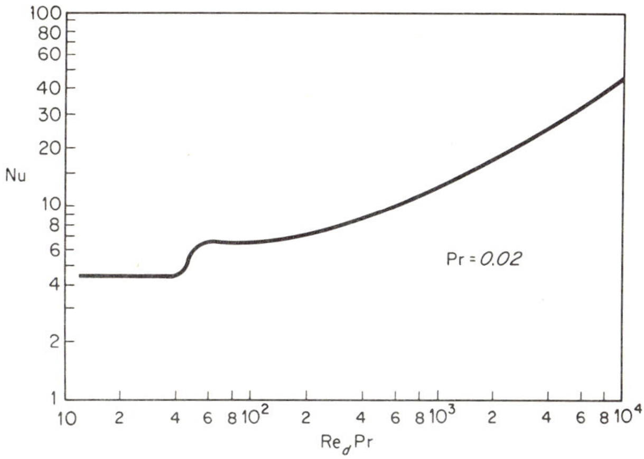

Fig. 11-2 Nusselt number for flow of a liquid metal through a tube with uniform heat flux q_w.

laminar and near-turbulent region deviate from each other and so far have not clearly indicated the effect of the transition to turbulence. Measurements in this region are difficult, and side effects often cannot be readily assessed. For example, it is frequently impossible to avoid superimposed free convection. Longitudinal conduction, as discussed before, may cause large errors. Opinion is also divided on how nonwetting of some liquid metals on solid surfaces influences heat transfer, and numerous experiments still have not completely clarified this point.

Boundary-layer flow is characterized by the fact that in a low Prandtl number fluid the thermal boundary layer is much thicker than the flow boundary layer. On this basis, the Nusselt number was calculated in Sec. 7-7 for the limiting case $\mathrm{Pr} \to 0$. The equation

$$\mathrm{Nu}_x = \frac{1}{\sqrt{\pi}} \sqrt{\mathrm{Re}_x \, \mathrm{Pr}} = 0.565 \sqrt{\mathrm{Re}_x \, \mathrm{Pr}} \qquad (11\text{-}8)$$

was obtained. Grosh and Cess[1] solved numerically the equation describing the temperature field in Sec. 7-7 and obtained values which were 7 to 12 percent smaller than the ones calculated from the equation above in a Prandtl number range between 0.005 and 0.025. They also showed that, for values $\mathrm{Re}_x \, \mathrm{Pr} > 50$, neglect of the longitudinal heat conduction is justified except in the immediate neighborhood of the plate's leading edge.

The heat transfer around the circumference of a cylinder with its axis normal to the flow direction can be calculated if the decrease of the velocity within the flow boundary layer is neglected. In an early paper

[1] R. J. Grosh and R. D. Cess, *Trans. ASME*, **80**:667–676 (1958).

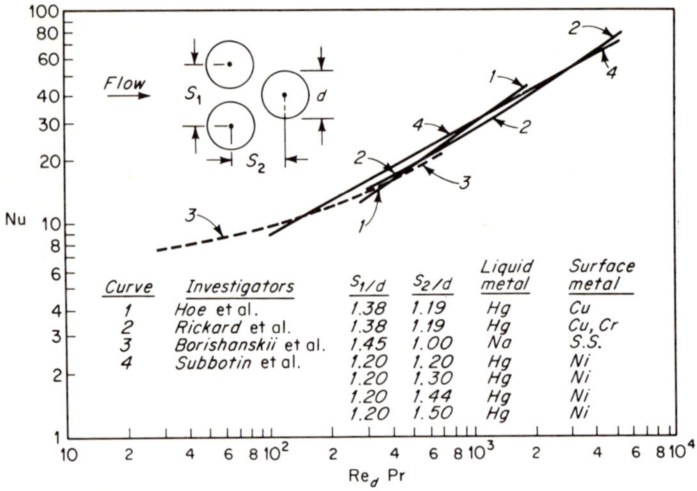

Fig. 11-3 Nusselt number for flow of a liquid metal through a bundle of staggered tubes. [*From O. E. Dwyer in Atomic Energy Commission, "Liquid Metals Handbook," Department of the Navy, Washington, D.C., 1952.*]

Boussinesq[1] showed that the energy equation applying to heat conduction in a two-dimensional inviscid flow field can be simplified when it is transformed from the geometric coordinates x, y to the flow potential φ and the stream function ψ as independent coordinates. It becomes

$$u_s \frac{\partial t}{\partial \varphi} = \alpha \left(\frac{\partial^2 t}{\partial \varphi^2} + \frac{\partial^2 t}{\partial \psi^2} \right) \tag{11-9}$$

If, as before, longitudinal conduction is neglected, the term $\partial^2 t/\partial \varphi^2$ becomes zero and the equation assumes the same form as that for a flat plate. The Nusselt number is then obtained by transforming the solution back to the geometric coordinates x and y. In this way Grosh and Cess calculated that the average Nusselt number for a single circular cylinder is

$$\overline{\mathrm{Nu}}_d = 1.015(\mathrm{Re}_d\ \mathrm{Pr})^{\frac{1}{2}} \tag{11-10}$$

The calculation assumes that the heat transport connected with the vortex motion in the back of the cylinder is negligibly small as compared with heat transport by conduction. Comparison with measurements indicates that this assumption is justified for values of the parameter Re Pr smaller than 500.†

[1] K. J. Boussinesq, Calcul du pouvoir refroidissant des courrants fluides, *J. Math.*, **1**:285 (1905).

† Much useful information on heat transfer of liquid metals is found in Atomic Energy Commission, *op. cit.*

Liquid metals are used as coolants for nuclear reactors. The geometries involved are mainly tube bundles with the liquid metal flowing in a direction parallel or normal to the tubes. Much analytical and experimental information has been collected on such configurations.[1]

Figure 11-3, presents, as an example, the results of measurements for a triangular rod assembly. It applies to a rod in the interior of the bundle. Circumferential heat conduction in the rod was small. The velocity in the Peclet number Pe = Re Pr is the value in the narrowest cross section. The agreement between the results obtained by the various investigators is quite satisfactory.

11-2 TRANSPIRATION AND FILM COOLING

Special cooling methods have been developed in an attempt to protect structural elements from the influence of a hot gas stream. The methods are applied in combustion chambers, in exhaust nozzles of rockets and jet engines, for gas-turbine blades, and in other technological applications.

Figure 11-4 presents sketches of these cooling methods. The upper left-hand sketch shows the standard *convection cooling,* in which the hot fluid is on one side and the coolant on the other side of the wall. The upper right-hand arrangement shows the *film-cooling process,* in which a stream of coolant is blown through a series of openings in a direction nearly tangential to the surface. In this way a layer is created which insulates the wall from the hot fluid. The coolant film is gradually destroyed by mixing with the hot fluid, so that its effectiveness decreases in the downstream direction. This disadvantage can be avoided by the

[1] Dwyer, *loc. cit.;* also R. P. Stein, Liquid Metal Heat Transfer *Advan. Heat Transfer,* **3**:101–174 (1966).

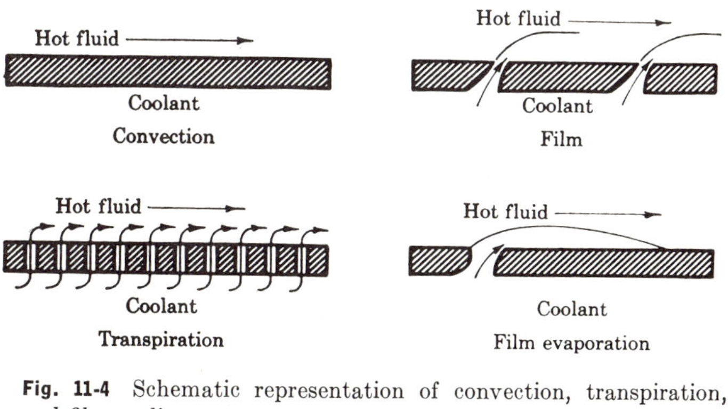

Fig. 11-4 Schematic representation of convection, transpiration, and film cooling.

process sketched in the lower left-hand part of the figure, which is generally referred to as *transpiration cooling*. In this method the wall is manufactured from a porous material, and the coolant is blown through the pores. The coolant film on the hot-fluid side is therefore continuously renewed, and the cooling effectiveness can be made constant along the surface. It has been assumed so far that the coolant and the hot fluid are both either gases or liquids. When the hot medium is a gas, the cooling effectiveness can be greatly increased by using a liquid as a coolant. This method is shown in the lower right-hand sketch of the figure. A liquid film is created on the hot-gas side of the wall, the liquid evaporates from its surface, and the heat absorbed in the evaporation process substantially increases the effectiveness of this cooling method, which may be called *evaporative film cooling*.

A very serious problem in the development of missiles, satellites, and spaceships is caused by the excessive heating of the skin of these vehicles during reentry through the atmosphere. A special method developed for this situation is *ablation cooling*. The skin is covered by a layer of a material which sublimes, melts, or decomposes when the temperature in the boundary layer increases by aerodynamic heating. The chemical process absorbs heat, and the mass flow of the generated gas away from the skin blocks the heat flux from the boundary layer toward the skin. In this way a very effective cooling is obtained.

In this section, consideration will be restricted to the processes within the boundary layer on the hot-gas side of the wall. From this viewpoint the processes in transpiration cooling, in evaporative film cooling, and in ablation cooling are basically similar since all are influenced by a continuous mass flow away from the surface. This group of cooling methods is therefore sometimes referred to as *mass-transfer cooling*. It will be discussed here under the heading *transpiration cooling*. Film cooling differs from transpiration cooling in that a local mass injection creates a cooling effect downstream from the point of injection. The processes are again similar in nature, whether the injection occurs through a slot in the wall, through a strip of a porous material, or by a strip of an ablating material. Evaporative film cooling creates a similar situation downstream from the location where the liquid film ends. All these processes will be considered under film cooling.

Transpiration cooling In this section, it will be assumed that the coolant is the same fluid as the mainstream or that both fluids differ very little in their properties. In addition, the properties will be considered constant.

The velocity field describing the laminar boundary layer with transpiration cooling has already been discussed for the above situation in Sec. 7-2, where it was pointed out that such boundary layers are self-similar for wedge-type flow and for the condition that the velocity v_w with which the coolant is ejected from the porous surface follows Eq. (7-20). Velocity profiles were presented in Fig. 7-3.

The temperature profiles in the laminar boundary layer are obtained by integration of the equation

$$\rho c_p \left(u \frac{\partial t^\circ}{\partial x} + v \frac{\partial t^\circ}{\partial y} \right) = k \frac{\partial^2 t^\circ}{\partial y^2} + \left(1 - \frac{1}{\mathrm{Pr}} \right) \mu \frac{\partial^2}{\partial y^2} \frac{u^2}{2} \tag{11-11}$$

with the boundary conditions

$$\begin{aligned} y &= 0 \qquad & t^\circ &= t_w \\ y &\to \infty \qquad & t^\circ &= t_s^\circ \end{aligned} \tag{11-12}$$

The differential equation and boundary conditions are the same as in Sec. 10-1. The velocity components u and v, however, must be introduced with the values as they apply to fluid ejection with the proper velocity v_w from the porous wall. The differential equation (11-11) is again linear, indicating that the procedure developed in Sec. 10-1 can be applied. The heat flux q_w at the wall surface is accordingly calculated with a heat-transfer coefficient h defined by

$$q_w = h(t_w - t_r) \tag{11-13}$$

with t_w denoting the wall-surface temperature and t_r the recovery temperature. The heat-transfer coefficient can be obtained by a solution of the homogeneous equation belonging to Eq. (11-11). Temperature profiles obtained in this way are presented in Fig. 11-5 for a self-similar

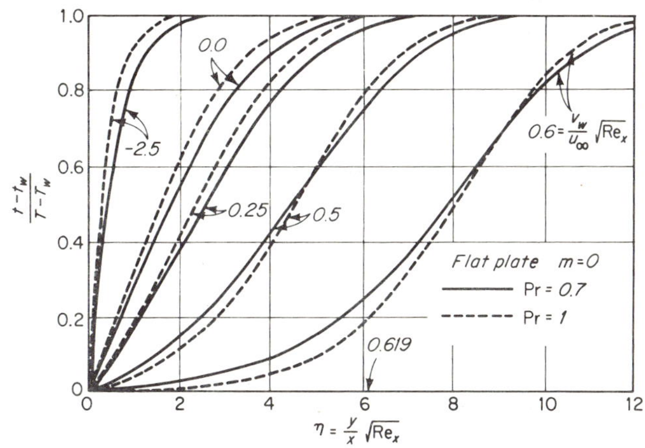

Fig. 11-5 Dimensionless temperature profiles for laminar flow over a flat plate. [*From J. P. Hartnett and E. R. G. Eckert, Trans. ASME,* **79**:*847* (*1957*).]

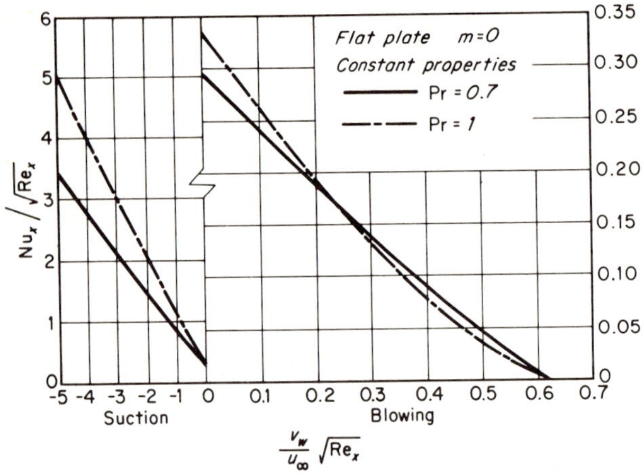

Fig. 11-6 Dimensionless local heat-transfer parameter as a function of suction or blowing rate for laminar flow over a flat plate. [*From J. P. Hartnett and E. R. G. Eckert, Trans. ASME,* **79**:*247* (*1957*).]

laminar boundary layer on a flat plate for the condition of a uniform wall temperature. The parameter on the curves, which are presented for two Prandtl numbers, is a dimensionless expression characterizing the blowing velocity v_w. It should be kept in mind that for the self-similar boundary layer the velocity v_w is proportional to $1/\sqrt{x}$ according to Eq. (7-20). The temperature gradient at the surface determines the heat flux q_w defined by the equation $q_w = -k\ (\partial t/\partial y)_w$ and the heat-transfer coefficient h. Figure 11-6 presents the parameter $\mathrm{Nu}_x/\sqrt{\mathrm{Re}_x}$, from which the heat-transfer coefficient can be calculated, using the definition $\mathrm{Nu}_x = hx/k$, for laminar flow over a flat plate, and Fig. 11-7 presents the same parameter for a laminar boundary layer in plane stagnation flow ($\beta = 1$).[1]

The recovery temperature t_r in Eq. (11-13) is obtained by a particular solution of the complete equation (11-11) corresponding to $q_w = 0$. Figure 11-8 presents the temperature recovery factor r for laminar flat-plate flow. The recovery temperature t_r is obtained from this recovery factor with the defining equation $r = (t_r - t_s)/(u_s^2/2c_p)$, in which t_s indicates the static stream temperature and u_s the velocity in the free stream.

Interesting information can be obtained from an energy balance on a control volume, as indicated in Fig. 11-9, which presents a cross section through the porous wall. The dashed line 1-2 may be located just outside the wall surface on the hot-fluid side. Coolant flows through this control

[1] For an extensive listing of solutions to self-similar boundary layers, including mass transfer at the wall; see C. F. Dewey and J. F. Gross, *Advan. Heat Transfer*, **4**:317–446 (1967).

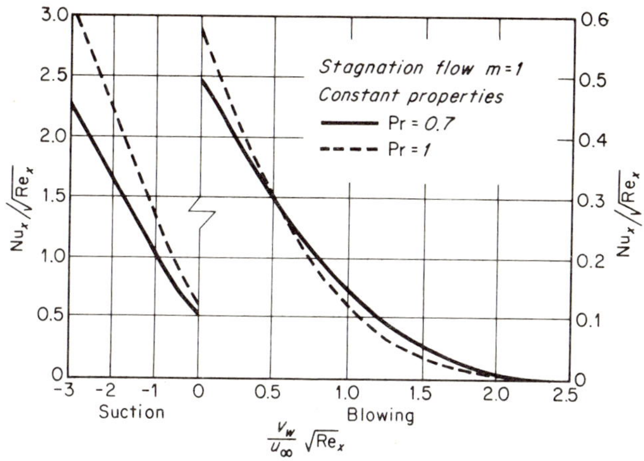

Fig. 11-7 Dimensionless local heat-transfer parameter for laminar plane stagnation flow. [*After J. P. Hartnett and E. R. G. Eckert.*]

surface with the velocity v_w. It moves toward the wall with the same velocity on the coolant entry side. A boundary layer will exist on this side of the wall, which under the influence of suction of fluid into the wall will be quite thin, as demonstrated in Sec. 7-2. The control surface 3-4 may be located outside of this boundary layer. The control volume is

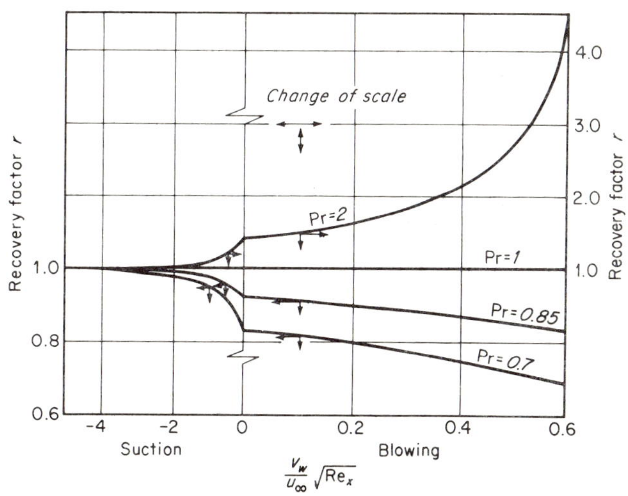

Fig. 11-8 Temperature recovery factor for laminar flow over a flat plate. [*From J. P. Hartnett and E. R. G. Eckert, Trans. ASME,* **79**:*247* (*1957*).]

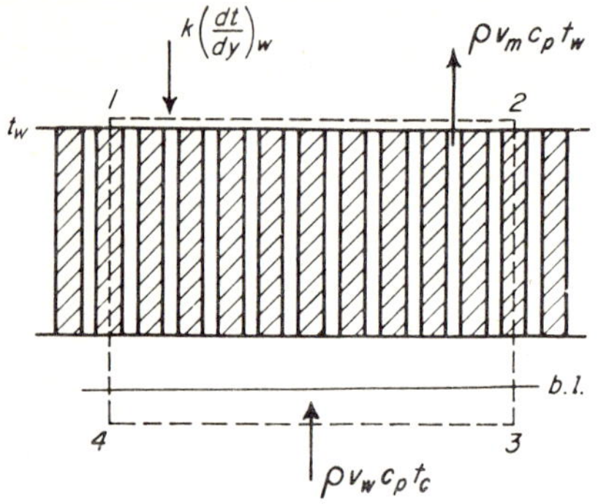

Fig. 11-9 Energy balance on a transpiration-cooled wall.

closed by a cylindrical control surface normal to planes 1-2 and 3-4. An energy balance has to include a heat flux $-k(\partial t/\partial y)_w$ and an enthalpy-convection flux $\rho v_w c_p t_w$ through plane 1-2. t_w denotes the fluid temperature at the wall surface. No heat will be transported by conduction through plane 3-4 because it is located outside the boundary layer, and only a convective enthalpy flux $\rho v_w c_p t_c$ has to be considered in this plane. t_c is the temperature with which the coolant is supplied. Neglecting heat fluxes within the wall in directions parallel to the surface and radiative heat transfer and assuming a steady state lead to the following form of the energy balance:

$$q_w = k\left(\frac{dt}{dy}\right)_w = \rho v_w c_p(t_w - t_c) \tag{11-14}$$

or with a heat-transfer coefficient h

$$h(t_r - t_w) = \rho v_w c_p(t_w - t_c) \tag{11-15}$$

For a self-similar boundary layer, the heat-transfer coefficient and the velocity v_w vary in the same way with x, and the recovery temperature is independent of x. For the flat plate, this can easily be checked with Figs. 11-6 and 11-8 and Eq. (7-20). The temperature t_w with which the fluid leaves the wall is then, according to Eq. (11-15), independent of x when the coolant is supplied with a constant temperature t_c.

Equation (11-15) offers also a convenient means of determining the heat-transfer coefficient h by experiments. The porous wall is usually manufactured in such a way that it contains a large number of very small and tortuous passages. The temperature of the solid surface on the ejection side is then practically equal to the fluid temperature t_w at that surface. A measurement of this temperature, of the coolant temperature t_c, and of the velocity v_w or the mass flux ρv_w is then sufficient to calculate h with Eq. (11-15). The recovery temperature is obtained by a measurement of the wall-surface temperature t_w for the condition $q_w = 0$. According to Eq. (11-14), this is the case when $t_w = t_c$.

It should be pointed out that some arbitrariness exists in the definition of a heat flux and a heat-transfer coefficient in transpiration cooling. Figure 11-9 indicates that a heat flux by conduction and an enthalpy transport by convection occur at the porous surface. The heat flux in the preceding discussion was defined as

$$q_w = -k\left(\frac{\partial t}{\partial y}\right)_w$$

The Nusselt numbers in Figs. 11-6 and 11-7 are based on this heat flux. Sometimes the convective energy transport is included in the heat-transfer coefficient in the literature, so that caution is necessary in using data on transpiration cooling.

The energy equation of a turbulent boundary layer with constant properties is again linear. Use of Eq. (11-13) to define a heat-transfer coefficient therefore takes care of the effect of viscous dissipation by introducing the recovery temperature, and the heat-transfer coefficient can be determined from Eq. (8-39) (Sec. 8-2), in which the dissipation term is excluded. When using this equation, we again consider a steady two-dimensional boundary layer on a flat plate. The velocity components $\bar{u}$ and $\bar{v}$ in this equation must be determined from an integration of the continuity and momentum equations (8-5) and (8-6) with the new boundary conditions

$$\begin{aligned} y &= 0 \qquad \bar{u} = 0 \qquad \bar{v} = v_w \\ y &\to \infty \qquad \bar{u} = \bar{u}_s \end{aligned} \tag{11-16}$$

To carry out an analysis we need assumptions regarding the turbulent-exchange parameters. A modified law of the wall has been obtained by Stevenson[1] in the following way. The flow field in the region close to the wall is approximated by the model of a simple shear flow in which $\partial\bar{u}/\partial x = 0$. The continuity equation then reduces to the statement $\bar{v} = \text{const} = v_w$, and the momentum equation is for the flat plate $(\partial\bar{p}/\partial x = 0)$

$$\mu\frac{d^2\bar{u}}{dy^2} - \rho\frac{d(\overline{u'v'})}{dy} - \rho\bar{v}_w\frac{d\bar{u}}{dy} = 0$$

A limiting relation for vanishing wall distance is obtained by assuming that the turbulent momentum transfer decreases toward zero at the wall surface provided the Reynolds number based on the ejection velocity v_w

[1] T. N. Stevenson, A Law of the Wall for Turbulent Boundary Layers with Suction or Injection, *College Aeron. Cranfield Rept. Aeron.* 166; *Brit. Aerons. Res. Council* 26,025, 1964.

and the pore diameter is sufficiently small. The above equation simplifies in this region to

$$\mu \frac{d^2\bar{u}}{dy^2} - \rho v_w \left(\frac{d\bar{u}}{dy}\right) = 0$$

Integration of this equation with the condition that at the wall

$$\tau_w = \mu \left(\frac{du}{dy}\right)_w$$

results in the following relation between the dimensionless wall distance and the dimensionless velocity in the boundary layer:

$$y^+ = \frac{v^*}{v_w} \ln\left(1 + \frac{v_w}{v^*} u^+\right) \tag{11-17}$$

The shear velocity $v^* = \sqrt{\tau_w/\rho}$ has again been introduced; u^+ denotes u/v^*, and y^+ stands for yv^*/v. In the outer portion of the wall layer, on the other hand, the laminar contribution to the shear can be neglected. If, in addition, Prandtl's mixing-length relation [Eq. (8-9)] is introduced with the assumption $l = \varkappa y$, the following relation is obtained:

$$\frac{2v^*}{v_w}\left[\left(\frac{v_w}{v^*} u^+ + 1\right)^{\frac{1}{2}} - 1\right] = \frac{1}{\varkappa} \ln y^+ + C \tag{11-18}$$

The constants in this equation are determined, so that the equation is identical with Eq. (8-13) for $v_w/v^* = 0$. Dvorak and Head[1] introduced an additional relation in which the right-hand side of Eq. (11-18) was replaced by

$$C_0 + C_1 \ln y^+ + C_2 \ln^2 y^+ + C_x \ln^3 y^+$$

to provide a smooth transition between Eqs. (11-17) and (11-18). The constants in this expression were determined accordingly. These equations together with a *velocity-defect law* to describe the velocity profile in the outer layer were sufficient for calculating the velocity field and, with the assumption of a constant turbulent Prandtl number, the temperature field and the heat transfer to the wall surface.[2]

The advent of the electronic computer has made it possible to attack the partial differential equations themselves with more general statements for the turbulent-exchange parameters.[3]

A convenient presentation of the results of such a calculation is obtained in a diagram in which the parameter $\rho v_w c_p/h_0$ is used as abscissa

[1] F. A. Dvorak and M. R. Head, Effect of Uniform Injection on Heat Transfer in the Constant Property Turbulent Boundary Layer, *ASME Preprint* 68-WA/HT-24, 1968.

[2] For a detailed discussion of this calculation, see *ibid.* and Stevenson, *op. cit.*

[3] Such a procedure is described in D. B. Spalding and S. V. Pantakar, "Heat and Mass Transfer in Boundary Layers," Morgan-Grampian, London, 1967.

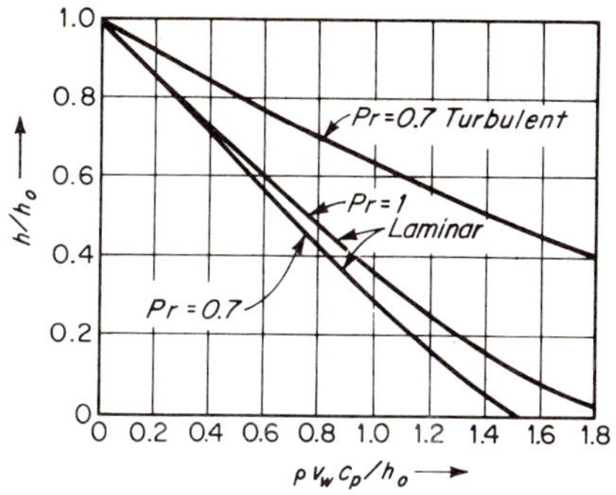

Fig. 11-10 Ratio of heat-transfer coefficient h with transpiration cooling to heat-transfer coefficient h_0 without transpiration for laminar and turbulent flow over a flat plate. [*After J. P. Hartnett and E. R. G. Eckert.*]

and h/h_0 as ordinate with h_0 denoting the heat-transfer coefficient as obtained for $v_w = 0$. The turbulent curve in Fig. 11-10 is based on an analysis by Rubesin[1] and on various experimental results; the two laminar curves are taken from Fig. 11-6. It can be observed that considerable reduction of the heat transfer from a hot gas to a wall surface can be obtained by transpiration cooling, especially for a laminar boundary layer. The curves in Fig. 11-10 are for design calculations often approximated by straight lines, a procedure which leads to simple equations describing the reduction in heat transfer by transpiration cooling.

Film cooling In this method a solid surface is protected from the influence of a hot gas stream by local release of a coolant from the surface. A cool layer is thus maintained for some distance downstream from the coolant ejection. The present discussion will be restricted to situations in which the coolant as well as the fluid in the mainstream are gases.

Figure 11-11 shows various ways of applying film cooling. In Fig. 11-11*a* the surface which is to be protected against the hot gas flowing over it is interrupted by a slot, and the coolant gas is ejected through the slot in a downstream direction. Figure 11-11*b* shows a similar arrangement but with a so-called *stepdown slot.* In Fig. 11-11*c* the coolant gas is ejected through a porous section in the wall and thus protects the solid portion downstream from the porous section. In Fig. 11-11*d* the coolant gas is generated from an ablating material and in Fig. 11-11*e* from a liquid film so that the generated vapor protects the downstream portion of the wall which is not covered by the liquid film. The ejection in Fig. 11-11*a* and *b* can be through slots which are continuous normal to the flow direction, or it can be through a series of interrupted slots or a series of holes. Similar continuous or interrupted ejection can be conceived with the arrangements presented in Fig. 11-11*c*

[1] M. Rubesin, *NACA Tech. Note* 3341, 1954.

and d. It has been pointed out that the effectiveness of film cooling decreases in the downstream direction, and in many applications new slots are provided at a certain distance from the first in order to renew the coolant layer.

One therefore has a large variety of geometric configurations in film-cooling devices to deal with. In addition, the film-cooling process depends on the dimensionless parameters describing the main flow as well as the coolant flow at the point of ejection. In the main flow, the characteristics of the boundary layer at the point of ejection are especially important; they are described, for instance, by a dimensionless expression for the boundary-layer thickness and by the state of turbulence within the boundary layer. Film cooling is often applied to situations with very large temperature differences, and the variation of properties throughout the flow field participating in the heat-transfer process can also become an important factor influencing the effectiveness of film cooling.

Thus the large number of parameters which must be considered requires a very extensive program if general quantitative information of the film-cooling process is desired. As a consequence, the effect of various parameters is not accurately known today. In the present discussion it will be assumed that the boundary layer at the point of ejection is a fully developed turbulent one, that the fluids involved are gases, and (for most of the discussion) that the temperature differences are sufficiently small for the thermodynamic and transport properties involved to be considered constant.

Continuous ejection The assumption of constant thermodynamic and transport properties which follows from the conditions listed above introduces a simplification, namely, that the shape of the temperature field and the effectiveness parameter, which will be introduced later, are the same whether the ejected fluid is colder or hotter than the fluid of the mainstream. This fact has been utilized in experiments which are easier to perform with a heated "coolant." Starting now from the consideration that the temperature field in the boundary layer (and especially the temperature of the wall downstream from the slot) is the main concern, one can argue that a simplified analysis can be based on a model which simulates the heat addition $m_s c_p t_s$ by the ejected stream m_s (per unit width) with temperature t_s into the boundary layer but neglects the influence of mass addition. This means that one replaces the mass ejection through the slot by a line heat source or sink arranged at the slot location.[1] Figure 11-12 then presents this model with a turbulent bound-

[1] E. Eckert, Transpiration and Film Cooling, *Heat Transfer Symp. Univ. Mich. 1952*, p. 195, 1953.

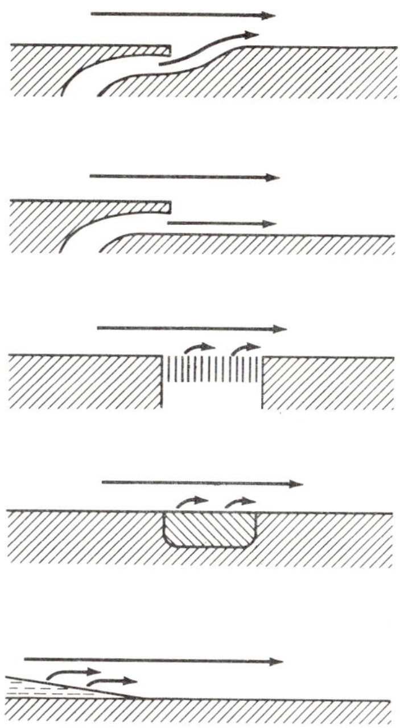

Fig. 11-11 Various film-cooling arrangements.

ary layer starting at a distance L upstream from the line heat source, whereas x measures the downstream distance from the source. Figure 11-12a indicates the heat-flux distribution along the surface, a heavy arrow representing the heat release by the line source and the area filled out with small arrows an additional heat release by a heat flux through the wall downstream from the source. As long as properties are considered constant, the energy equation of the turbulent boundary layer is linear. Therefore the law of superposition holds, and the situation sketched in Fig. 11-12a can be analyzed by superposition of the two heat-flux distributions sketched in Fig. 11-12b and c. The connection between heat flux and wall-surface temperature in Fig. 11-12c is conventionally described by a heat-transfer coefficient defined through the equation

$$q_w = h(t_w - t_e)$$

with the subscripts w and e denoting the wall surface and the free stream, respectively. The temperature assumed by an adiabatic wall downstream from the slot, as indicated in Fig. 11-12b, is denoted t_{aw} and referred to as adiabatic wall temperature. Superposition of the cases in Fig. 11-12c and b with proper boundary conditions leads (after a simple calculation) to the result that the wall temperature at the simultaneous

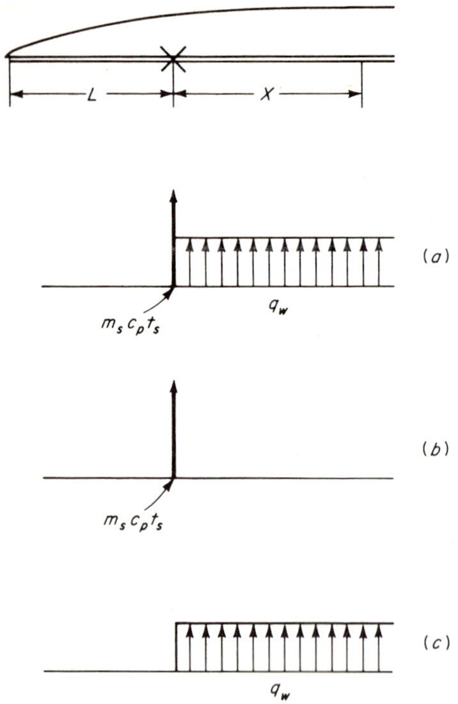

Fig. 11-12 Heat-source model of film cooling.

presence of the line heat source and the distributed heat sources, as indicated in Fig. 11-11*a*, is obtained from the equation

$$q_w = h(t_w - t_{aw}) \tag{11-19}$$

where h = same heat-transfer coefficient as for the Fig. 11-12*c* without film cooling
t_w = actual wall temperature
q_w = actual heat flux in Fig. 11-12*a*
t_{aw} = adiabatic wall temperature for Fig. 11-12*b*

This holds for any distribution of q_w along the coordinate x as long as it is the same in Fig. 11-12*a* and *c*. The validity of the superposition law for the actual film-cooling process has been checked experimentally. Figure 11-13 presents as an example the ratio of the heat-transfer coefficient h obtained with film cooling (corresponding to Fig. 11-12*a*) to the heat-transfer coefficient h_0 without film cooling (corresponding to Fig. 11-12*c*) plotted over the dimensionless distance x/s downstream from the slot for the film-cooling system of Fig. 11-11*a*.[1] The parameter on the

[1] J. P. Hartnett, R. C. Birkebak, and E. R. G. Eckert, "International Developments in Heat Transfer," pt. IV, p. 682, ASME, New York, 1961.

curves is the ratio of the density ρ_s times the exit velocity u_s from the slot to the external mass velocity $\rho_e u_e$ in the mainstream outside the boundary layer. The experiments were performed on a flat plate with a constant velocity u_e and a turbulent boundary layer arriving at the slot. It can be observed that, starting with a distance $x/s = 22$, the two heat-transfer coefficients differ by less than 10 percent as long as the ratio $\rho_s u_s/\rho_e u_e$ is smaller than 1. An approximation of this order is often acceptable in engineering calculations. Additional experiments confirming the superposition rule have been reported by Libby and Cresci.[1] This means that further discussion can now be restricted to the situation in which the wall downstream from the slot is adiabatic, since information on heat transfer to solid surfaces with the proper heat flux condition has already been presented.

For the adiabatic condition, some further information can be obtained from the heat-source model, again applying the law of superposition. A method[2] is available by which wall temperatures can be calculated for any arbitrary prescribed heat-flux distribution. This method was applied by Tribus and Klein to the heat-source model,[2] and the following relation was

[1] P. A. Libby and R. J. Cresci, *J. Aerospace Sci.*, **28**:51 (1961).

[2] M. Tribus and J. Klein, Forced Convection from Non-isothermal Surfaces, *Heat Transfer Symp. Univ. Mich., 1952,* p. 211, 1953.

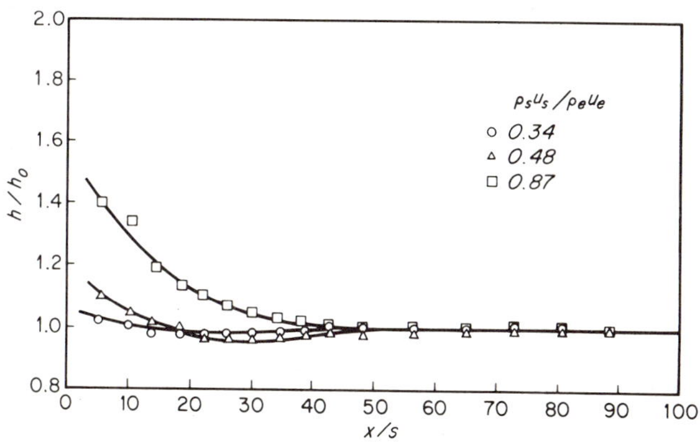

Fig. 11-13 Ratio of heat-transfer coefficient h with film cooling to heat-transfer coefficient h_0 without film cooling as a function of dimensionless distance x/s. *[From J. P. Hartnett, R. C. Birkebak, and E. R. G. Eckert, Velocity Distributions, Temperature Distributions, Effectiveness and Heat Transfer in Cooling of a Surface with a Pressure Gradient, "International Developments in Heat Transfer." pt. IV, pp. 682–689, ASME, New York, 1961.]*

obtained for a turbulent boundary layer on a flat plate and for a Prandtl number of 0.72:

$$\eta_0 = 5.76 \, \mathrm{Pr}^{\frac{2}{3}} \, \mathrm{Re}_s{}^{0.2} \left(\frac{\mu_c}{\mu_e}\right)^{0.2} \frac{c_{pc}}{c_{pe}} \left(\frac{\rho_e u_e x}{m_s}\right)^{-0.8} a \tag{11-20}$$

η_0 is called the film-cooling effectiveness. The slot Reynolds number Re_s is defined as m_s/μ_c with μ_c denoting the viscosity of the coolant. The parameter a describes the effect of the boundary-layer development upstream from the slot. The following relation is obtained by the same method:[1]

$$a = \left[1 - \left(\frac{L}{x + L}\right)^{0.9}\right]^{-0.8}$$

Sufficient experiments, however, are not yet available to judge how well this relation describes the influence of the upstream boundary layer, and it will be set equal to 1 for the following considerations. A differentiation between coolant properties (μ_c, c_{pc}) and mainstream properties (μ_e, c_{pe}) was achieved through the relation describing the heat flux of the line source in terms of the mass ejection, but it can be considered as only tentative.

Equation (11-20) applied to film cooling neglects the effect of mass ejection. One can therefore expect that it agrees best with reality for vanishing values of the dimensionless parameter $m_s/\rho_e u_e x$. This has been verified by the available experiments. A number of analyses have been published which include the effect of mass ejection in order to obtain an expression for the film-cooling effectiveness valid at larger values of this parameter. We discuss here the analysis by Goldstein and Haji-Sheikh,[2] which leads to a good representation of experimental results. The mass flow m within the boundary layer through a plane aa in Fig. 11-14 is composed of ejected fluid m_s and fluid m_e entering the boundary layer from the main flow according to

$$m = m_s + m_e$$

[1] E. R. G. Eckert and J. N. B. Livingood, *NACA Tech. Note* 3010, 1953.

[2] R. J. Goldstein and A. Hagi-Sheikh, *Proc. Japanese Soc. Mech. Engr., Semi-international Symp.*, **2**:213 (1967).

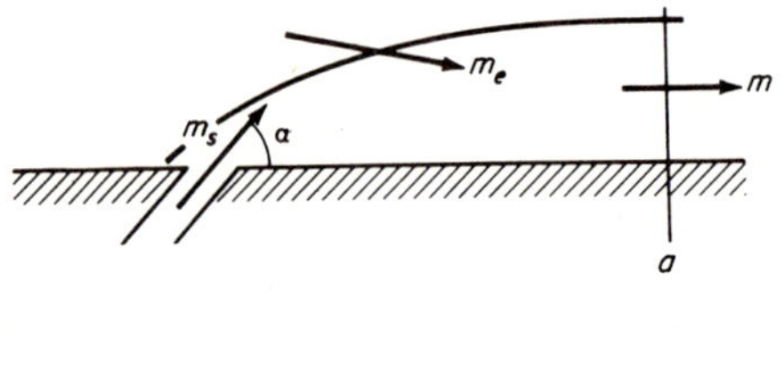

Fig. 11-14 Sketch of film-cooling arrangement.

The following energy balance must be fulfilled for steady state and an adiabatic surface when heat conduction through the plane aa is neglected:

$$c_{pc}m_s(t_c - t_e) = c_p m(\bar{t} - t_e)$$

With the expression

$$c_p m = c_{pc}m_s + c_{pe}m_e$$

and λ as a dimensionless parameter describing the average fluid temperature $\bar{t}$ according to the equation

$$\bar{t} - t_e = \lambda(t_{aw} - t_e)$$

one obtains the following expression for the film-cooling effectiveness:

$$\eta = \frac{1/\lambda}{1 + (c_{pe}/c_{pc})(m_e/m_s)} \tag{11-21a}$$

The parameter λ is now determined through the requirement that Eq. (11-20) be an asymptotic expression for this equation at vanishing values of the mass flow m_s. For this case Eq. (11-21a) can be simplified to

$$\eta_0 = \frac{c_{pc}m_s}{c_{pe}m_{e0}\lambda} \tag{11-21b}$$

The subscript zero indicates that this equation holds for vanishing values of the coolant mass flow m_s. The symbol m_{e0} denotes then the mass entering the boundary layer from the mainstream for zero injection, i.e., the mass flow entering a turbulent boundary layer on a solid surface. Expressions for this mass flow can be obtained from the literature;[1]

$$m_{e0} = \tfrac{7}{8}\rho_e u_e \delta_0$$

with the boundary-layer thickness δ_0 given by the expression

$$\frac{\delta_0}{x} = \frac{0.376}{(\mathrm{Re}_x)^{\frac{1}{5}}}$$

Introducing the last two equations into Eq. (11-21b) results in the following expression for the film-cooling effectiveness:

$$\eta_0 = \frac{3.04}{\lambda}(\mathrm{Re}_s)^{0.2}\left(\frac{\mu_c}{\mu_e}\right)^{0.2}\frac{c_{pc}}{c_{pe}}\left(\frac{\rho_e u_e x}{m_s}\right)^{-0.8}$$

and a comparison of this equation with Eq. (11-20) (with $a = 1$) indicates the following expression for the parameter λ:

$$\frac{1}{\lambda} = 1.9\ \mathrm{Pr}^{\frac{2}{3}}$$

[1] For instance, E. R. G. Eckert and R. M. Drake, Jr., "Heat and Mass Transfer," 2d ed., p. 144, McGraw-Hill, New York, 1959.

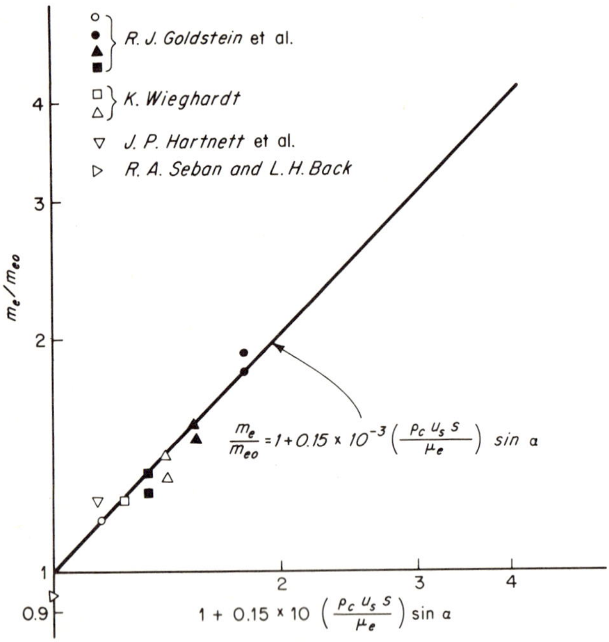

Fig. 11-15 Experimental results and straight-line correlation for the parameter m_e/m_{e0}. [*From R. J. Goldstein and A. Haji-Sheikh, Proc. Japan. Soc. Mech. Engr., Semiintern. Symp.*, **2**:*213* (*1967*).]

It remains to determine the ratio m_e/m_{e0}, which will be denoted by β. Goldstein and Haji-Sheikh obtained the expression

$$\beta = \frac{m_e}{m_{e0}} = 1 + 1.5 \times 10^{-4}\,\mathrm{Re}_s \frac{\mu_c}{\mu_e} \sin \alpha \qquad (11\text{-}22a)$$

from the available experimental information. In this equation, α denotes the angle under which the coolant is ejected from the surface, as indicated in Fig. 11-14. Figure 11-15 shows that this equation represents the available experimental results quite satisfactorily. The film-cooling effectiveness is then described by the equation

$$\eta = \frac{1.9\,\mathrm{Pr}^{\frac{2}{3}}}{1 + 0.329(\mathrm{Re}_s)^{-0.2} \dfrac{c_{pe}}{c_{pc}} \left(\dfrac{\mu_e}{\mu_c}\right)^{0.2} \left(\dfrac{\rho_e u_e x}{m_s}\right)^{0.8} \beta} \qquad (11\text{-}22b)$$

Figures 11-16 and 11-17 compare the film effectiveness obtained by this equation with experimental values for normal ejection of the coolant through a porous strip and for tangential injection through a stepdown slot, respectively. Similar agreement is obtained with experiments by Nishiwaki, Wieghardt, Hartnett, Eckert, and Birkebak.[1] It is still uncertain whether the influence of variable properties is expressed correctly in Eq. (11-22) because information from experiments with large temperature differences is insufficient to verify it.

Interrupted ejection From a stress standpoint, it is often preferable to eject the coolant through holes or through a series of short slots. This ejection method is more difficult to analyze because the velocity and temperature fields involved are three-dimensional. It is also found in experiments that the ejected coolant stream does not always flow along the surface but sometimes penetrates into the boundary layer or even partially through the boundary layer. The number of configurations with holes or slot arrangements is, of course, very large, and this makes it

[1] See Goldstein and Haji-Sheikh, *loc. cit.*, p. 567; also S. C. Kacker, B. R. Pai, and J. H. Whitelaw, The Prediction of Wall Jet Flows with Particular Reference to Film Cooling, *Progr. Heat Mass Transfer*, **2**:163 (1969).

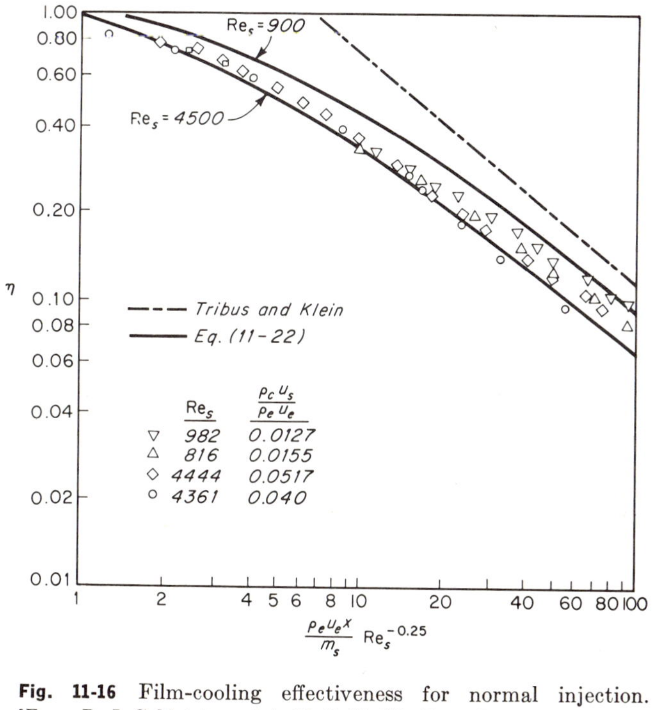

Fig. 11-16 Film-cooling effectiveness for normal injection. [*From R. J. Goldstein and A. Haji-Sheikh, Proc. Japan. Soc. Mech. Engr. Semiintern. Symp.* **2**:*213* (*1967*).]

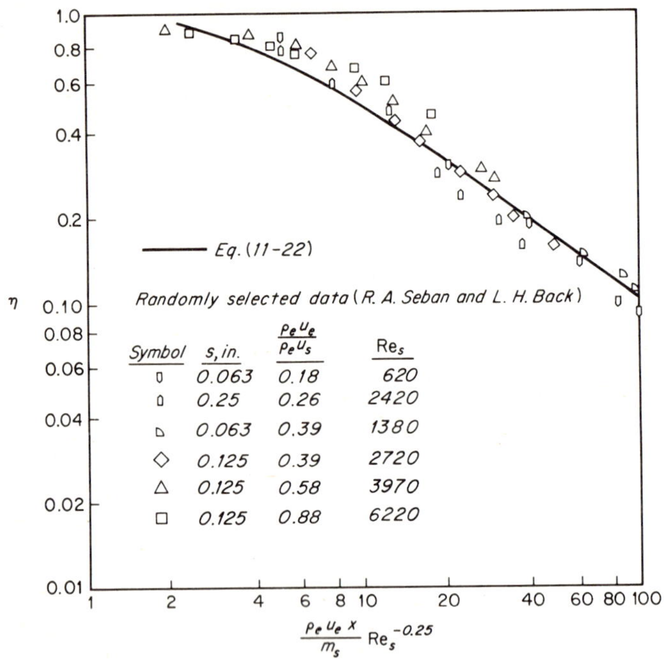

Fig. 11-17 Film-cooling effectiveness for tangential injection. [*From R. J. Goldstein and A. Haji-Sheikh. Proc. Japan. Soc. Mech. Engr. Semiintern. Symp.*, **2**:*213* (*1967*).]

desirable to generalize experimental results by an analytical model. The success of the line-source model as a representation of continuous ejection suggests using point heat sources or sinks at the proper locations within the film-cooled surface for the present situation.

A simple analytical expression is available which describes the temperature field created in a semi-infinite solid by a heat source of strength q moving along the surface in a straight line with a velocity u_e. The following equation describes the temperature difference $\theta = t - t_e$, with t denoting the local temperature in the solid:

$$\theta(x,y,z) = \frac{q}{2\pi\rho c_p \alpha r} e^{-(u_e/2x)(r-x)}$$

The coordinate system used for the point-source model is indicated in Fig. 11-18. The symbol

$$r = \sqrt{x^2 + y^2 + z^2}$$

[1] Eckert and Drake, *op. cit.*, p. 113.

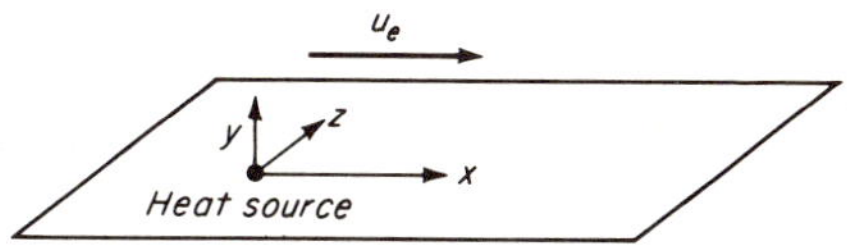

Fig. 11-18 Coordinate system for point-source model.

denotes the distance between the location in the solid with the temperature t and the heat source, and α denotes the thermal diffusivity. The temperature field on the surface of the solid in a coordinate system which moves with the heat source is shown in Fig. 11-19.

The equation above can also be used to describe the temperature field in a fluid moving with uniform velocity and homogeneous turbulence along a surface when a point source is located in the surface. The thermal diffusivity α in this case is to be replaced by the turbulent diffusivity ϵ. In addition, the equation can be simplified when the dimension x is large compared to y and z. With these changes, the temperature field is described by the equations

$$\frac{\theta(x,y,z)}{\theta(x,0,0)} = \exp\left\{-\frac{\mathrm{Pe}_t}{4}\left[\left(\frac{y}{x}\right)^2 + \left(\frac{z}{x}\right)^2\right]\right\} \tag{11-23a}$$

and

$$\theta(x,0,0) = \frac{1}{2\pi}\left(\frac{u_e x}{\nu}\right)^{-1}\frac{u_e q}{\rho c_p \epsilon \nu} \tag{11-23b}$$

In this equation the turbulent Peclet number Pe_t stands for the expression $u_e x/\epsilon$. In applying the point-heat-source model to film cooling, the

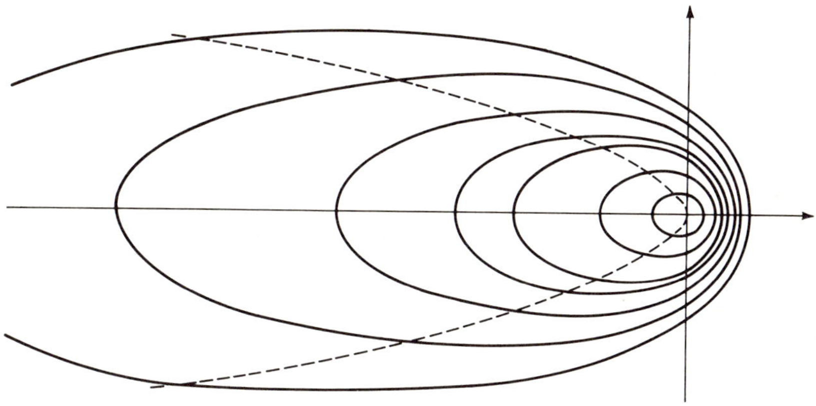

Fig. 11-19 Lines of constant temperature describing the temperature field created by a moving point source.

heat flux q issuing from the point source must be obtained from the equation

$$q = \rho c_p m_s (t_c - t_e) \tag{11-24}$$

Ramsey[1] carried out experiments which made it possible to investigate the applicability of the heat-source model. Air flowing over a flat plate with 60.7 m/s created a turbulent boundary layer, and 20 cm downstream from the start of the boundary layer, heated air was ejected from the surface through a hole with a diameter $D = 2.35$ cm in a direction normal to the surface. The two-dimensional temperature field in the locally adiabatic plate and the temperature field in the boundary layer were measured downstream from the hole. The velocity of the ejected airstream was varied. The following comparisons with the heat-source model are made for a blowing parameter $\rho_2 u_2/\rho_e u_e$ with a value 0.1. The indices 2 and e in this parameter refer to conditions at the exit of the hole and in the free stream, respectively. Figure 11-20 compares the measured temperature variation at the coordinate z with the one predicted by Eq. (11-23a). The distance z is normalized with the distance $z_{\frac{1}{2}}$ at which the ordinate parameter has the value 0.5. A value for the turbulent diffusivity ϵ was required to obtain the calculated curve in the

[1] J. W. Ramsey, The Interaction of a Heated Air Jet with a Deflecting Flow, Ph.D. thesis, University of Minnesota, June, 1969.

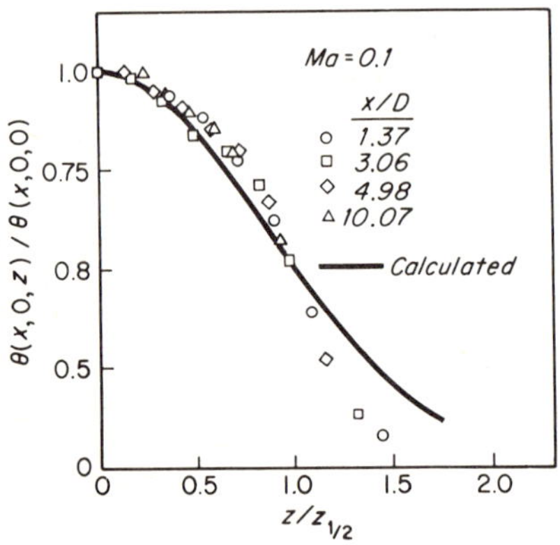

Fig. 11-20 Comparison of measured and predicted lateral wall-temperature distribution. [*From J. W. Ramsey, The Interaction of a Heated Air Jet with a Deflecting Flow, Ph.D. thesis, University of Minnesota, June, 1969.*]

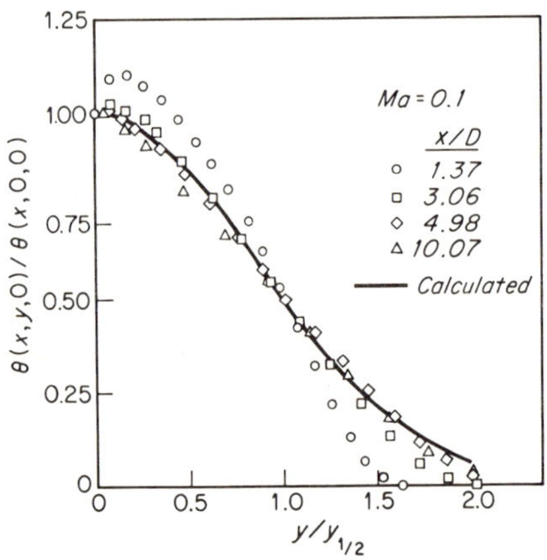

Fig. 11-21 Comparison of measured and predicted vertical temperature distributions. [*From J. W. Ramsey, The Interaction of a Heated Air Jet with a Deflecting Flow, Ph.D. thesis, University of Minnesota, June, 1969.*]

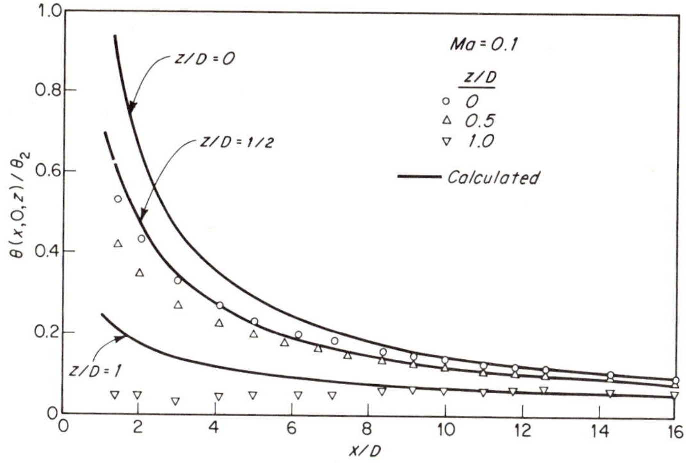

Fig. 11-22 Comparison of variation of measured and predicted wall-temperature distributions in downstream direction. [*From J. W. Ramsey, The Interaction of a Heated Air Jet with a Deflecting Flow, Ph.D. thesis, University of Minnesota, June, 1969.*]

figure. It was obtained by a least-squares approximation of the analytical values in Fig. 11-20 to the experimental ones. This method resulted in $\epsilon = 112\ \text{cm}^2/\text{s}$. Figure 11-21 presents a similar comparison of the vertical temperature variation within the boundary layer. The calculated curve was obtained with the same value of the turbulent diffusivity

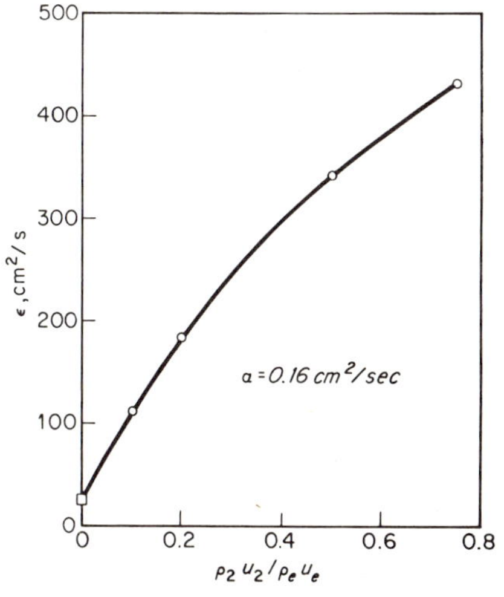

Fig. 11-23 Turbulent diffusivity for various injection rates. [*From J. W. Ramsey, The Interaction of a Heated Air Jet with a Deflecting Flow, Ph.D. thesis, University of Minnesota, June, 1969.*]

as Fig. 11-20. Figure 11-22 presents the variation of the adiabatic wall-surface temperature with downstream distance x and compares the measured values with the ones obtained from Eq. (11-23*b*), again with the aforementioned value of the diffusivity ϵ. The agreement is not too bad. It could be improved by use of a turbulent diffusivity which decreases with increasing distance x. Figure 11-23 presents values of the turbulent diffusivity ϵ for various values of the parameter $\rho_2 u_2/\rho_e u_e$. The turbulent diffusivity in the boundary layer at a distance $y = \delta/2$, according to the literature for the conditions of the experiment, has the value 26 cm^2/s. The curve in Fig. 11-23 extrapolates quite naturally to this value. The agreement between measured and predicted temperatures of the plate surface becomes poorer with increasing value of the parameter on $\rho_2 u_2/\rho_e u_e$, especially when the jet does not hug the surface but penetrates into the boundary layer or, under some conditions, even through the boundary layer. Such a condition was found to lead to very poor effectiveness values and should be avoided in a film-cooling arrangement.

Within the range of applicability, the heat-source model has considerable advantages because many different geometries can be analyzed simply by superposition of the temperature fields created by single point sources. The temperature fields created by a row or even by several rows of holes as well as by interrupted slots can be obtained in this way.

Wieghardt[1] and Malhotra and Cermak[2] have developed relations similar in shape to Eqs. (11-23*a*) and (11-23*b*) from experiments on turbulent spreading from a heat source or mass source, respectively, into a turbulent boundary layer. For the ejection of fluid through a hole, it is found, however, that the simple heat-source model leading to Eqs. (11-23*a*) and (11-23*b*) describes the measured temperature or concentration fields better than Wieghardt's or Malhotra's relation.

Laminar boundary layer The heat-source model is also useful for investigating film cooling in a laminar boundary layer. For the condition of continuous injection of a fluid at the start of the laminar boundary layer and for a situation in which viscous dissipation can be neglected, the solution employing the heat-source model has in fact already been discussed. It is presented in Fig. 7-9 by the curve for $\gamma = -0.5$. This curve has a horizontal tangent at $\eta = 0$ and therefore applies to an adiabatic wall condition. The effectiveness parameter η can be calculated from the consideration that the thermal energy removed by the heat sink (or the enthalpy defect of the coolant mass flow) must equal the enthalpy

[1] K. Wieghardt, *Z. Angew. Math. Mech.* **27**(11/12): 346–355 (1948).

[2] R. C. Malhotra and J. E. Cermak: *Intern. J. Heat Mass Transfer*, **7**:169–185 (1964).

Table 11-1

β	1.6	1	0	−0.1988	1	1
Pr	0.7	0.7	0.7	0.7	1	5
C	1.11	0.71	0.55	0.54	0.87	2.0

defect in the boundary layer at any cross section downstream from the ejection

$$m_s c_p(t_e - t_c) = \int_0^\infty \rho c_p(t_e - t)u\, dy$$

The integration must be performed at a constant value of x. With the notation in Sec. 7-2, the above equation transforms to

$$m_s(t_e - t_c) = \rho u_e \sqrt{\frac{2\nu x}{u_e}}\,(t_e - t_{aw}) \int_0^\infty \theta f'\, d\eta$$

and the effectiveness parameter is

$$\eta = C \frac{m_s}{\rho u_e x} \sqrt{\mathrm{Re}_x}$$

With introduction of a slot Reynolds number

$$\mathrm{Re}_s = \frac{m_s}{\mu}$$

the equation changes to

$$\eta = C \sqrt{\frac{m_s}{\rho u_e x}} \sqrt{\mathrm{Re}_s} \tag{11-25}$$

The constant C was obtained by a numerical integration based on the analysis of Levy[1] and is contained in Table 11-1. In reality the parameter η cannot assume values larger than 1, and Eq. (11-25) is therefore restricted to larger distances x from the location of ejection. It should give information on laminar film cooling to a first order of magnitude.

11-3 EXCHANGE PROCESSES IN RAREFIED GASES

In the conventional treatment of exchange processes (momentum, energy), the structure of the gas is considered to be continuous; that is to say, no consideration need be given to the molecular structure of the

[1] S. Levy, *J. Aeron. Sci.*, **19**:341 (1952).

gas. Flow and heat-transfer phenomena under these continuum conditions can be described adequately in terms of the Reynolds, Mach, Nusselt, and Prandtl numbers. However at small absolute pressures a gas partly loses its continuum characteristics, and important phenomena occur which are understandable only if account is taken of the molecular structure of the fluid. The study in *aerodynamic* terms of the flow and heat-transfer processes of rarefied gases began essentially with the pioneering paper of Tsien[1] in 1946. Although the field caught the imagination of many workers in the years that followed, and although the results have application in the broad field of aerospace technology, many fundamental questions remain to be answered by analysis and experiment.[2]

The tool for dealing with these low-density gas phenomena is the kinetic theory of gases first developed quantitatively by Daniel Bernoulli in 1738, and later more fully developed by Boltzmann and Maxwell in the nineteenth century. This theory considers the gas to consist of molecules which travel on straight paths, deviating only as a result of collisions with other molecules, walls, or surfaces. The average kinetic energy of the molecular aggregation determines the temperature of the gas. The collisions, at least in the simple theory, are considered to occur according to the laws of elastic impact. The statistically determined average length of the rectilinear paths between molecular collisions is called the *mean free-molecular path* λ. Since the velocity of each molecule changes upon collision, at any specific instance molecules of many different velocities are present. Most molecules have velocities near a velocity $\bar{v}$, the *mean molecular velocity*; i.e., there are few molecules of exceedingly high or exceedingly low velocities. The velocity distribution in a gas at equilibrium, calculated by Maxwell in 1859, is called the *maxwellian velocity distribution.*

One of the early successes of the kinetic theory of gases was that the theory predicted the astonishing fact that the dynamic viscosity and the thermal conductivity of ideal gases are independent of pressure. This means, for example, that a certain amount of heat is conducted through a stagnant layer of gas under given temperature conditions independent of the pressure of the gas. By means of simplified concepts we shall derive expressions for the viscosity and relate that property to the thermal conductivity, which was calculated in Chap. 2.

Kinetic theory explains the shear stresses in a flowing gas by the

[1] H. S. Tsien, Superaerodynamics, the Mechanics of Rarefied Gases, *Jr. Aeron. Sci.*, **13**:653–664 (1946).

[2] S. A. Schaaf, Mechanics of Rarefied Gases, in "Handbuch der Physik," vol. VIII/2, pp. 591–624, Springer-Verlag, Berlin, 1963. See also *Advan. Appl. Mech., Proc. Intern. Rarefied Gas Dynamics Symp. 1958–1969,* volumes which encompass most of the work in the field.

fact that molecules move back and forth between layers of gas flowing at different velocities. Thus a molecule from a low-velocity gas layer may jump into a layer of gas moving at a higher velocity, where after some collisions with other molecules its velocity is increased and the velocities of the other molecules in the collision process are decreased. In this way momentum is exchanged between the fluid layers. The same motion of the molecules causes an exchange of energy when there are temperature differences in the gas.

Consider the laminar flow of a gas parallel to a wall with a velocity gradient and a temperature gradient normal to the wall (Fig. 11-24). Molecules pass back and forth across the arbitrary plane *aa* parallel to the wall. During the time interval $d\tau$ molecules with velocity components v' normal to the plane *aa* will pass that plane when they are in a proximity $v'\,d\tau$ to the plane. The number of molecules per unit volume is n. The average velocity component toward plane *aa* of all molecules will be some fraction i' of the mean molecular velocity $\bar{v}$. Therefore all the molecules within the volume $i'\bar{v}\,d\tau$ will pass a unit area of the plane *aa*. The number of such molecules is $ni'\bar{v}\,d\tau$, and the number crossing a unit area per unit time from one side of *aa* is $i'n\bar{v}$. The molecules are characteristic of the flow velocity $\bar{u}$ from their last until their next collision. On the average, therefore, the molecules transport the flow velocity from a plane 1-1 to a plane 2-2 separated by a distance related to the molecular mean-free-path length λ. This distance we shall call $j\lambda$, where j cannot be much different from unity. A molecular stream of equal magnitude passes plane *aa* in the opposite direction and carries the flow velocity $\bar{u}'$ from plane 2-2 to plane 1-1. The momentum exchange resulting from this process is

$$i'n\bar{v}m(\bar{u} - \bar{u}')$$

where m is the mass of one molecule. When the velocity gradient at plane *aa* in the direction normal to that plane is du/dy, then $\bar{u} - \bar{u}'$ can

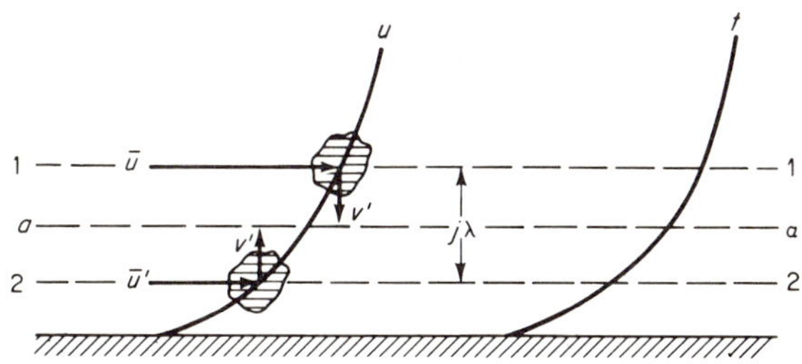

Fig. 11-24 Molecular momentum exchange.

be expressed as $j\lambda\, d\bar{u}/dy$. Therefore the exchange of momentum per unit area, or the *shear stress*, is

$$\tau = i'jnm\bar{v}\lambda \frac{d\bar{u}}{dy} = inm\bar{v}\lambda \frac{d\bar{u}}{dy}$$

the product $i'j$ being denoted by i. The viscosity μ is defined by the equation

$$\tau = \mu \frac{\partial \bar{u}}{\partial y}$$

from which, by comparison with the above expression for τ, we deduce that

$$\mu = inm\bar{v}\lambda$$

The product $nm = \rho$, the mass per unit volume. A thorough calculation for spherical molecules by Chapman and Enskog,[1] which accounted for the molecular velocity distribution, yielded for the numerical factor i the value 0.499 and thus for the dynamic viscosity

$$\mu = 0.499\rho\bar{v}\lambda \tag{11-26}$$

Kinetic theory shows that $\rho\lambda$ is independent of the pressure and further than the mean molecular velocity can be expressed as

$$\bar{v} = \sqrt{\frac{8RT}{\pi}}$$

where R is the gas constant for the gas in question. T is the absolute temperature; thus $\bar{v}$ is independent of the pressure. Therefore μ is independent of the pressure. Note that this result is not true for gases at very high pressures, where ultimately the viscosity depends upon the pressure.

When a thermal gradient dt/dy is present in the gas, thermal energy is exchanged by the molecules in much the same way as is momentum. If c_m is the specific heat per molecule, the heat exchange per unit area across the plane *aa* is

$$-q = in\bar{v}c_m\lambda \frac{dt}{dy}$$

Note that $nc_m = \rho c_v$, where c_v is the specific heat for constant volume, and that the thermal conductivity is defined by the equation

$$k = \frac{-q}{dt/dy}$$

[1] S. Chapman and T. G. Cowling, "The Mathematical Theory of Non-uniform Gases," 2d ed., Cambridge University Press, London, 1958.

Substituting these values into the expression for q, one obtains

$$k = i\bar{v}\lambda\rho c_v = \mu c_v \tag{11-27}$$

which agrees with the results given in Chap. 2.

Thus this simple theory relates mechanical and thermal properties which might have been suspected of being quite unrelated. Equation (11-27) yields acceptable order-of-magnitude values but does not agree with experimental data, as noted earlier. Chapman[1] introduced refinements in the theory which resulted in an expression for thermal conductivity that agrees with experimental data for monatomic molecules but not for more complex molecules, i.e., Eq. (2-26), repeated here for easy reference:

$$k = \tfrac{5}{2}\mu c_v \tag{11-28}$$

Eucken[2] developed an expression for the thermal conductivity for complex molecules

$$k = \tfrac{1}{4}(9\gamma - 5)\mu c_v \tag{11-29}$$

where $\gamma = c_p/c_v$ the ratio of specific heats. As noted in Chap. 2, Eq. (11-29) agrees surprisingly well with the experimental data for simple and complex molecules.

In the discussion of the exchange processes in rarefied gases that follows the various flow and thermal parameters must be viewed in terms of the results of kinetic theory.

From Eq. (11-29) one can write an expression for the *Prandtl number* on the basis of kinetic theory as

$$\text{Pr} = \frac{4\gamma}{9\gamma - 5} \tag{11-30}$$

which agrees reasonably well with measured values of the Prandtl number.

The *Reynolds number* can be interpreted from the kinetic theory as well,

$$\text{Re} = \frac{\rho UL}{\mu} = \frac{1}{0.499}\frac{U}{\bar{v}}\frac{L}{\lambda} \tag{11-31}$$

Thus the Reynolds number is the product of a velocity ratio and a length ratio. The velocity ratio is between the macroscopic flow velocity U and the mean molecular velocity $\bar{v}$; the length ratio is that between some significant dimension of the body L and the length of the mean free-molecular path λ.

[1] *Ibid.*

[2] A. Eucken, *Physik. Z.*, **14**:324 (1913).

A similar interpretation of the *Mach number* is also possible. The expression for the velocity of sound is

$$U_s = \sqrt{\gamma R T}$$

From the expression for the mean molecular velocity, therefore,

$$U_s = \bar{v}\sqrt{\frac{\gamma\pi}{8}}$$

and the Mach number becomes

$$\text{Ma} = \frac{U}{U_s} = \frac{U}{\bar{v}}\sqrt{\frac{8}{\gamma\pi}} \tag{11-32}$$

Thus the Mach number is related directly to the ratio of the macroscopic velocity U to the mean molecular velocity $\bar{v}$.

At ordinary pressures the intermolecular distances—even in gases, where the intermolecular distances are very large compared with liquids and solids—are still much smaller than the body dimensions with which we are usually concerned in heat-transfer calculations, and therefore the notion of a continuous gas is well fulfilled. At low pressures, with correspondingly low densities, the molecular mean free path becomes comparable with the body dimensions, and then the effect of molecular structure becomes a factor in flow and heat-transfer mechanisms.

The relative importance of effects due to the rarefaction of a gas can be indicated by a comparison of the magnitude of the mean free-molecular path in the gas with some significant dimension of the body. Hence, if L is some body dimension which is a characteristic dimension in the flow field, the effects of rarefaction phenomena on flow and heat transfer will become important as soon as the ratio λ/L can no longer be neglected. This ratio is dimensionless and is referred to as the *Knudsen number* Kn. The Knudsen number, which is thus of direct interest in the study of rarefied-gas flow and heat transfer, is expressible in terms of the Mach and Reynolds numbers

$$\text{Kn} = \frac{\lambda}{L} = \sqrt{\frac{\gamma\pi}{2}}\frac{\text{Ma}}{\text{Re}} \tag{11-33}$$

which can be evaluated approximately for a diatomic gas for $\gamma = 1.4$ to yield

$$\text{Kn} \approx 1.48\frac{\text{Ma}}{\text{Re}}$$

Tsien early proposed the division of fluid mechanics into various flow regimes based on the characteristic ranges of an appropriate Knudsen number. For flows of low viscosity, i.e., large Reynolds numbers, the characteristic dimension of greatest importance is the boundary-layer thickness δ. Since the boundary-layer thickness for laminar flow, for example, is related to the significant body dimension by the statement

$$\frac{\delta}{L} \sim \frac{1}{\sqrt{\text{Re}}}$$

the corresponding Knudsen number becomes

$$\text{Kn} = \frac{\lambda}{\delta} \sim \frac{\text{Ma}}{\sqrt{\text{Re}}} \tag{11-34}$$

Hence conventional gas dynamics will exist when $\text{Re} \gg 1$ and when the ratio $\text{Ma}/\sqrt{\text{Re}}$ is negligibly small. For slow, or *Stokes type*, flow, the significant flow dimension is the body dimension L. Thus the Knudsen number is based on L, to yield

$$\text{Kn} \sim \frac{\text{Ma}}{\text{Re}} \tag{11-35}$$

Thus continuum flow can be realized when Ma/Re is negligibly small. In flows for which the Reynolds number is only moderately large and for which the Mach number is large, the Knudsen number becomes of such a size that it no longer can be neglected, indicating the existence of rarefaction effects in the flow. Interest in the aerodynamic characteristics of low-density gases has arisen because of the growing importance of rockets, satellites, and other high-altitude, high-speed vehicles.

Extremely high-speed gas flows involve large variations in the gas temperature, particularly in regions of stagnation flows. These temperature variations are sufficiently great to cause dissociation and ionization of the gas and thus lead to measurable departures from ideal-gas behavior. In wind tunnels, on the other hand, gas is expanded from a reservoir, often at a relatively high temperature, to such low pressures and temperatures that condensation of the flowing gas may occur. These departures from perfect-gas behavior are known as *real-gas effects*. Work in low-density real gases has been undertaken seriously only within the last few years, and few results are yet available. The work to be discussed here will be confined to low-density, perfect-gas effects.

It is convenient to subdivide the field of rarefied-gas dynamics into several different flow regimes, depending upon greater or lesser rarefaction of the gas. The subdivision was suggested by Tsien and is best characterized by designated ranges in the Knudsen numbers (Fig. 11-25).

For extremely low densities the molecular mean free path λ is very

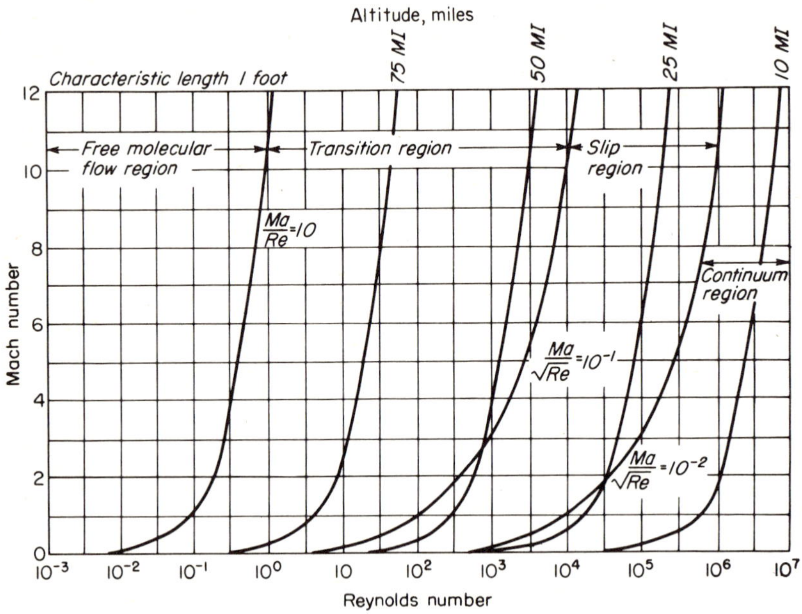

Fig. 11-25 Defined flow regimes of gas dynamics.

much larger than any significant body dimension L. In such a case the gas molecules striking and leaving the body surface do not collide with free-stream molecules until very far from the surface. As a first approximation in such a case the molecular distribution in the region away from the surface can be assumed to be undistorted, i.e., *maxwellian,* so that the flow near the body can be considered on the basis of the interaction between *free molecules* and the *surface.* This flow regime is named *free-molecule flow* or *Knudsen flow.* For such flows intermolecular collisions are of negligible importance. Free-molecule flow is characterized by large values of the Knudsen number.

In *continuum flow,* the state of ordinary gas dynamics, intermolecular collisions dominate the flow field, and a usual boundary condition at the interface between a fluid and a solid surface is that the fluid adjacent to the surface assumes both the velocity and temperature of the surface. The Knudsen number for continuum flow is small, approaching zero.

In *near-continuum flow,* where the Knudsen number is small but nonnegligible, one of the most interesting effects of gas rarefaction on the flow occurs. The gas adjacent to the surface, in contrast to its behavior in continuum flow, no longer reaches the velocity or temperature of the surface. The gas at the surface has a tangential velocity; it *slips* along the surface. The temperature of the gas at the surface is finitely different from the temperature of the surface; there is a *jump* in temperature between the surface and the adjacent gas. The *velocity-slip* and *temperature-jump* effects are related to the molecular mean-free-path lengths and parameters called *accommodation* and *reflection coefficients* which describe the statistical molecule-surface interaction. For such flows the departure from continuum flow is slight, and the phenomenon is called *slip flow*.

Flows for values of the Knudsen number between values for *slip flow* and *free-molecule flow* are called *transition flows* and are characterized by situations where intermolecular collision and collisions of gas molecules and surfaces are of more or less equal importance.

The demarcations between *slip*, *transition*, and *free-molecule-flow* regimes in terms of flow parameters such as Mach and Reynolds numbers are not simply defined, as the transition from the gas-dynamics continuum regime to the free-molecule-flow regime occurs gradually, showing no discontinuity experimentally; however, for convenience one can define the limits of the flow regimes in an arbitrary manner. Such limits may be

Continuum gas dynamics: $$\frac{\text{Ma}}{\sqrt{\text{Re}}} < 0.01$$

Free-molecule flow: $$\frac{\text{Ma}}{\text{Re}} > 10$$

The region between is occupied by the *slip-* and *transition-flow* regimes.

It should be noted here that the definitions given the limits of the flow regimes are approximate at best. For example, it will be shown later that the slip-flow regime is characterized by the fact that rarefaction effects come into the picture essentially through boundary conditions imposed upon continuum-flow equations and thus represent only small corrections to continuum-flow behavior.

The continuum flow in question, however, is not the ideal inviscid flow of classical fluid mechanics but a highly viscous flow. It may be understood from the relationships among the Knudsen, Mach, and Reynolds numbers that for rarefaction effects to occur for moderate values of the Mach number, the Reynolds numbers must be fairly low—certainly in the laminar-flow region and quite likely in the creeping-flow region of Stokes and Oseen. In these flow situations it is quite clear that viscous effects are of great importance. The fluid boundary layers, where such can be defined, increase in thickness, causing further displacement of the flow from surfaces. Skin-friction forces become quite large, and the effects of heat transfer affect the flow field much more. Heated boundary layers are thicker than cooled boundary layers, with consequent differences in skin-friction and displacement effects. The shape of the surface in the flow becomes of greater importance. Blunt objects are less sensitive to viscous effects than sharp slender objects.

It can be understood that the principal strong effects mentioned are far more characteristic of the near-continuum slip-flow region than are the weaker rarefaction effects. To define the region more carefully requires the additional specification of at least the surface-to-gas-temperature ratio and the body geometry.

The situation for near-free-molecule flow is not so complicated, insofar as the definition of the flow-regime limits are concerned, but the problems are just as formidable.

Kinetic theory The basic assumption in kinetic theory is that a gas can be considered as an aggregation of molecules moving freely with respect to each other, except for collisions. The molecules in the kinetic-theory treatment are generally considered to have only translational energies, i.e., no rotational or vibrational degrees of freedom.

The fundamental concept in kinetic theory is that of a molecular distribution function $f(\mathbf{x},\mathbf{v},\tau)$. The distribution function is the number density in phase space of molecules having a position $\mathbf{x}$, a velocity $\mathbf{v}$, at a time τ. The distribution function obeys the Maxwell-Boltzmann conservation equation

$$\frac{\partial f}{\partial \tau} + \sum_{i=1}^{3} v_i \frac{\partial f}{\partial x_i} = \iiint_{-\infty}^{+\infty} d\mathbf{v}' \iint d\omega \, [|\mathbf{v}' - \mathbf{v}| I(\mathbf{v}' - \mathbf{v}, \theta)(f_1 f_1' - f f')] \tag{11-36}$$

where $I(\mathbf{v}' - \mathbf{v}, \theta)$ is the differential cross section for the scattering of a molecule with velocity $\mathbf{v}'$ through an angle θ within the solid angle $d\omega$. The subscripts 1 represent the state of affairs after collision; i.e., velocities before collision may be $\mathbf{v}$, $\mathbf{v}'$ and after collision $\mathbf{v}_1$, $\mathbf{v}_1'$.

The left-hand side of Eq. (11-36) represents the rate of change in the distribution function $f(\mathbf{x},\mathbf{v},\tau)$ as the result of convection. The right-hand side is known as the *collision integral* and represents the rate of change in the distribution function as the result of molecular collisions. Equation (11-36) is valid in the general thermodynamic region characteristic of an ideal gas, a requirement generally satisfied in the gas flows of interest here.

If in Eq. (11-36), $f = f(\mathbf{v})$ only, an equilibrium solution is obtained such that

$$f = f^\circ = \frac{n}{(2\pi RT)^{\frac{3}{2}}} \exp\left[\frac{(v_1 - u_1)^2 + (v_2 - u_2)^2 + (v_3 - u_3)^2}{2RT}\right] \tag{11-37}$$

where n = number density of molecules
R = gas constant for the gas in question
T = absolute temperature
$\mathbf{u}$ = macroscopic gas velocity

The function f° is known as the *maxwellian velocity distribution.*

The nonequilibrium case is rather more complicated and by no means resolved. Most of the attempts made to resolve the problem are too formidable to be presented in a text of this scope, but some remarks are in order.

For all monatomic molecular models, and regardless of the magnitude of departure from equilibrium conditions, the five basic equations for the conservation of mass, momentum, and energy apply:

Mass:
$$\frac{\partial \rho}{\partial \tau} + \sum_{i=1}^{3} \frac{\partial(\rho u_i)}{\partial x_i} = 0 \tag{11-38}$$

Momentum:
$$\frac{\partial u_i}{\partial \tau} + \sum_{j=1}^{3} \left(u_j \frac{\partial u_i}{\partial x_j} - \frac{1}{\rho} \frac{\partial p_{ij}}{\partial x_j} \right) = 0 \tag{11-39}$$

Energy:
$$\frac{\partial p}{\partial \tau} + \sum_{i=1}^{3} \left(\frac{\partial(p u_i)}{\partial x_i} + \frac{2}{3} \frac{\partial q_i}{\partial x_i} - \frac{2}{3} \sum_{j=1}^{3} p_{ij} \frac{\partial u_i}{\partial x_j} \right) = 0 \tag{11-40}$$

where p_{ij} and q_i are moments of the distribution function and represent stresses or heat-flux components.

$$p_{ij} = m \iiint_{-\infty}^{+\infty} (v_i - u_i)(v_j - u_j) f \, dv_1 \, dv_2 \, dv_3 \tag{11-41}$$

$$q_i = \tfrac{1}{2} m \iiint_{-\infty}^{+\infty} (v_i - u_i)(\mathbf{v} - \mathbf{u})^2 f \, dv_1 \, dv_2 \, dv_3 \tag{11-42}$$

The conservation equations (11-38) to (11-40) are obtained from the Maxwell-Boltzmann equation (11-36) by utilizing the five collisional invariants, the mass, momenta, and energy of the molecule. The five equations are not completely determined, however, as the number of dependent unknowns is greater than 5. It remains for kinetic theory to provide the additional necessary relationships, which take the form of thermophysical properties, i.e., *viscosity*, *thermal conductivity*, and *specific heats.*

Certain specific results are available for the case of slight departure from equilibrium.[1] For the case of central force fields by which mole-

[1] Chapman and Cowling, *op. cit.;* J. O. Hirschfelder, C. F. Curtiss, and R. B. Bird, "Molecular Theory of Gases and Liquids," chap. 7, Wiley, New York, 1954; G. N. Patterson, "Molecular Flow of Gases," Wiley, New York, 1956.

cules repel each other with the inverse fifth power of the distance of separation, the rigorous theory yields

$$f = f^\circ \left[1 - \sum_{i=1}^{3} \sum_{j=1}^{3} \frac{p\delta_{ij} - p_{ij}}{2pRT} (v_i - u_i)(v_j - u_j) - \sum_{i=1}^{3} \frac{q_i}{pRT} (v_i - u_i) \frac{1 - (\mathbf{v} - \mathbf{u})^2}{5RT} \right] \tag{11-43}$$

from whence the Navier-Stokes equations and the Fourier heat-conduction equation for the case of the bulk viscosity $\mu' = -\frac{2}{3}\mu$ are, respectively,

$$p_{ij} = \delta_{ij} \left(-p - \tfrac{2}{3}\mu \sum_{k=1}^{3} \frac{\partial u_k}{\partial x_k} \right) + \mu \left(\frac{\partial u_i}{\partial x_j} + \frac{\partial u_j}{\partial x_i} \right) \tag{11-44}$$

$$q_i = -k \frac{\partial T}{\partial x_i} \tag{11-45}$$

where the coefficients μ and k are the coefficients of viscosity and thermal conductivity, as discussed in Chap. 2. Equations (11-44) and (11-45) are well known in the practice of conventional fluid mechanics, where they were derived in the same form from a macroscopic rather than a microscopic point of view.

Extensions of the foregoing calculations to the cases of finite departures from the equilibrium state have been attempted by many investigators, notable among them being the formulations of Burnett[1] and Grad.[2] The problem, while better understood, remains largely unsolved to date. The *Burnett* equations, for example, are of higher order in the derivatives than the *Navier-Stokes* equations, just as the *Navier-Stokes* equations are of higher order than the *Euler* equations for ideal inviscid fluids. The relative importance of these higher-order terms in the Burnett formulation, as compared, say, with the terms in the Navier-Stokes equations, is determined by the magnitude of the rarefaction parameters discussed earlier in the definition of the flow regimes. For example, the *Burnett stress term* is

$$\frac{\mu^2}{p} \frac{\partial u}{\partial x} \frac{1}{2} \left(\frac{\partial u}{\partial y} + \frac{\partial v}{\partial x} \right)$$

which divided by the *Navier-Stokes stress term*

$$\mu \left(\frac{\partial u}{\partial y} + \frac{\partial v}{\partial x} \right)$$

[1] D. Burnett, The Distribution of Molecular Velocities and the Mean Motion in a Non-uniform Gas, *Proc. London Math. Soc.*, **40**:382–400 (1935).

[2] H. Grad, On the Kinetic Theory of Rarefied Gases, *Commun. Pure Appl. Math.*, **2**:331–407 (1949).

yields

$$\frac{\mu}{p}\frac{\partial u}{\partial x} \sim \frac{\mu}{p}\frac{u}{L} \sim \frac{\mathrm{Ma}^2}{\mathrm{Re}}$$

Hence this term has an order of magnitude which depends on Mach and Reynolds numbers and which is not negligible in the slip-flow region. Other Burnett terms have a similar measure.

The foregoing notions are strictly applicable to a monatomic gas in which the molecules possess only the three translational degrees of freedom. For diatomic and polyatomic gases the distribution of internal energy over all degrees of freedom does not occur until the elapse of a *relaxation time* following any abrupt change of state in the gas. Internal energy is stored first in the translational degrees of freedom, and only after a sufficient number of collisions will energy be stored in the rotational and finally the vibrational degrees of freedom. The number of collisions required varies from a few in the case of air to a thousand or more in the case of carbon dioxide.

To treat problems involving atmospheric air (a fluid of obviously great interest) it is necessary to account for the rotational modes of nitrogen and oxygen, the principal components. At normal temperatures only the rotational states about the two axes of greatest moment of inertia are important.

The collision integral must be modified to include energy transfer from one state to the other in colliding molecules. Again for conditions of small departures from the equilibrium state, and if the relaxation time is small, the calculations result in the Navier-Stokes equations but with the inclusion of a *bulk-viscosity* term; i.e., the normal stresses in the gas are now expressed as

$$\sum_{i=1}^{3} -\frac{1}{3}p_{ii} = \rho RT + \varkappa \sum_{i=1}^{3} \frac{\partial u_i}{\partial x_i}$$

where $\varkappa = \mu' - \frac{2}{3}\mu$ and μ' is the *second coefficient of viscosity*.

This appears to be the correct expression for the normal stress in air at normal temperatures.[1] For other gases and for air at high temperatures there may be more than one relevant relaxation time, some of which may be long when compared with the translational relaxation time μ/p. A discussion of these features is beyond the scope of this book.

Gas-surface interaction The discussion thus far has avoided mention of the significant problem of the interaction of the molecules of a rarefied gas with solid surfaces, which (for internal flows) may be bounding surfaces such as tubes or conduits or (for external flows) the surfaces of

[1] L. Rosenhead, A Discussion of the First and Second Viscosity of Fluids, *Proc. Roy. Soc. London, Ser. A*, **226**(1164):1–65 (1954).

bodies submerged in the flow. The latter are perhaps the more interesting, certainly from the aerodynamics point of view.

While it is necessary to understand in some detail the kinetic state of the gas far from a surface, and therefore unaffected by the surface, it is within a few mean-free-path lengths of the surface that the most important processes take place. To establish any real understanding in this region requires a detailed knowledge of the physics of the interactions of molecules with the surface, in effect a joint distribution function which would indicate the number density of molecules reflected from a surface with velocity v_r from the number density of molecules incident upon the surface with velocity v_i. Such information does not yet exist.

To sidestep this lack of information Maxwell postulated an interaction model in which the incident molecules were either reflected *specularly*, i.e., regular reflection in which the normal component of velocity is reversed and the tangential component conserved, or *diffusely*, i.e., a random distribution of the reflected molecules. Maxwell introduced a single parameter f_s to indicate the fraction of the molecules undergoing diffuse reflection, such that $1 - f_s$ represents the fraction of molecules reflected specularly from the surface.

However, since both energy and momentum fluxes are important, there have evolved in more recent work specifications for three accommodation coefficients

$$\alpha_s = \frac{e_i - e_r}{e_i - e_w} \tag{11-46}$$

$$f_s = \frac{\tau_i - \tau_r}{\tau_i} \tag{11-47}$$

$$\sigma_s = \frac{p_i - p_r}{p_i - p_w} \tag{11-48}$$

where e_i, τ_i, p_i = fluxes of energy, tangential momentum, and normal momentum incident upon surface

e_r, τ_r, p_r = fluxes of energy, tangential momentum, and normal momentum reemitted from surface

e_w, p_w = fluxes reemitted from the surface as by a gas in complete maxwellian equilibrium with the surface

The quantities α_s, f_s, and σ_s are functions of the composition and temperature of the gas, the chemical state, physical structure, and temperature of the solid surface, and the velocity of the gas flow over the surface. Systematic investigations have been carried out on only a few of these parameters to date. Many investigations of α_s, rather fewer investigations of f_s, and almost no investigations of σ_s have been made. The available data are confined to incident energy levels of either a few hundredths of an electron volt or to high energies of 100 eV or more.

The range of energy of interest for applications in aerodynamics, 1 to 25 eV, has only recently been explored by means of flow-energized molecular beams.

Many of the direct macroscopic measurements of the accommodation coefficients have been made without careful consideration of either the chemical state of the solid surface or provision of a relevant flow regime.[1]

The concept of describing the gas-surface interaction in terms of the overall accommodation coefficients is useful only if these values are reasonably constant for a given surface and gas combination. Within the range of measurements available this seems to be the case for technical surfaces. Tables 11-2 and 11-3 show data on the thermal accommodation coefficients and on the tangential-momentum accommodation coefficients, respectively.

Slip flow In spite of, and in many ways because of, the great complications arising from the attempts to formulate the slip-flow problems, a considerable body of literature exists for slip-flow effects in fluid flow. This state of affairs is somewhat analogous to the treatment of the Navier-Stokes equations, for which no general solution and, indeed, only a few particular solutions exist. The equations are represented, however, by a vast literature of approximate solutions which make feasible the practical utilization of the ideas involved. The slip-flow studies are approximations in that the analyses have been based on the Navier-Stokes equations, usually for incompressible flow, with the slip-flow effects presumed to be concentrated in the contribution of additional terms in the tangential velocity and temperature boundary conditions, i.e., the *velocity-slip* and the *temperature-jump* conditions at the gas-surface interface. In spite of the broadness of the assumptions regarding the slip effects, the results have been checked experimentally to a surprising accuracy. Another interesting, and at the same time surprising conclusion, which has almost always been obtained in slip-flow analyses is that the results are found to correspond closely with free-molecule-flow results if the Knudsen numbers are taken as very large. Since slip flow is a phenomenon supposedly related to small Knudsen numbers, the exact significance of this correspondence to free-molecule-flow results at high Knudsen numbers is unknown; however, successful use of the correspondence has been made in semiempirical analysis to correlate experimental data over the entire flow regime from gas dynamics to free-molecule flow.

[1] J. P. Hartnett, A Survey of Accommodation Coefficients, *Advan. Appl. Mech. Rarefied Gas Dynamics*, suppl. 1, pp. 1–28, 1961; H. Y. Wachman, The Thermal Accommodation Coefficient: A Critical Survey, *J. Amer. Rocket Soc.*, **32**:2–12 (1962).

Table 11-2 Thermal accommodation coefficients α_s†

Gas	*Surface*	*Adsorbed gas*	*Temp.* °C	α_s	*Ref.*
H_2	Pt, bright			0.32	1
	Pt, black			0.74	1
O_2	Pt, bright			0.81	1
	Pt, black			0.93	1
N_2	Pt			0.50	2
	W			0.35	2
Air	Flat lacquer on bronze			0.88–0.89	3
	Polished bronze			0.91–0.94	3
	Machined bronze			0.89–0.93	3
	Etched bronze			0.93–0.95	3
	Polished cast iron			0.87–0.93	3
	Machined cast iron			0.87–0.88	3
	Etched cast iron			0.87–0.96	3
	Polished aluminum			0.87–0.95	3
	Machined aluminum			0.95–0.97	3
	Etched aluminum			0.89–0.97	3
He	W			0.025–0.057	4
	Ni, not flashed			0.20	4
	Ni, flashed			0.085	4
	W, flashed			0.17	5
	W, flashed			0.12	2
	W, not flashed			0.53	5
A	W, flashed			0.82	5
	W, flashed			0.46	5
	W, not flashed			1.00	5
He	H_2 adsorbed on W			0.041	6
	N_2 adsorbed on W			0.064	6
	W clean system			0.02	6
	W	None	30	0.0164	7
	W	None	−120	0.0130	7
	W	None	−196	0.0109	7
Ne	W	None	30	0.0412	7
	W	None	−120	0.0426	7
	W	None	−196	0.0495	7
Ar	W	None	30	0.271	7
	W	None	−120	0.300	7
	W	None	−196	0.549	7
Xe	W	None	30	0.773	7
	W	None	−120	0.878	7
	W	None	−196	0.942	7
He	K	None		0.083	8
Ar	K	None		0.444	8
He	K	None		0.042	8
Ar	K	None		0.41	8
He	W	0 (composite film)	32	0.107	6

Table 11-2 Thermal accommodation coefficients α_s† (continued)

Gas	*Surface*	*Adsorbed gas*	*Temp.* °C	α_s	*Ref.*
Ne	W	0 (composite film)	32	0.204	6
H_2	W	0	32	0.201	6
	W	None	32	0.105	6
He	W	N (monolayer)	32	0.040	6
Ne	W	N	32	0.117	6
H_2	W	N	32	0.642	6
CO_2	W	CO_2	32	0.990	6
He	Glass	None	29.3	0.31	7

† Compiled by F. C. Hurlbut; data from the following references: (1) M. Knudsen, "The Kinetic Theory of Gases," Methuen, London, 1934; (2) R. N. Oliver and M. Farber, Experimental Determination of Accommodation Coefficients as a Function of Temperature for Several Metals and Gases, Ph.D. dissertation, California Institute of Technology, 1950; (3) M. L. Wiedemann and P. R. Trumpler, Thermal Accommodation Coefficient, *Trans. ASME*, **68**:57–64 (1946); (4) J. K. Roberts, The Exchange of Energy between Gas Atoms and Solid Surfaces, *Proc. Roy. Soc. London, Ser. A*, **129**:146–151 (1930); **135**:192 (1932); **142**:518–524 (1933); (5) W. C. Michels, Accommodation Coefficients of Helium and Argon against Tungsten, *Phys. Rev.*, **40**:472 (1932); (6) H. Y. Wachman, The Thermal Accommodation Coefficient and Adsorption on Tungsten, Ph.D. dissertation, University of Missouri, 1957; (7) W. L. Silvernail, The Thermal Accommodation of Helium, Neon and Argon on Clean Tungsten from 77° to 303°K, Ph. D. dissertation, University of Missouri, 1954; (8) H. L. Petersen, The Accommodation Coefficients of Helium, Neon and Argon on Clean Potassium Surfaces from 77°K to 208°K, Ph.D. dissertation, University of Missouri, 1958.

Slip-flow analysis Since direct solutions for the flow and energy equations for the slip-flow regime are unavailable at the moment, the analytical attack on the problem has been made by using the continuum formulations for the flow and energy equations, relegating the rarefaction effects to the boundary conditions. Two definitive effects have been observed in slip-flow phenomena: (1) It was noted theoretically by Maxwell[1] and experimentally by Kundt and Warburg[2] that for rarefied-gas-flow conditions near a boundary, the flow velocity is not zero but slips along the surface with a finite velocity relative to the surface. (2) A temperature

[1] J. C. Maxwell, On the Condition to Be Satisfied by a Gas at the Surface of a Solid Body, "The Scientific Papers of James Clerk Maxwell," vol. 2, p. 704, Cambridge University Press, London, 1890.

[2] A. Kundt and E. Warburg, On the Friction and Thermal Conductivity in Rarefied Gases, *Phil. Mag.*, **50**:53 (1875).

Table 11-3 Tangential momentum accommodation coefficients f_s†

Gas	*Surface*	f_s	*Ref.*
Air	Machined brass	1.00	1
CO_2	Machined brass	1.00	1
Air	Old shellac	1.00	1
CO_2	Old shellac	1.00	1
Air	Hg	1.00	1
	Oil	0.90	1
CO_2	Oil	0.92	1
H_2	Oil	0.93	1
Air	Glass	0.89	1
He	Oil	0.87	1
Air	Fresh shellac	0.79	1
	Ag_2O	0.98	2
He	Ag_2O	1.00	2
H_2	Ag_2O	1.00	2
O_2	Ag_2O	0.99	2
Air	Oil on machined Al	0.90	3
N_2	Adsorbed gas on oil film on Al	0.8–0.93‡	4
O_2	Adsorbed gas on oil film on Al	0.8–0.95‡	4
Ar	Adsorbed gas on oil film on Al	0.8–0.95‡	4
Air	Adsorbed gas on oil film on Al	0.60–0.93‡	4
N_2	Glass	0.93–0.97¶	5

† Compiled by F. C. Hurlbut; data from the following references: (1) R. A. Millikan, Coefficients of Slip in Gases and the Law of Reflection of Molecules from Surfaces of Solids and Liquids, *Phys. Rev.*, **20**(3):217–238 (1923); (2) E. Blankenstein, Coefficients of Slip and Momentum Transfers in H_2, He, Air and O_2, *Phys. Rev.*, **22**:582 (1923); (3) S. F. Chiang, Drag on a Rotating Cylinder at Low Pressures, Ph.D. dissertation, University of California, Berkeley, 1951; (4) F. C. Hurlbut, Influence of Pressure History on Momentum Transfer in Rarefied Gas Flows, *Phys. Fluids*, **3**:4 (1960); (5) F. C. Hurlbut, Studies on Molecular Scattering at a Solid Surface, *J. Appl. Phys.*, **28**:844 (1957).
‡ The lower values depend on the time allowed for multilayer adsorption of test gas on the surface oil film.
¶ Computed from molecular-beam data.

discontinuity, or jump, was postulated by Poisson[1] during heat transfer from a surface to a rarefied gas, and this effect was found to exist for static systems by the experiments of Smoluchowski.[2]

[1] E. H. Kennard, "Kinetic Theory of Gases," pp. 162–181, McGraw-Hill, New York, 1938.

[2] M. von Smoluchowski, Über den Temperaturesprung bei Warmeleitung in Gasen, *Sitzber. Wien Akad.*, **107**:304 (1898); **108**:5 (1899); also *Anz. Akad. Wiss. Krakau ser. A*, **5**:129 (1910).

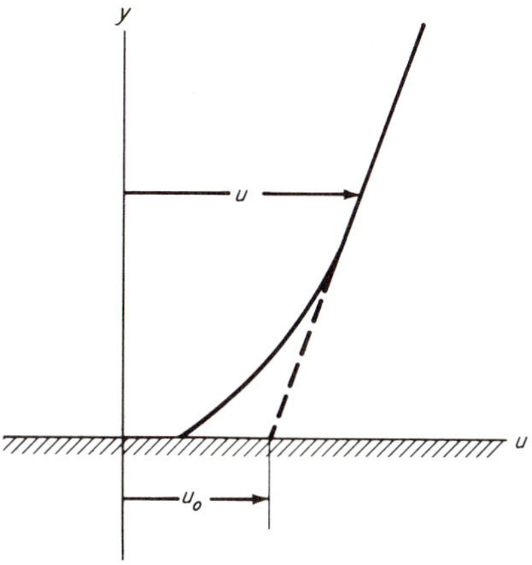

Fig. 11-26 Graphical representation of the slip-flow boundary conditions.

The velocity condition for slip flow adjacent to a surface can be treated in the following way. Several mean-free-path lengths away from the surface the gas is presumed to be in motion with a constant velocity gradient $\partial u/\partial y$ arising from a constant shear stress $\tau = \mu\, \partial u/\partial y$ (Fig. 11-26). The extrapolation of the macroscopic gas-velocity profile to the surface $y = 0$ yields u_0, the *slip velocity*. The distribution function far from the surface $y = 0$ is calculated to be

$$f = \frac{n}{(2\pi RT)^{\frac{3}{2}}} \exp\left[-\frac{(v_x - u)^2 + v_y{}^2 + v_z{}^2}{2RT}\right]\left[1 - \frac{1}{pRT}\,\mu\,\frac{\partial u}{\partial y}\,(v_x - u)v_y\right] \tag{11-49}$$

which is a special case of Eq. (11-43). Since the gas flow is steady, the shear stress τ is constant, and in terms of the extrapolated velocity profile the value of τ at the surface is the same as the value of τ far from the surface. Neglecting the effect of molecular collisions within the immediate vicinity of the wall, the momentum flux from the gas onto the surface is

$$\tau' = \int_{-\infty}^{+\infty}\int_{-\infty}^{0}\int_{-\infty}^{+\infty} mv_xv_yf\,dv_x\,dv_y\,dv_z \tag{11-50}$$

$$\tau' = \frac{1}{2}\,\mu\left(\frac{\partial u}{\partial y} + \frac{u_0}{\lambda}\right) \tag{11-51}$$

Of this flux a fraction f_s is transmitted; thus

$$\tau = \mu\,\frac{du}{dy} = \frac{1}{2}\,\mu f_s\left(\frac{du}{dy} + \frac{u_0}{\lambda}\right) \tag{11-52}$$

from which the slip velocity u_0 may be obtained

$$u_0 = \frac{2 - f_s}{f_s}\,\lambda\left(\frac{du}{dy}\right)_{y=0} \tag{11-53}$$

If temperature gradients exist in the direction of the flow, by a calculation similar to that above, Eq. (11-53) can be modified to read

$$u_0 = \frac{2 - f_s}{f_s} \lambda \left(\frac{du}{dy}\right)_{y=0} + 3\sqrt{\frac{RT}{8\pi}} \frac{\lambda}{T} \left(\frac{dT}{dx}\right)_{y=0} \tag{11-54}$$

where the second term on the right is that known as the *thermal creep*, which is of great importance in instrumentation for use in high-vacuum technology since the existence of temperature gradients can induce pressure gradients even when no flow exists.

For temperature gradients in the gas normal to the surface, calculations similar to the preceding ones lead to an expression for the temperature-jump condition at the surface

$$T_0 - T_w = \frac{2 - \alpha_s}{\alpha_s} \frac{2\gamma}{\gamma + 1} \frac{\lambda}{\mathrm{Pr}} \left(\frac{dT}{dy}\right)_{y=0} \tag{11-55}$$

where T_0 = temperature of gas at $y = 0$
T_w = temperature of wall at $y = 0$
γ = ratio of specific heats
Pr = Prandtl number of gas
α_s = thermal accommodation coefficient

Equations (11-53) to (11-55) are true rarefaction effects that appear to modify continuum-flow phenomena. From a series of definitive experiments in shock waves, where there are finite changes in the pressure and temperature through the shock thickness, there is strong evidence to support the use of the Navier-Stokes equations as representative of the near-continuum region, modified with the slip-flow boundary conditions, rather than the proposed Burnett, Grad, or Mott-Smith equations.[1] Even use of the Navier-Stokes equations is restricted to the treatment of those geometries or flow conditions for which solutions to the equations can be obtained. Some of the available results will be given here.

Flat plate Laminar slip flow over a flat plate at zero angle of attack can be treated in terms of the Rayleigh problem of an impulsively started plate. The inertia and viscous terms in the Navier-Stokes equations can be simplified in this case. Calculations will be presented first for the drag and then for the heat transfer for the same plate.

The flow can be considered in two dimensions, but for the impulsively started plate the convective inertia terms such as $u(\partial u/\partial x)$, and the viscous terms such as $\nu(\partial^2 u/\partial x^2)$ can be neglected, leaving for the

[1] L. Talbot and F. S. Sherman, Structure of Weak Shock Waves in a Monatomic Gas, *NASA Memo* 12-14-58 W, 1959; F. S. Sherman, A Low-density Wind Tunnel Study of Shock Wave Structure and Relaxation Phenomena in Gases, *NACA Tech. Note* 3298, 1955.

momentum equation

$$\frac{\partial u}{\partial \tau} = \nu \frac{\partial^2 u}{\partial y^2} \tag{11-56}$$

The left-hand term can be transformed into

$$\frac{\partial u}{\partial \tau} = \frac{\partial u}{\partial x}\frac{\partial x}{\partial \tau} + \frac{\partial u}{\partial y}\frac{\partial y}{\partial \tau}$$

and if a uniform velocity U is given to the plate in the x direction only, the above reduces, for $\partial y/\partial \tau = 0$, to

$$\frac{\partial u}{\partial \tau} = U\frac{\partial u}{\partial x}$$

and Eq. (11-56) becomes

$$U\frac{\partial u}{\partial x} = \nu \frac{\partial^2 u}{\partial y^2} \tag{11-57}$$

The appropriate boundary conditions are for $f_s = 1$, a reasonably satisfactory approximation,

$$u = U \qquad x = 0 \qquad \text{all } y$$

$$u = \lambda \frac{\partial u}{\partial y} \qquad x > 0 \qquad y = 0$$

A drag coefficient is defined as

$$C_D = \frac{\text{total drag on both sides of plate/unit width}}{\rho U^2 L/2}$$

where L is the length of the plate. The solution of Eq. (11-57) for the boundary conditions stated in terms of the defined drag coefficient becomes

$$C_D \,\text{Ma} = \frac{2.67}{X_1^2}\left(e^{X_1^2}\,\text{erfc}\, X_1 - 1 + \frac{2}{\sqrt{\pi}}X_1\right) \tag{11-58}$$

where $X_1 = \frac{2}{3}\sqrt{\text{Re}}/\text{Ma}$. A plot of Eq. (11-58) is shown in Fig. 11-27 in comparison with experimental data on the total drag of flat plates made in a low-density high-speed wind tunnel.[1]

The heat transfer from this plate can be calculated from the energy equation for the impulsively started plate.[2] In this instance one can use the same degree of approximation as in the treatment of the momentum equation neglecting the convective term in y and the heat-conduction term $k(\partial^2 t/\partial x^2)$ as being small in comparison with the term $k(\partial^2 t/\partial y^2)$.

[1] S. A. Schaaf, *Univ. Calif. Inst. Eng. Res. Rept.* HE-150-66, 1950; F. S. Sherman, *Univ. Calif. Inst. Eng. Res. Rept.* HE-150-70, 1950.

[2] R. M. Drake, Jr., and E. D. Kane, Heat Transfer Problems in High Speed Flow in Rarefied Gases, *Proc. Gen. Discussion Heat Transfer ASME-IME, London, 1951;* see also J. R. Stalder, G. Goodwin, and M. O. Creager, in *ibid.*

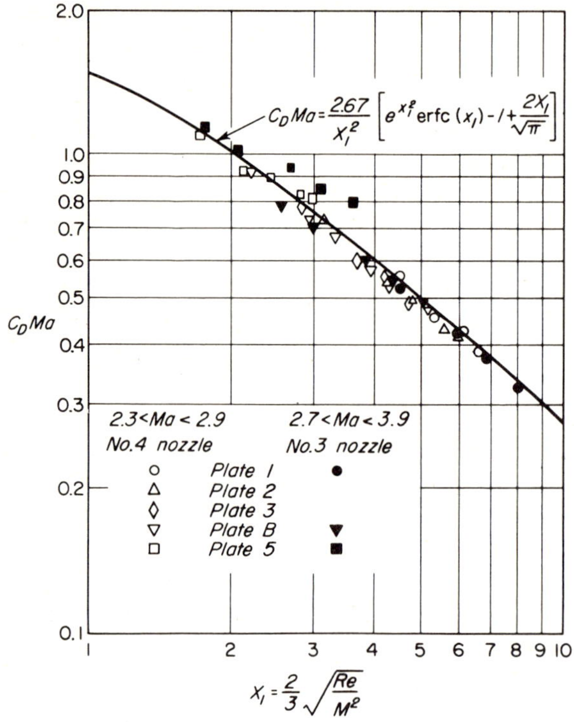

Fig. 11-27 Comparison of theory and experiment for flat-plate drag coefficient. [*Courtesy, F. S. Sherman*].

Thus for motion of the plate at constant velocity U in the x direction, the energy equation becomes

$$U \frac{\partial t}{\partial x} = \alpha \frac{\partial^2 t}{\partial y^2} \tag{11-59}$$

with the boundary conditions

$$t_{y=0} - t_w = \frac{2 - \alpha_s}{\alpha_s} \frac{2\gamma}{\gamma + 1} \frac{\lambda}{\Pr} \frac{dt}{dy} \qquad y = 0 \qquad x > 0$$

$$t = t_f \qquad x = 0 \qquad y \to \infty$$

For $\alpha_s = 0.8$ and $\gamma = 1.4$, values characteristic of air and technical surfaces, the solution of Eq. (11-59) becomes

$$\text{St Ma} = \frac{0.38}{X_2{}^2}\left(e^{X_2{}^2} \operatorname{erfc} X_2 - 1 + \frac{2}{\sqrt{\pi}} X_2\right) \tag{11-60}$$

where $X_2 = \sqrt{\text{Re Pr}/6.9\ \text{Ma}^2}$ and St is the Stanton number

$$\text{St} = \frac{\text{Nu}}{\text{Re Pr}} = \frac{h}{\rho U c_p}$$

There are no experimental data with which to compare Eq. (11-60); however, the results should be just as satisfactory as those of Eq. (11-58).

As one would expect, rarefaction effects show up at the high Mach numbers of hypersonic flow together with the complicated behavior of shock-wave interaction with the viscous boundary layers. The general picture is a formidable one analytically as well as experimentally and in the low-density flow regimes is represented by four hypersonic flow theories:[1] (1) weak-viscous-interaction theory; (2) strong-viscous-interaction theory; (3) strong viscous interaction with surface slip; and (4) free-molecule-flow theory. While a discussion of these theories, except the last, is beyond the scope of this book, for purposes of completeness the several results of the theories for flat-plate skin friction and heat transfer will be summarized here, and shown in comparison with existing experimental data.[2]

Flat plate, skin friction, hypersonic flow

1. Weak viscous interaction

$$\frac{C_f \sqrt{U_\infty x/\nu}}{\sqrt{C_*}} = 0.664\left(1 + \frac{3\beta^2 \bar{X}}{\gamma^2}\right) \tag{11-61}$$

2. Strong viscous interaction

$$\frac{C_f \sqrt{U_\infty x/\nu}}{\sqrt{C_*}} = 0.208\left[\left(1 + 2.5\frac{H_w}{H_0}\right)\bar{X}\right]^{\frac{1}{2}} \tag{11-62}$$

3. Strong viscous interaction with surface slip

$$\frac{C_f \sqrt{Uax/\nu}}{\sqrt{C_*}} = 0.208\left[\left(1 + 2.5\frac{H_w}{H_0}\right)\bar{X}\right]^{\frac{1}{2}}\left[1 - 0.85\left(\frac{\gamma - 1}{\gamma}\right)^{\frac{1}{2}}\frac{\sqrt{\bar{X}}}{\text{Ma}_\infty}\right] \tag{11-63}$$

4. Free-molecule flow

$$\text{Ma}_\infty C_f = \frac{f_s}{\text{Ma}_\infty}\sqrt{\frac{2}{\pi\gamma}} \tag{11-64}$$

[1] W. D. Hayes and R. F. Probstein, "Hypersonic Flow Theory," Academic, New York, 1959.

[2] J. E. Wallace and A. F. Burke, An Experimental Study of Surface and Flow Field Effects in Hypersonic Low Density Flow over a Flat Plate, *Advan. Appl. Mech.*, suppl. 3, vol. 1, pp. 487–507 (1965).

Flat-plate heat transfer, hypersonic flow

1. Weak viscous interaction

$$\frac{\text{St}\sqrt{U_\infty x/\nu}}{\sqrt{C_*}} = \frac{0.332(1 + \beta\bar{X})}{\sqrt{1 + 2\beta\bar{X}}} \tag{11-65}$$

2. Strong viscous interaction

$$\frac{\text{St}\sqrt{U_\infty x/\nu}}{\sqrt{C_*}} = 0.0788\left[\left(1 + 2.5\frac{H_w}{H_0}\right)\bar{X}\right]^{\frac{1}{2}} \tag{11-66}$$

3. Strong viscous interaction with surface slip

$$\frac{\text{St}\sqrt{U_\infty x/\nu}}{\sqrt{C_*}} = 0.0788\left(1 + 2.5\frac{H_w}{H_0}\right)^{\frac{1}{2}}\left(1 + 0\frac{\text{Ma}_\infty}{\sqrt{U_\infty x/\nu}}\right) \tag{11-67}$$

4. Free-molecule flow

$$\text{Ma}_\infty \text{St} = \frac{1}{\sqrt{2\pi\gamma}}\frac{\alpha_s}{1 - H_w/H_0}\left[1 + \frac{\gamma + 1}{\gamma(\gamma - 1)(\text{Ma}_\infty)^2}\left(1 - \frac{T_w}{T_0}\right)\right] \tag{11-68}$$

where $\dfrac{p}{p_\infty} = 1 + \beta\bar{X} + O(\bar{X})^2 \qquad T_* = \frac{1}{6}T_0 + \frac{1}{2}T_w$

$$\bar{X} = \frac{(\text{Ma}_\infty)^3\sqrt{C_*}}{\sqrt{U_\infty x/\nu}} \qquad \text{St} = \frac{q}{\rho_\infty U_\infty(H_0 - H_w)}$$

$$C_* = \frac{\mu_*}{\mu_\infty}\frac{T_\infty}{T_*} \qquad \gamma = \frac{c_p}{c_v}$$

H is the enthalpy and T is the absolute temperature of the flow. The subscripts w and 0 refer to wall and adiabatic wall states, respectively.

Comparison of these theories for the flat plate with experiment is shown by Figs. 11-28 and 11-29.[1]

The sphere Heat transfer from spheres has received some attention,[2] because the spherical shape lends itself well to experimental procedures and because the shape has considerable practical importance technically. The problem is complicated, however, by the fact that for the most part (unless the sphere diameter is very small) the Mach number must be

[1] The data in each case are reasonably well predicted by the hypersonic theory for values of $\bar{X} < 100$ whereas above that value the data are better represented by the free-molecule-flow theory, discussed later in this section.

[2] Drake and Kane, *loc. cit.;* R. M. Drake, Jr., and G. H. Backer, Heat Transfer from Spheres to a Rarefied Gas, *Trans. ASME,* **74**:7 (1952); T. D. Taylor, Heat Transfer from Single Spheres in a Low Reynolds Number Slip Flow, *Phys. Fluids,* **6**(7):987–992 (1963).

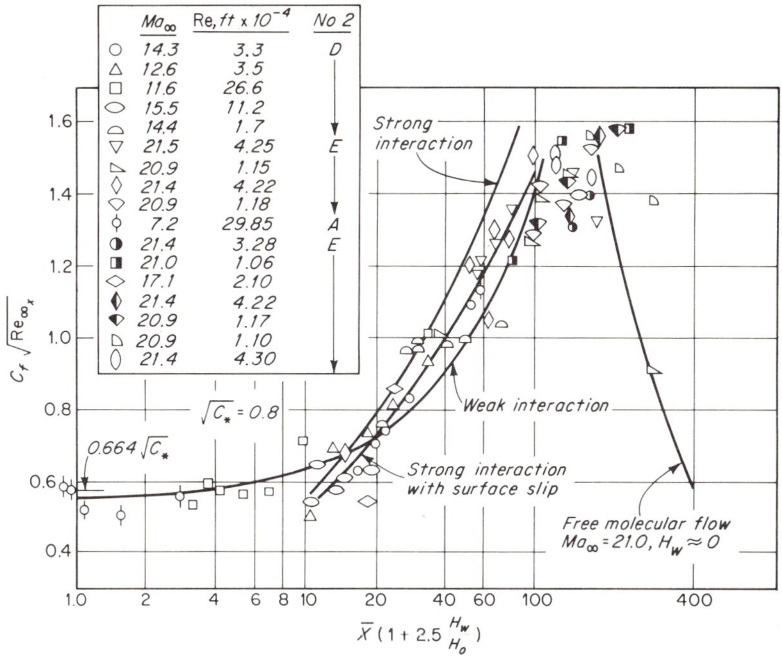

Fig. 11-28 Skin-friction distribution on a flat plate at zero angle of attack in hypersonic flow. [*J. E. Wallace and A. F. Burke, Advan. Appl. Mech., suppl. 3, vol. 1, pp. 487–507 (1965).*]

large if the Knudsen number is to be large enough to establish the flow as slip flow. Ordinarily this will mean a Mach number so large that the flow is supersonic. In such a case a shock wave will exist in front of the sphere (Fig. 11-30), and the conditions behind the shock wave must be considered before any sense can be made of the heat-transfer calculations. As one model, the sphere can be considered immersed in a supersonic rarefied-gas flow preceded by a *normal shock wave*, behind which can be calculated those gas states necessary to determine the heat transfer. As a second model one can consider a rarefied, subsonic flow obtainable at very low gas densities such that the shock wave is not present and therefore offers no real complication.[1] Kavanau has developed an expression for the rarefaction correction to the continuum solution for heat transfer from spheres that represents the experimental data within 10 percent. We shall consider this approach here.

[1] L. L. Kavanau and R. M. Drake, Jr., *Univ. Calif. Inst. Eng. Res. Rept.* HE-150-108, 1953; and L. L. Kavanau, Heat Transfer from Spheres to a Rarefied Gas in Subsonic Flow, *Trans. ASME*, **77**:5 (1955).

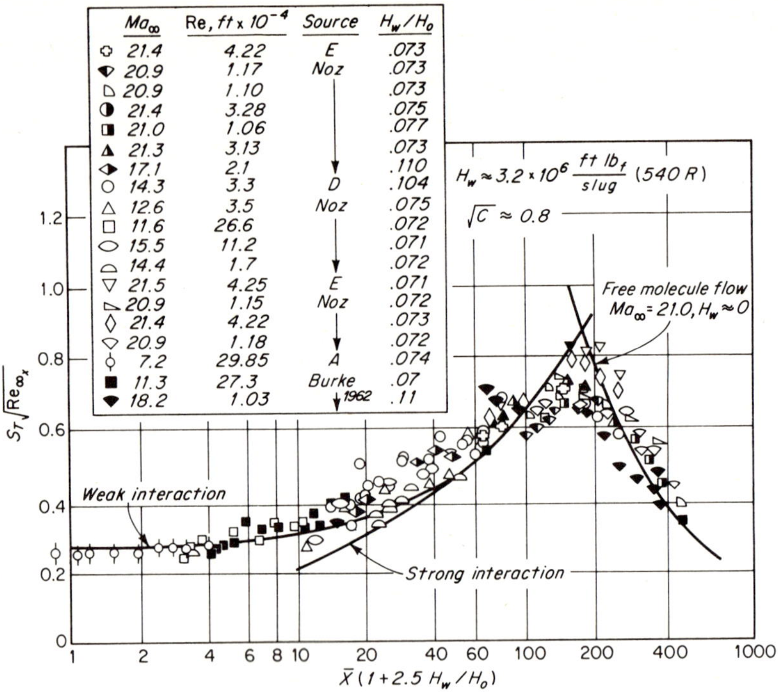

Fig. 11-29 Heat-transfer distribution on a flat plate at zero angle of attack in hypersonic flow. [*J. E. Wallace and A. F. Burke, Advan. Appl. Mech., suppl. 3, vol. 1, pp. 487–507 (1965).*]

If we consider the temperature jump in the slip-flow region as an effective thermal contact resistance at the gas-surface interface over and above the thermal resistance due to the viscous boundary layer, the heat-transfer coefficient at low densities can be determined to the first order by a correction to the continuum heat-transfer coefficient at the same Reynolds number.

Consider a surface in continuum flow so that heat is transferred between the surface and the stream. The surface temperature is uniform, constant, and equal to t_w°; the heat flux becomes

$$q = -k\left(\frac{\partial t}{\partial y}\right)_{y=0} = h^\circ(t_w^\circ - t_r^\circ) \tag{11-69}$$

where the superscript indicates the continuum state and t_r is the recovery temperature of the surface.

Now if the gas stream is rarefied while the Reynolds number is held constant, the additional resistance to heat transfer because of the temperature jump will appear, so that we can write a total convective

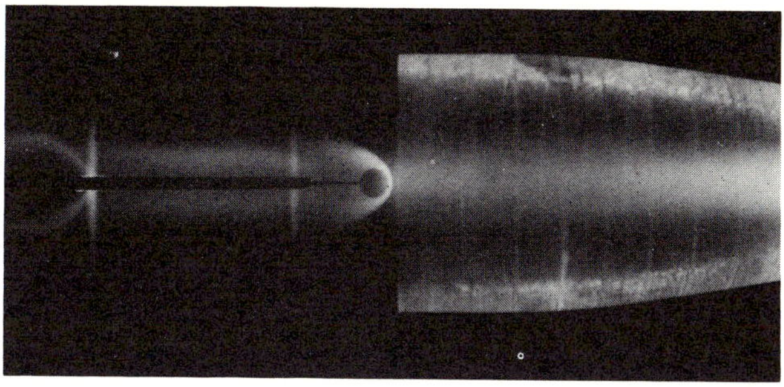

Fig. 11-30 Shock wave in front of sphere by nitrogen afterglow method. [*Courtesy of Low Pressure Research Project, University of California, Berkeley.*]

heat-transfer coefficient incorporating both the thermal boundary-layer and temperature-jump resistances by writing

$$q = h(t_w - t_r) \tag{11-70}$$

The first-order temperature jump can be written as

$$t_{y=0} - t_w = \xi \left(\frac{\partial t}{\partial y}\right)_{y=0} \tag{11-71}$$

where $t_{y=0}$ and $(\partial t/\partial y)_{y=0}$ are the temperature and temperature gradient in the layer of gas immediately adjacent to the surface. The quantity ξ is the temperature-jump distance defined from Eq. (11-55) as

$$\xi = 1.996 \frac{2 - \alpha_s}{\alpha_s} \frac{\gamma}{\gamma + 1} \frac{\lambda}{\text{Pr}}$$

or

$$\xi = \frac{\theta}{\text{Pr}} \lambda \tag{11-72}$$

The first-order temperature-jump condition requires that $t_r = t_r^\circ$. It is now assumed that q, the heat transferred between the surface and the gas, is the same with and without the temperature jump. This, in effect, assumes that Eq. (11-71) is valid when $t_{y=0} = t_w^\circ$. Combining Eqs. (11-71), (11-69), and (11-70) gives

$$\frac{h^\circ}{h} = 1 + \frac{\xi h^\circ}{k}$$

and substituting into this expression the value for ξ, we obtain with the aid of Eq. (11-33)

$$\frac{h^\circ}{h} = 1 + \frac{h^\circ L}{k} \frac{\text{Ma}}{\text{Re Pr}} \sqrt{\frac{\gamma \pi \theta^2}{2}}$$

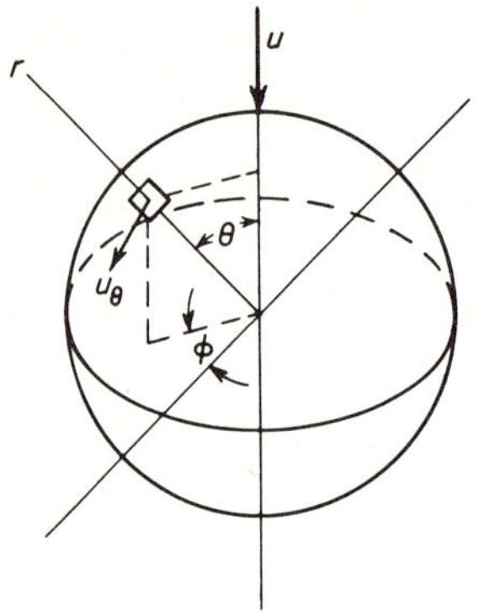

Fig. 11-31 Coordinate system for the sphere.

which can be written as

$$\frac{\text{Nu}^\circ}{\text{Nu}} = 1 + \sqrt{\frac{\gamma\pi\theta^2}{2}\frac{\text{Ma}}{\text{Re Pr}}}\,\text{Nu}^\circ \tag{11-73}$$

If Eq. (11-73) is evaluated for a diatomic gas with $\gamma = 1.4$ and an accommodation coefficient $a = 0.8$, we obtain approximately

$$\frac{\text{Nu}^\circ}{\text{Nu}} = 1 + 2.59\,\frac{\text{Ma}}{\text{Re Pr}}\,\text{Nu}^\circ$$

In order to interpret Eq. (11-73) it is necessary to have a solution for the continuum heat transfer from spheres. Such a solution has been obtained for an approximate velocity distribution about the sphere.[1] Consider the energy equation for a spherical coordinate system, neglecting the compressibility and dissipation of mechanical-energy terms, in terms of the geometry shown in Fig. 11-31

$$\rho c_p\left(u_r\frac{\partial t}{\partial r} + \frac{u_\theta}{r}\frac{\partial t}{\partial \theta} + \frac{u_\phi}{r\sin\theta}\frac{\partial t}{\partial \phi}\right) = k\left[\frac{1}{r^2}\frac{\partial}{\partial r}\left(r^2\frac{\partial t}{\partial r}\right) + \frac{1}{r^2\sin\theta}\frac{\partial}{\partial\theta}\left(\sin\theta\frac{\partial t}{\partial\theta}\right) + \frac{1}{r^2\sin^2\theta}\frac{\partial^2 t}{\partial\phi^2}\right] \tag{11-74}$$

As an approximation to the velocity slip condition let

$$u_r = u_\phi = 0 \qquad u_\theta = U$$

and from symmetry

$$\frac{\partial t}{\partial \phi} = \frac{\partial^2 t}{\partial \phi^2} = 0$$

[1] Drake and Kane, *loc. cit.;* also *Trans. ASME,* **74:**7 (1952).

and for small heat conduction in the θ direction compared with the convective flow in that direction,

$$k\frac{\partial t}{\partial \theta} = k\frac{\partial^2 t}{\partial \theta^2} = 0$$

Equation (11-74) thus reduces to the abbreviated form

$$\rho c_p \frac{U}{r}\frac{\partial t}{\partial \theta} = k\left[\frac{1}{r^2}\frac{\partial}{\partial r}\left(r^2\frac{\partial t}{\partial r}\right)\right] \tag{11-75}$$

Equation (11-75) can be solved for the boundary conditions

$$t = t_w \qquad r = r_0$$
$$t = t_f \qquad r \rightarrow \infty$$

by means of the Laplace transform[1] to yield for the average heat-transfer coefficient in terms of the Nusselt number

$$\overline{\text{Nu}} = \frac{h_{\text{av}}D}{k} = 2 + \frac{2}{\pi^2}\int_0^\infty \frac{(1 + e^{-\beta^2\pi})(1 + \beta^4)^{-1}\,d\beta}{[J_1^2(\alpha_1\beta) + Y_1^2(\alpha_1\beta)]\beta} \tag{11-76}$$

where D = diameter of sphere

$\alpha_1 = \sqrt{2\,\text{Re Pr}}$

$J_1(\alpha_1\beta)$, $Y_1(\alpha_1\beta)$ = Bessel functions of first and second kind

Equation (11-76) is plotted in Figs. 11-32 and 11-33 as the line $\text{Ma} \sim 0$. Using the combined results of Eqs. (11-73) and (11-76), we obtain the results as plotted in Fig. 11-33 in comparison with experimental heat-transfer data for small spheres. The constant 2.59 in Eq. (11-73) was adjusted empirically to a value 3.42 in order to represent the experimental data more accurately:

$$\overline{\text{Nu}} = \frac{\overline{\text{Nu}}^\circ}{1 + 3.42[\text{Ma}/(\text{Re Pr})]\,\overline{\text{Nu}}^\circ} \tag{11-77}$$

The traces in Figs. 11-32 and 11-33 representing heat transfer in free-molecule flow are the results of a kinetic-theory calculation made by Sauer.[2] The graphical results show the difference between taking the Reynolds number to zero as a result of a zero velocity and taking it to zero as a result of a zero density. The zero-velocity case for heat transfer from spheres becomes in the limit a case of radial conduction in a

[1] R. V. Churchill, "Modern Operational Mathematics in Engineering," McGraw-Hill, New York, 1944.

[2] F. M. Sauer, Convective Heat Transfer from Spheres in Free-molecule Flow, *J. Aeron. Sci.*, **18**(5):353 (1951).

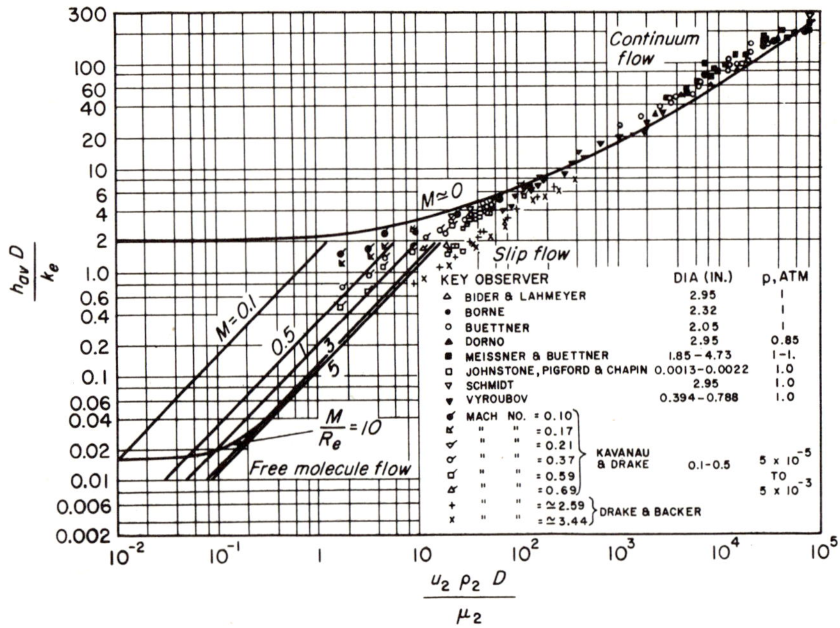

Fig. 11-32 Heat transfer from spheres (Reynolds number evaluated at condition 2 behind hypothetical normal shock for Ma > 1 and thermal conductivity evaluated at the equilibrium temperature). [*R. M. Drake, Jr., and G. H. Backer, Trans. ASME,* **74**:*7* (*1952*).]

stagnant gas, whereas the approach to the zero-density case leads to the free-molecule-flow regime and much lower heat-transfer coefficients.

An interesting feature of these rarefied-gas heat-transfer phenomena is shown in Fig. 11-34, a representation of the thermal recovery factor plotted in relation to the rarefaction parameter. The recovery factors are seen to increase at large values of the Knudsen number in slip flow and to approach a value greater than unity as predicted in the free-molecule-flow range. Another indication of this phenomenon, shown in Fig. 11-35, is the distribution of the thermal recovery factor through the boundary layer in low-density flow, supersonic nozzles. This energy distribution is typical of those found in high-speed gas flows, where the Prandtl number is less than unity. The fact that low-density flows have rather thick boundary layers makes this phenomenon more readily observable.

In the past few years there have been several investigations of the problem of heat transfer from spheres, particularly with respect to the search for an acceptable, broadly applicable theoretical determination.

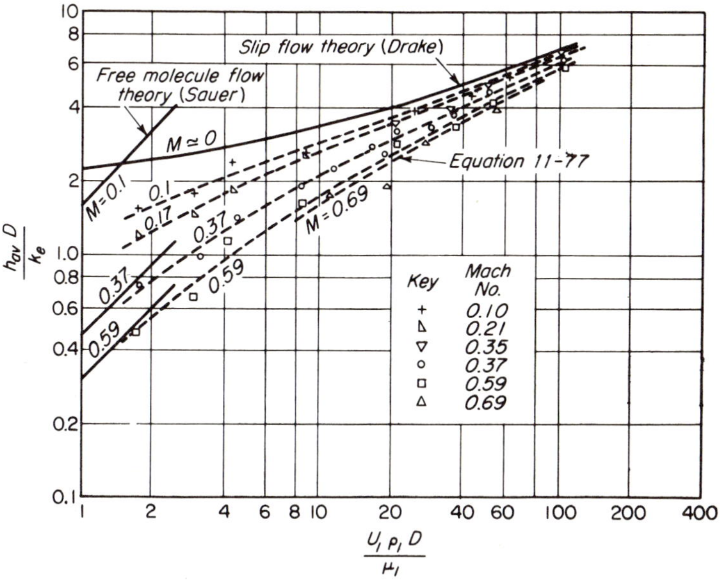

Fig. 11-33 Heat transfer from spheres at subsonic Mach numbers in a rarefied gas (Reynolds numbers evaluated at the free-stream condition 1 and thermal conductivity evaluated at the equilibrium temperature). [*L. L. Kavanau and R. M. Drake, Jr., Univ. Calif. Inst. Eng. Res. Rept. HE-150-108, 1953; also L. L. Kavanau, Trans. ASME,* **77**:*5* (*1955*).]

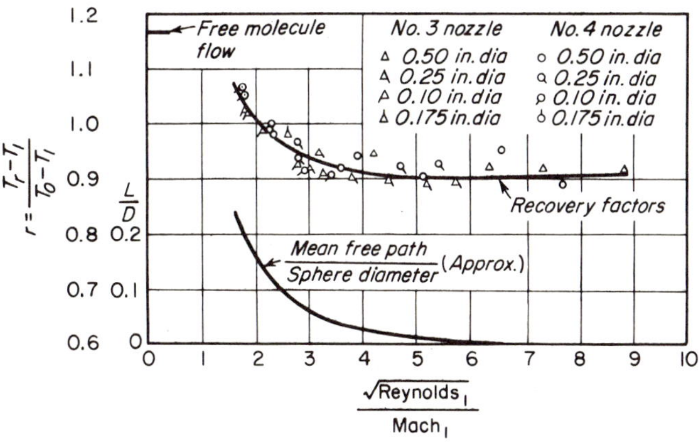

Fig. 11-34 Variation of thermal-recovery factor with rarefaction parameter: T_r = equilibrium temperature attained by high-conductivity sphere, T_0 = reservoir total temperature, T_1 = free-stream temperature. [*R. M. Drake, Jr., and G. H. Backer, Trans. ASME,* **74**:*7* (*1952*).]

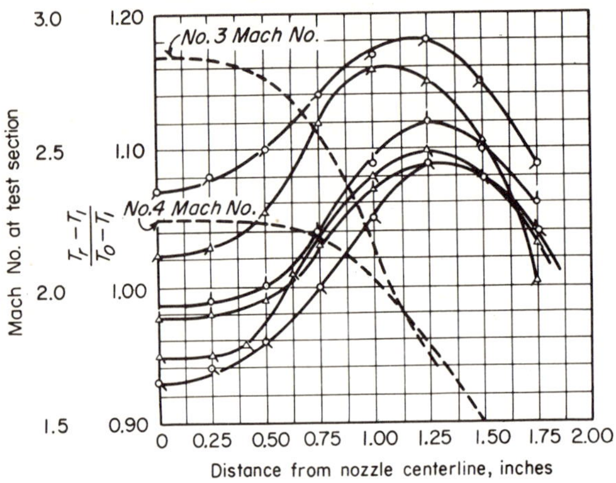

Fig. 11-35 Variation of thermal-recovery factor through laminar boundary layers in supersonic nozzles at very low pressures. T_r = recovery temperature attained by high-conductivity spherical probe, T_0 = the reservoir total temperature, T_1 = the free-stream temperature. Dashed lines relate to Mach number distribution radially in nozzle. Solid lines relate to recovery-factor distribution radially in the nozzle. Nozzle 3 is 13.34 cm in diameter; nozzle 4 is 13.97 cm in diameter. [*R. M. Drake, Jr. and G. H. Backer, Trans. ASME,* **74**:*7* (*1952*).]

Notable are the works of Takao,[1] Liu and Sugimura,[2] Cercignani and Pagani,[3] and Lees.[4]

The cylinder Heat transfer from cylinders in the slip-flow region produces effects quite similar to those described for spheres. An early slip-flow solution[5] was developed from the continuum form of the energy equation with temperature-jump boundary conditions such that the resulting equation for the Nusselt number for overall heat transfer from

[1] K. Takao, Heat Transfer from a Sphere in a Rarefied Gas, *Advan. Appl. Mech.*, suppl. 2, vol. 2, 102–111 (1963).

[2] C. Y. Liu and T. Sugimura, Rarefied Gas Flow over a Sphere at Low Mach Numbers, *Advan. Appl. Mech.* suppl. 5, vol. 1, pp. 789–796 (1969).

[3] C. Cercignani and C. D. Pagani, Variational Approach to Rarefied Flows in Cylindrical and Spherical Coordinates, *Advan. Appl. Mech.* suppl. 4, vol. 1, pp. 555–573 (1967).

[4] L. Lees, Kinetic Theory Description of Rarefied Gas Flows, *J. Soc. Indus. Appl. Math.* **13**(1):278–311 (1965).

[5] F. M. Sauer and R. M. Drake, Jr., Forced Convection Heat Transfer from Horizontal Cylinders in a Rarefied Gas, *J. Aeron. Sci.*, **20**(3):175–181 (1953).

the cylinder not only predicted with excellent correspondence the slip-flow heat-transfer experimental results in the range $0.1 < \text{Re} < 100$ and $0.0 < \text{Ma} \leq 1$ but for $\text{Kn} = 0$ accurately reflected the experimental data for the continuum-flow range, $0.01 < \text{Re} < 1{,}000$, $\text{Ma} \sim 0$.

Consider an infinite cylinder normal to a uniform gas flow (Fig. 11-36). The thermal-energy equation neglecting the terms representing compressibility and the dissipation of mechanical energy to heat may be written for steady conditions as

$$u \frac{\partial t}{\partial r} + \frac{v}{r} \frac{\partial t}{\partial \theta} + w \frac{\partial t}{\partial z} = \alpha \left(\frac{\partial^2 t}{\partial r^2} + \frac{1}{r} \frac{\partial t}{\partial r} + \frac{1}{r^2} \frac{\partial^2 t}{\partial \theta^2} + \frac{\partial^2 t}{\partial z^2} \right) \tag{11-78}$$

If the flow is two-dimensional, the conduction of heat in the z direction is zero; and if the radial velocity $u = 0$ and the tangential velocity $v = V$, a constant, Eq. (11-78) reduces to

$$\frac{V}{r} \frac{\partial \vartheta}{\partial \theta} = \alpha \left(\frac{\partial^2 \vartheta}{\partial r^2} + \frac{1}{r} \frac{\partial \vartheta}{\partial r} \right) \tag{11-79}$$

where $\vartheta = (t - t_w)/(t_r - t_w)$
t_w = constant surface temperature of cylinder
t_r = constant gas-recovery temperature

Boundary conditions applicable to both continuum and slip flow may be considered. In the continuum-flow case the fluid adjacent to the cylinder surface has the same temperature as the surface, and thus $\vartheta = 0$; while the fluid far away from the cylinder has the temperature t_r, and thus $\vartheta = 1$. The point $\theta = 0$, $r = r$, provides some difficulty as ϑ has there two values, 0 and 1. This is best seen if the problem is visualized in the $r\theta$ plane. However, at a distance along $\theta = 0$ somewhat greater than $r = r_1$, the cylinder surface, the onflow has the same characteristics as the free-stream flow; thus one concludes that for $\theta = 0$, $r = r_1$, $\vartheta = 1$.

For the slip-flow case, the above arguments apply except at the surface, $r = r_1$, where there exists a temperature jump between the surface

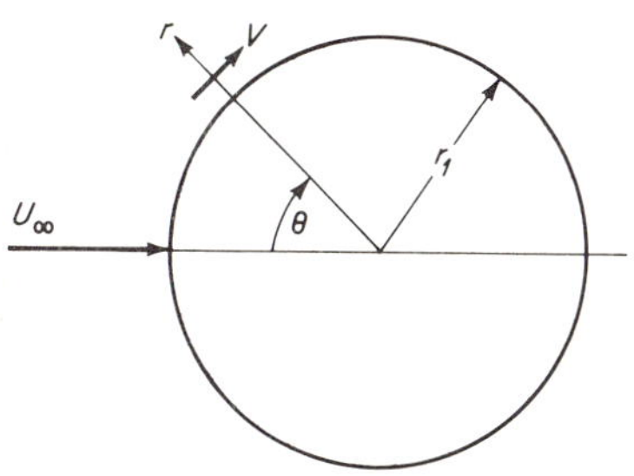

Fig. 11-36 Coordinate system for the flow about the circular cylinder.

and the gas, according to Eq. (11-55). The two sets of boundary conditions may be written as

Continuum flow:
$$\vartheta = 0 \qquad r = r_1 \qquad 0 < \theta < \pi$$
$$\vartheta = 1 \qquad r > r_1 \qquad \theta = 0$$

Slip flow:
$$\frac{\partial \vartheta}{\partial r} = L\vartheta \qquad r = r_1 \qquad 0 < \theta < \pi$$
$$\vartheta = 1 \qquad r > r_1 \qquad \theta = 0$$

where
$$\frac{1}{L} = 1.996 \frac{\gamma}{\gamma + 1} \frac{2 - \alpha_s}{\alpha_s} \frac{\lambda}{\Pr}$$

If a new independent variable $y = \sqrt{r}$ is introduced, Eq. (11-79) has the form

$$\frac{\partial^2 \vartheta}{\partial y^2} + \frac{1}{y} \frac{\partial \vartheta}{\partial y} - \frac{1}{K} \frac{\partial \vartheta}{\partial \theta} = 0 \tag{11-80}$$

where $K = \alpha/4V$ and the boundary conditions transform to

Continuum flow:
$$\vartheta = 0 \qquad y = \sqrt{r_1} \qquad 0 < \theta < \pi$$
$$\vartheta = 1 \qquad y > \sqrt{r_1} \qquad \theta = 0$$

Slip flow:
$$\frac{\partial \vartheta}{\partial y} = \xi\vartheta \qquad y = \sqrt{r_1} \qquad 0 < \theta < \pi$$
$$\vartheta = 1 \qquad y > \sqrt{r_1} \qquad \theta = 0$$

where
$$\xi = 2L\sqrt{r_1}$$

Equation (11-80) is identical with the equation for transient radial conduction of heat into an infinite region bounded internally by a circular cylinder. The temperature-jump boundary condition is identical with the radiation boundary condition. The solutions are well known,[1] and each result will be presented separately.

For the continuum boundary conditions the solution for Eq. (11-80) is

$$\vartheta = -\frac{2}{\pi} \int_0^\infty e^{-K\theta\beta^2} \frac{J_0(\beta\sqrt{r})Y_0(\beta\sqrt{r_1}) - Y_0(\beta\sqrt{r})J_0(\beta\sqrt{r_1})}{J_0{}^2(\beta\sqrt{r_1}) + Y_0{}^2(\beta\sqrt{r_1})} \frac{d\beta}{\beta} \tag{11-81}$$

where β is an arbitrary parameter of integration.

The local heat-transfer coefficient at the surface is given by

$$h = k\left(\frac{\partial \vartheta}{\partial r}\right)_{r=r_1} \tag{11-82}$$

Performing the necessary operations on Eq. (11-81) and using Eq. (11-82) results in

$$\mathrm{Nu} = \frac{hD}{k} = \frac{4}{\pi^2} \int_0^\infty \frac{e^{-K\theta\beta^2}}{J_0{}^2(\beta\sqrt{r_1}) + Y_0{}^2(\beta\sqrt{r_1})} \frac{d\beta}{\beta} \tag{11-83}$$

[1] H. S. Carslaw and J. C. Jaeger, "Conduction of Heat in Solids," 2d ed., pp. 334–336, Oxford University Press, New York, 1959.

The average heat-transfer coefficient is defined as

$$\mathrm{Nu}_{av} = \frac{D}{k}\frac{1}{\pi r_1}\int_0^{\pi} h(\theta) r_1\,d\theta = \frac{\bar{h}D}{k} \tag{11-84}$$

The evaluation of Eq. (11-84) using Eq. (11-83) results in the series

$$\frac{hD}{k} = 1.128\left(\frac{K\pi}{r_1}\right)^{-\frac{1}{2}} + 5.00 - 0.0941\left(\frac{K\pi}{r_1}\right)^{\frac{1}{2}} + 0.0625\frac{K\pi}{r_1} - 0.0588\left(\frac{K\pi}{r_1}\right)^{\frac{3}{2}} + 0.0677\left(\frac{K\pi}{r_1}\right)^2$$

for small values of $K\pi/r_1 = \alpha\pi/4Vr_1$.

If the tangential velocity is taken as $V = U_\infty/3$, a reasonable average velocity near the surface of the cylinder, the last equation becomes

$$\frac{K\pi}{r_1} = \frac{3\pi}{2\,\mathrm{Re}\,\mathrm{Pr}} \tag{11-85}$$

where $\mathrm{Re} = U_\infty D/\nu$. Substitution of this result into the series for $\mathrm{Re} > 300$ and $\mathrm{Pr} = 0.70$ yields

$$\frac{\bar{h}D}{k} = 0.500 + 0.435\left(\frac{U_\infty D}{\nu}\right)^{\frac{1}{2}} \tag{11-86}$$

which can be compared with the empirical result[1] based on the experimental results of many investigators

$$\frac{\bar{h}D}{k} = 0.32 + 0.43\left(\frac{U_\infty D}{\nu}\right)^{0.52} \tag{11-87}$$

Equations (11-86) and (11-87) represent the available data equally well, with Eq. (11-86) being the better in the region of smaller Reynolds numbers.

The same kind of solution can be effected for the slip-flow boundary conditions to yield for the local value of the Nusselt number

$$\frac{hD}{k} = \frac{4\xi^2}{\pi^2}\int_0^{\infty} \frac{e^{-K\theta\beta^2}}{[\beta J_1(\beta\sqrt{r_1}) + \xi J_0(\beta\sqrt{r_1})]^2 + [\beta Y_1(\beta\sqrt{r_1}) + \xi Y_0(\beta\sqrt{r_1})]^2}\frac{d\beta}{\beta} \tag{11-88}$$

where ξ may be seen to contain the temperature-jump effects.

Again, averaging Eq. (11-88) over $\theta = \pi$, one obtains

$$\frac{\bar{h}D}{k} = DL\left[1 - \frac{4}{3\sqrt{\pi}}DL\left(\frac{K\pi}{r_1}\right)^{\frac{1}{2}} + \frac{DL}{4}(1 + 2DL)\frac{K\pi}{r_1} + \cdots\right] \tag{11-89}$$

[1] W. H. McAdams, "Heat Transmission," 3d ed., McGraw-Hill, New York, 1954.

It can be shown from the slip-flow boundary conditions and values for $\alpha_s = 0.80$ and $\mathrm{Pr} = 0.70$ representative of air and technical surfaces that

$$DL = 0.267 \frac{\mathrm{Re}}{\mathrm{Ma}}$$

and, as before,

$$\frac{K\pi}{r_1} = \frac{3\pi}{2 \mathrm{Re} \mathrm{Pr}}$$

The graphical results of both Eqs. (11-86) and (11-89) are shown in Fig. 11-37. The correspondence of the simple theory and the experimental data is excellent for $\mathrm{Ma} < 2.0$. The free-molecule-flow solutions are shown in the figure for completeness.

As in the case of heat transfer from spheres, in the last several years much interest has been shown in the problem of heat transfer from cylinders, both theoretically and experimentally. The works of Dewey,[1] Baldwin, Sanborn, and Lawrence,[2] Vrebalovich,[3] Atassi and Brun,[4] and Atassi and Shen[5] are interesting and informative.

Transition flow In the pioneering paper in which Tsien[6] gave names to the flow regimes in rarefied-gas dynamics, he identified as *transition flow* the region where the dynamics of the flow is neither dominated by molecule-to-molecule encounters, as continuum flow is, nor by molecule-to-surface encounter, as free-molecule flow is. Transition-flow phenomena have attracted the attention of many investigators during the past few years, in both theoretical and experimental efforts. The theoretical studies are difficult and quite involved and as yet have not produced much in the way of meaningful results. The experimental efforts have produced a fair amount of good data encompassing the entire range from continuum to free-molecule flow for a number of simple geometries, i.e., the flat plate, the right-circular cylinder, and the sphere. Data for

[1] C. F. Dewey, Jr., A Correlation of Convective Heat Transfer and Recovery Temperature Data for Cylinders in Compressible Flow, *Intern. J. Heat Mass Transfer*, **8**:245–252 (1965); Hot Wire Measurements in Low Reynolds Number Hypersonic Flows, *J. Amer. Rocket Soc.*, **31**:1709–1718 (1961).

[2] L. V. Baldwin, V. A. Sandborn, and J. C. Lawrence, Heat Transfer from Transverse and Yawed Cylinders in Continuum, Slip and Free-molecule Air Flows, *J. Heat Transfer*, **82**:77–86 (1960).

[3] T. Vrebalovich, Heat Loss and Recovery Temperature of Fine Wires in Transonic Transitional Flow, *Advan. Appl. Mech.*, suppl. 4, vol. 2, pp. 1205–1219 (1967).

[4] H. Atassi and E. A. Brun, Heat Transfer from Wires in a Rarefied Low Speed Flow, *Advan. Appl. Mech.*, suppl. 4, vol. 2, pp. 1221–1234 (1967).

[5] H. Atassi and S. F. Shen, Hydrodynamic Description of the Rarefied Gas Flow over a Circular Cylinder in Steady Low Speed Motion and with Heat Transfer, *Advan. Appl. Mech.*, suppl. 5, vol. 1, pp. 805–810 (1969).

[6] Tsien, *loc. cit.*

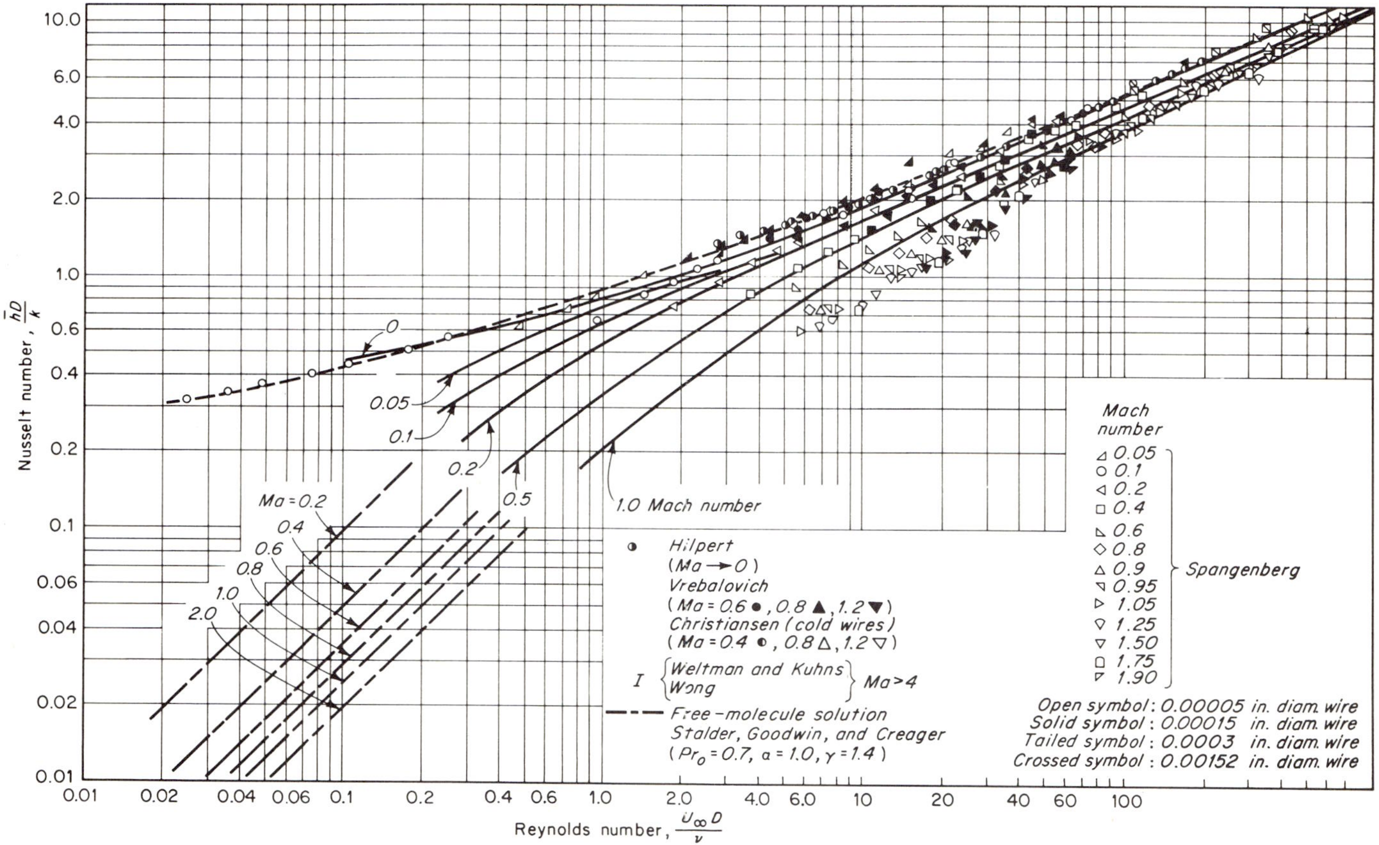

Fig. 11-37 Heat transfer for right-circular cylinders in air as a function of Reynolds and Mach numbers showing slip-flow effects.

both drag and overall heat transfer are available in the literature. Sherman[1] has surveyed the state of the experimental work and discussed some of the problems.

It is significant that out of these studies there has come what appears to be a universal correlation of subsonic data for the entire flow-regime spectrum. Originally suggested by Lees and Liu[2] and used more fully by Sherman, the correlation seems to have certain theoretical support.[3]

The correlation requires the knowledge in each case of three quantities: (1) $F(\text{Ma},\text{Re})$ the measured quantity in such dimensionless form that it becomes independent of Reynolds number in the free-molecule-flow limit; (2) $F_{FM}(\text{Ma})$, the limit of $F(\text{Ma},\text{Re})$ as $\text{Re} \to 0$ as a result of density reduction, a result usually available from theory; and, (3) $F_c(\text{Re})$, the limit of $F(\text{Ma},\text{Re})$ as $\text{Ma} \to 0$, which may be available either from theory or from experiments with viscous liquids. The correlation is represented as a plot of F/F_{FM} vs. F_c/F_{FM}. The function F may represent the skin-friction coefficient or the Stanton number.

In the correlation all data which are free from true rarefaction effects fall on a line $F/F_{FM} = F_c/F_{FM}$, and data for free-molecule-flow phenomena approach the line $F/F_{FM} = 1$. The correlation equation takes the form

$$\frac{F}{F_{FM}} = \frac{1}{1 + F_{FM}/F_c} \tag{11-90}$$

Equation (11-90) has been applied by Sherman to the experimental data for subsonic flat-plate skin-friction drag, for the drag of cylinders, and for subsonic heat transfer from spheres. The correlations are shown in Figs. 11-38 to 11-40. In Fig. 11-41 the data in the other three figures are plotted together, showing the rather universal nature of the correlation.

It can be seen from the plotted data that within the experimental accuracy claimed for them and the data smoothing inherent in logarithmic scale plotting of the variables, the Mach number dependence is eliminated.

The correlation shows, as it should, a reduction in momentum and energy exchange as the flow departs from a continuum state, as a consequence of the smaller gradients near the surface because of the velocity-

[1] F. S. Sherman, A Survey of Experimental Results and Methods for the Transition Regime of Rarefied Gas Dynamics, *Advan. Appl. Mech.*, suppl. 2, vol. 2, pp. 228–260 (1963).

[2] L. Lees and C. Y. Liu, Kinetic Theory Description of Conductive Heat Transfer from a Fine Wire, *Phys. Fluids*, **5**(10):1137–1148 (1962).

[3] Cercignani and Pagani, *loc. cit.;* Atassi and Shen, *loc. cit.*

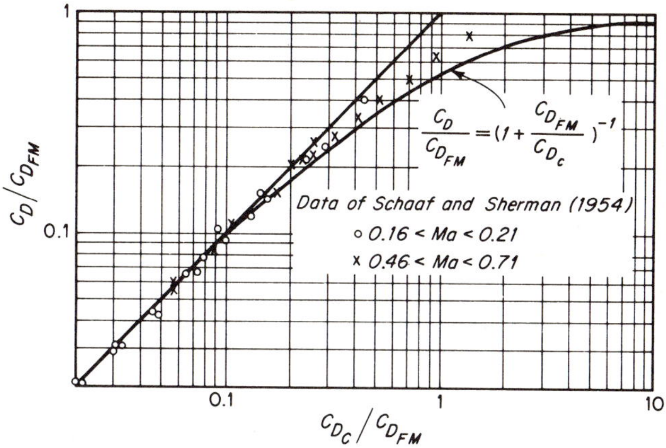

Fig. 11-38 Skin-friction drag of a flat plate in subsonic flow

slip and temperature-jump effects. Figure 11-41 demonstrates the lack of dependence on body shape, which can be rationalized inasmuch as for both creeping flow (Re $<$ 1) and free-molecule flow at low Mach numbers the body shape is known to be of small importance.

Recovery temperature Insofar as the heat-transfer coefficient is defined in terms of the adiabatic wall or equilibrium temperature of the body [Eq. (11-70), repeated here]

$$h = \frac{q}{t_w - t_r}$$

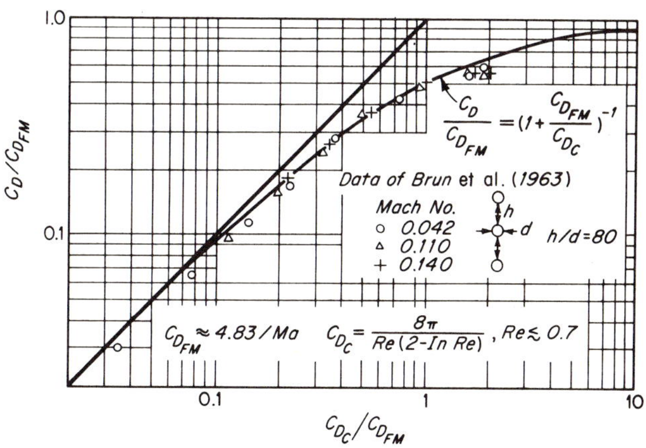

Fig. 11-39 Drag of a circular cylinder normal to a subsonic flow.

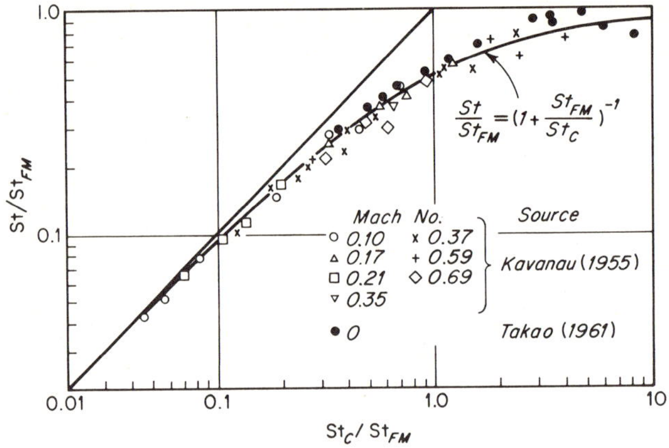

Fig. 11-40 Heat transfer from spheres in a subsonic flow.

it is necessary to have some knowledge of the recovery factor. Experimentally determined recovery factors for spheres were shown in Fig. 11-34 in terms of the recovery temperature or equilibrium temperature of the unheated, conducting sphere placed in the high-speed flow. Similar data exist for cylinders placed normally to the flow. Dewey[1] proposed a normalized recovery ratio which is quite successful in correlating

[1] Dewey, Hot Wire Measurements in Low Reynolds Number Hypersonic Flows, *J. Amer. Rocket Soc.*, **31**:1709–1718 (1961).

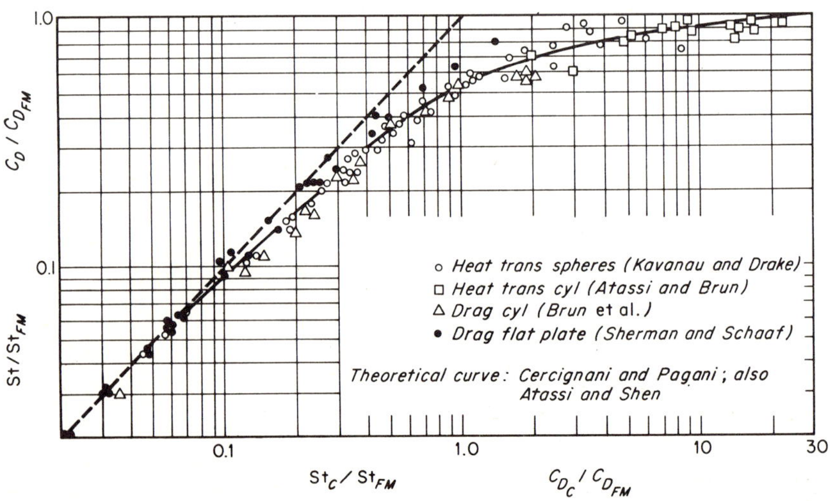

Fig. 11-41 Drag and heat transfer from plates, cylinders, and spheres in subsonic flow.

data on recovery temperatures of circular cylinders when plotted against the free-stream Knudsen number.

The normalized recovery ratio η_* is defined as

$$\eta_* = \frac{\eta - \eta_c}{\eta_f - \eta_c} \tag{11-91}$$

where η is the ratio of the cylinder recovery temperature to the free-stream stagnation temperature and the subscripts f and c denote free-molecule-flow and continuum high Reynolds number limits, respectively.

Experimental data on recovery temperatures from circular cylinders have been plotted on Fig. 11-42, which in spite of some data scatter provides a quite acceptable correlation. Dewey has provided as well empirical expressions for the values η_f and η_c:

$$\eta_c = 1 - 0.050 \frac{\mathrm{Ma}^{3.5}}{1.175 + \mathrm{Ma}^{3.5}} \tag{11-92}$$

$$\eta_f - \eta_c = 0.2167 \frac{\mathrm{Ma}^{2.80}}{0.8521 + \mathrm{Ma}^{2.80}} \tag{11-93}$$

Equation (11-92) is based on the data correlated by Morokovin,[1] while Eq. (11-93) is based on the calculations of Stalder, Goodwin, and Creager[2] for a diatomic gas $\gamma = 1.4$.

[1] M. V. Morokovin, Fluctuations and Hot Wire Anemometry in Compressible Flows, *AGARDograph* 24 (1956).

[2] J. R. Stalder, G. Goodwin, M. O. Creager, A Comparison of Theory and Experiment for High Speed Free-molecule Flow, *NACA Rept.* 1032 (1951) [supersedes *NACA Tech. Note* 2438 (1951)].

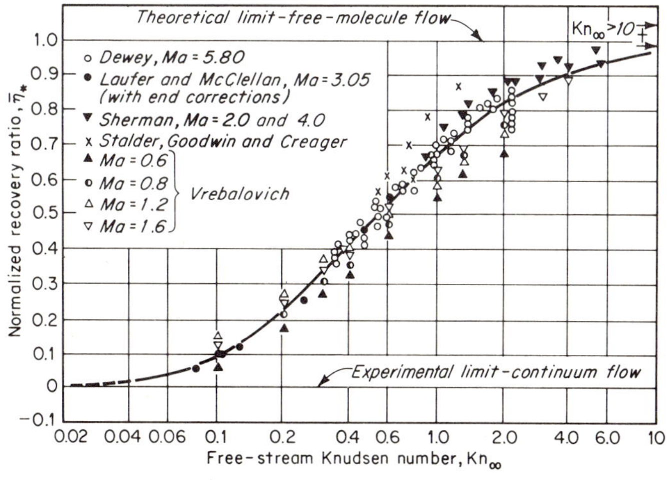

Fig. 11-42 Variation of normalized recovery temperatures as a function of the Knudsen number. [*C. F. Dewey, Jr., Intern. J. Heat Mass Transfer,* **8**:*245–252 (1965).*]

The solid line suggestive of all the data is represented by the equation

$$\eta_* = \frac{(Kn_\infty)^{1.193}}{0.4930 + (\text{Kn}_\infty)^{1.193}} \tag{11-94}$$

Free-convection heat transfer Rarefaction effects in the transition-flow region also occur in the absence of a forced bulk flow of the gas. In the continuum regime heat exchange without forced flow of the bulk fluid is known as *free-* or *natural-convection flow.* It is of interest to investigate the energy-exchange phenomenon in the case of small circular cylinders for Knudsen numbers corresponding to transition flow. In contrast to the relative wealth of data on forced-flow phenomena there are few data or analyses for the case of free-convection rarefied flow, although such analyses and data are important in terms of the design of vacuum instrumentation and thermal-property determination. The principal investigations have been those of Gordon,[1] Stathakis,[2] Tsubouchi and Sato,[3] Madden and Piret,[4] Kyte, Madden, and Piret,[5] and Drake, Smith, and Stathakis.[6]

The analysis presented here is based on an early model by Langmuir[7] and Rice[8] and modified by Madden and Piret[9] and Drake.[10]

Consider a circular cylinder of diameter D such that in the region $D/2 < r < D/2 + \lambda$ the heat transfer takes place by free-molecule flow; and beyond $r = D/2 + \lambda$ the heat transfer takes place by conduction through a stagnant film out to $r = b/2$; beyond $r = b/2$ the heat is trans-

[1] H. Gordon, Heat Transfer by Free-convection from Small Diameter Wires, Ph.D. dissertation, University of California, Berkeley, 1950.

[2] G. J. Stathakis, Heat Transfer by Free-convection to Gases at Low Pressures from a Small Diameter Hot Wire, M.S. thesis, University of California, Berkeley, 1954.

[3] T. Tsubouchi and S. Sato, Heat Transfer from Fine Wires and Particles by Natural Convection, *Chem. Eng. Progr.*, **56**:30 (1960).

[4] A. J. Madden and E. L. Piret, Heat Transfer from Wires to Gases at Sub-atmospheric Pressures under Natural Convection Conditions, *Proc. Gen. Discussion Heat Transfer ASME-IME, London, 1951.*

[5] J. R. Kyte, A. J. Madden, and E. L. Piret, Natural Convection Heat Transfer at Reduced Pressure, *Chem. Eng. Progr.*, **49**(12):653–662 (1953).

[6] R. M. Drake, Jr., C. G. Smith, and G. J. Stathakis, Free Convection Heat Transfer from Horizontal Cylinders to Gases at Transition Knudsen Numbers, in "Heat Transfer, Thermodynamics, and Education: Boelter Anniversary Volume," H. A. Johnson (ed.), McGraw-Hill, New York, 1964.

[7] I. Langmuir, Convection and Conduction of Heat in Gases and Liquids, *Phys. Rev.*, **34**:401 (1912).

[8] C. W. Rice, Free and Forced Convection of Heat in Gases and Liquids, *Trans. AIEE*, **42**:653–706 (1923).

[9] Madden and Piret, *loc. cit.*

[10] Drake, Smith, and Stathakis, *loc. cit.*

ferred by continuum free-convective flow. Let t_w be the temperature of the cylinder surface, t_λ be the gas temperature at $r = D/2 + \lambda$, and t_f be the temperature of the gas at $r = b/2$.

The heat loss from the cylinder by free-molecular-flow processes can be written,

$$Q = \frac{\pi D L \alpha_s \bar{k}_{w,\lambda}(\gamma + 1)(t_w - t_\lambda) \text{ Pr}}{4\gamma\lambda} \tag{11-95}$$

where $\bar{k}_{w,\lambda}$ is the average thermal conductivity in the region $D/2 \leq r \leq D/2 + \lambda$, and the other variables are as defined earlier.

The heat flow through the conduction layer may be expressed from the results from Chap. 3 as

$$Q = \frac{2\pi L \bar{k}_{\lambda,f}(t_\lambda - t_f)}{\ln [b/(D + 2\lambda)]} \tag{11-96}$$

The heat-transfer coefficient can be defined as

$$h = \frac{Q}{\pi D L(t_w - t_f)} \tag{11-97}$$

Equations (11-95) to (11-97) can be combined to form the expression for the Nusselt number

$$\frac{hD}{\bar{k}_{w,f}} = \frac{2}{8\dfrac{\gamma}{\gamma + 1}\dfrac{\lambda}{D}\dfrac{1}{\alpha_s \text{ Pr}}\dfrac{\bar{k}_{w,f}}{\bar{k}_{w,\lambda}} + \dfrac{\bar{k}_{w,f}}{\bar{k}_{\lambda,f}} \ln \dfrac{b}{D + 2\lambda}} \tag{11-98}$$

For small temperature differences the ratios of the thermal conductivities will be nearly unity; thus,

$$\text{Nu} = \frac{2}{8\dfrac{\gamma}{\gamma + 1}\dfrac{\text{Kn}}{\alpha_s \text{ Pr}} + \ln \dfrac{b/D}{1 + 2 \text{ Kn}}} \tag{11-99}$$

The Eucken relationship can be used for the Prandtl number (see Chap. 2) or Eq. (11-30)

$$\text{Pr} = \frac{4\gamma}{9\gamma - 5}$$

such that Eq. (11-99) becomes

$$\text{Nu} = \frac{2}{2\dfrac{9\gamma - 5}{\gamma + 1}\dfrac{\text{Kn}}{\alpha_s} - \ln (1 + 2 \text{ Kn}) + \ln \dfrac{b}{D}} \tag{11-100}$$

In Eq. (11-100) the first two terms in the denominator represent rarefaction effects, and the last one represents continuum heat conduction.

The last term can be related to continuum heat-flow effects in the following way.

A cylinder of diameter D with a surface temperature t_w losing heat by conduction through a layer of gas at temperature t_f of thickness b' may have the heat transfer stated as

$$Q = \frac{t_w - t_f}{(1/2\pi kL) \ln [(D + 2b')/D]} = h\pi DL(t_w - t_f) \qquad (11\text{-}101)$$

such that

$$\text{Nu} = \frac{hD}{k} = \frac{2}{\ln (1 + 2b'/D)} \qquad (11\text{-}102)$$

If $b'/D \ll 1$, the denominator can be expanded in a series. By omitting the higher-order terms in the series the expression for the Nusselt number becomes

$$\text{Nu} = \frac{D}{b'} \qquad (11\text{-}103)$$

Equation (11-103) is representative of the region treated by Hermann, and since that analysis yields for the average heat-transfer coefficient

$$\text{Nu} = 0.400\ (\text{Gr Pr})^{\frac{1}{4}} \qquad (11\text{-}104)$$

it is apparent that

$$\text{Nu} = \frac{D}{b'} = 0.400\ (\text{Gr Pr})^{\frac{1}{4}} \qquad (11\text{-}105)$$

The value b'/D is related to b/D of Eq. (11-100) simply as

$$\frac{b}{D} = \left(1 + \frac{2b'}{D}\right) = 1 + \frac{2}{0.400\ (\text{Gr Pr})^{\frac{1}{4}}}$$

which when combined with Eq. (11-100) yields

$$\text{Nu} = \frac{2}{\dfrac{2(9\gamma - 5)}{\gamma + 1}\dfrac{\text{Kn}}{\alpha_s} - \ln (1 + 2\ \text{Kn}) + \ln \left[1 + \dfrac{5}{(\text{Gr Pr})^{\frac{1}{4}}}\right]} \qquad (11\text{-}106)$$

The Grashof number can be related by means of the results from the kinetic theory of gases to the Knudsen number,

$$\text{Gr} = \frac{g\beta D^3\ \Delta T}{\nu^2}$$

where from kinetic theory,

$$\nu = 0.499 \sqrt{\frac{8RT}{\pi}}\ \lambda$$

such that

$$\text{Gr} = \left(\frac{\pi}{2}\frac{\Delta T}{T}\frac{g\beta D}{R}\right)\frac{1}{\text{Kn}^2}$$

which when introduced into Eq. (11-106) yields

$$\mathrm{Nu} = \frac{2}{\dfrac{2(9\gamma - 5)}{\gamma + 1}\dfrac{\mathrm{Kn}}{\alpha_s} - \ln(1 + 2\,\mathrm{Kn}) + \ln\left[1 + \dfrac{5\sqrt{\mathrm{Kn}}}{\left(\dfrac{\pi}{2}\dfrac{g\beta D}{R}\dfrac{\Delta T}{T}\mathrm{Pr}\right)^{\frac{1}{4}}}\right]} \tag{11-107}$$

A review of the high-pressure continuum data recently reported by Tsubouchi and Sato[1] indicates that the empirical expression

$$\mathrm{Nu} = 0.372(\mathrm{Gr\,Pr})^{0.27}$$

is superior to the Hermann result in representing the existing data; hence as a working result Eq. (11-107) can be modified to yield the expression

$$\mathrm{Nu} = \frac{2}{\dfrac{2(9\gamma - 5)}{\gamma + 1}\dfrac{\mathrm{Kn}}{\alpha_s} - \ln(1 + 2\,\mathrm{Kn}) + \ln\left[1 + \dfrac{5.38\,\mathrm{Kn}^{0.54}}{\left(\dfrac{\pi}{2}\dfrac{g\beta D}{R}\dfrac{\Delta T}{T}\mathrm{Pr}\right)^{0.27}}\right]} \tag{11-108}$$

A comparison of Eq. (11-108) with the presently available experimental data is shown for accommodation coefficients typical of air and helium on platinum in Fig. 11-43.

Free-molecule flow Momentum, normal and tangential, and energy exchange from a surface to a rarefied-gas stream in maxwellian equilibrium can be calculated entirely from fundamental notions of the kinetic theory of gases. Excellent treatises exist[2] on the theory in general, and specific applications of the theory to momentum and energy exchange from various geometries are available in the literature.[3]

The fundamental notion in using the kinetic theory in free-molecule flow is the neglect of molecule-to-molecule collisions. The interaction of molecules with surfaces in question is of importance, and the fluxes of mass, momentum, and energy incident upon and reflected or reemitted from an element of the surface can be treated as independent and non-

[1] Tsubouchi and Sato, *loc. cit.*

[2] Kennard, *op. cit.*; Chapman and Cowling, *op. cit.;* G. N. Patterson, "The Molecular Flow of Gases," Wiley, New York, 1956.

[3] A. K. Oppenheim, Generalized Theory of Convective Heat Transfer in a Free-molecule-Flow, *J. Aeron. Sci.*, **20**:49–58 (1953); J. K. Stalder and D. Jukoff, Heat Transfer to Bodies Travelling at High Speed in the Upper Atmosphere, *J. Aeron. Sci.*, **15**(7):381–391 (1948); F. M. Sauer, "Convective Heat Transfer from Spheres in Free-molecule-flow, *J. Aeron. Sci.*, **18**(5):353–354 (1951).

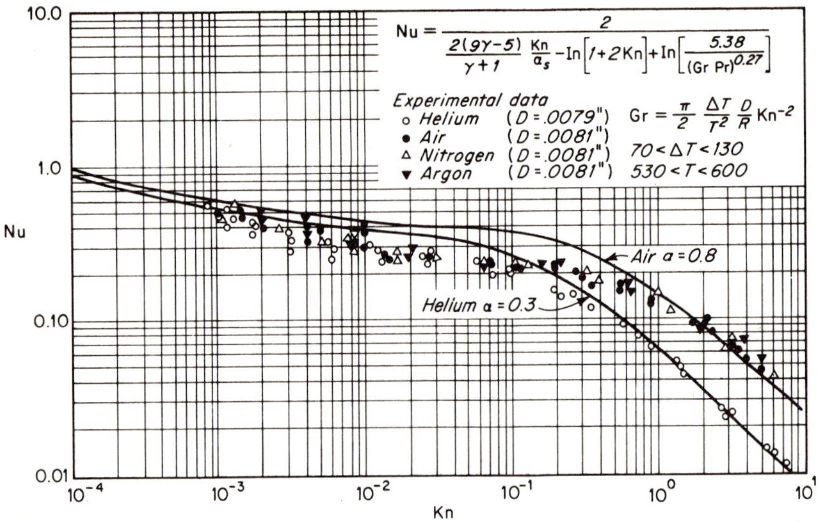

Fig. 11-43 Comparison of theoretical and experimental results for heat transfer from a horizontal cylinder to gases in the transition-flow regime. [*R. M. Drake, Jr., C. G. Smith, and G. J. Stathakis, in H. A. Johnson (ed.), "Heat Transfer, Thermodynamics, and Education: Boelter Anniversary Volume." Copyright 1964. McGraw-Hill Book Company. Used by permission.*]

interacting. The calculations for convex geometries are the simplest to undertake in that one avoids the need to account for the surface seeing itself and the interreflections of molecular fluxes. In this sense the calculations for free-molecule flow bear a strong resemblance to the calculations for radiant-energy exchange. In the case of convex geometries the molecular flux incident upon the surface has the characteristics of the free-stream flow, and molecules which have collided with the surface return to the free-stream flow with no further collision with the surface.

Consider a surface element arranged as shown in Fig. 11-44. The fluid in flow past the element is in maxwellian equilibrium, has a temperature T_0, a density ρ, and a macroscopic velocity U aligned in the xy plane at an angle θ with the surface element.

The distribution function is thus

$$f = \frac{\rho}{m(2\pi RT)^{3/2}} \exp\left[-\frac{(v_1 - U\sin\theta)^2 + (v_2 - U\cos\theta)^2 + v_3}{2RT}\right] \qquad (11\text{-}109)$$

where m = mass of molecule
R = gas constant
v_1, v_2, v_3 = components of molecular thermal velocity

The molecular flux N_i incident upon the surface area A is given by

$$N_i = \int_{-\infty}^{\infty} \int_{-\infty}^{\infty} \int_{0}^{\infty} f v_1 \, dv_1 \, dv_2 \, dv_3$$

which with the use of Eq. (11-109) yields

$$N_i = \frac{\rho}{m} \sqrt{\frac{RT}{2\pi}} \left[e^{-\eta^2} + \sqrt{\pi}\, \eta (1 + \operatorname{erf} \eta)\right] \tag{11-110}$$

where $\eta = S \sin \theta$ and S is the *molecular speed ratio,* defined as

$$S = \frac{U}{v_m} = \frac{U}{\sqrt{2RT}} = \sqrt{\frac{\gamma}{2}}\, \mathrm{Ma} \tag{11-111}$$

where $v_m = \sqrt{2RT}$ is the *most probable molecular velocity.*

If the surface is such that gas is neither adsorbed by the surface nor desorbed, the incident molecular flux is equal to the reemitted molecular flux, that is,

$$N_i = N_r \tag{11-112}$$

Technical materials usually have layers of adsorbed gases on the surfaces, and while these adsorbed gases present a problem, e.g., in vacuum systems in the maintenance of a high vacuum, the surface processes are too small to be of any importance in aerodynamic considerations. Equation (11-112) is thus a most reasonable representation of fact.

By similar calculations the incident fluxes of momentum normal and tangent to the surface element can be calculated:

$$p_i = \int_{-\infty}^{\infty} \int_{-\infty}^{\infty} \int_{0}^{\infty} m v_1^2 f \, dv_1 \, dv_2 \, dv_3 = \frac{\rho}{2\sqrt{\pi}} \left(\frac{U}{S}\right)^2 \left[\eta e^{-\eta^2} + \sqrt{\pi}\, (\tfrac{1}{2} + \eta^2)(1 + \operatorname{erf} \eta)\right] \tag{11-113}$$

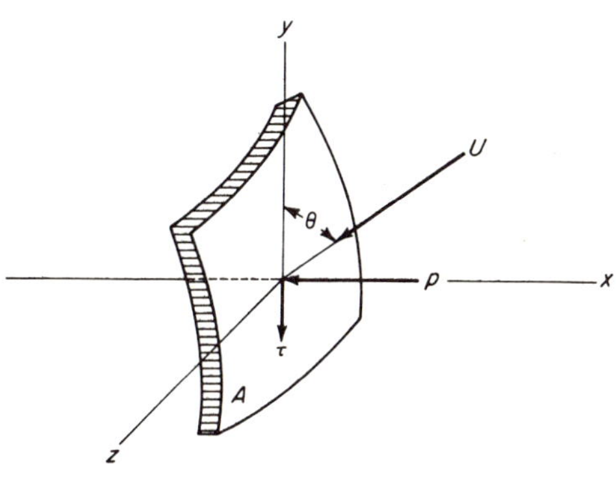

Fig. 11-44 Coordinate system relative to a surface for free-molecule-flow analysis.

and

$$\tau_i = -\int_{-\infty}^{\infty}\int_{-\infty}^{\infty}\int_{0}^{\infty} mv_1v_2f\,dv_1\,dv_2\,dv_3$$

$$= \frac{\rho U^2 \cos\theta}{2\sqrt{\pi}\,S}\,[e^{-\eta^2} + \sqrt{\pi}\,\eta(1 + \operatorname{erf}\eta)] \qquad (11\text{-}114)$$

In calculating the flux of energy incident upon the surface element a clear distinction must be made between the translational energy of the molecule and its internal energy. For monatomic molecules the energy is all in the translational mode; however, in diatomic gases, such as air, there are five degrees of freedom, at least at ordinary temperatures, three for translation and two internal degrees of freedom for rotation. Polyatomic molecules have additional internal degrees of freedom.

The total incident energy flux is thus

$$e_i = e_{i,\mathrm{tr}} + e_{i,\mathrm{int}} \qquad (11\text{-}115)$$

The translational component is given by

$$e_{i,\mathrm{tr}} = \int_{-\infty}^{\infty}\int_{-\infty}^{\infty}\int_{0}^{\infty} \tfrac{1}{2}mv^2v_1\,dv_1\,dv_2\,dv_3$$

$$= \rho RT\sqrt{\frac{RT}{2\pi}}\,[(S^2+2)e^{-\eta^2} + \sqrt{\pi}(S^2+\tfrac{5}{2})\eta(1+\operatorname{erf}\eta)] \qquad (11\text{-}116)$$

where $\quad v^2 = v_1^2 + v_2^2 + v_3^2$

The internal component of energy is calculated by use of the principle of equipartition of energy, which states that at equilibrium the internal energy of a gas is equally distributed over the internal degrees of freedom, each mode possessing $\frac{1}{2}RT$ units of energy per molecule.

The molecular degrees of freedom are related to the ratio of the specific heats of constant pressure and volume as

$$\frac{c_p}{c_v} = 1 + \frac{2}{j} \qquad j = \frac{2}{\gamma - 1}$$

where j is the total number of degrees of freedom of the molecule.

In every molecule there are three degrees of freedom for the translational modes; thus there are $j - 3$ internal degrees of freedom

$$j_{\mathrm{int}} = j - 3 = \frac{5 - 3\gamma}{\gamma - 1} \qquad (11\text{-}117)$$

Thus the internal energy associated with the incident molecular flow can be written

$$e_{i,\mathrm{int}} = \frac{5 - 3\gamma}{2(\gamma - 1)}\,mRTN_i$$

$$e_{i,\mathrm{int}} = \frac{\rho(5 - 3\gamma)}{\sqrt{\pi}(\gamma - 1)}\left(\frac{RT}{2}\right)^{\frac{3}{2}}[e^{-\eta^2} + \sqrt{\pi}\,\eta(1 + \operatorname{erf}\eta)] \qquad (11\text{-}118)$$

The results of Eqs. (11-116) and (11-118) are added as shown in Eq. (11-115) to give the expression of the total incident-energy flux.

The net pressure p, net shear stress τ, and net heat flux q with respect to the surface must include the effect of the reflected or reemitted molecules:

$$\begin{aligned} p &= p_i + p_r \\ \tau &= \tau_i - \tau_r \\ q &= e_i - e_r \end{aligned}$$

Making use of the accommodation coefficients from Eqs. (11-46), (11-47), and (11-48), one can write expressions for the reflected and reemitted molecules

$$p_r = (1 - \sigma_s)p_i + \sigma_s p_w \tag{11-119}$$

$$\tau_r = (1 - f_s)\tau_i \tag{11-120}$$

$$e_r = (1 - \alpha_s)e_i + \alpha_s e_w \tag{11-121}$$

where p_w and e_w represent the normal momentum and energy fluxes that would emanate from the surface as if from a gas in maxwellian equilibrium at the state of the surface.

The flux of normal momentum is represented by the expression

$$\begin{aligned} p_w &= \tfrac{1}{2}mv_mN_i = \tfrac{1}{2}m\sqrt{2\pi RT_w}\,N_i \\ &= \tfrac{1}{2}\rho RT_w\left[\frac{e^{-\eta^2}}{\sqrt{\pi}} + \eta(1 + erf\ \eta)\right] \end{aligned} \tag{11-122}$$

Again the energy flux must be treated in two parts, the translational and internal energy of the molecule. The translational energy of a molecule coming from the wall at temperature T_w is

$$e_{w,\mathrm{tr}} = 2N_imRT_w$$

The translational energy is distributed among three degrees of freedom; thus the internal modes account for the balance or

$$\begin{aligned} e_w &= 2N_imRT_w + (j - 3)\tfrac{1}{2}N_imRT_w \\ &= \frac{j+1}{2}mRT_wN_i \end{aligned}$$

which with Eqs. (11-110) and (11-116) becomes

$$e_w = \frac{\gamma + 1}{\gamma - 1}\rho\left(\frac{RT_w}{2}\right)^{\frac{3}{2}}\left[\frac{e^{-\eta^2}}{\sqrt{\pi}} + \eta(1 + \mathrm{erf}\ \eta)\right] \tag{11-123}$$

The expressions for the local momentum and energy exchange can be written, in terms of dimensionless quantities,

$$\frac{p}{\frac{1}{2}\rho U^2} = \frac{1}{S^2}\left\{\left(\frac{2-\sigma_s}{\sqrt{\pi}}\eta + \frac{\sigma_s}{2}\sqrt{\frac{T_w}{T_0}}\right)e^{-\eta^2} + \left[(2-\sigma_s)(\tfrac{1}{2}+\eta^2) + \frac{\sigma_s}{2}\eta\sqrt{\pi\frac{T_w}{T_0}}\right](1+\operatorname{erf}\eta)\right\} \tag{11-124}$$

$$\frac{\tau}{\rho U^2} = f_s\sqrt{1-\left(\frac{\eta}{S}\right)^2}\left[\frac{e^{-\eta^2}}{2S\sqrt{\pi}} + \frac{\eta(1+\operatorname{erf}\eta)}{2S}\right] \tag{11-125}$$

$$\frac{q}{\rho RT_0U} = \alpha_s\left\{\left[\frac{\gamma+1}{2(\gamma-1)}\frac{T_w}{T_0} - \left(S^2+\frac{\gamma}{\gamma-1}\right)\right]\left[\frac{e^{-\eta^2}}{2S\sqrt{\pi}} + \frac{\eta(1+\operatorname{erf}\eta)}{2S}\right] + \frac{1}{2}\frac{e^{-\eta^2}}{2S\sqrt{\pi}}\right\} \tag{11-126}$$

Of the last three equations, for the purposes intended for this discussion, we shall treat further only the equations for tangential-momentum and energy fluxes.

To calculate the overall shear and heat exchange from a body to a free-molecule flow, the following surface integrals may be defined:

$$\bar{G} = \frac{1}{A}\int_A \frac{e^{-\eta^2}}{2S\sqrt{\pi}}\,dA \tag{11-127}$$

$$\bar{F} = \frac{1}{A}\int_A \frac{\eta(1+\operatorname{erf}\eta)}{2S}\,dA \tag{11-128}$$

Furthermore, the following average values are of use:

$$\bar{\tau} = \frac{1}{A}\int_A \tau\,dA$$

$$\bar{q} = \frac{1}{A}\int_A q\,dA$$

Consider Eq. (11-125) for $\eta/S = \sin\theta = 0$, the flow U parallel to the surface,

$$\tfrac{1}{2}c_f = \frac{\bar{\tau}}{\rho U^2} = f_s(\bar{G}+\bar{F}) \tag{11-129}$$

and for the energy-flux expression, for all θ, integration over the surface area becomes

$$\frac{\bar{q}}{\rho RT_0U} = \alpha_s\left[\frac{\gamma+1}{2(\gamma-1)}\frac{T_w}{T_0}(\bar{G}+\bar{F}) - \left(S^2+\frac{\gamma}{\gamma-1}\right)(\bar{G}+\bar{F}) + \tfrac{1}{2}\bar{G}\right] \tag{11-130}$$

The heat transfer in free-molecule flow is aptly represented in terms of the Stanton number and the recovery temperature. For $\bar{q} = 0$, Eq.

(11-130) yields an expression for the recovery temperature T_r, where the recovery temperature is uniform over the body. That situation can exist when the thermal resistance due to heat conduction within the body is small in comparison with the thermal resistance due to the fluid convection on the surface. In such a case, the recovery-temperature ratio becomes

$$\frac{T_r}{T_0} = 2\frac{\gamma - 1}{\gamma + 1}\left(\frac{\gamma}{\gamma - 1} + S^2 - \frac{1}{2}\frac{\bar{G}}{\bar{G} + \bar{F}}\right) \tag{11-131}$$

The energy flux, the heat exchange between the body and the flowing gas, can be written in terms of the recovery temperature from Eqs. (11-130) and (11-131) as

$$\bar{q} = \alpha_s \frac{\gamma + 1}{2(\gamma - 1)} (\bar{G} + \bar{F})(T_w - T_r)\rho R U \tag{11-132}$$

and remembering that $R = c_p(\gamma - 1)/\gamma$, Eq. (11-132) becomes in terms of the Stanton number

$$\mathrm{St} = \frac{\bar{q}}{\rho c_p U(T_w - T_r)} = \frac{\bar{h}}{\rho c_p U} = \alpha_s \frac{\gamma + 1}{2\gamma} (\bar{G} + \bar{F}) \tag{11-133}$$

If the Eucken relationship is used for the Prandtl number, the Stanton number may as well be written

$$\mathrm{St} = \frac{\mathrm{Nu}}{\mathrm{Re\,Pr}} = \frac{9\gamma - 5}{4\gamma}\frac{\mathrm{Nu}}{\mathrm{Re}}$$

such that

$$\frac{\mathrm{Nu}}{\mathrm{Re}} = 2\alpha_s \frac{\gamma + 1}{9\gamma - 5} (\bar{G} + \bar{F})$$

The thermal recovery factor can be expressed in terms of Eq. (11-131) as

$$r = \frac{T_r - T_0}{T_T - T_0} = \frac{\gamma}{\gamma + 1}\left(2 + \frac{1}{S^2}\frac{\bar{F}}{\bar{G} + \bar{F}}\right) \tag{11-134}$$

where T_T is the total temperature (stagnation temperature) of the gas stream. The ratio of the total temperature to the free-stream temperature can be shown in terms of the molecular speed ratio S:

$$\frac{T_T}{T_0} = 1 + \frac{\gamma - 1}{2}\mathrm{Ma}^2 = 1 + \frac{\gamma - 1}{\gamma} S^2 \tag{11-135}$$

One of the more interesting results to come out of the theory, which has been confirmed by many experiments, is that the recovery temperature is greater than the total temperature of the gas stream. Equations like Eq. (11-131) can be written as

$$T_r = T_T f(S)$$

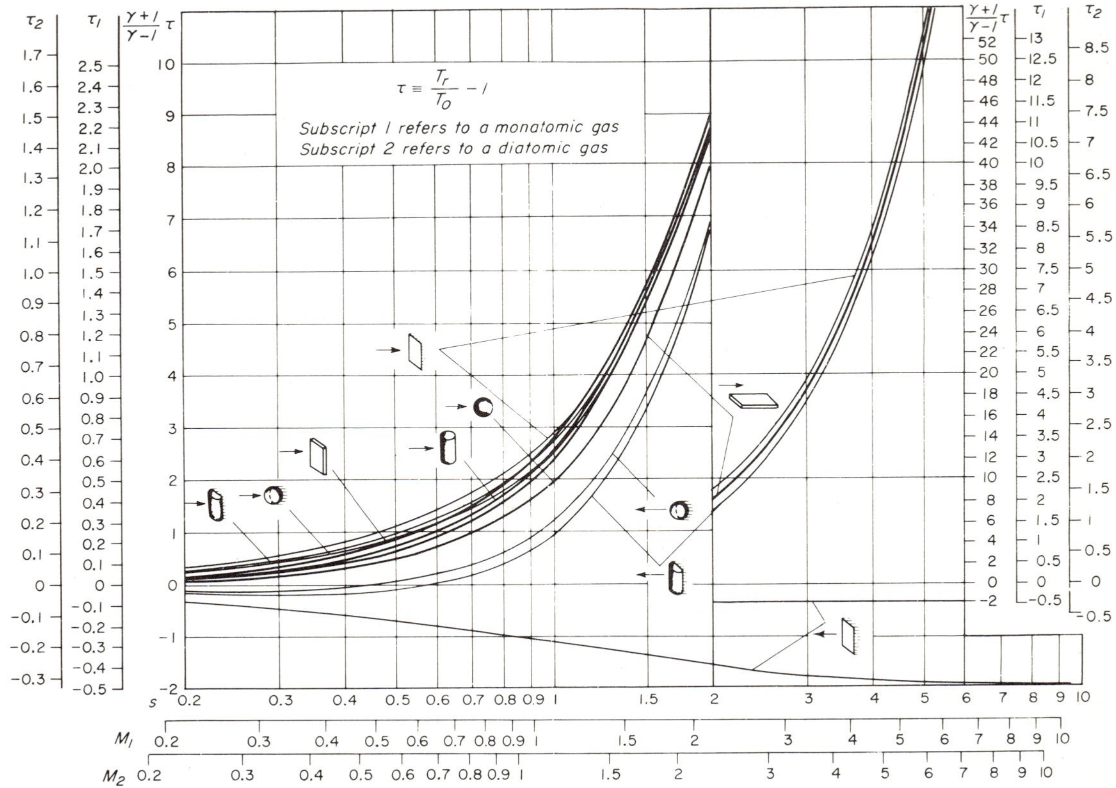

Fig. 11-45 Ratio of the wall-recovery temperature and the free-stream temperature for a flat plate, a sphere, and a circular cylinder in free-molecule flow. [*A. K. Oppenheim, J. Aeron. Sci.*, **20**:*1* (*1953*).]

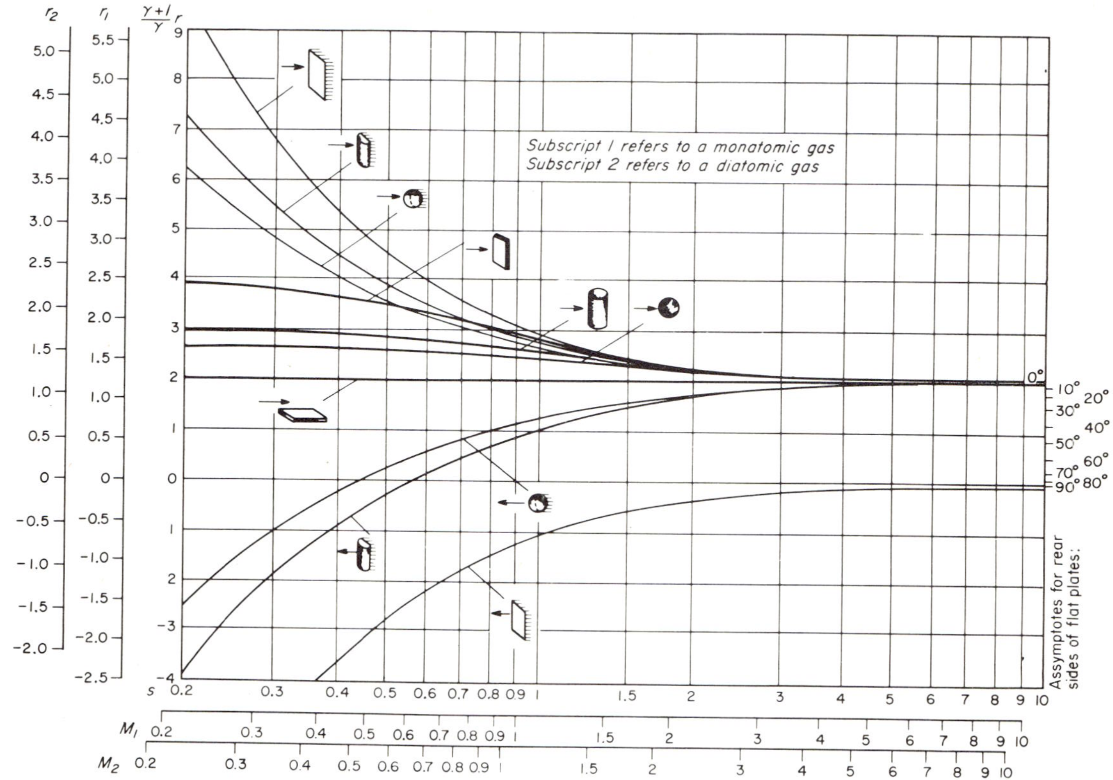

Fig. 11-46 Thermal-recovery factors for a flat plate, a sphere, and a circular cylinder in free-molecule flow. [*A. K. Oppenheim, J. Aeron. Sci.,* **20**:*1* (*1953*).]

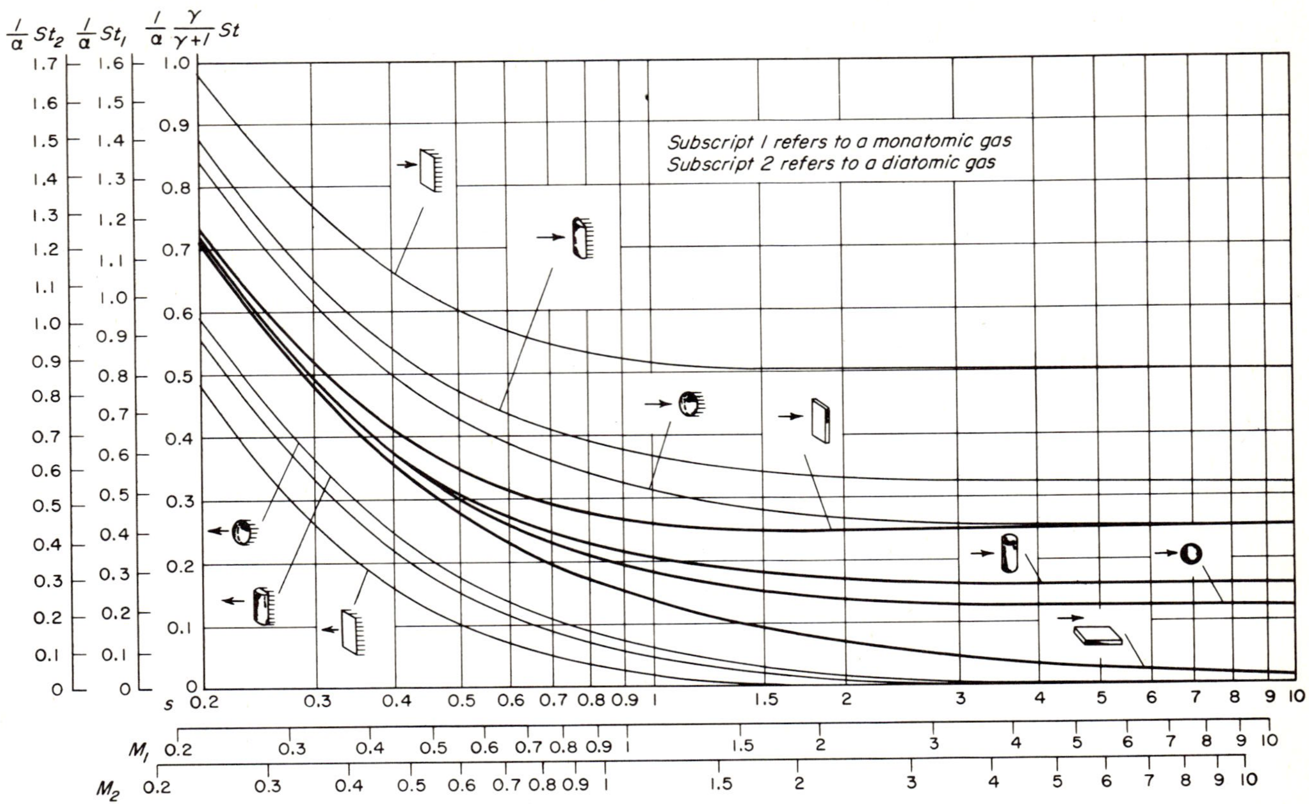

Fig. 11-47 Heat transfer from a flat plate, a sphere, and a circular cylinder in free-molecule flow. [*A. K. Oppenheim, J. Aeron. Sci.*, **20**:*1* (*1953*).]

where the quantity $f(S)$ is greater than unity. Comparison of Eqs. (11-131) and (11-135) confirm this statement.

Graphical displays of the results of Eqs. (11-131), (11-133) and (11-134) are shown in Figs. 11-45 to 11-47.

In addition to the shapes shown on the figures, other shapes can easily be deduced. For example, a flat plate inclined at an angle θ to the flow can be treated in terms of the flat plate normal to the flow. The modified surface-area integrals are

$$\bar{G}_\theta(S) = \frac{e^{-\eta^2}}{2S\sqrt{\pi}} = \frac{\eta}{S}\frac{e^{-\eta^2}}{2\eta\sqrt{\pi}} = \bar{G}_n(\eta)\sin\theta$$

and

$$\bar{F}_\theta(S) = \bar{F}_n(\eta)\sin\theta$$

The subscript θ indicates a flat plate at an angle of incidence θ with the onflow, while the subscript n indicates a flat plate arranged to be normal to the onflow. Thus

$$\mathrm{St}_\theta\,(S) = \mathrm{St}_n\,(\eta)\sin\theta$$

$$r_\theta(S) = \frac{2\gamma}{\gamma+1}\cos^2\theta + r_n(\eta)\sin^2\theta$$

It should be noted also that the flat-plate results apply to any surface making a constant angle with the flow; therefore, these results apply to wedges and cones in comparison with flat plates at angles of incidence with the flow, where the angles of incidence of the plate corresponds to half the opening angle of the wedge or cone. The zero-angle flat plate corresponds in condition to those occurring on a cylinder of any cross-sectional shape in axial flow.

12
Natural Convection

Flow and heat transfer under the influence of a body force will be considered in this chapter. Various body forces are encountered in engineering problems. The gravitational field surrounding the earth, the moon, and other planets creates body forces which are described by a gravitational acceleration g. To analyze flow in a rotating system, centrifugal forces and Coriolis forces have to be introduced into the momentum equations when they are written for a rotating coordinate system. Flow and heat transfer in gas turbines is an example of such a situation. Electric and magnetic fields can interact with a moving fluid when it is electrically conducting. For instance, boundary layers on the walls of magnetohydrodynamic power generators are strongly influenced by such forces.[1]

The discussion in this chapter will be restricted to body forces which are described by a constant body-force term g.

[1] E. R. G. Eckert and E. Pfender, Advances in Plasma Heat Transfer, *Advan. Heat Transfer*, **4**:229 (1967).

12-1 THE LAMINAR BOUNDARY–LAYER EQUATIONS

The laminar boundary-layer equations for a fluid with variable properties and under the presence of a body force were derived in Secs. 6-4 and 6-5. In this section we shall consider flow and heat transfer under the influence of the gravitational acceleration g. Velocities generated in this way are quite small, so that the effect of viscous dissipation can generally be neglected. More exactly, this means that the Eckert number Ec_0 in these equations is considered small. The equations will be written here in such a way that they apply to steady rotationally symmetric flow as well as to plane flow. With the notation in Fig. 12-1, the equations read

$$\frac{\partial}{\partial s}(\rho u r_0{}^n) + \frac{\partial}{\partial y}(\rho v r_0{}^n) = 0 \tag{12-1}$$

$$\rho\left(u\frac{\partial u}{\partial s} + v\frac{\partial u}{\partial y}\right) = \frac{\partial}{\partial y}\left(\mu\frac{\partial u}{\partial y}\right) - \frac{dp}{ds} + \rho g_s \tag{12-2}$$

$$\rho\left(u\frac{\partial i}{\partial s} + v\frac{\partial i}{\partial y}\right) = \frac{\partial}{\partial y}\left(k\frac{\partial t}{\partial y}\right) \tag{12-3}$$

The exponent n is set equal to 1 for rotationally symmetric flow and to zero for two-dimensional flow. The symbol g_s indicates the component of the gravitational acceleration in the s direction and the total-derivative symbols d in the expression for the pressure gradient dp/ds indicate that the pressure, according to the boundary-layer concept, is independent of y within the boundary layer. No distinction need be made between static and total enthalpy at the small velocities involved. Outside the boundary layer, the viscous effects are negligibly small, and the momentum equation (12-2) reduces along a streamline to

$$\frac{dp}{ds} = \rho_e g_s - \rho_e u_e \frac{du_e}{ds}$$

The index e denotes conditions outside the boundary layer. Introducing this expression into Eq. (12-2) results in

$$\rho\left(u\frac{\partial u}{\partial s} + v\frac{\partial u}{\partial y}\right) = \frac{\partial}{\partial y}\left(\mu\frac{\partial u}{\partial y}\right) + g_s(\rho - \rho_e) + \rho_e u_e \frac{du_e}{ds} \tag{12-4}$$

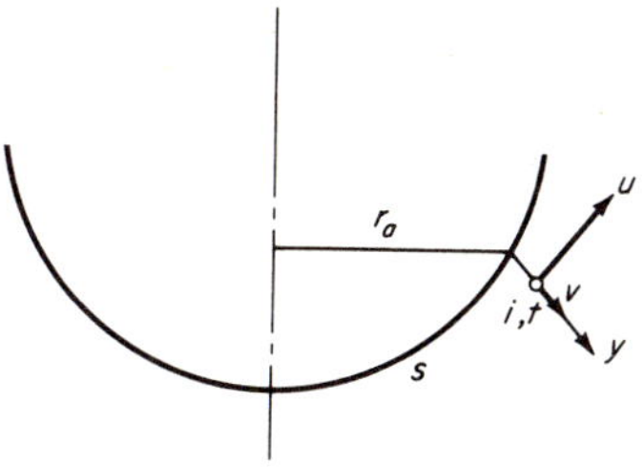

Fig. 12-1 Nomenclature for plane and rotationally symmetric free-convection flow.

For small temperature differences, the properties in this equation can be considered constant except for the term $\rho - \rho_e$. This approximation was introduced by Boussinesq.

So far the equations describe free convection regardless of the relation which determines the density variation within the boundary layer. In a one-component fluid, the density varies within the boundary layer in the y direction with temperature, and therefore the density difference in Eq. (12-4) will be expressed by a corresponding temperature difference. The following equation defines the thermal-expansion coefficient for a fluid:

$$\frac{1}{v}\left(\frac{\partial v}{\partial t}\right)_p = \beta$$

where v indicates the specific volume. Introducing the density ρ for the specific volume v results in

$$-\frac{1}{\rho}\left(\frac{\partial \rho}{\partial t}\right)_p = \beta$$

For small density differences, the last equation can be approximated by

$$\rho - \rho_e = \beta\rho(t_e - t) \tag{12-5}$$

The momentum equation now takes the form

$$\rho\left(u\frac{\partial u}{\partial s} + v\frac{\partial u}{\partial y}\right) = \frac{\partial}{\partial y}\mu\left(\frac{\partial u}{\partial y}\right) + \rho u_e\frac{du_e}{ds} - \beta\rho g_s(t - t_e) \tag{12-6}$$

At this point, it is convenient to consider which dimensionless variables influence the flow and heat transfer under the presence of body forces. This can be done by applying dimensional analysis. As in Sec. 9-1, the boundary-layer equations are for this purpose made dimensionless by referring all physical quantities to properly selected reference quantities. The velocity u_e was used for this purpose in Sec. 9-1. A reference length L will be used to make all length parameters dimensionless. Finally a temperature difference $t_w - t_e$ is selected for the temperatures.[1] The body force appears as an additional term in Eq. (12-6). When this equation is made dimensionless in the same way as Eq. (9-6), an additional parameter appears before the last term expressed in dimensionless quantities:

$$\frac{\beta g(t_w - t_e)L}{u_0^2}$$

No change occurs in the mass-continuity and the energy equation. However all three equations (12) are now interconnected. As a consequence,

[1] t_e is now considered constant.

the dimensionless parameters appearing in any of these equations influence the velocity field as well as the temperature field. The Nusselt number describing the film heat-transfer coefficient on a surface subjected to the flow will then be described by a function of the following form

$$\mathrm{Nu} = f\left(\mathrm{Re}, \mathrm{Pr}, \frac{\beta g(t_w - t_e)L}{u_0^2}\right)$$

The dimensionless coordinates on which the local Nusselt number depends additionally are omitted in this equation. The equation is somewhat inconvenient because the reference velocity u_0 appears in two of the parameters. This can be avoided when the velocity is removed from the parameter describing the influence of body forces by multiplying this parameter with the square of the Reynolds number

$$\frac{\beta g(t_w - t_e)L}{u_0^2}\,\frac{u_0^2 L^2}{\nu^2}$$

The parameter obtained in this way is called *Grashof number*

$$\mathrm{Gr} = \frac{\beta g(t_w - t_e)L^3}{\nu^2} \tag{12-7}$$

Another parameter, called the *Rayleigh number*, is being used more and more instead of the Grashof number. It is defined as

$$\mathrm{Ra} = \mathrm{Gr}\,\mathrm{Pr} = \frac{\beta g(t_w - t_e)L^3}{\nu\alpha} \tag{12-8}$$

It is of advantage to use the Rayleigh number because for many situations and for Prandtl numbers larger than 5 flow and heat transfer are either independent of, or only slightly dependent on, the Prandtl number when they are expressed in the form

$$\mathrm{Nu} = f(\mathrm{Re},\mathrm{Pr},\mathrm{Ra}) \tag{12-9}$$

A flow situation is called free- or natural-convection flow when it is created solely by body forces. In this case, then, no prescribed velocity is available which might be used to define a Reynolds number. As a consequence, the Reynolds parameter drops out of the dimensionless equations, and heat transfer, for instance, is described by an equation of the form

$$\mathrm{Nu} = f(\mathrm{Ra},\mathrm{Pr}) \tag{12-10}$$

12-2 THE FLAT PLATE AND THE HORIZONTAL CYLINDER

A solution to the system of Eqs. (12-1), (12-3), and (12-6) will now be obtained for a vertical plate in an infinitely large surrounding of constant temperature t_e with the boundary conditions

$$\begin{array}{llll} s = 0 & u = 0 & v = 0 & t = t_w \\ s \to \infty & u = 0 & & t = t_e \end{array} \tag{12-11}$$

The momentum equation (12-6) simplifies to

$$\rho\left(u\frac{\partial u}{\partial s} + v\frac{\partial u}{\partial y}\right) = \frac{\partial}{\partial y}\left(\mu\frac{\partial u}{\partial y}\right) - \beta\rho g_s(t - t_e) \tag{12-12}$$

when the velocity u_e is zero. The properties appearing in these equations will be considered constant, and the enthalpy i is correspondingly written as c_pt. The temperatures t_w and t_e are considered constant. Pohlhausen,[1] who was the first to solve this problem, found that the free-convection boundary layer on such a surface is again self-similar and that, accordingly, the above differential equations can be transformed to total differential equations by introduction of the following independent and dependent variables:

$$\eta = C\frac{y}{\sqrt[4]{x}} \qquad \theta = \frac{t - t_e}{t_w - t_e} \qquad \psi = 4\nu Cx^{\frac{3}{4}}\xi(\eta)$$

The constant in this equation is defined as

$$C = \sqrt[4]{\frac{g\beta(t_w - t_e)}{4\nu^2x^3}} = \frac{1}{2\sqrt{x^3}}\sqrt[4]{\mathrm{Gr}_x} = \frac{1}{2\sqrt{x^3}}\sqrt{\frac{\mathrm{Ra}_x}{\mathrm{Pr}}}$$

The velocity components expressed in these variables are

$$u = 4\nu x^{\frac{1}{2}}C^2\xi' \qquad v = \nu Cx^{-\frac{1}{4}}(\eta\xi' - 3\xi)$$

A prime indicates, again, differentiation with respect to η. The mass-continuity equation has been eliminated by introduction of a stream function ψ. The momentum and energy equations take the form

$$\xi''' + 3\xi\xi'' - 2\xi'^2 + \theta = 0$$
$$\theta'' + 3\,\mathrm{Pr}\,\xi\theta' = 0$$

The boundary conditions are

$$\begin{array}{llll} \eta = 0 & \xi = 0 & \xi' = 0 & \theta = 1 \\ \eta \to \infty & & \xi' = 0 & \theta = 0 \end{array}$$

Figures 12-2 and 12-3 present the velocity and the temperature profiles, respectively, as obtained by Pohlhausen from a numerical solution of the differential equations for air at atmospheric conditions. They are compared with results of measurements by Schmidt and Beckman,[2] which

[1] In E. Schmidt and W. Beckmann, *Tech. Mech. Thermodyn.* **1**:1 (1930).

[2] *Ibid.*

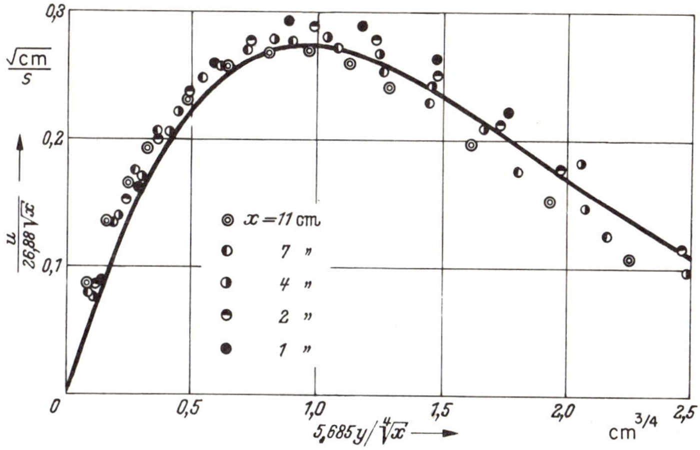

Fig. 12-2 Comparison between measured and calculated velocity profiles in free-convection flow of air on a vertical plate. [*Measured values after E. Schmidt and W. Beckmann; solid line, solution by E. Pohlhausen.*]

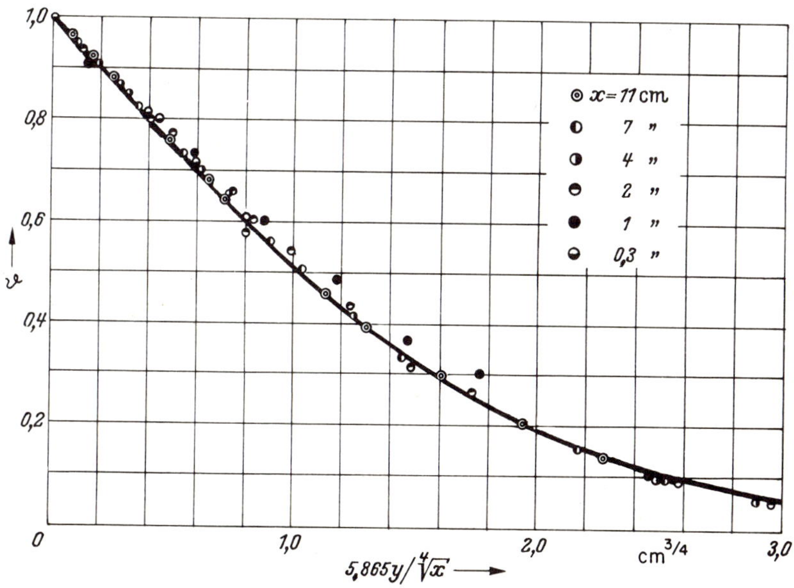

Fig. 12-3 Comparison between measured and calculated temperature profiles in free-convection flow of air on a vertical plate. [*Measured values after E. Schmidt and W. Beckmann; solid line, solution by E. Pohlhausen.*]

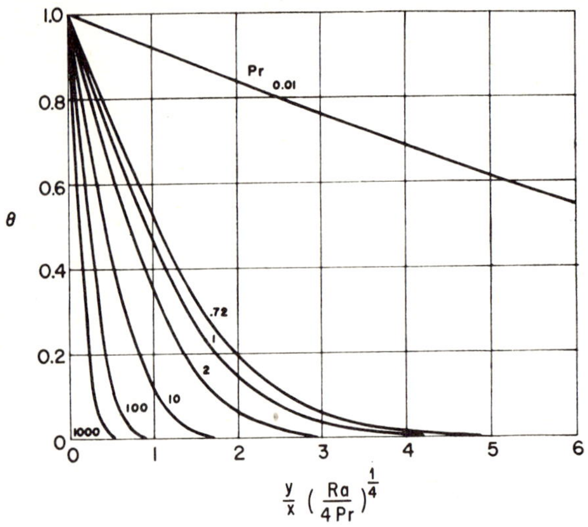

Fig. 12-4 Dimensionless temperature profiles for laminar free convection on a vertical plate. [*From S. Ostrach, NACA Tech. Note 2635, 1952.*]

are represented by the points in the figure. More recently, Ostrach[1] has solved the differential equations in a wide range of Prandtl numbers. Figures 12-4 and 12-5 present the temperature and velocity profiles he obtained. Local heat-transfer coefficients can again be found from the temperature gradients at the wall surface and can be presented by an equation of the form

$$\mathrm{Nu} = f(\mathrm{Pr}) \sqrt[4]{\mathrm{Ra}} \tag{12-13}$$

An approximate solution of the equations describing the present free-convection problem has been obtained by Squire.[2] He assumed for this purpose an appropriate shape of the velocity and temperature profiles and solved with their help the integrated momentum and energy equation. This analysis is described in Eckert.[3] The following equation results from this analysis

$$\mathrm{Nu} = 0.508 \frac{\mathrm{Pr}^{\frac{1}{4}}}{(0.952 + \mathrm{Pr})^{\frac{1}{4}}} \sqrt[4]{\mathrm{Ra}} \tag{12-14}$$

[1] S. Ostrach, *NACA Tech. Note* 2635, 1952.

[2] H. B. Squire and S. Goldstein, "Modern Developments in Fluid Dynamics," Oxford University Press, New York, 1938.

[3] E. R. G. Eckert, "Introduction to Heat and Mass Transfer," McGraw-Hill, New York, 1963.

The equation was found to agree well with experimental results. It will be observed that for large Prandtl numbers, the Nusselt number is a function of Rayleigh number only.

The equations for the vertical plate can also be used to describe heat transfer for free-convection flow over a flat plate inclined at an arbitrary angle against the vertical direction on the side which faces downward. The gravitational acceleration g must then be replaced by the component of this acceleration in the direction parallel to the plate surface. The condition for the flow on the upward-facing side of the plate is often different because an instability develops similar in nature to the Goertler instability on a concave surface in forced flow. A row of vortices with their axis parallel to the flow direction develops and creates convective energy transport, which adds to the overall heat transfer.[1] Analyses have also been performed for a large variety of geometries and boundary conditions. They are described, together with their solutions, in a survey paper by Ede.[2]

Figure 12-6 was obtained in a Zehnder-Mach interferometer with an adjustment which made all plates in this instrument parallel. The shadow in this photograph represents a vertical plate. The interference

[1] E. M. Sparrow, R. B. Husar, and R. J. Goldstein, Observations and Other Characteristics of Thermals, *J. Fluid Mech.*, **41**:793–800 (1970).

[2] A. J. Ede, Advances in Free Convection, in *Advan. Heat Transfer*, **4**:1 (1967).

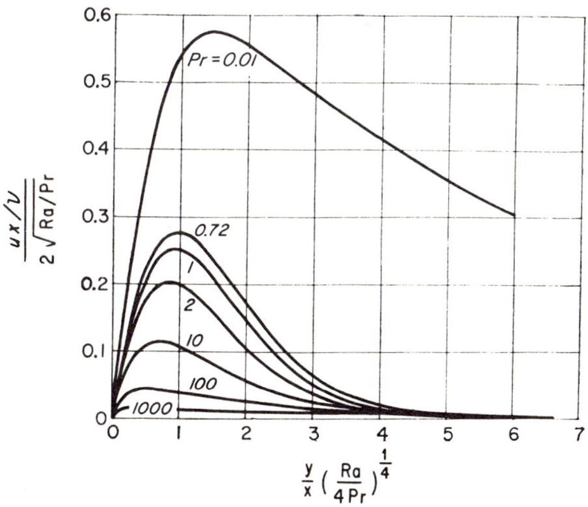

Fig 12-5 Dimensionless velocity profiles for laminar free convection on a vertical plate. [*From S. Ostrach, NACA Tech. Note 2635, 1952.*]

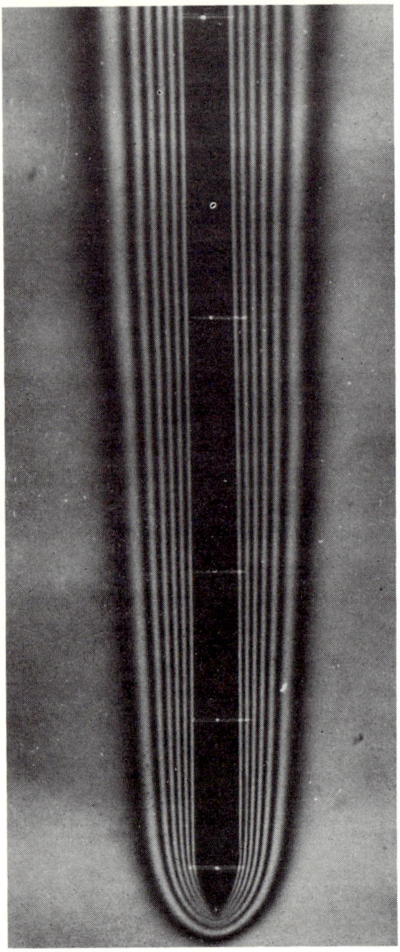

Fig. 12-6 Isotherms around a vertical plate in free-convection flow of air as recorded by an interferometer. [*Photograph by E. R. G. Eckert and E. Soehngen.*]

fringes surrounding this plate can be interpreted as lines of constant temperature. In this way the photograph gives a clear indication of the development of the boundary layer along such a plate. It can be observed that free-convection boundary layers are relatively thick. As a consequence, the analytical solution obtained from the boundary-layer equations does not describe the actual situation close to the lower end of the plate, where, according to the solution, the boundary layer would start with zero thickness. It is quite remarkable that the solution is nevertheless in good agreement with the actual situation a short distance downstream from the leading edge.

The results of an experimental study[1] on unsteady free convection

[1] R. J. Goldstein and E. R. G. Eckert, *Intern. J. Heat Mass Transfer*, **1**:208 (1960).

on a vertical heated plate will be discussed briefly because they exhibit features which are characteristic of unsteady convective-heat-transfer processes in general. The vertical plate was a metal foil 0.25 mm thick placed in water. An electric current was supplied beginning at a certain moment. This created a heat flux from the plate surface into the surrounding fluid which was locally uniform and suddenly increased stepwise from a value of zero to a constant value q_w because the heat capacity of the foil was very small.

Figure 12-7 presents the local heat-transfer coefficients plotted over the time with the distance x from the lower end of the plate as parameter. It can be observed that the heat-transfer coefficients are initially a function only of time and change through a short transition period to a condition where they are independent of time but dependent on the distance x. During the first period, the development of the boundary layer occurs in the same way as on a plate of infinite length. The streamlines and the isotherms are parallel to the plate surface, and, as a consequence, the heat flux into the fluid occurs by conduction only. The unsteady one-dimensional conduction equation therefore describes the heat transfer during this period. It is indicated in the figure by the heavy lines. The plate, at a certain location, becomes aware of the fact that it has a lower end only when fluid particles which have entered the boundary layer near the lower plate end arrive at this location. This, then, deter-

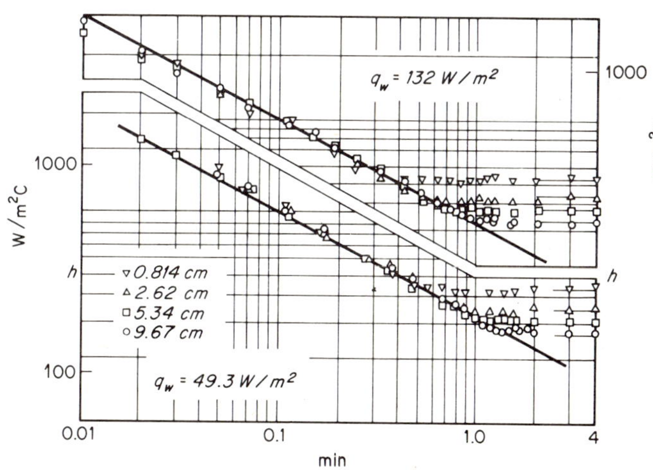

Fig. 12-7 Unsteady heat-transfer coefficient for free-convection flow of water on a vertical plate for a stepwise change in heat flux. [*From R. J. Goldstein and E. R. G. Eckert, Intern. J. Heat Mass Transfer,* **1**:*208 (1960).*]

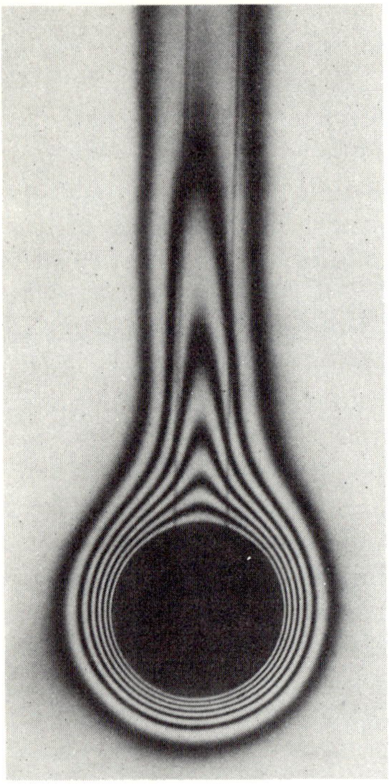

Fig. 12-8 Isotherms around a horizontal cylinder in free-convection flow of air as recorded by an interferometer. [*Photograph by E. R. G. Eckert and E. Soehngen.*]

mines in the figure the transition from the initial to the final steady condition.[1]

Figure 12-8 presents the field of isotherms around a horizontal cylinder the surface of which is kept at a higher temperature than the surrounding fluid. It can be observed that, at a Rayleigh number of 10^4, the thermal boundary layer has a thickness which cannot be considered as small compared to the cylinder diameter and that the boundary-layer simplifications therefore cannot be applied. An analytical solution to this problem can be obtained with the help of high-speed electronic computers. An earlier solution based on a strongly simplified model presented by Langmuir[2] was found to be in reasonably good agreement with experimental results.

12-3 THE TURBULENT BOUNDARY LAYER

Figure 12-9 is a photograph obtained in a Zehnder-Mach interferometer. It describes the temperature field for free-convection flow of air along a

[1] See survey paper by E. K. Kalinin and G. A. Dreitser, Unsteady Heat Transfer in Channels, *Advan. Heat Transfer*, **6**:367 (1970).

[2] I. Langmuir, *Phys. Rev.*, **34**:401 (1912), described in E. R. G. Eckert and R. M. Drake, "Heat and Mass Transfer," p. 318, McGraw-Hill, New York, 1959.

vertical plate of larger dimension than the one in Fig. 12-6. It can be observed that the boundary layer which is at first laminar and steady becomes unstable at a distance of approximately 20 in. Waves then appear which increase in amplitude as they travel downstream and finally transform into turbulence, which is at first fairly coarse but becomes finer-grained as the distance from the leading edge of the plate increases. The insert in Fig. 12-9 gives an idea of the structure of turbulence observed at large Rayleigh numbers. The plate, along which the boundary layer develops, is located just beyond the left-hand edge of the photograph. An instability Rayleigh number indicating the first appearance of the

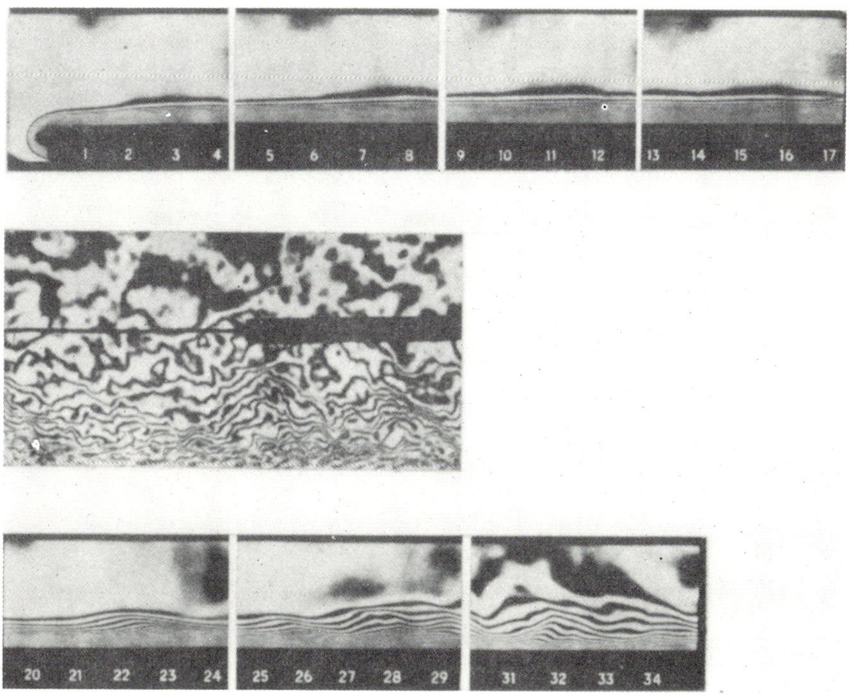

Fig. 12-9 Interference photographs of transition to turbulence in a free-convection boundary layer in air on a vertical plate. Numbers indicate distance in inches from the bottom. [*After E. R. G. Eckert and E. Soehngen*].

waves and a transition Rayleigh number describing the location where the disturbances have become irregular and statistical can be determined from such photographs and from measured local heat-transfer coefficients. The instability Rayleigh number has been determined in this way as having a value of approximately 10^9. It has not been as yet conclusively determined whether the transition is described solely by a Rayleigh number or whether the Prandtl number also has some influence. Thorough analytical and experimental studies on the stability of a free-convection boundary layer have been carried out by Gebhart.[1]

An early analysis of the turbulent boundary layer was performed by Eckert and Jackson.[2] The shape of the time-averaged velocity and temperature profiles has, for this purpose, been described by an appropriate equation based on available experimental information, and the integrated momentum and energy equations have been solved. The analysis is described in Eckert and Drake.[3] A more detailed analysis, which parallels the one described in Chap. 8 for forced flow, has recently been attempted by Kato, Nishiwaki, and Hirata.[4] A difficulty was caused by the fact that much less information on the turbulent transfer mechanism is available for free-convection boundary layers than for forced-convection flow. Correspondingly, the authors had to assume that the turbulent diffusivities are described by equations which have been established for forced flow.

The momentum equation written in time-averaged parameters has the form

$$\rho u \frac{\partial u}{\partial x} + \rho v \frac{\partial u}{\partial y} = \frac{\partial \tau}{\partial y} - \rho \beta g(t - t_e) \tag{12-15}$$

The equation is again simplified in the paper referred to by assuming that the boundary-layer flow near the wall surface can be approximated by a model in which the streamlines are parallel to the wall surface. Accordingly, Eq. (12-14) simplifies to

$$\frac{\partial \tau}{\partial y} - \rho g \beta(t - t_e) = 0$$

By integration, an equation can be obtained which describes the variation of the shear within the boundary layer

$$\tau = \tau_w - \int_0^y \rho g \beta(t - t_e)\, dy \tag{12-16}$$

[1] B. Gebhart, *J. Heat Transfer*, **91**:293 (1969).

[2] E. R. G. Eckert and T. W. Jackson, *NACA Rept.* 1015, 1951.

[3] *Op. cit.*

[4] H. Kato, N. Nishiwaki, and M. Hirata, *Intern. J. Heat Mass Transfer*, **11**:1117, (1968).

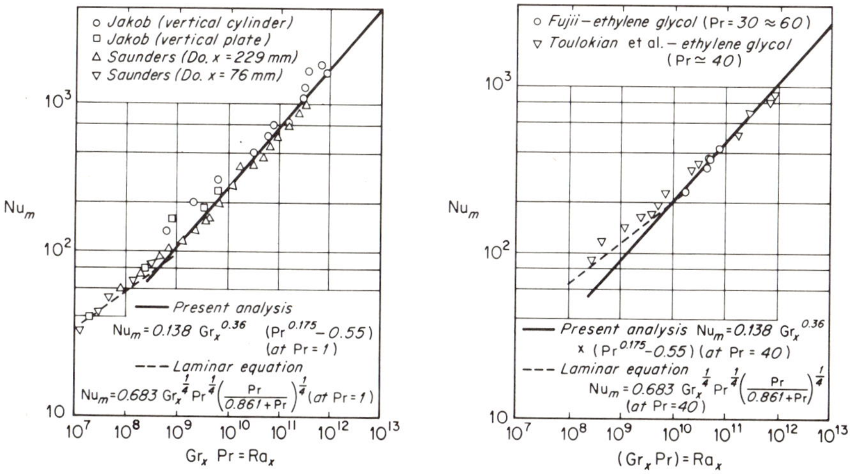

Fig. 12-10 Comparison of analytical and experimental Nusselt numbers for free-convection flow of fluids with Prandtl numbers around 1 and around 40. [*From H. Kato, N. Nishiwaki, and M. Hirata, Intern. J. Heat Mass Transfer*, **11**:*1117* (*1968*).]

Into this equation, then, is introduced an appropriate empirical relation describing the turbulent diffusivity of momentum, and the equation is transformed to the dimensionless variables (indicated by a superscript plus). The energy equation retains the form given in Chap. 8. The integrated momentum equation

$$\frac{\tau_w}{\rho} = \int_0^\delta g\beta(t - t_e)\, dy - \frac{d}{dx}\int_0^\delta u^2\, dy \tag{12-17}$$

and the corresponding integrated energy equation are then used to solve for the wall shear τ_w and the heat flux. The results of this analysis agree quite satisfactorily with experimental information. Figure 12-10 presents as an example the calculated Nusselt numbers as a function of the Rayleigh number and compares them with experimental information for fluids with a Prandtl number near 1 and 40.

12-4 FLUIDS ENCLOSED BETWEEN TWO PLANE WALLS[1]

In the preceding sections the objects on which heat transfer occurred were assumed to be located in a fluid of infinite extent and uniform tem-

[1] For a discussion of free convection in enclosed spaces see also S. Ostrach in "High Speed Aerodynamics and Jet Propulsion," vol. IV, sec. F, Princeton University Press, Princeton, N.J., 1964.

perature. In experiments, this is difficult to realize, and one has to worry about the effect of neighboring surfaces. Many engineering applications also present heat-transfer situations in which free convection occurs in closed spaces. The study of such situations is also important in geophysics. Free convection in a fluid enclosed between two parallel plane walls will be considered in this section as an example of free convection in an enclosed space.

Horizontal walls No free-convection currents occur in a fluid which is enclosed between two parallel horizontal plates as long as the temperature of the upper plate is higher than the temperature of the lower one. In this case heat flows from the upper to the lower plate, and the temperature in the fluid is constant in horizontal layers and increases in an upward direction. In a normal fluid for which the density decreases with increasing temperature, such a temperature field creates a situation in which less dense layers are located above denser ones, a stable situation which does not cause any convection currents. Heat will be transported by conduction only (not considering radiation), and the temperature profile will be linear. This situation can be disturbed only near the edge of the plates.[1]

The situation is different when a fluid is enclosed between two horizontal surfaces of which the upper surface is cooler than the lower one. Heat flows now through the fluid in the direction from the lower toward the upper surface, and, as a consequence, the fluid between the two plates assumes such temperatures that colder fluid layers are situated above warmer ones. For fluids whose density decreases with increasing temperature, this leads to an unstable situation. It does not give rise to convection currents as long as the Rayleigh number is low. However, when this parameter reaches a value around 1,700, based on distance and temperature difference of the plates, a peculiar free-convection flow pattern arises which can be observed in Fig. 12-11, in which the flow is made visible by tiny aluminum particles. The flow field has a cellular structure with more or less regular hexagonal cells. In the interior of these cells the flow moves upward, and along the rim of the cell it returns downward. This flow situation is maintained up to a Rayleigh number of around 47,000. Above this value, it changes to irregular turbulence.[2] The lower critical Rayleigh number at which the cellular flow pattern is established was calculated theoretically by Rayleigh (1916) after Benard first observed it (1900).

[1] A. V. Luikov, B. M. Berkovskii, and V. E. Fertman, Stimulation of Free Convection by Heating from Above, *Prog. Heat Mass Transfer*, **2**:77 (1969).

[2] R. V. Goldstein and T. Y. Chu, Thermal Convection in a Horizontal Layer of Air, *Progr. Heat Mass Transfer*, **2**:55 (1969).

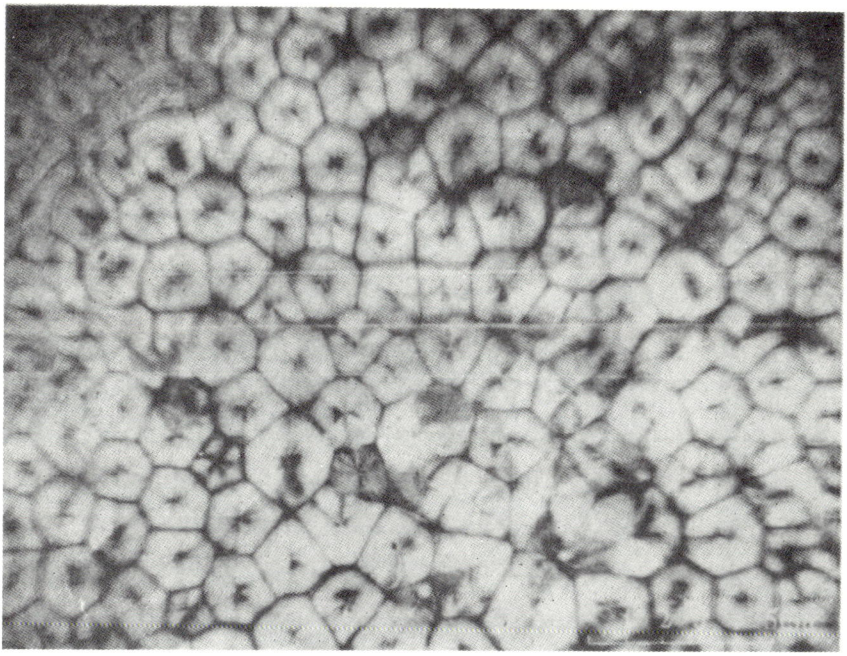

Fig. 12-11 Cellular flow pattern of free convection in horizontal air layers. [*After H. Siedentopf, from L. Prandtl, "Strömungslehre," Vieweg-Verlag, Brunswick, 1949.*]

Vertical walls When the fluid layer is enclosed between two vertical walls, it is expected that the height h of the walls will be an important geometric parameter as well as their distance d. In a dimensionless presentation, accordingly, the height-to-distance ratio appears in addition to the Rayleigh and Prandtl numbers.

The Zehnder-Mach interferometer can be used with advantage to study the temperature field in such a fluid layer. Figure 12-12 presents two interferograms obtained in a study[1] covering a large range of Rayleigh numbers. Figure 12-12*a* shows the temperature field characteristic of Rayleigh numbers below 1,000. The interference fringes are straight lines over the major portion of the temperature field, indicating that the temperature in the fluid decreases linearly along horizontal lines in the direction from the hot to the cold plate and that heat transfer is by conduction only. The field is disturbed at the lower and upper ends of the fluid layer. Figure 12-12*b* is an interferogram obtained at a large Rayleigh number. The temperature variation along horizontal lines is

[1] E. R. G. Eckert and W. O. Carlson, *Intern. J. Heat Mass Transfer,* **2**:106 (1961).

(*a*) (*b*)

Fig. 12-12 Interferograms presenting the temperature field for natural convection in air enclosed between a heated and a cooled vertical wall: (*a*) $h/d = 10$, $\mathrm{Ra}_d \approx 10^3$, (*b*) $h/d = 10$, $\mathrm{Ra}_d \approx 10^3$. [*From E. R. G. Eckert and W. O. Carlson, Intern. J. Heat Mass Transfer,* **2**:*106* (*1961*).]

now restricted to two boundary layers which develop along the heated and the cooled wall, whereas it is almost constant in the horizontal direction in the central layer. An evaluation of the interferograms, however, shows that the temperature increases almost linearly in an upward direction within this central portion.

The velocity field in the vertical fluid layer has been studied by flow visualization by Elder,[1] who found that the fluid rotates slowly at low values of the Reynolds number, moving upward along the heated plate and downward along the cooled plate with streamlines which are parallel to the vertical surfaces over the major portion of the fluid layer and which close near the upper and the lower end. This is in agreement with the observation that the heat transfer in the central portion of the fluid is by conduction only. At large Rayleigh numbers for which the temperature field is, for instance, illustrated by the interferogram of Fig. 12-12*b*, the fluid moves upward within the boundary layer adjacent to the hot wall and downward within the boundary layer adjacent to the cold wall. At first sight, one might expect the viscous forces exerted by the boundary layers on the central fluid to make it rotate. Why this is not the case can be understood from Fig. 12-13, which presents the results of a numerical solution of the conservation equations obtained by Elder on an electronic computer.[2] The lines in this figure are isotherms. The two boundary layers on the vertical walls can be clearly recognized, as well as the fact that the temperature in the central region increases vertically. It is especially interesting to observe that in the central region the temperature increases along horizontal lines in the direction from the hot toward the cold wall. This creates buoyancy forces which compensate the viscous forces exerted on the central region by the boundary layers and keeps the fluid in the central region stationary.

Starting at a Rayleigh number around 10^5, secondary flows in the

[1] J. W. Elder, *J. Fluid Mech.*, **23**:77–98 (1965).

[2] J. W. Elder, *J. Fluid Mech.*, **24**:823–843 (1966).

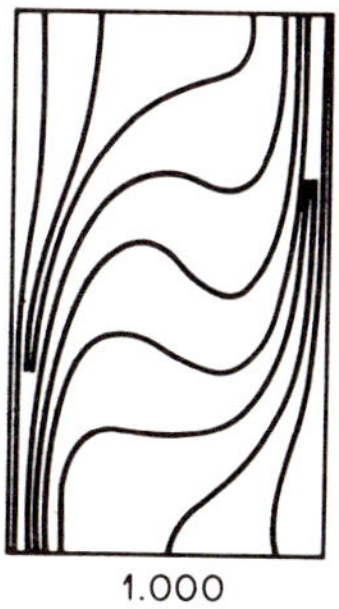

Fig. 12-13 Computer solution of the field of isotherms for natural convection in an enclosure with a heated and a cooled vertical wall: $Ra = 2 \times 10^4$, $Pr = 1$. [*From J. W. Elder, J. Fluid Mech.*, **23**:*77* (*1965*).]

form of a vertical row of vortices with horizontal axes develop,[1] with the number of vortices and their intensity increasing with increasing Rayleigh number. At Ra $\sim 10^9$ the flow becomes turbulent, first in the central region and then gradually spreading into the boundary layers. The average heat transfer from the hot to the cold surface at low Rayleigh numbers is described by Fourier's conduction equation and at higher Rayleigh numbers is determined by the thermal resistance of the two boundary layers arranged in series; the average heat-transfer coefficient for each boundary layer was found to differ from the heat-transfer coefficient of a single vertical plate in an infinite isothermal surrounding by not more than 20 percent.[2]

For engineering-design calculations, interest is mainly concentrated on a determination of the total heat flow through fluid layers of various geometries from a hot to a cold surface. It is surprising that the geometry exerts only a minor influence on this parameter. Kraussold[3] showed that the average heat flow per unit area as a first approximation can be represented by a single relation for cylindrical layers as well as for plane horizontal or vertical ones. The average heat flow q_m per unit time and area is conveniently characterized for such layers by an *equivalent heat conductivity* k_e, defined by

$$q_m = \frac{k_e}{b}(t_h - t_c)$$

where b = thickness of layer
t_h = temperature of hot surface
t_c = temperature of cold surface

For the cylindrical layer the heat flow q_m is based on the mean cross section of the layer (at radius r_m). Figure 12-14 shows the ratio k_e/k of the equivalent to the actual heat conductivity plotted vs. the Rayleigh number Ra_b based on the thickness b of the layer and on the temperature difference $t_h - t_c$. The experiments performed by Beckmann and Sellschopp on cylindrical layers and by the others on plane layers correlate quite well with the full line. For the range in which $k_e/k = 1$, heat transport through the layer takes place by conduction only. The deviation from $k_e/k = 1$ indicates the contribution of convection to the total heat flow. The correlation represented by the line can also be used for other geometries as a first approximation. According to Elder, it holds for any Prandtl number larger than 1 (gases can be included).

[1] J. W. Elder, *J. Fluid Mech.*, **23**:99–111 (1965).

[2] R. Eichhorn, Natural Convection in a Thermally Stratified Fluid, *Prog. Heat Mass Transfer*, **2**:41 (1969).

[3] H. Kraussold, *Forsch. Gebiete Ingenieurw.*, **2**:186–191 (1936).

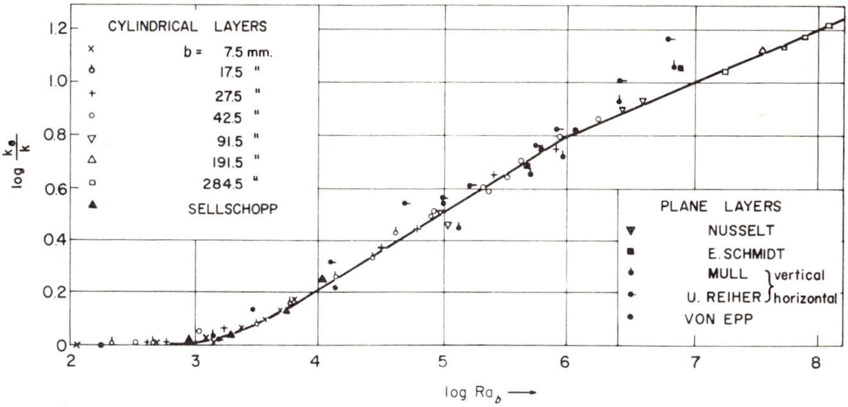

Fig. 12-14 Equivalent heat conductivity for free convection through fluid layers. *From H. Kraussold, Forsch. Gebiete Ingenieurw.*, **5**:*83* (*1934*).]

12-5 MIXED FREE AND FORCED CONVECTION

In Chaps. 7 to 11 on forced convection, the influence of body forces on flow and heat transfer has been neglected. Actually buoyancy forces are present in a normal fluid in which the density varies with temperature in any forced-convection flow occurring in the gravitational field of the earth, and it is of interest to know when they can be neglected and when they have to be accounted for. Such an investigation is made difficult by the large number of influencing parameters. In addition to Reynolds and Prandtl numbers, Rayleigh numbers as well as parameters describing the geometry of the boundaries and the orientation of the forced flow relative to the gravitational field are important. The situation which apparently has been investigated most intensively is forced flow through a circular pipe with its axis parallel to the direction of the gravitational acceleration and thus with the mean flow direction either parallel or opposite to it.

It is to be expected that for large Reynolds values (and correspondingly large flow velocities) and small Rayleigh numbers the influence of free convection on the heat transfer can be neglected. On the other hand, for large Rayleigh values and small Reynolds numbers free convection should be the dominating factor. This expectation is borne out in Fig. 12-15, which summarizes experimental and some analytical investigations on vertical tubes with various length-to-diameter ratios L/D and for fluids with various Prandtl numbers.[1] The limits between the forced-, free-,

[1] B. Metais and E. R. G. Eckert, Forced, Mixed, and Free Convection Regimes, *J. Heat Transfer*, **86**:295–296 (1964).

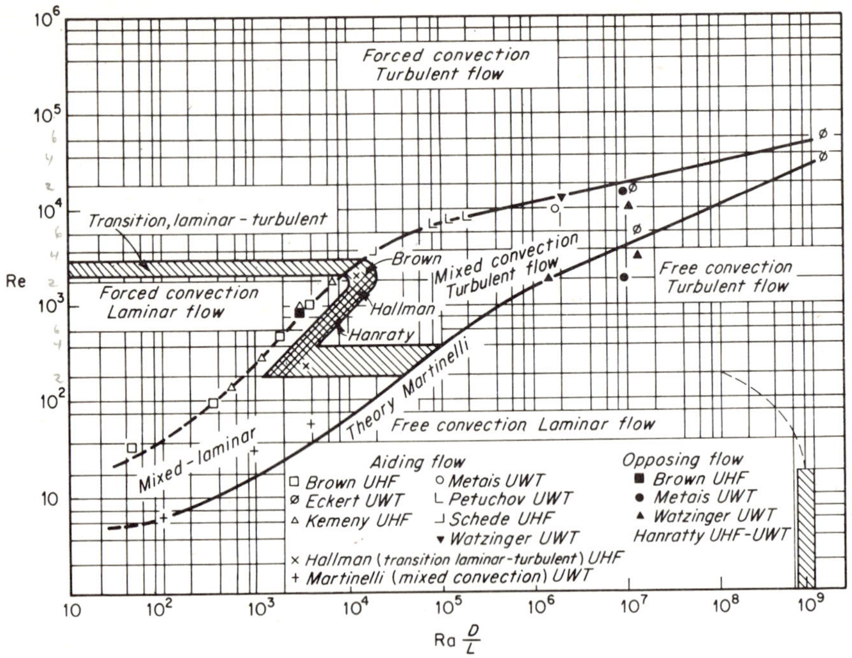

Fig. 12-15 Regimes of free, forced, and mixed convection for flow through vertical tubes. [*From B. Metais and E. R. G. Eckert, J. Heat Transfer,* **86**:*295* (*1964*).]

and mixed-convection regimes are defined in such a way that the actual heat flux under the combined influence of the forces does not deviate by more than 10 percent from the heat flux that would be caused in forced flow or in free-convection flow alone. It was found that the various experimental results could best be correlated by using the product Ra times D/L as abscissa. The figure contains the results of experiments with uniform heat flux (UHF) from the tube wall into the fluid as well as for uniform wall temperature (UWT). Most of the experiments were performed under the condition that the force caused by the pressure gradient determining the forced flow and the body force are parallel to each other near the tube wall. The results of a few experiments where the two forces are in opposite directions are also entered. The dashed area indicates the approximate range of transition from laminar to turbulent flow. The figure must be considered tentative, and more experiments are required to establish the limits between the various flow regimes accurately.

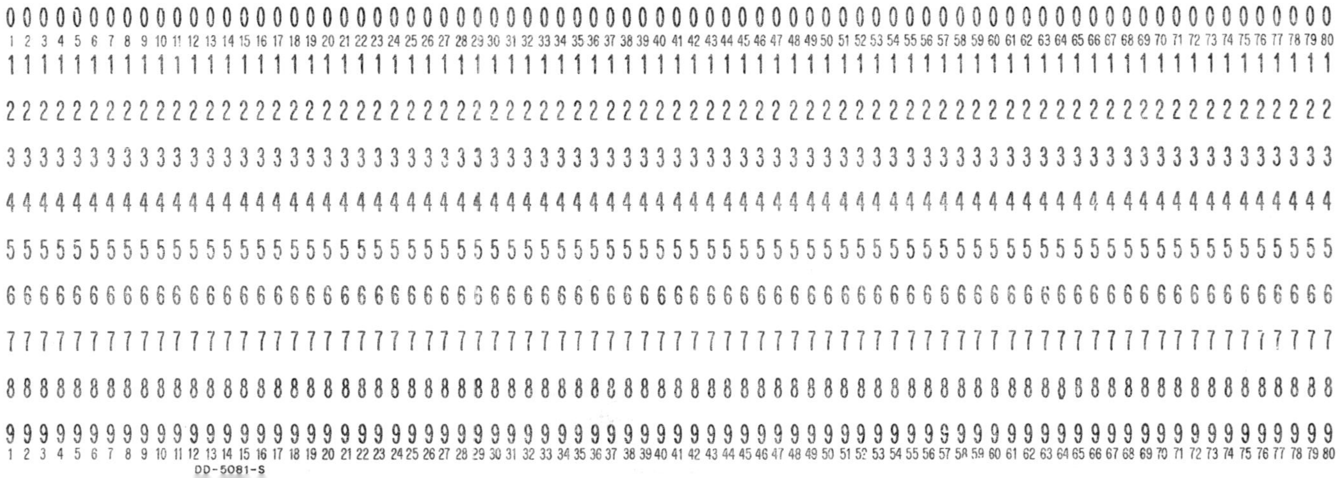
DD-5081-S

13
Condensation and Evaporation

Heat-transfer processes with a change of phase will be studied in this chapter. Depending on the direction of the heat flow, they occur as condensation or as evaporation, both of which are found in many technological applications.

Their analysis is made difficult by the fact that the interface between the vapor and the liquid often has an ill-defined shape. This occurs when the vapor condenses in the form of droplets or liquid evaporates in bubbles. Our knowledge of the formation of these droplets and bubbles is today not sufficient to develop an analysis of heat transfer starting from basic physical laws. As a consequence, our information is based mainly on experimental results, and the reader will find this reflected in the treatment of the boiling process particularly.

13-1 CONDENSATION

Condensation occurs when the walls of a container or a duct housing a vapor are cooled below the saturation temperature. The condensate on

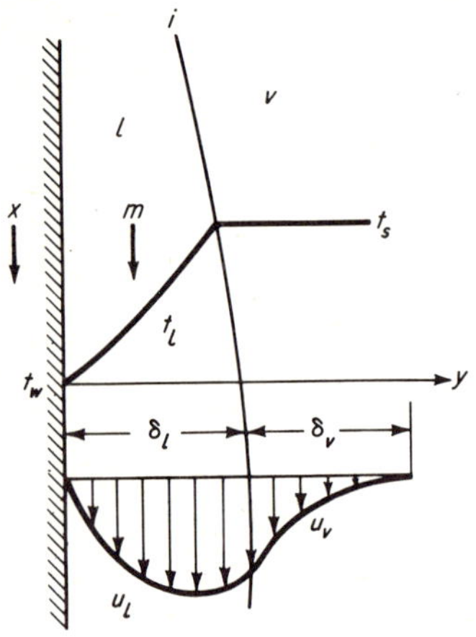

Fig. 13-1 Sketch of film condensation of a saturated vapor.

the walls appears in various forms, one being a continuous film created by condensation of the vapor on a vertical plate. Condensation of a saturated vapor in this form (film condensation) will now be discussed. By gravitational forces the film is made to flow down the wall, and in doing so, it drags a layer of vapor along with it even if the main body of the vapor is at rest. Qualitatively, the temperature and the velocity profiles in these layers have the shape sketched in Fig. 13-1, where δ_l indicates the thickness of the liquid condensate film and δ_v the thickness of the vapor boundary layer. In the film t_l and u_l denote, respectively, the temperature and the velocity parallel to the wall. u_v denotes the velocity in the vapor layer. In most applications, the thickness of both layers is found to be small compared to the dimension of the plate on which condensation occurs. Boundary-layer equations can therefore be used to describe the situation in the condensate film as well as in the vapor layer. For laminar two-dimensional steady state, the following equations describe the velocity and the temperature field:

$$\frac{\partial u}{\partial x} + \frac{\partial v}{\partial y} = 0 \tag{13-1}$$

$$\rho\left(u\frac{\partial u}{\partial x} + v\frac{\partial u}{\partial y}\right) = \mu\frac{\partial^2 u}{\partial y^2} - (\rho - \rho_e)g \tag{13-2}$$

$$\rho c_p\left(u\frac{\partial t}{\partial x} + v\frac{\partial t}{\partial y}\right) = k\frac{\partial^2 t}{\partial y^2} \tag{13-3}$$

The same set of equations applies for the liquid film as well as for the vapor layer, with the difference that no buoyancy forces occur within the vapor layer and the last term in Eq. (13-2) can therefore be dropped.

The following boundary conditions apply:

$$y = 0 \qquad u_l = 0 \qquad v_l = 0 \qquad t_l = t_w \tag{13-4a}$$
$$y \to \infty \qquad u = 0 \qquad t_v = t_s \tag{13-4b}$$

In addition, the following conditions must be satisfied at the interface:

$$y = \delta_l \qquad u_l = u_v \qquad t_l = t_s \qquad \mu_l \frac{\partial u_l}{\partial y} = \mu_v \frac{\partial u_v}{\partial y} \qquad k \frac{\partial t_l}{\partial y} = \frac{d\dot{m}}{\partial x} i_{lv} \tag{13-4c}$$

where t_s = saturation temperature
$\dot{m}$ = mass flux within the condensate film through a cross section at x
i_{lv} = heat of condensation

The mass flux $\dot{m}$ is connected with the velocity in the condensate film by the equation

$$\dot{m} = \int_0^{\delta_1} \rho_l u_l \, dy \tag{13-5}$$

The process of film condensation was first analyzed by Nusselt, who obtained a solution by simplifying the above system of equations. He neglected the convective transport in both layers and, in addition, considered the density of the condensate to be very large compared to the density of the vapor. In this way, Eqs. (13-2) and (13-3) simplify to

$$\mu_l \frac{\partial^2 u_l}{\partial y^2} = \rho_l g \tag{13-6}$$

$$k_1 \frac{\partial^2 t_l}{\partial y^2} = 0 \tag{13-7}$$

He also omitted the vapor film from his consideration, arguing that the viscosity of the liquid is much larger than the viscosity of the vapor. Thus the conditions at the interface given by Eq. (13-4c) simplify to

$$y = \delta_l \qquad \mu_1 \frac{\partial u_l}{\partial y} = 0 \qquad k_1 \frac{t_s - t_w}{\delta_1} = \frac{d\dot{m}}{dx} i_{lv} \tag{13-8}$$

In this way, he obtained for the thickness of the condensate film δ_l the relation

$$\delta_l = \sqrt[4]{\frac{4\nu k(t_s - t_w)x}{g\rho i_{lv}}} \tag{13-9}$$

and for the local Nusselt number

$$\mathrm{Nu}_x = \sqrt{\frac{g\rho i_{lv} x^3}{4\nu k(t_s - t_w)}} \tag{13-10}$$

The relations for the vertical plate can again be used for an inclined plate by replacing the gravitational acceleration g by its component parallel to the plate surface. The calculation was also extended to the case of a horizontal tube surrounded by vapor. For condensation of a vapor flowing through a duct, the velocity profile in Fig. 13-1 must be changed to take care of the forced-convection flow of the vapor.

A complete analysis of the system (13-1) to (13-3) of boundary-layer equations with the boundary conditions (13-4) has been performed by Koh, Sparrow, and Hartnett.[1] Condensation of a superheated vapor also creates a temperature boundary layer in the vapor.[2] All these processes are generally found to make very little alteration in the heat-transfer coefficient as it is described by Eq. (13-10). Considerably larger are the deviations of measured heat-transfer coefficients, which have been found to be by up to 20 percent smaller or up to 50 percent larger. These differences are believed to be caused by effects connected with the structure of the solid surfaces or with the behavior of the interface. A rough surface appears to retain the fluid, thus creating a thicker boundary layer, which will cause a smaller heat-transfer coefficient. The flow in the liquid film, on the other hand, may become unstable, creating surface waves which will increase heat transfer.

Especially large discrepancies were found between heat-transfer coefficients measured in the condensation of metal vapors and values calculated with the preceding equations. The measured values were found to be by up to orders of magnitude smaller. This was shown[3] to be due to a thermal resistance at the vapor-liquid interface, the magnitude of which can be calculated by the consideration that for a finite condensation rate the number of vapor molecules arriving at the interface must be larger than the number of molecules leaving the liquid surface. This surface resistance is negligibly small for normal fluids but can assume much larger values for metals.

The presence of even small amounts of noncondensable gases in the vapor also reduces heat transfer drastically. The condensation at the surface creates a continuous vapor flow toward the surface which carries the noncondensable gas along. This gas will accumulate near the interface until a steady state is obtained, when the diffusion and mass-transfer process away from the surface balances the convective transport toward it. The pressure in the vapor-gas mixture is now the sum of the partial

[1] J. C. Y. Koh, E. M. Sparrow, and J. P. Hartnett, *Intern. J. Heat Mass Transfer*, **2**:69–80 (1961).

[2] E. M. Sparrow and E. R. G. Eckert, *AIChE J.*, **7**:473–477 (1961).

[3] S. P. Sukhatme and W. M. Rohsenow, Heat Transfer during Film Condensation of a Liquid Metal Vapor, *ASME Paper* 65-HT-29, 1965.

pressures of the vapor and the gas. According to the discussion above, the partial pressure of the gas increases within the boundary layer in the direction toward the wall surface. The partial pressure of the vapor and the saturation temperature decrease correspondingly, thus reducing the driving temperature difference $t_s - t_w$ for the heat transfer through the condensate layer.[1] A quantitative investigation of these processes must rely on the methods discussed in Part IV, The Transfer of Mass.

The liquid film changes to turbulent flow when its thickness exceeds a certain value. Heat transfer through the turbulent film has been calculated only by approximate analyses. Such a calculation, for instance, was performed by Grigull.[2] The result shows that in the turbulent film the Nusselt number is a function of the dimensionless value in Eq. (13-10) and of the Prandtl number of the condensate. Since the equation for heat transfer in a turbulent film is rather complex, it is best presented in the form of a diagram (Fig. 13-2). In this figure, the Reynolds number of the film is plotted over the dimensionless value $kg^{\frac{1}{3}}\rho^{\frac{2}{3}}(t_s - t_w)x/(1.082 i_{lv}\mu^{\frac{5}{3}})$ with the Prandtl number as a parameter. The Reynolds number is based on the film thickness δ and the average velocity u_m in the film

$$\mathrm{Re}_\delta = \frac{u_m \delta}{\nu} \tag{13-11}$$

From this, the average heat-transfer coefficient $\bar{h}$ can be calculated by the fact that the heat flowing to the wall over the height x originates from the heat of condensation of the condensate passing through a cross section of the film at the point x

$$\bar{h}x(t_s - t_w) = \rho u_m i_{lv} \delta$$

Therefore

$$\mathrm{Re}_\delta = \frac{u_m \delta}{\nu} = \frac{\bar{h}x(t_s - t_w)}{i_{lv}\mu} \tag{13-12}$$

The Reynolds number calculated from Eq. (13-9) for laminar flow is

$$\mathrm{Re}_\delta = \frac{kg^{\frac{1}{3}}\rho^{\frac{2}{3}}x(t_s - t_w)}{1.082 i_{lv}\mu^{\frac{5}{3}}} \tag{13-13}$$

Which is represented in Fig. 13-2 by the dashed line. For turbulent flow, two bundles of lines are inserted in Fig. 13-2. One is calculated with a critical Reynolds number for transition to turbulence $\mathrm{Re}_c = 400$; the other with $\mathrm{Re}_c = 300$. The great number of inserted points representing experimental results indicates that $\mathrm{Re}_c = 300$ conforms somewhat better to reality. In the laminar region, Nusselt's theory is approximately verified by the experiments.

[1] W. J. Minkowycz and E. M. Sparrow, *Intern. J. Heat Mass Transfer,* **9**:1125 (1966).

[2] U. Grigull, *Forsch. Gebiete Ingenieurw.,* **13**:49–57, (1942).

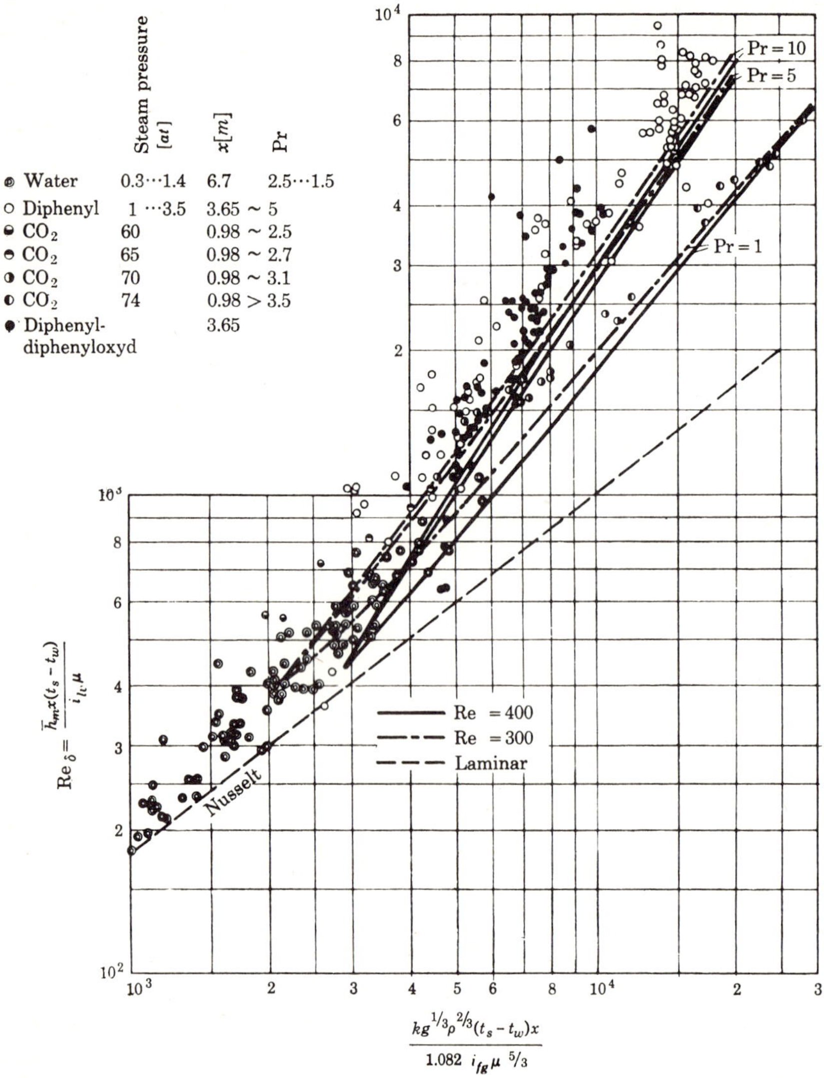

Fig. 13-2 Film condensation with laminar and turbulent flow. [*From U. Grigull, Forsch. Gebiete Ingenieurw.*, **13**:*49* (*1943*).]

Most experiments have confirmed Nusselt's liquid-film theory. Sometimes, however, much higher values have been measured. Schmidt, Schurig, and Sellschopp[1] showed that this is so because the vapor sometimes condenses not as a continuous film but as small droplets, which

[1] E. Schmidt, W. Schurig, and W. Sellschopp, *Tech. Mech. Thermodyn.*, **1**:53 (1930).

increase in size with time until they are so heavy that they run down the wall. In so doing they sweep their paths free of droplets. On these places the formation of tiny new droplets begins immediately. Figure 13-3 shows a photograph of this type of condensation. It can be easily understood that heat transfer for this type of condensation is much greater than for filmwise condensation. A great number of experiments have been conducted, which analyze the conditions under which the condensation will be filmwise or dropwise. Pure vapor on clean surfaces always condenses as a film. Impurities in the vapor or on the wall, especially of fatty acids, lead to condensation in the form of droplets. The formation of droplets is easier on rough surfaces than on smooth ones. On steel or aluminum tubes under ordinary conditions one must always expect filmwise condensation. Since it is difficult to predict with certainty the formation of dropwise condensation, it is recommended that calculations be made with the equation for film condensation.

13-2 EVAPORATION[1]

Evaporation of a liquid can occur in such a way that the vapor is produced at the liquid-vapor interface. This happens, for instance, when the heat of evaporation is transported into the fluid right at its surface, e.g., by radiation directed toward, and absorbed at, the surface of the liquid. This evaporation process is similar to the melting process discussed in the chapter on conduction and can be calculated by essentially the same methods (Fig. 13-4).

[1] See also A. Leppert and C. C. Pitt, Boiling, *Advan. Heat Transfer*, **1**:185 (1964).

Fig. 13-3 Dropwise condensation. [*E. Schmidt, W. Schurig, and W. Sellschopp, Tech. Mech. Thermodyn.* **1**:*53* (*1930*).]

In most engineering situations, heat is added to the liquid from a submerged solid surface, and the vapor is produced in the form of bubbles which start and grow at the heating surface, separating when they have reached a certain size and rising through the liquid. This mode of evaporation is called *boiling*. The complicated nature of the boiling process has long restricted our knowledge to a collection of empirical data. Recently, however, considerable progress has been made by intense experimental study and by development of models which sufficiently simplify the actual process so that dimensional analysis and analytical methods can be applied. There is good reason to expect that in the near future boiling heat transfer can be predicted in many cases by analysis just like other modes of heat exchange discussed in previous sections. In the meantime, the discussion must still be of a descriptive nature.

For a discussion of the essential features of a boiling process, let us consider evaporation of a liquid from a submerged heating surface. Let us assume that the liquid is originally well below the evaporation temperature. Then it can be observed that sometime after heat has been applied through the heating surface, vapor bubbles appear on that surface. These bubbles separate from the surface after having grown to a certain size and move into the bulk of the liquid. After a short path, however, they vanish again. Obviously they condense in the liquid, which in its bulk has not yet reached evaporation temperature. Only after the liquid has been heated sufficiently will the vapor bubbles travel all the way to the liquid surface and thus start a flow of vapor into the space above the surface. Both these processes have found engineering applications. The process in which the vapor is recondensed near the heating surface is called *local boiling* or *subcooled boiling*. The process in which the vapor bubbles penetrate the liquid surface is called *boiling with net evaporation*.

It has been established and can be readily understood that the boiling process is also influenced by the movement of the liquid near the heating surface. If this surface is submerged in a big container, movement of the liquid is generated only by free convection or by the stirring action of the bubbles. This type of boiling is classified as *pool boiling*. If, on the other hand, the boiling occurs on the walls of a tube through

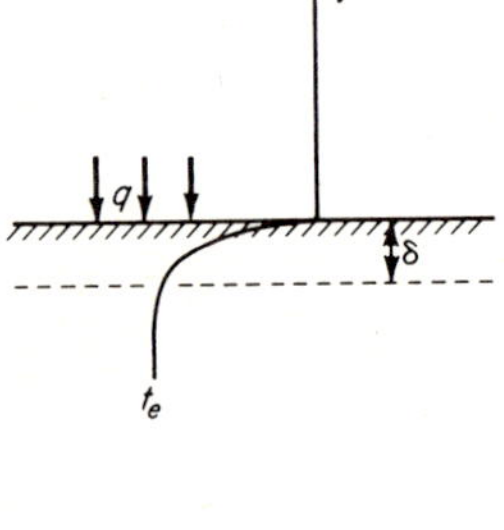

Fig. 13-4 Evaporation from an interface.

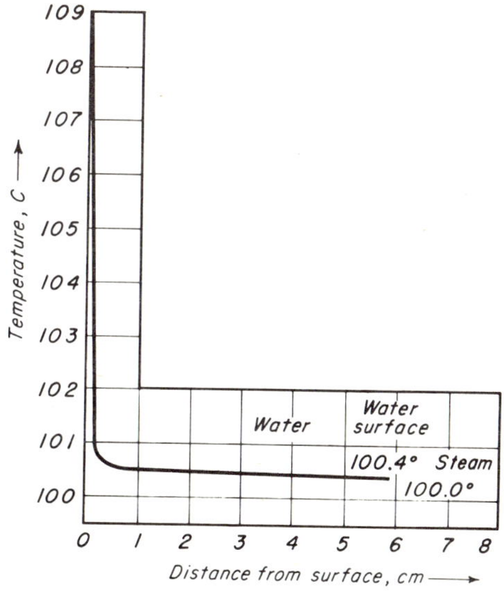

Fig. 13-5 Temperature profile in boiling water above a horizontal heating surface. [*After M. Jakob and W. Fritz, from M. ten Bosch, "Die Wärmeübertragung," 3d ed., Springer-Verlag, Berlin, 1936.*]

which the liquid flows with considerable velocity, this velocity makes itself felt and influences the bubble growth and separation. Boiling under such circumstances is called *forced-convection boiling.*

The most intensively investigated boiling process is *pool boiling with net evaporation.* This process will therefore be discussed in detail. The discussion will then be somewhat extended to local forced-convection boiling, and a few remarks will be devoted to forced-convection boiling with net evaporation.

Figure 13-5 presents the temperature field in boiling water measured for pool boiling above a horizontal heating surface with net evaporation. The space above the water surface was occupied by vapor and kept at atmospheric pressure (corresponding to a vapor temperature of 100 C). It is evident from the figure that the bulk of the water is at a fairly uniform temperature but slightly superheated (by 0.4 C). Near the heating surface, however, the water temperature rises in a thin layer with a very steep gradient to a surface temperature of 108.3 C. In its details the temperature field depends on the magnitude of the heat flux, pressure, geometry, and boiling liquid. Qualitatively, however, the features discussed here appear in the same way in all boiling processes.

For an understanding of this temperature condition and of the heat transfer connected with boiling, it is necessary to consider in some detail the process of bubble formation in a liquid.[1] For this purpose some con-

[1] F. Bosnjakovic, *Tech. Mech. Thermodyn.*, **1**:358 (1930).

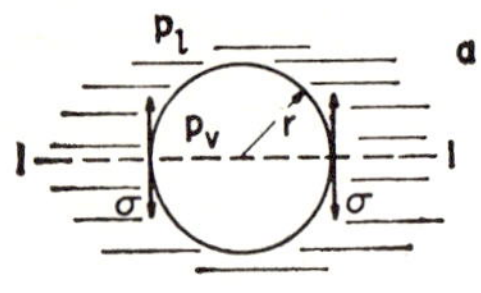

Fig. 13-6 Sketch of a vapor bubble in a saturated liquid.

sideration will first be given to a vapor bubble in thermal equilibrium within a liquid of uniform temperature. Under this condition, a bubble with a certain radius can be maintained, rather than causing it either to collapse by condensation of its vapor or to grow by evaporation from the liquid surface into its vapor space. In order to study the dynamic equilibrium of the bubble shown in Fig. 13-6, we imagine the bubble, with a radius r, to be cut into two halves and consider the equilibrium of one half. Forces which act on the surface within the plane 1-1 are the vapor pressure p_v acting from the inside of the bubble, the liquid pressure p_l from the outside of the bubble, and the surface tension within the surface of the bubble itself. If we denote this force per unit length of the surface by σ, then the force caused by surface tension is $2\pi r\sigma$. The vapor pressure and the liquid pressure generate a force in a direction normal to the plane 1-1 equal to the pressure difference times the cross-sectional area of the bubble. This gives the equation

$$\pi r^2(p_v - p_l) = 2\pi r\sigma$$

The pressure difference itself is

$$p_v - p_l = \frac{2\sigma}{r} \tag{13-14}$$

The thermal equilibrium requires that there be no temperature difference between the vapor in the bubble and the liquid surrounding it. The vapor temperature within the bubble must also be equal to the saturation temperature belonging to the pressure p_v.[1] The pressure in the liquid is smaller, and therefore the liquid must be superheated in order to allow a vapor bubble to be in equilibrium within it. The smaller the radius of the bubble, the greater the amount of superheat. If the equation were valid down to a radius equal to zero, this would mean that an infinitely large superheat is required to start a bubble within the liquid. Actually for very small bubble sizes considerations have to be made on the basis of

[1] Thermodynamics shows that the saturation pressure p_v of a vapor in equilibrium with a liquid at a certain temperature and separated by a plane interface is somewhat larger than the saturation pressure on a curved concave surface. This difference, however, is small compared with the pressure difference in Eq. (13-14) as long as the evaporation pressure is not close to the critical pressure of the boiling fluid.

the molecular structure rather than on a continuum basis, from which the above equation is derived. However, the equation makes it understandable that a large superheat is required to start the formation of new bubbles. The formation of new bubbles is easier when gases are contained in the form of small gas bubbles within the liquid. Such gases serve as nuclei for the bubble formation. The process of the formation of new bubbles is referred to as the *nucleation process.* A vapor bubble grows when the liquid temperature is larger than the saturation temperature corresponding to the pressure p_v. In this case, heat is conducted in the liquid toward the bubble surface, which is at saturation temperature. On the basis of a consideration of this conduction process, Forster and Zuber[1] were able to calculate this growth history of a bubble in good agreement with the results of measurements. Nucleation and growth of bubbles in a liquid occur in homogeneous nuclear reactors, where the heat of evaporation is supplied by the fission process of the fissionable atoms distributed in the liquid. The birth of new bubbles within the liquid is that part of the boiling process which cannot yet be predicted analytically.

If boiling occurs on a solid surface through which heat is transported into the liquid, the large superheat required for the growth of small bubbles is provided within the thin layer attached to the heating surface (Fig. 13-5). Places of especially large superheat will occur in grooves and pits within the surface, and it is very probable that the formation of new bubbles occurs at such places. This explains why in incipient boiling bubbles rise from a few spots on the heating surface only, the so-called *nucleation centers.* When the heat flux into the liquid gets larger, the superheat in the boundary layer adjacent to the heating surface also gets larger and conditions become conducive to bubble formation in smaller grooves or crevices: the number of nucleation centers increases. Experiments have shown the number of these centers to vary between 1 and 100 per square centimeter, depending on the heat-flux density.

It has also been established that a surface initially requires a smaller temperature difference to transmit a certain amount of heat into the boiling fluid and that the required temperature difference gradually increases with time. It is assumed that the gases which originally were adsorbed at the surface facilitate the bubble formation until they are gradually removed with the bubbles. The considerations discussed make it understandable that the microstructure of the surface as well as its condition as far as impurities or adsorbed gas films are concerned have a considerable influence on boiling. On the other hand, the shape of the surface should be of minor importance, since the thickness of the thermal bound-

[1] H. K. Forster and N. Zuber, *J. Appl. Phys.*, **25**:474 (1954).

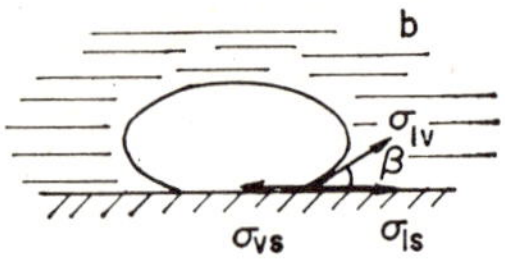

Fig. 13-7 Sketch of a vapor bubble on a surface.

ary layer on the surface is in most cases determined by the stirring action of the bubbles, a process concentrated in a comparatively small space.

Actually, the conditions for the dynamic equilibrium of a bubble attached to a solid surface are more complicated than for the bubble in the interior of the fluid. Figure 13-7 shows a sketch of a bubble attached to a surface. Under this condition three surface tensions have to be considered. One, σ_{lv}, is determined by the properties of the vapor and the liquid and is acting in the liquid-vapor interface. One, σ_{ls}, is determined by the properties of the solid surface and the liquid and is acting along the solid surface on the liquid side, and one, σ_{vs}, is determined by the properties of the vapor and the solid surface and is acting along the solid surface on the vapor side. The equilibrium in the horizontal direction requires the following relation among these surface tensions:

$$\sigma_{vs} - \sigma_{ls} = \sigma_{lv} \cos \beta \tag{13-15}$$

The angle β depends on the relative magnitude of the different surface tensions. It may be smaller or larger than 90°. In the first case the liquid is said to *wet* the surface. When the angle β is larger than 90°, the surface is considered to be *nonwetted.* It is frequently assumed that a thin vapor film exists between a liquid and a solid nonwetted surface even if no boiling occurs. It is understandable that the condition of surface wetting is of great importance in bubble formation. Excellent conditions with regard to wetting exist, for example, in boiling of benzene on a metal surface. The bubbles separate in the form of spheres with approximately 1 mm diameter. Water boiling on a clean metal surface generates bubbles of a somewhat elongated shape with a diameter of approximately 4 mm at 1 atm and 2 mm at 100 atm pressure at the time of separation. The angle β is 60 to 70°. Mercury does not wet steel surfaces. This causes the bubbles to grow larger (8 to 10 mm diameter) before they separate.

An interesting connection established between the bubble diameter d at separation and the frequency f of bubble formation is presented in Fig. 13-8. The frequency of bubble formation from a fixed nucleation center is very constant, but it varies somewhat from one center to another. The growth and separation of the bubbles from the heating surface is a dynamically violent process. Liquid is displaced from the surface as the bubble grows and rushes back when the bubble moves away. This produces a strong stirring action and makes it understandable that the thickness of the thermal bounday layer is little influenced by a bulk flow of the liquid over the surface. Inertia forces thus appear to dominate the bubble separation. Observations by Ellion[1] demonstrate this, showing that bubbles are propelled into the liquid with considerable velocity from the underside of a horizontal electrically heated ribbon in a direction opposite to that of the buoyancy force. For some time it was postulated that heat is transferred essentially in this process at first from the heating surface to the liquid and then from the liquid to the bubbles. Recent investigations, however, have demonstrated that a considerable fraction of the total heat flux from the surface to the bubble sometimes occurs through a very thin liquid layer between bubble and surface and through a dry spot within the bubble attachment area.[2]

Two definitions are found in the literature for a film heat-transfer coefficient in boiling. In one of them, the heat flux q at the heating surface is divided by the difference Δt between the surface temperature and the saturation temperature corresponding to the pressure in the liquid. The other definition uses the difference between the surface temperature and the fluid bulk temperature. This last definition agrees better with the one customarily used in convection without change of phase. The first one has an advantage in that the saturation temperature can be obtained more easily and accurately from a pressure measurement. It is

[1] M. E. Ellion, Ph.D. thesis, California Institute of Technology, 1953.

[2] R. W. Graham and R. C. Hendricks, *NASA Tech. Note* D-3943, 1967; R. R. Sharp, *NASA Tech. Note* D-1997 (1964).

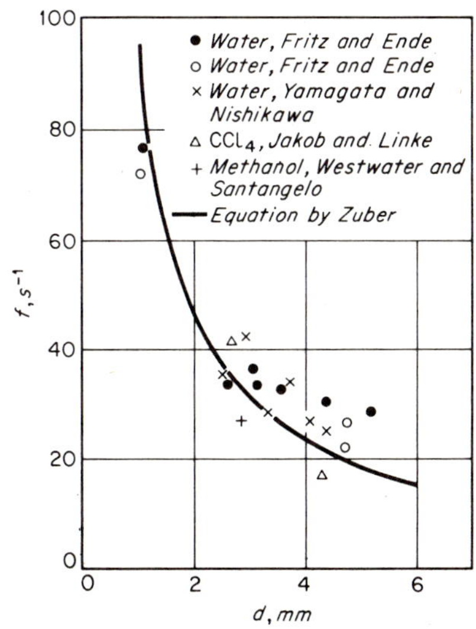

Fig. 13-8 Frequency of bubble formation as a function of bubble diameter at separation. [*From N. Zuber, U.S. Atomic Energy Comm. AECU-4439, 1959.*]

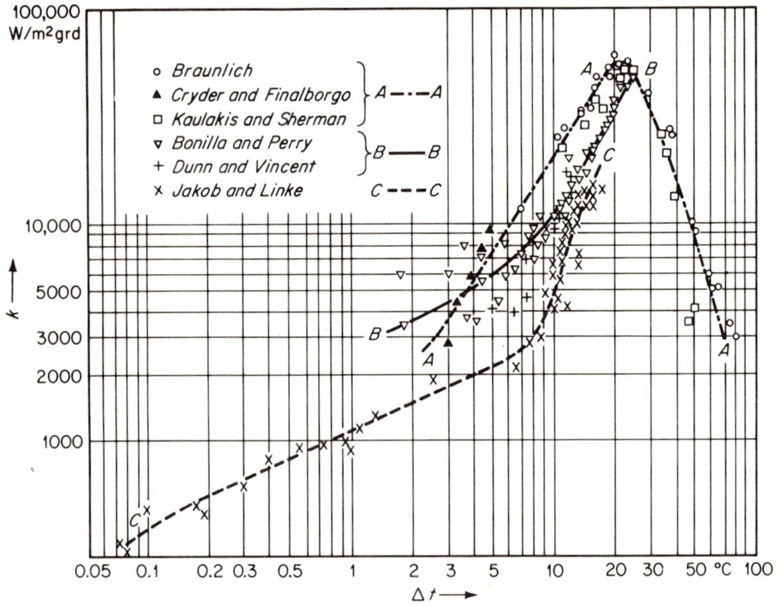

Fig. 13-9 Heat-transfer coefficient for water boiling on horizontal tubes (A) and on horizontal plates (B, C) at 1 atm. [*From W. H. McAdams, "Heat Transmission," 2d ed., McGraw-Hill Book Company. Used by permission. Copyright 1942.*]

therefore more frequently used. The difference between both heat-transfer coefficients, by the way, is small as becomes evident from Fig. 13-5. Figure 13-9 presents heat-transfer coefficients (according to the second definition) measured in pool boiling with net evaporation of water at atmospheric pressure. It can be observed that the heat-transfer coefficient increases at first at a moderate rate with the temperature difference Δt. At approximately $\Delta t = 5$ C, the rate of increase becomes considerably larger until at approximately $\Delta t = 20$ C a maximum is reached. Beyond this value of Δt, the heat-transfer coefficient drops off. Measurements which extended the temperature difference beyond the range presented in Fig. 13-9 showed that the heat-transfer coefficient reaches a minimum at a Δt of approximately 100 C and then starts rising again. Visual observation provided the explanation for this peculiar behavior.[1] It could be seen that in the region $0 < \Delta t < 5$ C bubbles occur only on a few selected spots of the heating surface and rise in widely separated columns. It was reasoned by Jakob that the small number of bubbles

[1] Described for the first time by S. Nukiyama, *J. Soc. Mech. Engr. Japan,* **37**:367 (1934).

can have only a minor influence on heat transfer and that the heat-transfer coefficient should be determined mainly by the free convection set up in the liquid by the temperature differences. Evaporation in this region is therefore called *free-convection boiling.* The number of spots on which bubbles start becomes larger with increasing Δt, and at a temperature difference beyond $\Delta t = 5$ C the heating surface is so densely populated with bubbles that their separation should cause a considerable stirring action in the fluid which increases the heat transfer. This is considered the reason for the steeper increase of the heat-transfer coefficient in the region 5 C $< \Delta t <$ 20 C, which is referred to as *nucleate boiling.* For the temperature region beyond 100 C, in which the heat-transfer coefficient rises again with increasing temperature difference, the evaporation rate has become so high that a continuous vapor blanket covers the surface and the vapor bubbles grow out of this blanket. Boiling in this region is called *film boiling.* The low conductivity of the vapor film through which the heat must be transported explains the lower values of the heat-transfer coefficient for this region. At the highest temperature differences the energy transport is increased by radiation. The fact that a vapor film covers the heating surface at high heat fluxes has been known for a considerable time. Every reader has probably made the observation that when a drop of water is sprayed onto a glowing surface, it does not touch the surface but floats on a vapor film and that the droplet takes a considerable time to evaporate. This is called the *Leidenfrost effect* after the man who reported and explained it in 1756.[1] The range 20 C $< \Delta t <$ 100 C, in which the heat-transfer coefficient decreases with increasing temperature difference, is a transition region in which part of the surface is covered by bubbles and part by a film. Evaporation in this region is referred to as *transition boiling* or *partial film boiling.* The difference between the various boiling regimes can be clearly observed in Fig. 13-10 which presents instantaneous photographs.

The heat flow q per unit surface area and unit time can be determined from Fig. 13-9 by multiplying the heat-transfer coefficient with the temperature difference Δt. When this is done, it is found that in the temperature range 20 C $< \Delta t <$ 100 C the heat flow q also drops with increasing Δt, as illustrated in Fig. 13-11. This has an important consequence in engineering applications, in many of which the heat flow q is the quantity controlled. Let us assume that the heat flux is adjusted to the value q_A in Fig. 13-11. The difference between the temperature of the heating surface and that of the liquid is Δt_A. Now when the heat flux is gradually increased, the temperature difference Δt rises steadily until the value q_B and with it Δt_B are reached. A slight increase of the heat flux

[1] J. G. Leidenfrost, "De aquae communis nonnullis qualitatibus tractatus," Duisburg, 1756.

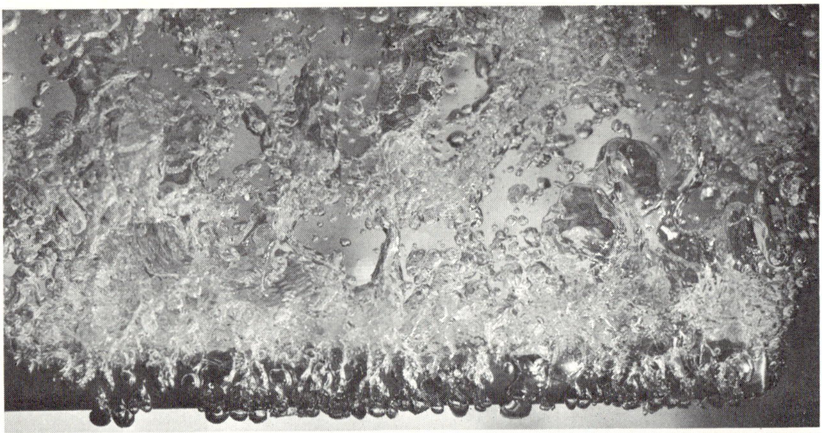

(a)

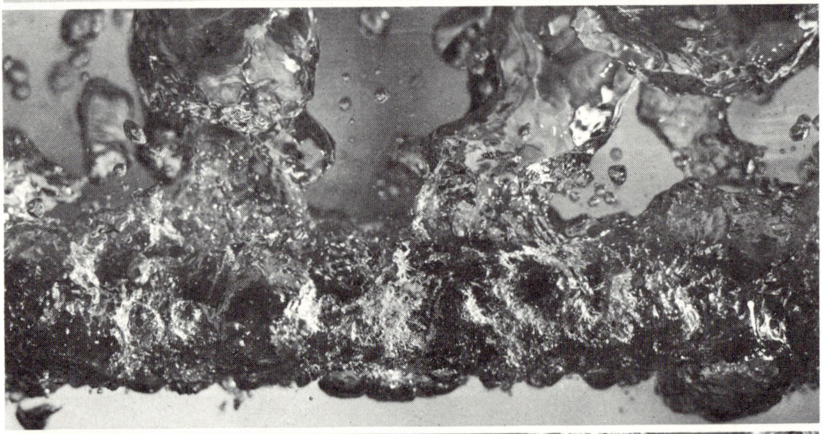

(b)

(c)

Fig. 13-10 Boiling of methanol on a horizontal tube with 9.5 mm diameter: (a) nucleate boiling, $\Delta t = 19$ C; (b) transition boiling, $\Delta t = 44$ C; (c) film boiling, $\Delta t = 82$ C. [*Photographs by J. W. Westwater.*]

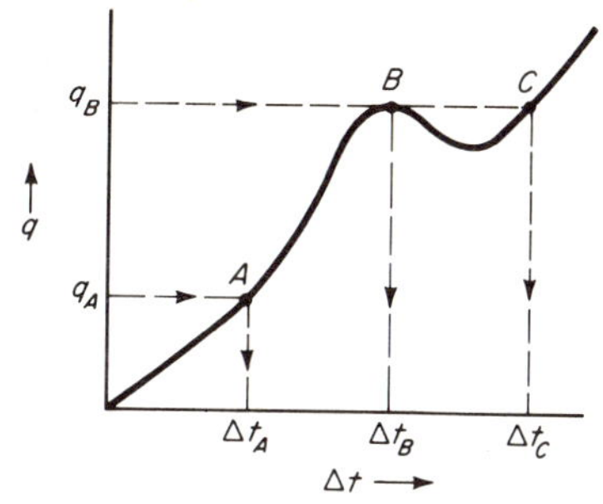

Fig. 13-11 Heat flux to a boiling liquid as function of the difference between surface temperature and saturation temperature.

beyond this value, however, makes the temperature rise until a new equilibrium condition slightly above point C is established. The corresponding temperature difference Δt_C is usually so high that the temperature of the heating surface is beyond the melting point of most metals, in which case the surface is destroyed. Point B is therefore referred to as the *burnout point* and the process as *boiling crisis*. The trend toward using high heat fluxes causes the engineer to design equipment so that it operates close to the burnout point, and the exact knowledge of this point therefore has great engineering importance.

For applications in which the heat flux q_w is prescribed, like nuclear reactors or devices which employ electric heating, it is more convenient

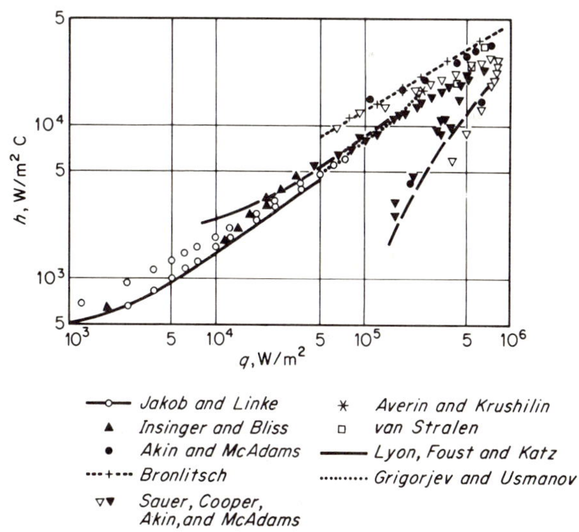

Fig. 13-12 Heat-transfer coefficient for water boiling on horizontal and vertical cylinders at 1.03 atm. [*From W. Fritz, Chem. Ingr.-Tech.*, **35**:*753* (*1963*).]

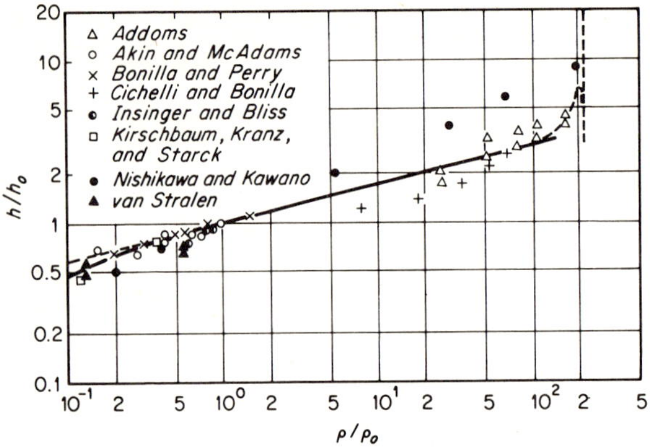

Fig. 13-13 Ratio of heat-transfer coefficient h at pressure p to heat-transfer coefficient h_0 at pressure $p_0 = 1.03$ atm. Heat-transfer coefficients are considered in function of heat flux. [*From W. Fritz, Chem. Ingr.-Tech.*, **35**:*753* (*1963*).]

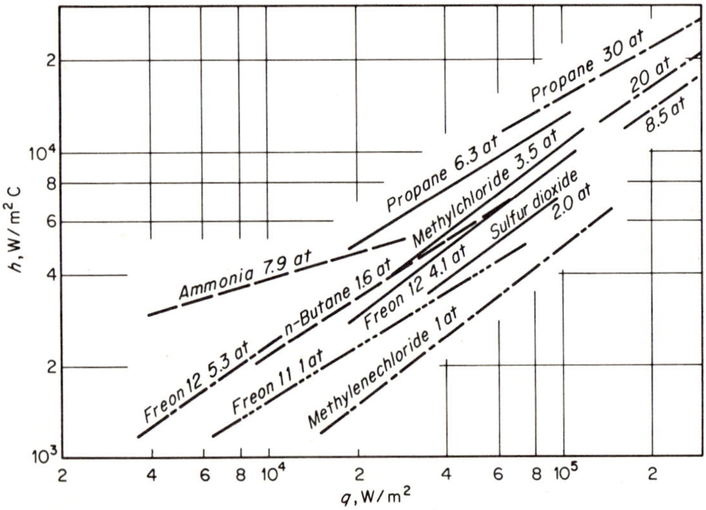

Fig. 13-14 Heat-transfer coefficient in boiling of refrigerants according to various investigators. [*From W. Fritz, Chem. Ingr.-Tech.*, **35**:*753* (*1963*).]

to have information on heat-transfer coefficients as function of the heat flux. Such a diagram is shown in Fig. 13-12, again for pool boiling of water at atmospheric pressure.[1]

With increasing pressure, the heat-transfer coefficients increase (probably because the bubbles separate earlier) and the frequency and with it the stirring action increase. Figure 13-13 demonstrates this effect.

Relations for heat-transfer coefficients in other liquids are reported in the literature and contained in various handbooks. Figure 13-14 presents such relations determined experimentally for refrigerants and Fig. 13-15 for liquid metals. The low values for mercury in this figure are due to the nonwetting character of this fluid. Small amounts of additives (K, Na, Mg) increase the heat-transfer coefficient of mercury considerably.

If the assumption is correct that heat transfer in the *free-convection*

[1] An excellent review paper by W. Fritz appears in *Chem. Ingr.-Tech.*, **35**:753 (1963).

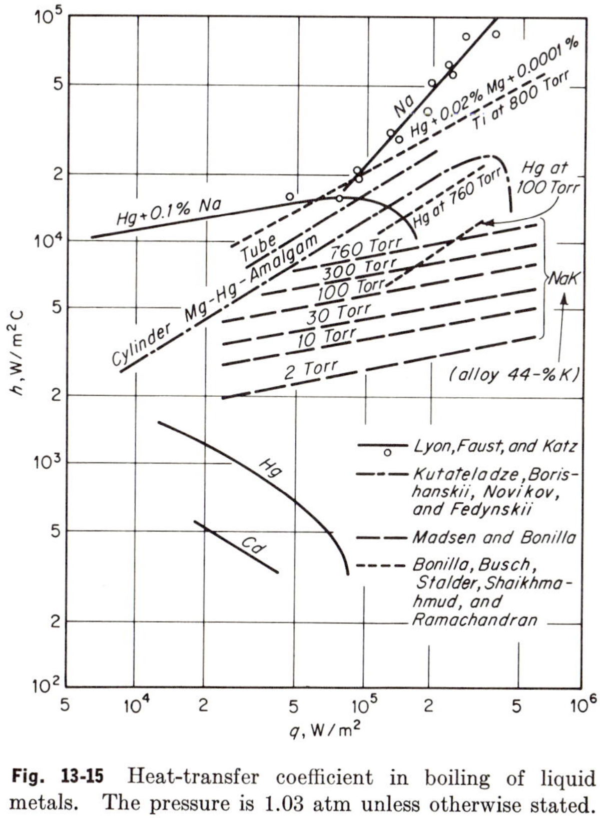

Fig. 13-15 Heat-transfer coefficient in boiling of liquid metals. The pressure is 1.03 atm unless otherwise stated. [*From W. Fritz, Chem. Ingr.-Tech.,* **35**:*753 (1963)*.]

boiling region is determined by the free-convection currents in the liquid, it must be possible to calculate the heat-transfer coefficient in this region by relations of the form

$$\mathrm{Nu} = f(\mathrm{Ra},\mathrm{Pr})_w \tag{13-16}$$

describing free-convection heat transfer. The shape of the function f will depend on the geometry of the heating surface. Jakob[1] has demonstrated that the known relations for free convection correlate measured heat-transfer coefficients for free-convection boiling satisfactorily.

In the *nucleate-boiling* region it has been assumed that the stirring action of the bubbles is the determining factor for the heat-transfer process. This implies that the nature of the heat transfer at the heating surface can be considered as one by forced convection in a liquid where the convection is supplied by the movement of the bubbles and that the heat-transfer coefficient should be described by a relation of the form $\mathrm{Nu} = f(\mathrm{Re},\mathrm{Pr})$. Attempts at a theoretical treatment of heat transfer in nucleate boiling have therefore concentrated on the effort to express the stirring action by the growth of the bubbles.[2] Semiempirical correlations have been presented by Rohsenow[2] and by Forster and Zuber.[2] Both start from the assumption that heat transfer from the heating surface to the stirred liquid should be presented by a correlation of the form $\mathrm{Nu} = f(\mathrm{Re},\mathrm{Pr})$. The length and velocity in these parameters are then related to bubble diameter, bubble velocity, and amount of vapor generated in the bubbles as a measure of bubble number. In this way Rohsenow obtained the following correlation:

$$\frac{c_l\,\Delta t}{i_{fg}} = a\left[\frac{q}{\mu_l i_{fg}}\sqrt{\frac{\sigma}{g(\rho_l - \rho_v)}}\right]^{1/3} (\mathrm{Pr}_l)^{1.7} \tag{13-17}$$

where Δt = difference between surface temperature and saturation temperature
i_{fg} = heat of evaporation
q = heat flux at heating surface
$c_l, \mu_l, \rho_l, \mathrm{Pr}_l$ = specific heat, viscosity, density, and Prandtl number of liquid at saturation temperature
ρ_v = vapor density at saturation temperature
σ = liquid-vapor surface tension at saturation temperature

The constant a depends on all the factors which influence nucleation. From an evaluation of available experimental results Rohsenow found values of a between 0.0027 and 0.013, and from their own experiments on heat transfer in forced-convection boiling with net evaporation Piret and

[1] M. Jakob, "Heat Transfer," Wiley, New York, 1949.

[2] W. M. Rohsenow, *Trans. ASME*, **74**:969 (1952); H. K. Forster and N. Zuber, *AIChE J.* **1**:531–535 (1955).

Isbin[1] determined values of the constant a between 0.0022 and 0.015. The wide range of this parameter indicates that considerable research is still required before a completely satisfactory correlation for nucleate boiling can be developed. The agreement between the constants a which have been measured in pool boiling and forced-convection boiling also indicates that in the nucleate-boiling range the stirring action of the bubbles is so great that the forced convection usually contributes little.

For heat transfer connected with *film boiling*, Bromley[2] developed a relation on the basis of a model which is essentially identical with Nusselt's film theory for condensation.[3] The vapor film adjacent to the heating surface is assumed to rise under the influence of buoyancy forces, and heat is transferred by conduction through this film. The resulting relation for a vertical wall is identical with Eq. (13-9) with the only difference that the density ρ in this equation must be replaced by the difference $\rho_l - \rho_v$ of the liquid and vapor density. Otherwise, the vapor properties have to be introduced into the equation. A relation for a horizontal tube analogous to Eq. (13-9) was found to agree with experimental results when a correction was applied to account for the fact that heat transfer by radiation through the vapor film increases the thickness of the vapor film and that the liquid exerts some frictional force on the moving vapor film.

It is of special importance to have a relation which predicts the peak nucleate heat flux occurring at the burnout point. Relying on previous work by Kutateladze[4] and Rohsenow and Griffith,[5] Zuber[6] was able to develop such an equation analytically by a stability consideration on the interface between the vapor film and the liquid under the condition that they move relative to each other. Considering the two fluids as inviscid and moving under the influence of gravity and surface tension, he obtained the following equation for the maximum heat flux q_B at burnout:

$$q_B = \frac{\pi}{24} i_{lv}\rho_v \left[\frac{\sigma g(\rho_l - \rho_v)}{\rho_v^2}\right]^{1/4} \left(\frac{\rho_l + \rho_v}{\rho_l}\right)^{1/2} \tag{13-18}$$

where i_{lv} = heat of evaporation
ρ_v = density of vapor
ρ_l = density of liquid
σ = vapor-liquid surface tension

[1] E. L. Piret and H. S. Isbin, *Chem. Eng. Progr.*, **50**:305 (1954).

[2] L. A. Bromley, *Chem. Eng. Progr.*, **46**:221–227 (1950).

[3] D. J. Jordan, Film and Transition Boiling, *Advan. Heat Transfer*, **5**:55 (1968).

[4] S. S. Kutateladze, *Izv. Akad. Nauk. SSSR, Otd. Tekhn. Nauk.* 529 (1951).

[5] W. M. Rohsenow and P. Griffith, *ASME and AIChE, Heat Transfer Symp. Louisville, Ky., March, 1955, Preprint 9.*

[6] N. Zuber, *Trans. ASME*, **80**:711–720 (1958).

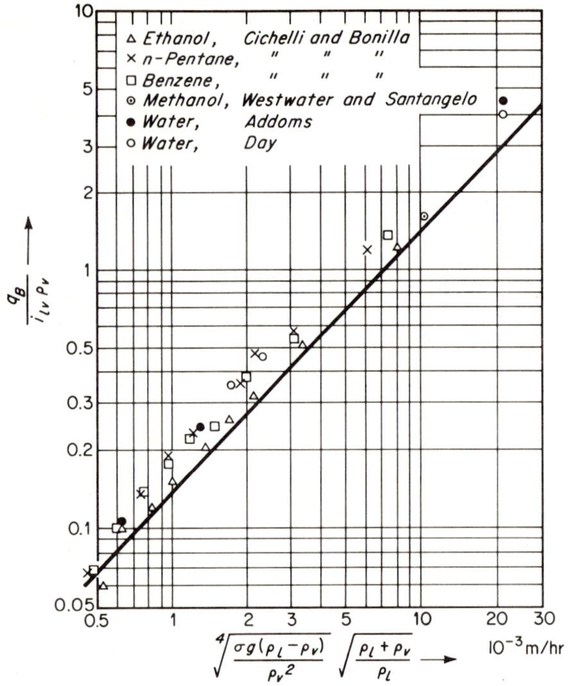

Fig. 13-16 Correlation for heat flux at burnout point and comparison with experimental results. [*From N. Zuber, ASME Paper 57-HT-4, 1957.*]

Figure 13-16 compares this equation in a dimensionless presentation with experimental results obtained by various investigations and with various fluids. The agreement, which can be considered as very satisfactory, appears to indicate that surface condition does not influence heat transfer at the burnout point. Zuber's relation is valid for pool boiling with net evaporation. Experiments have established that in local boiling the maximum heat flux is considerably higher (up to 4 times as large). Forced convection also increases the maximum heat flux.

The fact that heat-transfer coefficients from a surface into a boiling liquid are very large makes this process an effective one for keeping a wall cool. It is, for instance, used in the regenerative cooling of the walls of the combustion chambers in liquid-fuel rockets. In this application, one of the fuels is ducted around the wall of the combustion chamber. Usually the heat flow is so large that locally the saturation temperature of the fluid is exceeded and bubble formation occurs in the neighborhood of the wall. However, in the core of the coolant the temperature is lower than the evaporation temperature, and the bubbles collapse as soon as they separate from the surface and penetrate into the interior of the fluid. This heat-transfer process has to be classified as *forced-convection local boiling* and has been studied intensively in recent years. If, for instance, water is ducted through a pipe, and if the pipe is heated from the outside, the heat flow per unit area of the pipe wall follows the law indicated in Fig. 13-17. It increases at first with increasing difference between wall temperature and bulk temperature at the same rate as for forced flow without evaporation (*Colburn line*). As soon as bubble formation at the surface occurs, however, the increase in heat flow becomes very rapid and the temperature difference required is almost independent of the amount of heat flow. At the same time, it is found that various curves in the figure are obtained, depending on the magnitude of the saturation temperature. In a plot of heat flux q_w vs. the temperature difference $t_w - t_s$, on the other hand, these curves collapse nearly into a single one which gives heat-transfer coefficients very close to those measured in pool boil-

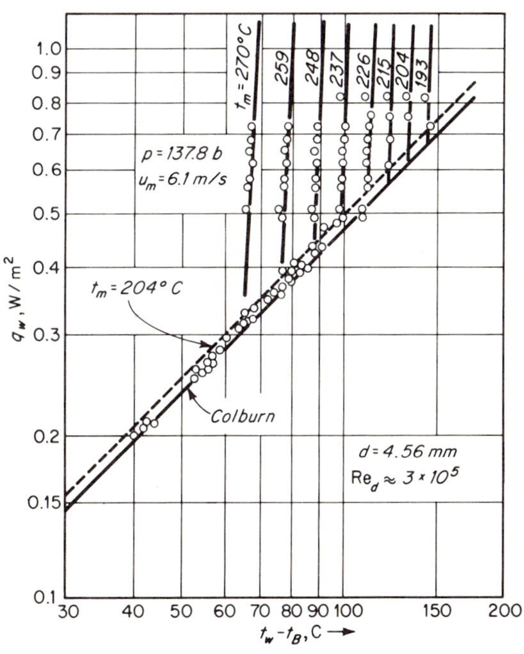

Fig. 13-17 Local boiling heat transfer for turbulent forced flow through a tube (d = 0.457 cm, Re = 3×10^5). [*From W. M. Rohsenow, Heat transfer with Evaporation, in "Heat Transfer: A symposium," University of Michigan Press, Ann Arbor, 1952.*]

ing, again demonstrating the dominant influence of the stirring action of the bubbles. If the abscissa in Fig. 13-17 is enlarged, it is found that the transition from forced convection without bubble formation to local boiling occurs less abruptly than appears in this figure but through a transition curve similar to that in Fig. 13-9 for the transition from free convection to nucleate boiling.

Special conditions arise when boiling occurs inside long tubes and when an appreciable portion of the liquid is converted to vapor. This occurs, for instance, in steam boilers and evaporators. The mixture of liquid and vapor is forced through the tubes by forced- or natural-convection flow. Pressure drop and heat transfer follow an involved and complicated pattern. This is an area which is closely related to two-phase flow, and for a more detailed discussion the reader is referred to texts on that subject.

part III

Thermal Radiation

III-1 INTRODUCTION

All substances, solid bodies as well as liquids and gases, at normal and especially at elevated temperatures emit energy in the form of radiation and are also capable of absorbing such energy. All heat-transfer processes are therefore more or less accompanied by a heat exchange by radiation. In some cases, radiative heat exchange is so small that it can be neglected or it is of such a nature that it is simply included in the conductive heat transfer. This is generally the case in solids and liquids. In gases, however, the radiative transport is so different in nature from the conductive and convective heat transport that it has to be considered separately. Heat exchange by radiation is the subject of Part III.

Calculations of heat transfer by conduction or convection in Parts I and II always started with an energy balance on a small volume element, and it turned out that it was sufficient for such a balance to know the conditions in the immediate neighborhood of the element. These conditions could be expressed by gradients of the state parameters involved, especially of the temperature, and as a consequence the energy balance could be written down as a differential equation. In radiative heat exchange the situation is usually more complicated. Let us consider, for example, that we want to study radiative heat exchange within the combustion chamber of a jet-engine combustor or a rocket. For this purpose, an energy balance on a small arbitrarily located volume element within the combustor has to be made. When the energy gain of the gas contained in this volume element is considered, the gas is found to absorb radiative energy which has been emitted from other gas elements or from elements of the solid walls even when these are at a considerable distance. Therefore, the energy balance depends not only on the conditions in the immediate neighborhood of the volume element under consideration, and the equation describing this balance is an integrodifferential equation. Such an equation, however, is quite difficult to solve, and practically all engineering calculations are based on simplifying assumptions. A second factor which makes radiative heat exchange more complicated than conductive or even convective heat transfer is connected with the fact that the radiative properties of the various substances encountered in engineering are more difficult to describe than, say, the heat conductivity, knowledge of which is required for calculation of a heat-conduction process.

Therefore, radiative properties must be discussed (Chaps. 15 and 16) before heat transfer by radiation, of main interest to engineers, can be treated. The basic concepts and relations for thermal radiation are covered in Chap. 14.

It is of interest to note that conditions existing in a rarefied gas, as described in Part II, Sec. 11-3, are quite similar to those for the radiative energy-exchange process. Heat conduction in a gas can be described by Fourier's differential equation as long as the mean free path of molecules is very small compared to the dimensions of the system involved. This is the case at normal pressures. With decreasing pressure, however, the mean free path increases, and there is a range of conditions under which the mean free path is of the same order of magnitude as the characteristic dimensions of the system. In this case, information at a certain location is obtained not only from the immediate neighborhood but also from regions farther removed from the point under consideration. As a consequence, one must again resort to integral equations or integrodifferential equations to set up a local energy balance. At the same time conditions encountered by the molecules as they impinge on solid walls enclosing the gas and as they are reflected from them become important. These conditions are described by accommodation coefficients, which by their nature are quite similar to reflectances used to describe the reflection characteristics of radiant beams impinging on the wall. Conditions again become simpler when the pressure is so small that collisions between molecules become rare compared with collisions of the molecules with the containing walls. This is the case when the mean-free-path length is large compared to the dimensions of the system. In rarefied-gas heat transfer, the various regimes which have just been described are characterized by a dimensionless parameter called the *Knudsen number*

$$\mathrm{Kn} = \frac{\Lambda}{L}$$

with Λ indicating the mean-free-path length and L indicating a characteristic dimension of the system. Fourier's differential equation can be used to describe a heat-conduction process in such a gas when Λ/L is very small compared to 1. Collisions between molecules, on the other hand, can be neglected when Λ/L is large compared to 1. The first regime is called *continuum regime,* the second one a *free-molecule regime.* In between is the *transition regime,* which is the most difficult to analyze. We shall find that similar regimes also exist in radiative energy exchange, where they are characterized by the value of the ratio of the mean-free-path length of the photons to a characteristic dimension of the system. The analogy between heat conduction in a rarefied gas and energy transport by radiation makes it possible to utilize analytical approaches which have been developed in one field for the other one as well. Under certain conditions, relations developed for one field can immediately be used in the other with proper reinterpretation of the symbols.

The radiative energy emitted from a substance increases very strongly with temperature. As a consequence, radiative heat exchange becomes large and finally dominating at high temperatures. Many processes in new engineering areas occur at high temperatures, and a knowledge of radiative heat transfer becomes very important for the design of the pertinent equipment. Nuclear power plants, gas turbines, and the various propulsion devices for aircraft, missiles, satellites, and space vehicles are examples. A fusion power plant with temperatures of millions of degrees will pose very severe cooling problems caused by radiative heat transfer.

14

Basic Concepts and Relations

Heat transfer by radiation can be thought of as transport of energy by photons (packages of energy) being released from excited molecules or atoms and traveling on straight paths until they are absorbed or scattered by some other atoms, or, alternatively, it can be visualized as energy transport in the form of electromagnetic waves. Parameters describing radiation are the photon or wave velocity c, the wavelength λ, and the frequency ν. In the concept of radiation as traveling photons, the energy E of a photon is connected with the frequency ν by Planck's relation $E = h\nu$, where h is Planck's constant ($h = 6.6256 \times 10^{-34}$ J s). Frequency and wavelength, interconnected by the equation $\lambda\nu = c$, are used interchangeably. Experimentalists prefer the wavelength because it can be measured more readily. In analytical work, the frequency is preferred because it does not change when a radiant beam travels from one medium into another, while wavelength and velocity generally undergo a sudden change at the interface. In spectroscopy, the wave number, defined as $1/\lambda$, is frequently used.

The velocity of thermal radiation in a vacuum is independent of frequency and has the value $c_0 = 2.997925 \times 10^8$ m/s. All media occurring in nature have wave velocities which are smaller and depend to some extent on frequency. These velocities c are connected with the velocity c_0 in vacuum by the equation $n_0 = c_0/c$, with n_0 indicating the refractive index. Air and gases generally have a refractive index which differs only in the fourth decimal from 1. Solids like glass or quartz and liquids like water have considerably larger refractive indices (of order 1.5). For these materials, the variation of n_0 with wavelength is also larger (for quartz, as an example, from 1.45 to 1.6). The refractive index is connected through Snell's law with the change in direction of a radiant beam traveling from vacuum through a smooth interface into the medium under consideration. It is $n_0 = (\sin \beta_0)/(\sin \beta)$, where β_0 indicates the angle between the approaching beam and the surface normal and β the angle between the refracted beam and the surface normal.

When a radiant beam travels from medium 1 into medium 2, the angles of the incident beam β_1, and of the refracted beam β_2 are connected by the equation

$$\frac{\sin \beta_2}{\sin \beta_1} = \frac{n_{10}}{n_{20}} = n_{12}$$

with n_{10} and n_{20} indicating the refractive indices of the two media toward a vacuum and n_{12} denoting the refractive index between the two media.

Electromagnetic waves are transverse waves. The state of polarization must be stated to describe a radiant beam completely. The laws of polarization will be discussed when their consideration for radiative heat transfer becomes important. The various electromagnetic waves differ by their wavelength λ in a wide range. The scale shown in Fig. 14-1 indicates how the waves are classified according to wavelength, measured in this figure in various scales, with 1 μm $= 10^{-6}$ m being a convenient scale for heat radiation. Molecular dimensions (diameter of a molecule)

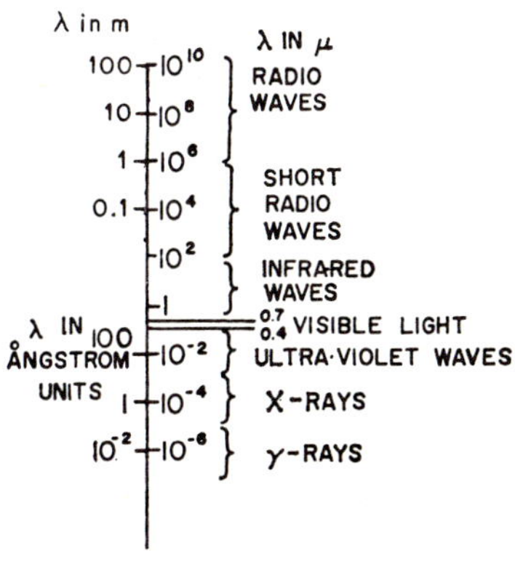

Fig. 14-1 Wavelengths of electromagnetic waves.

have numbers of order 1 when expressed in angstroms; 1 Å is equal to 10^{-10} m. The human eye is sensitive to rays in the range from 0.4 to 0.7 μm. Radiation emitted from heated bodies in many technological devices comprises rays with wavelengths 0.3 to 10 μm and more.

It can be observed that a large part of thermal radiation has wavelengths outside the visible range. The range with wavelengths larger than those of visible light is called *infrared;* that part with wavelengths smaller than those of light is referred to as *ultraviolet.*

14-1 RADIATION FIELD

The description of a radiation field developed in this section will be used in the following sections. It is based on the assumption that all characteristic dimensions of the system under consideration are large compared with the wavelengths of the radiation. This is the situation in almost all engineering applications with the possible exception of cryogenics, in which the wavelengths of the radiation are large as a consequence of the low temperatures.

A characteristic feature of the radiation process is that the energy transport occurs along straight lines as long as scattering or refraction is excluded. Consequently the following way of describing a radiation field is convenient. We focus attention at a point 0 arbitrarily located in the space under consideration and an area element dA arbitrarily located at this point (Fig. 14-2). We want to define the radiant energy flux per unit time through the area element dA in a direction characterized by the angle β toward the surface normal and by the azimuth angle γ. For this purpose we consider parallel pencils of radiation which have their apex in the area dA, all having a solid angle $d\omega$. One of these pencils with its apex at point 0 is indicated in Fig. 14-2. The radiant flux through the area element dA and contained within these pencils with solid angle $d\omega$ and within a frequency range $d\nu$ can then be expressed by

$$d\Phi = K_\nu \cos\beta \, dA \, d\omega \, d\nu \tag{14-1}$$

The parameter K_ν in this equation, which characterizes the intensity of the radiant flux at the selected point 0 and in the selected direction, is

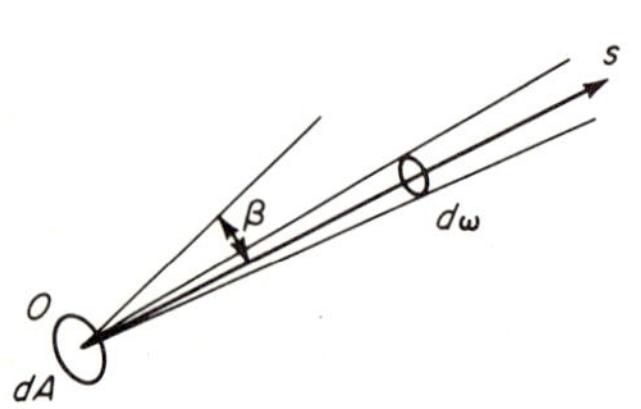

Fig. 14-2 Definition of intensity of radiation.

called the *monochromatic intensity of the radiant flux.* Using alternatively the wavelength to characterize the radiation, one can also write

$$d\Phi = K_\lambda \cos \beta \, dA \, d\omega \, d\lambda \tag{14-2}$$

The radiation field is described when the monochromatic flux intensities K_ν or K_λ are known for all points in the field in all directions and for all frequencies or wavelengths.

In a nonemitting and nonabsorbing medium, the intensity K_ν will be constant along each ray. K_ν increases or decreases along the radiant beam when the medium emits or absorbs radiation. In this case the radiant flux increases by the emission process in the specified direction, and the increase $d^2\Phi_e$ can be expressed by the equation

$$d^2\Phi_e = j_\nu \rho \, dV \, d\omega \, d\nu \tag{14-3}$$

in which j_ν is the *coefficient of emission,* or the energy emitted per unit time and per unit mass into a unit solid angle and within a unit frequency range. The notation $d^2\Phi_e$ indicates that the flux in Eq. (14-3) is by an order of magnitude smaller than the flux in Eq. (14-2). In the absorption process a certain fraction of the flux $d\Phi$ is absorbed along the path length ds, and this process is described by the equation

$$d^2\Phi_a = \varkappa_\nu \rho \, ds \, d\Phi \tag{14-4}$$

with $\varkappa_\nu$ indicating the *coefficient of absorption.* Some media scatter radiation. The intensity of the radiant beam is decreased by this process according to the equation

$$d^2\Phi_s = -\sigma_\nu \rho \, ds \, d\Phi \tag{14-5}$$

with σ_ν denoting the *coefficient of scattering.*

14-2 KIRCHHOFF'S LAWS

Kirchhoff was able to derive important relations between the radiation parameters discussed in the previous sections and others to be discussed by applying the laws of thermodynamics to systems of media in thermodynamic equilibrium. In discussing these laws, we shall essentially follow Planck's derivation.[1]

Consider, first, an isolated system with adiabatic walls very large in extent which contains a homogeneous and isotropic medium and is in thermodynamic equilibrium. Under this condition, the intensity $K_{\nu e}$ of the radiation fluxes crisscrossing the interior of the system must be uniform with regard to direction and location if we restrict ourselves to

[1] M. Planck, "The Theory of Heat Radiation," Dover, New York, 1959 (English translation of "Vorlesungen über die Theorie der Wärmestrahlung," Leipzig, 1923).

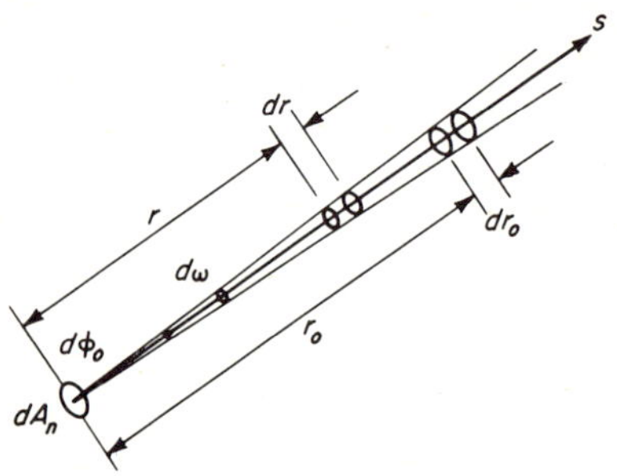

Fig. 14-3 Emission, absorption, and scattering of radiation.

regions far removed from the borders of the system. In order to derive a relation between this parameter and the coefficients of emission, scattering, and absorption, we study the radiative flux traveling through an area element dA_n within a pencil of solid angle $d\omega$, as indicated in Fig. 14-3. For simplicity, the surface normal of dA_n may coincide with the direction of the pencil of radiation. The monochromatic radiative flux $d\Phi_0$ traveling through dA_n in the direction indicated by the pencil is

$$d\Phi_0 = K_{\nu e}\, dA_n\, d\omega\, d\nu \tag{14-6}$$

A flux of equal magnitude travels through dA_n in the opposite direction, according to our statement above. We now trace the origin of this flux.

For this purpose, consider first the increase of the radiative flux traveling toward dA_n in the volume element $dV_0 = r_0^2\, d\omega\, dr_0$ at a distance r_0 from dA_n. The flux will increase by emission of radiation within the volume element and by the fact that radiation originally traveling in other directions will be scattered into the direction of the pencil. The emitted flux can be written according to Eq. (14-3)

$$d^2\Phi_{e,r_0} = j_\nu \rho r_0^2\, d\omega\, dr_0 \frac{dA_n}{r_0^2}\, d\nu$$

considering that the solid angle under which dA_n is seen from dV_0 is dA_n/r_0^2.

In order to determine the flux scattered into the pencil, one has to resort to a principle postulated by Helmholtz and well established since that time. Consider tracing a beam of radiation as it travels through a medium and is modified by absorption or scattering. Helmholtz' principle then states that the fractional decrease of the energy carried by a beam traveling in the opposite direction has exactly the same magnitude as the fractional decrease of the energy in the original beam. The principle can also be stated so that the coefficient of scattering and absorption has the same value whether radiant energy travels on a beam in a positive or a negative direction. In the situation considered here, the intensity of radiation is also independent of location and direction. Accordingly, the energy flux scattered into the pencil of radiation has the same value as the energy traveling within the beam in the opposite direction and being scattered out of the pencil. The energy scattered into the pencil is therefore described by Eq. (14-5) and, with the nomenclature indicated in Fig. 14-3, can be written as

$$d^2\Phi_{s,r_0} = \sigma_\nu \rho \, dr_0 \, K_{\nu e} \, dA_n \, d\omega \, d\nu$$

The total increase of energy by emission and scattering within the volume element is therefore

$$d^2\Phi_{r_0} = (j_\nu + \sigma_\nu K_{\nu e})\rho \, dA_n \, dr_0 \, d\omega \, d\nu$$

Not all this energy will arrive at the area element dA_n because part of it is either absorbed or scattered out of the pencil on its way. In order to calculate this decrease in energy, we shall consider the process occurring along a distance dr at the distance r from the area element. The decrease $d^3\Phi_r$ of the energy flux along the path length dr by absorption and scattering is

$$d^3\Phi_r = -\rho(\varkappa_\nu + \sigma_\nu) \, d^2\Phi_r \, dr$$

with $d^2\Phi_r$ indicating the flux that arrives at the location r. The above relation follows readily from Eqs. (14-4) and (14-5). Integration of this equation, with the condition that the energy flux is equal to $d^2\Phi_{r_0}$ at the distance r_0, results in

$$d^2\Phi_r = d^2\Phi_{r_0} e^{-\rho(\varkappa_\nu+\sigma_\nu)(r_0-r)}$$

The flux entering the beam in the volume element dV_0 and arriving at dA_n is therefore

$$d^2\Phi_0 = d^2\Phi_{r_0} e^{-\rho(\varkappa_\nu+\sigma_\nu)r_0}$$

An additional integration then gives the total flux $d\Phi_0$ which arrives within the pencil at dA_n:

$$\begin{aligned} d\Phi_0 &= \int_0^\infty d^2\Phi_0 = (j_\nu + \sigma_\nu K_{\nu e})\rho \, dA_n \, d\omega \, d\nu \int_0^\infty e^{-\rho(\varkappa_\nu+\sigma_\nu)r_0} \, dr_0 \\ &= \frac{j_\nu + \sigma_\nu K_{\nu e}}{\varkappa_\nu + \sigma_\nu} \, dA_n \, d\omega \, d\nu \end{aligned} \qquad (14\text{-}7)$$

For thermodynamic equilibrium this flux must be equal to the flux in the opposite direction given by Eq. (14-6). This results in the relation

$$K_{\nu e} = \frac{j_\nu + \sigma_\nu K_{\nu e}}{\varkappa_\nu + \sigma_\nu}$$

or

$$K_{\nu e} = \frac{j_\nu}{\varkappa_\nu} \qquad (14\text{-}8)$$

This relation was derived for a system which, as a whole, is in thermodynamic equilibrium. In such a system no net energy fluxes exist. As a rule engineering systems are not in thermodynamic equilibrium, and our task is to calculate the heat-transfer processes which are occurring. Equation (14-8) can be used for such a purpose when local thermodynamic equilibrium exists. This concept was introduced in our analysis of heat-convection processes and made it possible to use thermodynamic and transport properties like density, viscosity, and thermal conductivity in the conservation equations. For radiation processes, the concept of local thermodynamic equilibrium postulates that emission and absorption are determined by processes confined to a small volume element—small compared to the overall dimensions of the system under consideration. The emission and absorption are caused by molecular and atomic interactions and by interactions of these particles with photons. In a strict sense, the concept of local thermodynamic equilibrium then requires that the mean path lengths of molecules and atoms, as well as of photons, be of the order of magnitude of the dimension of this volume element. The first part of this condition is fulfilled when we can consider the medium as a continuum. A small photon beam length means that the local intensity of the radiant flux has the value $K_{\nu e}$. This condition of small photon beam length, however, can be relaxed when the emission process is primarily determined by molecular or atomic interactions and only to a negligible extent by photon absorption. This situation is still referred to as local thermodynamic equilibrium. Radiation under this condition is referred to as *thermal radiation*. Local thermodynamic equilibrium is closely approached in many engineering systems, and the discussion in the present text will be restricted to thermal radiation. For this situation, then, the ratio of the local coefficient of emission to the local coefficient of absorption is at any location equal to the intensity $K_{\nu e}$ which would exist if the whole system were in thermodynamic equilibrium with the parameters at the specific location.

We consider now conditions near an adiabatic wall (impermeable to radiation). A pencil of radiation with intensity $K_{\nu e}$ will maintain its magnitude until it intersects the wall. Since no net energy transfer can take place in a system in equilibrium, the flux of energy within the pencil in the opposite direction must have the same intensity, and it can be concluded that in a system in equilibrium the intensity is constant right up to the enclosing wall. We can now imagine successively putting other walls into our system with the resulting conclusion that the intensity of the radiative flux in a homogeneous and isotropic material is independent of the direction and location regardless of the size of the system as long as it is in thermodynamic equilibrium.

As a next step, we now consider two media, 1 and 2, separated by a smooth plane interface, as sketched in Fig. 14-4. We focus our attention on a beam which travels toward the interface within a pencil of opening

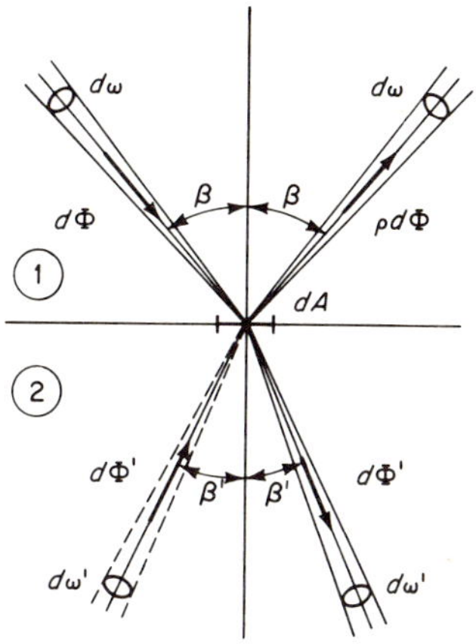

Fig. 14-4 Derivation of Kirchhoff's law for two adjoining media.

angle $d\omega$ inclined at an angle β toward the surface normal. The radiative flux arriving in this way at the area element dA is

$$d\Phi = K_{\nu e}\, dA \cos \beta \, d\omega \, d\nu$$

again assuming the system to be in thermodynamic equilibrium. Part of this flux is reflected within a pencil of the same solid angle $d\omega$ and at the same angle β because the specified interface reflects specularly. The magnitude of the reflected energy flux can be written as

$$\rho_\nu \, d\Phi$$

with ρ being called *reflectance* or *reflectivity*. The rest of the impinging flux penetrates the interface and continues in the second medium within a pencil of solid angle $d\omega'$ inclined at an angle β' toward the surface normal. According to Snell's law, the relation

$$\frac{\sin \beta'}{\sin \beta} = \frac{c'}{c} \tag{14-9}$$

exists. The total radiative flux within the pencil of opening angle $d\omega'$ is denoted $d\Phi'$. It will be composed of the flux $(1 - \rho_\nu)\, d\Phi$ and of the flux traveling within the dashed pencil toward the surface element dA and being reflected there. The flux traveling toward the surface element in medium 2 is also $d\Phi'$ because at equilibrium the intensity of radiation must be independent of location and direction in medium 2 and because the solid angles of the impinging and reflecting fluxes and the angles toward the surface normal are equal. From this we obtain

$$\rho_\nu' \, d\Phi' + (1 - \rho_\nu)\, d\Phi = d\Phi'$$

It has been found that the reflectances ρ_ν and ρ'_ν are of equal magnitude; therefore the equation simplifies to $d\Phi' = d\Phi$ with

$$d\Phi' = K'_{\nu e}\, dA \cos \beta' \, d\omega' \, d\nu$$

Setting the two fluxes $d\Phi$ and $d\Phi'$ equal to each other results in

$$K_{\nu e} = K'_{\nu e} \frac{\cos \beta' \, d\omega'}{\cos \beta \, d\omega} \tag{14-10}$$

The solid angles in a spherical coordinate system are

$$d\omega = \sin \beta \, d\beta \, d\varphi \qquad d\omega' = \sin \beta' \, d\beta' \, d\varphi$$

with $d\varphi$ indicating the azimuth angle. We therefore obtain for the ratio in the above equation

$$\frac{\cos \beta' \, d\omega'}{\cos \beta \, d\omega} = \frac{\sin \beta' \cos \beta' \, d\beta'}{\sin \beta \cos \beta \, d\beta}$$

The velocities c and c' of the radiation in the two media are independent of direction. A differentiation of Snell's law (14-9) therefore results in the relation

$$\frac{\cos \beta' \, d\beta'}{\cos \beta \, d\beta} = \frac{c'}{c}$$

Introducing this equation and Snell's law into Eq. (14-10) then gives

$$K_{\nu e} = K'_{\nu e} \frac{c'^2}{c^2}$$

Using the relation (14-8) finally gives Kirchhoff's law in the following form:

$$\frac{j_\nu}{\varkappa_\nu} c^2 = \frac{j'_\nu}{\varkappa'_\nu} c'^2 \tag{14-11}$$

This equation can be generalized to the statement that the ratio $(j_\nu/\varkappa_\nu)c^2$ is a universal function dependent on frequency and the local thermodynamic state but independent of the medium in which the radiation occurs. This is therefore a very useful relation between the various parameters describing thermal radiation. It should be remembered that it is not restricted to a situation of thermodynamic equilibrium but holds generally for thermal radiation.

Kirchhoff's law can also be expressed in a different set of parameters; one, describing what happens to a radiant beam at an interface, has already been introduced. It is the reflectance ρ_ν. In order to introduce the other parameters, we again consider two media, 1 and 2, within the closed system in thermodynamic equilibrium and a pencil of radiation with an infinitely small opening angle inclined in medium 1 at an angle β

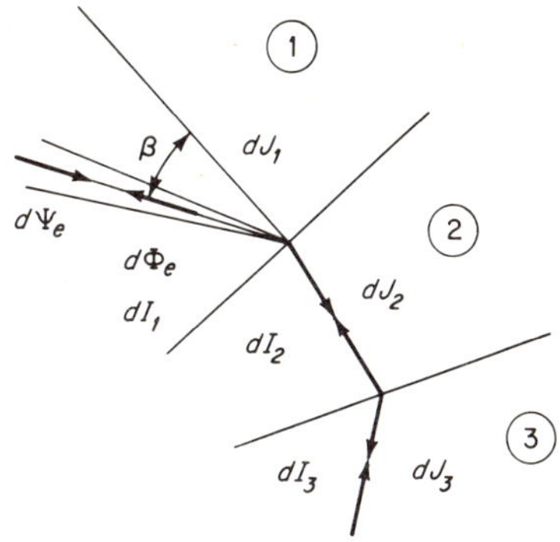

Fig. 14-5 Derivation of Kirchhoff's law.

toward the surface normal (Fig. 14-5). The radiant flux leaving the interface within this pencil is denoted $d\Phi_e$. This flux must have been generated in one of the n media of which the system may exist. Indicating the flux generated by emission in the individual media by dI_i, we can therefore write

$$d\Phi_e = dI_1 + dI_2 + dI_3 + \cdots + dI_n$$

Consider now the radiant flux $d\psi_e$ approaching the interface within the same pencil and trace the history of this flux as it is gradually spent by absorption in the various media, indicating the absorption in any one of these media by dJ_i. We can write

$$d\psi_e = dJ_1 + dJ_2 + dJ_3 + \cdots + dJ_n$$

From the fact that the system is in equilibrium, it follows that

$$d\Phi_e = d\psi_e$$

The fact that the system is in equilibrium and Helmholtz' principle require that

$$dI_1 = dJ_1,\ dI_2 = dJ_2,\ \ldots,\ dI_n = dJ_n$$

The absorptance $\alpha_{\nu 2}$ of medium 2 for the flux entering from medium 1 in the chosen direction is defined as

$$\alpha_{\nu 2} = \frac{dJ_2}{d\psi_e}$$

and, according to the statements above, it can also be written as

$$\alpha_{\nu 2} = \frac{dI_2}{d\Phi_e} \tag{14-12}$$

The flux arriving at the interface in medium 1 can be expressed by

$$d\Phi_e = K_{\nu e}\, dA \cos\beta\, d\omega\, d\nu$$

In an analogous manner, an intensity $i_{\nu e2}$ is defined for the flux being emitted from medium 2 into medium 1 within the pencil under consideration by

$$dI_2 = i_{\nu e2}\, dA \cos \beta \, d\omega \, d\nu$$

The above equation then reads

$$\alpha_{\nu 2} = \frac{i_{\nu e2}}{K_{\nu e}}$$

or

$$\frac{i_{\nu e2}}{\alpha_{\nu 2}} = K_{\nu e}$$

A universal relation independent of the nature of medium 1 can again be obtained by multiplying both sides of this equation by the square of the velocity c_1 in medium 1

$$\frac{i_{\nu e2}}{\alpha_{\nu 2}} c_1{}^2 = K_{\nu e} c_1{}^2 \tag{14-13}$$

The right-hand side of this equation is a universal relation according to Eq. (14-11). This, then, is the other form of Kirchhoff's law, stating that the intensity of radiation emitted into a medium from a neighboring one divided by the absorptance of the neighboring medium and multiplied by the square of the velocity c in the first medium is independent of the nature of both media and dependent only on frequency and the local thermodynamic state.

Of special interest is the intensity of radiation emitted by a medium with an absorptance $\alpha_\nu = 1$. Such a medium is called a *blackbody*. $\alpha_\nu = 1$ implies, according to Eq. (14-12), that no part of the radiation impinging on that medium is reflected or transmitted.

The intensity of emission of a blackbody may be denoted $i_{\nu b}$. From Eq. (14-13) it follows that

$$i_{\nu b} = K_{\nu e} \tag{14-14}$$

The same equation indicates that $i_{\nu b}c^2$ is again independent of the medium into which radiation is emitted. This of course implies, on the other hand, that the intensity of radiation emitted by a blackbody depends on the medium into which the radiation is emitted.

The value of the first form of Kirchhoff's law expressed by Eq. (14-11) is based on the fact that the parameters in this relation maintain their values for nonequilibrium situations as long as local equilibrium exists. This obviously is not the case in general for the parameters in the second form of Kirchhoff's law expressed by Eq. (14-13), because the intensity of the radiation generated by emission in medium 2 or generally in any medium will depend on the temperature distribution in that medium. There is, however, a large group of materials in which the energy emitted by a medium through its surface is generated in a very thin layer adjacent to the surface. This, for instance, is the case for metals, and other materials with the same characteristic will be discussed later. For such a material, the temperature differences in this thin layer behind the surface can be neglected, and the parameters $i_{\nu e}$ and α_ν again become parameters which can be connected with a surface temperature and which will with good approximation have the same values for a process in which a finite energy transfer occurs, as in thermodynamic equilibrium. Relation (14-13) then holds again generally regardless of whether thermodynamic equilibrium of the whole system exists or not.

14-3 RADIATION DENSITY AND RADIATION PRESSURE

These two parameters are useful in the analysis of thermal radiation. Since radiation moves with finite velocity, a finite amount of energy must be contained per unit volume in a radiation field. This quantity is called *radiation density.* We shall now establish the connection between the radiation density and the intensity K_ν of the radiant flux. For this purpose we focus attention on a location 1 in a radiation field (Fig. 14-6). We consider a control surface with the shape of a sphere of radius r surrounding point 1. The radiant energy dE passing through an area element dA_s of the control surface and traveling toward the area element dA at location 1 in a time element $d\tau$ is given by the relation

$$dE = K_\nu \, dA_s \frac{dA}{r^2} \, d\nu \, d\tau \tag{14-15}$$

since dA/r^2 is the solid angle under which the area element dA is seen from dA_s. The distance r between the two area elements is assumed sufficiently small to make the change in the energy flux from dA to dA_s by emission, absorption, or scattering negligible. Then the above equation also gives the flux which travels in the time element $d\tau$ through the

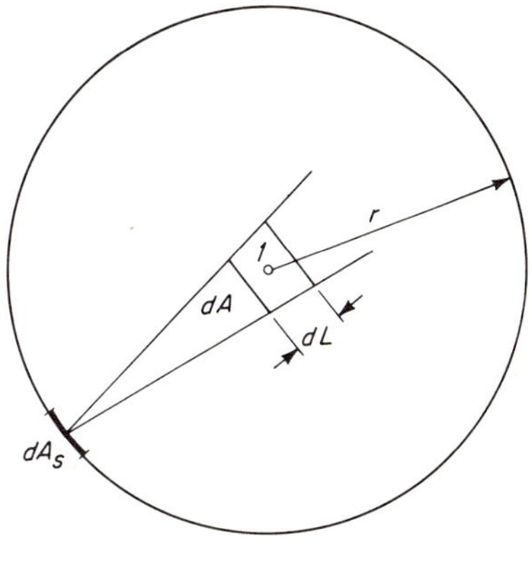

Fig. 14-6 Derivation of radiation density.

area element dA. This flux proceeds during the same time $d\tau$ through a distance dL given by the equation $dL = c\,d\tau$. The energy dE in Eq. (14-15) is therefore contained in a volume element with surface dA and height dL. Replacing the term $d\tau$ in the above equation by dL/c results in

$$dE = \frac{K_\nu\, dA_s\, dV\, d\nu}{cr^2}$$

where $dA\,dL$ has been replaced by dV. We now obtain the total flux contained in the volume element dV regardless of the direction in which it is traveling by considering the flux passing through the whole spherical shell. The total flux passing through this shell and contained in the volume element dV is then

$$\int_{\bigcirc} \frac{K_\nu\, dV\, d\nu}{cr^2}\, dA_s$$

The small circle underneath the integral sign indicates that the integration must be carried out over the surface of the sphere. The expression dA_s/r^2 under the integral sign is simply the solid angle $d\Omega$, under which the area element dA_s is seen from point 1. The monochromatic radiation density u_ν is obtained by dividing the last equation by dV and by $d\nu$

$$u_\nu = \frac{1}{c} \int_{\bigcirc} K_\nu\, d\Omega \tag{14-16}$$

In general, the directional dependence of K_ν must be known if one wants to calculate the radiation density. The radiation density will also generally vary from place to place in a radiation field.

For thermodynamic equilibrium, however, it has been shown in the previous section that the intensity $K_{\nu e}$ of the radiation field is independent of direction and location. In this case, the integration of Eq. (14-1) can easily be carried out and results in the relation

$$u_{\nu e} = \frac{4\pi}{c} K_{\nu e} \tag{14-17}$$

It has also been shown that the intensity $K_{\nu e}$ of the radiant flux for thermodynamic equilibrium is equal to the intensity of radiation $i_{\nu b}$ emitted from a blackbody. The radiation density can therefore also be expressed by

$$u_{\nu e} = \frac{4\pi}{c} i_{\nu b} \tag{14-18}$$

In 1865 James Clerk Maxwell (1831–1879) concluded from his electromagnetic theory that radiation reflected from an interface between two media exerts a pressure on that surface. The magnitude of this pressure will now be calculated under the assumption that the radiation travels in a vacuum and that it impinges on a totally reflecting ($\rho_\nu = 1$)

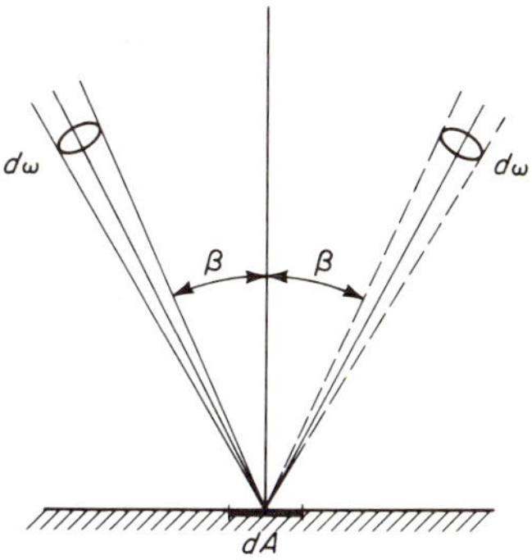

Fig. 14-7 Derivation of radiation pressure.

specular surface. The radiant flux $d\Phi$ impinging on an area element dA of such a surface within a pencil of solid angle $d\omega$ (Fig. 14-7) can be expressed by the equation

$$d\Phi = K_\nu \, dA \cos \beta \, d\omega \, d\nu$$

The flux arriving per unit area is therefore

$$\frac{d\Phi}{dA} = K_\nu \cos \beta \, d\omega \, d\nu$$

Einstein established the equivalence of energy E and mass m corresponding to the equation

$$E = mc_0^2 \tag{14-19}$$

The energy flux expressed by the preceding equation is therefore equivalent to a mass flux impinging per unit time on a unit area of the amount

$$\frac{d\dot{m}}{dA} = \frac{K_\nu}{c_0^2} \cos \beta \, d\omega \, d\nu$$

This flux is totally reflected into the dashed pencil with the same solid angle $d\omega$ and with the same angle β toward the surface normal. The change in momentum flux is therefore 2 times the normal component of the momentum flux arriving at the surface, and since it is a change in momentum flux per unit area, it must be equivalent to a pressure. Therefore the pressure caused by the radiant flux within the pencil of opening angle $d\omega$ is

$$dp_r = 2 \frac{K_\nu}{c_0} \cos^2 \beta \, d\omega \, d\nu$$

The radiation pressure due to all the radiation approaching the surface from all directions in space is obtained by an integration of the above equation over the whole hemispherical space and over all frequencies

$$p_r = \frac{2}{c_0} \int_{\bigtriangleup} K \cos^2 \beta \, d\omega \, d\nu \tag{14-20}$$

The term K in this equation is the intensity of the radiant flux integrated over all frequencies according to the equation

$$K = \int_0^\infty K_\nu \, d\nu \tag{14-21}$$

For a state of thermodynamic equilibrium, the integration can again be carried out since the equilibrium intensity K_e of the radiant flux is independent of direction. One obtains

$$p_r = \frac{4\pi K_e}{3c_0} \tag{14-22}$$

With Eq. (14-17) this relation can also be expressed as

$$p_r = \frac{u_e}{3} \tag{14-23}$$

The magnitude of the radiation pressure is, in general, very small. It is found, for instance, that the radiation pressure which the solar radiation exerts on the surface of our earth is 0.4 dyn/m^2 = 4×10^{-6} N/m^2. For a considerable time, therefore, experimental physicists attempted in vain to measure this pressure. The first who claimed to have demonstrated the existence of the radiation pressure was Crookes (1874), who detected the so-called radiometer effect by a device which is today available in many stores and is sometimes referred to as a "light mill." The radiometer effect, however, is not related to the radiation pressure. Only in 1900 was Lebedew able to measure the radiation pressure exerted by an arc lamp and to demonstrate that the above calculation gives the right amount. The radiation pressure has recently gained engineering importance in connection with space flight. The pressure exerted on some satellites has a measurable influence on their orbit. It has also been proposed as a means of propelling vehicles in space.

The force exerted by radiation on a surface has a direction normal to the surface as long as the reflection is specular and $\rho_\nu = 1$ because the component of the momentum flux parallel to the surface does not undergo any change in the reflection process. This is not necessarily the case in general, and a shear parallel to the surface may arise in addition to a pressure normal to the surface. One then refers to *radiation stress,* which can be calculated as soon as the directional dependence of the reflected radiation is known.

14-4 BLACKBODY RADIATION

A blackbody was defined in Sec. 14-2 as a medium which absorbs all the radiation impinging on its surface ($\alpha_\nu = 1$). At thermal equilibrium (at uniform temperature) such a body emits into a neighboring medium a radiant flux with the intensity $i_{\nu b}$. It was also found in Sec. 14-2 that this radiation intensity is equal to the intensity of the radiant flux existing in the neighboring medium for the condition of thermodynamic equilibrium

$$i_{\nu b} = K_{\nu e}$$

Kirchhoff's second law states that the product $i_{\nu b} c_1^2$ of the intensity of blackbody radiation times the wave velocity of radiation in the neighboring medium is a universal constant.

It remains to discuss the blackbody radiation intensity as a function of the parameters on which it depends. This will be done first for blackbody radiation into a vacuum, because the laws describing blackbody radiation have been derived analytically for such a condition. In the effort to derive this relation, Planck formulated his famous quantum theory, which became the origin of modern physics. He obtained the following relation for the monochromatic intensity from his quantum theory:

$$i_{\nu,b} = \frac{2h\nu^3}{c_0^2(e^{h\nu/\mathbf{k}T} - 1)} \tag{14-24}$$

Planck's constant h (6.6256×10^{-34} J s) and Boltzmann's constant $\mathbf{k}$ (1.38054×10^{-23} J K) in this equation are universal constants. The factor 2 appears in the equation because black radiation is unpolarized and the intensity of unpolarized radiation is 2 times the intensity of plane-polarized radiation. In order to convert this equation to the monochromatic intensity expressed as a function of wavelength, the following relations must be considered:

$$i_{\lambda,b}\, d\lambda = i_{\nu,b}\, d\nu \qquad d\lambda = -\frac{c_0}{\nu^2}\, d\nu \tag{14-25}$$

The second relation follows from $\lambda = c_0/\nu$, with c_0 being a constant. Planck's equation as function of wavelength is

$$i_{\lambda,b} = \frac{2hc_0^2}{\lambda^5(e^{hc_0/\mathbf{k}\lambda T} - 1)} = \frac{2C_1}{\lambda^5(e^{C_2/\lambda T} - 1)} \tag{14-26}$$

Numerical values of the constants C_1 and C_2 are listed together with some constants which will be introduced later in Table 14-1. Figure 14-8 presents curves of the intensity of monochromatic blackbody radiation into a vacuum as a function of wavelength for a number of temperatures. The wavelength range in which radiation is visible to the eye is indicated in the figure by the dashed area, and it can be recognized that for temperatures with which one is generally concerned in engineering, the bulk of the radiation occurs at wavelengths larger than the visible ones (in the infrared range). The wavelength distribution of the

Table 14-1 Radiation constants in the Planck, Wien, and Stefan-Boltzmann equations† and Table 14-2

C_1	0.18899×10^8 Btu μm^4/hr ft^2	0.59548×10^{-16} W m^2
C_2	25,898 μm R	1.43879 cm K
C_3	5216.0 μm R	0.28978 cm K
σ	0.17140×10^{-8} Btu/hr ft^2 R^4	5.6697×10^{-8} W/m^2 K^4
C_4	0.68790×10^{-13} Btu/hr ft^2 μm R^5	4.0950×10^{-6} W/m^3 K^5
C_5	0.84024×10^{-10} Btu/hr ft^2 R^4	0.27791×10^{-8} W/m^2 K^4

$$C_1 = hc_0^2 \qquad C_2 = \frac{hc_0}{k} \qquad C_3 = \frac{hc_0}{4.965114k}$$

$$C_4 = \frac{2C_1}{C_3^5(e^{4.965114} - 1)} \qquad C_5 = \frac{2k^4}{h^3c_0^2}$$

† According to *Natl. Bur. Std. U.S. News Bull.*, October, 1963.

radiation coming from the sun corresponds approximately to blackbody radiation at a temperature of 5600 K. At this temperature, approximately one-third of the total energy radiated is in the visible range.

The ratio of $i_{\lambda,b}/i_{\lambda,b,\max}$ on any isotherm is, according to Eq. (14-26),

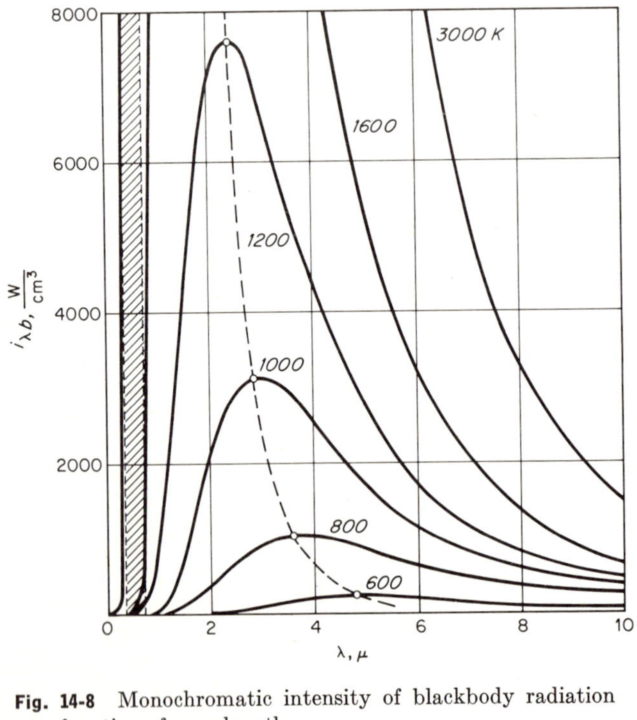

Fig. 14-8 Monochromatic intensity of blackbody radiation as a function of wavelength.

a function of λT only. The relation between the parameters is listed in Table 14-2. The maximum $i_{\lambda,b,\max}$ of the intensity occurs at a wavelength $\lambda_{\max}$, which is described by Wien's law

$$\lambda_{\max} T = C_3 \tag{14-27}$$

This relation as well as the next are immediately evident when both sides of Eq. (14-26) are divided by T^5. The value of the maximum itself can be calculated from the equation

$$i_{\lambda,b,\max} = C_4 T^5 \tag{14-28}$$

The constants C_3 and C_4 are also listed in Table 14-1. The intensity $i_{\lambda,b}$ can therefore easily be obtained from Tables 14-1 and 14-2.

In calculations of radiant heat transfer, one frequently needs the energy emitted by a blackbody in a certain wavelength range. For this reason, Table 14-2[1] contains a third column listing the value $i_{b,0-\lambda}/C_5 T^4$. The term $i_{b,0-\lambda}$ is defined by the equation

$$i_{b,0-\lambda} = \int_0^\lambda i_{\lambda,b}\, d\lambda \tag{14-29}$$

and expresses, therefore, the intensity of radiation emitted by a blackbody at temperature T in the wavelength range from zero to λ. The intensity emitted in the range between two wavelengths is obtained by subtraction of the corresponding values of $i_{b,0-\lambda}$ in Table 14-2. The constant C_5 is again found in Table 14-1. The total energy radiated by a blackbody into a unit solid angle in the whole wavelength range from 0 to ∞ is obtained by integration of Eq. (14-26) over all wavelengths. The result is

$$i_b = \int_0^\infty i_{\lambda,b}\, d\lambda = \frac{\sigma}{\pi} T^4 \tag{14-30}$$

The constant 5 is again listed in Table 14-1. The radiation e_b into all directions in space, called the *emissive power*, is found by a simple integration over the whole hemispherical space (Fig. 14-9)

$$e_b = i_b \int_{\triangle} \cos\beta\, d\omega = i_b \int_0^{2\pi} \int_0^{\pi/2} \cos\beta \sin\beta\, d\beta\, d\gamma = \pi i_b$$

Where γ indicates the azimuth angle. In this way the Stefan-Boltzmann law

$$e_b = \sigma T^4 \tag{14-31}$$

is obtained. The Stefan-Boltzmann constant σ is listed in Table 14-1.

For gases, the velocity c is practically equal to the velocity c_0 in vacuum; correspondingly, the above relations describe to a very good

[1] See also M. Czerny and A. Walther, "Tabellen der Bruchteilfunktionen zum Planckschen Strahlungsgesetz," Springer-Verlag, Berlin, 1961.

Table 14-2 Monochromatic intensities $i_{\lambda,b}$ and intensities $i_{b,0-\lambda}$ in the wavelength range between 0 and λ for black radiation in vacuum†

$\lambda T/C_2$	$i_{\lambda,b}/i_{\lambda,b,max}$	$i_{b,0-\lambda}C_5T^4$
.0600	.00350	.32176E-03
.0610	.00424	.40367E-03
.0620	.00509	.50239E-03
.0630	.00607	.62051E-03
.0640	.00719	.76087E-03
.0650	.00846	.92657E-03
.0660	.00990	.11210E-02
.0670	.01151	.13477E-02
.0680	.01332	.16106E-02
.0690	.01532	.19138E-02
.0700	.01754	.22617E-02
.0710	.01998	.26589E-02
.0720	.02265	.31104E-02
.0730	.02557	.36212E-02
.0740	.02875	.41966E-02
.0750	.03219	.48421E-02
.0760	.03591	.55635E-02
.0770	.03990	.63666E-02
.0780	.04419	.72575E-02
.0790	.04876	.82423E-02
.0800	.05364	.93274E-02
.0810	.05882	.10519E-01
.0820	.06431	.11824E-01
.0830	.07011	.13248E-01
.0840	.07622	.14799E-01
.0850	.08264	.16482E-01
.0860	.08937	.18305E-01
.0870	.09642	.20274E-01
.0880	.10377	.22396E-01
.0890	.11142	.24677E-01
.0900	.11938	.27123E-01
.0910	.12763	.29741E-01
.0920	.13618	.32537E-01
.0930	.14501	.35517E-01
.0940	.15412	.38688E-01
.0950	.16350	.42054E-01
.0960	.17314	.45622E-01
.0970	.18303	.49397E-01
.0980	.19317	.53385E-01
.0990	.20355	.57590E-01
.1000	.21415	.62018E-01
.1010	.22496	.66672E-01
.1020	.23598	.71558E-01
.1030	.24719	.76679E-01
.1040	.25858	.82041E-01
.1050	.27014	.87645E-01
.1060	.28185	.93496E-01
.1070	.29372	.99597E-01
.1080	.30571	.10595E+00
.1090	.31783	.11256E+00
.1100	.33006	.11943E+00
.1110	.34238	.12656E+00
.1120	.35479	.13395E+00
.1130	.36727	.14160E+00
.1140	.37982	.14952E+00
.1150	.39241	.15771E+00
.1160	.40505	.16616E+00
.1170	.41771	.17488E+00
.1180	.43038	.18387E+00
.1190	.44306	.19313E+00
.1200	.45573	.20266E+00
.1210	.46839	.21246E+00
.1220	.48102	.22252E+00
.1230	.49361	.23285E+00
.1240	.50616	.24345E+00
.1250	.51865	.25432E+00
.1260	.53107	.26544E+00
.1270	.54343	.27683E+00
.1280	.55569	.28849E+00
.1290	.56787	.30040E+00
.1300	.57995	.31256E+00
.1310	.59192	.32499E+00
.1320	.60378	.33766E+00
.1330	.61552	.35059E+00
.1340	.62713	.36376E+00
.1350	.63861	.37718E+00
.1360	.64995	.39084E+00
.1370	.66114	.40474E+00
.1380	.67218	.41887E+00
.1390	.68307	.43324E+00
.1400	.69380	.44784E+00
.1410	.70436	.46266E+00
.1420	.71476	.47770E+00
.1430	.72499	.49296E+00
.1440	.73503	.50844E+00
.1450	.74490	.52413E+00
.1460	.75459	.54003E+00
.1470	.76409	.55613E+00
.1480	.77341	.57242E+00
.1490	.78254	.58892E+00
.1500	.79147	.60561E+00
.1510	.80021	.62246E+00
.1520	.80676	.63954E+00
.1530	.81711	.65677E+00
.1540	.82527	.67418E+00
.1550	.83322	.69176E+00
.1560	.84098	.70951E+00
.1570	.84854	.72742E+00
.1580	.85590	.74549E+00
.1590	.86307	.76371E+00
.1600	.87003	.78209E+00
.1610	.87679	.80060E+00
.1620	.88336	.81926E+00
.1630	.88972	.83806E+00
.1640	.89569	.85699E+00
.1650	.9 167	.87605E+00
.1660	.9 764	.89523E+00
.1670	.91322	.91453E+00
.1680	.91861	.93395E+00
.1690	.92381	.95348E+00
.1700	.92882	.97312E+00
.1710	.93363	.99286E+00
.1720	.93826	.10127E+01
.1730	.94270	.10326E+01
.1740	.94696	.10527E+01
.1750	.95104	.10728E+01
.1760	.95493	.10930E+01
.1770	.95865	.11133E+01
.1780	.96218	.11337E+01
.1790	.96555	.11541E+01
.1800	.96874	.11746E+01
.1810	.97176	.11952E+01
.1820	.97461	.12158E+01
.1830	.97730	.12365E+01
.1840	.97982	.12572E+01
.1850	.98219	.12780E+01
.1860	.98439	.12989E+01
.1870	.98643	.13198E+01
.1880	.98833	.13407E+01
.1890	.99007	.13617E+01
.1900	.99166	.13827E+01
.1910	.99310	.14037E+01
.1920	.99440	.14248E+01
.1930	.99556	.14459E+01
.1940	.99657	.14670E+01
.1950	.99745	.14882E+01
.1960	.99820	.15093E+01
.1970	.99881	.15305E+01
.1980	.99929	.15517E+01
.1990	.99965	.15729E+01
.2000	.99988	.15940E+01
.2010	.99999	.16152E+01
.2020	.99998	.16364E+01
.2030	.99985	.16576E+01
.2040	.99961	.16788E+01
.2050	.99925	.17000E+01
.2060	.99878	.17212E+01
.2070	.99821	.17424E+01
.2080	.99753	.17635E+01
.2090	.99674	.17847E+01
.2100	.99586	.18058E+01
.2110	.99488	.18269E+01
.2120	.99380	.18480E+01
.2130	.99262	.18690E+01
.2140	.99136	.18901E+01
.2150	.99000	.19111E+01
.2160	.98856	.19321E+01
.2170	.98703	.19530E+01
.2180	.98542	.19739E+01
.2190	.98372	.19948E+01
.2200	.98195	.20156E+01
.2210	.98010	.20364E+01
.2220	.97818	.20572E+01
.2230	.97618	.20779E+01
.2240	.97411	.20986E+01
.2250	.97197	.21192E+01
.2260	.96977	.21398E+01
.2270	.96750	.21603E+01
.2280	.96516	.21808E+01
.2290	.96277	.22012E+01
.2300	.96031	.22216E+01
.2350	.94720	.23228E+01
.2400	.93283	.24224E+01
.2450	.91742	.25205E+01
.2500	.90112	.26169E+01
.2550	.88411	.27115E+01
.2600	.86653	.28043E+01
.2650	.84850	.28952E+01
.2700	.83014	.29842E+01
.2750	.81155	.30712E+01
.2800	.79283	.31562E+01
.2850	.77405	.32393E+01
.2900	.75529	.33204E+01
.2950	.73660	.33994E+01
.3000	.71805	.34765E+01
.3050	.69968	.35517E+01
.3100	.68152	.36249E+01
.3150	.66362	.36962E+01
.3200	.64599	.37656E+01
.3250	.62867	.38331E+01
.3300	.61168	.38989E+01
.3350	.59502	.39628E+01
.3400	.57871	.40251E+01
.3450	.56277	.40856E+01
.3500	.54719	.41444E+01
.3550	.53199	.42016E+01
.3600	.51716	.42572E+01
.3650	.50271	.43112E+01
.3700	.48864	.43638E+01
.3750	.47494	.44148E+01
.3800	.46161	.44645E+01
.3850	.44864	.45127E+01
.3900	.43604	.45596E+01
.3950	.42380	.46052E+01
.4000	.41190	.46495E+01
.4050	.40035	.46925E+01
.4100	.38914	.47344E+01
.4150	.37826	.47750E+01
.4200	.36770	.48146E+01
.4250	.35745	.48530E+01
.4300	.34751	.48904E+01
.4350	.33787	.49267E+01
.4400	.32852	.49620E+01
.4450	.31945	.49964E+01
.4500	.31066	.50298E+01
.4550	.30214	.50622E+01

Table 14-2 Monochromatic intensities $i_{\lambda,b}$ and intensities $i_{b,0-\lambda}$ in the wavelength range between 0 and λ for black radiation in vacuum† (continued)

.4600	.29387	.50938E+01
.4650	.28586	.51246E+01
.4700	.27810	.51544E+01
.4750	.27057	.51835E+01
.4800	.26327	.52118E+01
.4850	.25619	.52393E+01
.4900	.24933	.52661E+01
.4950	.24268	.52922E+01
.5000	.23624	.53176E+01
.5050	.22999	.53423E+01
.5100	.22393	.53664E+01
.5150	.21805	.53898E+01
.5200	.21235	.54126E+01
.5250	.20683	.54348E+01
.5300	.20147	.54565E+01
.5350	.19627	.54775E+01
.5400	.19123	.54981E+01
.5450	.18634	.55181E+01
.5500	.18160	.55376E+01
.5550	.17700	.55566E+01
.5600	.17253	.55751E+01
.5650	.16820	.55932E+01
.5700	.16400	.56108E+01
.5750	.15992	.56280E+01
.5800	.15596	.56447E+01
.5850	.15212	.56610E+01
.5900	.14838	.56769E+01
.5950	.14476	.56925E+01
.6000	.14124	.57076E+01
.6050	.13783	.57224E+01
.6100	.13451	.57369E+01
.6150	.13129	.57510E+01
.6200	.12816	.57647E+01
.6250	.12511	.57781E+01
.6300	.12216	.57912E+01
.6350	.11929	.58040E+01
.6400	.11650	.58165E+01
.6450	.11378	.58287E+01
.6500	.11115	.58407E+01
.6550	.10858	.58523E+01
.6600	.10609	.58637E+01
.6650	.10367	.58748E+01
.6700	.10131	.58857E+01
.6750	.09902	.58963E+01
.6800	.09679	.59067E+01
.6850	.09462	.59168E+01
.6900	.09251	.59267E+01
.6950	.09045	.59364E+01
.7000	.08845	.59459E+01
.7050	.08651	.59552E+01
.7100	.08461	.59642E+01
.7150	.08277	.59731E+01
.7200	.08097	.59818E+01
.7250	.07923	.59903E+01
.7300	.07752	.59986E+01
.7350	.07587	.60067E+01
.7400	.07425	.60147E+01
.7450	.07268	.60225E+01
.7500	.07115	.60301E+01
.7550	.06965	.60375E+01
.7600	.06820	.60449E+01
.7650	.06678	.60520E+01
.7700	.06540	.60590E+01
.7750	.06405	.60659E+01
.7800	.06274	.60726E+01
.7850	.06146	.60792E+01
.7900	.06021	.60856E+01
.7950	.05899	.60919E+01
.8000	.05780	.60981E+01
.8050	.05664	.61042E+01
.8100	.05551	.61101E+01
.8150	.05441	.61160E+01
.8200	.05333	.61217E+01
.8250	.05228	.61273E+01
.8300	.05126	.61328E+01
.8350	.05026	.61381E+01
.8400	.04928	.61434E+01
.8450	.04833	.61486E+01
.8500	.04739	.61537E+01
.8550	.04649	.61586E+01
.8600	.04560	.61635E+01
.8650	.04473	.61683E+01
.8700	.04388	.61730E+01
.8750	.04306	.61776E+01
.8800	.04225	.61821E+01
.8850	.04146	.61866E+01
.8900	.04069	.61909E+01
.8950	.03994	.61952E+01
.9000	.03920	.61994E+01
.9050	.03848	.62035E+01
.9100	.03778	.62075E+01
.9150	.03709	.62115E+01
.9200	.03642	.62154E+01
.9250	.03576	.62192E+01
.9300	.03511	.62230E+01
.9350	.03449	.62267E+01
.9400	.03387	.62303E+01
.9450	.03327	.62339E+01
.9500	.03268	.62374E+01
.9550	.03211	.62408E+01
.9600	.03154	.62442E+01
.9650	.03099	.62475E+01
.9700	.03045	.62507E+01
.9750	.02992	.62539E+01
.9800	.02941	.62571E+01
.9850	.02890	.62602E+01
.9900	.02841	.62632E+01
.9950	.02792	.62662E+01
1.0000	.02745	.62691E+01
1.0100	.02653	.62749E+01
1.0200	.02565	.62804E+01
1.0300	.02480	.62857E+01
1.0400	.02399	.62909E+01
1.0500	.02322	.62959E+01
1.0600	.02247	.63008E+01
1.0700	.02175	.63054E+01
1.0800	.02106	.63100E+01
1.0900	.02040	.63144E+01
1.1000	.01976	.63186E+01
1.1100	.01915	.63228E+01
1.1200	.01856	.63267E+01
1.1300	.01799	.63306E+01
1.1400	.01745	.63344E+01
1.1500	.01692	.63380E+01
1.1600	.01642	.63416E+01
1.1700	.01593	.63450E+01
1.1800	.01546	.63483E+01
1.1900	.01501	.63515E+01
1.2000	.01457	.63547E+01
1.2100	.01415	.63577E+01
1.2200	.01374	.63607E+01
1.2300	.01335	.63635E+01
1.2400	.01298	.63663E+01
1.2500	.01261	.63690E+01
1.2600	.01226	.63717E+01
1.2700	.01192	.63742E+01
1.2800	.01159	.63767E+01
1.2900	.01128	.63792E+01
1.3000	.01097	.63815E+01
1.3100	.01067	.63838E+01
1.3200	.01039	.63860E+01
1.3300	.01011	.63882E+01
1.3400	.00984	.63903E+01
1.3500	.00958	.63924E+01
1.3600	.00933	.63944E+01
1.3700	.00909	.63964E+01
1.3800	.00886	.63983E+01
1.3900	.00863	.64001E+01
1.4000	.00841	.64019E+01
1.4100	.00820	.64037E+01
1.4200	.00799	.64054E+01
1.4300	.00779	.64071E+01
1.4400	.00760	.64087E+01
1.4500	.00741	.64103E+01
1.4600	.00723	.64118E+01
1.4700	.00705	.64134E+01
1.4800	.00688	.64148E+01
1.4900	.00671	.64163E+01
1.5000	.00655	.64177E+01
1.5100	.00640	.64190E+01
1.5200	.00625	.64204E+01
1.5300	.00610	.64217E+01
1.5400	.00596	.64230E+01
1.5500	.00582	.64242E+01
1.5600	.00568	.64254E+01
1.5700	.00555	.64266E+01
1.5800	.00542	.64278E+01
1.5900	.00530	.64289E+01
1.6000	.00518	.64300E+01
1.6100	.00506	.64311E+01
1.6200	.00495	.64322E+01
1.6300	.00484	.64332E+01
1.6400	.00473	.64342E+01
1.6500	.00463	.64352E+01
1.6600	.00453	.64362E+01
1.6700	.00443	.64372E+01
1.6800	.00433	.64381E+01
1.6900	.00424	.64390E+01
1.7000	.00415	.64399E+01
1.7100	.00406	.64408E+01
1.7200	.00397	.64416E+01
1.7300	.00389	.64424E+01
1.7400	.00381	.64433E+01
1.7500	.00373	.64441E+01
1.7600	.00365	.64448E+01
1.7700	.00358	.64456E+01
1.7800	.00350	.64464E+01
1.7900	.00343	.64471E+01
1.8000	.00336	.64478E+01
1.8100	.00329	.64485E+01
1.8200	.00323	.64492E+01
1.8300	.00316	.64499E+01
1.8400	.00310	.64505E+01
1.8500	.00304	.64512E+01
1.8600	.00298	.64518E+01
1.8700	.00292	.64525E+01
1.8800	.00286	.64531E+01
1.8900	.00280	.64537E+01
1.9000	.00275	.64543E+01
1.9100	.00270	.64548E+01
1.9200	.00265	.64554E+01
1.9300	.00259	.64560E+01
1.9400	.00255	.64565E+01
1.9500	.00250	.64570E+01
1.9600	.00245	.64576E+01
1.9700	.00240	.64581E+01
1.9800	.00236	.64586E+01
1.9900	.00231	.64591E+01

† $i_{\lambda,b,\max} = C_4T^5$; constants C are listed in Table 14-1.

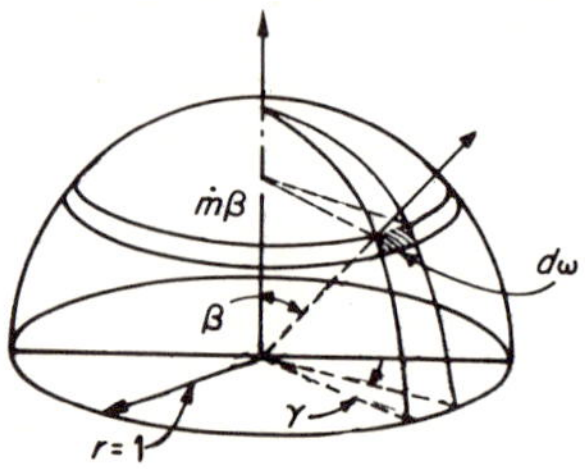

Fig. 14-9 Directional distribution of radiative flux.

approximation the behavior of radiation emitted from a blackbody into a gas. Solids and liquids, however, have refractive indices which are considerably larger than 1 and which also depend on wavelength. For such media, the monochromatic intensity of blackbody radiation is given by

$$i_{\nu,b} = \frac{2h\nu^3 n_0^2}{c_0^2(e^{h\nu/kT} - 1)} \tag{14-32}$$

which follows readily from Eq. (14-24). Wien's law cannot be derived for black radiation in such a medium, and the integration of Eq. (14-32) does not lead to the Stefan-Boltzmann law when n_0 is a function of ν.

Although no material existing in nature has exactly the properties of a blackbody, blackbody radiation can be produced for experimental purposes in the following way. According to Sec. 14-2, the intensity $K_{\nu e}$ of the radiant flux existing in a system in equilibrium is equal to the radiation intensity $i_{\nu b}$ emitted by a blackbody. The system in thermodynamic equilibrium in Fig. 14-10 exists of a cavity in a solid material which is heated and kept at a uniform temperature. Blackbody radiation then exists in the interior of the cavity and escapes through a small opening as indicated in the figure. The departure from equilibrium within the cavity can be minimized by making the hole small. The radiation escaping through the hole can in this way be made to approach blackbody radiation to any desired degree.

In the rest of this chapter, the derivation of the laws for blackbody radiation will be discussed. Historically, Boltzmann derived the fourth-power law in 1884 from classical thermodynamics. Wien then developed the displacement law (14-27) in 1896 on the same basis. For a derivation of the law describing the monochromatic radiation intensity, it was necessary to leave the ground of classical physics. Planck succeeded in deriving this law in 1900 through his quantum concept.

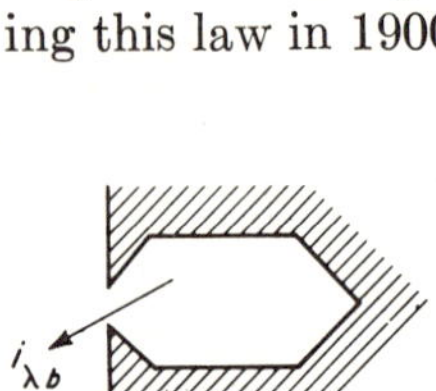

Fig. 14-10 Isothermal cavity as a blackbody.

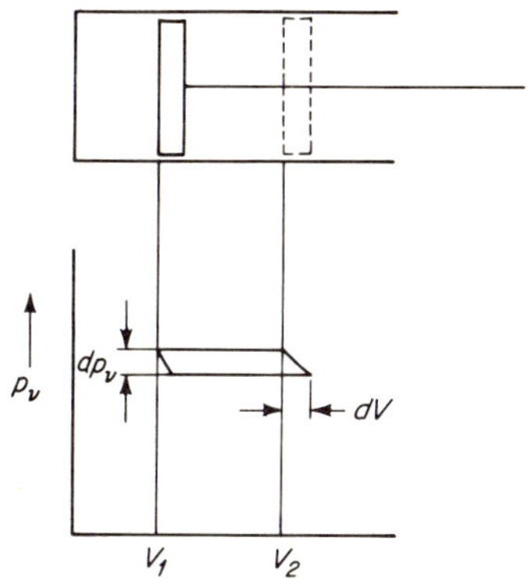

Fig. 14-11 Derivation of the Stefan Boltzmann law.

We shall first derive the Stefan-Boltzmann law. For this purpose imagine that the enclosure of uniform temperature which has been considered quite frequently now has the shape indicated in Fig. 14-11. It is a cylinder closed by a piston, the surface of which is assumed to be a perfect mirror ($\rho_\nu = 1$). The space within the cylinder may be assumed a perfect vacuum, and the cylinder surfaces are kept at a temperature T. Accordingly, the surfaces will radiate heat, and the interior of the cylinder is filled with heat radiation whose density is u_e. This density is a function of the temperature of the cylinder walls. It will also exert a pressure of the amount $p_r = u_e/3$ [Eq. (14-23)] on the wall of the piston. Now imagine the following cycle performed with the cylinder. The piston is moved slowly to the right so that the volume of the cylinder changes from V_1 to V_2. The temperature is kept constant during this time. This means that the radiation density is constant. Since the volume increases, the total amount of energy contained within the cylinder will increase during this period and has to be furnished by addition of heat Q from the surrounding to the cylinder walls.

The radiation pressure also stays constant during this period and will perform work on the piston of the magnitude

$$W = p_r(V_2 - V_1) = \frac{u_e}{3}(V_2 - V_1)$$

The internal energy contained within the space of the cylinder increases by an amount

$$\Delta U_e = u_e(V_2 - V_1)$$

The first law of thermodynamics states that the amount of heat added during this process must be

$$Q = \Delta U_e + W = \tfrac{4}{3}u_e(V_2 - V_1)$$

After the volume V_2 is reached, the cylinder is still moved on by an infinitely small amount dV, now without addition of heat. Correspondingly the density of radiation within the cylinder will decrease, and the temperature and radiation pressure will drop. This is indicated in the lower half of the diagram. After a temperature $T - dT$ is reached, the movement of the cylinder is reversed. The cylinder is moved to the left, at first in such a way that the temperature is kept constant and finally adiabatically, until the original state is reached again. The total work done on the cylinder during the cycle will be

$$dW = dp_r(V_2 - V_1) = \frac{du_e}{3}(V_2 - V_1)$$

The process which we consider is a reversible one if it is carried out slowly enough for the heat addition and extraction to occur without temperature drop. The cycle has been performed between two temperature limits T and $T - dT$, and according to the second law of thermodynamics, all reversible cycles performed between the two fixed temperature levels have the same thermal efficiency (dT/T), namely, the efficiency of a Carnot process. Accordingly, the thermal efficiency of the process considered here is

$$\eta = \frac{dT}{T} = \frac{dW}{Q} = \frac{du_e}{4u_e}$$

It follows that

$$\frac{du_e}{u_e} = 4\frac{dT}{T}$$

This differential equation can be integrated:

$$\ln u_e = 4 \ln T + \ln C$$

or

$$u_e = CT^4$$

According to Eq. (14-18) the last equation can also be written in terms of radiation intensity or emissive power of a blackbody and thus states the Stefan-Boltzmann law of radiation. Wien's law was derived in a similar manner.

Toward the end of the nineteenth century, there were a considerable number of attempts to predict the monochromatic emissive power theoretically. These, however, were unsuccessful in the sense that the calculations resulted in laws which were not in agreement with experimental results. Nevertheless, one of the predictions will be described since it leads to the step which Planck made in order to derive his law. This attempt was made by Rayleigh and Jeans (1900). It is based on the concept that in an enclosure of uniform temperature in which black radiation exists, waves of a large number of wavelengths travel through space, being in equilibrium with the surrounding walls by the continuous emission and absorption of energy by the molecules of which the wall material is composed. The simplest arrangement by which electromagnetic waves of a certain wavelength can be created is an arrangement of two atoms having a plus and minus electric charge and being interconnected by elastic forces so that they are able to perform oscillations. Such a harmonic oscillator will then be able to emit and absorb radiation. There must be a large number of such oscillators, with different frequencies, arranged along the wall of the enclosure to produce radiation of a blackbody. According to the law of equipartition of energy, derived from statistical mechanics, it is to be expected that the oscillators for each wavelength have, on an average, stored at a certain temperature the same amount of energy, namely, $\mathbf{k}T/2$ per degree of freedom. An oscillator stores kinetic energy as well as potential energy, and therefore its energy is

$$E = \mathbf{k}T$$

where $\mathbf{k}$ is the gas constant per molecule (Boltzmann's constant). If we knew, in addition, the number of oscillators for each wavelength, we would be able to calculate the energy contained in the arrangement of oscillators, and this again, according to the equipartition law, must be equal to the energy of the blackbody radiation which is in equilibrium with these oscillators. Rayleigh and Jeans in 1900 calculated the number dZ of waves in a frequency range between ν and $\nu + d\nu$ which are contained in a certain volume V from the concept that the waves must be standing waves with a whole number of waves within the confines of the volume. They derived the following equation,[1] assuming a vacuum within the enclosure:

$$dZ = \frac{8\pi\nu^2}{c_0^3} V \, d\nu$$

The radiation density $u_\nu \, d\nu$ of the radiation field in the volume V can then be obtained by multiplying the average energy of a specific oscillator which corresponds to the average energy of a wave with the number of such waves. This results in

$$Vu_\nu \, d\nu = E \, dZ = \frac{8\pi\nu^2}{c_0^3} V\mathbf{k}T \, d\nu$$

In experimental physics, the energy is usually expressed as a function of wavelength rather than of frequency. The equation connecting the wavelength with the frequency is

$$\nu\lambda = c_0$$

[1] For a derivation see, for instance, W. P. Allis and M. A. Herlin, "Thermodynamics and Statistical Mechanics," pp. 199–214, McGraw-Hill, New York, 1952.

with c_0 expressing the wave velocity in a vacuum. The amount of energy, therefore, can also be written

$$u_\lambda \, d\lambda = u_\nu \, d\nu$$

if the differential of the frequency is described by

$$d\nu = -\frac{c_0}{\lambda^2} \, d\lambda$$

Introducing those expressions into the above equation, neglecting the minus sign, and dividing by the volume V give for the monochromatic radiation density

$$u_\lambda = \frac{8\pi}{\lambda^4} \mathbf{k}T$$

This equation is the Rayleigh-Jean law for monochromatic black radiation. According to this law, the radiation will increase indefinitely when the wavelength goes toward zero, which is in contradiction to experiments. However, it was found that for large wavelengths, it describes the actual monochromatic black radiation quite well.

At this point, Planck stepped in with his bold idea that energy can be exchanged on a molecular basis not in any arbitrary quantity but only in multiples of a certain energy element ϵ. Therefore, the quantity $n\epsilon$, with n being a whole number, describes energy exchange on a molecular basis.

Planck showed that the average energy which an oscillator in equilibrium at the temperature T assumes on this basis is no longer described by the equipartition law as stated in the previous equation but by the relationship

$$E = \frac{\epsilon}{e^{(\epsilon/\mathbf{k}T)} - 1}$$

For small values of ϵ, the exponent in the denominator becomes small. In this case the e function can be developed into a series of which only the first two terms are considered. This procedure results in the following equation:

$$\lim_{\epsilon \to 0} E = \frac{\epsilon}{(1 + \epsilon/\mathbf{k}T) - 1} = \mathbf{k}T$$

For small values of ϵ, therefore, Planck's law of the energy of an oscillator transforms into the equipartition law. For increasing values of ϵ, however, the energy E drops off to smaller values and converges rapidly toward zero.

Planck specified the energy quantum additionally by assuming that it is proportional to the frequency of the oscillator according to the equation

$$\epsilon = h\nu \tag{14-29}$$

In this way he obtained the result that the energy drops off at large frequencies or at small wavelengths, which was necessary in order to bring the calculation into agreement with experimental results. The term h has been found to be a universal constant, the so-called *Planck's quantum.* If the new equation for the average energy stored in a linear oscillator is now introduced into the previous equation instead of $\mathbf{k}T$, the following equation is obtained for the energy density in a vacuum at a certain wavelength:

$$u_\lambda = \frac{8\pi h c_0}{\lambda^5(e^{hc_0/\lambda \mathbf{k}T} - 1)}$$

If in addition we change from the energy density to the radiation intensity, we obtain finally Planck's famous law for blackbody radiation intensity in a vacuum

$$i_{b\lambda} = \frac{2hc_0^2}{\lambda^5(e^{hc_0/\lambda \mathbf{k}T} - 1)}$$

This equation contains only one unknown constant, namely, the quantum h, which could be determined from the measurement of blackbody radiation intensity. The equation is in excellent agreement with the results of experimental determinations of the radiation intensity of a blackbody. Its derivation led to the concept that energy consists of finite particles, or quanta, in much the same way as matter, which can be broken up only into certain particles of finite size: molecules, atoms, electrons, and so on. This concept became the basis for the development of modern physics.

15

Radiation of Strongly Absorbing Media

Kirchhoff's laws, discussed in Sec. 14-2, established relations between radiation properties of a medium and the intensity of blackbody radiation. These relations will be briefly recapitulated here before starting with a discussion of the radiative properties of various materials. The first form of Kirchhoff's relations was concerned with radiation in the interior of a medium and established the following relationship between the coefficient of emission j_ν, the coefficient of absorption $\varkappa_\nu$, and the intensity $K_{\nu e}$ of the radiant flux for a system in thermal equilibrium:

$$\frac{j_\nu}{\varkappa_\nu} = K\nu_e$$

It was also found that the intensity of radiation in a system in thermodynamic equilibrium is equal to the intensity with which a blackbody emits radiation into this medium [Eq. (14-14)]. The above equation can therefore be written as

$$j_\nu = \varkappa_\nu i_{\nu b} \tag{15-1}$$

The same relation can also be written for the radiation intensity expressed in wavelengths instead of frequencies:

$$j_\lambda = \varkappa_\lambda i_{\lambda b} \tag{15-2}$$

Wavelength will be used in this chapter to characterize monochromatic radiation, since the discussion deals mainly with experimental information. It should be remembered that the parameters j_λ and $\varkappa_\lambda$ are functions of the local thermodynamic state in the medium and therefore have the same value for thermal radiation, regardless of whether the system as a whole is in equilibrium or not.

A second form of Kirchhoff's law concerned with the radiant flux at the interface between two media can be written in the form [Eqs. (14-13) and (14-14)]

$$i_\nu = \alpha_\nu i_{\nu b} \tag{15-3}$$

where i_ν = intensity of radiant beam emitted from neighborhood medium into medium 1
α_ν = absorptance of neighboring medium for radiation approaching interface from medium 1
$i_{\nu b}$ = radiation emitted by blackbody into medium 1

This last equation can again be written in radiation referred to wavelength instead of frequency

$$i_\lambda = \alpha_\lambda i_{\lambda b} \tag{15-4}$$

It will be observed that the two equations (15-1) and (15-3) or (15-2) and (15-4) are completely analogous in their form. The difference is that the first equation is concerned with radiation properties describing local conditions within a medium, whereas the second equation is concerned with properties at the interface between two media. Frequently a new term *emittance* is introduced by the equation

$$\epsilon_\lambda = \frac{i_\lambda}{i_{\lambda b}} \tag{15-5}$$

and Eq. (15-4) can then be written as

$$\epsilon_\lambda = \alpha_\lambda \tag{15-6}$$

expressing the fact that the monochromatic absorptance is equal to the monochromatic emittance. It has already been pointed out that these last parameters depend, in general, on the state of the medium which absorbs or emits the radiation and are different when this medium is in thermodynamic equilibrium or when the thermodynamic state changes locally. Only for strongly absorbing media can the praameters α_λ and ϵ_λ be uniquely related to the surface conditions, neglecting the differences in the thermodynamic state within the thin layer which contributes to absorption or emission.

A radiant beam incident on a medium is either reflected or absorbed in this medium or transmitted through it. The ratio of the reflected to the incident energy is called *reflectance* ρ_λ, the ratio of absorbed to incident energy is called *absorptance* α_λ, and the ratio of transmitted to incident energy is called *transmittance* τ_λ. According to this definition, the following relation exists between these parameters:

$$\rho_\lambda + \alpha_\lambda + \tau_\lambda = 1 \tag{15-7}$$

In a strongly absorbing medium, the transmitted portion is negligibly small, and the above equation reduces to

$$\rho_\lambda + \alpha_\lambda = 1 \tag{15-8}$$

Most solids absorb radiation entering through their surfaces very strongly, so that practically all the incident radiation is absorbed in a very thin layer below the surface. In metals, the thickness of this layer is a fraction of a micrometer. In electrically nonconducting materials, fractions of a millimeter are usually sufficient. This can be recognized from Tables 15-1 and 15-2. Exceptions are a few solid substances like glass, quartz, rock salt, and most liquids in the visible and near-infrared range. However, these substances also absorb very strongly in the bulk of the infrared, as indicated by the curve for optical glass in Fig. 15-1.

Table 15-1 Transmittance of metal foils for monochromatic radiation with wavelength λ†

	s, μm‡		
λ	*Silver*	*Gold*	*Platinum*
2.5		0.0272	0.0352
2.0		0.0238	0.0331
1.5	0.0222	0.0244	0.0309
1.2	0.0214	0.0249	0.0300
1.0	0.0230	0.0266	0.0284
0.8	0.0236	0.0281	0.0274
0.7	0.0233	0.0311	0.0263

† After E. Hagen and H. Rubens, *Ann. Phys.*, **8**:432 (1902).

‡ s denotes the thickness in micrometers within which the intensity of entering radiation has decreased to one-tenth of the original value.

Table 15-2 Transmittance of electric nonconductors for incident black radiation at temperature 95 C†

Substance	s, mm‡
Glass	0.3
Typewriter paper	0.06
Writing paper	0.07
Tracing paper	0.09
Mica	0.25

† After E. Eckert, *Forsch. Gebiete Ingenieurw.*, **9**:251 (1938).

‡ s denotes the thickness in millimeters within which the intensity of incident radiation is decreased to one-tenth of the original value.

Sulfarsenide glass has been developed especially for good transmittance in the infrared.

For strongly absorbing media, the following equations apply according to the preceding discussion:

$$\epsilon_\lambda = \alpha_\lambda \qquad \rho_\lambda + \alpha_\lambda = 1 \tag{15-9}$$

so that one monochromatic-radiation property determines the two others. It must be kept in mind, however, that this property depends on the

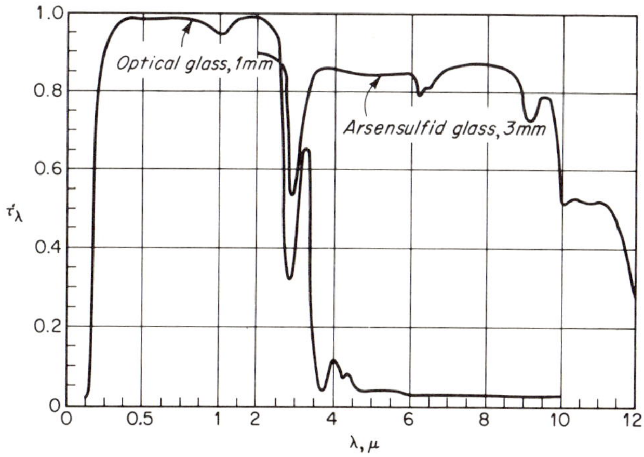

Fig. 15-1 Monochromatic transmittance of glass as a function of wavelength. [*From J. D'Ans and E. Lax, "Taschenbuch für Physiker und Chemiker," Springer-Verlag, Berlin, 1949*].

nature of both media separated by the interface, on the direction of the radiant beam, and also on its polarization. Radiation properties of solids and liquids are usually listed in the various handbooks for radiation into air. This is also true of the values presented in this and the following sections. Such values can be used with excellent approximation for radiation into any gas or into a vacuum, whereas they are different if one is concerned with radiation into a liquid or solid.

Gases are much weaker emitters and absorbers than solids and liquids. As a consequence, much thicker layers are involved in the radiative exchange process, and absorptances, emittances, and transmittances are useful radiation properties only as long as one is concerned with gas bodies of uniform thermodynamic state, e.g., uniform temperature and composition; otherwise one must use the emission, absorption, and scattering coefficients as defined at the beginning of this section. Gases are generally isotropic materials, and the coefficients mentioned are then independent of direction. Reflection of a radiant beam at the interface between two gases is very small and generally neglected. At the interface to a solid or liquid, the reflection process has to be considered.

15-1 RADIATION PROPERTIES OF STRONGLY ABSORBING MEDIA (SOLIDS AND LIQUIDS)

Emittance The emittance of surfaces has been defined by the equation

$$\epsilon_\lambda = \frac{i_\lambda}{i_{\lambda b}} \tag{15-10}$$

More exactly, this is the *monochromatic directional emittance* concerned with radiation leaving the surface in a specified direction. In addition to this emittance, a *hemispherical emittance* is often listed in reference books which is concerned with radiation emitted in all directions of the hemispherical space. It is defined as

$$\epsilon_{\lambda\frown} = \frac{e_\lambda}{e_{\lambda b}} \tag{15-11}$$

and can also be expressed as

$$\epsilon_{\lambda\frown} = \frac{\int_\frown i_\lambda \, dA \cos\beta \, d\omega \, d\lambda}{\int_\frown i_{\lambda b} \, dA \cos\beta \, d\omega \, d\lambda} = \frac{1}{\pi}\int_\frown \epsilon_\lambda \cos\beta \, d\omega$$

A special case of the directional emittance is the *normal emittance* $\epsilon_{b\perp}$, which is concerned with the radiation in the direction of the surface normal. Another distinction must be made between the *monochromatic emittance* and the *total emittance* referring to the total radiation regardless of wavelength. It is connected with the monochromatic emittance by the relation

$$\epsilon = \frac{\int_0^\infty \epsilon_\lambda i_{\lambda b}\, d\lambda}{\int_0^\infty i_{\lambda b}\, d\lambda} = \frac{\int_0^\infty \epsilon_\lambda i_{\lambda b}\, d\lambda}{\dfrac{\sigma}{\pi} T^4} \tag{15-12}$$

Electromagnetic theory shows that solids or liquids have a characteristically different behavior in the infrared, depending on whether they are electric conductors or nonconductors.

Metals (electric conductors) Metals as electric conductors have small emittance values, as can be seen in Table 15-3. The radiation properties of electric conductors are mainly determined by the free electrons. As a consequence, there exists a connection between the emittance and the specific electric resistance, the resistivity of a metal. The emittance increases with increasing electric resistivity. Because of the very thin layer beyond the surface which participates in the radiation, the emittance values depend very strongly on surface condition. A thin oxide layer, not visible to the eye, or a film of impurities or surface roughness changes the emittance considerably. Thicker oxide layers can increase the emittance by an order of magnitude, as various values listed in Table 15-3 indicate. The impossibility of describing the surface condition exactly accounts for the fact that emittance values listed in various reference books sometimes differ greatly.

The directional distribution of the total emittance for a few metals is shown in Fig. 15-2*a* and *b*. It will be observed that the directional emittance for metals increases with increasing angle β. Close to 90° the emittance decreases again and reaches the value zero at $\beta = 90°$. In Fig. 15-2*a* this occurs at such grazing angles that the decrease could not be measured. The directional distribution curves make it possible to calculate the ratio of the hemispherical to the normal emittance. Figure 15-3 presents this ratio, as obtained by electromagnetic theory for electric conductors and nonconductors, and values measured for a number of metal surfaces. The deviations between predicted and measured values are probably caused by surface effects, as discussed above. This figure is useful for converting normal emittances listed in Table 15-3 to hemispherical values. For larger values of the angle of the radiant beam measured toward the surface normal, the radiation is polarized in the sense that the intensity of the component whose electric field fluctuates parallel to the plane of incidence (the plane containing the surface normal and the radiant beam) is considerably larger than the intensity of the component

Table 15-3 Emittance $\epsilon_{\perp}$ in the direction of the surface normal for various materials at temperature t†

Surface	t, C	$\epsilon_{\perp}$
Metals:		
Aluminum:		
Bright rolled	170	0.039
	500	0.050
Polished, 98% pure	93	0.05
Aluminum paint	100	0.20–0.40
Bismuth, bright	80	0.340
Chrome, polished	150	0.058
Copper, polished	20	0.030
Lightly oxidized	20	0.037
Scraped	20	0.070
Black oxidized	20	0.78
Oxidized	131	0.76
Gold, polished	130	0.018
	400	0.022
Iron:		
Pure polished	94	0.06
Bright etched	150	0.128
Bright abrased	20	0.24
Red rusted	20	0.61
Hot-rolled	20	0.77
	130	0.60
Heavily crusted	20	0.85
Heat-resistant oxidized	80	0.613
Heat-resistant oxidized	200	0.639
Cast, bright	93	0.21
Oxidized	93	0.61
Black iron oxide	93	0.56
Lead:		
Pure polished	38	0.05
Gray oxidized	20	0.28
Mercury, pure clean	93	0.12
Molybdenum, polished	93	0.06
Nickel:		
Bright matte	100	0.041
Polished	100	0.045
Platinum:		
Pure polished	93	0.05
Black	93	0.93
Silumin, cast polished	150	0.186
Silver	20	0.020
Tin, bright tinned iron sheet	38	0.08
Tungsten, polished	93	0.04
Zinc:		
Matte	93	0.21
Gray oxidized	20	0.23–0.28
Alloys:		
Brass:		
Polished	38	0.05
Freshly rubbed with emery	38	0.21
Dull	38	0.22
Oxidized	38	0.46
Manganin, bright rolled	118	0.048

Table 15-3 Emittance $\epsilon_\perp$ in the direction of the surface normal for various materials at temperature t† (continued)

Surface	t, C	$\epsilon_\perp$
Paints:		
Varnish, dark glossy	38	0.89
Lacquer, clear on bright copper	38	0.07
White on bright copper, thin coat	38	0.85
White	100	0.925
Black matte	80	0.970
Enamel	20	0.85–0.95
Bakelite	80	0.935
Gold enamel	93	0.37
Red lead	100	0.93
Oil on polished iron, 0.0008 in. thick	38	0.22
0.0080 in. thick	38	0.81
Pigments:		
Acetylene soot	52	0.99
Camphor soot	52	0.98
Lampblack	52	0.94
Candle soot	52	0.95
Platinum black	52	0.91
Red (Fe_2O_3)	52	0.96
Green (Cr_2O_3)	52	0.95
White (Al_2O_3)	52	0.98
White (MgO)	52	0.97
Miscellaneous:		
Asbestos paper	260	0.94
Brick, mortar, plaster	20	0.93
Clay		
Fired	67	0.91
Corundum		
Emery rough	80	0.855
Glass	18	0.940
Hoar frost (0.1–0.2 mm thick)	38	0.98
Ice, smooth, water	0	0.966
Rough crystals	0	0.985
Limestone	38	0.95
Marble, white	38	0.95
Mica	38	0.75
Paper	95	0.92
Plaster of paris ($\frac{1}{2}$ mm)	38	0.91
Porcelain	20	0.92–0.94
Quartz, fused, rough	38	0.93
Refractory brick, ordinary	1100	0.59
White	1100	0.29
Dark chrome	1100	0.98
Rubber, grey, soft, rough	38	0.86
Sandstone	38	0.83
Tar paper	20	0.93
Velvet, black	38	0.97
Water (0.1 mm or more thick)	38	0.96
Waterglass	20	0.96
Wood, Beech	70	0.935
Oak, planed	38	0.90
Spruce, sanded	38	0.82

† From E. R. G. Eckert and R. M. Drake, "Heat and Mass Transfer," McGraw-Hill, New York, 1959, and J. R. Singham, *Intern. J. Heat Mass Transfer*, **5**:67 (1962).

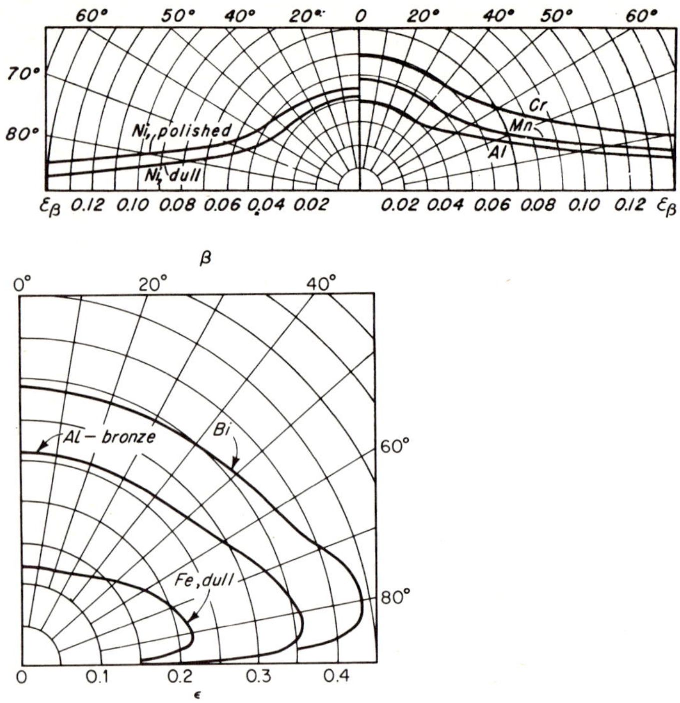

(b)

Fig. 15-2 Directional distribution of monochromatic emittance of metals. [*From E. Schmidt and E. Eckert, Forsch. Gebiete Ingenieurw.*, **6**:*175* (*1935*).]

with the electric field fluctuating normal to this plane. Figure 15-4 and Table 15-4 indicate the size of the emittance $\epsilon_{\lambda p}$ and $\epsilon_{\lambda n}$ for a surface with an emittance $\epsilon_\perp = 0.1$ according to electromagnetic theory.

Figure 15-5 presents values of monochromatic normal emittances as function of wavelength. The figure indicates that the monochromatic emittance for metals decreases in the infrared range in a regular way with increasing wavelength. This behavior is also predicted by electromagnetic theory. The titanium surface which has been heated for 303 hr at 810 K and the anodized aluminum surface behave differently as a consequence of the fact that these metal surfaces are covered by an oxide layer. Their behavior therefore approaches one which is characteristic for electric nonconductors. In the visible and near-infrared range, the behavior of the emittance for metal surfaces is quite irregular. Electromagnetic theory no longer applies in this range, and an estimate of the emittance

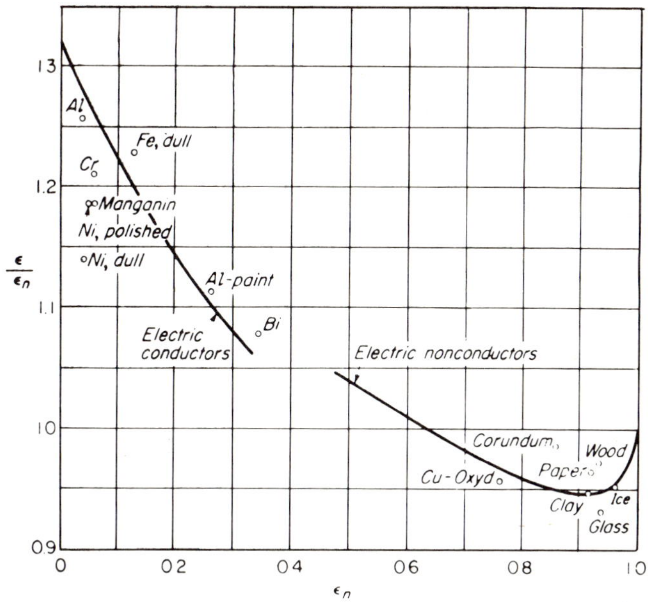

Fig. 15-3 Ratio of hemispherical to normal emittance of electric conductors and nonconductors as a function of normal emittance. [*From E. Schmidt and E. Eckert, Forsch. Gebiete Ingenieurw.*, **6**:*175* (*1935*).]

can better be made from the appearance of the surface to the eye of the observer when it reflects light. Partly as a consequence of this wavelength dependence and partly as a consequence of the fact that the monochromatic emittance depends on temperature, the total emittance increases with increasing surface temperature. This can be seen in Fig. 15-6. The increase in the normal emittance for the metals aluminum,

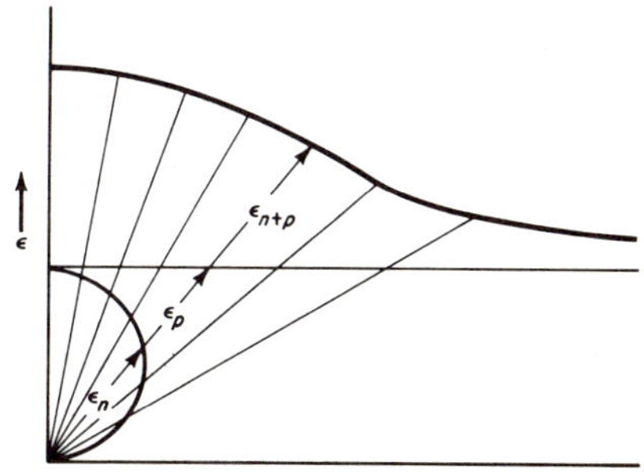

Fig. 15-4 Directional distribution of emittance of metals polarized normal and parallel to the plane of incidence.

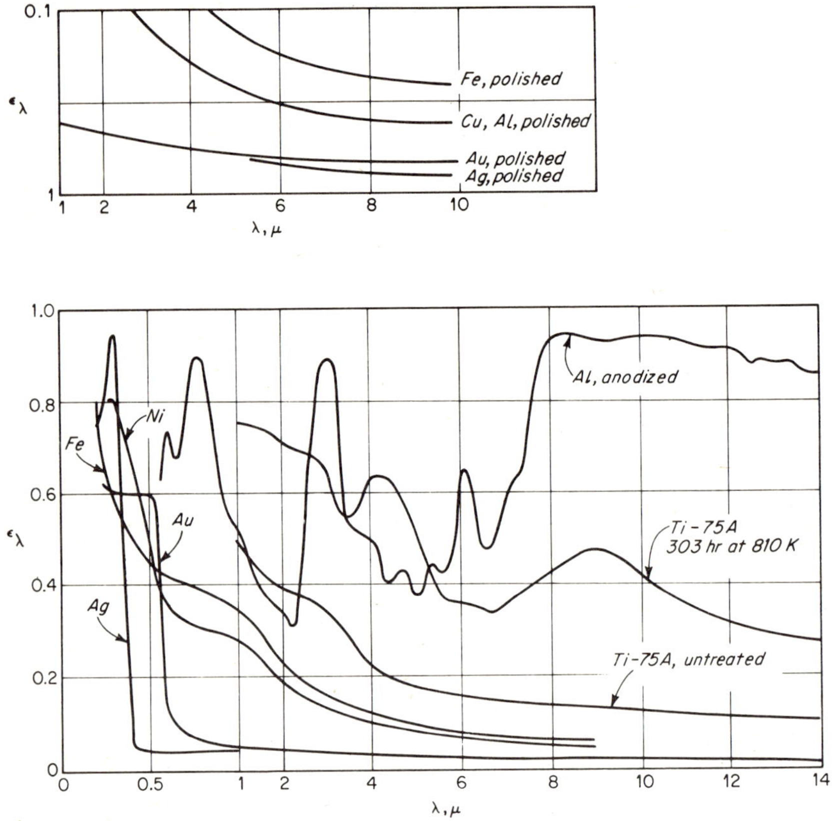

Fig. 15-5 Monochromatic normal emittance of metals as a function of wavelength. [*From J. D'Ans and E. Lax, "Taschenbuch für Physiker und Chemiker," Springer-Verlag, Berlin, 1949, and after J. T. Bevans, J. T. Gier, and R. V. Dunkle, Trans. ASME,* **80** (*1958*).]

tungsten, platinum, and molybdenum is quite regular and indicative of the general behavior of metals. The fact that the emittance of iron starts to increase at a faster rate beyond a temperature of 800 K must be attributed to the gradual formation of a thin oxide layer.

In conclusion of the discussion of emittance values for electric conductors, the main results obtained from electromagnetic theory may be briefly outlined.[1] Fresnel's relations connect the reflectance of an electric conductor with the refractive index n and the coefficient of absorp-

[1] For a more extensive discussion see, for instance, E. M. Sparrow and R. D. Cess, "Radiation Heat Transfer," Brooks/Cole, Belmont, Calif., 1966, or H. C. Hottel and A. F. Serafim, "Radiative Transfer," McGraw-Hill, New York, 1967.

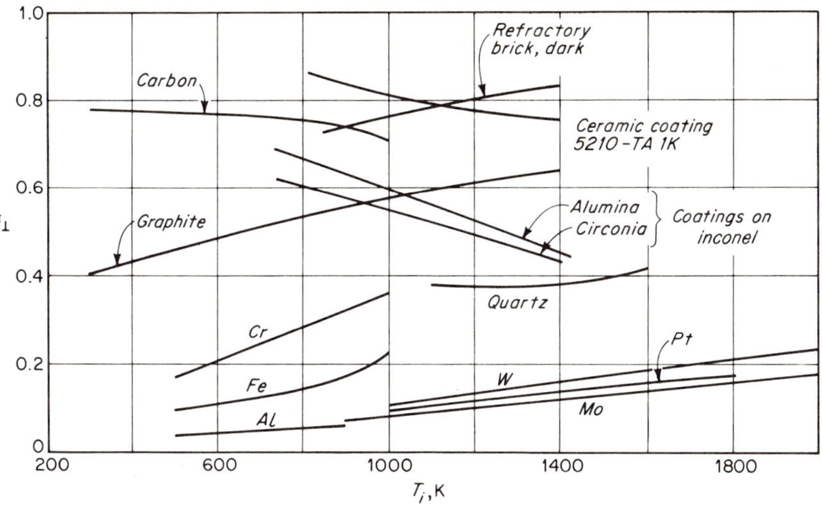

Fig. 15-6 Total normal emittance of various materials as a function of temperature. [*From various handbooks.*]

Table 15-4 Values of $\epsilon_{\lambda p}$ and $\epsilon_{\lambda n}$ as functions of β calculated from electromagnetic theory

β, deg	$\epsilon_{\lambda p}$	$\epsilon_{\lambda n}$
0	0.0500	0.0500
10	0.0491	0.0506
15	0.0482	0.0515
20	0.0469	0.0529
25	0.0453	0.0548
30	0.0433	0.0573
35	0.0410	0.0605
40	0.0384	0.0645
45	0.0355	0.0703
50	0.0323	0.0764
55	0.0289	0.0852
60	0.0252	0.0971
65	0.0214	0.1138
70	0.0173	0.1387
75	0.0131	0.1791
80	0.0088	0.2543
85	0.0044	0.4587
88°58′	$\epsilon_{\lambda n} = \epsilon_{\lambda n, \max} = 0.8182$	
90	0	0

tion $\varkappa$. For a radiant beam incident in a direction normal to the surface, for instance, the following equation applies:

$$\rho_\lambda = \frac{(n-1)^2 + \varkappa^2}{(n+1)^2 + \varkappa^2} \tag{15-13}$$

In metals, the coefficient of absorption $\varkappa$ is practically equal to the index of refraction n, and this latter value can be connected with the electric resistivity ρ. In this way the following equation is obtained for the reflectivity of a radiant beam incident with the angle β and polarized in a direction normal to the plane of incidence:

$$\rho_{\lambda n} = 1 - \epsilon_{\lambda n} = \frac{60\lambda/\rho - 2\sqrt{30\lambda/\rho}\cos\beta + \cos^2\beta}{60\lambda/\rho + 2\sqrt{30\lambda/\rho}\cos\beta + \cos^2\beta} \tag{15-14}$$

For a beam polarized in the plane of incidence, the equation is

$$\rho_{\lambda p} = 1 - \epsilon_{\lambda p} = \frac{60\lambda/\rho\cos^2\beta - 2\sqrt{30\lambda/\rho}\cos\beta + 1}{60\lambda/P\cos^2\beta + 2\sqrt{30\lambda/\rho}\cos\beta + 1} \tag{15-15}$$

The wavelength λ must be introduced into this equation in centimeters and the electric resistivity ρ in ohm-centimeters. Integration of the monochromatic intensities to obtain total reflectance values is best carried out numerically.[1] For this reason, an equation[1] will be presented only for the total normal emittance

$$\epsilon_\perp = 0.576\sqrt{\rho T} - 0.124\rho T \tag{15-16}$$

(ρ in ohm-centimeters and T in degrees Kelvin). For pure metals, the electric resistivity changes with temperature as

$$\rho = \rho_0 \frac{T}{273} \tag{15-17}$$

where ρ_0 is the resistivity at $T = 273$ K. This results in a temperature dependence of the emittance

$$\epsilon = 0.0348\sqrt{\rho_0}\,T - 0.00045\rho_0 T^2 \tag{15-18}$$

For good electric conductors, the first term in these equations is dominant.

Nonconductors Electrically nonconducting materials generally have large emittance values in the infrared. Such values are listed in Table 15-3 for materials of engineering importance. The distribution in the various directions in space is also different from that of metals, as can be seen in Fig. 15-7. Monochromatic emittances for a few nonconductors of engineering importance are presented in Fig. 15-8. It is found that in the infrared range almost all electric nonconductors have large values of monochromatic emittance, magnesium oxide being an exception. In

[1] E. Schmidt and E. Eckert, *Forsch. Gebiete Ingenieurw.*, **6**:175 (1935).

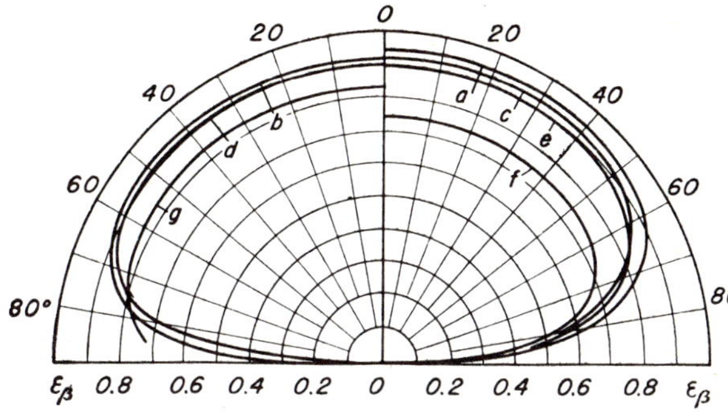

Fig. 15-7 Directional distribution of total emittance of electrically nonconducting materials. [*After E. Schmidt and E. Eckert, Forsch. Gebiete Ingenieurw.,* **6**:*175* (*1935*).]

the visible range and in the near infrared, the emittance values can again be estimated from their appearance to the eye. Bright surfaces have low emittance values, like the white tile in Fig. 15-8, whereas surfaces appearing dark to the eye have high emittance values in the visible and near infrared range. This behavior determines how the total emittance changes with temperature. For surfaces which appear bright to the eye, one can expect a decrease of ϵ with increasing temperature, whereas surfaces with a dark appearance will exhibit an increase or a small drop. Figure 15-6 verifies this rule.

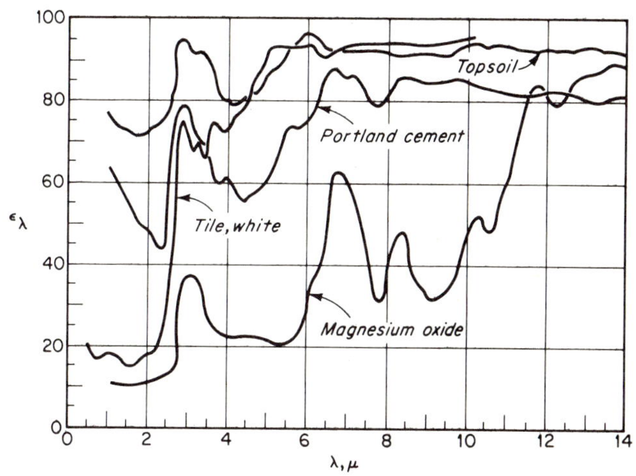

Fig. 15-8 Monochromatic normal emittance of electric nonconductors as function of wavelength. [*After W. Sieber, R. V. Dunkle, and J. T. Gier.*]

Electromagnetic theory can again be used to obtain information on emittances of electrically nonconducting materials which is useful as long as no experimental information is available. Fresnel's equation (15-13) is used to calculate the monochromatic reflectance. The absorption coefficient of electrically nonconducting materials is smaller than the corresponding value for metals; therefore the coefficient $\varkappa$ can be set approximately zero in Eq. (15-13) to obtain the reflectance for a beam incident normal to the surface. For arbitrary angles of incidence, the equation is

$$\rho_\lambda = 1 - \epsilon_\lambda = \frac{1}{2}\frac{\sin^2(\phi - \psi)}{\sin^2(\phi + \psi)}\left[1 + \frac{\cos^2(\phi + \psi)}{\cos^2(\phi - \psi)}\right] \tag{15-19}$$

The angle ϕ in this equation denotes the angle of incidence and ψ the angle at which the refracted beam leaves the interface. The following relation connects the two angles

$$\sin\phi = n \sin\psi \tag{15-20}$$

n denotes the index of refraction. The spatial distribution of the total emittance obtained from Eq. (15-19) by an integration over all wavelengths agrees well with measured distribution curves. The ratio of the normal to the hemispherical total emittance of electric nonconductors as obtained from the above equations is also inserted in Fig. 15-3. It can be observed that the agreement with measured values is not too bad.

Absorptance Monochromatic directional absorptance values are connected with the monochromatic directional emittance by Kirchhoff's relation

$$\alpha_\lambda = \epsilon_\lambda \tag{15-21}$$

Strictly speaking, this relation also holds only for a specified polarized radiant beam. Monochromatic, directional emittance values, as found in handbooks, can therefore be used to obtain absorptance values through Eq. (15-21). Unfortunately, information on emittance values of this kind is quite incomplete, and it is therefore of interest to determine under what conditions absorptance values can be obtained from hemispherical or total emittances. The directional absorptance is defined as the ratio of the absorbed energy flux to the energy flux impinging on a surface in a specified direction. Accordingly, the hemispherical absorptance is defined as the ratio of the absorbed energy flux integrated over all directions, or, in the form of an equation,

$$\alpha_{\lambda\bigtriangleup} = \frac{\int_{\bigtriangleup} \epsilon_\lambda K_{\lambda i}\, dA \cos\beta_i \, d\omega_i \, d\lambda}{\int_{\bigtriangleup} K_{\lambda i}\, dA \cos\beta_i \, d\omega_i \, d\lambda} \tag{15-22}$$

The index i refers to the incident beam, $K_{\lambda i}$ denotes the intensity of the incident flux, and Eq. (15-21) was used to replace the directional absorptance by the directional emittance. The terms dA and $d\lambda$ cancel out. The symbol underneath the integral sign indicates that the integration is to be performed over all directions in the hemispherical space. The equation demonstrates that the hemispherical absorptance depends on the directional distribution of $K_{\lambda i}$. Therefore no general tabulation of hemispherical absorptances is possible. Only when the intensity $K_{\lambda i}$ of the incident radiation is independent of direction can this value be taken out of the integrals, and the equation simplifies to

$$\alpha_{\lambda\bigtriangleup} = \frac{\int_{\bigtriangleup} \epsilon_\lambda \cos \beta_i \, d\omega_i}{\int \cos \beta_i \, d\omega_i} = \frac{1}{\pi} \int_{\bigtriangleup} \epsilon_\lambda \cos \beta_i \, d\omega_i = \epsilon_{\lambda\bigtriangleup} \tag{15-23}$$

This equation now demonstrates that for uniform irradiation the hemispherical absorptance is equal to the hemispherical emittance. Such a relation holds even when the irradiation occurs only over a solid angle which does not cover the whole hemisphere. If the irradiation is uniform over this solid angle, the average absorptance is equal to the average emittance, both averaged over the solid angle in question.

It is also of interest to investigate under what conditions total absorptance values for radiation integrated over all wavelengths can be obtained from total emittance values. To determine the total absorptance requires an integration of the absorbed energy flux $d\Phi_{\lambda a}$ and of the incident radiative energy flux over all wavelengths according to the equation

$$\alpha = \frac{\int_{\lambda=0}^{\infty} d\Phi_{\lambda a}}{\int_{\lambda=0}^{\infty} H_{\lambda i} \, dA \cos \beta_i \, d\omega_i \, d\lambda} \tag{15-24}$$

Introducing the definition of the monochromatic directional absorptance and using Kirchhoff's equation (15-21) results in

$$\alpha = \frac{\int_{\lambda=0}^{\infty} \epsilon_\lambda K_{\lambda i} \, d\lambda}{\int_{\lambda=0}^{\infty} K_{\lambda i} \, d\lambda} \tag{15-25}$$

Again the value of the total absorptance α depends on the way in which the intensity $K_{\lambda i}$ is distributed over the wavelength range, and no general tabulation of such values is possible.

Some statements, however, can be made for incident black radiation. In this case the above equation becomes

$$\alpha(T_s,T_i) = \frac{\int_{\lambda=0}^{\infty} \epsilon_\lambda(T_s) i_{\lambda bi}(T_i)\, d\lambda}{\int_{\lambda=0}^{\infty} i_{\lambda bi}(T_i)\, d\lambda} \tag{15-26}$$

In this equation the value of the monochromatic directional emittance ϵ_λ depends on surface temperature T_s; whereas the value for the incident monochromatic black radiation intensity $i_{\lambda bi}$ is a function of the temperature T_i which characterizes this radiation. Therefore the total absorptance α is now a function of the temperature of the surface as well as of the temperature of the incident black radiation. This distinguishes, in general, the total absorptance from the total emittance as defined by Eq. (15-12). The total absorptance will be equal to the total emittance for the following cases:

1. When the temperature T_i of the incident black radiation is equal to the temperature T_s of the absorbing surface.
2. When the monochromatic emittance ϵ_λ is independent of wavelength, because in this case the emittance can be moved out of the integral sign in the above equation. A body for which the emittance is independent of wavelength is call a *gray body*. For such a body $\alpha(T_s,T_i) = \epsilon(T_s)$.
3. When the monochromatic emittance is independent of temperature, the absorptance of a surface at temperature T_s for incident black radiation of temperature T_i is equal to the emittance of the surface at a temperature T_i $[\alpha(T_s,T_i) = \epsilon(T_i)]$. This follows from a comparison of Eqs. (15-26) and (15-12). The last-mentioned relation is quite useful for electric nonconductors in which the dependence of emittance on wavelength is much more pronounced than the dependence on temperature.

For metals, the variations of emittance with wavelength and temperature are of about the same magnitude. From electromagnetic theory then the following relation can be derived,[1] which is valid for incident black radiation:

$$\alpha(T_s,T_i) = \epsilon(\sqrt{T_sT_i}) \tag{15-27}$$

According to this relation, the absorptance of a surface at temperature T_s for incident black radiation of temperature T_i is equal to the total emittance of the surface at the temperature $\sqrt{T_sT_i}$. Relations derived from electromagnetic theory agree quite well with reality in the infrared

[1] E. Eckert, *Forch. Gebiete Ingenieurw.*, **7**:265–270 (1936).

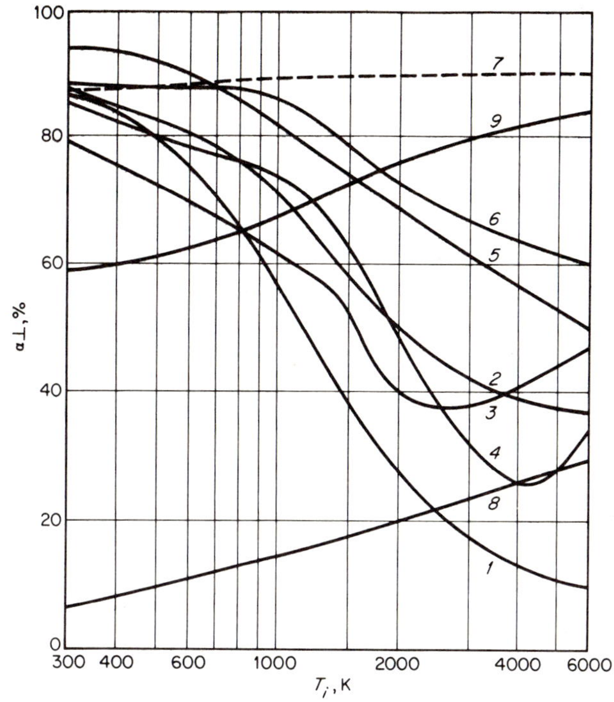

Fig. 15-9 Total normal absorptance of various materials for incident blackbody radiation of temperature T_i. [*From W. Sieber, Z. Tech. Phys.,* **22**:*130* (*1941*).]

region. Therefore the above relation can be used for radiation in the infrared wavelength range.

The integrals in the above equations are usually determined numerically or graphically. To obtain the absorptance for impinging black radiation, for instance, each ordinate of the pertinent curve in Fig. 14-8 is multiplied by the corresponding ϵ_λ (taken, for example, from Fig. 15-5 or 15-8). The resulting curve must be planimetered and divided by the area under the radiation curve in Fig. 14-8. In this way the absorptance (and reflectance) values in Fig. 15-9 were determined by Sieber.[1] In the curves, the difference between electric conductors (represented by aluminum) and nonconductors can be observed. For nonconductors, the absorptance decreases with temperature; for conductors, it increases. Industrial radiators have temperatures between 300 and 3000 K. For such radiation, the absorptance of nonconductors is greater than the

[1] W. Sieber, *Z. Tech. Phys.*, **22**:130–135 (1941).

values for conductors. The sun has an apparent temperature of 5600 K. At this temperature nonconductors with white surfaces absorb less of the radiation than metal surfaces. Only a few metals, e.g., silver, also have very small absorptance in the range of visible radiation. Silver has a value of $\rho = 0.96$ for the reflectance in this range.

Reflectance The reflectance in Eqs. (15-7) and (15-8) is concerned with the energy reflected on a surface regardless of how it is distributed in space. This follows from the derivation of the equations. In radiative exchange calculations, one is often concerned with the directional distribution of the reflected radiation; correspondingly, various definitions of reflectance are required. The reflectance in Eqs. (15-7) and (15-8) may be called directional-hemispherical (or angular-hemispherical) reflectance because it considers radiation incident in a specified direction being reflected into all directions in space. It is defined by the equation

$$\rho_{dh} = \frac{d\varphi_r}{d\varphi_i} = \frac{d\varphi_r}{K_i \cos \beta_i \, d\omega_i} \tag{15-28}$$

[$d\varphi = d\Phi/dA$; see Eq. (14-1)]. In describing the directional distribution of reflected radiation, it is of advantage to consider two extreme situations. Reflection is mirrorlike or specular if a radiant pencil approaching the surface with an angle β_i toward the surface normal is reflected within the incidence plane with the same angle and within an opening angle of the same magnitude. A perfectly diffuse surface is one which reflects an incident beam into all directions of space with an intensity independent of the direction of the reflected beam. A diffuse surface, in a general sense, is a surface which reflects an incident beam into all directions in space with an intensity dependent on the direction of the reflected beam. It is convenient to use two different definitions for the reflectance for specular and diffuse surfaces. The specular reflectance is the ratio of the intensity of the reflected radiation to the intensity of the incident radiation according to

$$\rho_s = \frac{K_r}{K_i} \tag{15-29}$$

The bidirectional, or biangular, reflectance used for a diffuse surface is defined as the ratio of the intensity of the radiation reflected in a certain direction to the specific energy flux $d\varphi_i$ of the incident radiation:

$$\rho_{bd} = \frac{dK_r}{d\varphi_i} = \frac{dK_r}{K_i \cos \beta_i \, d\varphi_i} \tag{15-30}$$

It will, in general, depend on the direction of incident radiation as well as on the direction of reflected radiation. The parameters in Eqs. (15-29) and (15-30) are defined such that their values are independent of the solid angle of the incident beam as long as this angle is small. The directional hemispherical reflectance can be obtained by an integration from the bidirectional reflectance values with the equation

$$\rho_{dh} = \int_{\bigcirc} \rho_{bd} \cos \beta_r \, d\omega_r \tag{15-31}$$

The integration has to be performed over all solid angles of the reflected rays.

The bidirectional reflectance depends on the azimuth angle as well as on the angle β_r, which is measured toward the surface normal. Its complete determination is therefore quite cumbersome, and few data exist for surfaces of engineering importance. The directional emittance and absorptance values as well as the directional hemispherical reflectance, on the other hand, are functions of the angle β_i alone as long as the surface is isotropic. For an unisotropic surface, like a grooved one, these values depend on the azimuth angle of the incident ray as well.

The directional hemispherical reflectance is difficult to measure, and it is often preferable to determine a reflectance value which may be called the *hemispherical directional reflectance.* This value is concerned with the energy reflected in a specific direction when the surface is irradiated from all directions in space. Its proper definition is

$$\rho_{hd} = \frac{\pi K_r}{\varphi_i} \tag{15-32}$$

It can be shown with the use of Helmholtz' theorem that this reflectance is equal to the directional hemispherical reflectance provided the irradiation has an intensity independent of direction.

The directional distribution of reflected radiation is presented for a few materials of engineering importance in Fig. 15-10*a* and *b*. Incident black radiation at approximately 280 C arrived at the surface within a cone with a 6° opening angle and at near normal incidence. The optics of the experimental setup was such that radiation reflected from a specular surface is recorded within an angular range of 11°. The reflection of the anodized aluminum surface can therefore be considered as perfectly specular; other surfaces indicate a gradual transition to diffuse radiation. The wood surface in Fig. 15-10*b* exhibits nearly perfect diffuse reflection.

The directional distribution of reflected radiation is mainly determined by the scale of the surface roughness and by the wavelength of the incident radiation. It is to be expected that a surface will act specularly when the ratio of the mean roughness dimension σ to the wavelength λ is small. It will act diffusely when this ratio is large. Experiments studying the directional distribution of monochromatic reflected radiation are

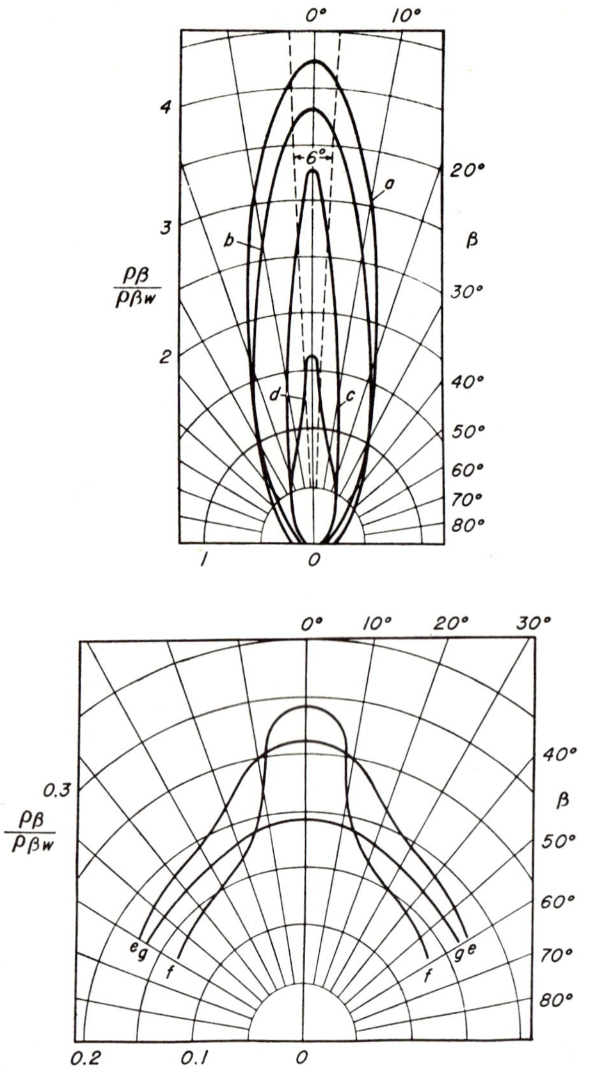

Fig. 15-10 Bidirectional reflectance of various materials: curve a = aluminum, anodized; curve b = aluminum paint; curve c = iron, scraped; curve d = black iron; curve e = copper oxide. [*From E. Eckert, Forsch. Gebiete Ingenieurw.*, **7**:*265 (1936)*.]

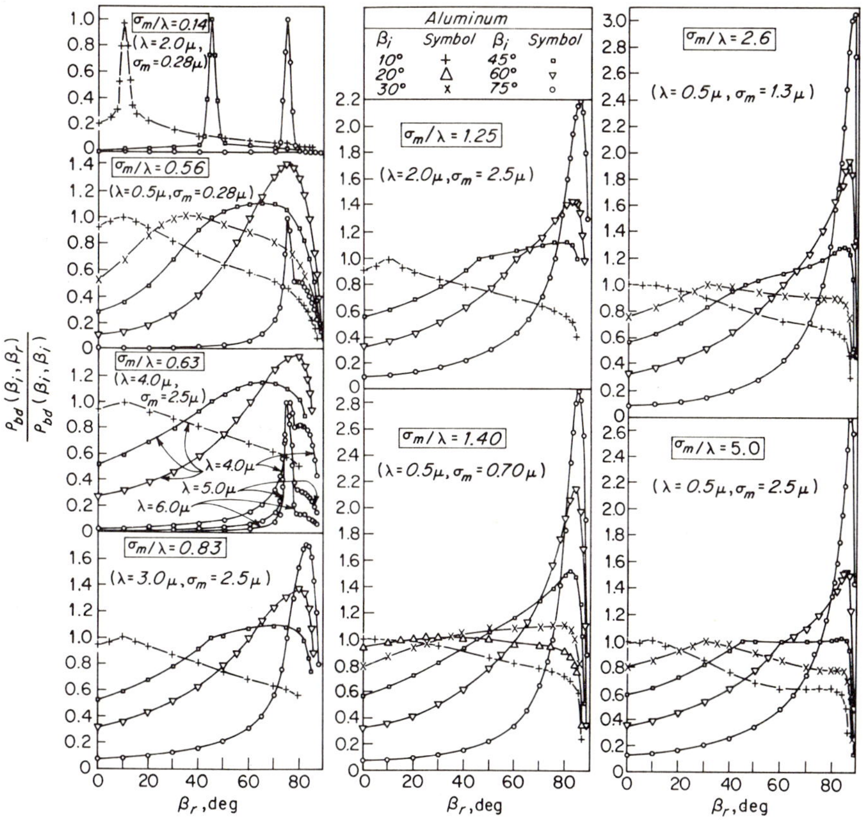

Fig. 15-11 Ratio of monochromatic bidirectional reflectance to monochromatic bidirectional reflectance in the direction of the specular ray for aluminum surfaces with varying roughness. [*From K. E. Torrance and E. M. Sparrow, J. Heat Transfer,* **87**:*283* (*1965*).]

therefore more revealing than measurements of total reflectances, as shown in Figs. 15-10.

Information on the influence of surface roughness on the bidirectional reflectance has recently been reported.[1,2] Figure 15-11 presents the results obtained for aluminum. Various samples with different roughness were obtained by grinding. Each of the diagrams in the figure

[1] R. C. Birkebak and E. R. G. Eckert, Effects of Roughness on Angular Distribution of Monochromatic Reflected Radiation, *J. Heat Transfer,* **87**:85 (1965).

[2] K. E. Torrance and E. M. Sparrow, Bi-angular Reflectance of an Electric Nonconductor as a Function of Wavelength and Surface Roughness, *J. Heat Transfer,* **87**:283 (1965).

presents the results for one of the samples. The parameter in the diagrams, σ_m/λ, is the ratio of the mean roughness size σ_m, which was measured mechanically by a small stylus, to the wavelength λ. The abscissa of the diagrams lists the angle β_r between the reflected ray and the surface normal. The ordinate lists the ratio of the bidirectional reflectance $\rho_{bd}(\beta_i,\beta_r)$ for an arbitrary value of β_r to the bidirectional reflectance $\rho_{bd}(\beta_i,\beta_i)$ for the situation when the angle of the reflected ray is equal to the angle of the incident ray ($\beta_r = \beta_i$). The various curves in each diagram are for different values β_i of the incident ray. All measurements were made in the plane of incidence. It can be observed that, for a roughness ratio $\sigma_m/\lambda = 0.14$, the aluminum surface acts almost like a specular surface, especially for larger values of β_i. With increasing value of the roughness ratio, the reflected radiation becomes more and more diffuse, and for a roughness ratio $\sigma_m/\lambda = 5.0$, the surface acts almost like an ideally diffuse one as long as the incident angle is not too large. One peculiarity can also be observed, namely, that the surfaces beginning with a roughness ratio 0.83 exhibit a peak in the intensity of the reflected radiation at an angle which is larger than the incidence angle. These peaks have been termed *off-specular peaks*. It can be observed, for instance, that the surface with the roughness ratio 1.40 has an off-specular peak at an angle of 85° for radiation incident at an angle of 75° and off-specular peaks at almost the same angle for the incidence angles 60 and 45°. The existence of off-specular peaks can be demonstrated analytically assuming a surface composed of a large number of plane facets oriented statistically in the various directions and reflecting specularly. An analysis[1] based on such a model of a rough surface and applying the laws of geometric optics predicts off-specular peaks in good agreement with measurements.

Bidirectional reflectances have also been measured on other metal surfaces with the result that the parameter determining the directional distribution of the reflected radiation is the roughness ratio σ/λ. This is to be expected since the reflection of radiation occurs very close to the surface. Measurements on magnesium oxide ceramic, an electrically nonconducting material, showed practically the same dependence of the bidirectional reflectance on the roughness ratio but one must be more careful in generalizing this result because the reflection in a nonconducting material occurs partially by internal reflections in the absorbing and refracting layer.

Radiation reflected specularly from a perfectly smooth surface of an electrically nonconducting material is partially polarized for incidence angles different from zero and is completely polarized for an incidence

[1] *Ibid.*

angle around 57° in a plane normal to the incidence plane. This is called *Brewster's angle.*

15-2 ENCLOSURES WITH A NONPARTICIPATING MEDIUM

With the knowledge of the radiation properties, we can now turn to the problem of analyzing energy exchange between surfaces in an enclosure filled with a medium not participating by emission, absorption, or scattering in the radiative transfer process. Engineering systems to which such calculations can be applied are, for instance, satellites or space probes, moving in a vacuum, so that radiation is the only means by which energy is transferred. The surfaces of the enclosure are assumed to be surfaces of materials which have been termed *strongly absorbing media.* These are media (solids or liquids) of such a nature that only a very thin layer below the surface participates in the radiative exchange. The problem of analyzing energy exchange in such an enclosure is of quite general usefulness because radiative exchange between a surface and its surroundings can be solved only if all the radiative energy which can arrive at the surface from any direction is accounted for. The enclosure may have windows or openings through which energy leaves or enters. In this case, however, the radiative flux entering the enclosure through the opening must be prescribed.

It will soon become apparent that the task of calculating the radiative energy exchange in an enclosure of somewhat complicated geometry with surfaces having radiation properties depending on direction and wavelength is formidable. For this reason we shall start with situations which are simplified by the fact that the surfaces emit with an intensity independent of direction and which have either perfectly diffuse or perfectly specular reflection. This leads to considerable simplification in the analysis, among other reasons as a consequence of the fact that the influence of the mutual spatial arrangement of the surfaces can be treated separately from the radiative-exchange calculation itself. The following subsection will discuss the *shape* or *angle factors* which determine the influence of the local arrangement of the surfaces involved.

The shape factor A term used extensively in calculations of radiant energy interchange is the shape factor or angle factor. In order to derive it, we consider two area elements, dA_1 and dA_2, with dimensions small compared to the distance s between them, as shown in Fig. 15-12. The radiation leaving the area dA_1 may be of such a nature that its intensity K_1 is independent of direction. It is, however, immaterial whether this radiative flux is produced by emission, reflection, or a combination of both processes and whether it is monochromatic or integrated over all

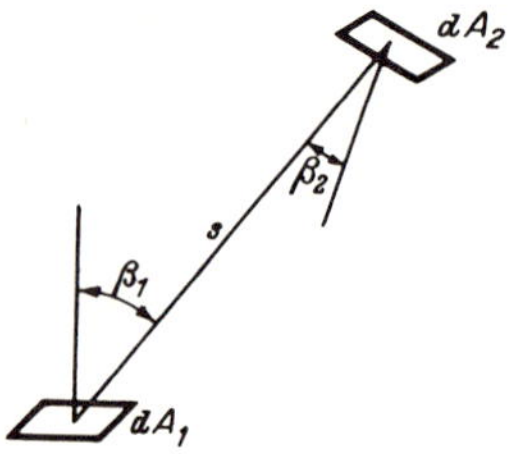

Fig. 15-12 Heat exchange by radiation of two small surfaces.

wavelengths. The flux leaving the area element per unit area and time in all directions in space may be denoted by B_1. The term *radiosity* has become customary for this quantity. The following relation exists between the quantities K_1 and B_1:

$$B_1 = \pi K_1 \tag{15-33}$$

since we assumed the intensity K_1 of the radiative flux to be independent of direction. This is the same relation which connects the emissive power and the intensity of emission for a blackbody, and it is also derived in the same way.

The angles between the two surface normals and s are denoted by β_1 and β_2, respectively. The heat flux leaving surface dA_1 per unit time is

$$d\Phi_1 = B_1\,dA_1 \tag{15-34}$$

That part of this heat flux which is incident on surface dA_2 can be written as

$$d^2\Phi_{dA_1-dA_2} = K_1\,dA_1 \cos\beta_1\,d\omega_1 \tag{15-35}$$

where $d\omega_1$ indicates the solid angle under which the surface element dA_2 is seen from dA_1. Expressing the solid angle through the other terms introduced above, one obtains alternatively

$$d^2\Phi_{dA_1-dA_2} = K_1\,dA_1 \cos\beta_1 \frac{\cos\beta_2\,dA_2}{s^2} \tag{15-36}$$

The shape factor of the area element dA_1 toward dA_2 is now defined as the ratio of the heat flux $d^2\Phi_{dA_1-dA_2}$ to the heat flux $d\Phi_1$

$$dF_{dA_1-dA_2} = \frac{\cos\beta_1\,d\omega_1}{\pi} = \frac{\cos\beta_1 \cos\beta_2\,dA_2}{\pi s^2} \tag{15-37}$$

It can be seen that the angle factor thus defined depends only on angular relationships describing the mutual position of the two area elements. This, however, depends on the condition that the intensity K_1 be independent of direction. Multiplication of the angle factor by the area element dA_1 results in

$$dA_1\, dF_{dA_1-dA_2} = \frac{\cos\beta_1 \cos\beta_2\, dA_1\, dA_2}{\pi s^2} \tag{15-38}$$

This equation is now symmetrical with respect to surfaces 1 and 2. The angle factor of surface dA_2 toward surface dA_1 is therefore

$$dF_{dA_2-dA_1} = \frac{\cos\beta_2 \cos\beta_1\, dA_1}{\pi s^2}$$

The following reciprocity relation results from the two preceding equations:

$$dA_1\, dF_{dA_1-dA_2} = dA_2\, dF_{dA_2-dA_1} \tag{15-39}$$

We shall see that such reciprocity relations are useful in the analysis of radiative-energy exchange.

The angle factor for a finite area A_2 is obtained by integration over all solid-angle elements or all area elements of A_2, according to the equation

$$F_{dA_1-A_2} = \frac{1}{\pi}\int_{\omega_1} \cos\beta_1\, d\omega_1 = \frac{1}{\pi}\int_{A_2} \frac{\cos\beta_1 \cos\beta_2}{s^2}\, dA_2 \tag{15-40}$$

This, then, expresses the angle factor of an area element dA_1 toward an area A_2 with dimensions of the same order of magnitude as the distance s—or alternatively that fraction of the radiant energy leaving an area element dA_1 which is incident on the area A_2—as long as the intensity K_1 of the energy flux leaving surface dA_1 is independent of direction. The reciprocity relation between the shape factor of dA_1 toward A_2 and of A_2 toward dA_1 again exists in the form

$$dA_1\, F_{dA_1-A_2} = A_2\, dF_{A_2-dA_1} \tag{15-41}$$

Relations for the angle factor described by Eq. (15-40) can be found by integration over the solid angle ω_1 or the area A_2. In some cases it is possible to simplify the integration process by converting the area integral in Eq. (15-40) into a line integral using Stokes' theorem.[1] Relations for angle factors of various shapes are also tabulated in textbooks and handbooks dealing with thermal radiation. A number of them are listed in Table 15-5.

As a last step, we shall derive the shape factor for two finite areas, A_1 and A_2. For this purpose, one additional restriction must be imposed, namely, that the radiosity B_1 be uniform over the area A_1. Starting with

[1] E. M. Sparrow, *J. Heat Transfer*, **85**:81–88 (1963).

Table 15-5†

RADIATING AREA ELEMENT

$$F_{1-2} = \frac{h}{2}\left[\frac{h^2+r^2+1}{\sqrt{(h^2+r^2+1)^2-4r^2}} - 1\right]$$

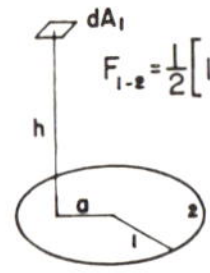

$$F_{1-2} = \frac{1}{2}\left[1 - \frac{h^2+a^2-1}{\sqrt{(h^2+a^2+1)^2-4a^2}}\right]$$

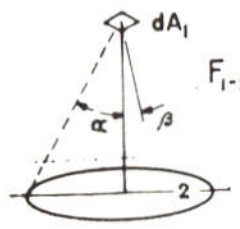

$$F_{1-2} = \sin^2\alpha \cos\beta$$

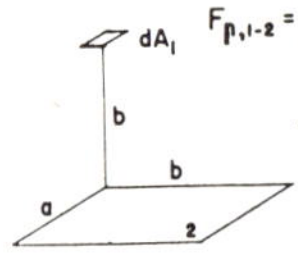

$$F_{p,1-2} = \frac{1}{2\pi}\left[\frac{a}{\sqrt{a^2+h^2}}\tan^{-1}\frac{b}{\sqrt{a^2+h^2}} + \frac{b}{\sqrt{b^2+h^2}}\tan^{-1}\frac{a}{\sqrt{b^2+h^2}}\right]$$

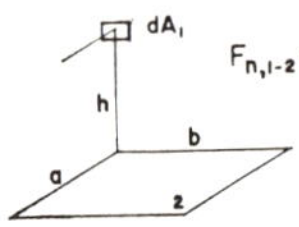

$$F_{n,1-2} = \frac{1}{2\pi}\left[\tan^{-1}\frac{b}{h} - \frac{h}{\sqrt{a^2+h^2}}\tan^{-1}\frac{b}{\sqrt{a^2+h^2}}\right]$$

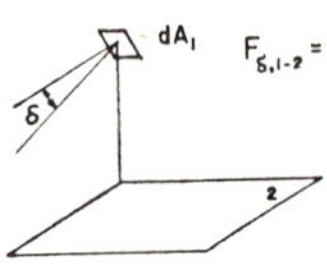

$$F_{\delta,1-2} = F_{p,1-2}\cos\delta + F_{n,1-2}\sin\delta$$

FINITE RADIATING AREA:

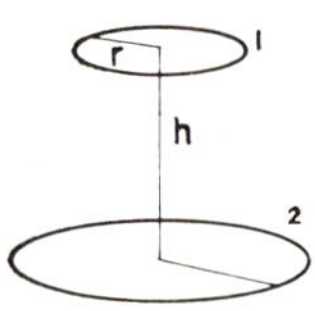

$$\bar{F}_{1-2} = \frac{1}{2}\left[\frac{h^2+1}{r^2} + 1 - \sqrt{\left(\frac{h^2+1}{r^2}+1\right)^2 - \frac{4}{r^2}}\right]$$

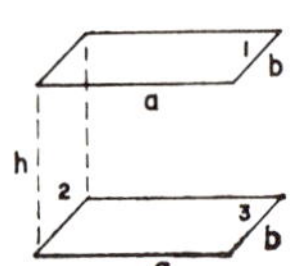

$$\bar{F}_{1-3} = \frac{2}{ab\pi}\left[b\sqrt{h^2+a^2}\tan^{-1}\frac{b}{\sqrt{a^2+h^2}} + a\sqrt{b^2+h^2}\tan^{-1}\frac{a}{\sqrt{b^2+h^2}} - bh\tan^{-1}\frac{b}{h} - ah\tan^{-1}\frac{a}{h} - \frac{h^2}{2}\ln\frac{(h^2+a^2+b^2)h^2}{(a^2+h^2)(b^2+h^2)}\right]$$

$$\bar{F}_{2-3} = \frac{1}{\pi}\left[\tan^{-1}\frac{b}{h} + \frac{a}{h}\tan^{-1}\frac{b}{a} - \frac{\sqrt{a^2+h^2}}{h}\tan^{-1}\frac{b}{\sqrt{a^2+h^2}} + \frac{h}{4b}\ln\frac{(a^2+b^2+h^2)h^2}{(a^2+b^2)(b^2+h^2)} + \frac{a^2}{4bh}\ln\frac{(a^2+b^2+h^2)a^2}{(a^2+b^2)(a^2+h^2)} - \frac{b}{4h}\ln\frac{(a^2+b^2+h^2)b^2}{(a^2+b^2)(h^2+b^2)}\right]$$

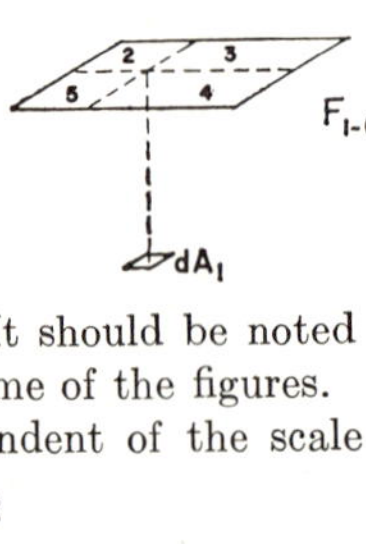

$$F_{1-(2+3+4+5)} = F_{1-2} + F_{1-3} + F_{1-4} + F_{1-5}$$

† It should be noted that one of the length dimensions has been set equal to 1 in some of the figures. This can always be accomplished since shape factors are independent of the scale of a configuration.

the shape factor A_1 toward A_2, we can write in terms of the energy flux leaving A_1 and arriving at A_2

$$F_{A_1-A_2} = \frac{\int_{A_1} dA_1 \int_{A_2} d^2\Phi_{dA_1-dA_2}}{B_1 A_1}$$

or

$$F_{A_1-A_2} = \frac{1}{A_1} \int_{A_1} F_{dA_1-A_2}\, dA_1 \tag{15-42}$$

The last equation indicates that we can interpret this shape factor as a mean value of $F_{dA_1-A_2}$ averaged over the surface A_1. Again, the reciprocity law is obtained in a simple way:

$$A_1 F_{A_1-A_2} = A_2 F_{A_2-A_1} \tag{15-43}$$

Another relation will be useful in the later analysis. When a surface A_1 is completely surrounded by n surfaces, one of which may be denoted by A_k, the relation

$$\sum_n F_{A_1-A_k} = 1 \tag{15-44}$$

follows easily from the definition of the shape factor as a ratio of energy fluxes. The numerical integration can be replaced by a graphical procedure. According to Eq. (15-40), the shape factor of a surface element dA_1 with respect to a finite surface A_2 is

$$F_{dA_1-A_2} = \frac{1}{\pi} \int \cos \beta_1 \, d\omega_1$$

The expression $\cos \beta_1 \, d\omega_1$ under the integral sign is the projection of the solid-angle element $d\omega_1$ on the plane of the radiating surface dA_1. The integral is the sum of the projections of all solid-angle elements $d\omega_1$, or the projection of the solid angle corresponding to the whole surface A_2 on the plane of dA_1. From this the following construction arises. We project by central projection the surface A_2 (denoted by 1 in Fig. 15-13) onto the surface of a hemisphere with the radius R. This surface 1′ must be projected once more by normal projection to the plane of the radiating surface dA_1. The area of this second projection 1″ divided by the area of the circle πR^2 gives the shape factor $F_{dA_1-A_2}$ as can be seen from the equation above. This graphical construction can be replaced by a simple mechanical integrator[1] or by optical projection.[2] For this purpose a point light source is arranged in the center of the hemisphere in Fig. 15-13. Area 1, the geometrical factor of which is to be determined, is made of

[1] H. Hottel, *Mech. Eng.*, **52** (1930); and V. H. Cherry, D. D. Davis, and L. M. K. Boelter, *Trans. Illum. Eng. Soc. N.Y.*, 1939.

[2] E. Eckert, *Z. Ver. Deut. Ingr.*, **79**:1495–1496 (1935).

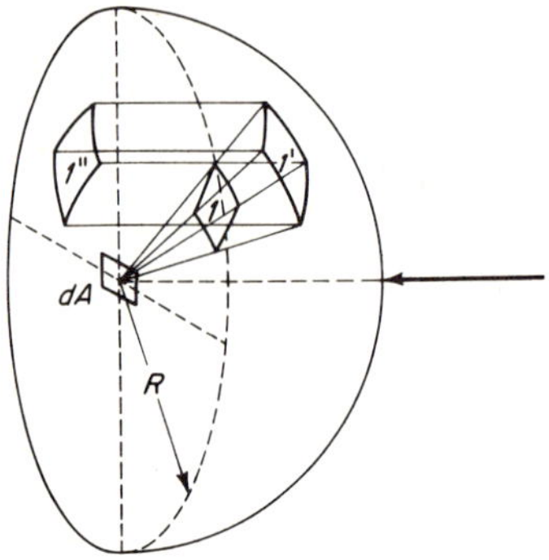

Fig. 15-13 Geometric interpretation of shape factor.

cardboard and arranged in the proper position. It is projected by the point light source as shadow on a milk-glass hemisphere. When this hemisphere is photographed from a great distance in the direction of the arrow, the ratio of the shadow of the cardboard area to the area of the circle representing the glass sphere is the shape factor. Figures 15-14*a* and 15-15 show two such photographs. Figure 15-14*a* determines the shape factor of the radiation cooling surfaces in the furnace of a steam boiler (Fig. 15-14*b*) as seen from the center of the stoker. Figure 15-15 determines the shape factor of the tungsten spiral as used in an electric light bulb. It was produced by positioning the model of the spiral a short distance from the point light source.[1]

From a study of the graphical construction shown in Fig. 15-13, analytical expressions can be derived for the shape factor of many geometries without any integration. An example will amplify this statement. The shape factor with which an area element dA irradiates a rectangle A may be determined when dA is arranged parallel to and below a corner of the rectangle A. Figure 15-16 shows this arrangement. Area A is divided into two triangles by the line AC, and the geometric construction is indicated for the triangle ABC. The primed figure $OB''C''$, which indicates the shape factor, is the projection of the sector $OB'C'$ of a great circle. The area of the sector $OB'C'$ is $\beta_1/2\pi$ times the area of the circle or $\beta_1/2$ (the radius of the sphere is chosen to be 1). The area of the sector $OB''C''$ is therefore $\beta_1/2 \sin \alpha_1$. The shape factor is this area divided by the area of the base circle, or $\beta_1/2\pi \sin \alpha_1$. In the same way the shape factor of the triangle ACD is found as $\beta_2/2\pi \sin \alpha_2$, and the shape factor of the rectangle is the sum of the shape factor of the two triangles, or

$$F_{1-2} = \frac{\beta_1 \sin \alpha_1 + \beta_2 \sin \alpha_2}{2\pi}$$

[1] *Ibid.*

(a)

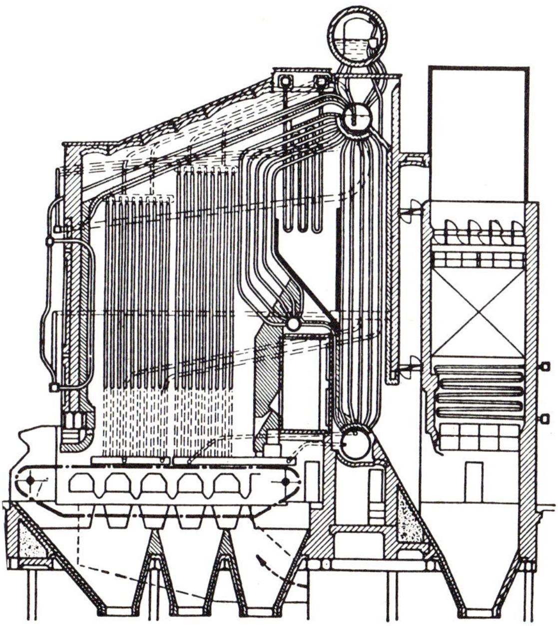

(b)

Fig. 15-14 (a) Photographic determination of shape factor of the radiation cooling surface in the furnace of the steam boiler in (b) as seen from the center of the stoker. [*From E. Eckert, Z. Ver. Deut. Ing.,* **79**:*1495* (*1935*).] (b) Steam boiler with radiation heating surface.

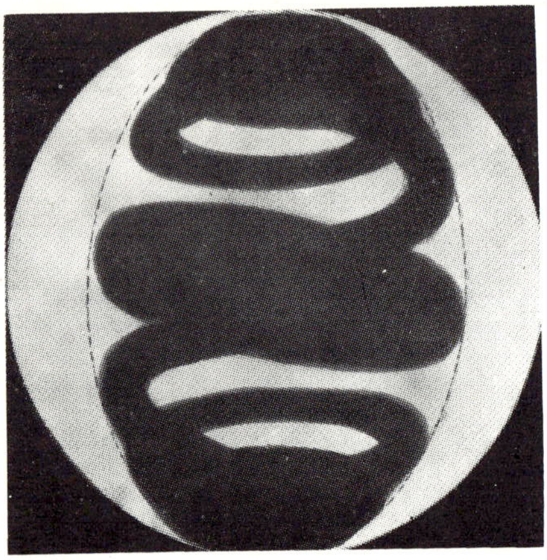

Fig. 15-15 Photographic determination of the shape factor of the tungsten coil in an electric bulb. [*From E. Eckert, Z. Ver. Deut. Ing.*, **79**:*1495* (*1935*).]

Replacement of the angles by the lengths a, b, and h results in the relation contained in Table 15-4 for this configuration.[1]

The manipulation of shape factors, indicated by Eqs. (15-43) and (15-44), sometimes referred to as *shape-factor algebra,* will be demonstrated in the following example (Fig. 15-17). A coil with radius r_c and surface area A_c is heated to a uniform temperature T_c, and as a consequence radiates energy; the task is to calculate the magnitude of the radiant energy flux which the coil emits into the surroundings. It will be assumed that the coil surface is black and that the medium in which the coil is arranged does not participate in the radiative exchange. The coil will also be assumed long enough to permit end effects to be neglected. The radiant flux which the coil sends into the surroundings must pass an imaginary cylindrical surface A_s with an arbitrary radius r_s. The problem can therefore be reformulated by asking what amount of the radiation emitted by the coil impinges on the surface A_s. This flux Φ_c can be obtained from the equation

$$\Phi_c = A_c F_{A_c-A_s} \sigma T_c^4 = A_s F_{A_s-A_c} \sigma T_c^4 \tag{15-45}$$

where the angle factor $F_{A_s-A_c}$ can be further expressed by the relation

$$F_{A_s-A_c} = \frac{1}{A_s} \int F_{dA_s-A_c}\, dA_s \tag{15-46}$$

The local shape factor $F_{dA_s-A_c}$ can easily be determined by the optical method, discussed before. A point light source located at the center of

[1] A program to calculate shape factors on a digital computer, written in Fortran IV language, is available: K. A. Toups, A General Computer Program for the Determination of Radiant-interchange Configurations and Form Factors CONFAC II, *ASTIA Doc.* AD440411, 1964.

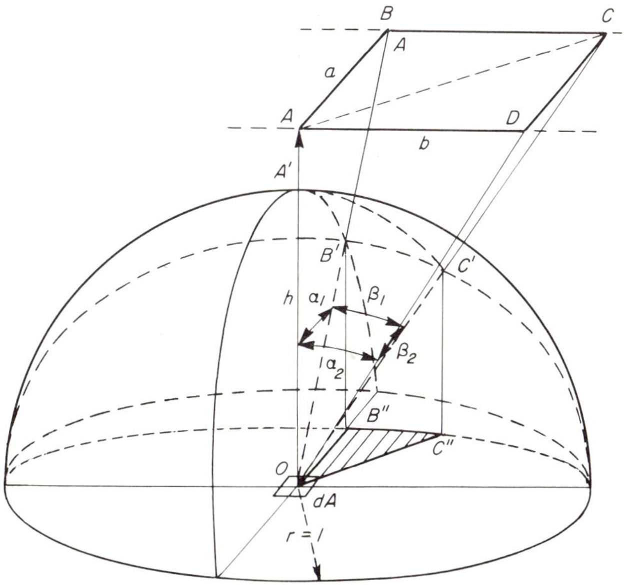

Fig. 15-16 Shape factor for radiative exchange between a rectangle and an area element.

the area element dA_s creates on a translucent hemisphere a shadow of the coil, as presented in Fig. 15-15. The ratio of the area of the shadow to the area of the illuminated surface is equal to the angle factor $F_{dA_s-A_c}$. The average angle factor required in Eq. (15-46) can be obtained by shifting the coil up and down through a distance s. The mean of several photographs of the kind as shown in Fig. 15-15 is therefore the average angle factor. It is found that in practice the local variation of the angle factor along the surface A_s is quite small as long as the radius r_s is not

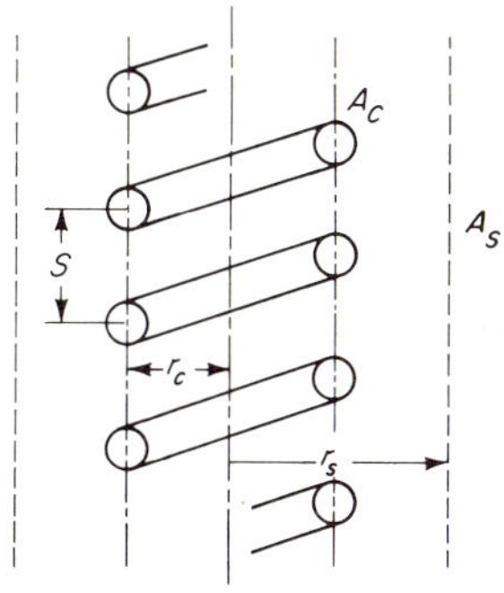

Fig. 15-17 Determination of shape factor for a coil.

too close to the radius r_c, so that a single photograph gives a sufficient approximation to the average angle factor, and with it the heat flux can be obtained from Eq. (15-45).

Enclosure with black surfaces We can now turn to the analysis of heat exchange within an enclosure for the situation where the medium inside the enclosure does not participate in the radiative process. As a first step, it will be assumed that the enclosure consists of black surfaces. The temperature is assumed constant over each surface, and the shape factors between any of the surfaces are assumed known.

We fix our attention at a surface i and calculate the net loss Q_i of this surface by the radiative energy exchange.

The radiative energy flux leaving surface i with area A_i is $A_i e_{bi}$, where e_{bi} denotes the blackbody emissive power according to the equation $e_{bi} = \sigma T_i^4$. The radiative energy flux leaving a surface k in the direction toward surface i is $A_k F_{k-i} e_{bk}$. All this heat arrives at surface i and is absorbed by it. The total radiative heat flux absorbed by the surface i is therefore

$$\sum_k A_k F_{k-i} e_{bk}$$

The net heat lost per unit time by surface i is therefore

$$Q_i = A_i e_{bi} - \sum_k A_k F_{k-i} e_{bk}$$

Using the reciprocity law for angle factors, the equation can be written

$$Q_i = A_i \left(e_{bi} - \sum_k F_{i-k} e_{bk} \right)$$

The sum of the angle factors for all surfaces surrounding surface i must be equal to 1.

$$\sum_k F_{i-k} = 1$$

Therefore the above relation can also be written

$$Q_i = A_i \Sigma F_{i-k} (e_{bi} - e_{bk}) \tag{15-47}$$

An equation of this form can be written for all n of the surfaces. In order to solve this system of equations, boundary conditions must be prescribed. Usually either the surface temperature T, and therefore the emissive power e_b, is known, or the net heat loss or gain Q of the surface is prescribed. The system of equations then can be used to calculate the net heat flux for the surfaces with known temperature or the temperature for the surfaces with known heat flux.

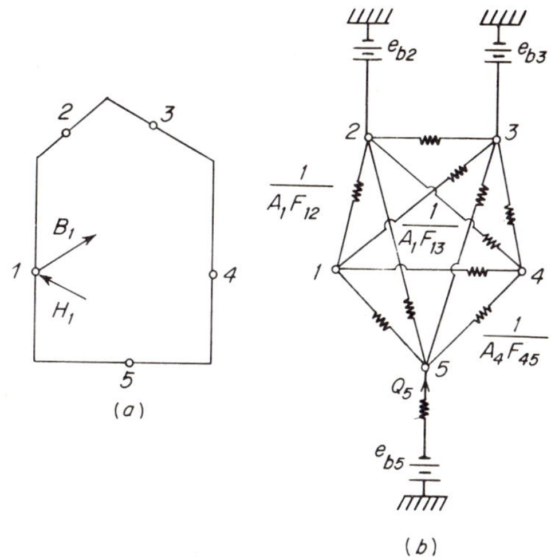

Fig. 15-18 Electrical analog of radiation in an enclosure with black walls.

Paschkis[1] and, more recently, Oppenheim[2] have pointed out that analogs in the form of an electric network can be devised for such a radiative exchange process. The network for the enclosure shown in Fig. 15-18*a* is sketched in Fig. 15-18*b*. The enclosure consists of five surfaces of which surfaces 2, 3, and 5 may be kept at a prescribed temperature whereas the heat loss of surfaces 1 and 4 may be equal to zero. A node in the analog corresponds to a surface in the enclosure. All nodes are interconnected by wires with resistances proportional to $1/A_iF_{i-k}$. The nodes corresponding to surfaces with known temperature are connected with electric potentials proportional to e_{bi}. The nodes corresponding to the adiabatic surfaces 1 and 4 are without external connection. The currents from the outside connections to the nodes are then proportional to the heat fluxes Q_i, and the electric potential of any of the nodes without an outside connection is proportional to the emissive power of the respective adiabatic surface. Actually an enclosure must be considered as a configuration in space, and all its surfaces must be included in the calculations. Figure 15-18 therefore considers a cylindrical enclosure infinite in height normal to the plane of drawing.

Enclosure with diffusely reflecting walls When the surfaces of the enclosure are nonblack, innumerable reflections occur within the enclosure. The calculations become quite involved if one attempts to trace the various rays through their reflections. For surfaces which emit radiation with an intensity independent of direction and which are perfectly

[1] V. Paschkis, *Electrotech. Maschinenbau*, **54**:617 (1936).

[2] A. K. Oppenheim, *Tran. ASME*, **78**:725 (1956).

diffuse reflectors, a simpler procedure is possible. In it one lumps all the radiation leaving a surface into one term, regardless of whether this radiative flux is generated by emission or reflection. The symbol B will be used to represent the total flux leaving a surface per unit area and time (B is chosen because this term is closely related to the brightness with which a surface appears to the eye). It has become customary to refer to it as *radiosity*. The heat flux arriving at a surface per unit area and time, regardless of how it has originated, will be denoted by H and will be referred to as *irradiance*. It has been pointed out that the terms B and H consider energy flux leaving or arriving at a surface in all directions in space, whereas the symbol K, used before, is connected with a certain direction.

We also assume that all the surfaces are gray. Then the following relations hold for surface i:

$$\rho_i + \alpha_i = 1 \qquad \alpha_i = \epsilon_i$$

We consider again an enclosure composed of n surfaces and fix our attention on surface i. The heat flux leaving the surface is A_iB_i. The heat flux arriving at the surface is A_iH_i. The net heat loss of the surface by radiation is therefore given by

$$Q_i = (B_i - H_i)A_i \tag{15-48}$$

The heat flux H_i arriving at the surface i must come from the surrounding surfaces. Accordingly, the following equation must hold:

$$H_i = \sum_k F_{i-k}B_k \tag{15-49}$$

The heat flux B_i leaving the surface finally must either have been emitted by the surface or been reflected on the surface. This is described by the equation

$$B_i = \epsilon_i e_{bi} + \rho_i H_i \tag{15-50}$$

Eliminating the term H_i from the three equations (15-48) to (15-50) and introducing $\rho_i = 1 - \epsilon_i$ gives

$$Q_i = \sum_k A_iF_{i-k}(B_i - B_k) \tag{15-51}$$

$$Q_i = A_i \frac{\epsilon_i}{1 - \epsilon_i} (e_{bi} - B_i) \tag{15-52}$$

Two equations of the same form can be written for each of the surfaces as compared with one equation for the enclosure with black surfaces. On the other hand, the terms B_i appear as additional unknown quantities, so that the system of equations can again be solved with the same boundary conditions as for the enclosure with black surfaces.

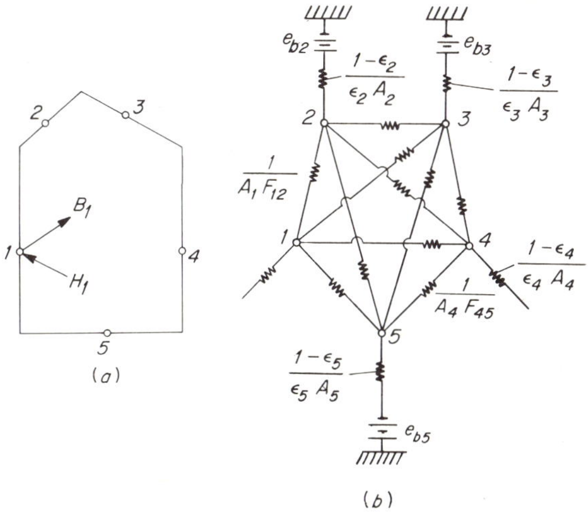

Fig. 15-19 Electrical analog of radiation in an enclosure with gray walls.

The electrical analog for the enclosure in Fig. 15-19*a* is presented in Fig. 15-19*b*. Equation (15-52) indicates that a resistance proportional to $(1 - \epsilon_i)/\epsilon_i A_i$ must be inserted into each connection of a node with an outside potential. An interesting conclusion can be drawn from this analog. Insertion of a resistance on the outside of a node without an outside connection (for instance, node 4) obviously has no influence on the potential of this node. This means that the temperature of an adiabatic wall is independent of the absorptance of this wall.

It will have occurred to the reader that the previous derivation rests on the assumption that all surfaces are irradiated uniformly over their areas or, in other words, that the terms H_i are constant over the individual surfaces. The subdivision of the enclosure into the n surfaces must therefore be made in such a way that this condition is reasonably approximated.

When the walls which exchange heat by radiation are nongray surfaces, Eqs. (15-51) and (15-52) have to be written for monochromatic radiation. The total amount of heat which a surface i loses is finally obtained by an integration over all wavelengths:

$$Q_i = \int_0^\infty Q_{\lambda i}\, d\lambda \tag{15-53}$$

The calculation may become a fairly involved trial-and-error procedure when the heat fluxes Q_i are prescribed on some of the surfaces. In many cases the calculation can be simplified when it is possible to subdivide

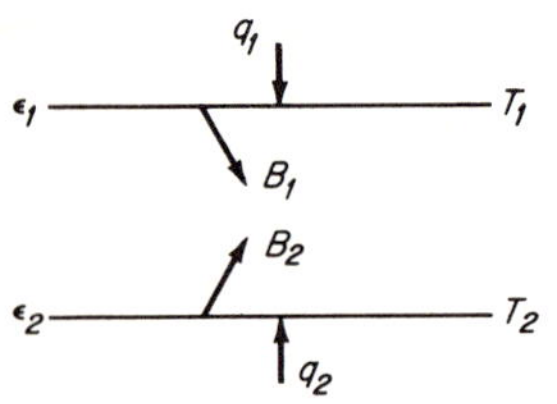

Fig. 15-20 Radiative exchange between two parallel surfaces.

the whole wavelength range into finite intervals $\Delta\lambda_i$ of such magnitude that the emittance ϵ_λ can be considered constant in each interval. The integral in Eq. (15-53) can then be replaced by a sum

$$Q_i = \Sigma Q_{\lambda i} \, \Delta\lambda_i \tag{15-54}$$

Two simple examples will demonstrate the procedure just discussed. As the first, consider an enclosure consisting of two parallel surfaces (Fig. 15-20) of infinite extent, with the temperatures T_1 and T_2 and the emittances ϵ_1 and ϵ_2. The blackbody emissive powers of the two surfaces are

$$e_{b1} = \sigma T_1{}^4 \qquad e_{b2} = \sigma T_2{}^4$$

The surfaces may again emit and reflect diffusely and may be gray surfaces, so that the following relations exist:

$$\alpha_1 = \epsilon_1 \qquad \alpha_1 + \rho_1 = 1$$
$$\alpha_2 = \epsilon_2 \qquad \alpha_2 + \rho_2 = 1$$

The net heat loss by radiation from the two surfaces per unit area may be q_1 and q_2, respectively. Equations (15-51) simplify to

$$q_1 = B_1 - B_2$$
$$q_2 = B_2 - B_1$$

and accordingly

$$q_1 = -q_2$$

Equations (15-52) read

$$q_1 = \frac{\epsilon_1}{1 - \epsilon_1} (e_{b1} - B_1)$$

$$q_2 = \frac{\epsilon_2}{1 - \epsilon_2} (e_{b2} - B_2)$$

From the equations above, the two radiosities B and the heat flux q_2 can be eliminated and the following equation results:

$$q_1 = \frac{1}{1/\epsilon_1 + 1/\epsilon_2 - 1} (e_{b1} - e_{b2}) \tag{15-55}$$

This relation is found in almost every textbook on radiative heat transfer. For nongray surfaces, Eq. (15-55) changes to

$$q_1 = \int_0^\infty \frac{e_{\lambda b1} - e_{\lambda b2}}{1/\epsilon_{\lambda 1} + 1/\epsilon_{\lambda 2} - 1} \, d\lambda \tag{15-56}$$

with $e_{\lambda b1}$ and $e_{\lambda b2}$ expressing the monochromatic black emissive power given by Planck's equation.

Two concentric spheres or cylinders whose surfaces reflect diffusely will be considered as the second example. It will again be assumed at first that the surfaces are gray. The surface of the inner cylinder or sphere will be denoted by A_1 and the surface of the outer cylinder or sphere by A_2. Equations (15-51) and (15-52) now take on the following form:

$$Q_1 = A_1 F_{1-2}(B_1 - B_2)$$
$$Q_1 = A_1 \frac{\epsilon_1}{1 - \epsilon_1} (e_{b1} - B_1)$$
$$Q_2 = A_2 F_{2-1}(B_2 - B_1) + A_2 F_{2-2}(B_2 - B_2)$$
$$Q_2 = A_2 \frac{\epsilon_2}{1 - \epsilon_2} (e_{b2} - B_2)$$

The following relations follow from shape-factor algebra:

$$F_{1-2} = 1 \qquad A_2 F_{2-1} = A_1 F_{1-2} = A_1 \qquad F_{2-2} = 1 - F_{2-1}$$

Utilizing these, the above equations simplify to

$$Q_2 = Q_1$$
$$q_1 = \frac{Q_1}{A_1} = \frac{e_{b1} - e_{b2}}{1/\epsilon_1 + A_1/A_2(1/\epsilon_2 - 1)} \tag{15-57}$$

The calculation for nongray surfaces starts with a consideration of monochromatic radiation. The total radiative heat exchange is finally obtained by integration over all wavelengths:

$$q_1 = \int_0^\infty \frac{e_{\lambda b1} - e_{\lambda b2}}{1/\epsilon_{\lambda 1} + A_1/A_2(1/\epsilon_{\lambda 2} - 1)} \, d\lambda \tag{15-58}$$

Equations (15-55) and (15-58) may be written in a general form

$$q = \epsilon_{1-2}(e_{b1} - e_{b2}) \tag{15-59}$$

with ϵ_{1-2} denoting a dimensionless interchange factor.

Enclosures with specularly reflecting walls As a next step in the analysis of radiative energy exchange in an enclosure, the condition in which one or more of the surfaces are specularly reflecting will be considered. First, however, we shall attempt to get some indication of the effect which specular or diffuse reflection has on the magnitude of the heat fluxes, which we do by considering a few simple geometries.

Energy exchange by radiation between two parallel, infinite surfaces has been analyzed in the preceding section. If one follows a ray emitted from one of the surfaces, one finds that it bounces back and forth between the two surfaces until it is finally absorbed. This happens, regardless of whether the surface reflects specularly or diffusely. As a consequence, Eq. (15-55) or (15-56) holds, regardless of the nature of reflection.

The energy exchange between two concentric spheres or cylinders with gray diffuse surfaces is given by Eq. (15-57). The radiative process is different when the outer surface A_2 reflects specularly. A ray originating at the inner surface is then alternately reflected on the outer and the inner surface, regardless of the direction in which it leaves. Rays leaving the outer surface can be separated into two groups: one which again is reflected back and forth between the two surfaces and another which never reaches the inner surface. From this it can easily be deduced that Eq. (15-55), or for nongray surfaces (15-56), gives the net heat transfer q_1 between the two surfaces per unit area of the inner surface. The dimensionless interchange factor in Eq. (15-59) has the form

$$\epsilon_{1-2} = \frac{1}{1/\epsilon_1 + A_1/A_2(1/\epsilon_2 - 1)}$$

for diffuse reflection and

$$\epsilon_{1-2} = \frac{1}{1/\epsilon_1 + 1/\epsilon_2 - 1}$$

for specular reflection. The spatial distribution of the reflection on the inner surface has obviously no influence on the interchange factor ϵ_{1-2}. Assuming the area ratio A_1/A_2 to be small compared to 1, the emittance of the inner surface ϵ_1 to be close to 1, and the emittance of the outer surface ϵ_2 to have the value 0.1 results in the value $\epsilon_{1-2} \approx 1$ for diffuse reflection and $\epsilon_{1-2} \approx 0.1$ for specular reflection on the outer surface. In this case, then, the heat fluxes for both situations obtained with Eq. (15-59) differ by an order of magnitude. Another example is provided by the infinitely deep cylindrical hole, as discussed on page 640ff. Figure 15-25 presents the apparent emittances of the surface of such a hole providing the reflection is diffuse. For specular reflection, the local apparent emittance ϵ_a is equal to 1, regardless of the value of the emittance ϵ of the material and of the distance x. This can be seen easily by following rays entering the hole from the outside. None of the rays is able to leave the hole again when the walls reflect specularly. According to Kirchhoff's laws, this means that the hole emits black radiation.

It is also clear, for instance, that the intensity of the energy flux coming from the sun, being reflected specularly from some part of the exterior of a satellite or spaceship and impinging on another part, will depend strongly on whether this second part is in the path of the specularly reflected radiation or not.

In order to develop the analysis for energy exchange in an enclosure with specularly reflecting surfaces, we shall start with the configuration shown in Fig. 15-21. Surfaces 1, 2, and 3 of the rectangular arrangement are assumed to be diffusely reflecting and surface 4 reflecting specularly. The configuration extends to a large distance normal to the plane of the drawing, so that end effects can be neglected. Diffuse reflection has the characteristic that a ray impinging on such a surface is reflected with a uniform intensity in all directions in space regardless of the direction from which the ray arrived. The previous history of the radiation is thus erased at each reflection process. As a consequence, reflected and emitted radiation can be combined into the radiosity term B. This is not the case on a specularly reflecting surface, and surface 4 will therefore be treated differently from the preceding analysis, whereas no change is required as far as surfaces 1, 2, and 3 are concerned. All surfaces involved may be gray surfaces, so that the total radiation for all wavelengths can be considered immediately. The following relations again apply to surface 1:

$$B_1 = \epsilon_1 e_{b1} + (1 - \epsilon_1)H_1 \tag{15-60}$$

$$\frac{Q_1}{A_1} = q_1 = B_1 - H_1 = \epsilon_1 e_{b1} - \epsilon_1 H_1 \tag{15-61}$$

$$q_1 = \frac{\epsilon_1}{1 - \epsilon_1}(e_{b1} - B_1) \tag{15-62}$$

The energy flux arriving at surface 1 from surface 2 will now be considered in two parts: (1) the flux which arrives directly and (2) the flux which arrives after reflection in the mirror 4. The first part can be expressed as

$$A_2F_{2-1}B_2 = A_1F_{1-2}B_2$$

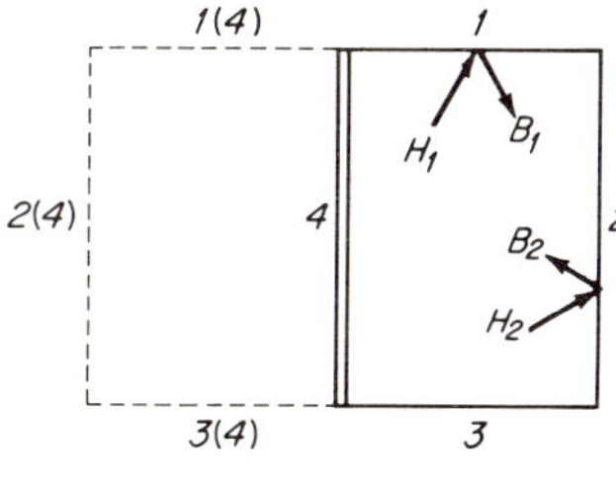

Fig. 15-21 Enclosure with one specularly reflecting wall.

In order to derive a relation for the second part, we make use of the law in optics stating that we can replace the rays coming from surface 2 and arriving at surface 1 after reflection in mirror 4 by the rays coming from the image of surface 2 as it is created by the mirror 4. This image is denoted by 2(4) in Fig. 15-21. The energy flux leaving surface 2 and arriving at surface 1 after reflection in mirror 4 is then $\rho_4 A_2 F_{2(4)-1}$, where $F_{2(4)-1}$ is the shape factor under which surface 1 is seen from the image 2(4) of surface 2 and ρ_4 denotes the reflectance of surface 4. The fluxes coming from surface 3 and arriving at surface 1 can be expressed in the same way. In this way the following equation is obtained:

$$A_1H_1 = A_2F_{2-1}B_2 + \rho_4A_2F_{2(4)-1}B_2 + A_3F_{3-1}B_3 + \rho_4A_3F_{3(4)-1}B_3 + \epsilon_4A_4F_{4-1}e_{b4}$$

The last term describes the radiative energy flux emitted by the mirror surface 4 and arriving at surface 1. Reciprocity relations apply to angle factors relating to images as well as to angle factors between real surfaces. As a consequence, the last equation can be simplified to

$$H_1 = F_{1-2}B_2 + \rho_4F_{1-2(4)}B_2 + F_{1-3}B_3 + \rho_4F_{1-3(4)}B_3 + \epsilon_4F_{1-4}e_{b4}$$

This relation is introduced into Eq. (15-61), to obtain

$$q_1 = B_1 - F_{1-2}B_2 - \rho_4F_{1-2(4)}B_2 - F_{1-3}B_3 - \rho_4F_{1-3(4)}B_3 - \epsilon_4F_{1-4}e_{b4} \tag{15-63}$$

Two analogous equations can be written for surfaces 2 and 3. The heat flux leaving the mirror surface 4 can finally be obtained by a flux balance through a control surface within the wall 4, just below the surface:

$$q_4 = \epsilon_4e_{b4} - \epsilon_4(F_{4-1}B_1 + F_{4-2}B_2 + F_{4-3}B_3) \tag{15-64}$$

In this way, seven equations are obtained, three each of the form of Eqs. (15-62) and (15-63) for the diffuse surfaces and one [Eq. (15-64)] for the specular surface. This results in a system of seven linear equations from which the three fluxes B and four additional quantities, either temperatures or net fluxes Q, can be calculated for the four surfaces.

In order to develop an electrical analog, Eq. (15-63) is written in the following form:

$$\frac{Q_1}{A_1} = F_{1-2}(B_1 - B_2) - \rho_4F_{1-2(4)}(B_1 - B_2) + F_{1-3}(B_1 - B_3) + \rho_4F_{1-3(4)}(B_1 - B_2) + \epsilon_4F_{1-4}(B_1 - e_{b4}) \tag{15-65}$$

That this equation is correct can be seen in two ways: First, the following relations between angle factors are evident from Fig. 15-21:

$$F_{1-2} + F_{1-3} + F_{1-2(4)} + F_{1-3(4)} = 1$$
$$F_{1-2(4)} + F_{1-3(4)} = F_{1-4}$$

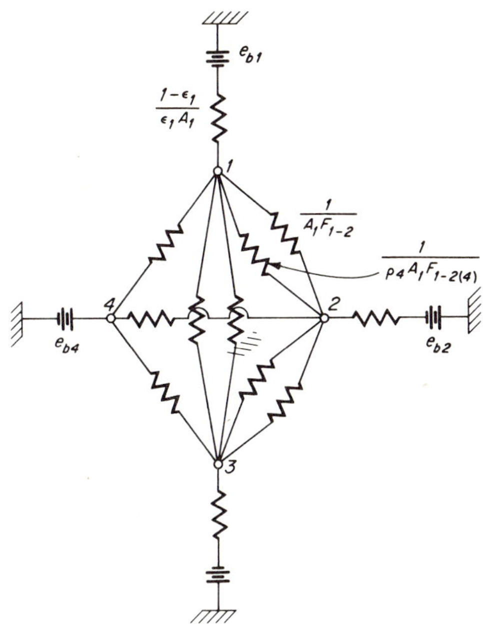

Fig. 15-22 Electrical analog to the enclosure of Fig. 15-21.

Introduction of these relations and reference to the gray-body relation $\epsilon = 1 - \rho$ transforms Eq. (15-63) into Eq. (15-65). The second proof considers the enclosure to be in thermal equilibrium; this means that all the surfaces have the same temperature. In such a system, the radiative fluxes B are isotropic and homogeneous ($B_1 = B_2 = B_3$), and the net energy fluxes Q are zero. From Eq. (15-63) then follows

$$1 = F_{1-2} + \rho_4 F_{1-2(4)} + F_{1-3} + \rho_4 F_{1-3(4)} + \epsilon_4 F_{1-4}$$

which, introduced into Eq. (15-63), changes it to Eq. (15-65). The electric network, as shown in Fig. 15-22, can be understood easily considering Eqs. (15-62) and (15-65) and the corresponding relations for the other surfaces.

The method can readily be extended to an enclosure with more than one specularly reflecting surface. Figure 15-23, for example, shows such a rectangular enclosure in which two surfaces, 3 and 4, reflect specularly. In this case double images like 1(4,3) must be included in the radiative flux balance. The rays shown in the enclosure demonstrate ways in which radiation leaving a point of surface 2 can arrive at a selected point on surface 1. Radiation can travel as indicated from 2 to 1 directly, or with a reflection on surface 3, or on surface 4, or finally with consecutive reflections on 3 and 4. The reflection or double reflec-

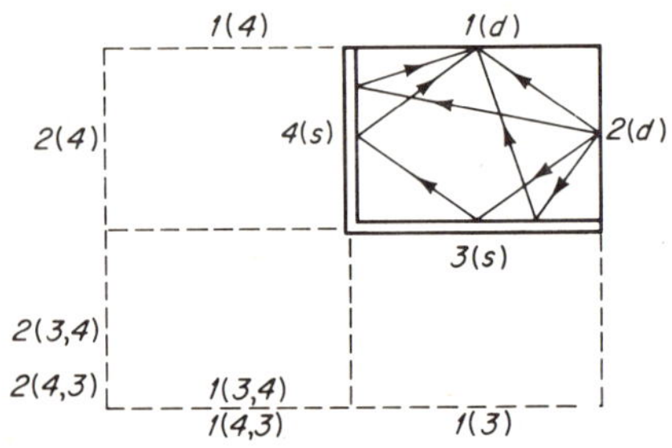

Fig. 15-23 Enclosure with two specularly reflecting walls.

tion can again be substituted by considering the radiation to come from the corresponding images or double images. The surface denoted by 1(4,3), for instance, is the image of surface 1 obtained by consecutive reflection on mirror surfaces 4 and 3. The contribution to the irradiation H_1 by the flux B_2 leaving surface 2 is thus described by the expression

$$B_2(F_{1-2} + \rho_3 F_{1-2(3)} + \rho_4 F_{1-2(4)} + \rho_3\rho_4 F_{1-2(3,4)})$$

A term with an angle factor $F_{1-2(4,3)}$ is missing in the above expression because the double image 2(4,3) cannot be seen from any point of surface 1 through the mirror surface 3. An expression similar to the one above describes the contribution to H_1 of the radiation leaving the surface 3. This arrangement also contains a new feature insofar as radiation leaving surface 1 will be reflected back to this surface on mirror 3. This flux per unit area of surface 1 is

$$\rho_3 B_1 F_{1-1(3)}$$

Radiation from surface 1 can also be returned to surface 1 by double reflection on mirrors 3 and 4. The angle factor describing this process is denoted by $F^*_{1-1(3,4)}$. The asterisk indicates that not all the double image 1(3,4) is seen from surface 1 through the mirror 4. A point at the left corner of surface 1 will see the whole of the double image through mirror 4, whereas a point at the right corner of 1 will just see one point, namely, the left corner of the double image 1(3,4) through the mirror 4. Accordingly, the local angle factor decreases as one moves on surface 1 from the left corner to the right corner. This has to be considered in deriving the average angle factor $F^*_{1-1(3,4)}$. Also to be considered is a path 1-4-3-1 for radiation originating at surface 1 and arriving at 1. The angle factors $F^*_{1-1(3,4)}$ and $F^*_{1-1(4,3)}$ can be seen to add up to $F_{1-1(3,4)}$ when the latter disregards the limitation of the finite size of mirror 4.

The analysis which was made for gray surfaces can easily be extended to nongray emitters by writing the relations for monochromatic radiation and integrating the results over all wavelengths.

The described procedure for specular reflection was first proposed by Eckert and Sparrow.[1] In recent publications, the calculation procedure for enclosures has been streamlined and adapted to electronic computers.[2] The linear character of the system of equations which describes the energy transfer suggests the possibility of obtaining new solutions by superposition. This can be utilized with advantage if a number of calculations have to be made for the same enclosure but with varying boundary conditions. In this case, one obtains a series of solutions imposing the condition that the blackbody emissive power on one of the surfaces be equal to 1 and zero on all the other surfaces. The net radiative heat flux $Q_i^{(1)}$ leaving surface i for the boundary conditions

$$e_{b1} = 1 \qquad e_{b2} = e_{b3} = e_{b4} = \cdots = 0$$

can be calculated in this way. In the same way, the heat flux for the boundary condition

$$e_{b1} = 0 \qquad e_{b2} = 1 \qquad e_{b3} = e_{b4} = \cdots = 0$$

is $Q_i^{(2)}$. It can then easily be seen that the heat flux leaving the surface i for the general condition that $e_{b1}, e_{b2}, e_{b3}, \ldots$ are the emissive powers of the various surfaces is obtained by the equation

$$Q_i = Q_i^{(1)}e_{b1} + Q_i^{(2)}e_{b2} + Q_i^{(3)}e_{b3} + \cdots$$

The use of the superposition principle for radiative exchange calculations has been introduced by Hottel.[3]

Enclosures with partially diffuse surfaces Many surfaces reflect impinging radiation in such a way that the intensity of the reflected flux is not uniform in the various directions but is also not confined to the specular reflection angle. Figure 15-10*a* and *b* presents examples of such surfaces. For an enclosure which contains such surfaces, an approximation to the radiant energy fluxes can be obtained by a combination of the procedures discussed in the two preceding sections. The total reflectance over the whole hemispherical space ρ then has to be expressed as the sum of a specular reflectance ρ_s and a diffuse reflectance ρ_d

$$\rho = \rho_s + \rho_d$$

Judgment is required in determining the two partial reflectances. The analysis is then performed separately for the specularly reflected energy

[1] E. R. G. Eckert and E. M. Sparrow, *Intern. J. Heat Heat Mass Transfer*, **83**:207–214 (1911).

[2] Sparrow and Cess, *op. cit.*

[3] H. C. Hottel, Radiant Heat Transmission, in W. H. McAdams, "Heat Transmission," 3d ed., McGraw-Hill, New York, 1954.

and for the diffusely reflected energy, and the two contributions are finally added up.

It is to be expected that radiative energy transfer in an enclosure is especially sensitive to the reflection characteristics of the surfaces when metal walls are involved. For certain geometries, it may even lead to a considerable error to approximate the emission characteristics of the surface by an intensity which is independent of direction or to neglect the fact that polarization occurs at large angles of the emitted radiation. A method which appears best suited to the inclusion of such features in an enclosure analysis is the Monte Carlo method, discussed below.

Integral equations for radiative exchange In the preceding sections it has been assumed that the walls of the enclosure consist of areas with locally uniform temperature. In addition, it was assumed that the radiosity and the irradiance are uniform over these areas. This is, in general, not strictly the case. One can, of course, reduce the inaccuracy caused by this assumption by dividing the enclosure surface into small areas. The number of linear equations to be solved increases in this way; however, with an electronic computer this does not pose a serious problem. If one carries the subdivision to infinitely small area elements, one ends up with an integral equation.

As an example, such an integral equation may be derived for a cylindrical hole of infinite depth drilled into a body of uniform temperature. The walls of the hole may be considered gray, and the emittance of the wall material denoted by ϵ. This property, together with the constant wall temperature, determines the radiative flux emitted from any location in the hole surface. The total flux B leaving any location in the hole will be larger because the emitted radiation is supplemented by radiation reflected on the surface. We can define a *local apparent emittance* of the wall

$$\epsilon_a(x) = \frac{B(x)}{e_b}$$

The radiative flux B as well as the apparent emittance must be considered as functions of the distance x from the end of the hole. The radiative flux arriving on a unit surface area at distance x may be denoted by H. The following equation relates the fluxes arriving and leaving such an area:

$$B = \epsilon e_b + (1 - \epsilon)H \tag{15-66}$$

In order to obtain an expression for the flux H, the radiative flux dH starting from an area element dA_2 and arriving at 1 will be expressed

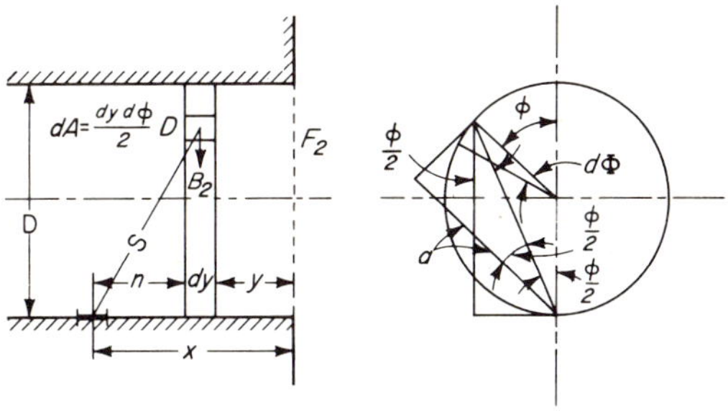

Fig. 15-24 Calculation of thermal radiation from cylindrical hole.

first. The coordinates indicating the position of dA_2 are y and ϕ. It follows that

$$dH = B_2\, dF_{1-2} = \epsilon_a(y)e_b \frac{\cos \beta_1 \cos \beta_2}{\pi s^2}\, dA_2 = \epsilon_a(y)e_b \frac{a^2\, dy\, D\, d\phi}{2\pi s^4}$$

The following relations are evident from Fig. 15-24:

$$a = D \cos^2 \frac{\phi}{2} \qquad s^2 = (x-y)^2 + D^2 \cos^2 \frac{\phi}{2}$$

After introduction of these, the above equation changes to

$$dH = \epsilon_a(y)e_b \frac{D^3 \cos^4 (\phi/2)}{\pi[(x-y)^2 + D^2 \cos^2 (\phi/2)]^2}\, d\frac{\phi}{2}\, dy$$

The total flux arriving at location 1 is obtained by integration over the periphery and over the length of the hole:

$$H = \frac{2D^3}{\pi} e_b \int_{y=0}^{\infty} \epsilon_a(y) \int_{\phi/2=0}^{\pi/2} \frac{\cos^4 (\phi/2)}{[(x-y)^2 + D^2 \cos^2 (\phi/2)]^2}\, d\frac{\phi}{2}\, dy$$

The integration over the angle ϕ can be carried out and with the following substitutions

$$\frac{x}{D} = \xi \qquad \frac{y}{D} = \eta$$

results in

$$H = e_b \int_0^{\infty} \epsilon_a(\eta) \left\{1 - |\xi - \eta| \frac{2(\xi-\eta)^2 + 3}{2[(\xi-\eta)^2 + 1]^{\frac{3}{2}}}\right\} d\eta \qquad (15\text{-}67)$$

The integration has to be carried out in two parts: from $y = 0$ to $y = x$ and from $y = x$ to $y =$ infinity. $|\xi - \eta|$ indicates that this expression

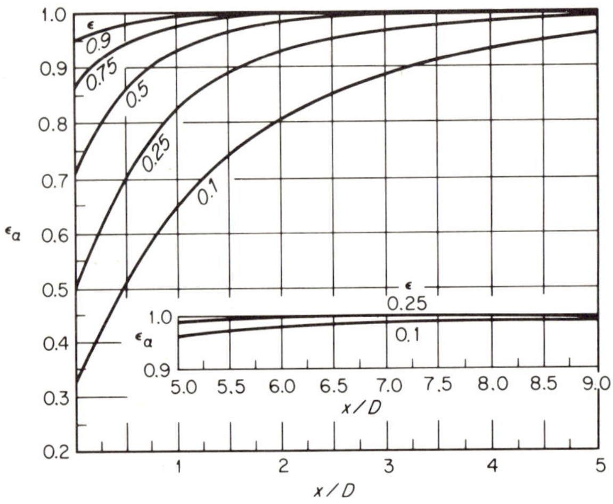

Fig. 15-25 Apparent emittance of the walls in a cylindrical hole. [*From E. M. Sparrow and L. V. Albers, J. Heat Transfer,* **82**:*253* (*1960*).]

must always be introduced as positive. Combination of Eqs. (15-66) and (15-67) gives

$$\epsilon_a(\xi) = \epsilon + (1 - \epsilon) \int_0^\infty \epsilon_a(\eta) \left\{ 1 - |\xi - \eta| \frac{2(\xi - \eta)^2 + 3}{2[(\xi - \eta)^2 + 1]^{\frac{3}{2}}} \right\} d\eta \qquad (15\text{-}68)$$

The unknown parameter ϵ_a appears in this equation not only on the left-hand side, where it has to be taken at the location ξ, but also on the right-hand side in the integral, where it has to be introduced at location η. Equation (15-68) is, therefore, an integral equation and in particular a homogeneous linear integral equation. A solution of this equation was obtained by Buckley[1] and more recently by Sparrow and Albers.[2] It is presented in Fig. 15-25, where it can be observed how the apparent emittance in the hole increases with increasing distance x. The figure permits one to determine the distance from the entrance of the hole required to have the apparent emittance approach, to any desired degree, the value 1 which holds for a blackbody.

Monte Carlo method The Monte Carlo method was originally developed for the analysis of the neutron flux in nuclear reactors. Its use for radiative-exchange calculations was proposed by Bird, Stewart, and

[1] H. Buckley, *Phil. Mag.*, **4**:753(1927), **6**:447 1928), **17**:576 (1934).

[2] E. M. Sparrow and L. V. Albers, *J. Heat Transfer*, **82**:253–255 (1960).

Lightfoot,[1] and the first analyses with this method were then performed by Howell and Perlmutter.[2] The method considers radiant energy as consisting of energy packets or energy particles and applies probability considerations to describe the wavelength and direction as well as the absorption, reflection, or scattering of each of these particles. Photons could actually be considered as such particles; however, their number is generally much too large and their energy content depends on wavelength, whereas for the present method it is of advantage to consider particles of equal energy. The analysis is performed with electronic computers, and the number of particles is selected according to the capability of the computer. Normally 10,000 to 50,000 particles are considered.

The method itself will now be described by a specific example, namely, radiative exchange in an enclosure as sketched in Fig. 15-26 with a nonparticipating medium inside the enclosure. The enclosure surface is subdivided into a large number of area elements ΔA. It is assumed that the monochromatic emittance ϵ_λ is a function of wavelength λ and of the angle β. The monochromatic bidirectional reflectance may also be known as a function of wavelength and the angles β and γ for the incident as well as for the reflected ray. The transmittance τ_λ is assumed zero, and the temperature T for each area element is prescribed. The task is to calculate the net heat loss or gain of each area element as it is caused by the radiative interchange. The radiative flux emitted by an area element ΔA_i is $\epsilon_{i\frown} e_{bi}\,\Delta A_i$, and the radiative flux emitted by all area elements of the enclosure is $\Phi = \Sigma \epsilon_{i\frown} e_{bi}\,\Delta A_i$. The energy content of each particle is then Φ/N if N denotes the number of particles selected by the analysis.

We now direct our attention to a specific area element ΔA and ask:

[1] R. B. Bird, W. E. Stewart, and E. N. Lightfoot, "Notes on Transport Phenomena," Wiley, New York, 1958.

[2] J. R. Howell and M. Perlmutter, *J. Heat Transfer*, **86**:116 (1969).

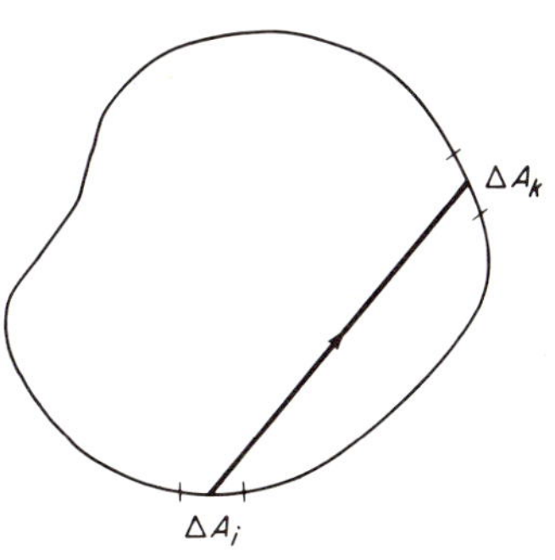

Fig. 15-26 Monte Carlo analysis of radiation in an enclosure.

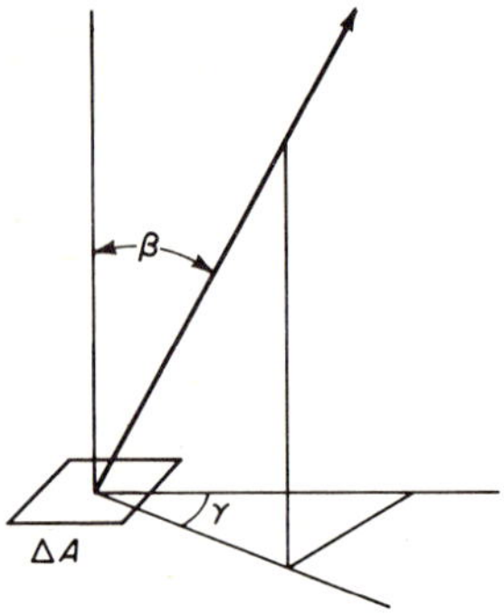

Fig. 15-27 Directional emission of radiation.

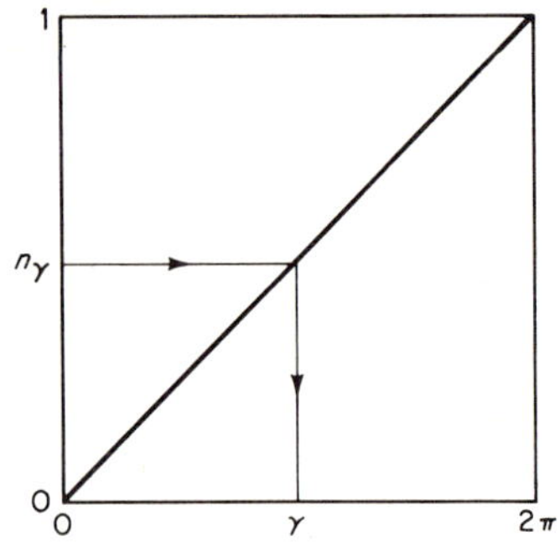

Fig. 15-28 Probability of angular emission with angle γ.

What is the probability that the particle is emitted from the area element in a specific direction characterized by the angles β and γ as shown in Fig. 15-27 and at a wavelength λ? As part of this question we shall first determine the probability that the particle leaves the area element within a plane containing the surface normal and oriented with angle γ. Obviously all angular positions have the same probability where the surface is isotropic, so that the probability that the particle leaves within an angle $d\gamma$ is $d\gamma/2\pi$. The probability n_γ that the particle leaves within the finite angle γ is therefore $n_\gamma = \int_0^\gamma d\lambda/2\pi = \lambda/2\pi$. In a plot with n_γ as ordinate and γ as abscissa, this integral is presented by a straight line, as shown in Fig. 15-28. This curve is called a *cumulative probability curve*, and it can be used for the present purpose in the following way. Consider a sequence of numbers which subdivide the range from zero to one in equal intervals and arbitrarily select one number. Such random-number subroutines are available on computers. This number can then be considered as giving the probability with which the energy particle leaves the area ΔA under the angle γ, which is found in Fig. 15-28 as that angle belonging to the value n_γ. The next step is to determine the probability that the energy packet leaves the area element in a direction belonging to a conical surface with the opening angle β. The energy flux leaving the area element within a solid angle $d\omega$ and a wavelength range $d\lambda$ is given by

$$d\varphi = \epsilon_\lambda i_{b\lambda} \cos\beta \, d\omega \, d\lambda = 2\pi\epsilon_\lambda i_{b\lambda} \cos\beta \sin\beta \, d\beta \, d\lambda$$

We shall assume that the dependence of the emittance ϵ_λ on wavelength and on the angle β has the form $\epsilon_\lambda = f_\lambda f_\beta$. The probability that the particle leaves the area element within the cone of angle β is then

$$n_\beta = \frac{\int_0^\beta \epsilon_\lambda i_{b\lambda} \sin\beta \cos\beta \, d\beta \, d\lambda}{\int_0^{\pi/2} \epsilon_\lambda i_{b\lambda} \sin\beta \cos\beta \, d\beta \, d\lambda} = \frac{\int_0^\beta f_\beta \sin\beta \cos\beta \, d\beta}{\int_0^{\pi/2} f_\beta \sin\beta \cos\beta \, d\beta}$$

The last term on the right-hand side of the equation was obtained by canceling out those terms which are constant in the integration. n_β again describes a cumulative probability curve, as shown in Fig. 15-29. The probability that the energy particle has a wavelength within the range from zero to λ is described by the equation

$$n_\lambda = \frac{\int_0^\lambda f_\lambda i_{b\lambda} \, d\lambda}{\int_0^\infty f_\lambda i_{b\lambda} \, d\lambda}$$

which is easily obtained in the same way as the probability n_β. It again describes a probability curve, as shown in Fig. 15-30. The probability that the energy particle under consideration leaves in a direction prescribed by the angles β and γ and at a wavelength λ is now found in the following way. A number in the series between zero and 1 is selected arbitrarily, and Fig. 15-28 determines the angle γ. A second number arbitrarily selected determines (through Fig. 15-29) the angle β, and a third number arbitrarily selected determines (through Fig. 15-30) the wavelength λ. In this way the direction in which the radiation particle leaves the area ΔA_i is determined, and from the geometry of the enclosure the area element ΔA_k is found on which the particle impinges. The next problem is to determine whether the particle is absorbed or reflected at this area element. The absorptance α_λ for this surface can be utilized in the probability analysis in the following way. A random number is

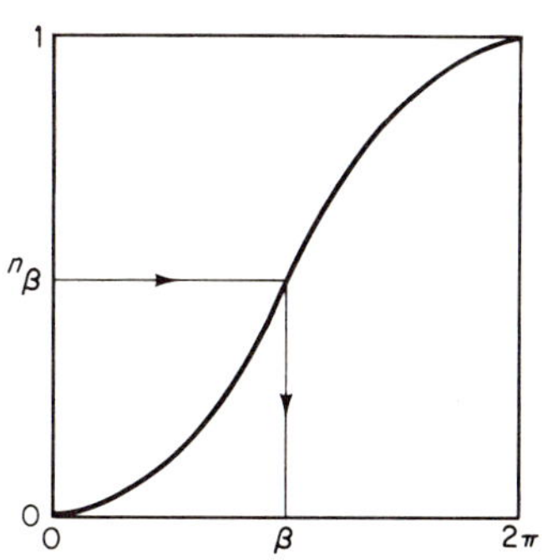

Fig. 15-29 Probability of angular emission with angle β.

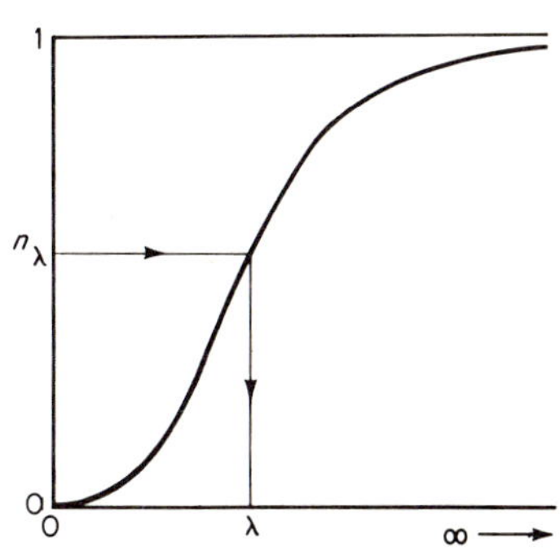

Fig. 15-30 Probability of wavelength distribution of emission.

again selected in the range zero to one and it is stipulated that the particle is absorbed when this number is smaller than the value of α_λ; otherwise it is reflected. A record is made of the fact that the particle is absorbed, and the calculation starts with a new particle emitted by the area element ΔA_i. When the particle is reflected, the calculation has to be continued to determine the direction in which it leaves the surface element ΔA_k. This is done with the help of cumulative probability curves constructed from the bidirectional reflectivity ρ_λ. The geometry of the enclosure then determines which new surface element is struck by the particle. The absorptance of that surface element determines whether the particle is absorbed or reflected. In the latter case, the calculation has to be continued until the particle is finally absorbed. All energy particles $N_i = (\epsilon_{i \frown} e_{bi}\,\Delta A_i/\Phi)N$ emitted by the area ΔA_i must be treated in this way, and the same calculation is performed for each of the area elements emitting radiation. As a result, one obtains the total number of energy particles N_i' which are absorbed by the area element ΔA_i, and the net heat loss of this area is given by the equation

$$Q_{\text{net},i} = \frac{\Phi}{N}\,(N_i - N_i')$$

In some cases the dependence of a radiation property on two parameters cannot be expressed in the form of a product, as was done in the preceding calculation. One must then determine the probability that two random effects simultaneously determine the probability. Procedures for this are available and are discussed, for instance, in a report by Kahn.[1] The Monte Carlo method requires a considerable effort for the solution of a problem, but it has the advantage of being able to account for very specialized conditions.[2] For instance, it has been extended to the analysis of radiative exchange in an enclosure filled with a medium which participates by emission, absorption, and scattering, and it has also been found useful for conductive- or convective-heat-transfer problems.

[1] H. Kahn, Applications of Monte Carlo, *U.S. Atomic Energy Comm. Rept.* AECU-3259, The RAND Corporation, 1954.

[2] J. R. Howell, Application of Monte Carlo to Heat Transfer Problems, *Advan. Heat Transfer*, **5**: (1968).

16

Radiation of Weakly Absorbing Media

16-1 RADIATION PROPERTIES OF WEAKLY ABSORBING MEDIA (GASES)

As a next step in the analysis of radiative exchange, we consider an enclosure filled with a medium that participates in this process. Since one usually encounters a gas as a participating medium in engineering systems, the discussion will concentrate on gases. Scattering has only a negligible effect on radiative exchange in a gas and will be neglected at first. In the early sections of Part III, the process of emission and absorption was characterized by a coefficient of emission j_λ and a coefficient of absorption κ_λ. For thermal radiation, these two coefficients are interrelated by Kirchhoff's equation

$$j_\lambda = \kappa_\lambda K_{\lambda e} = \kappa_\lambda i_{\lambda b} \tag{16-1}$$

so that knowledge of the coefficient of absorption is sufficient. This coefficient can be obtained from analyses based on certain models of the atomic structure and utilizing quantum mechanics; or it is obtained by measurement. Before attacking the analysis of radiative exchange in an

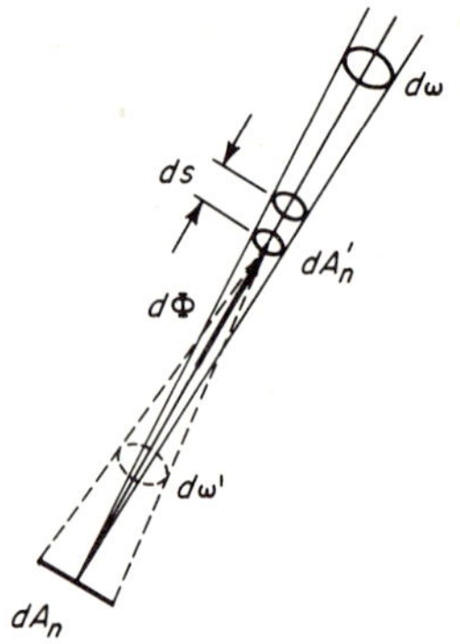

Fig. 16-1 Emission and absorption of radiation.

enclosure, it will be necessary to obtain information on this radiation property; and, as a first step, a method of measuring it will be discussed.

Emissivity, absorptivity From the equations in Chap. 14, describing the intensity of the radiative flux and the emission and absorption process, an equation can be derived which describes how the radiative flux varies along a pencil of radiation. This equation is referred to as the *transfer equation.* Figure 16-1 sketches a pencil of radiation passing normally through the area element dA_n. The radiative energy flux within this pencil with a solid angle $d\omega$ is

$$d\Phi = K_\lambda \, dA_n \, d\omega \, d\lambda$$

In general the flux will change along an infinitely small path length ds through the process of emission and absorption by the amount

$$d^2\Phi = dK_\lambda \, dA_n \, d\omega \, d\lambda$$

The increase in flux by emission is (because $dA_n' \, d\omega' = dA \, d\omega$)

$$d^2\Phi_e = \rho j_\lambda \, dA_n \, ds \, d\omega \, d\lambda$$

the decrease by absorption is

$$-d^2\Phi_a = -\rho \varkappa_\lambda K_\lambda \, dA_n \, d\omega \, d\lambda \, ds$$

Combining these three equations results in

$$dK_\lambda = \rho(j_\lambda - \varkappa_\lambda K_\lambda) \, ds \tag{16-2}$$

This is the transfer equation. For thermal radiation, it takes on the following form, considering Eq. (16-1):

$$dK_\lambda = \rho \varkappa_\lambda (i_{\lambda b} - K_\lambda) \, ds \tag{16-3}$$

Experiments to measure radiation properties of a gas or weakly absorbing liquid utilize essentially the equipment sketched in Fig. 16-2. The fluid fills a container a with two windows so that the radiative flux can be studied by an instrument b located on the outside. Adjacent to the second window is a blackbody c. Proper diaphragms and mirrors ensure that only radiant flux traveling within the shaded beam can enter the instrument b, where the arriving energy is measured. In order to obtain monochromatic values, the radiant pencil must be resolved with regard to wavelength before it is measured, and this is done in a spectroscope. The fluid filling the container is at uniform pressure and temperature, which means that the density as well as the absorption coefficient appearing in the transfer equation (and in a multicomponent mixture the concentration as well) is uniform along the beam over the length of the container. Provision is also made for the medium between the measuring instrument and the container on one side and the blackbody and the container on the other to play no part in the radiative exchange. The transfer equation can now be integrated along the path s by separation of variables

$$\frac{dK_\lambda}{i_{\lambda b} - K_\lambda} = \rho \varkappa_\lambda \, ds$$

and by integration

$$i_{\lambda b} - K_\lambda = Ce^{-\rho \varkappa_\lambda s} \tag{16-3a}$$

The constant C must be determined from the boundary condition that the flux entering the container at $s = 0$ is the flux $i_{\lambda b,c}$ coming from the blackbody c:

$$K_\lambda = i_{\lambda b,c}{}^{1} \qquad s = 0$$

[1] It is assumed that the energy reflected in the entry process is negligible, a good approximation for a gas. Otherwise $i_{\lambda b,c}$ must be interpreted as the intensity of the flux entering the medium.

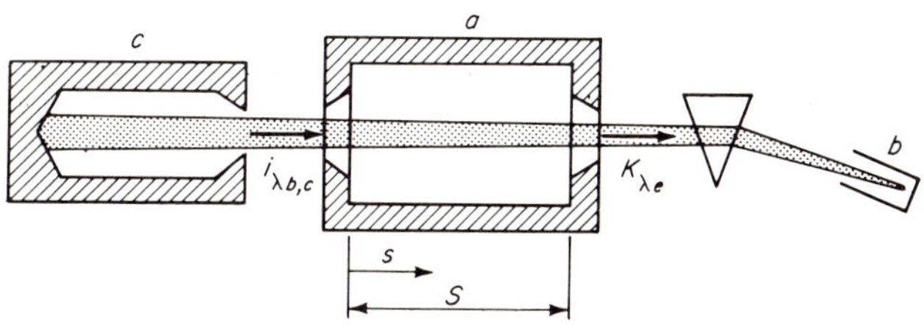

Fig. 16-2 Setup for measurement of radiation properties of a weakly absorbing medium.

The transfer equation then becomes

$$K_\lambda = i_{\lambda b}(1 - e^{-\rho \varkappa_\lambda s}) + i_{\lambda b,c} e^{-\rho \varkappa_\lambda s}$$

The radiation leaving the medium of dimension S in the beam direction and entering the measuring instrument is

$$K_{\lambda e} = i_{\lambda b}(1 - e^{-\rho \varkappa_\lambda S}) + i_{\lambda b,c} e^{-\rho \varkappa_\lambda S} \tag{16-4}$$

The first term in this equation obviously describes the effect of emission in the medium on the radiative flux, and the second describes the effect of absorption. In experiments one usually attempts to separate them. The emission process is separated when the temperature and consequently the radiative intensity $i_{\lambda b,c}$ of the blackbody c are made negligibly small relative to the intensity $i_{\lambda b}$ of blackbody radiation corresponding to the temperature of the medium. In this case, Eq. (16-4) simplifies to

$$K_{\lambda e} = i_{\lambda b}(1 - e^{-\rho \varkappa_\lambda S})$$

We can now define a monochromatic emissivity ϵ_λ of the medium as the ratio of the intensity of the radiative flux leaving the gas body at wavelength λ to the intensity of the radiative flux which would leave a blackbody at the same temperature

$$\epsilon_\lambda = 1 - e^{-\rho \varkappa_\lambda S} \tag{16-5}$$

The absorption process, on the other hand, can be isolated by an experiment in which the intensity of blackbody radiation $i_{\lambda b}$ corresponding to the medium temperature is negligibly small compared to the intensity of the radiation coming from the blackbody c. Equation (16-4) simplifies in this case to

$$K_{\lambda e} = i_{\lambda b,c} e^{-\rho \varkappa_\lambda S} \tag{16-6}$$

A transmissivity τ_λ of the medium may be defined as the ratio of the intensity of the radiant beam leaving the body to the intensity of the beam entering it. The absorptivity of the body is 1 minus the transmissivity[1]

$$\alpha_\lambda = 1 - \tau_\lambda = 1 - e^{-\rho \varkappa_\lambda S} \tag{16-7}$$

It can be seen that Kirchhoff's law $\alpha_\lambda = \epsilon_\lambda$ again holds for the parameters defined in this way. The coefficient of absorption $\varkappa_\lambda$ can, in principle, be determined from emission or absorption measurements utilizing Eq. (16-6) or (16-7). Actually, corrections have to be introduced, especially for gases, because spectroscopic measurements can be made only with a small but still finite wavelength interval $\Delta\lambda$.

As for solid surfaces, one can again define emissivities or absorp-

[1] α and τ are here referred to the intensity of the flux entering the medium.

tivities for the total radiative fluxes integrated over all wavelengths. The total emissivity, for instance, is then defined by the equation

$$\epsilon = \frac{\int_0^\infty \epsilon_\lambda i_{\lambda b}\, d\lambda}{\int_0^\infty i_{\lambda b}\, d\lambda}$$

The same restrictions which were discussed in Chap. 15 for solid surfaces apply to radiation of weakly absorbing media as well. Kirchhoff's law, for instance, does not hold generally for the total emissivities and absorptivities. The total emissivity is a function of the parameters describing the state of the radiating medium. The total absorptivity, on the other hand, depends in addition on the wavelength distribution of the radiative flux entering the body.

Gases are the most important representatives of a weakly absorbing medium from the viewpoint of engineering applications. Their radiation properties will therefore be discussed in the following section.

Radiation properties of gases Gases generally absorb and radiate energy by increasing or releasing some of the energy internally stored within the molecules. Such energy changes may be caused by changes in the rotation of the molecules or in the vibration of the atoms within the molecule, by changes in the electron orbits of the atoms, and finally by changes in the nuclear arrangements. They occur as quanta in finite steps, the amount of energy per step being smallest for changes in rotation and largest for changes in the nuclear arrangement. According to Planck's equation, $\Delta E = h\nu$, the radiation connected with small energy quanta ΔE has a small frequency or large wavelength. The larger the energy steps, the shorter the wavelength of the equivalent radiation. Changes in the energy state of the molecules are caused by collisions which have sufficient kinetic energy. The higher the temperature of the gas, the larger the energy quanta which can be released in the molecular collisions in measurable quantities and thus the shorter the wavelength of the released radiation. At temperatures in the range up to, say, 3000 C, as they occur in normal engineering applications, the bulk of radiation is connected with changes in the rotational or vibrational energy of the molecules.

The corresponding radiation has wavelengths in the range 1 to 30 μm (Fig. 14-8). Changes in the electron orbits move the radiation into the visible range. This radiation contributes, however, to the total energy essentially only at very high temperatures like that of the sun or a gaseous plasma. Changes in the nuclei finally shift the radiation to extremely short waves (Fig. 14-1). In this section, only infrared radiation will be considered. The atoms in their rotational or vibrational movement act as small transmitters of electromagnetic waves, since electric charges are usually connected with them.

Elementary gases, the molecules of which consist of atoms all of the same kind, have no free electric charges connected with the atom. Such gases as hydrogen, oxygen, and nitrogen emit practically no thermal radiation and are transparent to foreign infrared radiation. For engineering calculations, the radiation of carbon dioxide and of water vapor are most important, since these gases are good radiators and are present in combustion gases in high concentration. Carbon monoxide, sulfur dioxide, and methane also radiate well but are generally present only in small concentrations. Figures 16-3 and 16-4 show the absorption spectra of carbon dioxide and water vapor as measured spectroscopically. It can be seen that gases differ from solid and liquid bodies insofar as they radiate and absorb radiation only within certain wavelength ranges. For water vapor these ranges lie comparatively close together. The radiation taking place in the infrared is invisible. From the figures it can also be seen that gases, in contrast to solids and liquids, need great thickness to absorb the major part of incident radiation. Total emissivity values for gases are again defined by the equation

$$\epsilon = \frac{\int_0^\infty \epsilon_\lambda i_{b\lambda}\, d\lambda}{i_b}$$

For simple gases, such values can today be calculated applying certain models for the molecules and their vibration and rotation levels. For most gases, however, information on total emissivities has been obtained mainly by measurements in a test setup similar to the one shown in Fig. 16-2.

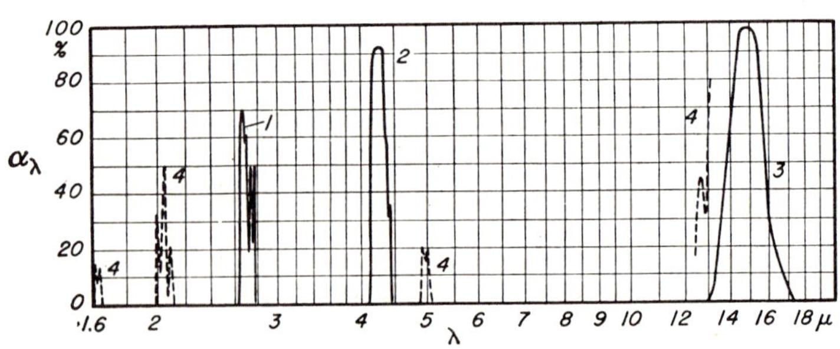

Fig. 16-3 Absorption bands of carbon dioxide: (1) 5-cm, (2) 3-cm, (3) 6.3-cm, and (4) 100-cm layers. [*Measured by E. F. Barker, G. Hertz, C. Schaefer, and B. Philipps, from E. Eckert, VDI-Forschungsh. 387, 1937.*]

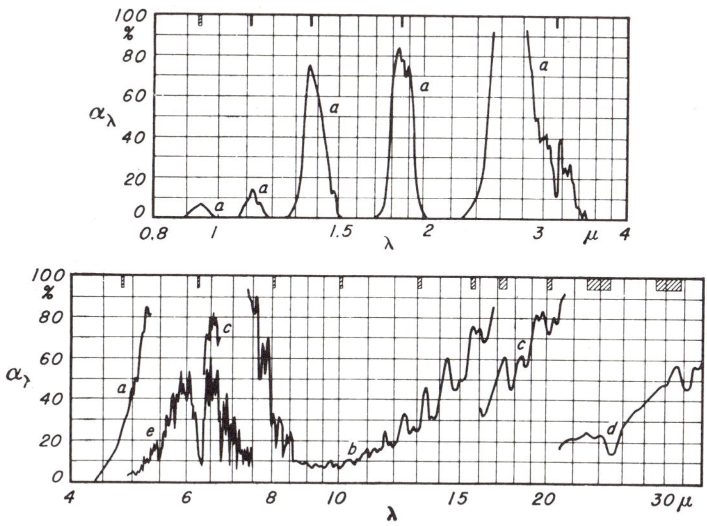

Fig. 16-4 Absorption bands of water vapor: (*a*) 127 C and 109 cm, (*b*) 127 C and 104 cm, (*c*) 127 C and 32.4 cm, (*d*) 81 C and 32.4 cm at 4 cm atm, (*e*) room temperature and 220 cm at 7 cm atm. [*From E. Schmidt, Forsch. Gebiete Ingenieurw.*, **3**:*57* (*1932*).]

A number of extensive experimental investigations were carried out in this field after Schack[1] emphasized the importance of gas radiation for heat transfer in furnaces and steam boilers. Measurements on carbon dioxide were made by Hottel and coworkers[2] and by Eckert.[3] Figure 16-5 shows the emissivity of carbon dioxide for different temperatures and thicknesses S and for a total pressure of 1 atm. Eckert made measurements at 1 atm with carbon dioxide–nitrogen mixtures of various partial pressures and thicknesses and found that the data correlate when plotted over the partial pressure p_p of carbon dioxide times thickness of the layer S, as shown in the figure. Hottel and Mangelsdorf used carbon dioxide–air mixtures. Both experimental results agree fairly well. Only the extrapolated values for higher temperatures show greater differences, Hottel's values being smaller. Since theoretical investigations show also that the emissivity decreases at higher temperatures, the curve for 1600 C in Fig. 16-5 is taken from Hottel's values.

The first measurements of water vapor were made by Schmidt,[4]

[1] A. Schack, *Z. Tech. Phys.*, **5**:267–278 (1924).

[2] H. C. Hottel and H. G. Mangelsdorf, *Trans. AIChE*, **31**:517–549 (1935); H. C. Hottel and V. C. Smith, *Trans. ASME*, **57**:463–470 (1935).

[3] E. Eckert, *VDI-Forschungsh.* 387, 1937.

[4] E. Schmidt, *Forsch. Gebiete Ingenieurw.*, **3**:57–70 (1932).

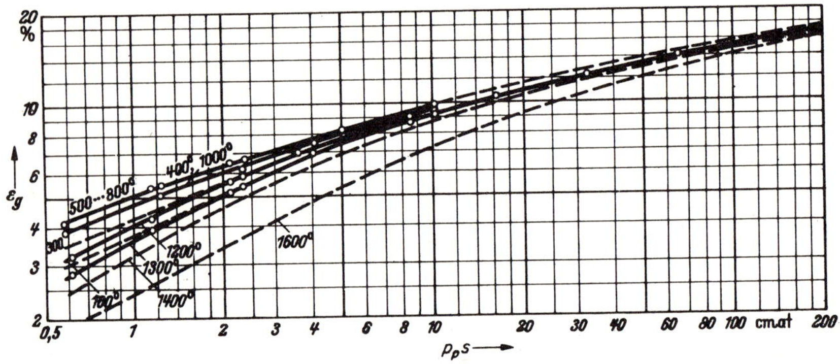

Fig. 16-5 Emissivity of carbon dioxide in mixture with nitrogen or air. S = thickness of gas layer; p_p = partial pressure of CO_2. Total pressure is 1 atm. Dashed lines indicate extrapolated values. [*From E. Eckert, VDI-Forschungsh. 387, 1937.*]

who investigated various thicknesses of pure water vapor. Experiments by Hottel and coworkers[1] with water-vapor–air mixtures of 22-in thickness agree with Schmidt's values at high partial pressures of the water vapor but are lower at low partial pressures. Measurements by Schmidt and Eckert[2] explained these differences and showed that the emissivity of this gas at a constant product of partial pressure of H_2O times thickness and at 1 atm total pressure is smaller for small partial pressures. This can be accounted for if the emissivity of pure water vapor is multiplied by a reduction factor less than 1 as soon as the partial pressure is less than 1 atm. Figure 16-6 shows the emissivity of pure water vapor according to the measurements of Schmidt for various temperatures and at atmospheric pressure. The small diagram gives the reduction factor f by which the values from the large diagram are to be multiplied when the partial pressure of water vapor is smaller than 1. This factor is a mean value derived from a paper by Schwiedessen,[3] who compared all measurements and extrapolated them to higher temperatures. They also take into account the results of new measurements.[4] The measurements by Schmidt and Eckert indicated a small dependence of the reduction factor on gas temperature, which has been neglected in the small diagram of Fig. 16-6.

[1] Hottel and Mangelsdorf, *loc. cit.*; Hottel and Smith, *loc. cit.*

[2] E. Schmidt and E. Eckert, *Forsch. Gebiete Ingenieurw.*, **8**:87 (1937).

[3] H. Schwiedessen, *Arch. Eisenhüttenw.*, **14**:9–14, 145–153, 207–210 (1940).

[4] H. C. Hottel and R. B. Egbert, *Trans. AIChE*, **38**:531–565 (1942); J. E. Eberhardt and H. C. Hottel, *Trans. AMSE*, **58**:185–193 (1936).

All measurements were carried out at a total pressure of the mixtures equal to 1 atm. At higher total pressures the radiation increases because the number of radiating molecules in a given thickness of the layer increases. However, the radiation of any molecule also increases somewhat with pressure. Information on this effect and on other gases is collected in the book by Hottel and Sarofim.[1]

In engineering calculations one often has to consider mixtures of water vapor and carbon dioxide with nonradiating gases. Generally, the monochromatic absorptivity of a mixture of two gases 1 and 2 with the absorption coefficients $\varkappa_{\lambda 1}$ and $\varkappa_{\lambda 2}$ is

$$\alpha_\lambda = 1 - e^{-\rho(\varkappa_{\lambda_1}+\varkappa_{\lambda_2})s}$$

assuming that the two gases do not influence each other in their absorption. The resulting absorptivity α_λ is therefore smaller than the sum of

[1] H. C. Hottel and A. F. Sarofim, "Radiative Transfer," McGraw-Hill, New York, 1967.

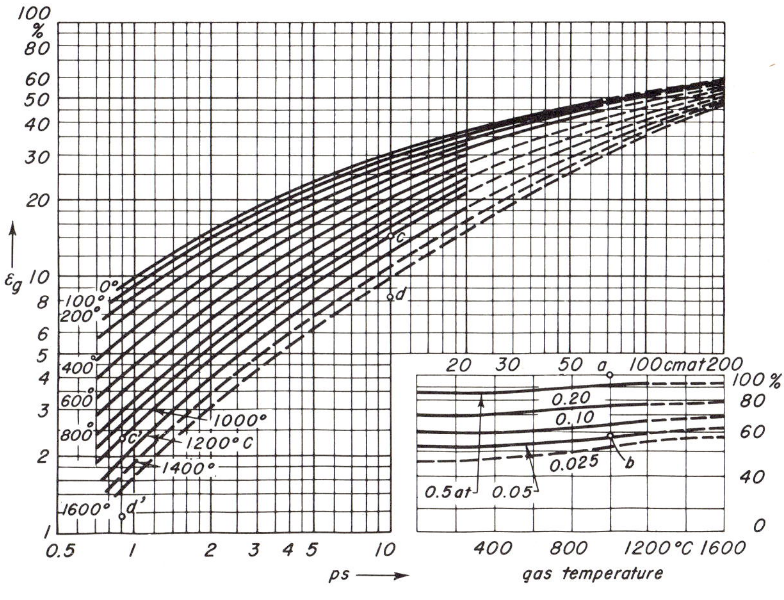

Fig. 16-6 Emissivity of water vapor in mixture with nitrogen or air. S = thickness of the gas layer; p_p = partial pressure of water vapor; total pressure p = 1 atm. When the partial pressure p_p is different from 1, the emissivities read in the big figure must be multiplied by f read from the small diagram. Dashed lines indicate extrapolated values. [*According to measurements by E. Schmidt, H. Hottel, H. Mangeladorf, and E. Eckert.*]

Table 16-1 Correction $\Delta\epsilon$ for simultaneous radiation of a water-vapor–carbon dioxide mixture†

$\epsilon_{CO_2+H_2C} = \epsilon_{CO_2} + \epsilon_{H_2O} - \Delta\epsilon$

CO_2, cm bar	H_2O, cm bar 5	30	100	K
5	0.000	0.014	0.034	400
	0.003	0.012	0.018	800
	0.007	0.020	0.035	1200
30	0.000	0.017	0.043	400
	0.008	0.024	0.032	800
	0.018	0.047	0.052	1200
100	0.000	0.017	0.043	400
	0.018	0.025	0.033	800
	0.030	0.049	0.054	1200

† From E. Eckert, *VDI-Forschungsh.* 387, 1937.

the absorptivities of the two gases individually. The same statement holds for the emissivity. Table 16-1 gives corresponding corrections $\Delta\epsilon$ to be subtracted from the sum of the individual emissivities of water vapor and carbon dioxide to obtain the radiation of a mixture of both gases.

Total absorptivities can be defined by an equation of the same form as the total emissivities with the blackbody intensity replaced by the intensity of the incident radiation. The difficulty recurs of the total absorptivity depending on the wavelength distribution of the incident radiation. For incident blackbody radiation, this distribution is a function of the temperature of the blackbody, and the total absorptivity accordingly depends on the temperature of the blackbody incident radiation as well as on the temperature of the gas. Relations approximating this dependence can be found in the book by Hottel and Sarofim.[1]

Total emissivities and absorptivities are sufficient for the calculation of radiative exchange in a gas-filled enclosure with black walls. For other surfaces, the reflected radiation depends in its wavelength character on monochromatic emissivities and absorptivities of the gas, and such an analysis should, in principle, start with a consideration of monochromatic radiative exchange. The difficulties involved in such an analysis, however, are so great that approximations generally have to be introduced, e.g., by the use of emissivities and absorptivities averaged

[1] *Ibid.*

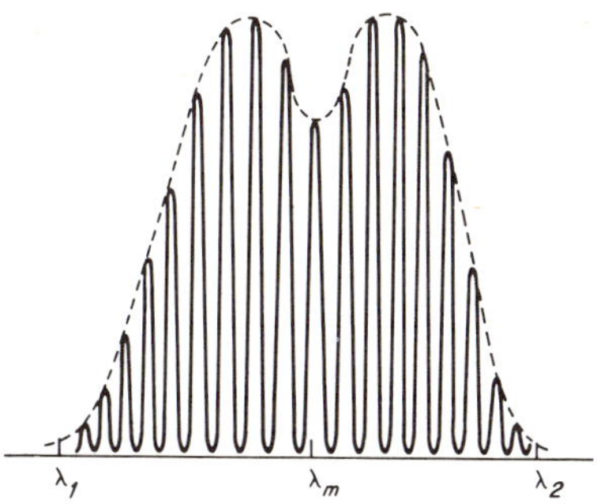

Fig. 16-7 Sketch of an absorption band of a gas.

over the individual bands which characterize gaseous radiation. With increasing resolution, the individual bands shown in Figs. 16-3 and 16-4 take on the shape sketched in Fig. 16-7. They consist of a large number of absorption lines because in the wavelength range of Figs. 16-3 and 16-4 the molecules change their vibrational and rotational energy simultaneously according to the equation

$$h\nu = (E_{v2} - E_{v1}) \pm (E_{r2} - E_{r1})$$

where E_v = vibrational energy level
E_r = rotational energy level
1, 2 = state before and after absorption of a photon

The change in the vibrational energy level determines the mean wavelength of the absorption band. A change in the rotational energy level shifts the frequency of the photons by small amounts toward larger or smaller values, depending on whether the change is positive or negative, and in this way creates the two groups of lines to the right and left of the mean wavelength in Fig. 16-7. The individual lines actually have a finite thickness and a shape similar to that of a probability curve. This effect, called *line broadening,* is caused by the natural uncertainty in the energy level connected with Heisenberg's uncertainty principle, with small changes in the energy level by collisions, and by the thermal motion of colliding particles. Correspondingly, one refers to natural, collision, and doppler broadening. An envelope of the band structure in Fig. 16-7 has a shape indicated by the dashed line and decreases only asymptotically toward zero at the two rims of the band. This causes difficulty in defining a mean band emissivity, which, in principle, would be given by the equation

$$\epsilon_b = \frac{\int_{\lambda_1}^{\lambda_2} \epsilon_\lambda i_{b\lambda}\, d\lambda}{\int_{\lambda_1}^{\lambda_2} i_{b\lambda}\, d\lambda}$$

The blackbody intensity changes, in general, only little over the wavelength range and can be canceled out of the equation so that it reads

$$\epsilon_b = \frac{\int_{\lambda_1}^{\lambda_2} \epsilon_\lambda \, d\lambda}{\lambda_2 - \lambda_1}$$

The difficulty in evaluating this equation is that the two wavelengths λ_2 and λ_1, which determine the borders of the band, must be selected somewhat arbitrarily. To avoid this, another parameter is defined with the numerator of this equation only. In addition, the integration is usually performed over the wave number $\omega = 1/\lambda$ instead of the wavelength. This leads to the integrated band emissivity

$$\epsilon_b = \int_{\omega_1}^{\omega_2} \epsilon_\omega \, d\omega$$

It must be kept in mind that the dimension of this parameter is 1 over length, whereas the emissivities defined before are dimensionless. An integrated band absorptivity can again be defined in the same way. The values for this parameter depend, in general, on the wavelength characteristics of the incident radiation and for blackbody incident radiation on its temperature. Band emissivities and absorptivities have been reported for various gases in the literature.[1] It is found that band emissivities and absorptivities do not increase exponentially with increasing path length, as described by Eqs. (16-5) and (16-7). The same statement holds for total gas emissivities and absorptivities. For the analysis, however, it is convenient to approximate them by an expression of the form

$$\epsilon_b = \Sigma a_i(1 - e^{-\rho \varkappa_\lambda S})$$

A small number of terms is generally sufficient for a good approximation.[2]

Analyses which take into account a local variation of the temperature or concentration of the radiating gas in the enclosure are based more conveniently on mass coefficients of absorption, for which the notation $\varkappa_\lambda$ has been introduced when dealing with monochromatic values. Mean values for the individual absorption bands, which are of advantage in such a calculation, are defined by the equation

$$S = \rho \int_{\omega_1}^{\omega_2} \varkappa_\omega \, d\omega$$

and values for such integrated band-absorption coefficients are again found in the literature for various gases.[3]

[1] D. K. Edwards, *J. Opt. Soc. Amer.*, **50**:617 (1965).

[2] Hottel and Sarofim, *op. cit.*, p. 247.

[3] More detailed information on radiation properties of gases are found in S. S. Penner, "Quantitative Molecular Spectroscopy and Gas Emissivities," Addison-Wesley, Reading, Mass., 1959; and C. L. Tien, *Advan. Heat Transfer*, **5**:257 (1968).

Various analyses have been reported in which the actual gas has been approximated by a gray gas. Definitions of mean absorption coefficients for those analyses will merely be listed here. Planck's mean absorption coefficient is defined as

$$\varkappa_P = \frac{\int_0^\infty \rho \varkappa_\lambda e_{b\lambda}\, d\lambda}{e_b}$$

and the Rosseland mean absorption coefficient as

$$\frac{1}{\varkappa_P} = \frac{\int_0^\infty \frac{1}{\rho \varkappa_\omega} \frac{de_{b\omega}}{dT}\, d\omega}{\int_0^\infty \frac{de_{b\omega}}{dT}\, d\omega}$$

The first coefficient is appropriate for optically thin gases and the second for optically thick ones.

In plasma technology and in the reentry process of satellites or space vehicles through the atmosphere, the gases involved are at a temperature of the order of 10,000 K. Radiation under this condition obtains its largest contribution from the visible or near-ultraviolet wavelength range. This radiation appears in the spectrogram as lines generated by transition between the various energy levels of the electrons in the atomic structure (bound-bound transition), or it is associated with the removal or capture of electrons by the atoms in the ionization process and appears then as continuum radiation.[1]

16-2 EQUIVALENT BEAM LENGTH

Calculations of radiative energy exchange in enclosures with a participating gas are often simplified in that the composition, pressure, and temperature of the gas are assumed as locally uniform. For such an analysis, the radiation parameters (emissivity, absorptivity, and transmissivity) defined in the preceding section are very useful. The following difficulty arises, however. For the analysis, the surface of the enclosure is broken up into a number of finite plane surfaces over which the temperature is assumed uniform. In order to calculate the radiative exchange between two such surfaces (for instance, i and k in Fig. 16-8) and the gas body, one must use in the equations for the radiation parameters some average of the beam lengths as they occur between the two surfaces. This *equivalent beam length* will be calculated in the present section. The concept of the equivalent beam length was introduced by Hottel and generalized by Dunkle.

[1] For a more detailed description and for tabulations of emissivities and absorptivities or absorption coefficients, the reaper is again referred to *ibid.*

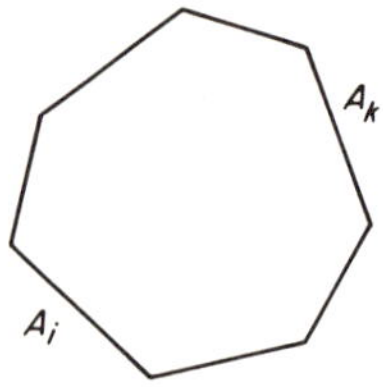

Fig. 16-8 Radiation analysis in an enclosure.

The radiative flux leaving surface A_i and arriving at surface A_k was written in Sec. 15-2 as $A_iB_iF_{i-k}$. In this equation B_i denotes the radiosity of surface A_i, and F_{i-k} indicates the shape factor of surface i toward surface k. The above expression assumes that the medium between the two surfaces does not participate in the radiation process. If this medium is an absorbing one, a fraction given by the absorptivity of the gas will be absorbed and will not reach surface A_k. The absorbed monochromatic energy flux will be written as

$$A_iB_{\lambda i}F_{i-k}\alpha_{\lambda g,i-k}$$

where $\alpha_{\lambda g,i-k}$ denotes an average absorptivity for all the rays in the various directions between surfaces A_i and A_k. A mean beam length can now be defined by the equation

$$\alpha_{\lambda g,i-k} = 1 - e^{-\rho\varkappa_\lambda L_e} \tag{16-8}$$

where L_e is an average value of the beam lengths between the two surfaces which may be called the equivalent beam length. For its calculation, we start with the monochromatic heat flux leaving a surface element dA_i in Fig. 16-9 in the direction of the area element dA_k and being absorbed by the gas between the two surfaces. This heat flux is

$$d^2\Phi_\lambda = K_{\lambda i}\, dA_i \cos\beta_i\, d\omega_i(1 - e^{-\rho\varkappa_\lambda S})$$

where $K_{\lambda i}$ is the monochromatic intensity of the heat flux. This intensity is connected with the radiosity by the expression $K_{\lambda i} = B_{\lambda i}/\pi$ provided the surface A_i emits and reflects diffusely. The heat flux leaving dA_i in the direction of the total area A_k and being absorbed by the gas is then obtained by integration over the total solid angle ω_i under which A_k is seen from dA_i:

$$d\Phi_\lambda = \frac{B_{\lambda i}\, dA_i}{\pi}\int_{\omega_i}(1 - e^{-\rho\varkappa_\lambda S})\cos\beta_i\, d\omega_i \tag{16-9}$$

On the other hand, this heat flux can be expressed with an equivalent beam length in the following way:

$$d\Phi_\lambda = B_{\lambda i}F_{dA_i-A_k}\, dA_i\,(1 - e^{-\rho\varkappa_\lambda L_e}) \tag{16-9a}$$

Setting the last two equations equal to each other results in the following equation, from which the equivalent beam length can be calculated:

$$1 - e^{-\rho\varkappa_\lambda L_e} = \frac{1}{\pi F_{dA_i - A_k}} \int_{\omega_i} (1 - e^{-\rho\varkappa_\lambda S}) \cos\beta_i \, d\omega_{\lambda i} \tag{16-10}$$

An approximation to the equivalent beam length which depends only on geometry and is usually sufficient for engineering purposes can be obtained in the following way. Equation (16-7) for the gas absorptivity can be simplified when the exponent $\rho\varkappa_\lambda S$ has a small value. In this case, the e function can be developed into a series, and higher-order terms can be neglected. This results in

$$\alpha_{\lambda g} = \rho\varkappa_\lambda S \tag{16-11}$$

A gas for which this approximation is valid is called a gas with vanishing self-absorption or an optically thin gas, because it has been demonstrated that the monochromatic gas absorptivity is equal to the monochromatic gas emissivity. Equation (16-11) for the emissivity, however, can also be obtained when the second term in the right-hand parentheses of Eq. (16-3) is neglected; this means, physically, when absorption of the radiant energy emitted by the gas is neglected within the gas body. Replacing the e function by the linear expression (16-11) on both sides of Eq. (16-10) gives

$$L_g = \frac{1}{\pi F_{dA_i - A_k}} \int_{\omega_i} S \cos\beta_i \, d\omega_i \tag{16-12}$$

This equation then describes a mean beam length for the energy flux from a diffusely radiating and reflecting surface dA_i toward the surface A_k

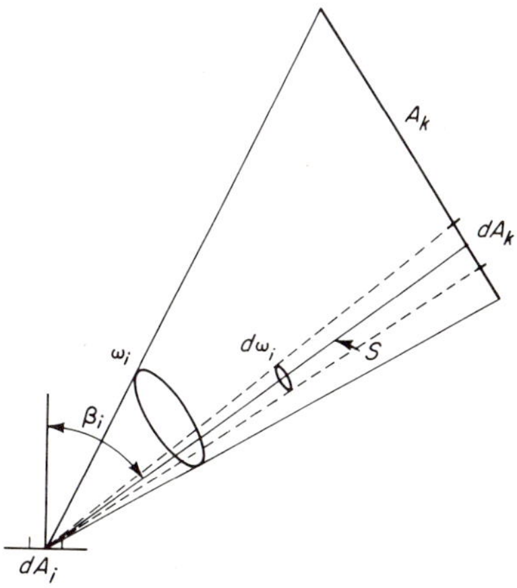

Fig. 16-9 Determination of geometric beam length.

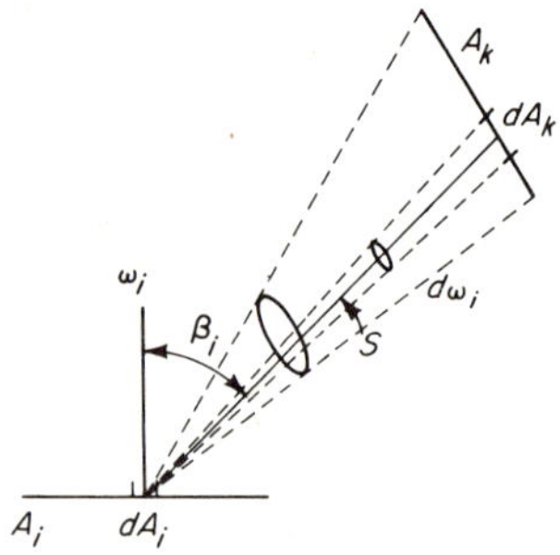

Fig. 16-10 Determination of geometric beam length.

being absorbed by a gas with a locally uniform coefficient of absorption and locally uniform density. This is now an equation which depends on geometry alone. Dunkle,[1] who introduced this parameter, calls it the *geometric beam length*. Dividing the equation by a characteristic dimension L of surface A_k, one obtains

$$\frac{L_g}{L} = \frac{1}{\pi F_{dA_i-A_k}} \int_{\omega_i} \frac{S}{L} \cos \beta_i \, d\omega_i \tag{16-13}$$

which now depends solely on angular relations. A simple geometric interpretation of the geometric mean beam length can be obtained when we recall that $\cos \beta_i \, d\omega_i$ is the projection $d\omega_i'$ of the solid angle $d\omega_i$ in Fig. 15-13 onto the plane of the area element dA_i. The geometric mean beam length is then given by

$$L_g = \frac{\int S \, d\omega_i'}{\omega_i'} \tag{16-14}$$

and can be interpreted as the mean distance between dA_i and A_k when equal increments $d\omega_i'$ are considered in the integration process.

In the analyses of an enclosure with a participating gas, we need equivalent beam lengths for radiant beams between two finite surfaces, for instance, A_i and A_k (Fig. 16-10). The radiative flux leaving surface A_i in the direction of A_k and being absorbed by the gas is obtained by an integration of Eq. (16-9):

$$\Phi_\lambda = \frac{B_{\lambda i}}{\pi} \int_{A_i} dA_i \int_{\omega_i} (1 - e^{-\rho \varkappa_\lambda S}) \cos \beta_i \, d\omega_i \tag{16-15}$$

On the other hand, the equivalent beam length is again defined by

$$\Phi_\lambda = (1 - e^{-\rho \varkappa_\lambda L_e}) B_{\lambda i} A_i F_{A_i-A_k} \tag{16-16}$$

[1] R. V. Dunkle, *J. Heat Transfer*, **86**:75 (1964).

For vanishing self-absorption one obtains from these two equations the relation for the geometric mean beam length

$$L_g = \frac{1}{\pi A_i F_{A_i - A_k}} \int_{A_i} dA_i \int_{\omega_i} S \cos \beta_i \, d\omega_i \tag{16-17}$$

Again the ratio of this mean beam length L_g to a characteristic dimension L of the configuration depends only on angular relations. Dunkle[1] has calculated this ratio for two configurations involving rectangles. Figures 16-11 and 16-12 present the result. To calculate the part of the energy leaving surface A_i which arrives at A_k one needs the transmissivity of the gas $\tau_{\lambda g, A_i - A_k}$; this is obtained from the relation

$$\tau_{\lambda g, A_i - A_k} = 1 - \alpha_{\lambda g, A_i - A_k} \tag{16-18}$$

The geometric mean beam length for radiation between an area A_i and the complete enclosure is

$$L_g = \frac{1}{\pi A_i} \int_{A_i} dA_i \int_{\triangle} S \cos \beta_i \, d\omega_i \tag{16-19}$$

The geometric beam length defined by Eq. (16-19) is useful in determining that portion of the heat flux leaving a surface A_i which is absorbed by an adjacent gas body. It can also be used to calculate the radiative flux emitted by the gas body and impinging on the surface area A_i.

[1] *Ibid.*

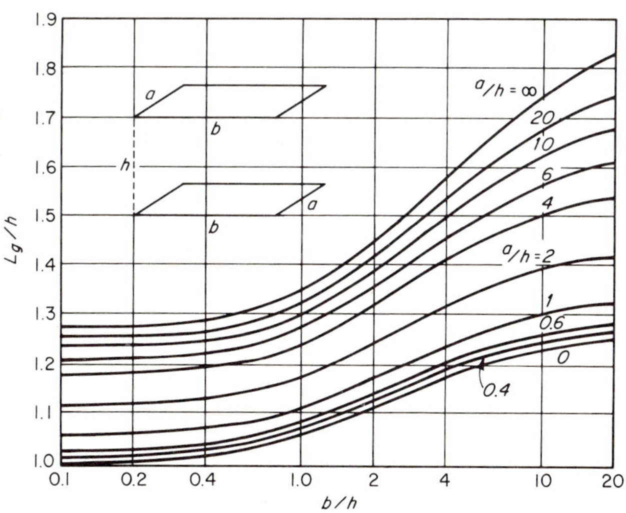

Fig. 16-11 Ratio of geometric mean beam length to distance h for two parallel rectangular surfaces. [*After R. V. Dunkle.*]

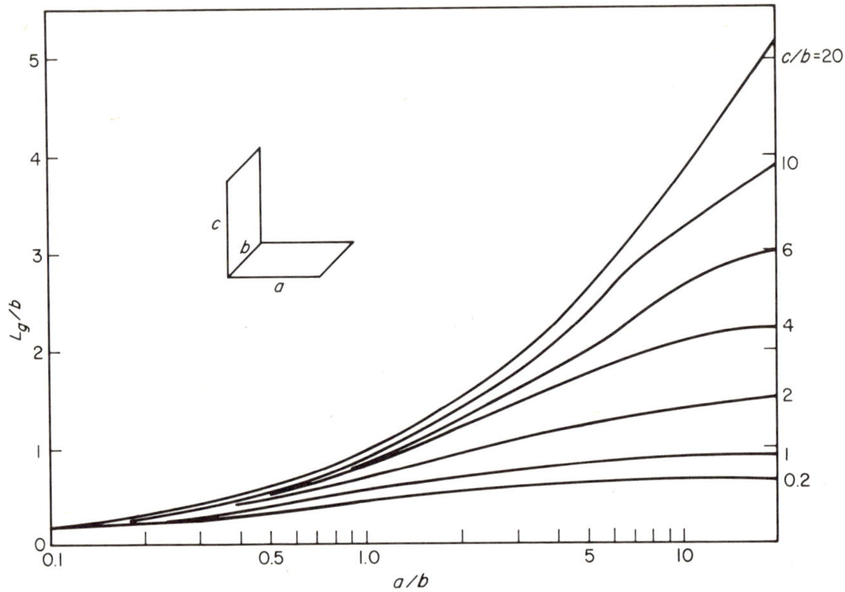

Fig. 16-12 Ratio of geometric mean beam length to dimension b for two rectangular surfaces normal to each other. [*After R. V. Dunkle.*]

Equivalent beam lengths for the same situations are defined by Eq. (16-10) when the integration is carried out over the whole hemispherical space. Values of this equivalent hemispherical beam length have been calculated by a number of investigators. Equation (16-10) shows that they are a function of the parameter $\rho\kappa_\lambda s$. Hottel, however, has shown that one can find a mean equivalent hemispherical beam length L_e in such a way that the ratio of the actual hemispherical emissivity ϵ_λ to the one calculated with the mean equivalent hemispherical beam length through the equation

$$\epsilon_{\lambda H} = 1 - e^{\rho\kappa_\lambda L_e}$$

has a minimum average deviation from 1 in the whole range of the parameter $\rho\kappa_\lambda s$. Equivalent beam lengths obtained in this way are given in Table 16-2. Figure 16-13 shows the ratio $\epsilon_\lambda/\epsilon_{\lambda H}$ for a number of geometries and therefore indicates to what degree an analysis with the mean equivalent beam length approximates the answer obtained from Eqs. (16-10). Of course, an additional approximation is involved when calculations are performed with the geometric beam length instead of the equivalent beam length. The difference between the answers obtained in these two ways will be discussed later.

A very simple relation can be obtained for the geometric mean

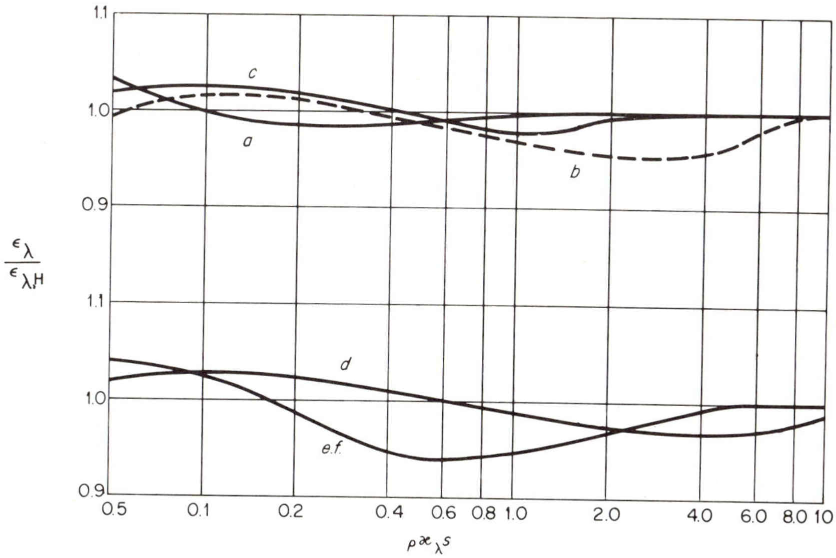

Fig. 16-13 Ratio of the mean hemispherical emissivity ϵ_λ of various gas bodies to the emissivity $\epsilon_{\lambda H}$ of a hemisphere into the center of its base. Curves a to f belong to the first six geometries listed in Table 16-2 in alphabetical order. The length s is the dimension d, r, or l, and $\epsilon_{\lambda H}$ is the emissivity of a hemisphere with radius L, as listed in the table. [*From E. Eckert, VDI-Forschungsh. 387, 1937.*]

beam length which describes the radiative flux emitted by a gas body into the whole surface A enclosing it. Figure 16-14 indicates such a gas body with the volume V and the surface area A. We first derive an expression to describe that part of the monochromatic radiation with wavelength λ originating in the volume element dV which arrives at the surface area dA_i. If the volume element is seen from the area dA_i under the solid angle $d\omega$, the heat radiated into the element dA_i per unit time can be written as

$$d^2\Phi_r = di_\lambda \, d\omega \, dA_i \cos \beta$$

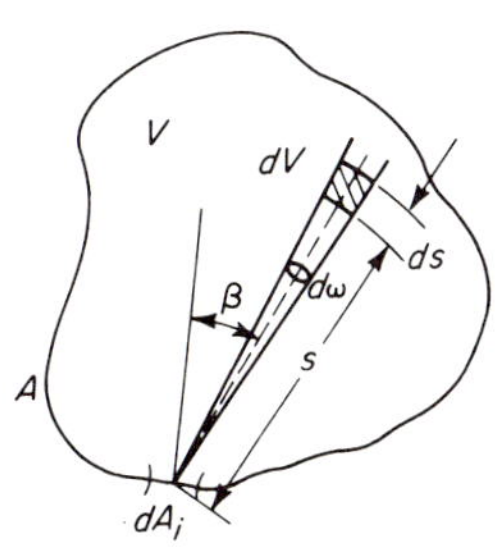

Fig. 16-14 Radiation in a weakly absorbing medium.

Table 16-2 Mean equivalent beam lengths l_f of gas volumes of different shape†

Shape of gas volume	L_e
Circular cylinder: height = diameter = d	
Irradiation into the center of the base	$0.77d$
Circular cylinder: height = ∞, diameter = d	
Irradiation into the convex surface	$0.95d$
Cylinder: height = ∞, base semicircle with radius r	
Irradiation into the center of the plane rectangular surface	$1.26r$
Sphere: diameter = d	
Irradiation into the surface	$0.65d$
Volume between two infinite planes separated by distance l	
Irradiation into the planes	$1.8l$
Circular cylinder: height = ∞, diameter = d	
Irradiation into the center of the base	$0.9d$
Tube bundle:‡	
In triangular arrangement	
$s = 2d$	$3.0(s - d)$
$s = 3d$	$3.8(s - d)$
In square arrangement	
$s = 2d$	$3.5(s - d)$
Cube: length = a	
Irradiation into each surface plane	$0.66a$

† According to H. C. Hottel and E. Eckert; from E. Eckert, *VDI-Forschungsh.* 387, 1937.
‡ s = distance between tube centers; d = tube diameter.

The term di_λ indicates the energy emitted by the volume element dV per unit area and solid angle in the direction toward A_i. With the emissivity

$$d\epsilon_\lambda = \rho\varkappa_\lambda\, ds$$

of the volume element with the thickness ds, this equation becomes

$$d^2\Phi_\lambda = i_{b\lambda}\, d\epsilon_\lambda\, d\omega\, dA \cos\beta = \frac{e_{b\lambda}}{\pi} \cos\beta\, \rho\varkappa_\lambda\, ds\, d\omega\, dA_i$$

The volume element itself can be expressed as

$$dV = s^2\, d\omega\, ds$$

Therefore, the radiative heat flux into the surface element becomes

$$d^2\Phi_\lambda = \frac{e_{b\lambda}}{\pi} \rho\varkappa_\lambda\, dV \frac{\cos\beta\, dA_i}{s^2}$$

The term $\cos\beta\, dA_i/s^2$ appearing in this equation can be interpreted as the solid angle $d\Omega$ under which the area dA_i is seen from the location of dV.

Therefore,

$$d^2\Phi_\lambda = \frac{e_{b\lambda}}{\pi} \rho\kappa_\lambda \, dV \, d\Omega$$

The total energy exchange between the volume element dV and the total surface area A of the gas body results from an integration of the above equation over the whole space surrounding dV:

$$d\Phi_\lambda = \frac{e_{b\lambda}}{\pi} \rho\kappa_\lambda \, dV \, 4\pi = 4\rho\kappa_\lambda e_{b\lambda} \, dV$$

By another integration, the monochromatic radiation emitted by the total volume V and arriving at its surface area A is obtained:

$$\Phi_\lambda = 4\rho\kappa_\lambda e_{b\lambda} V$$

For vanishing self-absorption, the heat flux can again be expressed by a geometric mean beam length with the equation

$$\Phi_\lambda = \rho\kappa_\lambda L_g e_{\lambda b} A$$

From the last two expressions one gets the simple relation

$$L_g = \frac{4V}{A} \tag{16-20}$$

It has been found that this expression of the geometric hemispherica beam length approximates the values for the mean equivalent hemispherical beam length in Table 16-2 satisfactorily for engineering calculations if the geometric beam length in Eq. (16-20) is multiplied by 0.9. This expression is therefore frequently used as a first approximation for the equivalent hemispherical beam length of various geometries.

So far the discussion has been based on the assumption that the absorption and emission process in a gas follows an exponential relation of the form as given in Eq. (16-8). On the other hand, it has been pointed out that band absorptivities or total absorptivities do not follow such an exponential law. The following procedure is recommended if one does not want to carry out the integration over the space coordinates in each individual case. One uses equivalent beam lengths or geometric beam lengths as defined in the preceding paragraphs and determines the gas emissivities, transmissivities, or absorptivities from equations or charts which express the actual variation of these parameters with the beam length S. For example, the total emissivities of carbon dioxide and water, are contained in Figs. 16-5 and 16-6. It can be shown that this procedure leads to approximations which are generally satisfactory for engineering calculations.

A simple example will demonstrate the calculation of geometric

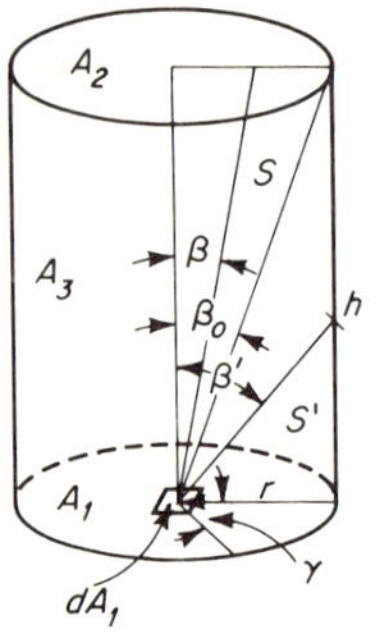

Fig. 16-15 Radiation in a weakly absorbing medium of cylindrical shape.

mean beam lengths. The enclosure is assumed to have the shape of a cylinder with circular cross section, as indicated in Fig. 16-15. The radius may be r and the height h. The geometric mean beam length will be calculated at first for radiation between dA_1 located at the center of the base and area A_2. A spherical coordinate system with the center at dA_1 may be used with the angles indicated in Fig. 16-15. Equation (16-13) changes to

$$L_g = \frac{1}{\pi F_{dA_1-A_2}} \int_0^{2\pi} \int_0^{\beta_0} S \cos\beta \sin\beta \, d\beta \, d\gamma$$

where γ is the azimuth angle. The beam length S can be written as $S = h/(\cos\beta)$, and the shape factor between the area element dA_1 and the area A_2 is $F_{dA_1-A_2} = \sin^2\beta_0$. This is easily seen by tracing the double projection illustrated in Fig. 15-16. Introducing these expressions and integrating over γ results in

$$L_g = \frac{2h}{\sin^2\beta_0} \int_0^{\beta_0} \sin\beta \, d\beta = 2h \frac{1 - \cos\beta_0}{\sin^2\beta_0}$$

The beam length necessary to calculate the emission of radiant energy from the cylindrical gas body into the area element dA_1 or the absorption of radiant energy coming from dA_1 by the gas body is given by the double integral

$$L_g = \frac{1}{\pi} \int_0^{2\pi} \int_0^{\pi/2} S \cos\beta \sin\beta \, d\beta \, d\gamma$$

Carrying out the integration results in the expression

$$L_g = 2(h + r - h\cos\beta_0 - r\sin\beta_0)$$

The geometric mean beam length which determines the emission of radiation from the gas body into the total surface of the enclosure finally, according to Eq. (16-20), is

$$L_g = \frac{4V}{A} = \frac{4\pi r^2 h}{2\pi r^2 + 2\pi r h} = r \frac{2}{r/h + 1}$$

For the special case where the height of the cylinder is equal to its diameter d, the following numerical values for the geometric mean beam lengths are obtained using the above equations:

For radiation between dA_1 and A_2:

$$L_g = 1.04d$$

For radiation between dA_1 and A_3:

$$L_g = 0.679d$$

For emission by the gas body into dA_1:

$$L_g = 0.76d$$

This value can be compared with the value $0.77d$ in Table 16-2.
For emission by the gas body into the total surface of the cylinder:

$$L_g = 0.666d$$

For the cylinder of infinite height, the geometric mean beam length is $L_g = d$ compared to $L_b = 0.9d$ in Table 16-2.

In the section on angle factors, reciprocity relations were obtained by the method referred to as angle-factor algebra. It is useful to obtain angle factors for new configurations from known ones. This angle-factor algebra can be extended to include the action of an emitting or absorbing gas. The following relations hold:

$$A_iF_{i-k}\alpha_{\lambda g,i-k} = A_kF_{k-i}\alpha_{\lambda g,k-i} \tag{16-21}$$

$$A_1F_{1-2}\alpha_{\lambda g,1-2} + A_1F_{1-3}\alpha_{\lambda g,1-3} = A_1F_{1-(2+3)}\alpha_{\lambda g,1-(2+3)} \tag{16-22}$$

$$A_i\epsilon_{\lambda g,i} = \sum_k A_iF_{i-k}\alpha_{\lambda g,i-k} \tag{16-23}$$

Equation (16-21) makes it possible to change the surface of reference for radiation between two surfaces. Equation (16-22) allows one to calculate gas absorptivity values for a composite area $2 + 3$ when the gas absorptivities for the two partial adjoining areas 2 and 3 are known. Equation (16-23) finally connects the gas absorptivity for radiation between pairs of surfaces to the gas emissivity of the total gas body in the enclosure.

16-3 ENCLOSURES WITH AN ABSORBING AND EMITTING MEDIUM

With the preparation afforded by the preceding section, we can now turn to the task of calculating heat transfer by radiation in an enclosure filled with an absorbing and emitting gas. It will be assumed that the walls of the enclosure emit radiation with an intensity which is independent of

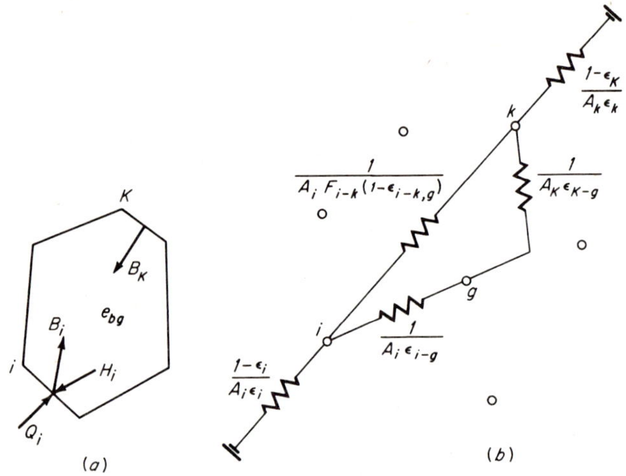

Fig. 16-16 (*a*) Enclosure containing a participating medium. (*b*) Electrical analog to the radiative exchange in the enclosure of (*a*).

direction and that they are diffuse reflectors and gray surfaces. The medium filling the enclosure will also be postulated as a gray gas ($\alpha = \epsilon$) with uniform temperature and concentration throughout the space inside the enclosure. Later some of these assumptions will gradually be relaxed.

We turn our attention to a surface i with the radiosity B_i, the irradiation H_i, and the net heat loss by radiation Q_i (Fig. 16-16*a*). The following equations connect these quantities (see page 630):

$$Q_i = A_i(B_i - H_i)$$
$$B_i = \epsilon_i e_{bi} + (1 - \epsilon_i)H_i$$

Elimination of H_i results in

$$Q_i = A_i \frac{\epsilon_i}{1 - \epsilon_i} (e_{bi} - B_i) \tag{16-24}$$

The irradiation H_i is now being connected with the radiosity of the surrounding surfaces and the emission from the gas. The radiative flux leaving surface k in the direction of surface i is

$$A_iF_{i-k}B_k$$

Not all of this radiation will arrive at surface i because part of it is absorbed in the gas. The radiative flux arriving is

$$A_iF_{i-k}(1 - \alpha_{i-k,g})B_k$$

where $\alpha_{i-k,g}$ denotes the average absorptivity for all rays between surfaces i and k.

The radiative flux emitted by the gas and arriving at i is

$$A_i \epsilon_{i,g} e_{bg}$$

where $\epsilon_{i,g}$ indicates the emissivity of the gas body into area i. Finally the radiative flux leaving i is

$$A_i B_i$$

From these components the net radiative heat loss of surface i can be obtained:

$$\begin{aligned} Q_i &= A_i[B_i - \Sigma F_{i-k}(1 - \alpha_{i-k,g})B_k - \epsilon_{i,g}e_{bg}] \\ &= A_i \epsilon_{i,g}(B_i - e_{bg}) + \Sigma A_i F_{i-k}(1 - \alpha_{i-k,g})(B_i - B_k) \end{aligned} \tag{16-25}$$

The net heat loss of the gas caused by the radiative exchange is

$$Q_g = \sum_i A_i \epsilon_{i,g}(e_{bg} - B_i) \tag{16-26}$$

Equations (16-24) to (16-26) are sufficient to calculate the unknown parameters in this problem—the heat loss for surfaces with known temperature, the temperature for adiabatic surfaces, and the heat loss or the temperature of the gas in the enclosure.

The electric analog to the enclosure sketched in Fig. 16-16*a* is shown in Fig. 16-16*b*. Only the connections between nodes i, k, and g are inserted. Node g corresponds to the enclosed gas. Analogous connections have to be made between the other nodes. Node g is either connected with a battery giving the potential e_{bg} when the temperature of the gas is prescribed or is left floating when no heat sources or sinks are contained in the gas. For the first situation, the current from the battery into node g is proportional to the heat loss of the gas. In the second situation, the potential at node g is proportional to the emissive power of a blackbody at the gas temperature.

Referring to the section on gas radiation, it is evident that real gases differ considerably from a gray one. For a calculation of radiative heat exchange in an enclosure filled with a nongray gas, Eqs. (16-24) to (16-26) must be applied to monochromatic radiation. The total heat loss is then found by an integration of the monochromatic heat flux over all wavelengths. This calculation becomes very cumbersome, and an approximation to the real conditions is therefore desirable. The concept of a selectivity gray gas[1] is useful for this purpose. To approximate a real gas by a selectively gray one, the whole wavelength range is divided into two portions, one comprising the wavelength ranges in which the gas absorbs and the other comprising the rest. The calculation in the first range is then made in exactly the same manner as for a gray gas,

[1] Eckert, *op. cit.*

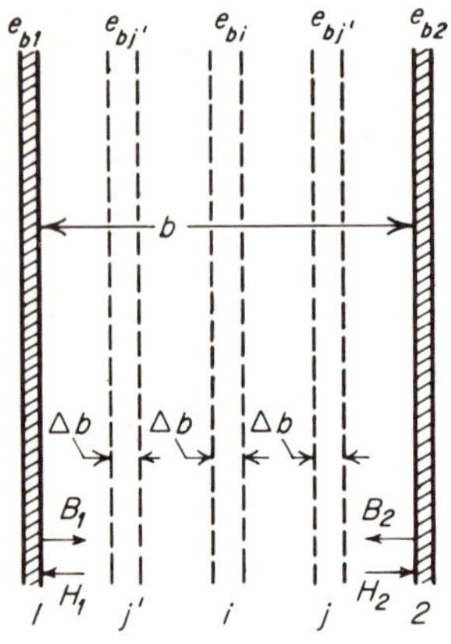

Fig. 16-17 Emitting and absorbing medium within two parallel walls.

assuming that in this range or in the individual bands of this range the absorptivity in independent of wavelength. For the second part, the calculation method developed for a nonradiating gas in the preceding section is used. Another method, which approximates a real gas even better, is described by Hottel and Sarofim.[1]

In many cases the assumption that the temperature of the gas body in the enclosure is uniform is not realistic. An obvious improvement of the calculation procedure is obtained when the gas body is divided into a number of parts and the temperature and concentration are allowed to vary from one part to the next. As an example, the geometry in Fig. 16-17 will be investigated.

An absorbing and emitting gray gas of locally uniform composition is enclosed between two infinite parallel walls 1 and 2. The distance between the walls is b. For our calculation, we shall divide the gas body into n layers of equal thickness Δb. The temperature within each of these layers and on both solid walls is assumed constant.

For the calculation of the radiative heat exchange, the radiation parameters of the gas layers and of the walls must be determined. The following values are needed and will be assumed independent of temperature. The solid walls will be postulated as diffusely emitting and reflecting.

Let ϵ_w = hemispherical emittance of walls
ϵ_g = hemispherical emissivity of gas layer with thickness Δb
$\tau_{gw,k}$ = hemispherical transmissivity of a gas body consisting of k layers for wall radiation
$\tau_{gg,k}$ = hemispherical transmissivity of gas body consisting of k layers for gas radiation emitted from one layer

[1] Hottel and Sarofim, *op. cit.*, p. 247.

$\alpha_{gw,k}$ = ratio of radiative flux absorbed in gas layer with thickness Δb to radiative flux leaving solid wall with k gas layers in between

$\alpha_{gg,k}$ = ratio of radiative flux absorbed in gas layer to radiative flux emitted from another gas layer (both with thickness Δb) with k gas layers in between

The following expressions describe these radiation parameters:

$$\epsilon_g = 2 \int_0^{\pi/2} (1 - e^{-a/(\cos\beta)}) \sin\beta \cos\beta \, d\beta$$

$$\tau_{gw,k} = 2 \int_0^{\pi/2} e^{-ka/(\cos\beta)} \sin\beta \cos\beta \, d\beta$$

$$\tau_{gg,k} = \frac{2}{\epsilon_g} \int_0^{\pi/2} (1 - e^{-a/(\cos\beta)}) e^{-ka/(\cos\beta)} \sin\beta \cos\beta \, d\beta$$

$$\alpha_{gg,k} = \frac{2}{\epsilon_g} \int_0^{\pi/2} (1 - e^{-a/(\cos\beta)})^2 \, e^{-ka/(\cos\beta)} \sin\beta \cos\beta \, d\beta$$

with the abbreviation $a = \rho\varkappa \, \Delta b$. It can easily be seen that the following relation exists between the radiation parameters:

$$\tau_{gg,k}\epsilon_g = \alpha_{gw,k}$$

The radiative flux arriving at wall 1 will be called H_1, and the radiative flux leaving this wall will be called B_1. Again Eq. (16-24) applies to the present situation. It remains to express the net heat loss q_1 per unit area of wall 1 by the arriving and the leaving radiative fluxes and to describe the arriving flux by the various parts of which it consists. An arbitrary layer is denoted by i. The following equation holds:

$$q_1 = B_1 - \left(\tau_{gw,n}B_2 + \sum_{i=1}^{i=n} \epsilon_g\tau_{gg,i-1}e_{bi}\right)$$

This equation can now be written for the situation where the two walls and the enclosed gas are all at the same temperature. From Kirchhoff's law, it is known that in an enclosure of uniform temperature the radiative flux through any arbitrary plane is equal to the radiative flux coming from a black surface and that the net heat loss by radiation of wall 1 is zero. The above equation gives for this special situation

$$1 - \tau_{gw,n} - \sum_{i=1}^{i=n} \epsilon_g\tau_{gg,i-1} = 0$$

With this relation, the preceding equation can be transformed to

$$q_1 = \tau_{gw,n}(B_1 - B_2) + \sum_{i=1}^{i=n} \epsilon_g\tau_{gg,i-1}(B_1 - e_{bi}) \tag{16-27}$$

A corresponding equation holds for wall 2.

As a next step, the radiative loss of the gas layer i will be calculated. This quantity is determined by the radiation emitted from the gas layer through both its surfaces, by the radiative flux which comes from the two walls and is absorbed by the gas layer, and by the radiation originating in the two gas bodies on both sides of the layer and being absorbed in the layer. This energy balance can be expressed by the equation

$$q_i = 2\epsilon_g e_{bi} - \alpha_{gw,i-1}B_1 - \alpha_{gw,n-i}B_2 - \sum_{j=i+1}^{j=n} \epsilon_g \alpha_{gg,j-i-1} e_{bj} - \sum_{j'=1}^{j'=i-1} \epsilon_g \alpha_{gg,i-j'-1} e_{bj'}$$

Changing this equation for the condition of a uniform temperature results (with the same consideration as above) in the relations

$$2\epsilon_g = \alpha_{gw,i-1} + \alpha_{gw,n-i} + \sum_{j=i+1}^{j=n} \epsilon_g \alpha_{gg,j-i-1} + \sum_{j'=1}^{j'=i-1} \epsilon_g \alpha_{gg,i-j'-1}$$

and

$$q_i = \alpha_{gw,i-1}(e_{bi} - B_1) + \alpha_{gw,n-i}(e_{bi} - B_2) + \sum_{j=i+1}^{j=n} \epsilon_g \alpha_{gg,j-i-1}(e_{bi} - e_{bj}) + \sum_{j=1}^{j=i-1} \epsilon_g \alpha_{gg,i-j'-1}(e_{bi} - e_{bj'}) \quad (16\text{-}28)$$

Equations (16-24), (16-27), and (16-28), together with two corresponding equations for wall 2, completely describe the problem. It can easily be checked that the number of equations is equal to the number of unknowns.

The electrical analog which corresponds to these equations is shown in Fig. 16-18. The vertical row of nodes 1 to n corresponds to the n gas layers. The necessary connections are indicated between node i and the nodes 1, 2, j, and j', respectively. The network would have to be completed by making all the other connections in the same way. The temperature of the various gas layers can be determined by internal heat sources or sinks. In this case the temperature of the layers has to be considered as prescribed. Accordingly, the potential of nodes 1 to n is fixed by appropriate batteries. The temperature of some or all gas layers may also be determined solely by the radiative heat exchange. In this case, the corresponding nodes are left floating, and the potentials of the nodes indicate their blackbody emissive power and thus their temperature. Finally, the heat-generation rate in the individual layers may be prescribed. The corresponding electric currents (q_i, q_j) then have to be generated by adjustment of the potential of the corresponding batteries.

For a nongray gas, the calculation must be performed first for monochromatic radiation with a succeeding integration over all wave-

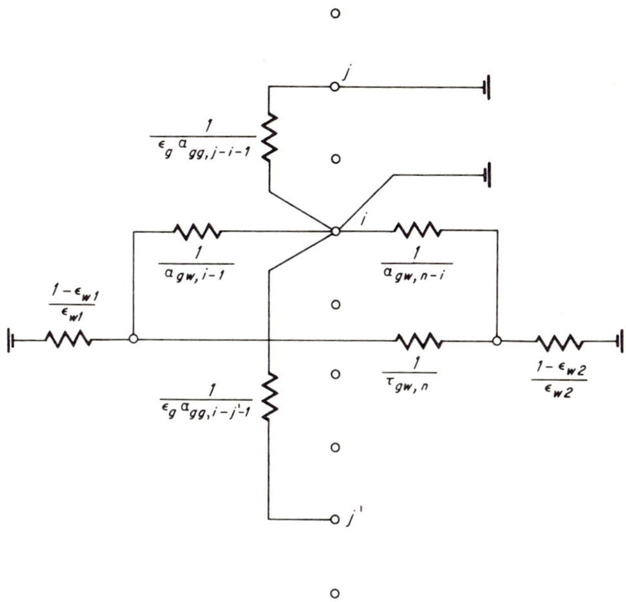

Fig. 16-18 Electrical analog of the radiation in the enclosure of Fig. 16-17.

lengths. This procedure can again be simplified by the concept of a selectively gray gas.[1]

The number of surfaces or partial volumes into which an enclosure is subdivided can be increased at will, and thus the model on which the analysis is based can be made to approach the actual situation more and more accurately. The number of equations to be solved increases proportionately; this, however, does not cause a serious obstacle when a digital electronic computer is used. One can also proceed to differential areas and volumes and then end up with a system of integral equations. These are usually so involved that solutions have to be obtained electronically. With a digital computer, this means that the integral equations have to be converted back to finite-difference equations. Monte Carlo methods also prove a very useful approach to problems involving an enclosure with an emitting-absorbing and possibly scattering medium.

In recent engineering applications, situations have arisen which require the study of radiative heat exchange in fairly simple geometries. On vehicles reentering through the atmosphere, a hot layer exists between the shock and the surface of the vehicle, composed of an inviscid layer

[1] For a thorough discussion of these methods see Hottel and Sarofim, *op. cit.*

and a viscous boundary layer. This region is heated to extremely high temperatures, so that knowledge of the heat transfer from these layers into the surface of the vehicle by radiation becomes important. The simple geometry of the radiating-absorbing layer makes it possible to analyze the radiation process in such a way that parameters like the temperature distribution, concentration distribution within the boundary layer, and the directional characteristics of the radiation properties can be accounted for realistically. Methods for the analysis of such problems are discussed in the following section.

16-4 RADIATIVE ENERGY EXCHANGE IN A MEDIUM WITH NONUNIFORM TEMPERATURE

In the preceding sections, the analysis of energy exchange by radiation in a system with an absorbing and emitting gas was simplified by the assumption that the temperature and composition of the gas are uniform. In this and the following sections, this simplification will be dropped. In order to calculate the temperature field in such a medium, two equations are needed, the radiative transfer equation and an energy balance in differential form, just as the calculation of the temperature field in a conducting medium starts with Fourier's heat-conduction equation and with an energy balance.

The energy-transfer equation for an emitting, absorbing, and scattering medium is obtained by an extension of Eq. (16-2) to include the effect of scattering and has the following form:

$$\frac{dK_\lambda}{ds} = \rho \dot{j}_\lambda - (\varkappa_\lambda + \sigma_{\lambda,s})\rho K_\lambda + \frac{\rho}{4\pi}\oint \sigma_{\lambda,bd} K'_\lambda \, d\omega' \qquad (16\text{-}29)$$

The equation describes the change in the intensity K_λ of the radiative flux traveling in the direction s along an infinitely small distance ds (Fig. 16-19). The first term on the right-hand side describes the increase in flux by emission of radiation. The second term deals with the decrease in the flux caused by absorption ($\varkappa_\lambda$) and by scattering of energy into different directions. The symbol $\sigma_{\lambda,s}$, a directional-spherical scattering coefficient, describes the amount of energy scattered out of the direction s into all directions of the spherical space. The last term on the right side of the equation deals with the energy which is scattered into the direction s from other directions. The symbol K'_λ denotes the radiative energy flux in the direction s', and the scattering coefficient $\sigma_{\lambda,bd}$ is a bidirectional parameter similar to the one introduced in the description of reflection from a surface. Also called the *scattering function*, it describes the process in which radiation originally traveling in the direc-

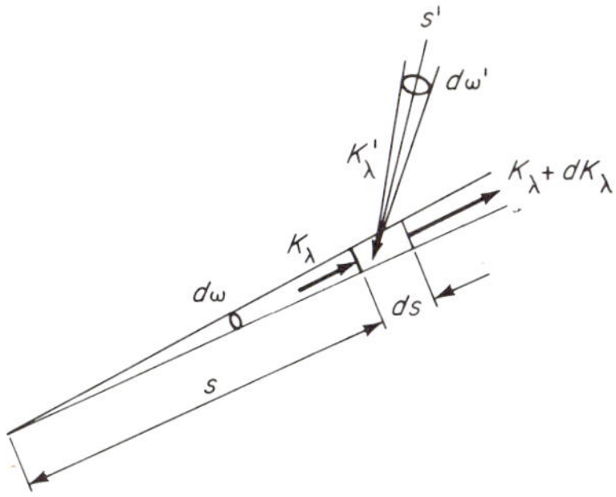

Fig. 16-19 Emission, absorption, and scattering of radiation.

tion s' is scattered into the direction s. The circle under the integral sign indicates that the integration is to be performed over the complete spherical space. The factor $1/4\pi$ enters this term through the definition of $\sigma_{\lambda,bd}$. For isotropic scattering, the scattering coefficient can be moved out of the integral, and for thermal radiation Kirchhoff's relation

$$j_\lambda = \varkappa_\lambda i_{b\lambda}$$

can again be introduced. This simplifies the transfer equation to the form

$$\frac{dK_\lambda}{ds} = \rho\varkappa_\lambda(i_{\lambda b} - K_\lambda) - \rho\sigma_{\lambda,s}K_\lambda + \rho\,\frac{\sigma_{\lambda,bd}}{4\pi}\oint K'_\lambda\,d\omega' \qquad (16\text{-}30)$$

For a system in thermal equilibrium, K'_λ is independent of direction, and the last term on the right-hand side of the equation reduces to $\rho\sigma_{\lambda,bd}K_\lambda$. This term and the preceding one must also be equal, so that $\sigma_{\lambda,s} = \sigma_{\lambda,bd}$. This is the reason why the factor $1/4\pi$ was included in Eq. (16-29).

The energy balance on a volume element describes the fact that a volume element in an emitting, absorbing medium loses energy by emission and gains energy by absorption. It may, in addition, gain energy by energy sources within the volume element or by energy transfer into the volume element in any form except by radiation, e.g., by thermal conduction or convection. All the energy terms not connected with radiative transfer will be combined into one term and will be denoted by Φ per unit mass and time. The energy balance then takes on the form per unit volume

$$\rho\int_0^\infty d\lambda \oint j_\lambda\,d\omega - \rho\int_0^\infty d\lambda \oint \varkappa_\lambda K_\lambda\,d\omega = \rho\Phi \qquad (16\text{-}31)$$

Scattering need not be considered in this equation, because it only changes the direction of the radiative fluxes passing through the volume element. For thermal radiation and for an isotropic material, in which the coefficient $\varkappa_\lambda$ is independent of direction, the energy-balance equation simplifies to

$$4\pi \int_0^\infty \varkappa_\lambda i_{\lambda b}\, d\lambda = \int_0^\infty \varkappa_\lambda\, d\lambda \int_{\bigcirc} K_\lambda\, d\omega + \Phi \tag{16-32}$$

Information on the scattering function can be obtained by analysis for uniform spherical particles with the help of Maxwell's equations describing electromagnetic waves. The analysis, first performed by Gustav Mie, leads to fairly complicated mathematical expressions.[1] The scattering function is found to depend on the parameter $\alpha = \pi D/\lambda$ and the refractive index n. D is the diameter of the scattering particle. Scattering, by the way, leads to polarization, as does reflection on solid surfaces. The analysis simplifies greatly for the condition $\alpha \ll 1$, and the corresponding solution was obtained by Lord Rayleigh. It is described by the relation

$$\frac{\sigma_{\lambda,bd}}{\sigma_{\lambda,s}} = \frac{3}{4}(1 + \cos^2 \beta)$$

where β is the angle between the incident and the scattered ray. The scattering coefficient $\sigma_{\lambda,s}$ was found to increase inversely proportional to the fourth power of the wavelength. This explains why the sky looks blue during the day and why the sun takes on a reddish color when it dips below the horizon. Rayleigh scattering is generally not important in engineering systems because long path lengths are required to decrease the intensity of a beam by an appreciable amount. For the condition $\alpha \gg 1$, scattering can be investigated on the basis of geometric optics. The situation becomes involved when particles which are not uniform in size or not spherical are involved in the scattering process. Information on scattering coefficients must then be obtained primarily by experiments. Results for clouds of particles of various materials are contained in a publication by Love and Beattie.[2] For situations in which the shape of the scattering function is not known, scattering is often assumed to be isotropic, corresponding to the relation $\sigma_{\lambda,bd}/\sigma_{\lambda,s} = 1$. It is easily checked that the scattering coefficient has the dimension length squared divided by mass. It is therefore sometimes called scattering cross section and compared with the projected area of the particles per unit mass.

In the following discussion, scattering will again be omitted. In this case the transfer equation (16-30) simplifies to

$$\frac{dK_\lambda}{ds} = \rho\varkappa_\lambda(i_{\lambda b} - K_\lambda) \tag{16-33}$$

[1] A review of the method and its results is contained in H. C. Van De Hulst, "Light Scattering by Small Particles," Wiley, New York, 1957.

[2] T. J. Love and J. F. Beattie, Experimental Determination of Thermal Radiation Scattering by Small Particles, *USAF Aerospace Res. Lab. Rept.* ARL 65-110, June, 1965.

a first-order total differential equation which can be integrated by multiplying both sides with the integrating factor

$$\exp\left(\int_0^s \rho\kappa_\lambda \, ds\right)$$

The result is

$$K_\lambda(s) = K_\lambda(0) \exp\left(-\int_0^s \rho\kappa_\lambda \, ds\right) + \int_0^s \rho\kappa_\lambda i_{\lambda b}(s') \exp\left(-\int_{s'}^s \rho\kappa_\lambda \, ds\right) ds' \tag{16-34}$$

This equation considers the change in the intensity K_λ of the radiative flux in the direction s (Fig. 16-20) along a finite path length s. $K_\lambda(0)$ denotes the flux intensity at $s = 0$. The first term on the right-hand side of the equation then expresses the fact that this intensity decreases by absorption in its travel through the distance s. The second term, on the other hand, describes the increase in intensity along the path length s by emission of radiative energy. $i_{\lambda b}(s')$ indicates the blackbody intensity at the location s', and the density ρ and coefficient of absorption κ_λ appearing before $i_{\lambda b}(s')$ are the values at the same location. In this way the equation is physically self-evident. It can easily be checked that it changes to the form of Eq. (16-4) when the temperature, pressure, and composition of the gas are independent of location (ρ, κ_λ, $i_{\lambda b}$ are constant). The integral $\int \rho\kappa_\lambda \, ds$ is the *optical path length.*

The use of these equations will now be described in a simple example. Consider a layer of a gray gas with two parallel planes as boundaries (Fig. 16-21). The lower boundary may be a black surface with temperature T_0 and the blackbody intensity $i_{b0} = (\sigma/\pi)T_0^4$, whereas the upper boundary transmits all radiation perfectly (one might imagine a nonparticipating gas in the region beyond this boundary). No radiation enters the gas body through its upper boundary, and no heat sources exist within the gas. The problem is to determine the temperature in the

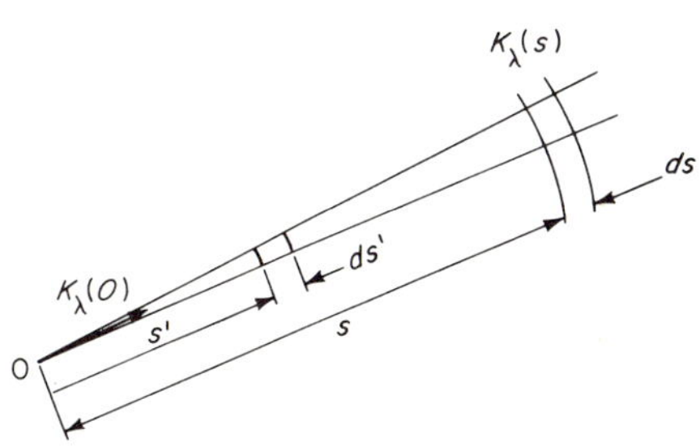

Fig. 16-20 Emission and absorption of radiation.

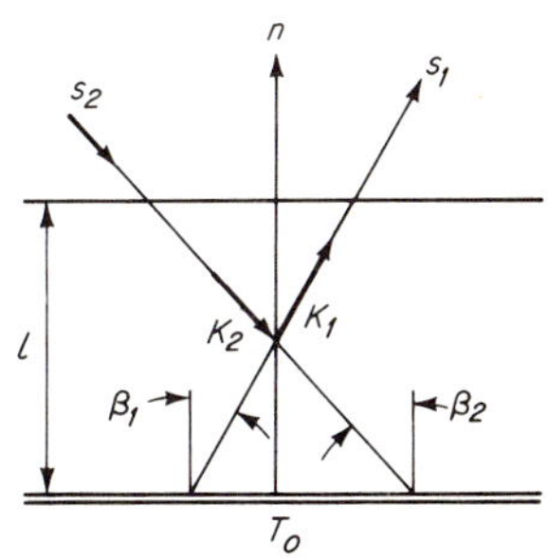

Fig. 16-21 Radiation in a gaseous layer.

gas as it is established by radiative energy exchange. This temperature will be a function of the coordinate n normal to the boundaries when the gas body is unlimited in horizontal direction and when the surface temperature T_0 is locally constant. The density of the gas and its coefficient of absorption may be functions of temperature and therefore functions of the coordinate n. The intensity of the radiative flux traveling toward the upper boundary along the direction s_1, characterized by its angle β_1 toward the surface normal, is K_1. The intensity of the radiative flux traveling toward the lower boundary of the gas layer in the direction s_2, characterized by the angle β_2 measured toward the surface normal, is K_2. The transfer equations for both fluxes are obtained from Eq. (16-33), in which the subscript λ can be dropped for a gray gas. The differential ds_1 can be expressed by

$$ds_1 = \frac{dn}{\cos \beta_1}$$

Introducing this into the transfer equation gives

$$\cos \beta_1 \frac{dK_1}{dn} = \rho\varkappa(i_b - K_1)$$

We can now introduce an optical path length τ by the following equation defining its differential:

$$d\tau = \rho\varkappa \, dn$$

The optical path length itself is then

$$\tau = \int_0^n \rho\varkappa \, dn$$

when it is counted from the lower surface of the layer ($n = 0$). The transfer equation then simplifies to

$$\cos \beta_1 \frac{dK_1}{d\tau} = i_b - K_1 \tag{16-35}$$

In the same way the transfer equation for the radiative flux K_2 is

$$\cos \beta_2 \frac{dK_2}{d\tau} = K_2 - i_b \tag{16-36}$$

when the following relation is introduced:

$$-ds_2 = \frac{dn}{\cos \beta_2}$$

The boundary conditions are

$$\begin{aligned} \tau &= 0 \qquad & K_1 &= i_{b0} \\ \tau &= \tau_l \qquad & K_2 &= 0 \end{aligned} \tag{16-37}$$

where τ_l denotes the integral

$$\tau_l = \int_0^l \rho\varkappa \, dn \tag{16-38}$$

The energy balance Eq. (16-32) also simplifies for a gray gas to the form

$$4\pi i_b = \int_{\triangle} K_1 \, d\omega_1 + \int_{\triangledown} K_2 \, d\omega_2 \tag{16-39}$$

The three equations (16-35), (16-36), and (16-39) together with the boundary conditions [Eq. (16-37)] are best solved numerically by an iteration process. One assumes a temperature variation $T(\tau)$ which also specifies the blackbody intensity $i_b(\tau)$, the coefficient of absorption $\varkappa(\tau)$, and the density $\rho(\tau)$. The equation $l = \int_0^{\tau_l} \frac{d\tau}{\rho\varkappa}$ then also determines the boundary τ_l. The two transfer equations can now be solved for each angle β_1 and β_2, respectively, resulting in the flux intensities K_1 and K_2. When these flux intensities are entered into the right-hand side of Eq. (16-39), a new approximation of the blackbody intensity i_b is obtained. The process can be repeated until sufficient convergence is obtained. An approximation, referred to as the *Milne-Eddington approximation*, has been developed to simplify such an analysis. It is based on certain assumptions of the angular distributions of the flux intensities K.

Calculations of this form were originally made by astrophysicists interested in the temperature distribution in the outer layer (photosphere) of stars in which radiative energy exchange determines the temperature. Similar situations are also encountered in the reentry process where emission and absorption in the hot gas layer between the shock wave and the object are important contributors to energy transfer.

The results of such an analysis performed by Usiskin and Sparrow[1] will now be discussed. The analysis is concerned with radiative energy transfer through a gas layer between two parallel black walls having emissive powers e_{b1} and e_{b2}, respectively. Heat sources are again assumed absent in the gas, the local temperature of which is characterized by the corresponding blackbody emissive power e_{bg}. The absorption coefficient $\varkappa$ and the density ρ are assumed constant. Figure 16-22 presents the results. Two dashed curves are added as limiting cases. The figure looks very similar to one which describes heat conduction in a rarefied gas. The parameter characterizing the temperature field in such a gas is the ratio of the mean-free-path length Λ of the molecules to the characteristic dimension L. This ratio Λ/L is called the *Knudsen number*. Continuum conduction exists in such a gas when the Knudsen number is very small, which is the case at normal pressures. The temperature then

[1] C. M. Usiskin and E. M. Sparrow, *Intern. J. Heat Mass Transfer*, **1**:28 (1960).

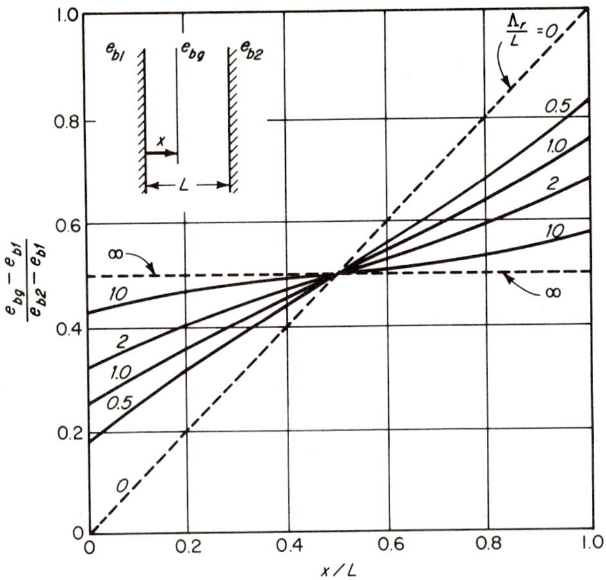

Fig. 16-22 Emissive power in a radiating and absorbing gray gas confined between two black parallel walls. [*After C. M. Usiskin and E. M. Sparrow.*]

drops from the value at surface 1 linearly through the gas to the value at surface 2. An exactly analogous condition obviously exists in a radiating gas layer when the parameter $1/\rho\varkappa L$ is small, as can be observed in Fig. 16-22. One can define the reciprocal of $\rho\varkappa$ as a photon mean path length Λ_r and a Knudsen number Λ_r/L for radiation. When this Knudsen number is very small compared to 1, the blackbody emissive power of a gas drops linearly from the value e_{b1} to the value e_{b2}. The reason for this analogy between continuum conduction and radiation at small Knudsen numbers is that a fluid element at a certain location obtains direct information by traveling molecules or photons from its immediate surroundings only. Rosseland[1] has demonstrated the analogy between continuum conduction and radiation at small Knudsen numbers and has found that the net energy transfer q per unit area and time can in this case be described by a vector equation

$$\mathbf{q} = -\tfrac{4}{3}\Lambda_r \text{ grad } e_b \tag{16-40}$$

which is completely analogous to Fourier's heat-conduction equation

$$\mathbf{q} = -k \text{ grad } T$$

[1] S. Rosseland, "Astrophysik und atom-theoretische Grundlagen," pp. 41–44, Springer-Verlag, Berlin, 1931.

This means that solutions for a heat-conduction problem can be immediately transferred to a corresponding radiative-transfer problem when the thermal conductivity k in the conduction solution is replaced by the radiative conductivity $\frac{4}{3}\Lambda_r$ and the temperature by the blackbody emissive power. A short derivation of Rosseland's equation will be given for the case of a unidirectional variation of temperature in direction x (Fig. 16-23). According to Sec. 14-2, the intensity of the radiant flux approaches the blackbody intensity with vanishing photon path length. Accordingly, the energy flux q per unit area of an area element dA in a plane at the location x_0 is given by the integral

$$q = \int_0 i_b \cos\beta \, d\omega$$

The intensity of blackbody radiation will vary with x and can be approximated by the expression

$$i_b = i_{b0} + (x - x_0)\frac{di_b}{dx}$$

for a small distance $x - x_0$. For a radiant beam in the direction S, this distance is connected with the photon path lengths by the equation

$$x - x_0 = \Lambda_r \cos\beta$$

Inserting the last two expressions into the above integral results in

$$q = 2\pi \int_0^{\pi} \left(i_{b0} - \Lambda_r \cos\beta \frac{di_b}{dx} \right) \cos\beta \sin\beta \, d\beta$$

or

$$q = -2\pi \int_{-1}^{1} \Lambda_r \frac{di_b}{dx} \cos^2\beta \, d(\cos\beta) = -\frac{4\pi}{3}\Lambda_r \frac{di_b}{dx}$$

Replacement of the intensity i_b by the emissive power e_b gives finally

$$q = -\tfrac{4}{3}\Lambda_r \frac{de_b}{dx}$$

an expression which is identical with Eq. (16-40).

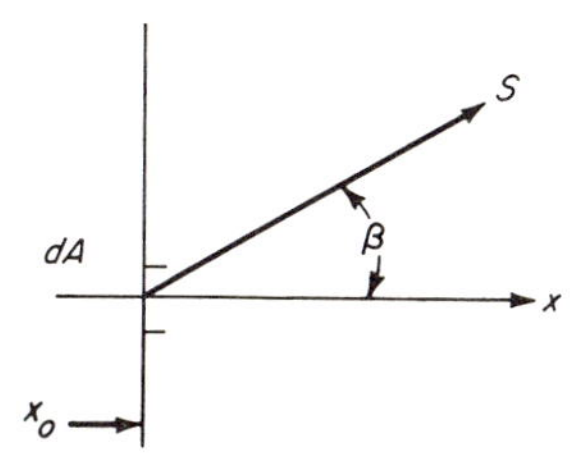

Fig. 16-23 Derivation of Rosseland's law.

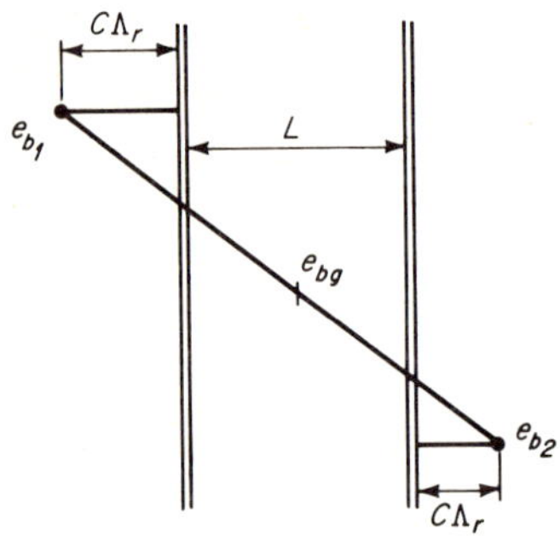

Fig. 16-24 Radiation in a gaseous layer bordered by two parallel black walls.

The other limiting case indicated by the horizontal dashed line in Fig. 16-22 describes a situation in which the ratio of the radiation mean free path Λ_r to the characteristic length L is very large. This is the situation already discussed as vanishing self-absorption. In this case, radiation emitted by the gas can travel to the two black walls with negligible absorption in the intervening gas layers. As a consequence, the radiative-exchange process is the same for the various gas layers independent of their location, and the temperature in the gas must then also be independent of the distance x.

The full curves between the two limiting cases in Fig. 16-22 are almost straight lines, again suggesting an analogy to the conduction process in a rarefied gas which for moderately small Knudsen numbers can still be described by Fourier's equation provided a temperature jump is introduced at the two surfaces.[1] In the same way, a *radiation jump* can be used to obtain a simple relation which in good approximation describes the energy exchange by radiation between the two walls. Figure 16-24 is a simplified sketch of the variation of the blackbody emissive power in the gas layer. The straight line representing this variation reaches the emissive power e_{b1} and e_{b2} at the distance $C\Lambda_r$ from the wall surface. The energy flux from the hot to the cold wall per unit area and time can then be written as

$$q = \tfrac{4}{3}\Lambda_r \frac{e_{b1} - e_{b2}}{L + 2C\Lambda_r} \tag{16-41a}$$

For $\Lambda_r/L \gg 1$ the photons can travel almost freely between the two black walls, and the heat flux is correspondingly

$$q = e_{b1} - e_{b2} \tag{16-41b}$$

[1] E. R. G. Eckert, Similarities between Energy Transport in Rarefied Gases and by Thermal Radiation, in W. Ibele (ed.) "Modern Developments in Heat Transfer," Academic, New York, 1963, p. 159.

The constant C can now be determined in such a way that Eq. (16-41*a*) transforms into Eq. (16-41*b*) for $\Lambda_r/L \gg 1$:

$$\tfrac{4}{3}\Lambda_r = 2C\Lambda_r$$

This procedure, suggested by Probstein,[1] results in the equation

$$q = \frac{\frac{4}{3}}{L/\Lambda_r + \frac{4}{3}}(e_{b1} - e_{b2}) \tag{16-42}$$

which has been found to approximate the results in Fig. 16-22 very well. Unfortunately such simple expressions do not fit the actual situation as well for other geometries or for gases with internal heat sources.

In real gases, the situation is complicated by the fact that the absorption coefficient, and with it the photo mean-free-path length Λ_r, changes strongly with wavelength. As a consequence, some wavelength regions are in the domain for which Rosseland's approximation applies, other regions in the domain where the gas can be considered as having vanishing self-absorption, and others in domains that lie in between. Nevertheless, the approximations which apply to the two extremes are often useful in the analysis of radiative exchange in real gases. In certain cases, the whole wavelength range, for example, can be subdivided into regions $\Delta\lambda$ in which one or the other approximation applies.

16-5 RADIATION OF FLAMES

From the emissivity charts for carbon dioxide and water vapor, the radiation of combustion gases after completion of combustion can be calculated, e.g., the heat transfer by radiation within the tube bundles (convective surface) of water-tube boilers. The radiation of flames is usually considerably greater than the radiation as computed from the carbon dioxide and water molecules present.

Most fuels burn with a luminous flame. The yellow glow of a flame comes from the hydrocarbons in the fuel which are gradually broken down in the flame, forming molecules with an increasing percentage of carbon. Carbon particles occur, which glow in the flame and give it the yellowish color. These particles also emit considerable thermal radiation. All bituminous fuels, e.g., wood, lignite, and the younger coals, burn in this way with luminous flames. Only anthracite, coke, and some gases (generator gas, hydrogen, blast-furnace gas) have nonluminous flames. The faint bluish shine which these flames emit, called *chemoluminosity*, arises from the chemical reactions within the gaseous components. It is not connected, however, with an energy flux of any importance as has

[1] R. F. Probstein, *AIAA J.*, **1**:1202 (1963).

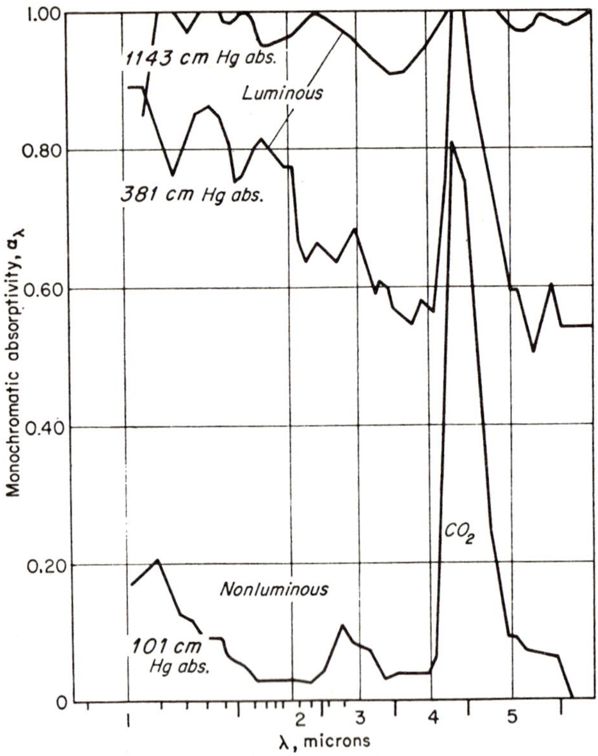

Fig. 16-25 Monochromatic absorptivity of luminous and nonluminous benzene-air flames. [*From E. C. Miller, A. E. Blake, R. M. Schirmer, G. D. Kittredge, and E. H. Fromm, Radiation from Laboratory Scale Jet Combustor Flames, Phillips Petroleum Company, Res. Rept., 1526–56R U.S. Navy Contract 52-(32-C), 1956.*]

been proved by several investigations. The emission of a luminous flame depends on the number of carbon particles, which varies greatly with the conditions under which the combustion occurs. It is influenced, for instance, by the mixing of air and the combustible gases and by the temperatures of both components. As long as these conditions cannot be predicted in a furnace, there is no possibility of calculating the radiation of a flame exactly.

Figure 16-25 shows an example of the results of absorptivity measurements on a benzene-air flame of approximately 3200 F. The flame filled a tubular combustor of 2-in diameter. The measurement was made along a beam in the direction of the tube diameter 5 in from the fuel nozzle. The fuel-air ratio was approximately 0.01. Three pressure levels were maintained: 40, 150, and 450 in Hg abs. It can be observed in the figure that with increasing pressure the flame becomes gradually luminous and the luminous radiation fills the empty valleys more and more between the absorption bands caused by gaseous radiation. The wavelength range shown in the figure is the important one for the specific flame temperature. Blackbody radiation at this temperature has its maximum at 2 μm wavelength and drops to below 10 percent of the maximum at 1 and 5 μm.

For a first approximation, the flame is often considered as a blackbody, and the fact that this is not entirely true is corrected for by an empirical factor p smaller than 1. Assuming that the flame fills all the interior of the furnace, this gives for the net heat flow Q per unit time to the furnace wall the equation

$$Q = p\epsilon_w \sigma A(T_f^4 - T_w^4) \tag{16-43}$$

where ϵ_w = emissivity (equal to absorptivity) of furnace walls
T_f = absolute temperature of flame
T_w = absolute wall temperature
A = area of wall

The empirical factor p depends mainly on the fuel and on the size of the furnace. For large furnaces p is nearly unity. In water-cooled furnaces of steam boilers, in addition to the radiation from the flame, the cooling surface receives radiation from the refractory-lined walls and

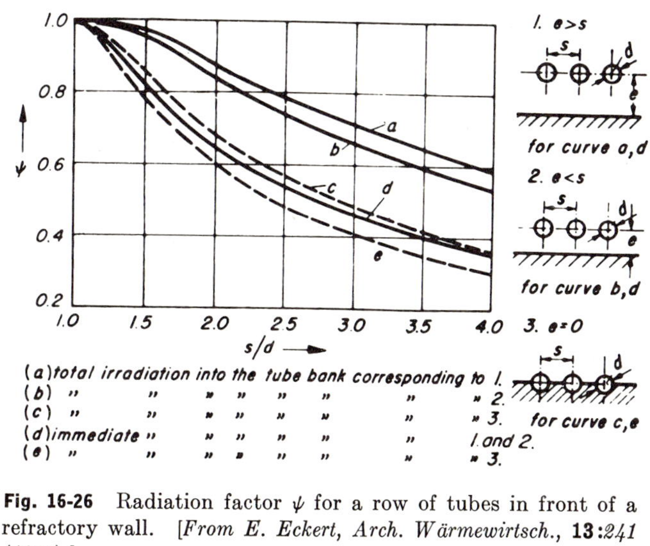

Fig. 16-26 Radiation factor ψ for a row of tubes in front of a refractory wall. [*From E. Eckert, Arch. Wärmewirtsch.*, **13**:*241* (*1932*).]

from the fuel layer on the stokers and grates. The factor p therefore is greater in furnaces in which only a small part is water-cooled than in furnaces whose walls are entirely lined with cooling surfaces. In furnaces of usual size p lies between 0.6 and 1. If the net radiative heat flow to a cooling surface consisting of a row of tubes before a refractory wall is to be calculated with Eq. (16-43), the question arises of which area is to be used for A. According to investigations of this situation made by Hottel[1] and Eckert,[2] the area A is determined by multiplying the area of the furnace walls before which the tubes are arranged by a factor ψ, represented in Fig. 16-26. The ψ values were calculated with the methods discussed in Chap. 15.

[1] H. C. Hottel, *Mech. Eng.*, **52**:699 (1930); *Trans. ASME*, **53**:265 (1931).
[2] E. Eckert, *Hauptverein Deut. Ingr. Mitt.*, **1931**:483–486; *Arch. Wärmewirtsch.*, **13**:241 (1932).

17
Combined Transfer Processes; Temperature Measurement

17-1 RADIATION AND CONDUCTION

In many engineering applications, energy transfer by radiation occurs simultaneously with conduction and convection. In this case, the energy equation (16-32) has to be combined with the energy-conservation equation in Part II in the way that the heat-source term Φ in Eq. (16-32) has to be expressed by the terms in the conservation equation (6-28). A detailed discussion of analyses of this kind is available.[1] In this section only combined conduction and radiation will be discussed for a situation corresponding to the one presented in Fig. 16-22. Accordingly, a conducting, emitting, and absorbing gray gas is enclosed between two parallel black walls with temperatures T_{w1} and T_{w2}. The process is steady, and the properties are assumed constant. Equation (16-32) with the source term Φ replaced by the expression

$$k \frac{d^2T}{dx^2}$$

[1] E. M. Sparrow and R. D. Cess, "Radiation Heat Transfer," Brooks/Cole, Belmont, Calif., 1966, and *Advan. Heat Transfer*, **1**:1 (1964); **3**:176 (1966).

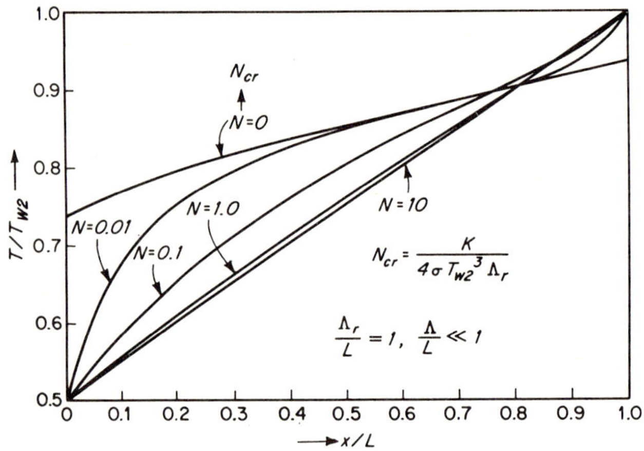

Fig. 17-1 Temperature variation through a gray radiating, absorbing, and heat-conducting gas enclosed between two parallel black walls. [*From R. Viskanta and R. J. Grosh, Heat Transfer by Simultaneous Conduction and Radiation in an Absorbing Medium, ASME Paper 61-SA-34, 1961.*]

and Eq. (16-33) with proper boundary conditions describe this process. An analysis of this situation has been published.[1] The term

$$N_{cr} = \frac{k}{4\sigma T_{w2}{}^3\Lambda_r}$$

appears as the parameter on the curves. It can be interpreted as presenting essentially the ratio of the mean-free-path length Λ for conduction to the mean-free-path length Λ_r of the photons because the thermal conductivity k is proportional to the molecular free path length. The curve labeled $N_{cr} = 10$ is almost a straight line connecting the two wall temperatures, indicating that the heat-conduction process dominates. Figure 17-2, which was obtained from Fig. 17-1, lists blackbody emissive powers instead of temperatures on the ordinate. Now the curve labeled $N_{cr} = 0$ has almost become a straight line which is identical with the corresponding line in Fig. 16-22. The full lines in this figure are transferred from Fig. 17-1. The dashed line obviously would give the result for a parameter N_{cr} which is small but not zero. It is interesting to observe that the variation of the blackbody emissive power and consequently of the temperature in the bulk of the medium is now determined by the radiation process, whereas heat conduction becomes effective in two thin "boundary layers" only, removing the temperature jump which

[1] R. Viskanta and R. J. Grosh, Heat Transfer by Simultaneous Conduction and Radiation in an Absorbing Medium, *ASME Paper* 61-SA-34, 1961.

would exist if the energy transfer were solely by radiation. This situation obviously offers an opportunity of simplifying the analyses formulating a boundary-layer approach to the heat-conduction process. In general, the shape of the temperature field is determined by a number of dimensionless parameters, namely, the ratio of the photon path length to the thickness L of the medium, Λ_r, which in the two figures was assumed to have the value 1; the ratio of the molecular path length to the thickness L, which in the figure is estimated to be small compared to 1 according to continuum conduction; the ratio of the two wall temperatures, which in the figures has the value 0.5; and the parameter N_{cr}.

Of special interest is the magnitude of the heat flux occurring in such combined conduction and radiation transfer. Viskanta and Grosh calculated this heat flux in the range

$\dfrac{T_{w1}}{T_{w2}}$	from 0.1 to 0.5
T_{w1}	from 400 to 2000 K
N_{cr}	from 0.1 to 1
$\dfrac{L}{\Lambda_r}$	from 0.1 to 1

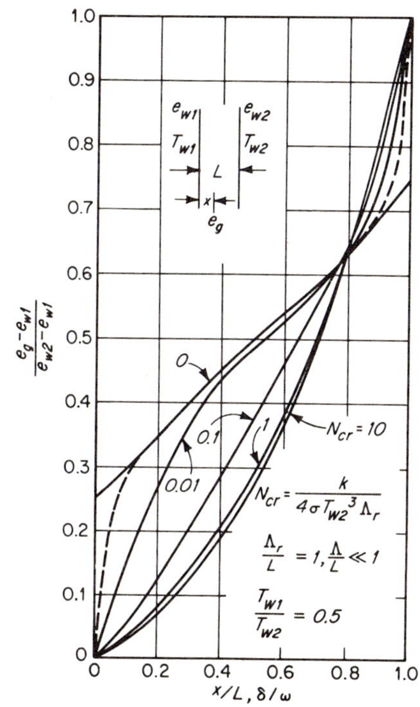

Fig. 17-2 Emissive power in a radiating, absorbing, and heat-conducting gas confined between two parallel walls. [*After R. Viskanta and R. J. Grosh.*]

If one compares the heat fluxes obtained in this analysis with those obtained by calculating radiation and conduction separately and adding the two terms, one finds that in the range of parameters stated above the actual heat flux is 3 to 14 percent larger than the heat flux obtained by superposition of the conduction and radiation processes.

With Rosseland's approximation, the calculation of a combined conductive and radiative heat transfer becomes especially simple. Equation (16-40) changes to

$$q = -\tfrac{16}{3}\Lambda_r \sigma T^3 \frac{dT}{dx} \tag{17-1}$$

when the Stefan-Boltzmann law is introduced. This expression, then, can be combined with Fourier's equation with the result

$$q = -(k + \tfrac{16}{3}\Lambda_r \sigma T^3)\frac{dT}{dx}$$

which indicates that the effect of radiation can be combined with that of conduction by introducing an effective thermal conductivity which is given by the term in parentheses in the above equation. Conductivity values obtained by measurements actually contain, in general, the radiation term. It can be expected that Eq. (16-40) is a valid approximation for a larger range of photon path length than Eq. (17-1), because the linearization of the emissive power with respect to temperature has not yet been introduced. Either of the two equations describes a combined radiative and conductive transfer process with sufficient accuracy in most unorganic liquids.

In organic liquids, however, the mean photon path length cannot be considered short, as was shown in a remarkable experimental investigation by Schödel and Grigull.[1] The temperature field was studied in horizontal liquid layers enclosed between two plates and subject to a downward heat flux. Figure 17-3 shows a photograph obtained with a Zehnder-Mach interferometer. A 15-mm liquid layer of carbon tetrachloride was enclosed between two silver plates with an emittance value 0.05. The temperature difference between the surfaces of the two plates was 0.89 C, and the average temperature of the liquid was 25 C. The interferometer was adjusted in such a way that the interference fringes are vertical in the central part of the layer. The deviation of the actually measured fringes from a vertical line thus indicates the effect of a photon path length which is of the same order as the thickness of the liquid layer. The fringes have a shape similar to the dashed line in Fig.

[1] G. Schödel and U. Grigull, "Heat Transfer for 1970," vol. III, p. 546, Elsevier, Amsterdam, 1970.

Fig. 17-3 Interference photograph showing the temperature field for combined conduction and radiation in a layer of carbon tetrachloride liquid. [*Photograph by G. Schödel and U. Grigull.*]

17-2. The emissive power used in Fig. 17-2 can be replaced by the temperature because the temperature difference is very small compared to the temperature level. The investigation demonstrated that a number of other organic liquids have also a photon path length which cannot be considered as small.

The results presented in the analyses in this section and Sec. 16-5 give an interesting insight into radiative transfer processes in a gray medium,[1] but they should not be taken as presenting good approximations to conditions in a radiating and absorbing gas because in such a medium the radiation properties and, correspondingly, the photon path length depend strongly on wavelength. An analysis based on a band approximation is much more appropriate for this situation.

17-2 RADIATION ERROR IN TEMPERATURE MEASUREMENTS

Temperature measurements in a solid or fluid by an immersed probe (thermometer, thermocouple, etc.) are subject to systematic errors, and an estimate of their magnitude should be made in each installation. Such errors are caused by heat conduction within the stem of the thermometer and by the heat capacity of the instrument when the temperature changes with time.[2] An additional error arises when temperatures are measured in a nonparticipating or weakly absorbing medium, e.g., a gas, because the thermometer exchanges heat with surrounding solid surfaces by radiation. This radiation error will be calculated in the present section.

A thermometer is assumed to be positioned in a nonradiating gas of

[1] See also R. Viscanta, Radiation Transfer and Interaction of Convection with Radiation Heat Transfer, *Advan. Heat Transfer*, **3**:176 (1966).

[2] E. R. G. Eckert, "Introduction to Heat and Mass Transfer," McGraw-Hill, New York, 1963.

temperature T_g in an enclosure, the walls of which have a temperature T_w. A heat balance, describing the heat transferred per unit surface area and time from the gas to the thermometer by convection and from the thermometer to the walls by radiation, gives

$$h_c(T_g - T_t) = \epsilon_t \sigma (T_t^4 - T_w^4)$$

where T_t = thermometer temperature
h_c = convective-heat transfer coefficient
ϵ_t = emittance of thermometer surface

The right-hand term describes the heat loss by radiation, because the thermometer surface area is small compared with the area of the surrounding walls. The radiation error $T_t - T_g$ can be calculated from this equation. A simplification is possible when the difference between the thermometer temperature and the wall temperature is small compared to either temperature. The right-hand term in the above equation can then be linearized and a heat-transfer coefficient h_r introduced, resulting in the equations

$$h_c(t_g - t_t) = h_r(t_t - t_w) \tag{17-1a}$$

$$h_r = 4\epsilon_t \sigma T_w^3 \tag{17-1b}$$

The relative error E is

$$E = \frac{t_g - t_t}{t_g - t_w} = \frac{1}{1 + h_c/h_r} \tag{17-2}$$

If, for instance, the air temperature is measured by a mercury thermometer in a heated room with walls which are cooler than the air, heat exchange between air and thermometer is by free convection. Convective-heat-transfer coefficients are then of the order of 10 W/m² C. Radiative-heat-transfer coefficients at temperatures near atmospheric temperature are found to be of order 2.5 W/m² C. The relative radiation error then becomes approximately 20 percent! This means that no accurate temperature measurement is possible with an ordinary thermometer under these circumstances.

The radiation error can be substantially reduced by a radiation shield. Figure 17-4 is a sketch of a thermometer arranged to measure the gas temperature t_g in an enclosure with walls of temperature t_w. The shield s must be designed so that it protects the temperature-sensing element of the thermometer from the radiation coming from the walls but does not inhibit convective heat transfer at the thermometer surface. For instance, a short, hollow cylinder open on both ends as indicated in Fig. 17-4 is a desirable configuration. For a calculation of the radiation error, values referring to the thermometer have the index t, and values pertaining to the shield the index s. A_s is the surface area of one side

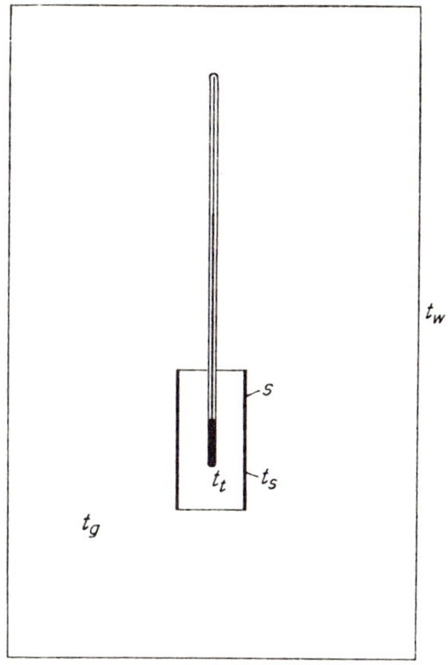

Fig. 17-4 Calculation of the radiation error of a thermometer with a radiation shield.

of the shield and A_t the thermometer surface area. Heat balances can be written for the thermometer and the shield.

$$h_{r,t}(t_t - t_s) = h_{c,t}(t_g - t_t)$$
$$2A_s h_{c,s}(t_g - t_s) + A_t h_{r,t}(t_t - t_s) = A_s h_{r,s}(t_s - t_w)$$

Elimination of t_s and calculation of the relative error results in

$$E = \frac{t_g - t_t}{t_g - t_w} = \frac{h_{r,t}}{h_{r,t} + h_{c,t} + 2\frac{h_{r,t} + h_{c,s}}{h_{r,s}} + 2\frac{h_{c,s}h_{c,t}}{h_{r,s}} + \frac{A_t}{A_s}\frac{h_{r,t} + h_{c,t}}{h_{r,s}}}$$

The last term in the denominator can usually be neglected, since $A_t/A_s \ll 1$. The relative error simplifies then to

$$E = \frac{t_g - t_t}{t_g - t_w} = \frac{1}{(1 + h_{c,t}/h_{r,t})[1 + 2(h_{c,s}/h_{r,s})]} \tag{17-3}$$

The symbols are the same as for the unshielded thermometer. Equation (17-3) shows that all radiation terms must be kept small in order to reduce the error. This can be done by using a material of low emittance for the outside of the shield. The inside of the shield should be blackened in order to avoid reflection of wall radiation onto the thermometer. If possible, the surface of the thermometer should be given a coating of low emittance.

Under extreme conditions, e.g., when measuring gas temperatures in furnaces, shielding is not sufficient to reduce radiation errors, and one

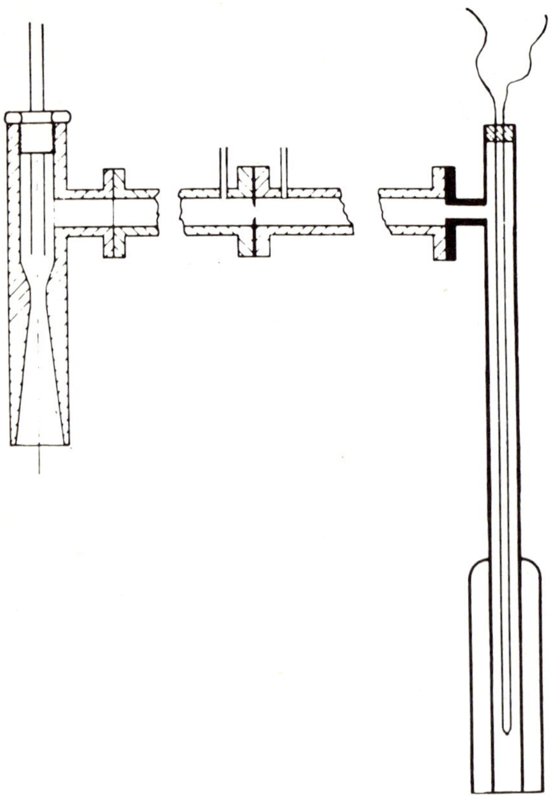

Fig. 17-5 Suction thermometer.

must then try to increase the convective heat transfer by decreasing the diameter of the thermometer (very thin wire thermocouples) or by artificially increasing the gas velocity past the thermometer. This can be done by sucking the gas through the radiation shield. Such a thermometer, called a *suction thermometer*, is shown in Fig. 17-5.

A special situation arises when the total temperature is to be measured in a high-velocity, high-temperature gas stream. The diffusor thermometer (Fig. 10-8) is the most common instrument for this purpose. Three errors—conduction, incomplete temperature recovery, and radiation—have to be considered in a steady state, and in addition, heat-capacity errors arise for unsteady temperatures. By multiple shielding and careful optimization of conditions with respect to all errors satisfactory instruments can be designed.[1]

[1] F. D. Werner and R. E. Keppel, An Improved Multiple Shield Gas Temperature Probe, *Proc. Third Midwestern Conf. Fluid Mechanics, 1953,* pp. 463–478.

17-3 PYROMETRY

All thermometers have an upper limit to which they can be used. A platinum–platinum-rhodium thermocouple, for instance, can measure temperatures up to 1500 C. Other metal combinations are known which can be used up to 2500 C, but they are difficult to handle. Beyond these limits temperature is measured by the intensity of emitted radiation, a method known as *pyrometry*. With this technique, the international temperature scale is extended beyond the temperature of melting gold 1063 C by measurement of the radiation intensity of a black surface. Pyrometers have an additional advantage in that the measurement can be made from a distance. They are therefore used even for lower temperatures and, when properly designed and applied, are very accurate instruments.

The most commonly used methods for measuring the temperatures of the surfaces of solids or liquids will be discussed. Two classes of instruments are in existence. One class measures the total (integrated) radiation emitted from a surface calorimetrically. The second class compares by observation with the eye the monochromatic radiation intensity of the object in the visible-wavelength range with a known monochromatic radiation intensity.

Each pyrometer must be calibrated by viewing with it objects having a known radiation intensity. These are, for instance, created by immersing blackbodies (ceramic pieces with cavities) in baths of molten metal with a known temperature. In its use the instrument has one basic disadvantage, which is connected with the fact that the radiation intensity of a solid or liquid surface depends not only on its temperature but also on its emissivity. Either the emissivity of the surface has to be known, or it has to be increased artificially to the value 1 when the true temperature is to be determined from a pyrometer reading. Otherwise, an apparent temperature is read which does not agree with the true temperature. A number of differently defined apparent temperatures are used in pyrometry. The *total radiation temperature* T_t is the temperature of a blackbody when it emits the same total radiation intensity as the surface toward which the pyrometer is directed. From this definition follows the relation between T_t and the true surface temperature T:

$$\sigma T_t^4 = \epsilon\sigma T^4 \qquad \text{or} \qquad T = \frac{T_t}{\sqrt[4]{\epsilon}} \tag{17-4}$$

with ϵ indicating the emittance of the surface. For a metal with an emittance $\epsilon = 0.05$, the apparent total radiation temperature is only approximately half the true temperature.

The *apparent monochromatic temperature* T_λ is the temperature of a blackbody which at a defined wavelength in the visible range (usually at $\lambda_r = 0.665\ \mu m$) has the same radiation intensity as the surface under consideration. According to this definition,

$$\frac{\lambda_r^{-5}}{e^{C_2/\lambda_r T_2} - 1} = \epsilon_{\lambda_r} \frac{\lambda_r^{-5}}{e^{C_2/\lambda_r T} - 1}$$

At temperatures occurring in engineering applications, the first term in the denominator of the above equation is usually large compared with 1, and the 1 can be neglected without causing an error which would exceed the obtainable accuracy. This equation then can be transformed into

$$\frac{1}{T} = \frac{1}{T_\lambda} + \frac{\lambda_r}{C_2} \ln \epsilon_{\lambda_r} \tag{17-5}$$

where ϵ_λ denotes the monochromatic emittance of the surface. The equation shows that the apparent monochromatic temperature is always lower than the true temperature. The same is true for the total radiation temperature. For metals which have especially low emittances ϵ_{λ_r} is usually higher than ϵ. The monochromatic temperature then is closer to the true temperature than the total radiation temperature.

The *color temperature* T_c is defined as the temperature at which a blackbody has the same ratio of radiation intensities at two defined wavelengths in the visible range as the surface in question. Usually the wavelengths of red light ($\lambda_r = 0.665\ \mu m$) and of green light ($\lambda_g = 0.544\ \mu m$) are used. Broadly speaking, the color temperature is that temperature at which a radiating blackbody appears to the eye to have the same color as the surface under consideration. From the definition it follows that

$$\frac{1}{T} = \frac{1}{T_c} + \frac{\lambda_r \lambda_g}{(\lambda_r - \lambda_g)C_2} \ln \frac{\epsilon_{\lambda_g}}{\epsilon_{\lambda_r}} \tag{17-6}$$

The color temperature for a gray surface for which the emittance does not vary with wavelength ($\epsilon_{\lambda_g} = \epsilon_{\lambda_r}$) is therefore equal to the true temperature. For most surfaces the color temperature comes closer to the true temperature than the other apparent temperatures.

Figures 17-6 and 17-7 show the two most commonly used pyrometers. The total radiation pyrometer (Fig. 17-6) uses a front gold-plated mirror *a* to concentrate the radiation coming from a surface *s* on a thermopile *b* and measures the electric potential generated in the thermopile by a millivoltmeter *c*. Diaphragms *d* prevent stray radiation from impinging on the thermopile. This pyrometer therefore measures the apparent total radiation temperature. The optical pyrometer with vanishing filament (Fig. 17-7) is essentially a telescope. Its objective *a* when directed on a radiating object *b* creates an image *c* in the plane

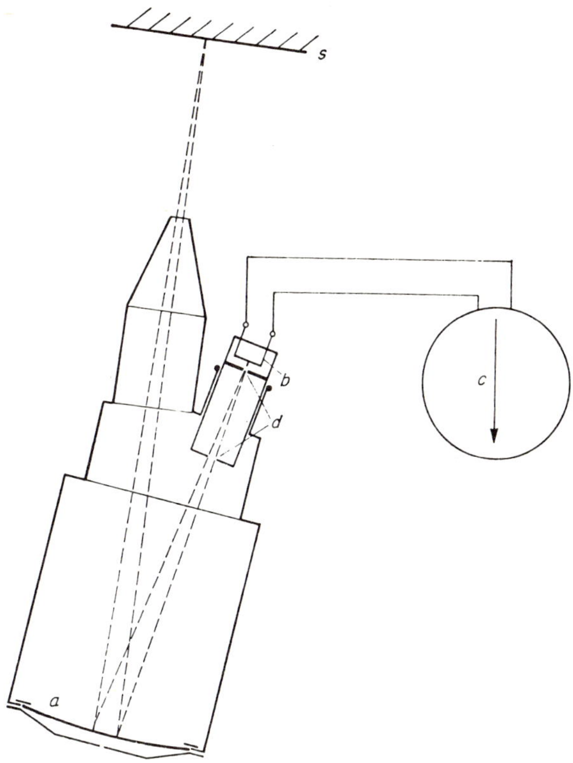

Fig. 17-6 Pyrometer for measuring total radiation. [According to *E. Schmidt.*]

where the filament d of a small lamp is located. Through the ocular e and a color filter f, the eye observes the image c of the object and the lamp filament simultaneously. The temperature of the filament is changed by adjustment of the resistance g in the circuit which heats the lamp. It is adjusted until the radiation intensity of the filament is identical with the apparent intensity of the image. The heating current for the filament

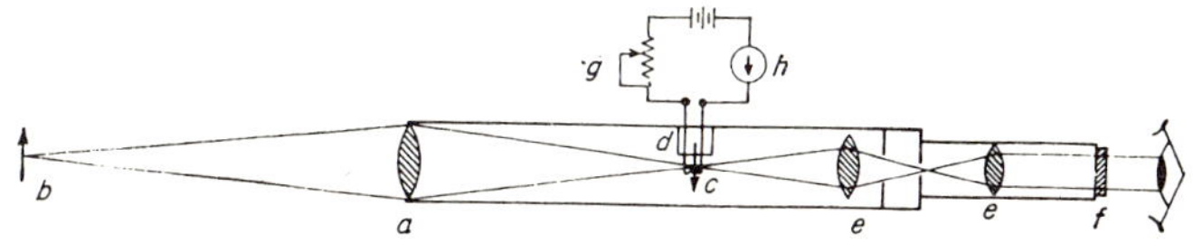

Fig. 17-7 Optical pyrometer.

read on instrument h is, when calibrated, an indication of the radiation intensity of the object. By a smoked glass the apparent intensity can be adjusted to that value for which the human eye is most sensitive. Usually the color filter is a red filter ($\lambda_r = 0.665\ \mu$m), and the instrument reads the apparent monochromatic temperature. The color temperature can be measured when the pyrometer is additionally equipped with a green filter λ_g. The following relation connects the color temperature T_c with the temperature T_{λ_r} read with the red filter and the value T_{λ_g} read with the green filter

$$\frac{1}{T_c} = \frac{1/\lambda_r T_{\lambda_r} - 1/\lambda_g T_{\lambda_g}}{1/\lambda_r - 1/\lambda_g} \tag{17-7}$$

To obtain the true temperature by pyrometry, either the emittance values of the surface must be known or the emittance must be made to approach the value 1. This can often be done by drilling a hole into the surface and by viewing the bottom of the hole with the pyrometer.

Temperatures in flames or radiating gases are most commonly measured with the *line-reversal method,* based on Kirchhoff's law. The gas is made luminous at the desired location by introducing traces of sodium. When the gas is viewed by a spectroscope, the sodium lines can be observed as bright-yellow lines. A blackbody is arranged behind the gas or flame. It produces a continuous spectrum in the spectroscope. The sodium lines appear as bright lines on this background when the gas temperature is higher than that of the blackbody and as dark lines when the blackbody temperature is higher. The lines disappear when gas and blackbody have the same temperature, because in this case the gas absorbs just as much of the background radiation as it radiates (Kirchhoff's law). A measurement of the blackbody temperature then reveals the gas temperature at the location containing the sodium traces which is viewed by the pyrometer.

The physical processes on which this method is based can be traced by referring to Eq. (16-4), which describes the intensity of a radiant beam as affected by emission from a gas with a blackbody temperature $i_{\lambda b}$ and by absorption in the gas of background radiation with a blackbody intensity $i_{\lambda b,c}$. In the wavelength region λ' in which the gas is nonabsorbing, the following relation exists: $K_{\lambda',e} = i_{\lambda' b,c}$. Equation (16-4) describes the intensity $K_{\lambda,e}$ for monochromatic radiation at a wavelength λ in which the gas absorbs. The background radiation is now adjusted in such a way that $K_{\lambda e} = K_{\lambda' e}$. In addition, the blackbody intensity $i_{\lambda' b,c}$ can be considered with good accuracy equal to $i_{\lambda b,c}$ when the two wavelengths λ and λ' are close to each other. Equation (16-4) then becomes

$$K_{\lambda e} = i_{\lambda b}(1 - e^{-\rho \kappa_\lambda S}) + K_{\lambda e} e^{-\rho \kappa_\lambda S}$$

from which follows $K_{\lambda e} = i_{\lambda b} = i_{\lambda b,c}$. This indicates that the gas temperature is equal to the temperature of the blackbody. The method determines an average temperature in the gas along the path length S along which the gas has been made absorbing and emitting by sodium addition.

At very high temperatures of order 10,000 K, measurements are performed through spectroscopy. Considerable knowledge of atomic physics is required to evaluate such measurements, and the reader is referred to the literature in this field.[1]

[1] H. R. Griem, "Plasma Spectroscopy," McGraw-Hill, New York, 1964.

18
Solar Radiation

A large energy source is available to us in the form of radiation coming from the sun, and considerable efforts are being made to find ways by which this energy can be utilized effectively. For instance, the possibility of heating homes by this source is being studied at present at various research centers. This section is devoted to a brief discussion of solar radiation. The sun radiates very nearly like a "black" circular disk with a temperature of 5600 K. Rays drawn from a point on the earth toward two opposite points on the circumference of the sun include an angle of 32 minutes or 0.00931 rad. As a consequence of the high temperature, the maximum radiation intensity is found at 0.5 μm wavelength and approximately one-third of the radiation occurs in the visible-wavelength range, the rest in the infrared up to approximately 3 μm. Part of the solar radiation which is directed toward the earth is absorbed, reflected, or refracted away by the atmosphere, and the rest reaches the surface of the earth. In a yearly average the earth absorbs approximately 43 percent of the radiation coming from the sun (27 percent directly and 16 percent as diffuse sky radiation), 42 percent is reflected or refracted back to space from clouds and the air and reflected from the earth's surface, and 15 percent is absorbed in the atmosphere.

The amount of solar radiation which impinges on a unit area of a surface normal to the sun's radiation and situated outside the atmosphere does not depend on the location on the earth or on the time of the day and is therefore often called the *solar constant.* It changes somewhat during the year, however, because it depends on the distance from the sun. Its amount is 1,450 W/m^2 s in January and 1,370 W/m^2 s in July. Solar absorbers will often have a position which is fixed relative to the earth's surface. Solar radiation impinges on such a surface under an angle which changes during the day and the year. If absorption of radiation in the atmosphere is neglected, the amount of radiation impinging on a unit area of such a surface can readily be calculated from the angle between the surface normal and the solar rays. Figure 18-1 presents the results of such a calculation for a horizontal surface and for a vertical one facing south as a function of geographic latitude and time of the year. Two facts are interesting to note. During the winter months, when solar energy is needed for heating, a vertical surface in the range of latitudes covered by the United States receives a considerably larger energy flux than a horizontal one. This energy flux also does not vary appreciably with geographic latitude.

Actually this amount of energy is not available because part of the

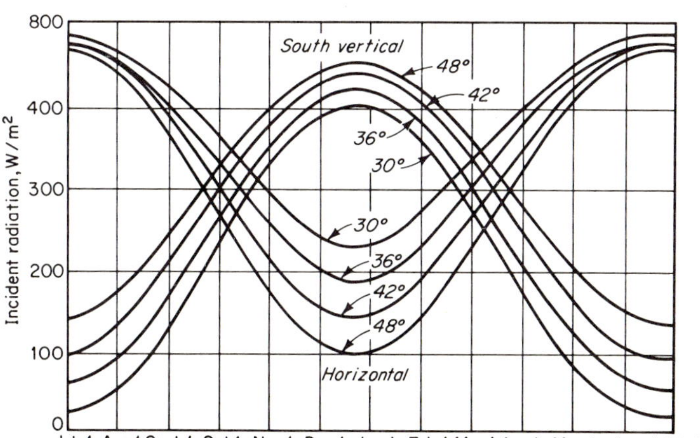

Fig. 18-1 Daily solar irradiation on a horizontal and south-facing vertical surface at various north latitudes outside the atmosphere. [*From R. C. Jordan and J. L. Threlkeld, Trans. Amer. Soc. Heating Ventilating Engrs.,* **60**:*177* (*1954*).]

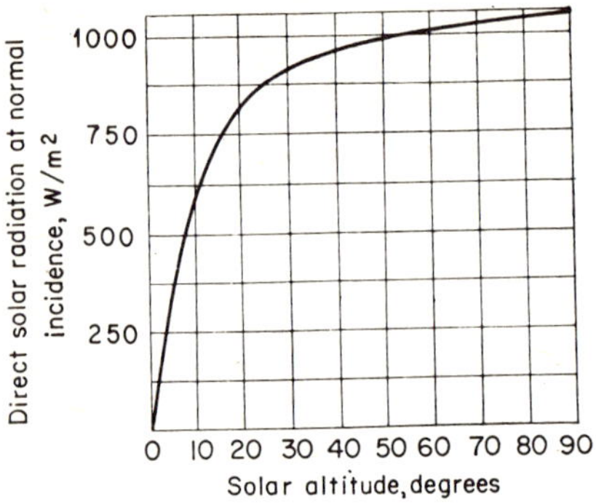

Fig. 18-2 Direct solar irradiation on a surface normal to the sun's rays at the earth's surface on a cloudless winter day. [*From R. C. Jordan and J. L. Threlkeld, Trans. Amer. Soc. Heating Ventilating Engrs.*, **60**:*177* (*1954*).]

solar radiation is absorbed in the atmosphere or reflected and refracted back into space. Absorption is caused in clear air mainly by ozone and water vapor. Additional absorption is caused by dust particles, which, of course, depend considerably on location. The amount by which solar radiation is reduced by absorption depends also on the length of travel of the sun rays through the atmosphere or on the angle between the sun rays and the earth's surface (solar altitude). Figure 18-2 gives, according to investigations by Moon,[1] the solar radiation incident on a surface normal to the sun rays and located on the earth's surface for representative conditions of the absorbing constituents in the atmosphere on a cloudless winter day.

Part of the sun's radiation reaches the earth as diffuse sky radiation. The amount of this radiation varies widely. On a cloudless day approximately 50 to 100 W/m^2 s of sky radiation impinges on a horizontal surface.

On cloudy days the solar radiation drops to very low values (1 percent and less of radiation on clear days).

The amount of solar radiation which a surface absorbs depends on its absorptance. It has to be kept in mind that solar radiation has short wavelengths and that the absorptance for such radiation may be considerably different from the absorptance for longwave radiation, which can, for instance, be obtained through use of Table 15-3. Figure 15-9 shows, for instance, that the white surface 1 has considerably smaller absorptances than the aluminum surface for solar radiation. Surfaces which are to be protected against the sun's radiation should be given a white, nonmetallic surface coat, e.g., roofs of cars. For solar absorbers one is interested in having a surface which has a high absorptance for solar radiation but a low emittance for longwave radiation in order to minimize heat losses by radiation from the surface. Most surfaces with high

[1] P. Moon, *J. Franklin Inst.*, **230**:583–617 (1940).

absorptance have the opposite trend (see Fig. 15-18). The National Physical Laboratory in Israel was successful[1] in developing a surface with the desired properties by applying a thin black layer of a nonmetallic material on a metal surface. Shortwave radiation is absorbed in the layer, whereas longwave radiation is transmitted and reflected from the metal surface.

Solar energy has the advantage that it is readily available; on the other hand, its flux density is too small for many purposes. In such cases, collectors have to be used which concentrate the energy on the target. Parabolic mirrors are generally used for this purpose. We shall determine what flux density can be obtained by the use of such a mirror, which is shown in Fig. 18-3.[2] Solar radiation approaches it with an angle $\alpha = 32$ min. The bundle of rays approaching the mirror along its axis is reflected in itself. At a distance f from the apex, which is equal to the focal length of the mirror, it creates an image of the sun of diameter $d = f\alpha$ and with an area $\pi f^2\alpha^2/4$. Onto this area is also collected radiation which is incident on the other parts of the mirror surface. There are, however, certain losses connected with this process, because not all the incoming rays are concentrated on the sun's image. A bundle incident on point 1 of the surface is, for instance, reflected as a cone with the opening angle α, the axis of which goes through the focal point. The plane of the sun's image cuts out of this cone an elliptical area

$$\frac{\alpha^2\rho^2}{\cos\theta}\frac{\pi}{4}$$

[1] *Sci. Amer.*, **194**:102 (1956).

[2] N. K. Hiester, T. E. Tietz, E. Loh, and P. Duwez, *Jet Propulsion*, **27**:507–513 (1957).

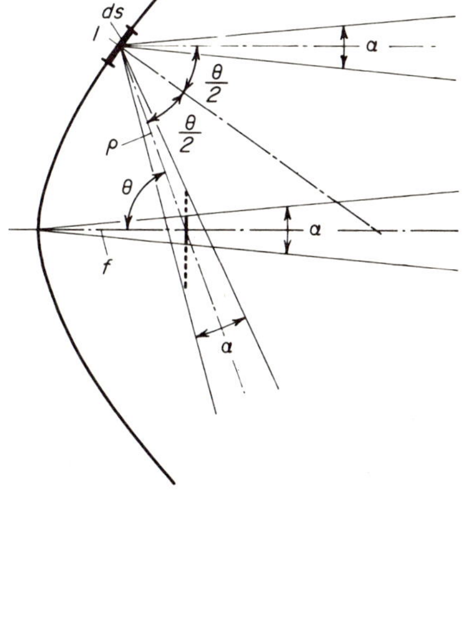

Fig. 18-3 Parabolic mirror for a solar furnace.

The ratio of the area of the sun's image to this area indicates the ratio of the radiation impinging on the sun's image after reflection at point 1 to the radiation incident at point 1.

$$\frac{f^2}{\rho^2} \cos \theta \tag{18-1}$$

The energy reflected under ideal conditions from a ring-shaped area dA of the mirror with the length ds in the meridional plane is

$$dq = C\, dA \cos \frac{\theta}{2}$$

when C is the solar constant. With the expressions

$$ds = \frac{\rho\, d\theta}{\cos(\theta/2)} \qquad dA = 2\pi\rho \sin\theta\, ds = 2\pi\rho^2 \sin\theta \frac{d\theta}{\cos(\theta/2)}$$

which are evident from the figure, the reflected radiation becomes

$$dq = C 2\pi\rho^2 \sin\theta\, d\theta$$

The amount of this radiation collected at the sun's image is found by multiplying this expression by Eq. (18-1). The total energy collected on the sun's image by the mirror is then found by integration:

$$Q = 2\pi C f^2 \int_0^\theta \sin\theta \cos\theta\, d\theta = \pi C f^2 \sin^2\theta$$

Up to this point it has been assumed that none of the energy approaching from the sun is lost in the collection process. In reality, a reduction of this energy will be caused by absorption in the atmosphere and by the fact that the mirror surface will have reflectance smaller than 1 and will also suffer losses caused by imperfect geometry. These three factors will be accounted for by two efficiency terms η_a and η_r, respectively. The heat flux collected at the image of the sun is

$$Q = \pi C \eta_a \eta_r f^2 \sin^2\theta \tag{18-2}$$

The flux density, or energy arriving per unit area, is obtained by dividing Eq. (18-2) by the area of the sun's image:

$$q = \frac{4}{\alpha^2} C \eta_a \eta_r \sin^2\theta$$

This equation can be written in a different way which is more readily understandable. The solar constant can be expressed as

$$C = \frac{\alpha^2}{4} \sigma T_s^4 \tag{18-3}$$

in which T_s indicates the blackbody temperature of the sun (5600 K). The term $\alpha^2/4$ is the shape factor of the sun from the location of the mirror. With this equation the heat flux density can be expressed as

$$q = \sigma T_s^4 \eta_a \eta_r \sin^2 \theta$$

When the mirror is made with an opening angle of 60°, and when the values $\eta_a = 0.7$ and $\eta_r = 0.8$ are introduced, values which can be met in real installations, the heat-flux density is 42 percent of the value σT_s^4 or 42 percent of the heat-flux density from a black surface at the temperature of the sun (2,150 kW/m^2 s). It can be seen that high flux densities can be obtained by a solar collector.

It is also of interest to calculate to what temperature T_m a target located at the position of the sun's image can be heated by a solar collector under optimum conditions. A target will obviously assume its highest temperature when heat losses from its back surfaces are prevented and when no heat conduction or convection occurs on the front surface, which is exposed to the solar radiation. Under these circumstances the heat loss will be by radiation alone. Equating this heat loss with the heat flux arriving from the sun gives

$$\epsilon \sigma T_m^4 = \alpha \sigma T_s^4 \eta_a \eta_r \sin^2 \theta$$

For a gray surface the absorptance α is equal to the emittance ϵ, and the temperature of the object becomes

$$T_m = T_s \sqrt[4]{\eta_a \eta_r} \sin^{\frac{1}{2}} \theta \tag{18-4}$$

With the same values as before, $\theta = 60°$, $\eta_a = 0.7$, $\eta_r = 0.8$, one finds that the maximum temperature of the object is equal to 80 percent of the temperature of the sun, or 4500 K.

part **IV**

The Transfer of Mass

In many engineering activities, especially in chemical engineering, processes are utilized which are classified as *mass-transfer processes*. This term refers to a situation where, in a multicomponent mixture with local concentration differences, one or several components are transported from one location to another relative to the mixture. This transport may, for instance, be effected by diffusion or by turbulent mixing or generally by the combined action of diffusion, turbulence, and convection. From an engineering point of view the mass transport through an interface between various media or phases of the same medium is of special importance. The interface or surface on which mass transfer occurs can be that between a gas and a solid or a liquid or that between a liquid and a solid. It may be predetermined in its shape by the specific situation, e.g., the surface of a pond from which evaporation occurs, or it may be determined by the physical process itself, e.g., evaporation of a liquid in droplet form or of gas and vapor bubbles in a liquid. The discussion in this book is not concerned with a determination of the shape of the interphase but considers the mass transfer at the interface only.

19

Conservation Laws

19-1 PARAMETERS DESCRIBING A MIXTURE

A study of papers in the field of mass transfer is made difficult by the fact that various parameters are used to characterize the composition of the mixture. For this reason the various parameters used will be discussed in this section. We assume, for this purpose, that a mixture is contained in the volume $\mathcal{V}$ with the component masses $m_1, m_2, m_3, \ldots m_n$. The mass of an arbitrary component will be denoted by m_m. The total mass of the mixture is

$$m = \sum_1^n m_m \tag{19-1}$$

The mixture density is

$$\rho = \frac{m}{\mathcal{V}}$$

In the same way one can define a component density

$$\rho_m = \frac{m_m}{\mathcal{V}} = c_m \tag{19-2}$$

The component density is often referred to as *concentration* and is then denoted by c_m. The following equation connects the component density or concentration with the mixture density:

$$\sum_1^n \rho_m = \sum_1^n c_m = \rho \tag{19-3}$$

One also defines a mass fraction of component m as

$$w_m = \frac{m_m}{m} \tag{19-4}$$

The sum of all component mass fractions must have the value 1:

$$\sum_1^n w_m = 1 \tag{19-5}$$

A mixture is sometimes also characterized by the number of moles of the various components. This number for component m is

$$N_m = \frac{m_m}{M_m} \tag{19-6}$$

where M_m denotes the molecular weight of this component. The number of moles per unit volume, the mole density, is

$$n_m = \frac{N_m}{\mathcal{V}} \tag{19-7}$$

for which the following relation exists:

$$\sum_1^n n_m = n \tag{19-8}$$

where n denotes the total number of moles of the mixture per unit volume.

Physicists extend the term to other particles, like atoms, ions, and electrons, and use the term particle density. The particle density of molecules, for instance, is obtained by multiplication of the mole density by Avogadro's number, the number of molecules per mole:

$$n_{p,m} = 6.03 \times 10^{23} n_m$$

with $n_{p,m}$ indicating the number of molecules per kilomole. The mole fraction of component m is defined as

$$\omega_m = \frac{N_m}{N} \tag{19-9}$$

with the relation

$$\sum \omega_m = 1 \tag{19-10}$$

In a mixture of ideal gases, the composition is often expressed by the partial pressures. For such a mixture of ideal gases which do not react chemically, Dalton's law applies: where different gases are present in the same space, every gas fills the whole available space as if the other gases were not present. The pressure which any of the component gases creates in this way according to the ideal-gas law, is called its *partial pressure.* According to this law, the following relation describes the partial pressure of component m:

$$p_m \mathcal{V} = m_m R_m T = \frac{m_m}{M_m} \mathcal{R} T = N_m \mathcal{R} T$$

where R_m denotes the gas constant for the component m and $\mathcal{R}$ denotes the universal gas constant. From the relation

$$\Sigma p_m = p \tag{19-11}$$

it follows for the total pressure of the mixture that

$$p\mathcal{V} = mRT = N\mathcal{R}T$$

with the gas constant R of the mixture $R = \Sigma w_m R_m$.

Parameters introduced before to describe the composition of a mixture can also be expressed by partial pressures through the relations

$$\rho_m = \frac{p_m}{R_m T}$$

$$w_m = \frac{p_m R}{p R_m}$$

$$n_m = \frac{p_m}{\mathcal{R} T}$$

$$\omega_m = \frac{p_m}{p}$$

which follow readily from the equations given above.

The relations above are used to describe the composition of a mixture as long as the composition is uniform within the volume $\mathcal{V}$. They can also be used to describe the local composition in a mixture with locally varying composition when the volume $\mathcal{V}$ is made sufficiently small.

In a moving multicomponent fluid with mass transfer, the individual components of the mixture move with different velocities. The velocity of a one-component substance was defined in Chap. 6 as the mass flow rate per unit area divided by the local density; consequently, the velocity v of a multicomponent mixture is defined as the mass flow rate of the mixture through a specific area per unit of this area divided by the mixture density. It is therefore given by the equation

$$v_i = \frac{1}{\rho} \sum_{1}^{n} \rho_m v_{mi} \tag{19-12}$$

with v_{mi} indicating the velocity of component m. In the following discussion the mass flow per unit time and unit area of component m will be needed through an area which moves with the mixture; this means with the velocity v_i. This specific mass flow rate will be denoted as j_{mi}. The following equation then describes this vector quantity

$$j_{mi} = \rho_m(v_{mi} - v_i) \tag{19-13}$$

The following relation holds:

$$\sum_1^n j_{mi} = 0 \tag{19-14}$$

The terms defined by Eqs. (19-12) and (19-13) are convenient for mass-transfer calculations when the composition of a mixture is expressed either in concentrations or in mass fractions. Sometimes analyses are performed using moles to denote the composition. In this case it is convenient to use a *mole-averaged velocity*

$$v_i^* = \frac{1}{n} \sum_1^n n_m v_{mi}$$

For clarity, the velocity defined by Eq. (19-12) is then referred to as the *mass-averaged velocity.*

In this book mass-transfer calculations will be based on mass fractions because this definition of the composition lends itself more readily to an extension of the relations obtained for one-component fluids to multicomponent mixtures. The mass-averaged velocity, for instance, is then defined in the same way as the velocity of a one-component fluid which means that the continuity equation (6-8) retains its validity. It will also be found that analogies between heat- and mass-transfer situations can more readily be established.

19-2 CONSERVATION LAWS FOR MULTICOMPONENT FLUIDS

The continuity equation for a one-component fluid was derived in Chap. 6 as Eq. (6-8). The same derivation for a component m in a multicomponent mixture results in

$$\frac{\partial \rho_m}{\partial \tau} + \frac{\partial}{\partial x_i} \rho_m v_{mi} = \psi_m \tag{19-15}$$

The left-hand side of the equation is completely analogous to Eq. (6-8). On the right-hand side, however, a term ψ_m appears, expressing the fact that mass of component m may be generated by some chemical reaction. In a combustion process, for instance, mass of carbon dioxide may be generated within the volume element under consideration. ψ_m therefore indicates a source or, if negative, a sink for mass m_m. The continuity equation for the mixture can now be obtained by summing up the n equations of the form (19-15) with the result

$$\frac{\partial}{\partial \tau} \Sigma \rho_m + \frac{\partial}{\partial x_i} \Sigma \rho_m v_{mi} = \Sigma \psi_m$$

With Eqs. (19-3) and (19-12), this equation transforms into

$$\frac{\partial \rho}{\partial \tau} + \frac{\partial}{\partial x_i} \rho v_i = 0 \tag{19-16}$$

because no generation or destruction of the total mass occurs as long as nuclear processes are excluded (mass-conservation law). It is seen that in this way one comes back to Eq. (6-8).

Introduction of the specific mass flow rate j_{mi} as defined by Eq. (19-13) into the continuity equation (19-15) leads to the relation

$$\frac{\partial \rho_m}{\partial \tau} + \frac{\partial}{\partial x_i} \rho_m v_i + \frac{\partial}{\partial x_i} j_{mi} = \psi_m$$

This equation can further be transformed by introduction of the mass fraction utilizing the equation

$$w_m = \frac{\rho_m}{\rho}$$

with the result

$$\rho \frac{\partial w_m}{\partial \tau} + \rho v_i \frac{\partial w_m}{\partial x_i} + w_m \left(\frac{\partial \rho}{\partial \tau} + \frac{\partial}{x_i} \rho v_i \right) + \frac{\partial}{\partial x_i} j_{mi} = \psi_m$$

The expression in parentheses is equal to zero, according to Eq. (19-16). Introducing, in addition, the substantial derivative defined by Eq. (6-4) leads to the continuity equation for the component m

$$\rho \frac{D w_m}{d\tau} + \frac{\partial}{\partial x_i} j_{mi} = \psi_m \tag{19-17}$$

This form of the continuity equation will generally be used in the following calculations.

n equations of the form (19-17), or alternately $n - 1$ equations of this form together with Eq. (19-5), with the momentum equation and the energy equation describe the composition, the flow field, and the temperature field in a moving multicomponent fluid. To solve the equations, however, one needs relations which connect the energy flux ϵ_i and the mass fluxes j_{mi} with the other parameters appearing in the equations. This is accomplished by the so-called constitutive relations, which will be developed in the following chapter.

20
Constitutive Equations

20-1 FLUXES AND FORCES

The constitutive equations establish relations between the fluxes involved in transfer processes and the respective driving potentials. Equations of this kind have already been introduced in Part II. Fourier's equation (6-32), for instance, describes the relation between the energy flux ϵ_i and the temperature gradient as driving potential. The relations connecting the viscous stresses with the deformation velocities, for instance Eq. (6-18) for a newtonian fluid, are of the same kind. It was mentioned that viscous stresses are caused in gases by the momentum transport of the molecules through the surfaces of the control volume. They can therefore be interpreted as momentum fluxes. Fick's law is a well-known relation between the mass fluxes caused by diffusion and the composition gradients. This law will be discussed in the following section.

When heat and mass transfer occur simultaneously in a moving fluid, the relations between the fluxes and the driving potentials are of a more intricate nature. It is found that diffusional mass fluxes are not only created by composition gradients but also by pressure gradients and by body forces as long as different forces act on the various fluid components. In addition, irreversible thermodynamics shows that any of the driving potentials present can cause any of the fluxes mentioned above, with the restriction that the driving potentials referred to in irreversible thermodynamics as forces and the respective fluxes must both be of the same nature; this means that they must both be either tensors or vectors. Table 20-1 gives the forces involved in a combined heat- and mass-transfer process and the possible fluxes and indicates which forces cause which respective fluxes. The blanks result because the corresponding forces and fluxes are of a different nature. Entries in the main diagonal from upper left to lower right refer to processes which have already been discussed. The laws describing these processes are mentioned. Fourier's law, for instance, describes the relation between an energy flux ϵ_i and a temperature gradient. It is found, however, that an energy flux can also be generated by composition gradients, pressure gradients, or body forces. The energy flux caused by a composition gradient was discovered in 1873 by Dufour and is correspondingly often referred to as the *Dufour* effect; it is also called the *diffusion-thermo effect.* On the other hand, mass fluxes can also be created by temperature gradients, as was established by Soret. This is the thermal-diffusion effect. In general, the thermal-diffusion and the diffusion-thermo effects are of a smaller order of magnitude than the effects described by Fourier's or Fick's laws and are often neglected in heat- and mass-transfer processes. There are, however, exceptions. The thermal-diffusion effect, for instance, has been utilized for isotope separation, and in mixtures between gases with very light molecular weight (H_2, He) and of medium molecular weight (N_2, air)

Table 20-1 Fluxes and forces according to irreversible thermodynamics

	Forces		
Fluxes	*Deformation velocities (tensor)*	*Temperature gradients (vector)*	*Composition gradients, pressure gradients, body forces (vector)*
Viscous stresses p_{ij} (tensor)	Newton's law, or equivalent		
Energy flux ϵ_i (vector)		Fourier's law	Diffusion thermo (Dufour effect)
Mass fluxes j_{mi} (vector)		Thermal diffusion (Soret effect)	Fick's law and extension

the diffusion-thermo effect was recently found to be of a magnitude such that it cannot be neglected.

Generally both these effects will be omitted in the following discussions.

20-2 ISOTHERMAL DIFFUSION

In discussing Fick's law in some detail, a two-component mixture will be assumed in the absence of body forces. The mixture may be at rest ($v_i = 0$) and at a uniform temperature. Accordingly, the pressure will also be uniform. Mass fluxes by diffusion are then created by composition gradients only.

Equimass diffusion The two components of the mixture will be denoted by the indices 1 and 2. The mass flux of component 1 per unit area and time through a plane at rest so that the mixture mass flux is zero is described by the following form of Fick's equation:

$$\dot{j}_{1i} = -\rho D_{12} \frac{\partial w_1}{\partial x_i} \tag{20-1}$$

where D_{12} is the mass-diffusion coefficient of component 1 in the mixture and ρ is the mixture density. In the same way the mass flux of component 2 through the same plane can be written

$$\dot{j}_{2i} = -\rho D_{21} \frac{\partial w_2}{\partial x_i} \tag{20-2}$$

The following relations hold, according to the preceding section:

$$w_1 + w_2 = 1 \qquad \dot{j}_{1i} + \dot{j}_{2i} = 0$$

From the first of these relations follows

$$\frac{\partial w_1}{\partial x_i} + \frac{\partial w_2}{\partial x_i} = 0$$

These equations, when introduced into Fick's relation (20-1), result in the statement

$$D_{12} = D_{21} = D \tag{20-3}$$

The indices on the mass-diffusion coefficient are no longer necessary and will be omitted.

Equimolal diffusion The relations above describe the mass fluxes through a plane arranged in the mixture in such a way that no mixture mass flux occurs through it. It can readily be understood, however, that a flux of moles of the mixture will, in general, be present through this plane. To show this, we denote the flux of moles per unit area and time through this plane by $\dot{n}_{1i}$ and $\dot{n}_{2i}$, respectively. These mole fluxes are connected with the mass fluxes by the equations

$$\dot{n}_{1i} = \frac{\dot{j}_{1i}}{M_1} \qquad \dot{n}_{2i} = \frac{\dot{j}_{2i}}{M_2} = -\frac{\dot{j}_{1i}}{M_2}$$

The mole flux of the mixture through this plane is therefore

$$\dot{n}_i = \dot{n}_{1i} + \dot{n}_{2i} = \dot{j}_{1i}\left(\frac{1}{M_1} - \frac{1}{M_2}\right)$$

There must be another plane which moves through the mixture in such a way that the flux of mixture moles through that plane becomes zero. The velocity v_i^* of this plane was given in the preceding section. The flux of moles through that plane is given by a different form of Fick's law:[1]

$$j_{1i}^* = -nD\frac{d\omega_1}{dx_i} \tag{20-4}$$

In an ideal gas the mole density n of the mixture is uniform in a fluid at constant pressure and temperature. Fick's relation for the mole flux of component 1 can then be written as

$$j_{1i}^* = -nD\frac{d}{dx_i}\left(\frac{n_1}{n}\right) = -D\frac{dn_1}{dx_i} \tag{20-5}$$

Alternately it can also be expressed by partial pressures as the driving potential

$$j_{1i}^* = -\frac{D}{\mathcal{R}T}\frac{dp_1}{dx_i} \tag{20-6}$$

Diffusion through an arbitrary plane We now consider the mass fluxes through a plane through which the mixture moves with an arbitrary velocity v_i. Figure 20-1 sketches this situation, assuming the flow to be in the vertical y direction, and 1-1 indicates the plane though which the mixture flows with a velocity v. A plane moving with the fluid, the position of which originally coincided with plane 1-1, may be in position 2-2 after a time element $d\tau$. The distance between the two planes is $v\,d\tau$,

[1] For the analysis connecting Eqs. (20-1) and (20-4) for ideal-gas mixtures, see E. R. G. Eckert, "Introduction to Heat and Mass Transfer," pp. 287–290, McGraw-Hill, New York, 1963.

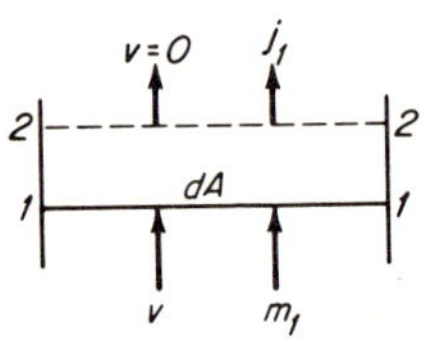

Fig. 20-1 Derivation of mass-transfer equation.

and the mass of component 1 passing through plane 2-2 in the time interval $d\tau$ is $j_1\,dA\,d\tau$. The mass passing through plane 1-1 in the same time interval is $\dot{m}_1\,dA\,d\tau$, where $\dot{m}_1$ is the mass flux of component 1 per unit area and time through plane 1-1. The mass passing through 1-1 must either have passed through 2-2 or be contained in the volume between both planes. Therefore

$$\dot{m}_1\,dA\,d\tau = j_1\,dA\,d\tau + \rho w_1\,dA\,v\,d\tau$$

Canceling the common factors and generalizing to three dimensions results in the following equation for the mass flux of component 1 per unit area and time through a plane with an arbitrary velocity v_i:

$$\dot{m}_{1i} = j_{1i} + \rho w_1 v_i \tag{20-7}$$

The mass flux of component 2 is

$$\dot{m}_{2i} = j_{2i} + \rho w_2 v_i = -j_{1i} + \rho w_2 v_i$$

The mixture mass flux is

$$m_i = m_{1i} + m_{2i} = \rho v_i \tag{20-8}$$

Diffusion through a semipermeable plane A plane is called semipermeable when the mass flux of one of the components is zero. Such a plane is, for instance, the surface of water which evaporates into an adjoining airstream because water but no air passes through the surface. We shall assume that

$$\dot{m}_{2i} = 0$$

The velocity v_i of the mixture is then obtained from the above equation for $\dot{m}_{2i}$

$$v_i = \frac{j_{1i}}{\rho w_2} = \frac{j_{1i}}{\rho(1 - w_1)}$$

and the mass flux of component 1 is, according to Eq. (20-7),

$$m_{1i} = \frac{1}{1 - w_1} j_{1i} = \dot{m}_i = \rho v \tag{20-9}$$

Enthalpy transport by diffusion A diffusion process generates an enthalpy transport even when the temperature is locally uniform. This transport through an arbitrary plane is given by

$$\dot{q}_{Di} = \dot{m}_{1i} i_1 + \dot{m}_{2i} i_2 \tag{20-10}$$

with i_1 and i_2 denoting the respective enthalpies of the components. Such an enthalpy transport occurs even through a plane through which the mixture mass flow is zero ($v_i = 0$). Its magnitude is

$$\epsilon_{Di} = j_{1i}(i_1 - i_2) \tag{20-11}$$

We shall call it *energy transport by interdiffusion.* The enthalpy transport through an arbitrary plane can be expressed in a different way by introducing Eq. (20-7) and the corresponding equation for m_{2i} into Eq. (20-10):

$$\begin{aligned} \dot{q}_{Di} &= (j_{1i} + \rho w_1 v_i)i_1 + (j_{2i} + \rho w_2 v_i)i_2 \\ &= \epsilon_{Di} + \rho v_i i \end{aligned} \tag{20-12}$$

where $i = w_1 i_1 + w_2 i_2$ is the enthalpy of the mixture. A correction is necessary for liquids with appreciable values of the heat of mixing. For these, the mixture enthalpy is not equal to the sum of the component enthalpies, as was assumed in Eqs. (20-10) and (20-12).

For the semipermeable plane, Eq. (20-10) simplifies to

$$\dot{q}_{Di} = \dot{m}_i i_1 = \rho v_i i_1 \tag{20-13}$$

Equation (20-12) retains its form:

$$\dot{q}_{Di} = \epsilon_{Di} + \rho v_i i \tag{20-14}$$

The only difference in the right-hand term of Eq. (20-13) and in the second right-hand term in Eq. (20-14) is caused by the enthalpies. In Eq. (20-13) the enthalpy of component 1 appears and in Eq. (20-14) the enthalpy of the mixture. This difference will become important in later discussions.

In the following sections, where heat- and mass-transfer processes with chemical reactions will be considered, the enthalpies will be interpreted as absolute enthalpies including heats of formation.

The scheme in Table 20-1 indicates that in any originally isothermal mixture, diffusion thermo creates an energy flux which will then create temperature differences. It has been mentioned that this effect is quite small in many engineering situations. It has not been included in this chapter. Equations (20-7) to (20-14) maintain their validity even when thermal-diffusion and diffusion-thermo effects are included.[1]

[1] A summary of recent investigations on mass- and heat-transfer situations in which the diffusion-thermo effect becomes important is contained in E. R. G. Eckert, Diffusion Thermo Effects in Mass Transfer Cooling, *Proc. Fifth U.S. Natl. Congr. Appl. Mech.*, 1966, p. 639.

20-3 NONISOTHERMAL MOVING TWO-COMPONENT FLUID

The expressions for the various fluxes are quite involved when all the driving potentials listed in Table 20-1 are present simultaneously. They are listed here for a two-component fluid.

$$j_{1,i} = \underbrace{-\rho D \frac{\partial w_1}{\partial x_i}}_{\text{Fick's diffusion}} + \underbrace{\frac{M_1 M_2 D w_1}{p} \frac{\partial p}{\partial x_i}}_{\text{pressure diffusion}} + \underbrace{\frac{M_1 M_2 w_1 w_2 D}{RT} (g_{2i} - g_{1i})}_{\text{body-force diffusion}} \underbrace{- \frac{\rho D w_1 w_2 \alpha}{T} \frac{\partial T}{\partial x_i}}_{\text{thermal diffusion}} \tag{20-15}$$

$$\epsilon_i = \underbrace{-k \frac{\partial t}{\partial x_i}}_{\text{Fourier conduction}} + \underbrace{j_{1i}(i_1 - i_2)}_{\text{interdiffusional convection}} - \underbrace{\alpha RT \frac{M^2}{M_1 M_2} j_{1,i}}_{\text{diffusion thermo}} \tag{20-16}$$

For a derivation of these equations the reader is referred to the literature.[1]

The pressure-diffusion term becomes important in swirling flow. It is negligible in normal boundary-layer flow because the pressure variation in the direction normal to the flow, where the important diffusional mass fluxes occur, is negligible in this situation. The body-force diffusion takes on finite values when different body forces act on the two components. This is the case in plasma technology, where the fluid interacts with electric and magnetic forces.[2] Thermal diffusion becomes important in special situations only, and the same is true of the diffusion-thermo term in Eq. (20-16).

Viscous stresses are caused by deformation velocities alone because of their tensorial nature, but a correction has to be applied to Eq. (6-18) for a newtonian fluid when it is to be used for a two-component mixture because the velocities in that equation are mass-averaged velocities whereas viscous stresses are connected with momentum fluxes. Kármán, who pointed this out, derived the following equation for a newtonian gas:

$$p_{ij} = \mu \left(\frac{\partial v_i}{\partial x_j} + \frac{\partial v_j}{\partial x_i} \right) - \mu' \frac{\partial v_i}{\partial x_i} \delta_{ij} - \rho(w_1 v'_{1i} v'_{1j} + w_2 v'_{2i} v'_{2j}) \tag{20-17}$$

The last term on the right-hand side is called the *diffusional momentum transport* p_{Dij}. The velocities in this term are the difference between the

[1] S. Chapman and T. G. Cowling, "The Mathematical Theory of Non-uniform Gases," Cambridge University Press, Cambridge, 1953; J. O. Hirschfelder, C. F. Curtiss, and R. B. Bird, "Molecular Theory of Gases and Liquids," Wiley, New York, 1954; R. B. Bird, W. E. Stewart, and E. N. Lightfoot, "Transport Phenomena," Wiley, New York, 1960.

[2] E. R. G. Eckert and E. Pfender, Advances in Plasma Heat Transfer, *Advan. Heat Transfer*, **4**:229 (1967).

component velocities and the mass-averaged mixture velocity according to the equation

$$v'_{1i} = v_{1i} - v_i \qquad v'_{1j} = v_{1j} - v_j$$

The diffusional momentum transport is again negligibly small in most situations. For boundary-layer flow, Lees calculated the ratio of this term p_{Dij} to the largest term $\mu\, \partial u/\partial y$ in the rest of the equation to be of the order $1/(\mathrm{Re}_{x0}\, \mathrm{Sc}_0)$. The Schmidt number Sc_0 is defined as the ratio of a reference kinematic viscosity ν_0 to a reference mass diffusivity D_0.

Except for special situations in which the terms discussed above become important, the various fluxes can be represented in a two-component mixture by the following equations:

$$\dot{j}_{1,i} = -\rho D \frac{\partial w_1}{\partial x_i} \tag{20-18}$$

$$\epsilon_i = -k \frac{\partial t}{\partial x_i} - \rho D(i_1 - i_2) \frac{\partial w_1}{\partial x_i} \tag{20-19}$$

$$p_{ij} = \mu \left(\frac{\partial v_i}{\partial x_j} + \frac{\partial v_j}{\partial x_i} \right) - \mu' \frac{\partial v_i}{\partial x_i} \delta_{ij} \tag{20-20}$$

The treatment of multicomponent mixtures becomes much more difficult, mainly because Eq. (20-18) has to be replaced by the relation

$$\dot{j}_{mi} = -\rho D_{me} \frac{\partial w_m}{\partial x_i}$$

in which the effective diffusion coefficients D_{me} are connected[1] in a very involved way with the binary coefficients, values for which are reported in the literature. The discussions in the following chapters will be concerned mainly with two-component mixtures.

In the recent literature, heat and mass transfer in multicomponent fluid mixtures is frequently treated by approximating the multicomponent mixture by a two-component mixture. For this purpose, the various components are arranged in two groups in such a way that the molecular weights of the components in each of the groups differ less than the molecular weights of the two groups. The total mixture is then considered as a two-component fluid consisting of the two groups. Dissociation and recombination in a high-temperature gas boundary layer, for instance, is analyzed such that one group consists of the molecules and the other group of the atoms. This considerably simplifies the determination of the diffusion coefficients involved. A few references using this approach are listed below.[2]

[1] Chapman and Cowling, *op. cit.*; Hirschfelder, Curtiss, and Bird, *op. cit.*; and Bird, Stewart, and Lightfoot, *op. cit.*

[2] H. Hoshizaki, *J. Amer. Rocket Soc.*, **32**:1544 (1962); J. A. Fay and N. H. Kemp, *AIAA J.*, **1**:2741 (1963); A. Pallone and W. van Tassel, *Phys. Fluids*, **6**:983 (1963).

21
Boundary-layer Equations

The conservation equations developed in the previous chapters of this book also apply for a multicomponent fluid undergoing a combined heat- and mass-transfer process. Section 6-1 contains the conservation equations for mixture mass, Sec. 6-2 developed the momentum equation, and Sec. 6-3 the energy equation. In connection with the energy equation one can raise the point that a term should be added to describe the heat generation by a chemical process which the fluid mixture within the volume element may undergo. This, however, is not necessary when the mixture enthalpy i is interpreted as including the chemical formation energy in addition to the sensible internal energy. The continuity equation (19-17) has to be written for $n - 1$ of the components when Eq. (19-5) is used in addition.

Boundary-layer simplifications have already been made to the mixture mass, the momentum, and the energy equations in Sec. 6-4. The same procedure as applied to the energy equation can be used to simplify the component continuity equations. In this way one obtains the following set of equations describing flow and heat and mass transfer for a two-dimensional boundary layer:

$$\frac{\partial \rho}{\partial \tau} + \frac{\partial(\rho u)}{\partial x} + \frac{\partial(\rho v)}{\partial y} = 0 \tag{21-1}$$

$$\rho \frac{\partial u}{\partial \tau} + \rho u \frac{\partial u}{\partial x} + \rho v \frac{\partial u}{\partial y} = -\frac{\partial p}{\partial x} + \frac{\partial}{\partial y}\left(\mu \frac{\partial u}{\partial y}\right) + \rho g_x \tag{21-2}$$

$$\frac{\partial p}{\partial y} = 0 \tag{21-3}$$

$$\rho \frac{\partial i^\circ}{\partial \tau} + \rho u \frac{\partial i^\circ}{\partial x} + \rho v \frac{\partial i^\circ}{\partial y} - \frac{\partial p}{\partial \tau} + \frac{\partial \epsilon_y}{\partial y} = \frac{\partial}{\partial y}\left(\mu u \frac{\partial u}{\partial y}\right) + \rho u g_x \tag{21-4}$$

$$\rho \frac{\partial w_m}{\partial \tau} + \rho u \frac{\partial w_m}{\partial x} + \rho v \frac{\partial w_m}{\partial y} + \frac{\partial j_{my}}{\partial y} = \psi_m \tag{21-5}$$

There are $n - 1$ equations of the form (21-5). It may be recalled that one of the essential simplifications leading to the boundary-layer equations is the fact that only one term need be maintained in each of the expressions for the energy and mass fluxes, namely, ϵ_y and j_{my}. For the mass flux, this is permissible when the term $\text{Re}_0\ \text{Sc}_0$ is very large. The mass-source term ψ_m in Eqs. (21-5) is of the same order as the terms on the left-hand side of the equation when the expression $\psi_0 L/(\rho_0 V_0\, \Delta w_{m0})$ is of order 1. ψ_0 indicates a reference mass source selected arbitrarily from among those occurring in the process under consideration. Δw_{m0} denotes a characteristic mass-fraction difference of component m, again arbitrarily selected. The rest of the symbols were defined in Sec. 6-4.

Boundary conditions to this set of differential equations must be stated to describe a specific problem entirely.

A simplification of the boundary-layer equations is possible in a situation where the multicomponent fluid is premixed, so that local composition differences are caused by chemical processes within the boundary layer only. Figure 21-1*a* shows an example of such a situation as it occurs in the reentry of satellites or space vehicles through the atmosphere. An airstream with a temperature of order 10,000 K flows over a solid surface with a temperature of say 1000 K. As a consequence, the oxygen and the nitrogen in the approaching airstream will be partially

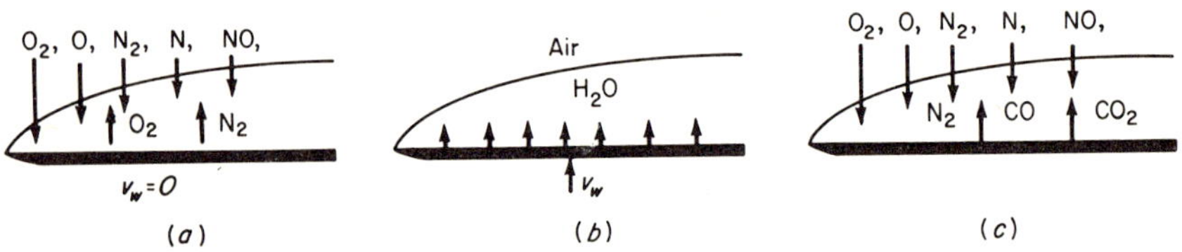

Fig. 21-1 Chemical reactions and mass transfer in boundary-layer flow.

dissociated, and nitric oxide will also be present, so that the following components will be found: O_2, O, N_2, N, and NO. The temperature in the boundary layer will decrease in the direction toward the wall surface; correspondingly, recombination occurs within the boundary layer. The local composition within this layer will depend on the speed of the chemical recombination process relative to the speed with which the various components diffuse into each other. Two extreme cases are usually considered. In one the chemical processes are very fast compared to the diffusion processes. The gas mixture is then in chemical equilibrium at any location within the boundary layer. This means that the local composition is described by a thermodynamic relation as a function of the temperature and pressure existing within the boundary layer. The component continuity equations (21-5) are then replaced by this thermodynamic relation. The energy equation (21-4) can also be simplified. Equation (20-15) will be considered for this purpose. Generalized to a multicomponent mixture, it reads

$$\epsilon_y = -k_f \frac{\partial t}{\partial y} - \rho \Sigma D_{me} i_m \frac{\partial w_m}{\partial y} \tag{21-6}$$

The concentration is now a function of local temperature and only indirectly of the location within the boundary layer. The equation can therefore be written

$$\epsilon_y = -\left(k_f + \rho \Sigma D_{me} i_m \frac{\partial w_m}{\partial t}\right) \frac{\partial t}{\partial y} = -k_e \frac{\partial t}{\partial y} \tag{21-7}$$

The whole term within the parentheses is again a function of temperature and pressure and can be calculated through thermodynamic relations. It is referred to as the *equilibrium conductivity* k_e. Introducing this term into Eq. (21-4) results in the relation

$$\rho \frac{\partial i^\circ}{\partial \tau} + \rho u \frac{\partial i^\circ}{\partial x} + \rho v \frac{\partial i^\circ}{\partial y} - \frac{\partial p}{\partial \tau} = k_e \frac{\partial t}{\partial y} + \frac{\partial}{\partial y}\left(\mu u \frac{\partial u}{\partial y}\right) + \rho u g_x \tag{21-8}$$

To distinguish the conductivity in Eq. (21-6) from the equilibrium conductivity, it has been identified by the index f and will be referred to as *frozen conductivity*. It describes the heat conduction in a process in which the composition of the gas mixture is locally uniform. Equations (21-1) to (21-3) and (21-8) now describe the combined heat- and mass-transfer process. The form of these equations is the same as for a one-component mixture, and the analysis can proceed as in Part II.

The difference between the frozen and equilibrium heat conductivity is often very large. Figure 21-2 presents both conductivities for air at atmospheric pressure as a function of temperature. It will be observed that the frozen conductivity increases more or less regularly

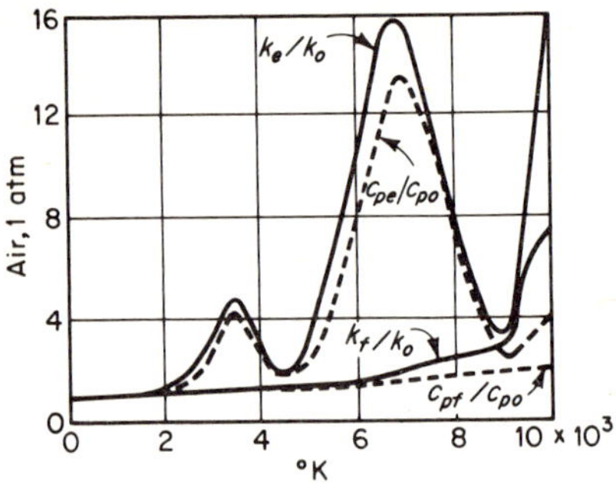

Fig. 21-2 Thermal conductivity and specific heat for thermodynamic equilibrium condition (k_e, c_{pe}) and for frozen condition (k_f, c_{pf}). All values are made dimensionless by the properties calculated from simple kinetic theory (k_o, c_{po}). [*From F. C. Hansen, Approximations for the Thermodynamic and Transport Properties of High-temperature Air, NACA Tech. Note* 4150, *1958*.]

up to a temperature of approximately 9000 K. The equilibrium conductivity exhibits two pronounced peaks in this temperature region. The first peak is caused by the fact that molecular oxygen starts dissociating at around 2000 K and the energy transport by interdiffusion of atoms and molecules increases the conductivity. The dissociation process is practically complete at a temperature around 5000 K, but now the dissociation of nitrogen molecules into atoms starts. This causes the second peak in the curve for the equilibrium conductivity. The increase in both conductivity values at temperatures above 9000 K is caused by ionization. The electrons created by this process are very mobile because they are so small, and thus may increase both conductivity values. Figure 21-2 also shows curves for specific heats at constant pressure. In a multicomponent fluid one must again distinguish between a specific heat at the frozen condition and a specific heat for chemical-equilibrium processes. The second value includes chemical energy converted into internal energy as a consequence of the change in composition.

A second extreme situation in the heat- and mass-transfer process is one in which the chemical reactions occur very slowly compared to the diffusion processes. One refers to such a boundary-layer situation as a *frozen boundary layer*. Whether local composition differences occur in such a boundary layer depends on the characteristics of the wall. If the wall is composed of a catalytic material, recombination will occur at the surface so that in the example of Fig. 21-1*a* the gas close to the surface consists of O_2 and N_2 only. Interdiffusion of the molecules and atoms will then determine the local composition within the boundary layer, a process which will be considered in the following chapters. If the wall material is completely noncatalytic, the composition of the gas will be uniform within the boundary layer and the same as in the free stream.

The mass-transfer process just considered is characterized by the fact that the wall does not participate in the chemical process except as a catalyst; consequently no mass flow leaves or enters the wall surface, and $v_w = 0$. There are other mass-transfer processes in which a finite mass flow at the wall surface exists, a simple example being shown in Fig. 21-1*b*. Air moves over a surface from which water evaporates. A finite mass flow of water in the form of vapor leaves the wall surface, and the velocity v_w has a finite value. In condensation of water vapor out of an airstream at a surface, on the other hand, the velocity v_w is negative. An example of a mass-transfer process with chemical reactions and with a finite mass flux leaving the surface is shown in Fig. 21-1*c*. In contrast to Fig. 21-1*a*, the wall is assumed to be of carbon material; consequently at a sufficiently high temperature the oxygen in the boundary layer will combine with carbon to form CO or CO_2, and the consumption of carbon creates a mass flow leaving the surface.

The following chapters will discuss processes described by the set of boundary-layer equations (21-1) to (21-4) considering at first a fluid with constant properties in forced laminar flow and in the absence of chemical reactions. Following this, an example including chemical reactions will be discussed. Turbulent flow of a multicomponent fluid with constant properties will be taken up next, and finally gas mixtures with variable properties and with chemical reactions will be considered.

22
Forced Convection

22-1 LAMINAR FLOW, CONSTANT-PROPERTY FLUID

Without chemical reactions In this section heat and mass transfer will be analyzed for two-dimensional boundary-layer flow to a two-component fluid with constant properties. This is interpreted to mean that variations in the mixture properties as they appear in Eqs. (21-1) to (21-5) can be neglected and that, in addition, the interdiffusional convection appearing in Eq. (20-15) is small compared to the Fourier conduction term in this equation. Coupled effects like thermal diffusion will also be neglected, and it is assumed that body forces and chemical reactions are absent or negligible. A situation with no body forces is referred to as *forced convection.* If the velocities are sufficiently small, the kinetic energy can be neglected compared to the differences in enthalpy pertinent to the problem. The continuity and momentum equations for this situation are Eqs. (7-1) to (7-3) in Part II. In the energy equation (21-4), the difference between static and total enthalpy vanishes, and it is convenient to introduce the temperature t by the equation $i = c_p t$. In addition, the terms $\partial p/\partial \tau$ and $\partial/\partial y(\mu u\, \partial u/\partial y)$ are of a small order of magnitude and can be neglected. The body-force term also drops out. This leads to the equation

$$\frac{\partial t}{\partial \tau} + u\,\frac{\partial t}{\partial x} + v\,\frac{\partial t}{\partial y} = \alpha\,\frac{\partial^2 t}{\partial y^2} \tag{22-1}$$

The conservation equation of component 1 takes on the form

$$\frac{\partial w_1}{\partial \tau} + u\,\frac{\partial w_1}{\partial x} + v\,\frac{\partial w_1}{\partial y} = D\,\frac{\partial^2 w_1}{\partial y^2} \tag{22-2}$$

when the mass flux is introduced through Eq. (20-14). The following boundary conditions apply:

$$y = 0 \qquad u = 0 \qquad v = v_w \qquad t = t_w \qquad w_1 = w_{1w} \tag{22-3}$$

$$y = \infty \qquad u = u_s \qquad t = t_s \qquad w_1 = w_{1s} \tag{22-4}$$

The velocity v_w at the surface on which the boundary layer develops may have a finite value or a value zero depending on the specific mass-transfer situation, as has been discussed with the aid of Fig. 21-1.

The temperature t_s and the mass fraction w_{1s} in the stream outside the boundary layer are constant when they are uniform in the approaching stream. The temperature t_w and the mass fraction w_{1w}, however, may be prescribed as varying along the surface or may be determined as variable by a combined heat- and mass-transfer process.

The assumption of a constant-property fluid greatly simplifies the analysis and understanding of the heat- and mass-transfer process. Temperature and composition do not appear in Eqs. (7-1) to (7-3), which means that the flow process and the development of the velocity boundary layer is the same in an isothermal one-component fluid. Any solution obtained in Part II for the proper boundary conditions can immediately be applied to the heat- and mass-transfer situation. The temperature field described by Eq. (22-1) and the mass-fraction field described by Eq. (22-2) depend on the flow field because they contain the velocity components u and v. They are, however, independent of each other. This means that the temperature field in a combined heat- and mass-transfer process is exactly the same as the temperature field in a one-component fluid subject to heat transfer with proper boundary conditions. A corresponding statement can also be made for the composition field. A coupling between the two fields can of course occur in the sense that the mass-transfer process may determine the velocity v_w.

In Part II it was established that the temperature field ϑ in a heat-transfer process at forced convection can be represented as function of a Reynolds number, the velocity ratio v_w/u_s, and the Prandtl number of the fluid when represented in a dimensionless form:

$$\vartheta = \frac{t - t_s}{t_w - t_s} = f\left(\mathrm{Re}, \frac{v_w}{u_s}, \mathrm{Pr}\right) = f\left(\frac{u_s L}{\nu}, \frac{v_w}{u_s}, \frac{\nu}{\alpha}\right) \tag{22-5}$$

An inspection of Eqs. (22-1) and (22-2) immediately allows us to establish the shape of the mass-fraction field for the corresponding boundary conditions. The only difference in the two equations is that the temperature t in the first equation is replaced by w_1 in the second equation and, correspondingly, the thermal-diffusion coefficient α is replaced by the mass-diffusion coefficient D. This means that the same replacement in Eq. (22-5) for the temperature field must lead to the equation

$$\varphi_1 = \frac{w_1 - w_{1s}}{w_{1w} - w_{1s}} = f\left(\frac{u_s L}{\nu}, \frac{v_w}{u_s}, \frac{\nu}{D}\right) = f\left(\mathrm{Re}, \frac{v_w}{u_s}, \mathrm{Sc}\right) \tag{22-6}$$

describing the dimensionless mass-fraction field. The functional relationship in the equations describing the two fields is the same. One need only replace ϑ and Pr in a known equation describing the temperature field by φ_1 and Sc to obtain the equation for the mass-fraction field of a comparable situation. Any solution, steady or unsteady, found in Part II or elsewhere in the literature for a heat-transfer situation under the conditions set at the beginning of this section can thus readily be used to obtain the solution for the corresponding mass-transfer situation.

For engineering purposes, one is usually less interested in the shape of the temperature or composition profiles than in the values of the heat transfer or mass transfer at the surface, both values being conventionally described by heat- or mass-transfer coefficients respectively. It is convenient to define a heat-transfer coefficient with the energy flux ϵ_w according to the equation

$$\epsilon_w = h(t_w - t_s) \tag{22-7}$$

It must be kept in mind, however, that this definition is not generally accepted and that equations defining the heat-transfer coefficient with different heat fluxes are used in the mass-transfer literature. The energy flux ϵ_w is given by the first term in Eq. (20-19) because the second term in this equation was assumed to be negligible for a constant-property fluid. Introduction of Eq. (22-5) leads to the relation

$$\epsilon_w = -k\left(\frac{\partial t}{\partial y}\right)_w = -k(t_w - t_s)\left(\frac{\partial \theta}{\partial y}\right)_w$$

The Nusselt number is then

$$\mathrm{Nu} = \frac{hL}{k} = -L\left(\frac{\partial\theta}{\partial y}\right)_w \tag{22-8}$$

In an analogous manner, a mass-transfer coefficient h_M will be defined with the mass flux j_{1i} according to the equation

$$j_{1w} = \rho h_M(w_{1w} - w_{1s}) \tag{22-9}$$

From Eqs. (20-18) and (22-6) it follows that

$$j_{1w} = -\rho D\left(\frac{\partial w_1}{\partial y}\right)_w = -\rho D(w_{1w} - w_{1s})\left(\frac{\partial \varphi_1}{\partial y}\right)_w$$

A dimensionless parameter characterizing the mass-transfer coefficient is the Sherwood number

$$\mathrm{Sh} = \frac{h_M L}{D} = -L\left(\frac{\partial\varphi_1}{\partial y}\right)_w \tag{22-10}$$

From the similarity between the dimensionless temperature and composition fields, it follows that similarity also exists between the two derivatives appearing in Eqs. (22-8) and (22-10). This has the consequence that a relation describing the Sherwood number in a mass-transfer process is obtained from the relation for the Nusselt number in a corresponding heat-transfer process by replacing the Nusselt number by the Sherwood number and the Prandtl number by the Schmidt number:

$$\mathrm{Sh} = f\left(\mathrm{Re}, \frac{v_w}{u_s}, \mathrm{Sc}\right) \tag{22-11}$$

As an example of heat transfer at a surface with finite v_w, Fig. 11-6 presented Nusselt numbers describing local heat-transfer coefficients for steady, two-dimensional, laminar flow over a flat plate for two Prandtl numbers. The diagram can be reinterpreted as presenting local Sherwood Sh numbers for fluids with the two Schmidt numbers Sc of 0.7 and 1.

In many mass-transfer processes with moderate mass-fraction differences, it is found that the temperature profile is little influenced by the velocity v_w and has the same shape as for flow over a solid surface ($v_w = 0$). Figure 22-1 indicates, for instance, that this applies as long as the parameter $(v_w/u_s)\sqrt{\mathrm{Re}_x}$ is smaller than about 0.025. In this case any solutions for local or average heat-transfer coefficients connected with flow over a solid surface can be used to obtain mass-transfer coefficients for corresponding mass-transfer situations. This also applies, of course, to mass-transfer processes with $v_w = 0$, like the one shown in Fig. 21-1*a*.

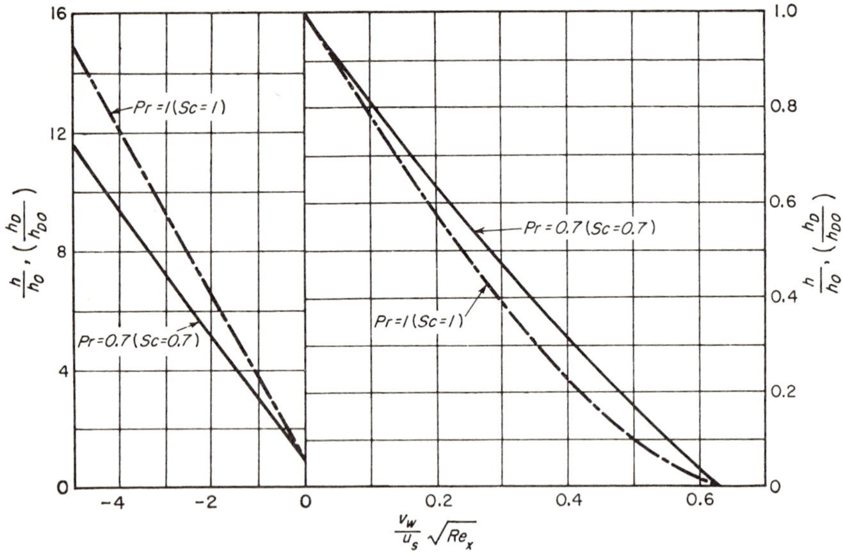

Fig. 22-1 Heat- and mass-transfer parameters for laminar flow over a flat plate. 0 indicates heat- and mass-transfer coefficients with vanishing concentration gradients. Pr belongs to h and Sc to h_d. [*From J. P. Hartnett and E. R. G. Eckert, Trans. ASME,* **79**:*247* (*1957*).]

As an example, the relation (7-70) was derived in Part II to describe the Nusselt number for heat transfer in boundary-layer flow over a flat plate:

$$\mathrm{Nu} = \frac{hx}{k} = 0.332\sqrt{\mathrm{Re}_x}\,\sqrt[3]{\mathrm{Pr}} \qquad \text{for Pr} > 0.6$$

The Sherwood number describing mass transfer from a flat plate into a laminar boundary layer is then

$$\mathrm{Sh} = \frac{h_M x}{D} = 0.332\sqrt{\mathrm{Re}_x}\,\sqrt[3]{\mathrm{Sc}} \qquad \text{for Sc} > 0.6$$

It must be kept in mind that the equation for the Nusselt number holds for a uniform wall temperature. Correspondingly, the equation for the Sherwood number requires a uniform mass fraction w_1 along the plate surface.

For many two-component gas mixtures, the ratio Pr/Sc has values not far from 1. It is therefore of advantage to consider heat and mass transfer in a fluid where this ratio is equal to 1. The ratio of the two parameters is referred to as the *Lewis number*. It is a dimensionless transport property according to the relation

$$\mathrm{Le} = \frac{\mathrm{Pr}}{\mathrm{Sc}} = \frac{D}{\alpha} = \frac{\rho c_p D}{k} \tag{22-12}$$

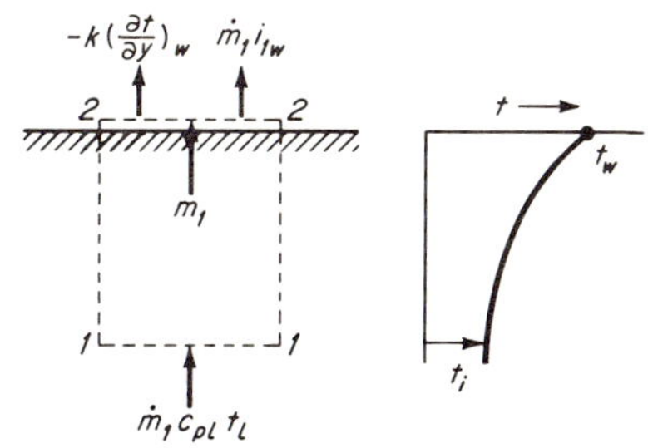

Fig. 22-2 Energy balance for ablation cooling.

Equation (22-11) shows that for a gas with Le = 1, the Sherwood number is simply equal to the Nusselt number:

$$\frac{hL}{k} = \frac{h_M L}{D}$$

or

$$h_M = \frac{h}{\rho c_p} \tag{22-13}$$

This relation, utilized by Lewis[1] in 1927, is often referred to as the *Lewis relation.*

Example 22-1 The skin of high-speed vehicles is often protected against destruction by frictional heat in the atmosphere through ablation cooling. The skin is covered by a material, e.g., a plastic, which sublimes under the influence of the heat transfer through the boundary layer to the skin. This process will be investigated in a simplified manner considering the gas mixture of the air with the gases generated by the ablation process as a constant-property two-component fluid. An energy balance will be performed on a control volume indicated by dashed lines in Fig. 22-2. The surface of the ablating material recedes in time, and it is assumed that the control volume moves with the receding surface. Accordingly, a mass flux $\dot{m}_1$ per unit time and area occurs through both control surfaces 1-1 and 2-2, as indicated in the figure. The ablating material originally has the temperature t_i. With the start of the heating process, the temperature of the solid will at first be raised close to the surface only. This temperature increase will penetrate into the interior of the material as time proceeds, and finally a quasi-steady temperature field will be established of the form sketched in Fig. 22-2 when the heating rate is constant in time. The location of the control surface 1-1, may be selected such that the temperature at this surface remains at the original temperature t_i. When the assumption is made that no energy transport occurs in the solid material parallel to its surface, an energy balance on the material moving through the control volume results in

$$-k\left(\frac{\partial t}{\partial y}\right)_w + \dot{m}_1 i_{1w} = \dot{m}_1 c_{p1} t_i$$

where 1 = ablating material
w = conditions at wall
c_{p1} = specific heat of solid ablating material at temperature t_i

[1] W. K. Lewis, *Trans. AIChE,* **20**:9 (1927).

This equation follows from Eq. (20-13) when a heat-conduction term is added. The first term on the left side of the equation can be expressed by a heat-transfer coefficient h. The equation thus takes the form

$$h(t_s - t_w) = \dot{m}_1(i_{1w} - c_{p1}t_i)$$

The mass flux $\dot{m}_1$ is given for a semipermeable wall by Eq. (20-9) and can again be expressed by a mass-transfer coefficient h_M

$$\dot{m}_1 = \frac{j_{1w}}{1 - w_{1w}} = \frac{h_M}{1 - w_{1w}}(w_{1w} - w_{1s})$$

Combining the last two equations results in the relation

$$h(t_s - t_w) = h_M \frac{w_{1w}}{1 - w_{1w}}(i_{1w} - c_{p1}t_i)$$

For a gas mixture with a Lewis number of 1, the equation simplifies to

$$\rho c_p(t_s - t_w) = \frac{w_{1w}}{1 - w_{1w}}(i_{1w} - c_{p1}t_i)$$

At a given pressure in the airstream along the surface, the enthalpy i_{1w} of the ablated gas is a function of the wall-surface temperature t_w. In the same way there will be a thermodynamic relation connecting the partial pressure (and at a given total pressure the mass fraction w_{1w}) with the wall-surface temperature t_w. The only unknown term in the above equation is then the wall-surface temperature t_w. In order to calculate the ablation rate $\dot{m}_1$, one needs a relation for the mass-transfer coefficient h_M. Figure 22-1 shows that for laminar boundary-layer flow over a flat plate the mass-transfer coefficient decreases almost linearly with the blowing parameter. One thus obtains the following approximation describing the mass-transfer coefficient:

$$h_M = h_{M0} - 0.7v_w = h_{M0} - 0.7\frac{\dot{m}_1}{\rho}$$

The term h_{M0} indicates the mass-transfer coefficient for $v_w = 0$. Introducing this value into the above equation for the mass flow rate results in the equation

$$\dot{m}_1 = \left(h_{M0} - 0.7\frac{\dot{m}_1}{\rho}\right)\frac{w_{1w}}{1 - w_{1w}}$$

which can be solved for the ablation rate $\dot{m}_1$. For laminar boundary-layer flow over a flat plate, for instance, the mass-transfer coefficient h_{M0} is obtained from the relation

$$\frac{h_{M0}x}{D} = 0.332\sqrt{\text{Re}_x}\sqrt[3]{\text{Pr}}$$

keeping in mind that the Lewis number was assumed to have the value 1.

With chemical reactions Analyzing a combined heat- and mass-transfer process in a multicomponent fluid with chemical reactions in the boundary layer can best be done for a specific situation. A constant-property fluid flows with a steady two-dimensional laminar boundary layer over a flat plate. The process is subject to the restrictions in this

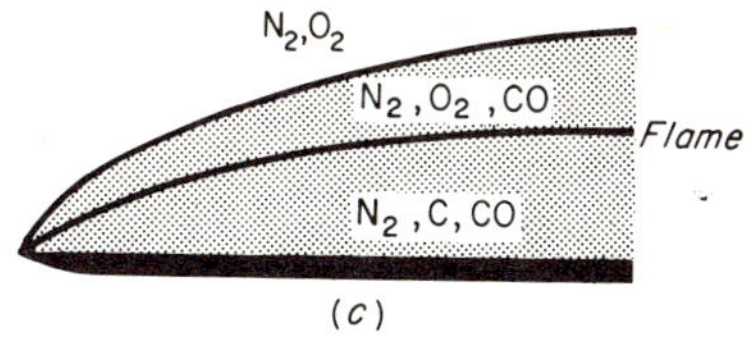

Fig. 22-3 Combustion of a carbon surface in a laminar airstream.

section. Figure 22-3 illustrates the process under consideration. Air may flow over a solid carbon surface and the temperature conditions may be such that the oxygen in the airstream reacts with the carbon and forms carbon monoxide, CO. It will be assumed that the chemical reaction is very fast compared to the diffusion processes. Under this condition combustion occurs within a thin sheet which will be simplified as a flame surface. The following gas components are considered to occur within the boundary layer: N_2, O_2, C, CO. This certainly simplifies the actual situation, in which more atomic species and other molecular compounds would be present. It should, however, act as a first approximation of the actual situation. Another approach to the problem, which is restricted to gases but includes effects which are neglected here, will be discussed in Sec. 22-2. The temperature of the carbon surface is influenced by heat conduction in the solid material and by heat radiation. This temperature, together with the pressure within the boundary layer, will determine the gas composition at the surface and the ablation rate. For the present analysis, however, it is advantageous to consider either the ablation rate or the velocity v_w as the independent variable.

In a constant-property fluid, the flow process which determines the velocity field in the boundary layer is completely independent of the heat- and mass-transfer situation and can be calculated by the methods discussed in Part II. The heat- and mass-transfer processes are also independent of each other. We shall start our consideration with the mass transfer. Figure 22-4 presents a qualitative picture of the variation of the various components to be expected within the boundary layer.

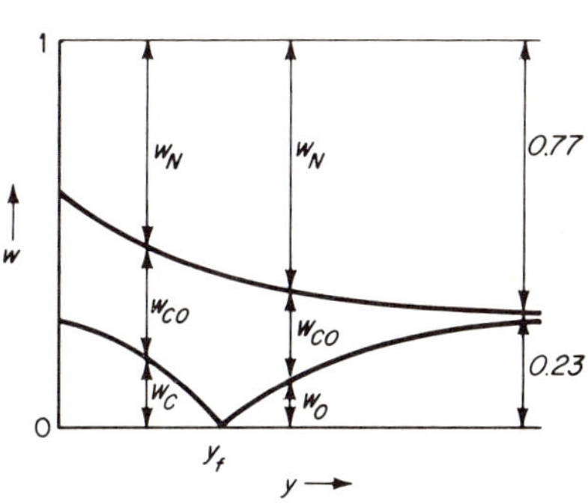

Fig. 22-4 Sketch of mass-fraction profiles for the combustion process in Fig. 22-3.

Since we assume that the reaction equilibrium is very close to complete combustion, all the carbon and oxygen is converted at the flame front by a stoichiometric chemical reaction to CO, no oxygen is found in the inner part of the boundary layer, and no carbon in the outer part of the boundary layer. The air outside the boundary layer consists of the mass fractions $w_{N,s} = 0.77$ and $w_{O,s} = 0.23$ of N_2 and O_2, respectively. The shape of the mass-fraction profiles within the boundary layer can be calculated from the diffusion equations. Three such equations will be used of the form

$$u \frac{\partial w_m}{\partial x} + v \frac{\partial w_m}{\partial y} = D \frac{\partial^2 w_m}{\partial y^2} \qquad m = \text{N, O, C} \tag{22-14}$$

with the index m denoting the three chemical species N, O, and C, respectively. All the diffusion coefficients have the same value according to our constant-property assumption. In addition, the relation

$$w_N + w_O + w_C + w_{CO} = 1 \tag{22-15}$$

holds with the boundary conditions

$$\begin{array}{llll} y = 0 & w_N = w_{N,w} & w_O = 0 & w_C = w_{C,w} \\ y = y_f & & w_O = w_C = 0 & \\ y \to \infty & w_N = 0.77 & w_O = 0.23 & w_C = 0 \end{array} \tag{22-16}$$

The mass fraction of the nitrogen must pass smoothly through the flame front, as indicated in Fig. 22-4, because the nitrogen does not participate in the chemical reaction. The boundary conditions stated above would be sufficient to calculate the various mass-fraction profiles were it not for the fact that the mass fractions of nitrogen and carbon are not known at the wall surface. In addition, the location y_f of the flame front is not known. This means that three additional relations must be supplied. Two of them are obtained from the fact that the carbon surface is impermeable to nitrogen and carbon monoxide:

$$\begin{aligned} \dot{m}_{N,w} &= -\rho D \left(\frac{\partial w_N}{\partial y}\right)_w + \rho w_N v_w = 0 \\ \dot{m}_{CO,w} &= -\rho D \left(\frac{\partial w_{CO}}{\partial y}\right)_w + \rho w_{CO} v_w = 0 \end{aligned} \tag{22-17}$$

Since in our system no diffusion equation for CO has been used, it is of advantage to replace the second equation above by a mass balance for the carbon. Adding the last two equations and considering Eq. (22-15) results in

$$\rho D \left(\frac{\partial w_C}{\partial y}\right)_w + (1 - w_C) v_w = 0 \tag{22-18}$$

A third relation is obtained from the consideration that carbon and oxygen have to arrive at the flame front in the mass ratio required for stoichiometric combustion:

$$\left(\frac{\partial w_C}{\partial y}\right)_f = \frac{4}{3}\left(\frac{\partial w_O}{\partial y}\right)_f \tag{22-19}$$

No convective transport appears in this equation because the mass fraction of both elements is zero at the flame front.

In order to proceed with the solution, we shall assume that the boundary layer is caused by air flow at constant velocity over a plane carbon surface. The boundary layers under this condition are self-similar, as discussed in part II, and the partial differential equations (22-14) can be transformed into total differential equations by introduction of the following dimensionless parameters (see page 284):

$$\eta = y\sqrt{\frac{u_s}{2\nu x}} \qquad f_w = -v_w\sqrt{\frac{x}{2u_s\nu}} \tag{22-20}$$

In addition, we shall describe the mass-fraction profiles by the dimensionless parameters

$$\varphi_N = \frac{w_N - w_{N,w}}{w_{N,s} - w_{N,w}} \qquad \varphi_O = \frac{w_O}{w_{O,s}} \qquad \varphi_C = \frac{w_{C,w} - w_C}{w_{C,w}} \tag{22-21}$$

The boundary-layer equations (22-14) transform in this way to equations of the form

$$\varphi_m'' + \mathrm{Sc}\, f\varphi_m' = 0 \qquad m = \mathrm{N, O, C} \tag{22-22}$$

with the boundary conditions

$$\begin{array}{llll} \eta = 0 & \varphi_N = 0 & \varphi_O = 0 & \varphi_C = 0 \\ \eta = \eta_f & & \varphi_O = 0 & \varphi_C = 1 \\ \eta \to \infty & \varphi_N = 1 & \varphi_O = 1 & \varphi_C = 1 \end{array} \tag{22-23}$$

A prime again indicates differentiation with respect to η. Equations (22-22) can now be readily solved with the boundary conditions stated in (22-23), but even this is unnecessary because the form of Eqs. (22-22) is the same as the form of the equation analyzed in Sec. 11-3. The similarity between heat and mass transfer allows us, furthermore, to interpret the temperature profiles in Fig. 11-5 as mass-fraction profiles. We can therefore select the proper mass-fraction profile for the prescribed Schmidt number Sc and blowing parameter f_w from Fig. 11-5 and modify by a change of scale corresponding pieces of this curve such that the boundary conditions (22-23) are fulfilled. This is sketched in Fig. 22-5 with an assumed value of η_f.

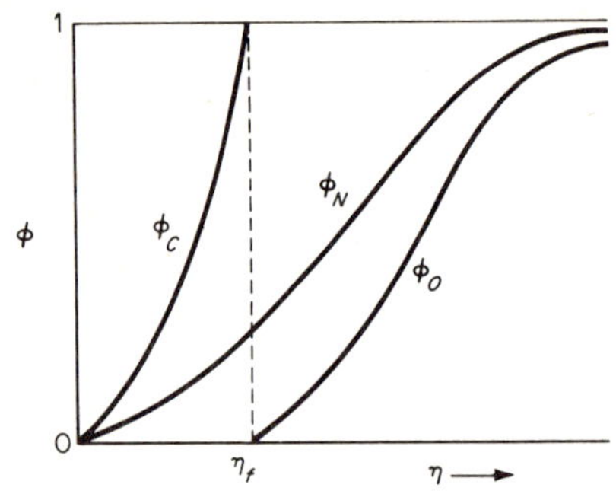

Fig. 22-5 Sketch of profiles of dimensionless mass fraction in the combustion process in Fig. 22-3.

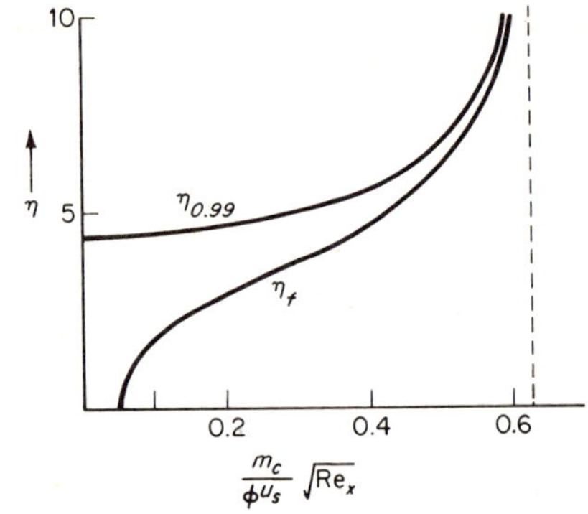

Fig. 22-6 Dimensionless distance η_f of the flame front from the carbon surface and dimensionless boundary-layer thickness $\eta_{0.99}$ as a function of the rate of combustion for the process in Fig. 22-3.

Equations (22-17) to (22-19), transformed to the new variables, result in

$$(w_{N,s} - w_{N,w})\varphi'_{N,w} = -2w_{N,w}f_w \text{ Sc} \tag{22-24}$$

$$w_{C,w}\varphi'_{C,w} = 2(1 - w_{C,w})f_w \text{ Sc} \tag{22-25}$$

$$w_{C,w}\varphi'_{C,f} = \tfrac{4}{3}w_{O,s}\varphi'_{O,f} \tag{22-26}$$

The derivatives of the dimensionless mass-fraction parameters φ appearing in these equations can be obtained from Fig. 11-5. The mass fraction $w_{C,w}$ can now be calculated from Eq. (22-25), and it can be checked whether Eq. (22-26) is satisfied. If not, the position η_f of the flame front must be changed until it is satisfied. This determines the correct value for $w_{C,w}$. The mass fraction $w_{N,w}$ of nitrogen at the carbon surface is then obtained from Eq. (22-24). With this information we can now actually calculate the mass-fraction profiles as sketched in Fig. 22-4. Another iteration may be necessary if the calculated mass fraction of the carbon at the wall surface does not agree with the value prescribed by the thermodynamic relation connecting the mass fraction of gaseous carbon in a sublimation process with the temperature of the carbon surface and the pressure in the boundary layer.

Figure 22-6 presents the results of a numerical[1] analysis for the combustion process considered in this section. The dimensionless boundary-

[1] E. R. G. Eckert and J. P. Hartnett. *Z. Angew. Math. Phys.*, **11**(b):259 (1958); J. P. Hartnett and E. R. G. Eckert, *1958 Heat Transfer Fluid Mech. Inst., Stanford, California*, p. 54.

layer thickness $\eta_{0.99}$, defined as that wall distance at which the velocity u is 99 percent of the free-stream velocity u_s, and the dimensionless distance η_f of the flame front are plotted over the ablation parameter f_w. The calculation results in $\eta_f = 0$ at $f_w = 0.05$. Only at f_w values larger than 0.05 is the flame lifted off the carbon surface. At values smaller than 0.05 the oxygen mass fraction at the surface will have finite values. Increasing values of f_w mean increasing surface temperatures according to the thermodynamic relation $w_{C,w} = f(t_w,p)$.

The temperature condition within the boundary layer can now be calculated. An equation of the form (22-22) with φ replaced by ϑ and Sc by Pr again describes the temperature profile within the boundary layer on the flat plate. We must write this equation separately for the temperature in the inner and in the outer layer, corresponding to the dimensionless parameters

$$\vartheta_i = \frac{t_i - t_w}{t_f - t_w} \qquad \vartheta_o = \frac{t_0 - t_f}{t_s - t_f} \tag{22-27}$$

and with the boundary conditions

$$\begin{array}{lll} \eta = 0 & \vartheta_i = 0 & \\ \eta = f & \vartheta_i = 1 & \vartheta_a = 0 \\ \eta \rightarrow \infty & \vartheta_a = 1 & \end{array} \tag{22-28}$$

The profiles of the dimensionless parameters ϑ are obtained piecemeal from Fig. 11-5 with the proper value of the Prandtl number and of the blowing parameter f_w. These profiles are shown in Fig. 22-7. To calculate the temperatures themselves we write an energy balance for a thin layer containing the flame front which describes the fact that the

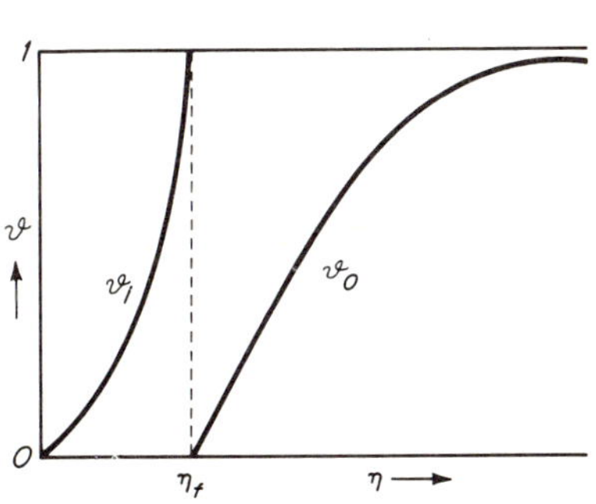

Fig. 22-7 Sketch of dimensionless temperature profiles for the combustion process in Fig. 22-3.

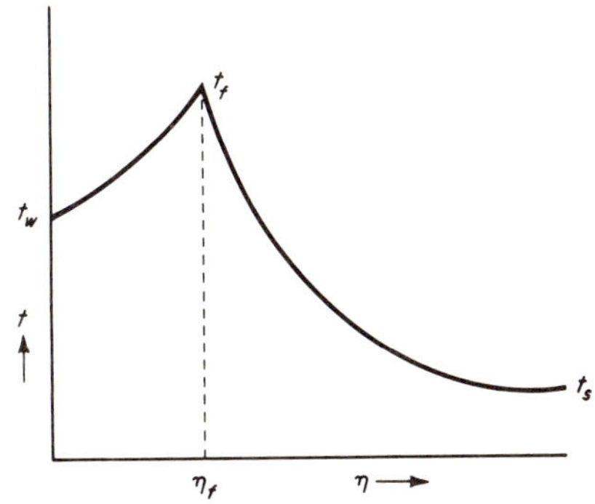

Fig. 22-8 Sketch of temperature profile within the boundary layer for the combustion process in Fig. 22-3.

heat conduction from the inner layer to the flame front plus the heat generated by the combustion process must be equal to the heat conducted away in the outer layer:

$$-k\left(\frac{\partial t_i}{\partial y}\right)_f - \rho D\left(\frac{\partial w_C}{\partial y}\right)_f H = -k\left(\frac{\partial t_a}{\partial y}\right)_f \tag{22-29}$$

where H is the heat of combustion of carbon into CO per unit mass of carbon. The convective enthalpy transport drops out of this equation because the temperature is equal on both sides of the flame front. In the dimensionless parameters, the above equation is

$$(t_f - t_w)\vartheta'_{if} - (t_s - t_f)\vartheta'_{af} = \frac{\text{Le}}{c_p} H w_{C,w}\varphi'_{C,f} \tag{22-30}$$

The equation can be solved for the only unknown value, the flame temperature t_f, and the temperature profile can now be obtained as sketched in Fig. 22-8. The energy flux ϵ_w at the wall surface finally is obtained from the equation

$$\epsilon_w = -k\left(\frac{\partial t_i}{\partial y}\right)_w = -k(t_f - t_w)\vartheta'_{iw} \tag{22-31}$$

It is of interest to calculate the ratio of the heat flux ϵ_w to the heat flux ϵ_{w0} which occurs for the same flow conditions and for the same wall and stream temperatures t_w and t_s without combustion, assuming a gas mixture with a Lewis number equal to 1. A straightforward calculation leads to the result

$$\text{Le} = 1 \qquad \frac{\epsilon_w}{\epsilon_{w0}} = \frac{t_s - t_w + \frac{3}{4}(H/c_p)w_{O,s}}{t_s - t_w}$$

The equation can also be expressed in enthalpies when the numerator and denominator are multiplied by the specific heat c_p:

$$\frac{\epsilon_w}{\epsilon_{w0}} = \frac{i_s^* - i_w}{i_s - i_w} \tag{22-32}$$

where $i_s^* = c_p t_s + \frac{3}{4} H w_{O,s}$ = enthalpy of air outside boundary layer including the chemical energy converted by combustion of the oxygen in the air to carbon monoxide

$i_s = c_p t_s$ = enthalpy of the air outside the boundary layer without this chemical energy

$i_w = c_p t_w$ = enthalpy of the gas at the carbon surface

The position of the flame front does not enter into this relation at all, which indicates that the relation (22-32) may have a more general validity than is evident right now. That it does will become clear in a later section.

22-2 TURBULENT FLOW, CONSTANT-PROPERTY FLUID

The turbulent boundary-layer equations were developed in Sec. 8-1. It was found that the continuity equation (8-5) has the same form as the corresponding equation for laminar flow when the velocity components in this equation are interpreted as time-averaged values, averaged over a time sufficiently long for the turbulent fluctuations to be suppressed. The momentum equation (8-6) is enlarged to include the action of turbulent shear as well as of laminar shear. The component of the shear term which appears in the boundary-layer equation has the form

$$p_{xy} = p_{yx} = \mu \frac{\partial \bar{u}}{\partial y} - \rho \overline{u'v'} = \rho(\nu + \epsilon_m) \frac{\partial \bar{u}}{\partial y} \tag{22-33}$$

with the term $\rho u'v'$ describing the turbulent shear. In order to proceed with a turbulent boundary-layer analysis, a turbulent diffusivity ϵ_m was introduced, and the turbulent shear was written as $\rho\epsilon_m \, \partial\bar{u}/\partial y$. Experimental information must be used to relate the turbulent diffusivity ϵ_m to the time-averaged velocity, to the wall distance, and to other parameters involved. Similarly, it was found in Sec. 8-2 [Eq. (8-39)] that now the energy flux ϵ_y also consists of two terms, a term describing Fourier's conduction law using the time-averaged temperature gradient and a term describing enthalpy transport by the turbulent-mixing movements. This lead to the equation

$$\epsilon_y = -k \frac{\partial \bar{t}}{\partial y} + \rho c_p \overline{t'v'} = -\rho c_p \left(\frac{\nu}{\mathrm{Pr}} + \frac{\epsilon_m}{\mathrm{Pr}_t}\right) \frac{\partial \bar{t}}{\partial y} \tag{22-34}$$

An analogous treatment of the component continuity equation leads to the following equation for the mass flow j_m of component m:

$$j_{my} = -\rho D \frac{\partial \overline{w_m}}{\partial y} + \rho \overline{w_m' v'} = -\rho \left(\frac{\nu}{\mathrm{Sc}} + \frac{\epsilon_m}{\mathrm{Sc}_t}\right) \frac{\partial \overline{w_m}}{\partial y} \tag{22-35}$$

The turbulent Prandtl number Pr_t in Eq. (22-34) expresses the ratio of the turbulent diffusivity ϵ_m for momentum to the turbulent diffusivity ϵ_H for heat. The turbulent Schmidt number Sc_t in Eq. (22-35) is defined as the ratio of the turbulent diffusivity ϵ_m for momentum to the turbulent diffusivity ϵ_M describing the mass transfer process. In addition, a turbulent Lewis number Le_t is defined as the ratio of the turbulent Prandtl number to the turbulent Schmidt number:

$$\mathrm{Pr}_t = \frac{\epsilon_m}{\epsilon_H} \qquad \mathrm{Sc}_t = \frac{\epsilon_m}{\epsilon_M} \qquad \mathrm{Le}_t = \frac{\mathrm{Pr}_t}{\mathrm{Sc}_t} \tag{22-36}$$

It was mentioned in Sec. 8-2 that in heat-transfer analysis the turbulent Prandtl number can often be considered as a constant but that newer experiments have indicated that the turbulent Prandtl number in boundary-layer or tube flow varies with wall distance and this variation also depends on the laminar Prandtl number. Available experimental evidence, however, as well as theoretical considerations, point to the fact that the turbulent Lewis number has the value 1. It can then readily be seen, by introducing Eqs. (22-34) and (22-35) into the boundary-layer energy and component-mass equations, respectively, that the equations become similar when, in addition, the laminar Lewis number is equal to 1. Therefore, mass-transfer relations obtained from heat-transfer coefficients through the Lewis relation expressed by Eq. (22-13) approximate conditions in turbulent flow quite well even when the laminar Lewis number deviates from 1, because the turbulent contribution to the energy and to the mass flux is generally more important than the laminar contribution. When this approximation is not satisfactory, the rule obtained from the similarity of the two boundary-layer equations still applies: a heat-transfer relation for turbulent flow can be turned into a mass-transfer relation for an analogous mass-transfer situation by replacing the Nusselt number with the Sherwood number and the laminar Prandtl number with the laminar Schmidt number.

22-3 GASES WITH VARIABLE PROPERTIES

The discussion in the preceding sections is most suitable for liquids in which the property variation is usually small, for which, however, the Prandtl and the Lewis numbers involved can have values far removed from 1. The opposite situation often exists in gases, especially when chemical reactions are involved. The temperature differences existing within the boundary layer are then quite large, and this also causes a large variation in the properties. On the other hand, it is found that the Prandtl number and the Lewis numbers in gas mixtures have values which do not vary as much and are comparatively close to the value 1. The Prandtl number for air, for instance, varies between 0.74 and 0.6 in the temperature range from 20 to 10.000 C. Lewis numbers for gas mixtures have, in general, values between 1 and 1.4 unless lightweight gases like hydrogen are involved. The procedure discussed in this section is adapted to heat- and mass-transfer processes in gases by allowing variation of the thermodynamic and transport properties involved but considering an ideal-gas mixture with a Prandtl number and Lewis number equal to 1.

In this section it is of advantage to use the enthalpy i of the gas mixture in the analysis instead of the temperature and accordingly to express the conduction term in the energy equation (21-6) by an enthalpy gradient instead of a temperature gradient. In a boundary layer, the pressure is constant on any normal to the surface, and the mixture enthalpy i changes throughout the boundary layer with the temperature and mass fraction of the components. Accordingly, in the relation

$$i = i(t, w_m, p)$$

only the first two terms are considered variable. In addition, the mixture enthalpy of a gas is equal to the sum of the enthalpies of the components involved:

$$i = \Sigma i_m w_m$$

The enthalpies i_m will be defined by the equation

$$i_m = \int_0^T c_{pm}\, dT + i_{0m}$$

with i_{0m} denoting the heat of formation of this component from the elements (for instance $CO_2 = C + O_2$), where the enthalpy of the elements contains an arbitrary constant. The total differential of the enthalpy i can then be written

$$di = \frac{\partial i}{\partial t}\, dt + \sum \frac{\partial i}{\partial w_m}\, dw_m = c_{pf}\, dt + \sum i_m\, dw_m$$

The term $\partial i / \partial t$ is the change of the enthalpy of the mixture per unit temperature variation under the condition that the composition stays constant. It is therefore denoted by c_{pf} and called the *frozen specific heat at constant pressure*. Replacing the temperature gradient in the heat-conduction term of Eq. (21-6) with the enthalpy gradient results in the equation

$$\epsilon_y = -\frac{k_f}{c_{pf}} \frac{\partial i}{\partial y} - \rho \Sigma \left[D_m \left(1 - \frac{1}{\mathrm{Le}_{mf}} \right) i_m \frac{\partial w_m}{\partial y} \right] \qquad (22\text{-}37)$$

Where Le_{mf} denotes the Lewis number for component m in the mixture under frozen conditions. In the next step, the static enthalpy i in Eq. (22-37) can be converted to the total enthalpy i° utilizing the relation

$$di^\circ = di + u\, du$$

Introducing, in addition, the frozen Prandtl number Pr_f leads to the final relation for the energy flux

$$\epsilon_y = -\frac{\mu}{\mathrm{Pr}_f} \frac{\partial i^\circ}{\partial y} + \frac{\mu}{\mathrm{Pr}_f} u \frac{\partial u}{\partial y} - \rho \sum \left[D_m \left(1 - \frac{1}{\mathrm{Le}_{mf}} \right) i_m \frac{\partial w_m}{\partial y} \right]$$

Introduction of this equation into Eq. (21-4) leads to the following form of the energy equation for a two-dimensional boundary layer:

$$\rho \frac{\partial i^\circ}{\partial \tau} + \rho u \frac{\partial i^\circ}{\partial x} + \rho v \frac{\partial i^\circ}{\partial y} - \frac{\partial p}{\partial \tau} = \frac{\partial}{\partial y} \left(\frac{\mu}{\mathrm{Pr}_f} \frac{\partial i^\circ}{\partial y} \right) - \frac{\partial}{\partial y} \left[\mu \left(1 - \frac{1}{\mathrm{Pr}_f} \right) \mu \frac{\partial u}{\partial y} \right]$$
$$+ \frac{\partial}{\partial y} \left\{ \rho \sum \left[D_m \left(1 - \frac{1}{\mathrm{Le}_{mf}} \right) i_m \frac{\partial w_m}{\partial y} \right] \right\} + \rho u g_x \qquad (22\text{-}38)$$

It was pointed out that the enthalpy in this equation must be interpreted as including heats of formation.

The equation simplifies radically when a gas is considered for which the frozen Prandtl and Lewis numbers all have the value 1. In addition, we shall restrict ourself to forced convection and steady flow and accordingly drop the corresponding terms in the energy equation. This leads to the equation

$$\rho u \frac{\partial i^\circ}{\partial x} + \rho v \frac{\partial i^\circ}{\partial y} = \frac{\partial}{\partial y}\left(\mu \frac{\partial i^\circ}{\partial y}\right) \tag{22-39}$$

In many mass-transfer processes involving gases, it is found that the properties involved vary much more as a consequence of the large temperature differences than as a consequence of the local variations in composition. Neglecting therefore the variation with composition, one can observe that the temperature field is fully described by the continuity and momentum equations (6-43) and (6-44) and by the energy equation (22-39). This system of equations contains no parameter depending on the local composition and can therefore be solved independently of the mass-transfer equations. The heat-transfer coefficient is also independent of the local variation of composition if it is defined using a difference in total enthalpy as the driving potential instead of temperature difference according to the equation

$$\epsilon_{y,w} = -\mu \left(\frac{\partial i^\circ}{\partial y}\right)_w = h_i(i_w - i_s^\circ) \tag{22-40}$$

The fact that the enthalpies contain arbitrary constants does not influence the heat-transfer coefficient since the constants drop out when the ratio $di^\circ/(i_w - i_s^\circ)$ is formed. Heat-transfer coefficients themselves can be obtained from Sec. 10-2 or 10-3. It is, of course, advantageous to introduce the properties into the equations describing the heat-transfer coefficient at the mean composition of the gas mixture.

Equations describing the mass-fraction fields of the various components are of the form of Eq. (21-5). It can be observed that the mass-source term ψ_m in these equations is the one term which prevents similarity between these equations and the energy equation (22-39). This term can be removed[1] by introducing a species mass fraction $\tilde{w}_m$. The species mass fraction is defined as the mass fraction of a chemical species regardless of the fact that chemical compounds in this species occur. In the situation sketched in Fig. 22-3, for instance, the species

[1] The following discussion follows essentially L. Lees, Convective Heat Transfer with Mass Addition and Chemical Reactions, *Combustion and Propulsion, Third AGARD Colloq.* pp. 451–498; reprinted in J. P. Hartnett (ed.), "Recent Advances in Heat and Mass Transfer," pp. 161–207, McGraw-Hill, New York, 1961.

mass fraction $\bar{w}_O$ is connected with the component mass fractions w_{O_2} and w_{CO} by the equation

$$\bar{w}_O = w_{O_2} + \tfrac{4}{7}w_{CO}$$

No chemical species is created or destroyed in a chemical process, nuclear processes excluded. The component mass-conservation equations therefore assume the form

$$\rho u \frac{\partial \bar{w}_m}{\partial x} + \rho v \frac{\partial \bar{w}_m}{\partial y} = \frac{\partial}{\partial y}\left(\mu \frac{\partial \bar{w}_m}{\partial y}\right) \tag{22-41}$$

which is now similar to the energy equation (22-39), so that the analogy between mass-fraction fields and the total enthalpy field is established. A mass-transfer coefficient h_M is defined by equation

$$j_{mw} = -\mu \left(\frac{\partial \bar{w}_m}{\partial y}\right)_w = h_M(\bar{w}_{mw} - \bar{w}_{m0}) \tag{22-42}$$

and the Lewis relation between the heat-transfer coefficient h_i and the mass-transfer coefficient h_M takes the form

$$h_M = h_i \tag{22-43}$$

In the present analysis it is of advantage to describe the transfer coefficients in a dimensionless form by Stanton numbers St instead of Nusselt numbers or Sherwood numbers, according to the definitions

$$\mathrm{St}_i = \frac{h_i}{\rho_s u_s} \qquad \mathrm{St}_M = \frac{h_M}{\rho_s u_s}$$

The Lewis relation in Stanton numbers is

$$\mathrm{St}_M = \mathrm{St}_i$$

The method of solving a combined heat- and mass-transfer process with chemical reactions can now be demonstrated by an example which again considers an airstream over a flat carbon plate where a steady two-dimensional laminar boundary layer exists and the temperature is high enough to cause chemical reactions. In contrast to the treatment of the same problem in the preceding section, a considerable number of chemical conversions will be admitted, according to the scheme

$$\begin{array}{lll} 2O \rightleftarrows O_2 & C + O \rightleftarrows CO & C + \tfrac{1}{2}O_2 \rightleftarrows CO \\ 2N \rightleftarrows N_2 & C + N \rightleftarrows CN & C + \tfrac{1}{2}N_2 \rightleftarrows CN \end{array}$$

It will also be assumed that the free stream of air has a temperature sufficiently high for the oxygen and nitrogen to exist in atomic as well as molecular form. This is indicated in Fig. 22-9. All the oxygen may be consumed at the surface, so that only the components CO, CN, and C exist in the gas immediately at the surface. The Prandtl and

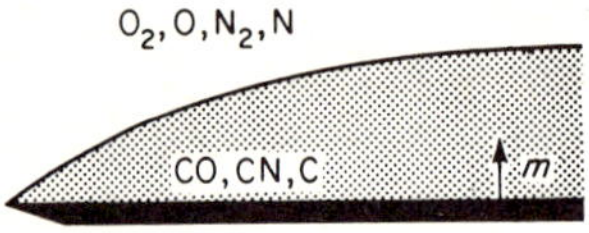

Fig. 22-9 Dissociation and combustion in laminar boundary-layer flow of air over a carbon surface.

Lewis numbers will be approximated by the value 1. The species mass fractions of oxygen, nitrogen, and carbon are described by the following equations:

$$\begin{aligned} \tilde{w}_O &= w_O + w_{O_2} + \tfrac{4}{7} w_{CO} \\ \tilde{w}_N &= w_N + w_{N_2} + \tfrac{7}{13} w_{CN} \\ \tilde{w}_C &= w_C + \tfrac{3}{7} w_{CO} + \tfrac{6}{13} w_{CN} \end{aligned} \tag{22-44}$$

A relation for the species mass fraction of oxygen at the surface is obtained from the consideration that the surface is impermeable to oxygen:

$$m_{\tilde{O},w} = -\rho D \left(\frac{\partial \tilde{w}_O}{\partial y} \right)_w + \dot{m} \tilde{w}_{O,w} = 0$$

Introducing a mass-transfer coefficient and considering the fact that this coefficient is equal to the heat-transfer coefficient h_i through the Lewis relation transforms this equation to

$$h_i(\tilde{w}_{O,w} + \tilde{w}_{O,s}) + m\tilde{w}_{O,w} = 0 \tag{22-45}$$

It was shown in Sec. 11-3 that the heat-transfer coefficient decreases approximately linearly with the mass flow $\dot{m}$ from the surface into the boundary layer for laminar boundary-layer flow over a flat plate. It can therefore be described by the relation

$$h_i = h_{i0} - 0.7\dot{m}$$

in which h_{i0} denotes the heat-transfer coefficient based on enthalpies at a mass flow rate $\dot{m} = 0$. Dividing the above equation by $\dot{m}$ leads to the relation

$$\frac{h_i}{\dot{m}} = \frac{h_{i0}}{\dot{m}} - 0.7 = B \tag{22-46}$$

Introducing this parameter B into Eq. (22-45) results in

$$(B + 1)\tilde{w}_{O,w} - B\tilde{w}_{O,s} = 0$$

The species mass fractions are now replaced by the component mass fractions, with the result

$$\tfrac{4}{7}(B + 1)w_{CO,w} = B(w_{O,s} + w_{O_2,s})$$

The mass fraction of CO at the wall surface is thus obtained as

$$w_{CO,w} = \frac{7B}{4(B+1)}(w_{O,s} + w_{O_2,s}) = \frac{7B}{4(B+1)} 0.23 \tag{22-47}$$

In the same way the mass fraction of CN at the wall surface is obtained from the consideration that the wall is also impermeable to nitrogen:

$$w_{\mathrm{CN},w} = \frac{13B}{7(B+1)} 0.77 \tag{22-48}$$

Ablation of carbon from the surface provides the mass flux $\dot{m}$, according to the relation

$$\dot{m}_\mathrm{C} = \dot{m} = -\rho D \left(\frac{\partial w_\mathrm{C}}{\partial y}\right)_w + \dot{m}\tilde{w}_{\mathrm{C},w}$$

Introduction of the heat-transfer coefficients leads to

$$\dot{m} = h_i(\tilde{w}_{\mathrm{C},w} - \tilde{w}_{\mathrm{C},s}) + \dot{m}\tilde{w}_{\mathrm{C},w} \tag{22-49}$$

With the parameter B one obtains

$$\tilde{w}_{\mathrm{C},w} = \frac{1}{B+1}$$

keeping in mind that no carbon exists in the airstream outside the boundary layer. The mass fraction of the gaseous component carbon then is described by the relation

$$\tilde{w}_{\mathrm{C},w} = w_{\mathrm{C},w} - \tfrac{3}{7}w_{\mathrm{CO},w} - \tfrac{6}{13}w_{\mathrm{CN},w} \tag{22-50}$$

In this way the composition of the gas at the wall surface has been determined. The ablation rate of the carbon can also be obtained from Eq. (22-49).

Calculations of this form must be made in the analysis of the ablation cooling process. In this case it is especially important to know the heat flux to the surface as influenced by the ablation process. The heat flux within the boundary layer entering the surface is

$$q_w = \epsilon_w + \dot{m}i_w = h_i(i_w - i_s^\circ) + \dot{m}(w_{\mathrm{CO},w}i_{\mathrm{CO},w} + w_{\mathrm{CN},w}i_{\mathrm{CN},w} + w_{\mathrm{C},w}i_{\mathrm{C},w}) \tag{22-51}$$

All terms in this equation are known when the total enthalpy i_s° in the free stream and, with the surface temperature t_w, the mixture enthalpy $i_w = i_w^\circ$ at the wall surface are prescribed. The first will be known from the conditions in the free stream; the second is determined by heat conduction into the interior of the solid material by the heat release through the evaporation process and by the effects of thermal radiation and must be obtained from an energy balance including the ablating material as well as the boundary layer. It may have been observed that in the preceding analysis no assumption has been made about the rate of the homogeneous chemical reactions within the boundary layer. The heat and mass transfer is therefore the same whether this rate is fast—leading to chemical equilibrium within the boundary layer—or slow—leading to a chemically frozen condition. Only the reaction rate of the heterogeneous reactions at the surface influences the heat and mass transfer at the surface. In the preceding example, the reaction rate of C with O and N was assumed fast. Otherwise finite mass fractions of O and N would exist at the surface.

Correction formulas for Le $\neq$ 1 and Pr $\neq$ 1 It has been mentioned that the Prandtl and Lewis numbers of gas mixtures do not deviate strongly from the value 1 as long as very lightweight gases like hydrogen or helium are excluded. This suggests using relations for an ideal gas with Pr = 1 and Le = 1 as a first approximation and applying corrections for the deviations of the parameters mentioned from the value 1. Couette flow of a two-component gas mixture has been analyzed by Monaghan, Crabtree, and Woods.[1] Zero mass release from either of the two walls bounding the Couette flow was specified. The temperature of the moving wall was assumed sufficiently high for dissociation of the gas to occur. Local chemical equilibrium was postulated. The heat flux from the stationary wall into the fluid could be expressed by the following equation:

$$q_w = \frac{k}{c_p b}\left[(i_w - i_s^\circ) + (1 - \mathrm{Pr})\frac{u_s^2}{2} + (1 - \mathrm{Le})i_D(w_{as} - w_{aw})\right] \qquad (22\text{-}52)$$

where b = distance of two walls bounding Couette flow
w = stationary wall
s = wall moving in its own plane with velocity u_s
i_D = heat of dissociation per unit mass of atomic species
w_{as}, w_{aw} = the mass fractions of atomic species at two wall surfaces

The form of this equation suggests a generalization which will describe conditions in boundary-layer flows

$$\epsilon_w = h_i\left[(i_w - i_s^\circ) + (1 - r_v)\frac{u_s^2}{2} + (1 - r_{\mathrm{ch}})i_D(w_{as} - w_{aw})\right] \qquad (22\text{-}53)$$

The equation defines two recovery factors, r_v, describing the recovery of kinetic energy, and r_{ch}, describing the recovery of chemical energy. The index s now refers to conditions at the outer border of the boundary layer. h_i is the heat-transfer coefficient for a gas with Pr = Le = 1. Without chemical reactions and with zero mass release from the surface, the relations describing the recovery factor for kinetic energy in boundary layers are well known. They are approximated by the equation

$$r_v = \sqrt{\mathrm{Pr}} \qquad (22\text{-}54)$$

[1] R. J. Monaghan, L. F. Crabtree, and B. A. Woods, Features of Hypersonic Heat Transfer, *Advan. Aeron. Sci. Proc. First Intern. Congr. Aeron.* vol. 1, *Sci. Madrid Sept. 8–13, 1958,* pp. 287–313, 1959.

for laminar flow and

$$r_v = \sqrt[3]{\text{Pr}} \qquad (22\text{-}55)$$

for turbulent flow. The effect of dissociation in a laminar air boundary layer at a three-dimensional stagnation point has been studied by Fay, Riddell, and Kemp.[1] The equation

$$r_{\text{ch}} = \text{Le}^{0.52} \qquad (22\text{-}56)$$

was found to approximate the results of this analysis well for chemical equilibrium and the equation

$$r_{\text{ch}} = \text{Le}^{\frac{2}{3}} \qquad (22\text{-}57)$$

for chemically frozen conditions within the boundary layer and complete recombination on a cold catalytic wall.

Conditions for boundary-layer flow over a surface from which mass is released have been studied to a lesser extent. Lees[2] provides the following estimate for the chemical recovery factor: $r_{\text{ch}} = \text{Le}^n$, with n having values between $\frac{2}{3}$ and 1 for moderate blowing at the surface and frozen catalytic wall for laminar or turbulent boundary layers. The expression $i_D(w_{as} - w_{aw})$ can be replaced by

$$\Sigma i_{\text{ch},m}(w_{ms} - w_{mw})$$

in which $i_{\text{ch},m}$ indicates the enthalpy change of component m created in the chemical process.

[1] J. A. Fay, F. R. Riddell, and N. H. Kemp, Stagnation Point Heat Transfer in Dissociated Air Flow, *Jet Propulsion,* **27**:672–674 (1957).

[2] Lees, *lop. cit.*

23
Free Convection

Free-convection flows can be caused in mass-transfer processes when the density of the fluid changes with composition. Heat and mass transfer in free convection will be considered in this chapter for a two-component fluid and for boundary-layer flow under the influence of gravitational acceleration g.

23-1 ISOTHERMAL MASS TRANSFER

In Chap. 12 dealing with free convection of a one-component fluid under the influence of temperature differences, the effect of the buoyancy force was expressed through a thermal expansion coefficient β. In the same way an expansion coefficient for mass transfer can be defined through the equation

$$\frac{1}{v}\left(\frac{\partial v}{\partial w_1}\right)_p = \gamma$$

where v denotes the specific volume of the mixture. This expansion coefficient can also be expressed with the mixture density ρ:

$$-\frac{1}{\rho}\left(\frac{\partial \rho}{\partial w_1}\right)_p = \gamma \tag{23-1}$$

In Chap. 12 the buoyancy force term in the boundary-layer momentum equation led through dimensional analysis to a dimensionless parameter which was formulated either as a Grashof number [Eq. (12-7)] or a Rayleigh number [Eq. (12-8)]. In a completely analogous derivation, one obtains a Grashof number

$$\mathrm{Gr}_M = \frac{\gamma g(w_{1w} - w_{1e})L^3}{\nu^2} \tag{23-2}$$

or a Rayleigh number

$$\mathrm{Ra}_M = \mathrm{Gr}_M\,\mathrm{Sc} = \frac{\gamma g(w_{iw} - w_{ie})L^3}{\nu D} \tag{23-3}$$

for mass transfer. The analogy between the boundary-layer equations describing heat transfer in Chap. 12 and the equations for isothermal mass transfer leads to the conclusion that there exists for every heat-transfer situation a corresponding mass-transfer situation with analogous boundary conditions and with the same geometry. For such an analogous mass transfer, the Sherwood number can be obtained by simply replacing in the corresponding relation for heat transfer the Nusselt number by the Sherwood number, the Rayleigh number Ra by the Rayleigh number Ra_M for mass transfer, and the Prandtl number Pr by the Schmidt number:

$$\mathrm{Nu} = f(\mathrm{Ra}, \mathrm{Pr}) \tag{23-4}$$

This analogy holds for turbulent as well as for laminar flow.

23-2 SIMULTANEOUS HEAT AND MASS TRANSFER

In many cases, a mass-transfer process is inherently connected with heat transfer. In such a situation the density may vary with temperature as well as with composition. For small density differences, this can be expressed by the equation

$$d\rho = \frac{\partial \rho}{\partial t}\,dt + \frac{\partial \rho}{\partial w_1}\,dw_1 = -\rho(\beta\,dt + \gamma\,dw_1) \tag{23-5}$$

Correspondingly, two body-force terms appear in Eq. (12-6). They have the form

$$\beta \rho g(t - t_e) + \gamma \rho g(w_1 - w_{1e})$$

Dimensional analysis performed on the new system of equations leads to the result that the Nusselt number or the Sherwood number, describing heat transfer and mass transfer, respectively, is now a function of the Rayleigh numbers for heat transfer, the Rayleigh numbers for mass transfer, the Prandtl number, and the Schmidt number. This is expressed by the relation

$$\left.\begin{matrix}\mathrm{Nu}\\ \mathrm{Sh}\end{matrix}\right\} = f(\mathrm{Ra},\mathrm{Ra}_M,\mathrm{Pr},\mathrm{Sc}) \tag{23-6}$$

Because of the large number of parameters, considerable experimentation or analysis is required to obtain solutions to such combined heat- and mass-transfer problems.

appendix A

Dimensions and Units

The laws of physics and those of heat transfer are most effectively presented in mathematical language. The equations used in physics, however, differ from the ones which we use in mathematics in that they deal with physical quantities. Such quantities have *dimensions* (length, velocity, energy) and are measured in certain *units* (centimeters, centimeters per second, feet, feet per second). A physical quantity is known only when the unit in which it is measured is stated and a number is given indicating how often the unit is contained in the quantity. Each symbol in a physical equation stands, therefore, for the product of a number with a unit ($V = 3$ ft/s). We distinguish between basic units (centimeters, feet, seconds) and derived units. The latter are composed of groups of basic units (centimeters per second, feet per second). The concepts of dimension and unit should be clearly distinguished. A physical quantity has only one dimension but can be expressed in different units. The dimension *length,* for instance, can be measured in meters, centimeters, feet, inches, or miles. It is possible to convert each of the different units for the same dimension into any one of the others. For instance

$$1 \text{ ft} = 30.48 \text{ cm}$$
$$1 \text{ hr} = 3{,}600 \text{ s}$$

It is never possible to convert one dimension into another. It is of advantage to keep the number of basic units small. Different systems of units are distinguished by which units are basic.

The physical system of units is based on three units: the centimeter (cm) as the unit for the dimension length, the gram (g) as the unit for the dimension mass, and the second (s) as the unit for the dimension time. Sometimes the Celsius degree (C)[1] is used as a fourth basic unit. All other units are formed from these basic ones as derived units. The number of basic units is in no way prescribed to us by nature. The degree Celsius (C), for instance, can be expressed by the basic units centimeters and grams with the use of the laws of thermodynamics. The number of basic units can, in principle, be reduced still further. For instance, it would be possible to eliminate the gram as a basic unit for the mass by the use of the general gravitational law. The force with which two masses m_1 and m_2 at a distance r attract each other is expressed by Newton's gravitational law, usually written

$$F = K\frac{m_1 m_2}{r^2} \tag{A-1}$$

This expression contains a universal constant K, which in the physical system of units has the value $K = 6.685 \times 10^{-8}\ \text{cm}^3/\text{g s}^2$. The constant is omitted when the equation is used to derive a unit for the mass, and the equation is written as

$$F = \frac{m_1 m_2}{r^2}$$

To eliminate the force, this equation is combined with Newton's law, written, for instance, for the mass m_1 subject to an acceleration a

$$F = m_1 a$$

Combination of both equations results in

$$m_2 = ar^2$$

[1] The notation °C is often used. In derived units it may, however, lead to errors, since the symbol for degree may be misinterpreted as a power (ft^{2}°C). Therefore, in this book the letter C alone is used as symbol for the Celsius degree (and correspondingly F, K, R for degree Fahrenheit, Kelvin, and Rankine).

It can be seen that the right-hand side of the equation has the dimension cm^3/s^2. Since different terms appearing in a physical equation must have the same dimension, we have obtained in this way a dimension in which mass can be measured. It is customary, however, to retain the gram as the basic unit for mass in addition to the others mentioned above.

The *engineering system of units,* which in the United States is based on English units, has grown under the influence of engineering needs and has actually never developed into such a universally accepted system as the physical system of units. For instance, there is no complete agreement on the number or kind of basic units used. Usually the engineering system is defined in such a way that basic units are used for the dimensions length (feet), force (pounds force), time (seconds), and temperature (degrees Fahrenheit). Other units, however, are used continuously besides the ones listed above, e.g., inches, hours, and degrees Rankine. Moreover, quite often in the engineering literature the pound is used to represent a mass and a force simultaneously.

The 1960 Eleventh General Conference on Weights and Measures defined and officially recommended the *Système International d'Unités* (SI system), sometimes also referred to as the MKSA system. It is based on the following basic units:

m = meter (for length)
kg = kilogram (for mass)
s = second (for time)
A = ampere (for electric current)
K = Kelvin (for temperature)
cd = candela (for light flux)

All physical quantities can be measured in derived units based on these basic units. For convenience, certain derived units are given special names and symbols. The following ones are of importance in the context of this book. Energy is measured in joules (1 J = 1 m^2 kg/s^2) or in kilojoules (1 kJ = 1,000 J) and power in watts (1 W = 1 m^2 kg/s^3) or kilowatts (1 kW = 1,000 W). Force is measured in Newtons (1 N = 1 kg m/s^2) and pressure in bars

$$(1 \text{ bar} = 10^5 \text{ N/m}^2 = 10^5 \text{ kg/m s}^2, 1 \text{ atm} = 0.98066 \text{ bar})$$

The relations between the SI units and the corresponding units in the English system are listed in Table A-1. It is convenient to remember that the units for energy in both systems are almost equal in size

$$(1 \text{ kJ} = 0.9478 \text{ Btu})$$

Multiples of the units are formed with certain prefixes. The important ones are:

10^9 = giga (G)
10^6 = mega (M)
10^3 = kilo (k)
10^2 = hecto (h)
10 = deka (da)
10^{-1} = deci (d)
10^{-2} = centi (c)
10^{-3} = milli (m)
10^{-6} = micro (μ)
10^{-9} = nano (n)

For instance, 1 km = 10^3 m, 1 μm = 10^{-6} m. The SI system has many advantages. It combines in a natural way, for instance, electrical units with those used in mechanics and thermodynamics. A number of countries have already adopted it by law. Great Britain is presently converting to it. It is to be hoped that it will also be used more and more in the United States. The National Bureau of Standards and the National Aeronautics and Space Agency have already adopted it. Numerical calculations in this text are based on this system.

General adoption of the SI system will greatly simplify numerical calculations; however, one must expect a period of transition in which the various systems will be used simultaneously. Because of the diversified applications of heat transfer, papers on this subject are found in journals of physics as well as of engineering. As a consequence, the necessity arises for everyone working in this field continuously to convert values from one system of units to the other. The possibility of errors in such calculations, which is great, can be substantially reduced if the following rules are adhered to:

1. Always remember that a physical quantity is described only when the unit is stated as well as the number indicating how often the unit is contained in the quantity. If, therefore, a physical equation is used to make numerical calculations, for each symbol not only a number but also the corresponding unit should be introduced. Some examples in the following paragraphs will indicate the application of this rule.
2. It must be kept in mind that only quantities of the same kind, which means with the same dimension, can be compared (added, subtracted). As a consequence, all terms in a physical equation must have the same dimension.
3. The pound as the unit of mass or the unit of force must be clearly dis-

tinguished when calculating in the English system of units. In particular it must be clarified in each instance whether property values (density ρ, specific heat c, etc.) are based on mass or weight. In the United States they are usually based on mass, whereas on the European continent an engineering system is occasionally still used which bases properties on a unit standard weight, defined as the force which a mass of 1 kg experiences under the influence of the standard gravitational acceleration. In the following examples the pound as a unit of force will be written lb_f and as unit of mass lb.

The whole problem of systems of units loses much of its significance when these rules are consistently applied, and conversion from one system to another becomes almost automatic. The following examples will explain these rules.

Example A-1 If a substance expands reversibly under constant pressure p by the volume ΔV, the first law of thermodynamics relates the heat Q added in this process to the change in internal energy ΔU and the mechanical work performed in the expansion. The following equation describing this law is found in many textbooks:

$$Q = \Delta U + \frac{p\,\Delta V}{J} \tag{A-2}$$

J is the mechanical equivalent of heat. Assume that the values for the different quantities appearing on the right-hand side of the equation are known (ΔU = 3 Btu, p = 1,000 lb_f/ft^2, ΔV = 2 ft^3) and that the heat added in this process is to be calculated. Remembering that J = 778.26 ft lb_f/Btu and introducing the values for the symbols on the right-hand side of Eq. (A-2) lead to the expression

$$Q = 3 \text{ Btu} + \frac{1{,}000 \text{ lb}_f/\text{ft}^2 \times 2 \text{ ft}^3}{778.26 \text{ ft lb}_f/\text{Btu}} = 5.57 \text{ Btu}$$

Actually the factor J appearing in the first law of thermodynamics as written above is only a conversion factor from one unit for the dimension work or energy into another, and it is not necessary to have a special symbol for it in a physical equation. We usually do not indicate for other physical quantities which unit is to be used. The expression for the first law, in the process considered here, can then be written

$$Q = \Delta U + p\,\Delta V \tag{A-3}$$

We shall show that the calculation procedure which applies the rules stated above leads to exactly the same result for the heat Q added in the process. Introducing the prescribed values into the right-hand side of the equation gives

$$Q = 3 \text{ Btu} + 1{,}000 \text{ lb}_f/\text{ft}^2 \times \text{ft}^3 = 3 \text{ Btu} + 2{,}000 \text{ ft lb}_f$$

This time the two terms on the right-hand side of the equation appear in different units. In order to add the terms, one term must be converted to the units of the other. Since it is customary to express heat in Btu, the conversion will be made on the second term. Introducing the conversion factor from foot-pounds force into Btu, the calculation can be continued:

$$1 \text{ ft lb}_f = \frac{1}{778.26} \text{ Btu}$$

$$Q = 3 \text{ Btu} + 2{,}000 \frac{1}{778.26} \text{ Btu} = 5.57 \text{ Btu}$$

The following example will show that the numerical calculation is simple in the SI system. We consider the process described by Eq. (A-3) with the values $\Delta U = 3$ kJ, $p = 0.1$ bar, $\Delta V = 0.5$ m^3.

$$\begin{aligned} Q &= \Delta U + p\,\Delta U = 3 \text{ kJ} + 0.1 \text{ bar} \times 0.5 \text{ m}^3 \\ &= 3 \text{ kJ} + 10^4 \text{ N/m}^2 \times 0.5 \text{ m}^3 \\ &= 3 \text{ kJ} + 5 \times 10^3 \text{ N m} \\ &= 3 \text{ kJ} + 5 \text{ kJ} = 8 \text{ kJ} \end{aligned}$$

Example A-2 In flow of an inviscid incompressible fluid, Bernoulli's equation holds:

$$p_T = p_{st} + \rho \frac{v^2}{2} \qquad \text{(A-4)}$$

It states that the sum of the static pressure p and the dynamic pressure $\rho v^2/2$ at each location in the fluid is equal to the total pressure $p°$. Let us assume that the following values are prescribed:

$$p = 2 \text{ lb}_f/\text{ft}^2 \qquad v = 50 \text{ ft/s} \qquad \rho = 0.1 \text{ lb/ft}^3$$

and that the total pressure is to be calculated. The calculation proceeds in the same way as in the first example, introducing the values for the symbols on the right-hand side of the equation

$$p° = 2\frac{\text{lb}_f}{\text{ft}^2} + 0.1\frac{\text{lb}}{\text{ft}^3}\frac{2{,}500}{2}\frac{\text{ft}^2}{\text{s}^2} = 2 \text{ lb}_f/\text{ft}^2 + 125 \text{ lb/ft s}^2$$

Again, it is seen that the two terms on the right-hand side, which must have the same dimension, are expressed in different units. Therefore, one of the units must be converted into the other. For this conversion, we must remember the definition of the pound force: one pound force is that force which the mass of one pound exerts in a gravitational field with the standard gravitational acceleration 32.174 ft/s^2. This definition can be written as an equation

$$1 \text{ lb}_f = 1 \text{ lb} \times 32.174 \text{ ft/s}^2 = 32.174 \text{ lb ft/s}^2$$

The equation is used in the above expression to replace the lb:

$$p° = 2\frac{\text{lb}_f}{\text{ft}^2} + \frac{125}{32.174}\frac{1}{\text{ft s}^2}\frac{\text{lb}_f \text{ s}^2}{\text{ft}} = 5.89 \text{ lb}_f/\text{ft}^2$$

Bernoulli's equation is sometimes found in the literature in the following form:

$$p° = p + \rho\frac{v^2}{2g_c} \qquad \text{(A-5)}$$

The symbol g_c in this equation is the conversion factor

$$g_c = 32.174 \text{ lb ft/lb}_f \text{ s}^2$$

The calculation is made in the same way as above with the following results:

$$p^\circ = 2\,\frac{\text{lb}_f}{\text{ft}^2} + 0.1\,\frac{\text{lb}}{\text{ft}^3}\,\frac{2.500\ \text{ft}^2}{2}\,\frac{1}{\text{s}^2}\,\frac{1}{32.174}\,\frac{\text{lb}_f\ \text{s}^2}{\text{lb ft}} = 5.89\ \text{lb}_f/\text{ft}^2$$

It can again be seen that either one of Eqs. (A-4) or (A-5) for Bernoulli's law leads to the same value for the total pressure.

Example A-3 A third example deals with temperature measurements in a high-speed gas flow. In this book it has been explained that two different kinds of temperature can be distinguished in such a flow: a total temperature T° and a static temperature T. Both are related by the following equation when the specific heat c_p is considered constant:

$$T^\circ = T + \frac{v^2}{2c_p} \tag{A-6}$$

In this equation the specific heat c_p is based on a unit mass. The calculation proceeds in the following way when again the values for the terms on the right-hand side are prescribed: $T = 100$ F, $v = 1{,}000$ ft/s, $c_p = 0.240$ Btu/lb F

$$\begin{aligned} T^\circ &= 100\ \text{F} + \frac{10^6}{2}\ \text{ft}^2/\text{s}^2 \times \frac{1}{0.240}\ \text{lb F/Btu} \\ &= 100\ \text{F} + 2.08 \times 10^6\ \text{lb ft}^2\ \text{F/s}^2\ \text{Btu} \end{aligned}$$

Obviously the second term must have the dimension of temperature. This suggests replacement of the lb in the numerator by lb_f to obtain an expression for the mechanical work:

$$\begin{aligned} T^\circ &= 100\ \text{F} + 2.08 \times 10^6\ \text{ft}^2\ \text{F/s}^2\ \text{Btu} \times \frac{1}{32.174}\ \text{lb}_f\ \text{s}^2/\text{ft} \\ &= 100\ \text{F} + 6.46 \times 10^4\ \text{lb}_f\ \text{ft F/Btu} \end{aligned}$$

Finally the Btu can be converted to ft lb_f:

$$\begin{aligned} T^\circ &= 100\ \text{F} + 6.46 \times 10^4\ \text{lb}_f\text{ft F/Btu} \times \frac{1}{778.26}\,\frac{\text{Btu}}{\text{ft lb}_f} \\ &= 100\ \text{F} + 83\ \text{F} = 183\ \text{F} \end{aligned}$$

In the SI system with the data ($T = 50$ C, $v = 300$ m/s, $c_p = 0.574$ kJ/kg C) the calculation proceeds in the following way:

$$\begin{aligned} T^\circ &= 50\ \text{C} + \frac{9 \times 10^4}{2}\ \text{m}^2/\text{s}^2 \times \frac{1}{0.574}\ \text{kg C/kJ} \\ &= 50\ \text{C} + 7.85 \times 10^4\,\frac{\text{N m C}}{\text{kJ}} \end{aligned}$$

and since, 1 N m = 1 J,

$$T^\circ = 128.5\ \text{C}$$

These examples have demonstrated that conversion factors like the mechanical equivalent of heat J and the gravitational conversion factor g_c can be omitted from physical equations when numerical calculations are made following the three stated rules. In the equations in this book, therefore, such conversion factors were not included. Calculations can easily be performed in any system of units. The SI system has the advantage that only powers of 10 occur in the conversion of units within this system. An extensive tabulation of conversion factors is available.[1] The most useful of the physical constants and conversion factors are shown in Tables A-1 to A-3.

Table A-1 Physical constants†

Constant	*Value*	*SI*
Speed of light in vacuum	2.997925×10^{8}	m/s
Avogadro constant	6.02252×10^{23}	$mole^{-1}$
Planck constant	6.6256×10^{-34}	J s
Gas constant	8.3143	J/K mole
Normal volume of a perfect gas	2.24136×10^{-2}	m^3/mole
Boltzmann constant	1.38054×10^{-23}	J/K
First radiation constant ($2\pi hc^2$)	3.7405×10^{-16}	W m^2
Second radiation constant	1.43879×10^{-2}	m K
Wien displacement constant	2.8978×10^{-3}	m K
Stefan-Boltzmann constant	5.6697×10^{-8}	W/m^2 K^4

† From E. A. Mechtly, The International System of Units, *NASA* SP-7012, 1964.

[1] E. A. Mechtly, The International System of Units, *NASA* SP-7012, 1964.

Table A-2 Conversion factors†

To convert from—	*To—*	*Multiply by—*
Acceleration		
ft/s²	m/s²	−01‡ 3.048*
Area		
ft²	m²	−02 9.290304*
in²	m²	−04 6.4516*
mile² (U.S. statute)	m²	+06 2.589988110336*
Density		
lb_m/in^3	kg/m³	+04 2.7679905
lb_m/ft^3	kg/m³	+01 1.6018463
slug/ft³	kg/m³	+02 5.15379
Energy		
Btu (International Steam Table)	J	+03 1.05504
cal (International Steam Table)	J	+00 4.1868
eV	J	−19 1.60210
ft lb_f	J	+00 1.3558179
kcal (International Steam Table)	J	+03 4.1868
kW hr	J	+06 3.60*
Force		
dyne	N	−05 1.00*
kg_f	N	+00 9.80665*
lb_f (avoirdupois)	N	+00 4.4482216152605*
poundal	N	−01 1.38254954376*
Length		
Å	m	−10 1.00*
ft	m	−01 3.048*
in	m	−02 2.54*
light-year	m	+15 9.46055
m	wavelengths Kr 86	+06 1.65076373*
mile (U.S. statute)	m	+03 1.609344*

† Selected from E. A. Mechtly, The International System of Units, *NASA* SP-7012, 1964.

‡ These figures indicate the power of 10 by which the number to the right has to be multiplied, for example, $-01 \rightarrow 10^{-1}$. Values which are exact by NBS standards are followed by an asterisk.

Table A-2 Conversion factors† (continued)

To convert from—	*To—*	*Multiply by—*
Mass		
lb_m (avoirdupois)	kg	−01 4.5359237*
Power		
ft lb_f/s	W	+00 1.3558179
hp (550 ft lb_f/s)	W	+02 7.4569987
Pressure		
atm	N/m^2	+05 1.01325*
bar	N/m^2	+05 1.00*
mm Hg (0°C)	N/m^2	+02 1.333224
psi (lb_f/in^2)	N/m^2	+03 6.8947572
torr (0°C)	N/m^2	+02 1.33322
Speed		
ft/s	m/s	−01 3.048*
mph (U.S. statute)	m/s	−01 4.4704
Temperature		
Celsius	K	K = C + 273.15
Fahrenheit	K	$K = \frac{5}{9}(F + 456.67)$
Rankine	K	$K = \frac{5}{9}R$
Thermal conductivity		
Btu in/ft^2 s F	J/m s K (W/m K)	+02 5.1887315
Viscosity		
stoke	m^2/s	−04 1.00*
ft^2/s	m^2/s	−02 9.290304*
lb_m/ft s	N s/m^2	+00 1.4881639
poise	N s/m^2	−01 1.00*
Volume		
ft^3	m^3	−02 2.8316846592*
gal (U.S. liquid)	m^3	−03 3.785411784*
oz (U.S. fluid)	m^3	−05 2.95735295625*

Table A-3 Conversion factors in heat transfer

Heat:
- 1 kJ = 0.94783 Btu_{IT}
- 1 kJ = 0.23884 $kcal_{IT}$

Heat flux:
- 1 W = 3.4121 Btu/hr
- 1 W = 0.85984 kcal/hr

Heat flux per unit area:
- 1 W/m^2 = 0.3170 Btu/ft^2 hr
- 1 W/m^2 = 0.85984 $kcal/m^2$ hr

Heat-transfer coefficient:
- 1 W/m^2 C = 0.1761 Btu/ft^2 hr F
- 1 W/m^2 C = 0.85984 $kcal/m^2$ hr C

Heat per unit mass:
- 1 kJ/kg = 0.4299 Btu/lb
- 1 kJ/kg = 0.23884 kcal/kg

Specific heat:
- 1 kJ/kg C = 0.23884 Btu/lb_m F
- 1 kJ/kg C = 0.23884 kcal/kg C

Thermal conductivity:
- 1 W/m C = 0.5778 Btu/ft hr F
- 1 W/m C = 0.85984 kcal/m hr C

Thermal diffusivity:
- 1 m^2/s = 10.7639 ft^2/s
- 1 m^2/s = 3,600 m^2/hr

Dynamic viscosity:
- 1 kg/m s = 0.006720 lb_m/ft s
- 1 kg/m s = 10 g/cm s = 10 poises

Kinematic viscosity:
- 1 m^2/s = 10.7639 ft^2/s
- 1 m^2/s = 10^4 cm^2/s = 10^4 stokes

appendix **B**

Thermophysical Properties

Appendix B is intended primarily to support the text and not necessarily to fulfill the requirements of a handbook; therefore, the accent is on brevity and simplicity. The reader or researcher desiring more detailed or additional material should consult any of the excellent and extensive publications listed.[1]

Tables B-1 and B-3 were taken from Koch [9], with the kind permission of the author. Some supplementary information has been added to Table B-1 with correction of the values of thermal conductivity in consideration of newer data. Table B-2 presents data on the thermal conductivity of metals and alloys at low temperatures from Lees [16]. Table B-3 gives properties of a number of liquids in the saturated state. Excluding the region in the neighborhood of the critical state, the variation of the listed properties with pressure is small, so that the table has a universal value. The viscosity is given as the kinematic viscosity.

Table B-4, which presents the properties of several important gases

[1] See References, p. 791–792.

at atmospheric pressure, was computed largely on the basis of [14] except for the data on helium and ammonia and supplementary data for air and hydrogen. The supplementary data for air, which were taken from [28], [29], [38], and [39], are considered the best information available at this time.

The data for helium were taken from [10], and [30] to [32]. The data for ammonia came from [9] except for the thermal conductivity, which came from [17]. The supplementary data for hydrogen were taken from [7] and [17].

Excluding again the region of the critical state, the property values for other pressures can be obtained from those given in the following manner. The density can be determined from the ideal-gas equation of state $\rho = p/RT$. From this it follows that at any temperature the density is $\rho = \rho_0(p/p_0)$, where $p_0 = 10.1325 \times 10^4$ N/m^2 and ρ_0 is the density as given in Table B-4 for the temperature in question.

Furthermore, the specific heat c_p varies only a little with pressure for a fairly wide range of pressure. This pressure independence is true also for the thermal conductivity k, the dynamic viscosity μ, and therefore for the Prandtl number Pr. The kinematic viscosity ν and the thermal diffusivity α are inversely proportional to the density; therefore, for a given temperature, they are inversely proportional to the pressure. In this way, property values for other conditions within the ideal-gas range can be found from the ones listed in the table.

The thermal and transport properties of gas mixtures require additional considerations. The properties of primary interest are the density, specific heat, viscosity, and thermal conductivity for binary gas mixtures. The density of the mixture presents no difficulty, nor does the specific heat, since these values for nonreactive mixtures can be determined easily from the component properties. For example, the specific heat of such a mixture is given by

$$c_{p_m} = wc_{p_1} + (1 - w)c_{p_2} \tag{B-1}$$

where w is the mass fraction of component 1.

The thermal conductivity and the viscosity present somewhat more difficulty. Computation of either of these properties in simple ratio or as molal averages of the components may lead to values that are quite unsatisfactory when compared with experimental measurements. Fortunately, calculation methods of reasonable simplicity exist, which depend on a knowledge of the properties of the pure components and which provide reasonably accurate results. A fairly comprehensive comparison of the presently available calculation methods with existing experimental data is given by Tsederberg [24]. Of the more acceptable methods

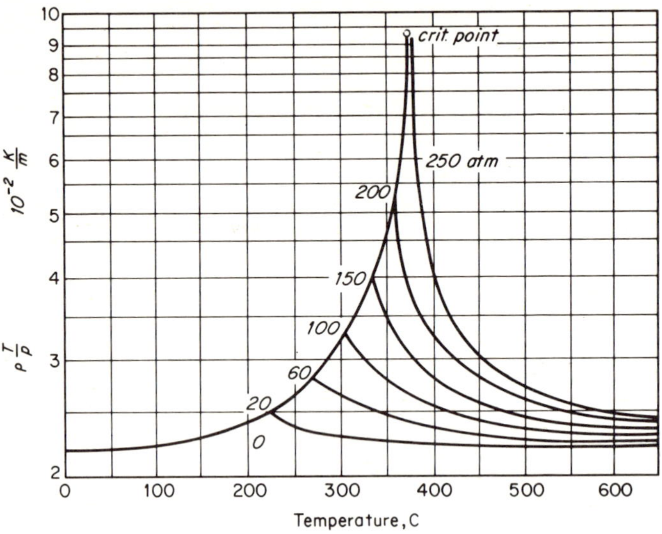

Fig. B-1 The density ρ of water vapor showing the dependence on pressure and temperature.

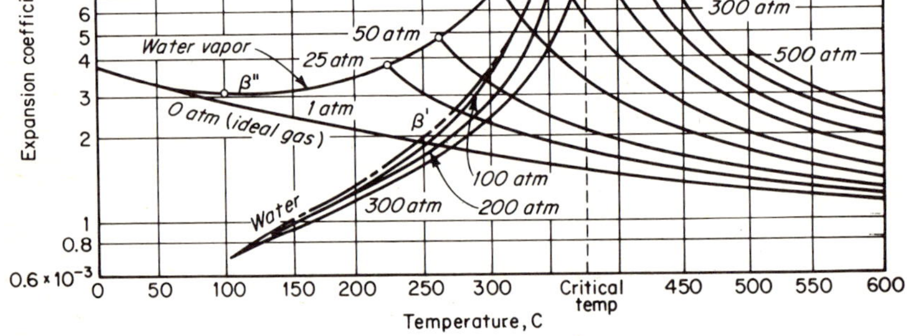

Fig. B-2 The expansion coefficient $\beta = \frac{1}{v}\left(\frac{\partial v}{\partial T}\right)_p$ of water and water vapor as a function of the pressure and temperature.

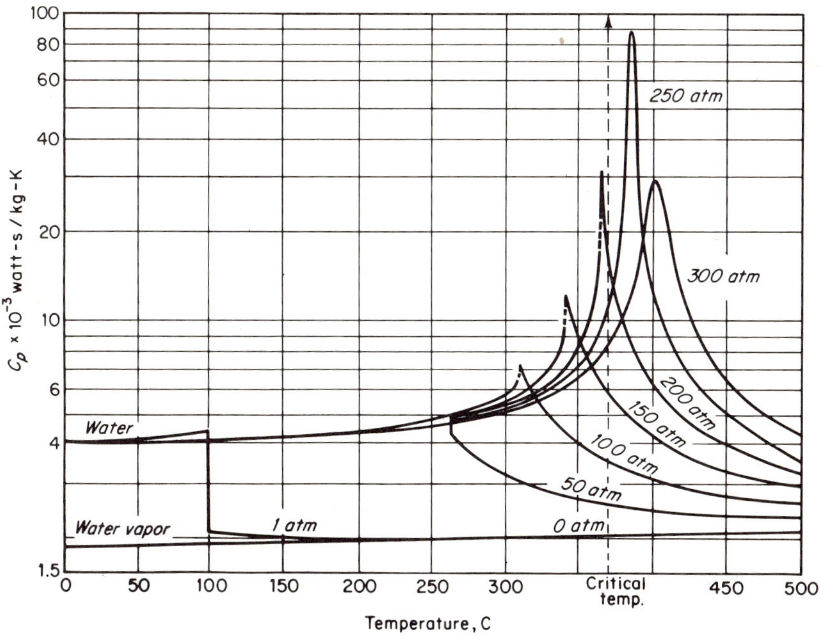

Fig. B-3 The constant-pressure specific heat c_p of water and water vapor as a function of pressure and temperature.

Wilke [33] gives the following expression for the viscosity of a binary mixture of gases:

$$\mu_m = \frac{\mu_1}{1 + (x_2/x_1)\phi_{12}} + \frac{\mu_2}{1 + (x_1/x_2)\phi_{21}} \tag{B-2}$$

where

$$\phi_{ij} = \frac{[1 + (\mu_i/\mu_j)^{\frac{1}{2}}(M_j/M_i)^{\frac{1}{4}}]^2}{(4/\sqrt{2})(1 + M_i/M_j)^{\frac{1}{2}}}$$

where M_i, M_j = molecular weights of the component pure gases
x_1, x_2 = mole fractions
μ_i, μ_j = viscosities of pure components

Lindsay and Bromley [34] present a method based on the Wassiljewa [35] computation method for thermal conductivities which appears satisfactory; it results in the relationship

$$k_m = \frac{k_1}{1 + A_1(x_2/x_1)} + \frac{k_2}{1 + A_2(x_1/x_2)} \tag{B-3}$$

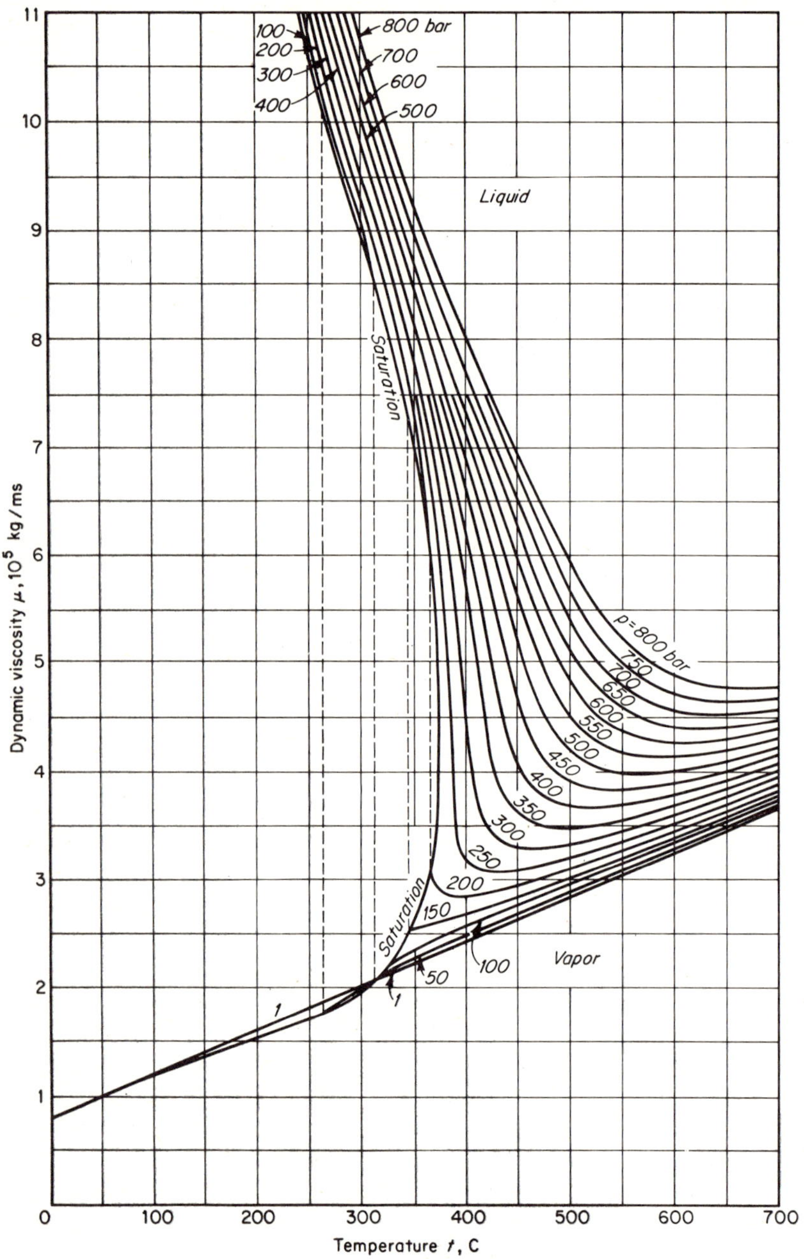

Fig. B-4 The dynamic viscosity μ of water and water vapor as a function of pressure and temperature. [*U. Grigull, F. Mayinger, and J. Bach, Wärme- und Stoffübertragung,* **1**:*15–34* (*1968*).]

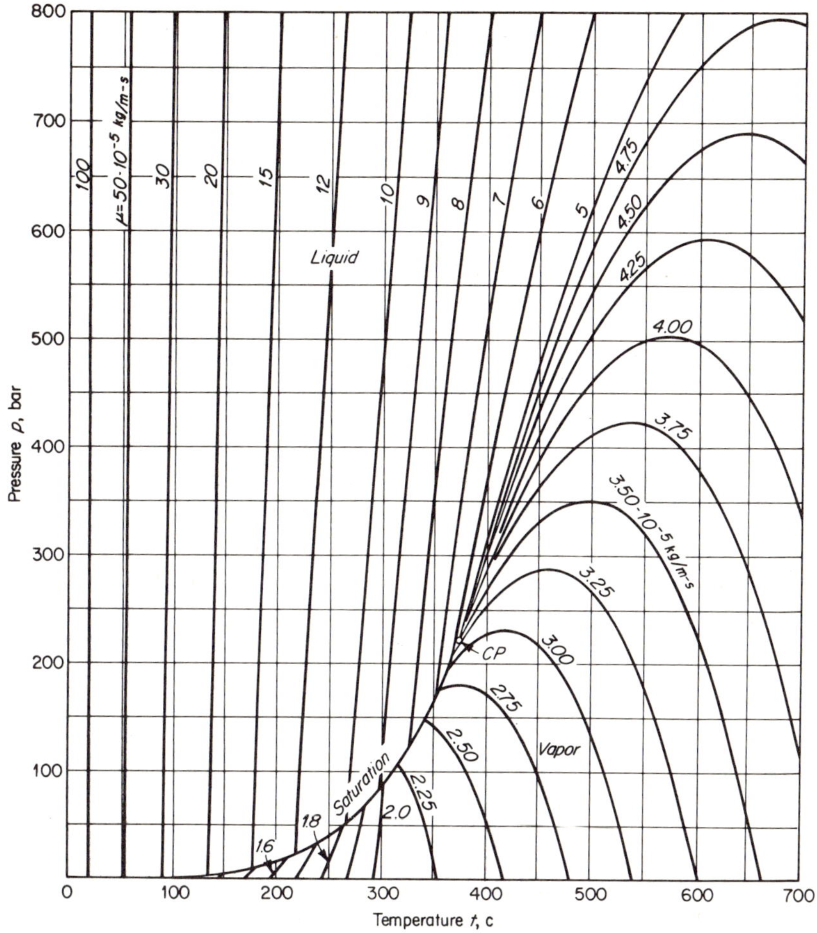

Fig. B-5 Pressure-temperature diagram showing the dynamic viscosity μ as a parameter. [*U. Grigull, F. Mayinger, and J. Bach, Wärme- und Stoffübertragung,* **1**:*15–34* (*1968*).]

where the Wassiljewa constants A_1 and A_2 are

$$A_1 = \tfrac{1}{4}\left\{1 + \left[\frac{\mu_1}{\mu_2}\left(\frac{M_2}{M_1}\right)^{\frac{3}{4}} \frac{1 + S_1/T}{1 + S_2/T}\right]^{\frac{1}{2}}\right\}^2 \frac{1 + S_{12}/T}{1 + S_1/T} \tag{B-4a}$$

$$A_2 = \tfrac{1}{4}\left\{1 + \left[\frac{\mu_2}{\mu_1}\left(\frac{M_1}{M_2}\right)^{\frac{3}{4}} \frac{1 + S_2/T}{1 + S_1/T}\right]^{\frac{1}{2}}\right\}^2 \frac{1 + S_{12}/T}{1 + S_2/T} \tag{B-4b}$$

where μ = viscosity of pure component
M = molecular weight
T = absolute temperature, K
S = Sutherland constant

$S_{12} = \sqrt{S_1 S_2}$ except for gases containing water vapor or ammonia, in which case $S_{12} = 0.733\sqrt{S_1 S_2}$. Where Sutherland constants are not available, $S = 1.5T_B$ is the value recommended. T_B is the normal boiling temperature of the component in question.

The general effect of mixing two gases of different molecular weights is that the resulting thermal conductivity is lower than that calculated by molal averaging. Mixtures of gases containing water vapor or ammonia have thermal conductivities higher than those values predicted from molal averages. Lenoir [17] has extended the calculations shown here

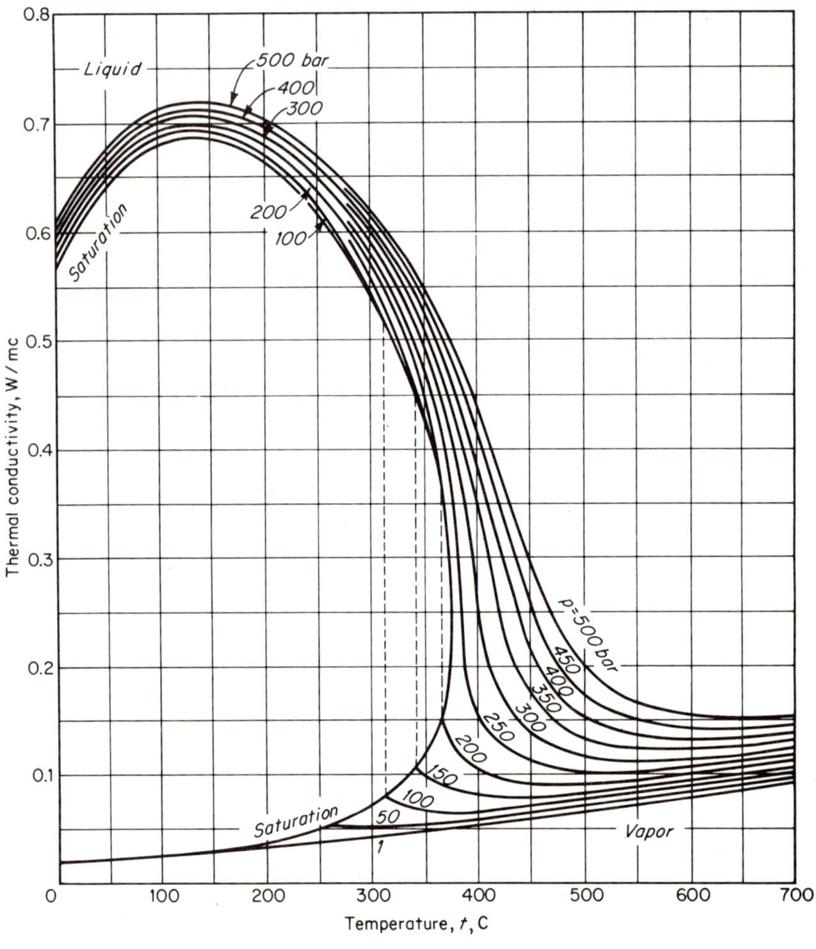

Fig. B-6 The thermal conductivity k of water and water vapor as a function of pressure and temperature. [*U. Grigull, F. Mayinger, and J. Bach, Wärme- und Stoffübertragung,* **1**:*15–34* (*1968*).]

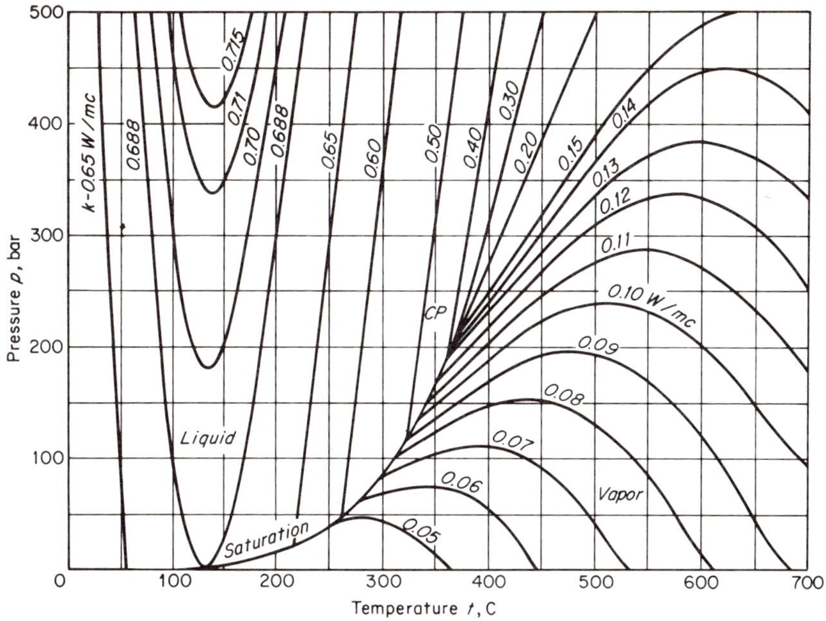

Fig. B-7 Pressure-temperature diagram showing the thermal conductivity k as a parameter. [*U. Grigull, F. Mayinger, and J. Bach, Wärme- und Stoffübertragung,* **1**:*15–34 (1968).*]

to multicomponent mixtures and has calculated a number of Wassiljewa constants from the above equations. Lenoir also gives thermal-conductivity values for many technically useful gases at atmospheric pressure.

The dependence of the property values on pressure and temperature in the neighborhood of the critical state is shown for water and water vapor in Figs. B-1 to B-9. A similar dependence exists for other materials, and therefore deviations from the values calculated for the ideal-gas state can at least be determined approximately from the fact that the percentage variation of any value in the region of the critical state has the same order of magnitude for all materials at corresponding locations with respect to their critical state, i.e., equal value of p/p_c and T/T_c, where p_c and T_c represent the critical pressure and temperature, respectively. Figures B-1 to B-3 were prepared from steam tables.

Table B-5 is presented to give typical values of nonmetallic substances and stems from several sources, notably [2] and [5]. Table B-5 has been supplemented by Table B-6, giving thermal conductivities of insulating materials over their useful temperature range. These data were also taken from [2].

Table B-1 Property values

Metal	Properties at 20 C				Thermal conductivity k, W/m K									
	ρ, kg/m³	c_p Ws/kg K	k, W/m K	α, m²/s	−100 C −148 F	0 C 32 F	100 C 212 F	200 C 392 F	300 C 572 F	400 C 752 F	600 C 1112 F	800 C 1472 F	1000 C 1832 F	1200 C 2192 F
Aluminum:														
Pure	2,707	0.896×10^3	204	8.418×10^{-5}	215	202	206	215	228	249				
Al-Cu (Duralumin) 94–96 Al, 3–5 Cu, trace Mg	2,787	0.883	164	6.676	126	159	182	194						
Al-Mg (Hydronalium) 91–95 Al, 5–9 Mg	2,611	0.904	112	4.764	93	109	125	142						
Al-Si (Silumin) 87 Al, 13 Si	2,659	0.871	164	7.099	149	163	175	185						
Al-Si (Silumin, copper bearing) 86.5 Al, 1 Cu	2,659	0.867	137	5.933	119	137	144	152	161					
Al-Si (Alusil) 78–80 Al, 20–22 Si	2,627	0.854	161	7.172	144	157	168	175	178					
Al-Mg-Si 97 Al, 1 Mg, 1 Si, 1 Mn	2,707	0.892	177	7.311		175	189	204						
Lead	11,373	0.130	35	2.343	36.9	35.1	33.4	31.5	29.8					
Iron:														
Pure	7,897	0.452	73	2.034	87	73	67	62	55	48	40	36	35	36
Wrought iron (C H 0.5 %)	7,849	0.46	59	1,626		59	57	52	48	45	36	33	33	33
Cast iron (C ≈ 4 %)	7,272	0.42	52	1.703										
Steel (C max ≈ 1.5 %)														
Carbon steel C ≈ 0.5 %	7,833	0.465	54	1.474		55	52	48	45	42	35	31	29	31
1.0 %	7,801	0.473	43	1.172		43	43	42	40	36	33	29	28	29
1.5 %	7,753	0.486	36	0.970		36	36	36	35	33	31	28	28	29
Nickel steel Ni ≈ 0 %	7,897	0.452	73	2.026										
10 %	7,945	0.46	26	0.720										
20 %	7,993	0.46	19	0.526										
30 %	8,073	0.46	12	0.325										
40 %	8,169	0.46	10	0.279										
50 %	8,266	0.46	14	0.361										
60 %	8,378	0.46	19	0.493										
70 %	8,506	0.46	26	0.666										
80 %	8,618	0.46	35	0.872										
90 %	8,762	0.46	47	1.156										
100 %	8,906	0.448	90	2.276										

Invar Ni = 36 %	8,137	0.46	10.7	0.286										
Chrome Steel Cr = 0 %	7,897	0.452	73	2.026	87	73	67	62	55	48	40	36	35	36
1 %	7,865	0.46	61	1.665		62	55	52	47	42	36	33	33	
2 %	7,865	0.46	52	1.443		54	48	45	42	38	33	31	31	
5 %	7,833	0.46	40	1.110		40	38	36	36	33	29	29	29	
10 %	7,785	0.46	31	0.867		31	31	31	29	29	28	28	29	
20 %	7,689	0.46	22	0.635		22	22	22	22	24	24	26	29	
30 %	7,625	0.46	19	0.542										
Cr-Ni (chrome-nickel): 15 Cr, 10 Ni	7,865	0.46	19	0.526										
18 Cr, 8 Ni (V2A)	7,817	0.46	16.3	0.444		16.3	17	17	19	19	22	26	31	
20 Cr, 15 Ni	7,833	0.46	15.1	0.415										
25 Cr, 20 Ni	7,865	0.46	12.8	0.361										
Ni-Cr (nickel-chrome): 80 Ni, 15 Cr	8,522	0.46	17	0.444										
60 Ni, 15 Cr	8,266	0.46	12.8	0.333										
40 Ni, 15 Cr	8,073	0.46	11.6	0.305										
20 Ni, 15 Cr	7,865	0.46	14.0	0.390		14.0	15.1	15.1	16.3	17	19	22		
Cr-Ni-Al: 6 Cr, 1.5 Al, 0.55 Si (Sicromal 8(	7,721	0.490	22	0.594										
24 Cr, 2.5 Al, 0.55 Si (Sicromal 12)	7,673	0.494	19	0.501										
Manganese steel Mn = 0 %	7,897	0.494	73	1.863										
1 %	7,865	0.46	50	1.388										
2 %	7,865	0.46	38	1.050		38	36	36	36	35	33			
5 %	7,849	0.46	22	0.637										
10 %	7,801	0.46	17	0.483										
Tungsten steel W = 0 %	7,897	0.452	73	2.026										
1 %	7,913	0.448	66	1.858										
2 %	7,961	0.444	62	1.763		62	59	54	48	45	36			
5 %	8,073	0.435	54	1.525										
10 %	8,314	0.419	48	1.391										
20 %	8,826	0.389	43	1.249										
Silicon steel Si = 0 %	7,897	0.452	73	2.026										
1 %	7,769	0.46	42	1.164										
2 %	7,673	0.46	31	0.888										
5 %	7,417	0.46	19	0.555										

Table B-1 Property values (continued)

Metal	Properties at 20 C				Thermal conductivity k, W/m K									
	ρ, kg/m^3	c_p Ws/kg K	k, W/m K	α, m^2/s	−100 C −148 F	0 C 32 F	100 C 212 F	200 C 392 F	300 C 572 F	400 C 752 F	600 C 1112 F	800 C 1472 F	1000 C 1832 F	1200 C 2192 F
Copper:														
Pure	8,954	0.3831×10^3	386	11.234×10^{-5}	407	386	379	374	369	363	353			
Aluminum bronze 95 cu, 5 Al	8,666	0.410	83	2.330										
Bronze 75 Cu, 25 Sn	8,666	0.343	26	0.859										
Red Brass 85 Cu, 9 Sn, 6 Zn	8,714	0.385	61	1.804		59	71							
Brass 70 Cu, 30 Zn	8,522	0.385	111	3.412	88		128	144	147	147				
German silver 62 Cu, 15 Ni, 22 Zn	8,618	0.394	24.9	0.733	19.2		31	40	45	48				
Constantan 60 Cu, 40 Ni	8,922	0.410	22.7	0.612	21		22.2	26						
Magnesium:														
Pure	1,746	1.013	171	9.708	178	171	168	163	157					
Mg-Al (electrolytic) 6–8 % Al, 1–2 % Zn	1,810	1.00	66	3.605		52	62	74	83					
Mg-Mn 2 % Mn	1,778	1.00	114	6.382	93	111	125	130						
Mg-Mn 2 % Mn	1,778	1.00	114	6.382	93	111	125	130						
Molybdenum	10,220	0.251	123	4.790	138	125	118	114	111	109	106	102	99	92
Nickel:														
Pure (99.9 %)	8,906	0.4459	90	2.266	104	93	83	73	64	59				
Impure (99.2 %)	8,906	0.444	69	1.747		69	64	59	55	52	55	62	67	69
Ni-Cr 90 Ni, 10 Cr	8,666	0.444	17	0.444		17.1	18.9	20.9	22.8	24.6				
80 Ni, 20 Cr	8,314	0.444	12.6	0.343		12.3	13.8	15.6	17.1	18.9	22.5			
Silver:														
Purest	10,524	0.2340	419	17.004	419	417	415	412						
Pure (99.9 %)	10,524	0.2340	407	16.563	419	410	415	374	362	360				
Tungsten	19,350	0.1344	163	6.271		166	151	142	133	126	112	76		
Zinc, pure	7.144	0.3843	112.2	4.106	114	112	109	106	100	93				
Tin, pure	7,304	0.2265	64	3.884	74	65.9	59	57						

Table B-2 Conductivities of metals at low temperatures†
Thermal conductivity k, W/m K

Material	*State or composition*	−170 C −274 F	−160 C −256 F	−150 C −238 F	−125 C −193 F	−100 C −148 F	− 75 C −103 F	−50 C −58 F	−25 C 13 F	0 C 32 F	18 C 64.4 F
Copper	Pure soft drawn	467	453	441	419	408	402	395	391	388	384
Silver	0.999 pure	417	419	419	443	422	421	419	419	412	408
Zinc	Pure, redistilled cast	117.3	116.7	115.8	113.2	113.5	112.7	112.7	112.7	112.7	112.7
Cadmium	Pure, redistilled cast	100.6	100.2	99.7	98.0	96.7	95.2	94.3	93.1	91.7	90.9
Aluminum	0.99 Al	220	215	213	206	206	206	208	209	209	211
Tin	Pure cast	81.7	80.5	79.3	76.3	73.7	72.0	70.3	68.7	67.2	65.8
Lead	Pure	38.9	38.6	38.1	37.4	36.5	35.7	35.7	35.1	35.1	34.8
Iron	0.994 Fe, 0.0015 Mn, 0.0013 Ni, 0.003 C	63.3	63.7	64.0	64.6	63.7	62.8	62.0	61.6	61.6	61.6
Nickel	0.99 Ni	53.7	54.0	54.3	55.2	56.1	56.9	57.5	58.2	58.7	58.7
Steel	Approx. 0.01 C	47.4	47.4	47.4	47.4	47.8	48.1	48.6	48.6	48.6	48.1
Brass	0.7 Cu, 0.3 Zn	73.4	75.8	77.9	83.8	89.3	94.2	98.5	102.3	106.4	108.9
German Silver	0.62 Cu, 0.15 Ni, 0.22 Zn	17.7	18.0	18.5	18.9	19.7	20.6	21.5	22.7	23.5	24.7
Manganin	0.84 Cu, 0.04 Ni, 0.12 Mn	14.26	14.68	14.68	15.51	16.36	17.19	18.0	19.2	20.9	21.8
Lipowitz alloy	0.50 Bi, 0.25 Pb, 0.14 Sn, 0.11 Cd	17.8	17.8	17.8	17.8	17.8	18.0	18.0	18.5	18.5	18.5

† C. H. Lees, *Phil. Trans. Roy. Soc. London, Ser. A*, **208**:381 (1908).

Because of the importance of molten metals in contemporary power-generation systems, Table B-7 has been added, giving the properties of several molten metals and molten-metal alloys. The data on which the table was formulated were taken from [36].

Table B-8 gives the critical states of several technically important elements and compounds. These data were taken from [37].

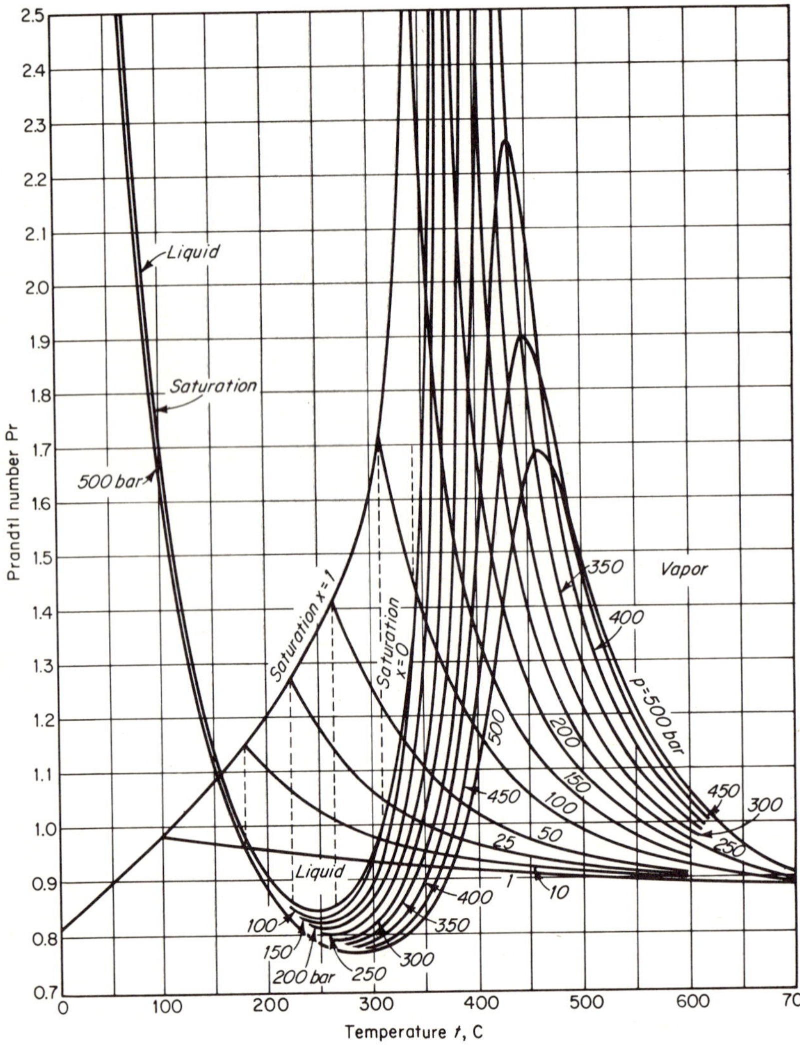

Fig. B-8 The Prandtl number Pr of water and water vapor as a function of pressure and temperature. [*U. Grigull, F. Mayinger, and J. Bach, Wärme- und Stoffübertragung,* **1**:*15–34* (*1968*).]

Table B-3 Property values of fluids in a saturated state

t, C	ρ, kg/m^3	c_p, Ws/kg K	ν, m^2/s	k, W/m K	α, m^2/s	Pr	β, K^{-1}
Water, H_2O							
0	1,002.28	4.2178×10^3	1.788×10^{-6}	0.552	1.308×10^{-7}	13.6	
20	1,000.52	4.1818	1.006	0.597	1.430	7.02	0.18×10^{-3}
40	994.59	4.1784	0.658	0.628	1.512	4.34	
60	985.46	4.1843	0.478	0.651	1.554	3.02	
80	974.08	4.1964	0.364	0.668	1.636	2.22	
100	960.63	4.2161	0.294	0.680	1.680	1.74	
120	945.25	4.250	0.247	0.685	1.708	1.446	
140	928.27	4.283	0.214	0.684	1.724	1.241	
160	909.69	4.342	0.190	0.680	1.729	1.099	
180	889.03	4.417	0.173	0.675	1.724	1.004	
200	866.76	4.505	0.160	0.665	1.706	0.937	
220	842.41	4.610	0.150	0.652	1.680	0.891	
240	815.66	4.756	0.143	0.635	1.639	0.871	
260	785.87	4.949	0.137	0.611	1.577	0.874	
280.6	752.55	5.208	0.135	0.580	1.481	0.910	
300	714.26	5.728	0.135	0.540	1.324	1.019	
Ammonia, NH_3							
−50	703.69	4.463×10^3	0.435×10^{-6}	0.547	1.742×10^{-7}	2.60	
−40	691.68	4.467	0.406	0.547	1.775	2.28	
−30	679.34	4.476	0.387	0.549	1.801	2.15	
−20	666.69	4.509	0.381	0.547	1.819	2.09	
−10	653.55	4.564	0.378	0.543	1.825	2.07	
0	640.10	4.635	0.373	0.540	1.819	2.05	
10	626.16	4.714	0.368	0.531	1.801	2.04	
20	611.75	4.798	0.359	0.521	1.775	2.02	2.45×10^{-3}
30	596.37	4.890	0.349	0.507	1.742	2.01	
40	580.99	4.999	0.340	0.493	1.701	2.00	
50	564.33	5.116	0.330	0.476	1.654	1.99	
Carbon dioxide, CO_2							
−50	1,156.34	1.84×10^3	0.119×10^{-6}	0.0855	0.4021×10^{-7}	2.96	
−40	1,117.77	1.88	0.118	0.1011	0.4810	2.46	
−30	1,076.76	1.97	0.117	0.1116	0.5272	2.22	
−20	1,032.39	2.05	0.115	0.1151	0.5445	2.12	
−10	983.38	2.18	0.113	0.1099	0.5133	2.20	
0	926.99	2.47	0.108	0.1045	0.4578	2.38	
10	860.03	3.14	0.101	0.0971	0.3608	2.80	
20	772.57	5.0	0.091	0.0872	0.2219	4.10	14.00×10^{-3}
30	597.81	36.4	0.080	0.0703	0.0279	28.7	

Table B-3 Property values of fluids in a saturated state (continued)

t, C	ρ, kg/m^3	c_p, Ws/kg K	ν, m^2/s	k, W/m K	α, m^2/s	Pr	β, K^{-1}
Sulfur dioxide, SO_2							
−50	1,560.84	1.3595 × 10^3	0.484 × 10^{-6}	0.242	1.141 × 10^{-7}	4.24	
−40	1,536.81	1.3607	0.424	0.235	1.130	3.74	
−30	1,520.64	1.3616	0.371	0.230	1.117	3.31	
−20	1,488.60	1.3624	0.324	0.225	1.107	2.93	
−10	1,463.61	1.3628	0.288	0.218	1.097	2.62	
0	1,438.46	1.3636	0.257	0.211	1.081	2.38	
10	1,412.51	1.3645	0.232	0.204	1.066	2.18	
20	1,386.40	1.3653	0.210	0.199	1.050	2.00	1.94 × 10^{-3}
30	1,359.33	1.3662	0.190	0.192	1.035	1.83	
40	1,329.22	1.3674	0.173	0.185	1.019	1.70	
50	1,299.10	1.3683	0.162	0.177	0.999	1.61	
Methyl chloride, CH_3Cl							
−50	1,052.58	1.4759 × 10^3	0.320 × 10^{-6}	0.215	1.388 × 10^{-7}	2.31	
−40	1,033.35	1.4826	0.318	0.209	1.368	2.32	
−30	1,016.53	1.4922	0.314	0.202	1.337	2.35	
−20	999.39	1.5043	0.309	0.196	1.301	2.38	
−10	981.45	1.5194	0.306	0.187	1.257	2.43	
0	962.39	1.5378	0.302	0.178	1.213	2.49	
10	942.36	1.5600	0.297	0.171	1.166	2.55	
20	923.31	1.5860	0.293	0.163	1.112	2.63	
30	903.12	1.6161	0.288	0.154	1.058	2.72	
40	883.10	1.6504	0.281	0.144	0.996	2.83	
50	861.15	1.6890	0.274	0.133	0.921	2.97	
Dichlorodifluoromethane (Freon), CCl_2F_2							
−50	1,546.75	0.8750 × 10^3	0.310 × 10^{-6}	0.067	0.501 × 10^{-7}	6.2	2.63 × 10^{-3}
−40	1,518.71	0.8847	0.279	0.069	0.514	5.4	
−30	1,489.56	0.8956	0.253	0.069	0.526	4.8	
−20	1,460.57	0.9073	0.235	0.071	0.539	4.4	
−10	1,429.49	0.9203	0.221	0.073	0.550	4.0	
0	1,397.45	0.9345	0.214	0.073	0.557	3.8	
10	1,364.30	0.9496	0.203	0.073	0.560	3.6	
20	1,330.18	0.9659	0.198	0.073	0.560	3.5	
30	1,295.10	0.9835	0.194	0.071	0.560	3.5	
40	1,257.13	1.0019	0.191	0.069	0.555	3.5	
50	1,215.96	1.0216	0.190	0.067	0.545	3.5	
Eutectic calcium chloride solution, 29.9% $CaCl_2$							
−50	1,319.76	2.608 × 10^3	36.35 × 10^{-6}	0.402	1.166 × 10^{-7}	312	
−40	1,314.96	2.6356	24.97	0.415	1.200	208	
−30	1,310.15	2.6611	17.18	0.429	1.234	139	
−20	1,305.51	2.688	11.04	0.445	1.267	87.1	
−10	1,300.70	2.713	6.96	0.459	1.300	53.6	

Table B-3 Property values of fluids in a saturated state (continued)

t, C	ρ, kg/m³	c_p, Ws/kg K	ν, m²/s	k, W/m K	α, m²/s	Pr	β, K⁻¹
Eutectic calcium chloride solution, 29.9% $CaCl_2$ (continued)							
0	1,296.06	2.738×10^3	4.39×10^{-6}	0.472	1.332×10^{-7}	33.0	
10	1,291.41	2.763	3.35	0.485	1.363	24.6	
20	1,286.61	2.788	2.72	0.498	1.394	19.6	
30	1,281.96	2.814	2.27	0.511	1.419	16.0	
40	1,277.16	2.839	1.92	0.523	1.445	13.3	
50	1,272.51	2.868	1.65	0.535	1.468	11.3	
Glycerin, $C_3H_5(OH)_3$							
0	1,276.03	2.261×10^3	0.00831	0.282	0.983×10^{-7}	84.7×10^3	
10	1,270.11	2.319	0.00300	0.284	0.965	31.0	
20	1,264.02	2.386	0.00118	0.286	0.947	12.5	0.50×10^{-3}
30	1,258.09	2.445	0.00050	0.286	0.929	5.38	
40	1,252.01	2.512	0.00022	0.286	0.914	2.45	
50	1,244.96	2.583	0.00015	0.287	0.893	1.63	
Ethylene glycol, $C_2H_4(OH_2)$							
0	1,130.75	2.294×10^3	57.53×10^{-6}	0.242	0.934×10^{-7}	615	
20	1,116.65	2.382	19.18	0.249	0.939	204	0.65×10^{-3}
40	1,101.43	2.474	8.69	0.256	0.939	93	
60	1,087.66	2.562	4.75	0.260	0.932	51	
80	1,077.56	2.650	2.98	0.261	0.921	32.4	
100	1,058.50	2.742	2.03	0.263	0.908	22.4	
Engine oil (unused)							
0	899.12	1.796×10^3	0.00428	0.147	0.911×10^{-7}	47,100	
20	888.23	1.880	0.00090	0.145	0.872	10,400	0.70×10^{-3}
40	876.05	1.964	0.00024	0.144	0.834	2,870	
60	864.04	2.047	0.839×10^{-4}	0.140	0.800	1,050	
80	852.02	2.131	0.375	0.138	0.769	490	
100	840.01	2.219	0.203	0.137	0.738	276	
120	828.96	2.307	0.124	0.135	0.710	175	
140	816.94	2.395	0.080	0.133	0.686	116	
160	805.89	2.483	0.056	0.132	0.663	84	
Mercury, Hg							
0	13,628.22	0.1403×10^3	0.124×10^{-6}	8.20	42.99×10^{-7}	0.0288	
20	13,579.04	0.1394	0.114	8.69	46.06	0.0249	1.82×10^{-4}
50	13,505.84	0.1386	0.104	9.40	50.22	0.0207	
100	13,384.58	0.1373	0.0928	10.51	57.16	0.0162	
150	13,264.28	0.1365	0.0853	11.49	63.54	0.0134	
200	13,144.94	0.1570	0.0802	12.34	69.08	0.0116	
250	13,025.60	0.1357	0.0765	13.07	74.06	0.0103	
315.5	12,847	0.134	0.0673	14.02	81.5	0.0083	

Table B-4 Property values of gases at atmospheric pressure

T, K	ρ kg/m^3	c_p, Ws/kg K	μ, kg/ms	ν, m^2/s	k, W/m K	α, m^2/s	Pr
Air							
100	3.6010	1.0266 $\times 10^3$	0.6924 $\times 10^{-5}$	1.923 $\times 10^{-6}$	0.009246	0.02501 $\times 10^{-4}$	0.770
150	2.3675	1.0099	1.0283	4.343	0.013735	0.05745	0.753
200	1.7684	1.0061	1.3289	7.490	0.01809	0.10165	0.739
250	1.4128	1.0053	1.488	9.49	0.02227	0.13161	0.722
300	1.1774	1.0057	1.983	15.68	0.02624	0.22160	0.708
350	0.9980	1.0090	2.075	20.76	0.03003	0.2983	0.697
400	0.8826	1.0140	2.286	25.90	0.03365	0.3760	0.689
450	0.7833	1.0207	2.484	28.86	0.03707	0.4222	0.683
500	0.7048	1.0295	2.671	37.90	0.04038	0.5564	0.680
550	0.6423	1.0392	2.848	44.34	0.04360	0.6532	0.680
600	0.5879	1.0551	3.018	51.34	0.04659	0.7512	0.680
650	0.5430	1.0635	3.177	58.51	0.04953	0.8578	0.682
700	0.5030	1.0752	3.332	66.25	0.05230	0.9672	0.684
750	0.4709	1.0856	3.481	73.91	0.05509	1.0774	0.686
800	0.4405	1.0978	3.625	82.29	0.05779	1.1951	0.689
850	0.4149	1.1095	3.765	90.75	0.06028	1.3097	0.692
900	0.3925	1.1212	3.899	99.3	0.06279	1.4271	0.696
950	0.3716	1.1321	4.023	108.2	0.06525	1.5510	0.699
1000	0.3524	1.1417	4.152	117.8	0.06752	1.6779	0.702
1100	0.3204	1.160	4.44	138.6	0.0732	1.969	0.704
1200	0.2947	1.179	4.69	159.1	0.0782	2.251	0.707
1300	0.2707	1.197	4.93	182.1	0.0837	2.583	0.705
1400	0.2515	1.214	5.17	205.5	0.0891	2.920	0.705
1500	0.2355	1.230	5.40	229.1	0.0946	3.262	0.705
1600	0.2211	1.248	5.63	254.5	0.100	3.609	0.705
1700	0.2082	1.267	5.85	280.5	0.105	3.977	0.705
1800	0.1970	1.287	6.07	308.1	0.111	4.379	0.704
1900	0.1858	1.309	6.29	338.5	0.117	4.811	0.704
2000	0.1762	1.338	6.50	369.0	0.124	5.260	0.702
2100	0.1682	1.372	6.72	399.6	0.131	5.715	0.700
2200	0.1602	1.419	6.93	432.6	0.139	6.120	0.707
2300	0.1538	1.482	7.14	464.0	0.149	6.540	0.710
2400	0.1458	1.574	7.35	504.0	0.161	7.020	0.718
2500	0.1394	1.688	7.57	543.5	0.175	7.441	0.730
Helium							
3		5.200 $\times 10^3$	8.42 $\times 10^{-7}$		0.0106		
33	1.4657	5.200	50.2	3.42 $\times 10^{-6}$	0.0353	0.04625 $\times 10^{-4}$	0.74
144	3.3799	5.200	125.5	37.11	0.0928	0.5275	0.70
200	0.2435	5.200	156.6	64.38	0.1177	0.9288	0.694
255	0.1906	5.200	181.7	95.50	0.1357	1.3675	0.70
366	0.13280	5.200	230.5	173.6	0.1691	2.449	0.71
477	0.10204	5.200	275.0	269.3	0.197	3.716	0.72
589	0.08282	5.200	311.3	375.8	0.225	5.215	0.72
700	0.07032	5.200	347.5	494.2	0.251	6.661	0.72
800	0.06023	5.200	381.7	634.1	0.275	8.774	0.72
900	0.05286	5.200	413.6	781.3	0.298	10.834	0.72

Table B-4 Property values of gases at atmospheric pressure (continued)

T, K	ρ, kg/m^3	c_p, Ws/kg K	μ, kg/ms	ν, m^2/s	k, W/m K	α, m^2/s	Pr
Hydrogen							
30	0.84722	10.840 $\times$ 10^3	1.606 $\times$ 10^{-6}	1.895 $\times$ 10^{-6}	0.0228	0.02493 $\times$ 10^{-4}	0.759
50	0.50955	10.501	2.516	4.880	0.0362	0.0676	0.721
100	0.24572	11.229	4.212	17.14	0.0665	0.2408	0.712
150	0.16371	12.602	5.595	34.18	0.0981	0.475	0.718
200	0.12270	13.540	6.813	55.53	0.1282	0.772	0.719
250	0.09819	14.059	7.919	80.64	0.1561	1.130	0.713
300	0.08185	14.314	8.963	109.5	0.182	1.554	0.706
350	0.07016	14.436	9.954	141.9	0.206	2.031	0.697
400	0.06135	14.491	10.864	177.1	0.228	2.568	0.690
450	0.05462	14.499	11.779	215.6	0.251	3.164	0.682
500	0.04918	14.507	12.636	257.0	0.272	3.817	0.675
550	0.04469	14.532	13.475	301.6	0.292	4.516	0.668
600	0.04085	14.537	14.285	349.7	0.315	5.306	0.664
700	0.03492	14.574	15.89	455.1	0.351	6.903	0.659
800	0.03060	14.675	17.40	569	0.384	8.563	0.664
900	0.02723	14.821	18.78	690	0.412	10.217	0.676
1000	0.02451	14.968	20.16	822	0.440	11.997	0.686
1100	0.02227	15.165	21.46	965	0.464	13.726	0.703
1200	0.02050	15.366	22.75	1107	0.488	15.484	0.715
1300	0.01890	15.575	24.08	1273	0.512	17.394	0.733
1333	0.01842	15.638	24.44	1328	0.519	18.013	0.736
Oxygen							
100	3.9918	0.9479 $\times$ 10^3	7.768 $\times$ 10^{-6}	1.946 $\times$ 10^{-6}	0.00903	0.023876 $\times$ 10^{-4}	0.815
150	2.6190	0.9178	11.490	4.387	0.01367	0.05688	0.773
200	1.9559	0.9131	14.850	7.593	0.01824	0.10214	0.745
250	1.5618	0.9157	17.87	11.45	0.02259	0.15794	0.725
300	1.3007	0.9203	20.63	15.86	0.02676	0.22353	0.709
350	1.1133	0.9291	23.16	20.80	0.03070	0.2968	0.702
400	0.9755	0.9420	25.54	26.18	0.03461	0.3768	0.695
450	0.8682	0.9567	27.77	31.99	0.03828	0.4609	0.694
500	0.7801	0.9722	29.91	38.34	0.04173	0.5502	0.697
550	0.7096	0.9881	31.97	45.05	0.04517	0.6441	0.700
600	0.6504	1.0044	33.92	52.15	0.04832	0.7399	0.704
Nitrogen							
100	3.4808	1.0722 $\times$ 10^3	6.862 $\times$ 10^{-6}	1.971 $\times$ 10^{-6}	0.009450	0.025319 $\times$ 10^{-4}	0.786
200	1.7108	1.0429	12.947	7.568	0.01824	0.10224	0.747
300	1.1421	1.0408	17.84	15.63	0.02620	0.22044	0.713
400	0.8538	1.0459	21.98	25.74	0.03335	0.3734	0.691
500	0.6824	1.0555	25.70	37.66	0.03984	0.5530	0.684
600	0.5687	1.0756	29.11	51.19	0.04580	0.7486	0.686
700	0.4934	1.0969	32.13	65.13	0.05123	0.9466	0.691
800	0.4277	1.1225	34.84	81.46	0.05609	1.1685	0.700
900	0.3796	1.1464	37.49	91.06	0.06070	1.3946	0.711
1000	0.3412	1.1677	40.00	117.2	0.06475	1.6250	0.724
1100	0.3108	1.1857	42.28	136.0	0.06850	1.8591	0.736
1200	0.2851	1.2037	44.50	156.1	0.07184	2.0932	0.748

Table B-4 Property values of gases at atmospheric pressure (continued)

T, K	ρ, kg/m³	c_p, Ws/kg K	μ, kg/ms	ν, m²/s	k, W/m K	α, m²/s	Pr
Carbon dioxide							
220	2.4733	0.783 × 10³	11.105 × 10⁻⁶	4.490 × 10⁻⁶	0.010805	0.05920 × 10⁻⁴	0.818
250	2.1657	0.804	12.590	5.813	0.012884	0.07401	0.793
300	1.7973	0.871	14.958	8.321	0.016572	0.10588	0.770
350	1.5362	0.900	17.205	11.19	0.02047	0.14808	0.755
400	1.3424	0.942	19.32	14.39	0.02461	0.19463	0.738
450	1.1918	0.980	21.34	17.90	0.02897	0.24813	0.721
500	1.0732	1.013	23.26	21.67	0.03352	0.3084	0.702
550	0.9739	1.047	25.08	25.74	0.03821	0.3750	0.685
600	0.8938	1.076	26.83	30.02	0.04311	0.4483	0.668
Carbon monoxide							
220	1.55363	1.0429 × 10³	13.832 × 10⁻⁶	8.903 × 10⁻⁶	0.01906	0.11760 × 10⁻⁴	0.758
250	0.8410	1.0425	15.40	11.28	0.02144	0.15063	0.750
300	1.13876	1.0421	17.843	15.67	0.02525	0.21280	0.737
350	0.97425	1.0434	20.09	20.62	0.02883	0.2836	0.728
400	0.85363	1.0484	22.19	25.99	0.03226	0.3605	0.722
450	0.75848	1.0551	24.18	31.88	0.0436	0.4439	0.718
500	0.68223	1.0635	26.06	38.19	0.03863	0.5324	0.718
550	0.62024	1.0756	27.89	44.97	0.04162	0.6240	0.721
600	0.56850	1.0877	29.60	52.06	0.04446	0.7190	0.724
Ammonia, NH_3							
220	0.3828	2.198 × 10³	7.255 × 10⁻⁶	1.90 × 10⁻⁵	0.0171	0.2054 × 10⁻⁴	0.93
273	0.7929	2.177	9.353	1.18	0.0220	0.1308	0.90
323	0.6487	2.177	11.035	1.70	0.0270	0.1920	0.88
373	0.5590	2.236	12.886	2.30	0.0327	0.2619	0.87
423	0.4934	2.315	14.672	2.97	0.0391	0.3432	0.87
473	0.4405	2.395	16.49	3.74	0.0467	0.4421	0.84
Steam (H_2O vapor)							
380	0.5863	2.060 × 10³	12.71 × 10⁻⁶	2.16 × 10⁻⁵	0.0246	0.2036 × 10⁻⁴	1.060
400	0.5542	2.014	13.44	2.42	0.0261	0.2338	1.040
450	0.4902	1.980	15.25	3.11	0.0299	0.307	1.010
500	0.4405	1.985	17.04	3.86	0.0339	0.387	0.996
550	0.4005	1.997	18.84	4.70	0.0379	0.475	0.991
600	0.3652	2.026	20.67	5.66	0.0422	0.573	0.986
650	0.3380	2.056	22.47	6.64	0.0464	0.666	0.995
700	0.3140	2.085	24.26	7.72	0.0505	0.772	1.000
750	0.2931	2.119	26.04	8.88	0.0549	0.883	1.005
800	0.2739	2.152	27.86	10.20	0.0592	1.001	1.010
850	0.2579	2.186	29.69	11.52	0.0637	1.130	1.019

Table B-5 Property values of selected nonmetals

Material	t, C	ρ, kg/m^3	c_p, Ws/kg K	k, W/m K	α, m^2/s
Aerogel, silica	120	136.2		0.022	
Asbestos	−200	469.3		0.074	
	0	469.3		0.156	
	0	576.7	0.816×10^3	0.151	
	100	576.7	0.816	0.192	
	200	576.7		0.208	
	400	576.7		0.223	
	−200	696.8		0.156	
	0	696.8		0.234	
Brick, dry	20	1,762–1,810	0.84	0.38–0.52	$0.028–0.034\times10^{-5}$
Bakelite	20	1,273.5	1.59	0.232	0.0114
Cardboard, corrugated				0.064	
Clay	20	1,457.7	0.88	1.279	0.101
Concrete	20	1,906–2,307	0.88	0.81–1.40	0.049–0.070
Coal, anthracite	20	1,201–1,506	1.26	0.26	0.013–0.015
Powdered	30	737	1.30	0.116	0.013
Cotton	20	80	1.30	0.059	0.194
Cork, board	30	160		0.043	
Expanded scrap	20	44.9–118.5	1.88	0.036	0.015–0.044
Ground	30	150.6		0.043	
Diatomaceous earth	38	320.4		0.062	
	871	320.4		0.142	
Earth, coarse gravelly	20	2,050	1.84	0.52	0.0139
Felt, wood	30	330.0		0.05	
Fiber, insulating board	21	237.1		0.048	
Red	20	1,289.5		0.47	
Glass plate	20	2,707	0.8	0.76	0.034
Glass, borosilicate	30	2,227		1.09	
Wool	20	200.2	0.67	0.040	0.028
Granite				1.7–4.0	
Ice	0	913	1.93	2.22	0.124
Marble	20	2,499–2,707	0.808	2.8	0.139
Rubber, hard	0	1,198.2		0.151	
Sandstone	20	2,162–2,307	0.71	1.63–2.1	0.106–0.126
Silk	20	57.7	1.38	0.036	0.044
Wood, oak radial	20	609–801	2.39	0.17–0.21	0.0111–0.0121
Fir (20% moisture) radial	20	416.5–421.3	2.72	0.14	0.0124

Table B-6 Variation of thermal conductivities of insulating materials with temperature†

Material	*Mean temperature*										*Limiting-use temperature*	
	37.8 C 100 F	93.3 C 200 F	148.9 C 300 F	204.4 C 400 F	260 C 500 F	315.6 C 600 F	426.7 C 800 F	537.8 C 1000 F	815.6 C 1500 F	1093.3 C 2000 F	C	F
Asbestos (577 kg/m³) laminated asbestos felt	0.168	0.190	0.202	0.209	0.213	0.216	0.225					
Approx. 146 laminations/m	0.057	0.064	0.069	0.076	0.083						371.1	700
Approx. 73 laminations/m	0.078	0.087	0.095	0.104	0.112						260	500
Corrugated asbestos (14.6 plies/m)	0.087	0.100	0.119								148.9	300
85% magnesia (208 kg/m³)	0.059	0.062	0.066	0.069							315.6	600
Diatomaceous earth, asbestos and bonder	0.078	0.081	0.085	0.087	0.092	0.095	0.104	0.112			871.1	1600
Diatomaceous earth, brick	0.093	0.097	0.100	0.104	0.109	0.112	0.119	0.126			871.1	1600
Diatomaceous earth, brick	0.220	0.225	0.230	0.237	0.242	0.247	0.260	0.273	0.305		1093.3	2000
Diatomaceous earth, brick	0.222	0.227	0.234	0.241	0.247	0.256	0.268	0.282	0.317	0.351	1371.1	2500
Diatomaceous earth, powder (density, 288 kg/m³)	0.067	0.073	0.076	0.083	0.088	0.093	0.106	0.118				
Rock wool....................	0.052	0.059	0.067	0.076	0.087	0.099						

† L. S. Marks, "Mechanical Engineers' Handbook," 5th ed. Copyright 1951. McGraw-Hill Book Company. Used by permission.

Table B-7 Properties of liquid metals†

Metal, composition, and melting point	t, C	ρ, kg/m³	c_p, Ws/kg K	ν, m²/s	k, W/m K	α m²/s	Pr
Bismuth (271.1 C)	316	10,011	0.1444×10^3	1.617×10^{-7}	16.4	1.138×10^{-5}	0.0142
	538	9,739	0.1545	1.133	15.6	1.035	0.0110
	760	9,467	0.1645	0.8343	15.6	1.001	0.0083
Lead (327.2 C)	371	10,540	0.159	2.276	16.1	1.084	0.024
	482	10,412	0.155	1.849	15.6	1.223	0.017
	704	10,140		1.347	14.9		
Mercury (−38.9) see Table B-3 Table B-3							
Potassium (63.9 C)	149	807.3	0.80	4.608	45.0	6.99	0.0066
	427	741.7	0.75	2.397	39.5	7.07	0.0034
	704	674.4	0.75	1.905	33.1	6.55	0.0029
Sodium (97.8 C)	93	929.1	1.38	7.516	86.2	6.71	0.011
	371	860.2	1.30	3.270	72.3	6.48	0.0051
	704	778.5	1.26	2.285	59.7	6.12	0.0037
NaK(56/44, 19 C)	93	887.4	1.130	6.522	25.6	2.552	0.026
	371	821.7	1.055	2.871	27.5	3.17	0.0091
	704	740.1	1.043	2.174	28.9	3.74	0.0058
NaK(22/78, −11.1 C)	93	849.0	0.946	5.797	24.4	3.05	0.019
	399	775.3	0.879	2.666	26.7	3.92	0.0068
	760	690.4	0.883	2.118			
PbBi(44.5/55.5, 125 C)	149	10,524	0.147		9.05	0.586	
	371	10,236	0.147	1.496	11.86	0.790	0.189
	649	9,835		1.171			

† Atomic Energy Commission, "Liquid Metals Handbook," 23 ed., Department of the Navy, Washington, D.C., 1952.

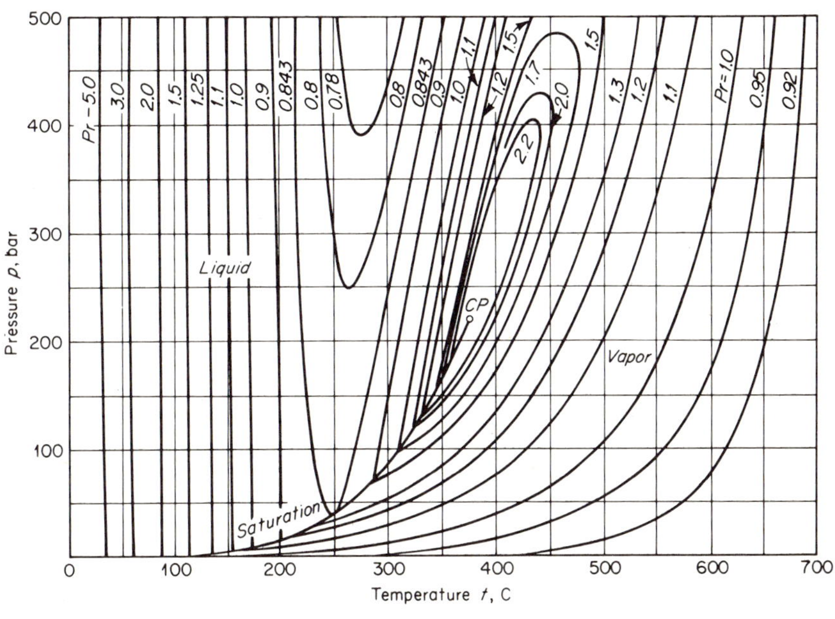

Fig. B-9 Pressure-temperature diagram showing the Prandtl number Pr as a parameter. [*U. Grigull, F. Mayinger, and J. Bach, Wärme- und Stoffübertragung,* **1**:*15–34* (*1968*).]

Table B-8 Critical-state conditions†

Material	T_c, C	P_c, N/m^2	V_c, m^3/kg
Air	−140	3,771 × 10^3	0.287 × 10^{-2}
Alcohol, methyl	240	7,977	0.368
Ethyl	243.3	6,391	0.362
Ammonia	132.2	11,301	0.425
Argon	−122.2	4,861	0.187
Butane	152.8	3,647	0.443
Carbon dioxide	31.1	7,398	0.218
Carbon monoxide	−138.9	3,544	0.318
Carbon tetrachloride	282.8	4,557	0.181
Chlorine	143.9	7,715	0.175
Ethene	32.2	4,944	0.474
Ethylene	9.4	5,847	0.456
Helium	−267.8	228	1.442
Hexane	235	2,992	0.425
Hydrogen	−240	1,296	3.221
Methane	−82.2	4,640	0.618
Methyl chloride	143.3	6,667	0.268
Neon	−228.9	2,696	0.206
Nitric oxide	−93.9	6,584	0.194
Nitrogen	−147.2	3,392	0.325
Octene	296.1	2,496	0.425
Oxygen	−118.9	5,033	0.231
Propane	95.6	4,357	0.443
Sulfur dioxide	157.2	7,874	0.194
Water	373.9	22,105	0.312

† "Handbook of Chemistry and Physics," 39th ed., Chemical Rubber Publishing Company, Cleveland, Ohio, 1957–1958.

Table B-9 gives some available diffusion coefficients and the corresponding Schmidt numbers. The supplementary data shown here were taken largely from the data reported in [37].

Table B-10 and B-11 give emissivities for thermal radiation from selected sources which are indicated in the table. The agreement of data on emissivities between investigators is in general not good, but it is believed that the data shown are reliable.

Tables B-12 and B-13 values of surface tension of various liquids, the data for which were taken completely from data reported in [37].

Table B-9 Coefficients of diffusion D†

Water, H_2O, in air: D, $(\text{m}^2/\text{s}) = 2.302 \dfrac{p_0}{p}\left(\dfrac{T}{T_0}\right)^{1.81} \times 10^{-5}$

Ammonia, NH_3, in air: $D(\text{m}^2/\text{s}) = 1.969 \dfrac{p_0}{p}\left(\dfrac{T}{T_0}\right)^{1.81} \times 10^{-5}$

$p_0 = 0.98 \times 10^5$ N/m² $\quad T_0 = 256$ K

Diffusing material	*Medium of diffusion*	*Temp.,* C	*Diffusion coefficient,* m²/s	*Schmidt no.* Sc = ν/D
Solids:				
Cu	Al	462.2	8.45×10^{-10}	
Hg	Pb	177.2	1.95×10^{-10}	
	Pb	197.2	5.02×10^{-10}	
Liquids:				
HCl	H_2O	0.0	2.23×10^{-6}	0.81
NH_3	H_2O	4.0	1.21×10^{-6}	1.5
Gases:				
NH_3	Air	0.0	2.165×10^{-5}	0.634
CO_2	Air	0.0	1.198×10^{-5}	1.14
	H_2	18.0	6.048×10^{-5}	0.158
Hg	N_2	18.4	$325. \times 10^{-5}$	0.00424
O_2	Air	0.0	1.533×10^{-5}	0.895
	N_2	11.8	2.025×10^{-5}	0.681
H_2	Air	0.0	5.472×10^{-5}	0.250
	O_2	14.0	7.748×10^{-5}	0.182
H_2	N_2	12.5	7.376×10^{-5}	0.187
H_2O	Air	8.0	2.062×10^{-5}	0.615
	Air	16.0	2.815×10^{-5}	0.488
C_6H_6	Air	0.0	0.7497×10^{-5}	1.83
	CO_2	0.0	0.5268×10^{-5}	1.37
	H_2	0.0	2.936×10^{-5}	3.26
CS_2	Air	19.9	0.8807×10^{-5}	1.68
Ether	Air	19.9	0.7692×10^{-5}	1.93
Ethyl alcohol	Air	0.0	1.013×10^{-5}	1.36
	Air	40.4	1.180×10^{-5}	1.45

† "Handbook of Chemistry and Physics," 39th ed., Chemical Rubber Publishing Company, Cleveland, Ohio, 1957–1958; and other sources.

Table B-10 Emissivities ϵ_n of the radiation in the direction of the normal to the surface and ϵ of the total hemispherical radiation for various materials for the temperature t†‡

Surface	t, C	ϵ_n	ϵ
Gold, polished	130	0.018	
	400	0.022	
Silver	20	0.020	
Copper, polished	20	0.030	
Lightly oxidized	20	0.037	
Scraped	20	0.070	
Black oxidized	20	0.78	
Oxidized	131	0.76	0.725
Aluminum, bright rolled	170	0.039	0.049
	500	0.050	
Aluminum paint	100	0.20–0.40	
Silumin, cast polished	150	0.186	
Nickel, bright matte	100	0.041	0.046
Polished	100	0.045	0.053
Manganin, bright rolled	118	0.048	0.057
Chrome, polished	150	0.058	0.071
Iron, bright etched	150	0.128	0.158
Bright abrased	20	0.24	
Red rusted	20	0.61	
Hot rolled	20	0.77	
	130	0.60	
Hot cast	100	0.80	
Heavily rusted	20	0.85	
Heat-resistant oxidized	80	0.613	
	200	0.639	
Zinc, gray oxidized	20	0.23–0.28	
Lead, gray oxidized	20	0.28	
Bismuth, bright	80	0.340	0.366
Corundum, emery rough	80	0.855	0.84
Clay, fired	70	0.91	0.86
Lacquer, white	100	0.925	
Red lead	100	0.93	
Enamel, lacquer	20	0.85–0.95	
Lacquer, black matte	80	0.970	
Bakelite lacquer	80	0.935	
Brick, mortar, plaster	20	0.93	
Porcelain	20	0.92–0.94	
Glass	90	0.940	0.876
Ice, smooth, water	0	0.966	0.918
Rough crystals	0	0.985	
Waterglass	20	0.96	
Paper	95	0.92	0.89
Wood, beech	70	0.935	0.91
Tarpaper	20	0.93	

† From measurements by E. Schmidt and E. Eckert.

‡ For metals, the emissivities rise with rising temperature, but for nonmetallic substances (metal oxides, organic substances) this rule is sometimes not correct. Where the exact measurements are not given, take for bright metal surfaces an average ratio $\epsilon/\epsilon_n = 1.2$ and for other substances with smooth surfaces $\epsilon/\epsilon_n = 0.95$; for rough surfaces use $\epsilon/\epsilon_n = 0.98$.

Table B-11 Emissivities as functions of wavelength

Wavelength, μ	0.50	0.60	0.95	1.8	2.1	3.6	4.4	5.4	8.8	9.3	Authority
Aluminum:											
Polished	....	....	0.26	....	0.17	0.08	....	0.05	....	0.04	Coblentz
Oxidized	....	....	....	....	....	0.18	....	0.12	....	0.11	Randolph and Overholser
Duralumin	....	0.53	....	....	....	....	....	....	....	....	Coblentz
Chromium	....	0.49	0.43	....	0.36	0.26	....	0.17	....	0.08	Coblentz and Hulbert
Copper:											
Polished	....	....	0.26	....	0.17	0.18	....	0.05	....	0.04	Coblentz, E. Schmidt
Oxidized	....	....	....	....	....	0.77	....	0.83	....	0.87	Coblentz
Gold, polished	....	....	....	....	0.03	0.03	....	0.02	....	0.02	Burgess and Foote
Steel, polished	....	0.45	0.37	....	0.23	0.14	....	0.10	....	0.07	Randolph and Overholser
Iron:											
Polished	....	0.45	0.35	....	0.22	0.13	....	0.08	....	0.06	Coblentz
Cast, oxidized	....	....	....	....	....	0.76	....	0.66	....	0.63	Overholser
Galv., new	0.66	0.66	0.67	0.42	....	....	....	....	....	0.23	Beckett
Galv., dirty	0.89	0.89	0.89	0.90	....	....	....	....	....	0.28	Beckett
Galv., white-washed	0.24	0.21	0.22	0.37	....	....	....	....	....	....	Beckett
Magnesium	....	0.30	0.26	....	0.23	0.18	....	0.13	....	0.07	Coblentz and Hulbert
Monel	....	0.43	0.29	....	0.16	0.10	....	....	....	....	Coblentz
Molybdenum	0.55	....	0.43	....	0.18	0.11	....	0.08	....	0.06	Coblentz and Hulbert
Brass:											
Polished	....	....	....	....	....	....	....	....	....	0.05	E. Schmidt
Oxidized	....	....	....	....	....	....	....	....	....	0.61	Randolph and Overholser
Paint:*											
Lampblack	....	0.97	0.97	....	....	....	0.97	....	0.96	....	Coblentz
Red	....	0.74	0.59	....	....	....	0.70	....	0.96	....	Coblentz
Yellow	0.39	0.30	....	....	....	....	0.59	....	0.95	....	Coblentz
White	0.18	0.14	0.16	....	....	....	0.77	....	0.95	....	Coblentz, E. Schmidt
Aluminum	....	0.45	....	....	....	....	....	....	....	....	Coblentz
Silver, polished	0.11	....	0.04	....	0.03	0.03	....	0.02	....	0.01	Coblentz
Platinum black	0.97	....	0.97	....	0.97	0.97	0.96	0.96	0.93	0.93	Coblentz
White paper	....	0.28	0.25	....	....	0.82	....	....	....	0.95	Coblentz
White marble	....	0.47	....	....	....	0.93	....	....	....	0.95	Coblentz
Graphite	0.78	....	0.73	....	0.64	0.54	....	0.49	....	0.41	Coblentz and Hulbert

* Various paints are indistinguishable as regards thermal radiation for temperatures below about 200 F (i.e., black, dead black, dark green, white, black enamel, white enamel).

Table B-12 Surface tension of liquids in contact with their vapor

Substance	*Temp.*, C	*Surface tension* N/m
Acetic acid	10	29.1×10^{-3}
	20	28.1
	50	25.1
Acetone	0	26.5
	20	24.0
	40	21.4
Ammonia	11	23.6
	34	18.4
Bromine	20	41.8
Carbon dioxide	20	1.172
	−11	9.21
Carbon tetrachloride	20	27.2
	100	17.42
	200	6.60
Ethyl alcohol	10	23.8
	20	22.9
	30	22.1
Ethyl ether	20	17.19
	50	13.57
Hydrazine	25	92.5
Hydrogen peroxide	18	76.9
Methyl alcohol	50	20.3
Methyl ether	−10	16.52
	−40	21.2
Naphthalene	127	29.1
Tetrachloroethylene	20	31.9
Toluene	10	27.9
	20	28.8
	30	27.7

Table B-13 Surface tension of water in contact with air

Temp., C	*Surface tension*, N/m
0	78.2×10^{-3}
10	74.8
20	73.4
30	71.6
40	70.2
50	68.5
60	66.7
70	64.9
80	63.2
100	59.3

REFERENCES FOR PROPERTY VALUES

1. "International Critical Tables," McGraw-Hill, New York, 1926–1930.
2. Baumeister, T., and L. S. Marks (eds.): "Mechanical Engineers' Handbook," 6th ed., McGraw-Hill, New York, 1958.
3. D'Ans, V., and E. Lax: "Taschenbuch für Chemiker and Physiker," Springer-Verlag, Berlin, 1943.
4. Henning, F.: "Wärmetechnische Richtwerte," VDI, Berlin, 1938.
5. McAdams, W. H.: "Heat Transmission," 3d ed., McGraw-Hill, New York, 1954.
6. Landolt, H.: "Landolt-Bornstein Physikalischchemische Tabellen," Springer-Verlag, Berlin, 1923.
7. Keenan, J. H., and J. Kaye: "Gas Tables," Wiley, New York, 1945.
8. Keenan, J. H., and F. G. Keyes: "Thermodynamic Properties of Steam," Wiley, New York, 1936.
9. Koch, B.: "Grundlagen des Wärmeaustausches (Stoffwerte)," Dissen T. W. Verlaganstalt H. Bücke und Söhne, 1950.
10. Akin, S. W.: The Thermodynamic Properties of Helium, *Trans. ASME,* **72**(6): 751–757 (1950).
11. Kowalczyk, L. S.: Thermal Conductivity and Its Variability with Temperature and Pressure, *Trans. ASME,* **77**(7):1021–1036 (1955).
12. Wooley, H. W., R. B. Scott, and F. G. Brickwedde: Compilation of Thermal Properties of Hydrogen in its Various Isotopic and Ortho-Para Modifications, *J. Res. Natl. Bur. Std.,* **41** (1948).
13. Masi, J. F.: Survey of Experimental Determinations of Heat Capacities of Ten Technically Important Gases, *Trans. ASME,* **76**(7):1067–1074 (1954).
14. Hilsenrath, J., et al.: Tables of Thermal Properties of Gases, *Natl. Bur. Std. U.S. Circ.* 564, 1955.
15. Keyes, F. G., and D. J. Sandell, Jr.: New Measurements of the Heat Conductivity of Steam and Nitrogen, *Trans. ASME,* **72**(6):767–778 (1950).
16. Lees, C. H.: Conductivities of Metals and Alloys at Low Temperatures, *Phil. Trans. Roy. Soc. London, Ser. A,* **208**:381 (1908).
17. Lenoir, J. M.: Thermal Conductivity of Gases at Atmospheric Pressure, *Univ. Arkansas Eng. Exptl. Sta. Bull.* 18, 1953.
18. Coblentz, W. W.: The Diffuse Reflecting Power of Various Substances, *Natl. Bur. Std. U.S. Bull.,* **9**:283–324 (1913).
19. Randolf, C. F., and M. J. Overholzer: *Phys. Rev.,* **2**:144 (1915).
20. Hulbert, E. O.: *Astrophys. J.,* **42**:205 (1915).
21. Burgess, G. K., and P. D. Foote: *Natl. Bur. Std. U.S. Bull.,* **11**:41 (1915–1916); **12**:83 (1915–1916).
22. Schmidt, E.: *Gesundh. Ingr.,* **20** (1927).
23. Westenberg, A. A.: A Critical Survey of the Major Methods for Measuring and Calculating Dilute Gas Properties, *Advan. Heat Transfer* **3**:253–302 (1966).
24. Tsederberg, N. V.: "Thermal Conductivity of Gases and Liquids," M.I.T., Cambridge, Mass., 1965.
25. Sato, T., and T. Minamiyama: Viscosity of Steam at High Temperatures and Pressures, *Intern. J. Heat Mass Transfer,* **7**:199–209 (1964).
26. Tye, R. R.: "Thermal Conductivity," vols. I and II, Academic, New York, 1969.
27. Grigull, V., F. Mayinger, and J. Bach: Viskosität, Wärmeleitfähigkeit und Prandtl-Zahl von Wasser und Wasserdampf, *Wärme- u Stoffübertragung,* **1**:15–34 (1968).
28. Stops, D. W.: Effect of Temperature on the Thermal Conductivity of Gases, *Nature,* **164**:966–967 (1949).
29. Glassman, I., and C. Bonilla: *Chem. Engr. Progr. Symp.,* **49(5)** (1953).

30. Ubbink, J. B., and W. J. de Haas: Thermal Conductivity of Gaseous Helium, *Physica* **10**:465–467 (1943).
31. Keesom, W. H.: "Helium," Elsevier, Amsterdam, 1942.
32. Kelley, K. K.: *U.S. Bur. Mines Bull.* 371, 1934.
33. Wilke, C. R.: *J. Chem. Phys.*, **18**:517 (1950).
34. Lindsay, A. L., and L. A. Bromley: *Ind. Eng. Chem.*, **42**:1, 508 (1950).
35. Wassiljewa, A.: *Physik Z.*, **5**:737 (1904).
36. Atomic Energy Commission: "Liquid Metals Handbook," 2d ed., Department of the Navy, Washington, D.C. 1952.
37. "Handbook of Chemistry and Physics," 39th ed., Chemical Rubber, Cleveland, Ohio, 1957–1958.
38. Poferl, D. J., R. A. Svehla, and K. Lewandowski: Thermodynamic and Transport Properties of Air and the Combustion Products of Natural Gas and of ASTM-A-1 Fuel with Air, *NASA Tech. Note* D-5452, 1969.
39. Lee, J. S., and P. J. Bobbitt: Transport Properties at High Temperatures of CO_2—N_2—O_2—Ar Gas Mixtures for Planetary Entry Applications, *NASA Tech. Note* D-5476, 1969.

Name Index

Subject Index